Climate Studies
Introduction to Climate Science

D0164639

Joseph M. Moran

Education Program

American Meteorological Society
Boston, MA

The American Meteorological Society
Education Program

The American Meteorological Society (AMS), founded in 1919, is a **scientific and professional** society. Interdisciplinary in its scope, the Society actively promotes the development and dissemination of information on the atmospheric and related oceanic and hydrologic sciences. AMS has more than 13,000 professional members from more than 100 countries and over 175 corporate and institutional members representing 40 countries.

The Education Program is the initiative of the American Meteorological Society fostering the teaching of the atmospheric and related oceanic and hydrologic sciences at the precollege level and in community college, college and university programs. It is a unique partnership between scientists and educators at all levels with the ultimate goals of (1) attracting young people to further studies in science, mathematics and technology, and (2) promoting public scientific literacy. This is done via the development and dissemination of scientifically authentic, up-to-date, and instructionally sound learning and resource materials for teachers and students.

AMS Climate Studies, a new component of the AMS education initiative, is an introductory undergraduate Climate Science course offered partially via the Internet in partnership with college and university faculty. **AMS Climate Studies** provides students with a comprehensive study of the principles of Climate Science while simultaneously providing classroom and laboratory applications focused on the rapidly evolving interdisciplinary field of Climate Science.

Developmental work for **AMS Climate Studies** and the companion **DataStreme: Earth's Climate System** was supported by the National Aeronautics and Space Administration under Grants Number NNX09AP58G and NNX08AN53G. Any opinions, findings, and conclusions or recommendations expressed in this material are those of the author and do not necessarily reflect the views of the National Aeronautics and Space Administration.

Climate Studies: Introduction to Climate Science/ Joseph M. Moran. — 1st edition
ISBN-10: 1-878220-04-7
ISBN-13: 978-1878220-04-2
Copyright © 2010 by the American Meteorological Society

Published by the American Meteorological Society
45 Beacon Street, Boston, MA 02108

Cover photographs
top right: Pack Ice in Southern Spitsbergen, Norway, © Radius Images/Corbis
bottom right: Smokestacks, © Fotosearch
bottom left: Electric generating windmills, Highway 206, OR, © John Dittli/Larry Ulrich Stock Photography, Inc

BRIEF CONTENTS

CONTENTS

CHAPTER 3   PLANETARY ENERGY BUDGET IN EARTH'S CLIMATE SYSTEM   67

CHAPTER 4   THERMAL RESPONSE OF THE CLIMATE SYSTEM   101

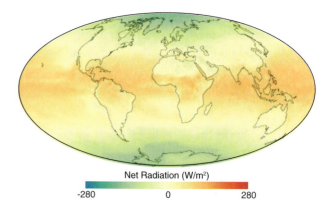

Net Radiation (W/m²)
-280   0   280

CHAPTER 5   WATER IN EARTH'S CLIMATE SYSTEM   133

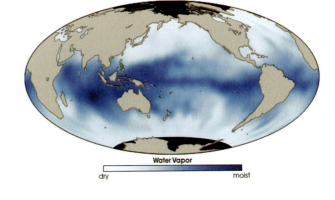

Water Vapor
dry   moist

CHAPTER 6   GLOBAL ATMOSPHERIC CIRCULATION   169

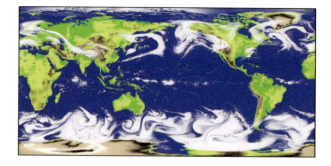

CHAPTER 7   ATMOSPHERIC CIRCULATION AND REGIONAL CLIMATES   205

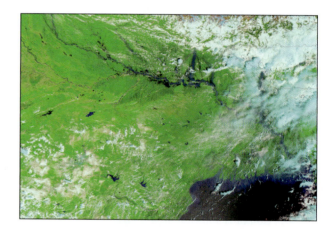

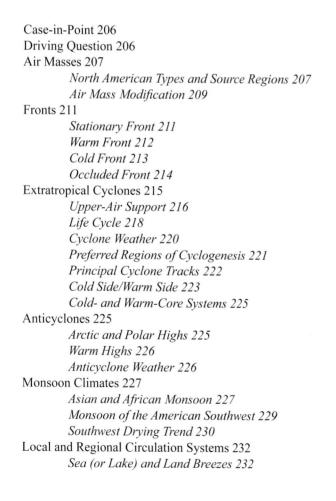

CHAPTER 8   CLIMATE AND AIR/SEA INTERACTIONS   245

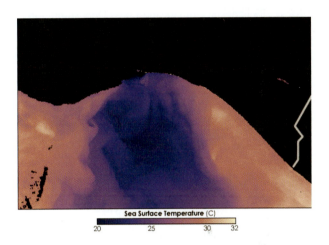

Sea Surface Temperature (C)
20   25   30   32

CHAPTER 9   THE CLIMATE RECORD: PALEOCLIMATES   277

CHAPTER 10   INSTRUMENT-BASED CLIMATE RECORD AND CLIMATOLOGY OF SEVERE WEATHER   309

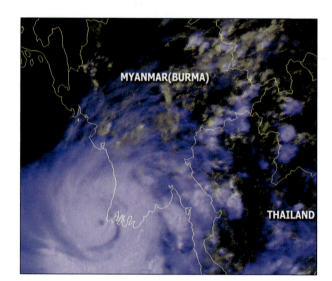

CHAPTER 13   CLIMATE CLASSIFICATION   399

CHAPTER 14   RESPONDING TO CLIMATE CHANGE   423

CHAPTER 15   CLIMATE CHANGE AND PUBLIC POLICY   453

APPENDIX I   CONVERSION FACTORS   473

APPENDIX II   MILESTONES IN THE HISTORY OF CLIMATE SCIENCE   475

GLOSSARY   483

INDEX   509

PREFACE

Welcome to *Climate Studies*! You are about to embark on an exciting study of climate science. The purpose of this book is to provide you with background information on Earth's climate system, the scientific principles that govern climate, climate variability and climate change with the implications for society. Also covered are risk management strategies aimed at countering negative impacts of global climate change. *Climate Studies* was developed by the Education Program of the American Meteorological Society (AMS) with support and assistance provided by the National Aeronautics and Space Administration (NASA).

Climate Studies is closely aligned with the essential principles of climate science identified in a recent climate literacy initiative by the National Oceanic and Atmospheric Administration (NOAA) and the American Association for the Advancement of Science (AAAS). The ultimate goal is to promote climate science literacy at all levels of formal and informal education where climate science literacy is defined as "an understanding of the climate's influence on you and society and your influence on climate." The NOAA/AAAS initiative emphasizes that climate science literacy is part of science literacy. That is, "each essential principle is supported by fundamental concepts comparable to those underlying the National Science Education Standards and the AAAS Benchmarks for Science Literacy."

Climate Studies is guided by new findings of learning science that redefine what it means to be proficient in science. According to the National Research Council of the National Academies, Board on Science Education (2007), students who are proficient in science "(1) know, use, and interpret scientific explanations of the natural world, (2) generate and evaluate scientific evidence and explanations, (3) understand the nature and development of scientific knowledge, and (4) participate productively in scientific practices and discourse." These strands of proficiency are learning goals in *Climate Studies* that address the knowledge and reasoning skills essential for students to be proficient in climate science and participate as climate science literate citizens.

Climate Studies may serve as a stand-alone textbook in an undergraduate college course on climatology, climate change, weather and climate, or atmospheric science. No prior course work in meteorology or atmospheric science is assumed. *Climate Studies* also serves as the reference book for *AMS Climate Studies*, a turnkey course package developed, licensed, and nationally implemented by AMS. Each of the first 12 chapters corresponds to one week of the *AMS Climate Studies* course. A companion *Investigations Manual* plus course website provide students with twice-weekly investigations on climate science partially delivered via the Internet. The course can be offered in face-to-face, blended, and totally online instructional environments.

The course package, *AMS Climate Studies*, follows learning science in providing strategically designed "student encounters with science that take place in real time and over a period of months and years (e.g., *learning progressions*)." *AMS Climate Studies* seeks to engage learners in exploring their world by investigating meaningful questions. Investigations have printed and electronic components that make use of climate information/data available via the course website. Investigations engage participants in observation, prediction, data analysis, inference, and critical thinking. Application of information-age technology enables participants to develop their ability to retrieve and analyze real-world data and share interpretations. Throughout the course, participants assemble learning materials for assessment purposes.

This book consists of 15 chapters. Chapters are organized so that concepts build logically one upon the other so that the components of Earth's climate system emerge as highly interactive and subject to fundamental laws. Topics covered include the components of Earth's climate system (Chapter 1), observing and modeling Earth's climate system (Chapter 2), Earth's radiation budget and the greenhouse effect (Chapter 3), heat transfer mechanisms and controls of temperature (Chapter 4), humidity, clouds, and precipitation (Chapter 5), how the planetary-scale circulation influences climate (Chapter 6), synoptic and regional climatology (Chapter 7), atmosphere-ocean interactions and climate (Chapter 8), reconstructing the climate past (Chapter 9), trends in the instrument-based climate record and the climatology of severe weather systems (Chapter 10), natural agents and mechanisms responsible for climate change (Chapter 11), and the influence of human activity on climate change (Chapter 12). The final three chapters cover climate classification systems (Chapter 13), climate change mitigation, adaptation, and geoengineering (Chapter 14), and international and national aspects of public policy directed at reducing anthropogenic emissions of heat-trapping gases (Chapter 15). Accompanying all chapters are investigations in the separate *Investigations Manual*.

Each chapter opens with a *Case-in-Point*, an authentic, relevant, and real-life event or issue that highlights or applies one or more of the main concepts or principles covered in the chapter. In essence, the Case-in-Point previews the chapter and is intended to engage reader interest early on. For example, Chapter 4 opens with a summary of the reasons why Death Valley, CA, has the hottest and driest climate of any place in North America. The Case-in-Point is followed by a sample *Driving Question*, a broad-based query that links chapter concepts and provides a central focus for that chapter's topics. Chapter content is science-rich and informs additional driving questions. Each chapter closes with a list of *Basic Understandings* and *Enduring Ideas*, as well as questions for *Review* and *Critical Thinking*. Two *Essays* at the end of each chapter address in some depth specific topics that complement or supplement a concept covered in the chapter narrative. Examples include *Asteroids, Climate Change, and Mass Extinctions*, *Contrails and Climate Change*, *Cloud Forests and Climate Change*, *Permafrost and Climate Change*, and *The Extreme Climate of Mount Washington, NH*. All terms bold-faced in the narrative are defined in the *Glossary* at the back of the book. Appendixes cover unit conversions and milestones in the history of climate science.

Development work ultimately culminating in the national dissemination of *AMS Climate Studies* was supported by NASA. For this support and encouragement we are very grateful to Jack Kaye, Associate Director of Research of the Earth Science Division within NASA's Science Mission Directorate, and Ming-Ying Wei, Manager of Education Programs, NASA's Office of Earth Science. We acknowledge with much appreciation the assistance of the following NASA scientists who reviewed sections of *Climate Studies*: Michele M. Rienecker, Global Modeling and Assimilation Office, NASA Goddard Space Flight Center; David H. Rind, NASA Goddard Institute for Space Studies; and Norman G. Loeb, NASA Langley Research Center.

Climate Studies was a team effort involving atmospheric scientists, climate specialists, and educators at all levels from K-12 through college. The principal author of the textbook is Joseph M. Moran, Associate Director

of the American Meteorological Society's Education Program and Professor Emeritus of Earth Science at the University of Wisconsin-Green Bay. Much of the initial design of the course scope and sequence was inspired by the insightful comments and suggestions of Kathleen V. Schreiber, Millersville University, and Samantha W. Kaplan, the University of Wisconsin-Stevens Point. We also acknowledge with much thanks the active encouragement and interest of Frank Niepold of NOAA's Climate Program Office.

Book development benefited greatly from suggestions and critical reviews provided by James A. Brey, Ira W. Geer, Bernard A. Blair, Elizabeth W. Mills, Thomas P. Kiley, Jr., Emily E. Ruwe, Emily Gracey Miller, and Katie L. O'Neill of the AMS Education Program, Robert S. Weinbeck of SUNY College at Brockport and the AMS Education Program, and Paul A.T. Higgins of the AMS Policy Program. Edward J. Hopkins of the University of Wisconsin-Madison is singled out for his exceptionally thorough analysis of science content. Thanks also to Peter Crane, former Director of Education at the Mount Washington Observatory, for his critical review of the Essay on Mount Washington's climate.

Special thanks are extended to the many college and K-12 educators who enthusiastically gave of their valuable time and efforts as participants in the pilot testing phase of all course materials. We are very grateful to Lisa Bastiaans, Nassau Community College; William Blanchard, Georgian Court University; Margaret F. Boorstein, C.W. Post Campus of Long Island University; William Buckler, Youngstown State University; Toni DeVore, Ohio Valley University; Kelly Esslinger, Arizona Western College; Thomas Gill, University of Texas at El Paso; Jerry Griffith, The University of Southern Mississippi; Jacquelyn Hams, Los Angeles Valley College; Don Hellstern, Eastfield College; Steve LaDochy, California State University, Los Angeles; Julie Lambert, Florida Atlantic University; Michael Leach, New Mexico State University at Grants; Joan Lindgren, Florida Atlantic University; Pedro Ramirez, California State University, Los Angeles; Richard Schultz, Elmhurst College; David Travis, University of Wisconsin-Whitewater; Gail Wyant, Cecil Community College; and Patrick Wyant, Cecil Community College. We acknowledge with gratitude the following K-12 educators who served as Local Implementation Team leaders for the pilot testing of DataStreme Earth's Climate System: Samuel Wine (Kansas), Freida Blink (California and Nevada), Craig Wolter (Minnesota), Mark Mettert (Indiana), Thomas Kelly (Michigan), Carol Hildreth (New York), Roy Chambers (Oregon), Bruce Smith (Wisconsin), and Vicky Peterson (Minnesota). We thank Marsha Rich (New Hampshire) for her help with the Essay on Mount Washington climate, and Carol Hildreth, Toni DeVore, and Bruce Smith for their thorough reviews of the learning materials.

With his usual enthusiasm, skill, and dedication, Bernard A. Blair of the AMS Education Program tackled the sometimes daunting task of formatting the book, seeing to the many details and daily challenges of guiding the project from initial concept to final product. Along the way, the manuscript benefitted greatly from technical editing by Elizabeth W. Mills of the AMS Education Program. Norman J. Frisch of Brockport, NY, excelled at converting line drawings into final art. Unless otherwise indicated, Joseph M. Moran supplied photographs.

A special note concerns the use of units in *Climate Studies*. Generally, the International System of Units (abbreviated SI, for Systéme International d'Unitès) is employed with equivalent English or other units following in parentheses. Exceptions are units used by convention or convenience in the atmospheric sciences or the user community (e.g., knots, calories, millibars). Also, the equivalence between units is given in context; that is, where general estimates are used, approximate values are shown in all units. Conversion factors are given in Appendix I. Note also that dates adhere to the system whereby BCE is Before the Common Era and CE is the Common Era.

James A. Brey, Ph.D.
Director, AMS Education Program

CHAPTER 1

CLIMATE SCIENCE FOR TODAY'S WORLD

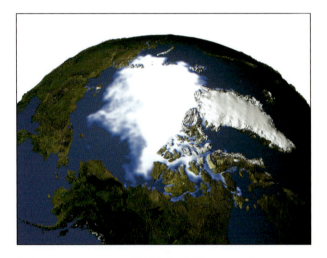

Dwindling arctic sea-ice. [NASA Earth Observatory]

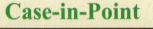

Case-in-Point

Today's much discussed proposition that human activity can contribute to climate change is not new. In fact, during much of the 18th and 19th centuries, debate raged among natural scientists over whether deforestation and cultivation of land in America were responsible for changing the climate. In 1650, prior to colonization, tall forests blanketed most of what is now the eastern United States, but over the subsequent 200 years, settlers cleared the forests over much of New England, the mid-Atlantic region, and parts of the Midwest. By about 1920, almost all of the tall forests were gone as the land was converted to farms, towns, and cities.

     Among the earliest proponents of a possible link between land clearing and climate change was Benjamin Franklin (1706-1790), a man of many talents and interests. In 1763, Franklin wrote that by clearing the woods, colonists exposed the once shaded soil surface to more direct sunshine thereby absorbing more heat. Hence, snow

melted more quickly. Thomas Jefferson (1743-1826), third President of the United States, shared Franklin's view that deforestation and cultivation of the soil ameliorated the climate. Others claimed that these landscape changes caused winters to be less severe and summers to be more moderate. However, Franklin and Jefferson also recognized that many years of instrument-based weather observations would be needed to firmly establish a link between deforestation and climate change.

     Prior to the end of the 18th century, Noah Webster (1758-1843), author of the first American dictionary, weighed in on the climate change debate. According to Webster, most proponents of a warming climate based their arguments largely on anecdotal information and faulty memories of what the weather had been like many years prior. While rejecting the idea of a large-scale warming trend, Webster believed that deforestation and cultivation of land in America had caused the climate to become seasonally more variable because cleared land would be hotter in summer and colder in winter.

Until the early decades of the 19th century, most information on climate was qualitative, consisting of pronouncements by various authorities or the memories of the elderly. In the second half of the 19th century with the increasingly widespread availability of thermometers and other weather instruments along with establishment of regular weather observational networks operated by the U.S. Army Medical Department and the Smithsonian Institution, quantitative climate data became available for analysis. Those data failed to show an unequivocal relationship between deforestation, cultivation, and climate change.

Today, climate scientists remain intrigued by the possible influence of land use patterns on climate. Vegetation is an important component of the climate system (e.g., slowing the wind, transpiring water vapor into the atmosphere, absorbing sunlight and carbon dioxide for photosynthesis). It is reasonable to assume that transformation of forests to cropland would affect these and other processes that influence the climate. Unlike their early predecessors in the climate/land use debate, today's climate scientists have access to regional climate models to predict the role played by changes in land use patterns on climate. These computerized numerical models simulate the interactions between vegetation and atmosphere taking into account biological and physical characteristics of the land.

Driving Question:
What is the climate system and why should we be concerned about climate and climate change?

We are about to embark on a systematic study of climate, climate variability, and climate change. Earth is a mosaic of many climate types, each featuring a unique combination of physical, chemical, and biological characteristics. Differences in climate distinguish, for example, deserts from rainforests, temperate regions from glacier-bound polar localities, and treeless tundra from subtropical savanna. We will come to understand the spatial and temporal (time) variations in climate as a response to many interacting forcing agents or mechanisms both internal and external to the *planetary system*. At the same time we will become familiar with the scientific principles and basic understandings that underlie the operations and interactions of those forcing agents and mechanisms. This is **climate science**, the systematic study of the mean state of the atmosphere at a specified location and time period as governed by natural laws.

Our study of climate science provides valuable insights into one of the most pressing environmental issues of our time: global climate change. We explore the many possible causes of climate change with special emphasis on the role played by human activity (e.g., burning fossil fuels, clearing vegetation). A thorough grounding in climate science enables us to comprehend the implications of anthropogenic climate change, how each of us contributes to the problem, and how each of us can be part of the solution to the problem.

Our primary objective in this opening chapter is to begin constructing a framework for our study of climate science. We begin by defining climate and showing how climate relates to weather, as the state of the atmosphere plays a dominant role in determining the global and regional climate. The essential value in studying climate science stems from the ecological and societal impacts of climate and climate change. Climate is the ultimate environmental control that governs our lives; for example, what crops can be cultivated, the supply of fresh water, and the average heating and cooling requirements for homes.

By its very nature, climate science is interdisciplinary, drawing on principles and basic understandings of many scientific disciplines. We recognize climate as a system in which Earth's major subsystems (i.e., atmosphere, hydrosphere, cryosphere, geosphere, and biosphere) individually and in concert function as controls of climate. Linking these subsystems are biogeochemical cycles (e.g., global carbon cycle, global water cycle, global nitrogen cycle), pathways for transfer of climate-sensitive materials (e.g., greenhouse gases, atmospheric particulates) and energy and energy transfers among Earth-bound reservoirs. This chapter closes with the *climate paradigm*, a rudimentary theoretical framework that encapsulates the basic ingredients of our study of climate science.

Defining Climate

The study of climate began with the ancient Greek philosophers and geographers. Climate is derived from the Greek word *klima* meaning "slope," referring to the variation in the amount of sunshine received at Earth's surface due to the regular changes in the Sun's angle of inclination upon a spherical Earth. This was the original basis for subdividing Earth into different climate zones. Parmenides, a philosopher and poet who lived in the mid 5th century BCE, is credited with devising the first climate classification scheme. His classification consists of a latitude-bounded five-zone division of Earth's surface based on the intensity of sunshine: a torrid zone, two temperate zones, and two frigid zones (Figure 1.1). According to Parmenides, the torrid zone was uninhabitable because of heat and the frigid zones were uninhabitable because of extreme cold.

Hippocrates (*ca.* BCE 460-370), considered the founder of medicine, authored the first climatography, *On Airs, Waters, and Places*, about BCE 400. (A *climatography* is a graphical, tabular or narrative description of the climate.) Aristotle who adopted Parmenides' climate classification, followed in about BCE 350 with *Meteorologica*, the first treatise on meteorology, which literally means the study of anything from the sky. Strabo (*ca.* BCE 64 – CE 24), author of the 17-volume treatise *Geographica*, noted

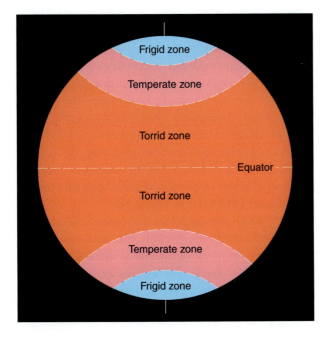

FIGURE 1.1
Parmenides developed the first global climate classification scheme in the mid 5th century BCE.

that climate zones correspond to temperature differences as well as amount of sunshine. He was the first to observe that temperature varied with both latitude and altitude. In addition, Strabo attributed local variations in climate to topography and land/water distribution.

CLIMATE VERSUS WEATHER

Weather and climate are closely related concepts. According to an old saying, *climate is what we expect and weather is what actually happens*. In this section, we describe the relationship between weather and climate and focus on two complementary working definitions of climate: an *empirical definition* that is based on statistics and a *dynamic definition* that incorporates the forces that govern climate. The first <u>describes</u> climate whereas the second seeks to <u>explain</u> climate.

Everyone has considerable experience with the weather. After all, each of us has lived with weather our entire life. Regardless of where we live or what we do, we are well aware of the far-reaching influence of weather. To some extent, weather dictates our clothing, the price of orange juice and coffee in the grocery store, our choice of recreational activities, and even the outcome of a football game. Before setting out in the morning, most of us check the weather forecast on the radio or TV or glance out the window to scan the sky or read the thermometer. Every day we gather information on the weather through our senses, the media, and perhaps our own weather instruments. And from that experience, we develop some basic understandings regarding the atmosphere, weather, and climate.

Weather is defined as the state of the atmosphere at some place and time, described in terms of such variables as temperature, humidity, cloudiness, precipitation, and wind speed and direction. Thousands of weather stations around the world monitor these weather variables at Earth's surface at least hourly every day. A place and time must be specified when describing the weather because the atmosphere is dynamic and its state changes from one place to another and with time. When it is cold and snowy in Boston, it might be warm and humid in Miami and hot and dry in Phoenix. From personal experience, we know that tomorrow's weather may differ markedly from today's weather. *If you don't like the weather, wait a minute* is another old saying that is not far from the truth in many areas of the nation. **Meteorology** is the study of the atmosphere, processes that cause weather, and the life cycle of weather systems.

While weather often varies from one day to the next, we are aware that the weather of a particular locality tends to follow reasonably consistent seasonal variations, with temperatures higher in summer and lower in winter. Some parts of the world feature monsoon climates with distinct rainy and dry seasons. We associate the tropics with warmer weather and seasonal temperature contrasts that are less than in polar latitudes. In fact, experienced meteorologists can identify readily the season from a cursory glance at the weather pattern (atmospheric circulation) depicted on a weather map. These are all aspects of climate.

An easy and popular way of summarizing local or regional climate is in terms of the averages of weather elements, such as temperature and precipitation, derived from observations taken over a span of many years. In this empirically-based context, **climate** is defined as weather (the state of the atmosphere) at some locality averaged over a specified time interval. Climate must be specified for a particular place and period because, like weather, climate varies both spatially and temporally. Thus, for example, the climate of Chicago differs from that of New Orleans, and winters in Chicago were somewhat milder in the 1980s and 1990s than in the 1880s and 1890s.

In addition to average values of weather elements, the climate record includes extremes in weather. Climatic summaries typically tabulate extremes such as the coldest, warmest, driest, wettest, snowiest, or windiest day, month or year on record for some locality. Extremes are useful aspects of the climate record if only because what has happened in the past can happen again. For this reason, for example, farmers are interested in not only the average rainfall during the growing season but also the frequency of exceptionally wet or dry growing seasons. In essence, records of weather extremes provide a perspective on the variability of local or regional climate.

In 1935, delegates to the International Meteorological Conference at Warsaw, Poland, standardized the averaging period for the climate record. Previously it was common practice to compute averages for the entire period of station record even though the period of record varied from one station to another. This practice was justified by the erroneous assumption that the climate was static. By international convention, average values of weather elements are computed for a 30-year period beginning with the first year of a decade. (Apparently, selection of 30 years was based on the Brückner cycle, popular in the late 19th century and consisting of alternating episodes of cool-damp and warm-dry weather having a period of nearly 30 years. However, the Brückner cycle has been discredited as a product of statistical smoothing of data.) At the close of the decade, the averaging period

is moved forward 10 years. Current climatic summaries are based on weather records from 1971 to 2000. Average July rainfall, for example, is the simple average of the total rainfall measured during each of thirty consecutive Julys from 1971 through 2000.

Selection of a 30-year period for averaging weather data may be inappropriate for some applications because climate varies over a broad range of time scales and can change significantly in periods much shorter than 30 years. For example, El Niño refers to an inter-annual variation in climate involving air/sea interactions in the tropical Pacific and weather extremes in various parts of the world (Chapter 8). The phenomenon typically lasts for 12 to 18 months and occurs about every 3 to 7 years. For some purposes, a 30-year period is a short-sighted view of climate variability. Compared to the long-term climate record, for example, the current 1971-2000 averaging period was unusually mild over much of the nation.

In the United States, 30-year averages are computed for temperature, precipitation (rain plus melted snow and ice), and degree days and identified as *normals*. Averages of other climate elements such as wind speed and humidity are derived from the entire period of record or at least the period when observations were made at the same location. Other useful climate elements include average seasonal snowfall, length of growing season, percent of possible sunshine, and number of days with dense fog. Tabulation of extreme values of weather elements is usually also drawn from the entire period of the observational record.

Climatic summaries (e.g., *Local Climatological Data)* are available in tabular formats for major cities (along with a narrative description of the local or regional climate) as well as climatic divisions of each state. The National Oceanic and Atmospheric Administration's (NOAA's) **National Weather Service** is responsible for gathering the basic weather data used in generating the nation's climatological summaries. Data are processed, archived, and made available for users by NOAA's **National Climatic Data Center (NCDC)** in Asheville, NC.

While the empirical definition of climate (in terms of statistical summaries) is informative and useful, the *dynamic definition* of climate is more fundamental. It addresses the nature and controls of Earth's climate together with the causes of climate variability and change operating on all time scales. Climate differs from season to season and with those variations in climate, the array of weather patterns that characterize one season differs from the array of characteristic weather patterns of another season. (As mentioned earlier, this explains why an experienced meteorologist can deduce the season from the weather pattern.) The status of the **planetary system** (that is, the Earth-atmosphere-land-ocean system) determines (or selects) the array of possible weather patterns for any season. In essence, this status constitutes boundary conditions (i.e., forcing agents and mechanisms) such as incoming solar radiation and the albedo (reflectivity) of Earth's surface. Hence, in a dynamic context, climate is defined by the boundary conditions in the planetary system coupled with the associated typical weather patterns that vary with the seasons. For example, the higher Sun's path across the local sky and the longer daylight length in Bismarck, ND during July increase the chance of warm weather and possible thunderstorms, whereas lower Sun angles and shorter daylight duration during January would mean colder weather and possible snow.

Climatology, the subject of this book, is the study of climate, its controls, and spatial and temporal variability. Climatology is primarily a field science rather than a laboratory science. The field is the atmosphere and Earth's surface where data are obtained by direct (*in situ*) measurement by instruments and remote sensing, mostly by sensors flown aboard Earth-orbiting satellites (Chapter 2). Nonetheless, laboratory work is important in climatology; it involves analysis of climate-sensitive samples gathered from the field (Figure 1.2). For example, analysis of glacial ice cores, tree growth rings, pollen profiles, and deep-sea sediment cores enables climatologists to reconstruct the climate record prior to the era of weather instruments (Chapter 9).

FIGURE 1.2
The thickness of annual tree growth rings provides information on past variations in climate, especially the frequency of drought.

The only scientific experiments routinely conducted by climate scientists involve manipulation of numerical climate models. Usually these global or regional models are used to predict the climatic consequences of change in the boundary conditions of Earth's climate system. Furthermore, climatology is an interdisciplinary science that reveals how the various components of the natural world are interconnected. For example, the composition of the atmosphere is the end product of many processes where gases are emitted (e.g., via volcanic eruptions) or absorbed (e.g., gases dissolving in the ocean). The composition of the atmosphere, in turn, affects the ocean, living organisms, geological processes, and climate.

THE CLIMATIC NORM

Traditionally, the **climatic norm**, or normal, is equated to the average value of some climatic element such as temperature or precipitation. This tradition sometimes fosters misconceptions. For one, "normal" may be taken to imply that the climate is static when, in fact, climate is inherently variable with time. Furthermore, "normal" may imply that climatic elements occur at a frequency given by a Gaussian (bell-shaped) probability distribution, although many climatic elements are non-Gaussian.

Many people assume that the *mean* value of a particular climatic element is the same as the *median* (middle value); that is, 50% of all cases are above the mean and 50% of all cases fall below the mean. This assumption is reasonable for some climatic elements such as temperature, which approximates a simple Gaussian-type probability distribution (Figure 1.3A). Hence, for example, we might expect about half the Julys will be warmer and half the Julys will be cooler than the 30-year mean July temperature. On the other hand, the distribution of some climatic elements, such as precipitation, is non-Gaussian, and the mean value is not the same as the median value (Figure 1.3B). In a dry climate that is subject to infrequent deluges of rain during the summer, considerably fewer than half the Julys are wetter than the mean and many more than half of Julys are drier than the mean. In fact, for many purposes the median value of precipitation is a more useful description of climate than the mean value as extremes (*outliers*) are given less weight.

For our purposes, we can think of the climatic norm for some locality as encompassing the total variation in the climate record, that is, both averages plus extremes. This implies, for example, that an exceptionally cold winter actually may not be "abnormal" because its mean temperature may fall within the expected range of variability of winter temperature at that location.

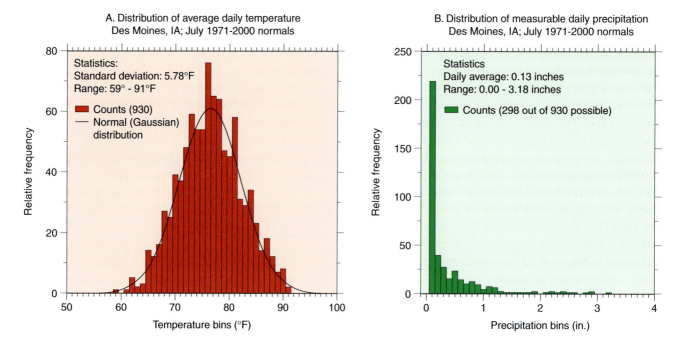

FIGURE 1.3
Distribution of average daily temperature for the month of July in Des Moines, IA, for 1971-2000 (A). Distribution of measurable daily precipitation for the month of July in Des Moines, IA, for 1971-2000 (B). [Courtesy of E.J. Hopkins]

HISTORICAL PERSPECTIVE

Early observers kept records of weather conditions using primitive instruments or qualitative descriptions, jotting them down in journals or diaries. In North America, the first systematic weather observations were made in 1644-1645 at Old Swedes Fort (now Wilmington, DE). The observer was Reverend John Campanius (1601-1683), chaplain of the Swedish military expedition. Campanius had no weather instruments, however. He wrote in his diary qualitative descriptions of temperature, humidity, wind, and weather. Campanius returned to Sweden in 1648 but fifty years passed before his grandson published his weather observations.

Long-term instrument-based temperature records began in Philadelphia in 1731; Charleston, SC, in 1738; and Cambridge, MA, in 1753. The New Haven, CT, temperature record began in 1781 and continues uninterrupted today.

On 2 May 1814, James Tilton, M.D., U.S. Surgeon General, issued an order that marked the first step in the eventual establishment of a national network of weather and climate observing stations. Tilton directed the Army Medical Corps to begin a diary of weather conditions at army posts, with responsibility for observations in the hands of the post's chief medical officer. Tilton's objective was to assess the relationship

between weather and the health of the troops, for it was widely believed at the time that weather and its seasonal changes were important factors in the onset of disease. Even well into the 20th century, more troops lost their lives to disease than combat. Tilton also wanted to learn more about the climate of the then sparsely populated interior of the continent.

The War of 1812 prevented immediate compliance with Tilton's order. In 1818, Joseph Lovell, M.D., succeeded Tilton as Surgeon General and issued formal instructions for taking weather observations. By 1838, 16 Army posts had recorded at least 10 complete (although not always successive) years of weather observations. By the close of the American Civil War, weather records had been tabulated for varying periods at 143 Army posts. In 1826, Lovell began compiling, summarizing, and publishing the data and for this reason Lovell, rather than Tilton, is sometimes credited with founding the federal government's system of weather and climate observations.

In the mid-1800s, Joseph Henry (1797-1878), first secretary of the Smithsonian Institution in Washington, DC, established a national network of volunteer observers who mailed monthly weather reports to the Smithsonian. The number of citizen observers (mostly farmers, educators, or public servants) peaked at nearly 600 just prior to the American Civil War. Henry knew the value of rapid communication of weather data and realized the potential of the newly invented electric telegraph in achieving this goal. In 1849, Henry persuaded the heads of several telegraph companies to direct their telegraphers in major cities to take weather observations at the opening of each business day and to transmit these data free of charge to the Smithsonian. Henry supplied thermometers and barometers (for measuring air pressure). Availability of simultaneous weather observations enabled Henry to prepare the first national weather map in 1850; later he regularly displayed the daily weather map for public viewing in the Great Hall of the Smithsonian building. By 1860, 42 telegraph stations, mostly east of the Mississippi River, were participating in the Smithsonian network.

The success of Henry's Smithsonian network and another telegraphic-based network operated by Cleveland Abbe (1838-1916) at the Mitchell Astronomical Observatory in Cincinnati, OH, persuaded the U.S. Congress to establish a telegraph-based storm warning system for the Great Lakes. In the 1860s, surprise storms sweeping across the Great Lakes were responsible for a great loss of life and property from shipwrecks. President Ulysses S. Grant (1822-1885) signed the Congressional resolution into law on 9 February 1870 and the network, initially composed of 24 stations, began operating on 1 November 1870 under the authority of the U.S. Army Signal Corps. Although the network was originally authorized for the Great Lakes, in 1872, Congress appropriated funds for expanding the storm-warning network to the entire nation. The network soon encompassed stations previously operated by the Army Medical Department, Smithsonian Institution, U.S. Army Corps of Engineers, and Cleveland Abbe. With the expansion of telegraph service nationwide, the number of Signal Corps stations regularly reporting daily weather observations reached 110 by 1880.

On 1 July 1891, the nation's weather network was transferred from military to civilian hands in the new U.S. Weather Bureau within the U.S. Department of Agriculture, with a special mandate to provide weather and climate guidance for farmers. Forty-nine years later, aviation's growing need for weather information spurred the transfer of the Weather Bureau to the Commerce Department. Many cities saw their Weather Bureau offices relocated from downtown to an airport, usually in a rural area well outside the city. In 1965, the Weather Bureau was reorganized as the National Weather Service (NWS) within the *Environmental Science Services*

Administration (ESSA), which became the **National Oceanic and Atmospheric Administration (NOAA)** in 1971.

Today, NWS Forecast Offices operate at 122 locations nationwide. NWS and the Federal Aviation Administration (FAA) operate nearly 840 automated weather stations, many at airports, which have replaced the old system of manual hourly observations. This **Automated Surface Observing System (ASOS)** consists of electronic sensors, computers, and fully automated communications ports (Figure 1.4). Twenty-four hours a day, ASOS feeds data to NWS Forecast Offices and airport control towers. Nearly 1100 additional automatic weather stations, which are funded by other federal and state agencies, supply hourly weather data from smaller airports.

FIGURE 1.4
The National Weather Service's Automated Surface Observing System (ASOS) consists of electronic meteorological sensors, computers, and communications ports that record and transmit atmospheric conditions (e.g., temperature, humidity, precipitation, wind) automatically 24 hours a day.

FIGURE 1.5
This NWS Cooperative Observer Station is equipped with maximum and minimum recording thermometers housed in a louvered wooden instrument shelter. Nearby is a standard rain gauge. Instruments are read and reset once daily by a volunteer observer.

In addition to the numerous weather stations that provide observational data primarily for weather forecasting and aviation, another 11,700 cooperative weather stations are scattered across the nation (Figure 1.5). These stations, derived from the old Army Medical Department and Smithsonian networks, are staffed by volunteers who monitor instruments provided by the National Weather Service. The principal mission of member stations of the **NWS Cooperative Observer Network** is to record data for climatic, hydrologic, and agricultural purposes. Observers report 24-hr precipitation totals and maximum/minimum temperatures based on observations made daily at 8 a.m. local time; some observers also report river levels. Traditionally, observers mailed in monthly reports or telephoned their reports to the local NWS Weather Forecast Office; more recently they enter that data into a computer which formats and transmits data to computer workstations in the NWS *Advanced Weather Interactive Processing System (AWIPS)*.

Climate and Society

Probably the single most important reason for studying climate science is the many linkages between climate and society. For one, climate imposes constraints on social and economic development. For example, the abject poverty of North Africa's Sahel in large measure is due to the region's subtropical climate that is plagued by multi-decadal droughts (Chapter 5). In other regions, climate provides resources that are exploited to the advantage of society. For example, some climates favor winter or summer recreational activities (e.g., skiing, boating) that attract vacationers and feed the local economy. Severe weather (e.g., tornadoes, hurricanes, floods, heat waves, cold waves, and drought) can cause deaths and injuries, considerable long-term disruption of communities, property damage, and economic loss. The impact of Hurricane Katrina on the Gulf Coast is still being felt many years after that weather system made landfall (August 2005).

Regardless of a nation's status as developed or developing, it is not possible to weather- or climate-proof society to prevent damage to life and property. In the agricultural sector, for example, the prevailing strategy is to depend on technology to circumvent climate constraints. Where water supply is limited, farmers and ranchers routinely rely on irrigation water usually pumped from subsurface aquifers (e.g., the High Plains Aquifer in the central U.S.) or transferred via aqueducts and canals from other watersheds. Because of consumers' food preferences and for economic reasons, this strategy is preferred to matching crops to the local or regional climate (e.g., dry land farming). Other strategies include construction of dams and reservoirs to control runoff and genetic manipulation to breed drought resistant crops. Although these strategies have some success, they have limitations and often require tradeoffs. For example, many rivers around the world lose so much of their flow to diversions (mostly for irrigation) that they are reduced to a trickle or completely dry up prior to reaching the sea at least during part of the year. Consider, for example, the Colorado River.

By far, the nation's most exploited watershed is that of the Colorado River, the major source of water for the arid and semi-arid American Southwest. The Colorado River winds its way some 2240 km (1400 mi)[1] from its headwaters in the snow-capped Rocky Mountains of Colorado to the Gulf of California in extreme northwest Mexico (Figure 1.6). Along the river's course, ten major dams and reservoirs (e.g., Lake Mead behind Hoover Dam, Lake Powell behind Glen Canyon Dam) regulate its flow. Water is diverted from the river to irrigate about 800,000 hectares (2 million acres) and meet the water needs of 21 million people. Governed by the Colorado River Compact, aqueducts and canals divert water for use in 7 states and northern Mexico. A

---

[1] For unit conversions, see Appendix I.

FIGURE 1.6
The nation's most exploited watershed is that of the Colorado River, the major source of water for the arid and semi-arid American Southwest. The Colorado River winds its way from its headwaters in the snow-capped Rocky Mountains of Colorado to the Gulf of California in extreme northwest Mexico.

714-km (444-mi) aqueduct system transfers water from the Colorado River to Los Angeles and the irrigation systems of California's Central and Imperial Valleys. The Central Arizona Project, completed in 1993, diverts Colorado River water from Lake Havasu (behind Parker Dam) on the Arizona/California border to the thirsty cities of Phoenix and Tucson. Where its channel finally enters the sea, watershed transfers and evaporation have so depleted the river's discharge that water flows in the channel only during exceptionally wet years.

Compounding the constraints of climate on society is the prospect of global climate change. The scientific evidence is now convincing that human activity is influencing climate on a global scale with significant consequences for society. As we will see in much greater detail later in this book, burning of fossil fuels (coal, oil, natural gas) and clearing of vegetation is responsible for a steady build-up of atmospheric carbon dioxide (CO_2) and enhancement of Earth's greenhouse effect. This enhancement is exacerbated by other human activities that are increasing the concentration of methane (CH_4) and nitrous oxide (N_2O), also greenhouse gases. The consequence is global warming and alteration of precipitation patterns.

In addition, certain human activities are making society and ecosystems more vulnerable to climate change. An **ecosystem** consists of communities of plants and animals that interact with one another, together with the physical conditions and chemical substances in a specific geographical area. Deserts, tropical rain forests, and estuaries are examples of natural ecosystems. Most people live in highly modified terrestrial ecosystems such as cities, towns, farms, or ranches. For example, the human population of the coastal zone is rising rapidly putting more and more people at risk from rising sea level. The 673 coastal counties of the U.S. represent 17% of the nation's land area but have three times the nation's average population density. The population of Florida's coastal counties increased 73% between 1980 and 2003. A consequence of global warming is sea level rise (due to melting glaciers and thermal expansion of sea water); higher sea level, in turn, increases the hazards associated with storm surges (rise in water level caused by strong onshore winds in tropical and other coastal storms). These hazards include coastal flooding, accelerated coastal erosion, and considerable damage to homes, businesses, and infra-structure (e.g., roads, bridges).

With the human population growing rapidly in many areas of the globe, more people are forced to migrate into marginal regions, that is, locales that are particularly vulnerable to excess soil erosion (by wind or water) or where barely enough rain falls or the growing season is hardly long enough to support crops and livestock. These are typically boundaries between ecosystems, known as **ecotones**. Ecotones are particularly vulnerable to climate change in that even a small change in climate can spell disaster (e.g., crop failure and famine).

An important consideration regarding weather and climate extremes (hazards) is **societal resilience**, that is, the ability of a society to recover from weather- or climate-related or other natural disasters. For example, if climate change is accompanied by a higher frequency of intense hurricanes in the Atlantic Basin, there is even greater urgency for a coordinated preparedness plan that would minimize the impact of landfalling hurricanes especially on low-lying communities along the Gulf Coast. These preparations must involve investment in appropriately designed infra-structure that will reduce flooding and allow for the quick evacuation of populations that find themselves in harm's way.

Assessment of societal resilience to climate-related hazards requires understanding of the regional bias of severe weather events. The climate record indicates that although tornadoes have been reported in all states, they are most frequent in the Midwest (tornado alley).

Hurricanes are most likely to make landfall along the Gulf and Atlantic Coasts, but are rare along the Pacific Coast. Droughts are most common on the High Plains whereas forest fires are most frequent in the West.

Our understanding of the potential impact of climate and climate change on society requires knowledge of (1) the structure and function of Earth's climate system, (2) interactions of the various components of that system, and (3) how human activities influence and are influenced by these systems. We begin in the next section with an overview of Earth's climate system.

The Climate System

What is the climate system and, more fundamentally, what is a system? A **system** is an entity whose components interact in an orderly manner according to the laws of physics, chemistry, and biology. A familiar example of a system is the human body, which consists of various identifiable subsystems including the nervous, respiratory, and reproductive systems, plus the input/output of energy and matter. In a healthy person, these subsystems function internally and interact with one another in regular and predictable ways that can be studied based upon analysis of the energy and mass budgets for the systems. Extensive observations and knowledge of a system enable scientists to predict how the system and its components are likely to respond to changing internal and external conditions. The ability to predict the future state(s) of a system is important, for example, in dealing with the complexities of global climate change and its potential impacts on Earth's subsystems and society.

The 1992 **United Nations Framework Convention on Climate Change** defines Earth's **climate system** as the totality of the atmosphere, hydrosphere (including the cryosphere), biosphere and geosphere and their interactions. In this section, we examine each subsystem, its composition, basic properties, and some of its interactions with other components of the climate system. The view of Planet Earth in Figure 1.7, resembling a "blue marble," shows all the major subsystems of the climate system. The ocean, the most prominent feature covering more than two-thirds of Earth's surface, appears blue. Clouds obscure most of the ice sheets (the major part of the cryosphere) that cover much of Greenland and Antarctica. The atmosphere is made visible by swirling storm clouds over the Pacific Ocean near Mexico and the middle of the Atlantic Ocean. Viewed edgewise, the atmosphere appears as a thin, bluish layer. Land (part of the geosphere) is mostly green because of vegetative cover (biosphere).

FIGURE 1.7
Planet Earth, viewed from space by satellite, appears as a "blue marble" with its surface mostly ocean water and partially obscured by swirling masses of clouds. [Courtesy of NASA, Goddard Space Flight Center]

ATMOSPHERE

Earth's **atmosphere** is a relatively thin envelope of gases and tiny suspended particles surrounding the planet. Compared to Earth's diameter, the atmosphere is like the thin skin of an apple. But the thin atmospheric skin is essential for life and the orderly functioning of physical, chemical and biological processes on Earth. While a person can survive for days without water or food, a lack of atmospheric oxygen can be fatal within minutes. Air density decreases with increasing altitude above Earth's surface so that about half of the atmosphere's mass is concentrated within about 5.5 km (3.4 mi) of sea level and 99% of its mass occurs below an altitude of 32 km (20 mi). At altitudes approaching 1000 km (620 mi), Earth's atmosphere merges with the highly rarefied interplanetary gases, hydrogen (H_2) and helium (He).

Based on the vertical temperature profile, the atmosphere is divided into four layers (Figure 1.8). The **troposphere** (averaging about 10 km or 6 mi thick) is where the atmosphere interfaces with the hydrosphere, cryosphere, geosphere, and biosphere and where most weather takes place. In the troposphere, the average air temperature drops with increasing altitude so that it is usually colder on mountaintops than in lowlands (Figure 1.9). The troposphere contains 75% of the atmosphere's mass and 99% of its water. The **stratosphere** (10 to 50 km or 6 to 30 mi above Earth's surface) contains the *ozone*

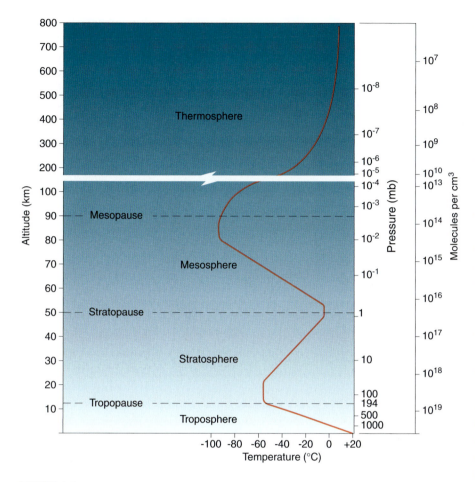

FIGURE 1.8

Based on variations in average air temperature (°C) with altitude (scale on the left), the atmosphere is divided into the troposphere, stratosphere, mesosphere, and thermosphere. Scales on the right show the vertical variation of atmospheric pressure in millibars (mb) (the traditional meteorological unit of barometric pressure) and the number density of molecules (number of molecules per cm³). [Source: US Standard Atmosphere, 1976, NASA, and U.S. Air Force]

FIGURE 1.9

Within the troposphere, the average air temperature decreases with increasing altitude so that it is generally colder on mountain peaks than in lowlands. Snow persists on peaks even through summer.

shield, which prevents organisms from exposure to potentially lethal levels of solar ultraviolet (UV) radiation. Above the stratosphere is the **mesosphere** where the average temperature generally decreases with altitude; above that is the **thermosphere** where the average temperature increases with altitude but is particularly sensitive to variations in the high energy portion of incoming solar radiation.

Nitrogen (N_2) and oxygen (O_2), the chief atmospheric gases, are mixed in uniform proportions up to an altitude of about 80 km (50 mi). Not counting water vapor (with its highly variable concentration), nitrogen occupies 78.08% by volume of the lower atmosphere, and oxygen is 20.95% by volume. The next most abundant gases are argon (0.93%) and carbon dioxide (0.038%). Many other gases occur in the atmosphere in trace concentrations, including ozone (O_3) and methane (CH_4) (Table 1.1). Unlike nitrogen and oxygen, the percent volume of some of these trace gases varies with time and location.

In addition to gases, minute solid and liquid particles, collectively called **aerosols**, are suspended in the atmosphere. A flashlight beam in a darkened room reveals an abundance of tiny dust particles floating in the air. Individually, most atmospheric aerosols are too small to be visible, but in aggregates, such as the multitude of water droplets and ice crystals composing clouds, they may be visible. Most aerosols occur in the lower atmosphere, near their sources on Earth's surface; they derive from wind erosion of soil, ocean spray, forest fires, volcanic eruptions, industrial chimneys, and the exhaust of motor vehicles. Although the concentration of aerosols in the atmosphere is relatively small, they participate in some important processes. Aerosols function as nuclei that promote the formation of clouds essential for the global water cycle. Some aerosols (e.g., volcanic dust, sulfurous particles) affect the climate by interacting with incoming solar radiation and dust blown out over the tropical Atlantic Ocean from North

TABLE 1.1
Gases Composing Dry Air in the Lower Atmosphere (below 80 km)

Gas	% by Volume	Parts per Million
Nitrogen (N_2)	78.08	780,840.0
Oxygen (O_2)	20.95	209,460.0
Argon (Ar)	0.93	9,340.0
Carbon dioxide (CO_2)	0.0388	388.0
Neon (Ne)	0.0018	18.0
Helium (He)	0.00052	5.2
Methane (CH_4)	0.00014	1.4
Krypton (Kr)	0.00010	1.0
Nitrous oxide (N_2O)	0.00005	0.5
Hydrogen (H)	0.00005	0.5
Xenon (Xe)	0.000009	0.09
Ozone (O_3)	0.000007	0.07

Africa may affect the development of tropical cyclones (hurricanes and tropical storms).

The significance of an atmospheric gas is not necessarily related to its concentration. Some atmospheric components that are essential for life occur in very low concentrations. For example, most water vapor is confined to the lowest kilometer or so of the atmosphere and is never more than about 4% by volume even in the most humid places on Earth (e.g., over tropical rainforests and seas). But without water vapor, the planet would have no water cycle, no rain or snow, no ocean, and no fresh water. Also, without water vapor, Earth would be much too cold for most forms of life to exist.

Although comprising only 0.038% of the lower atmosphere, carbon dioxide is essential for photosynthesis. Without carbon dioxide, green plants and the food webs they support could not exist. While the atmospheric concentration of ozone (O_3) is minute, the chemical reactions responsible for its formation (from oxygen) and dissociation (to oxygen) in the stratosphere (mostly at altitudes between 30 and 50 km) shield organisms on Earth's surface from potentially lethal levels of solar UV radiation.

The atmosphere is dynamic; the atmosphere continually circulates in response to different rates of heating and cooling within the rotating planetary system. On an average annual basis, Earth's surface experiences net radiational heating (absorbing more incident radiation than it emits), but the atmosphere undergoes net radiational cooling (to space). Also, net radiational heating occurs in the tropics, while net radiational cooling characterizes higher latitudes. Variations in heating and cooling rates give rise to *temperature gradients*, which are differences in temperature from one location to another. In response to temperature gradients, the atmosphere (and ocean) circulates and redistributes heat within the climate system. Heat is conveyed from warmer locations to colder locations, from Earth's surface to the atmosphere and from the tropics to higher latitudes. As discussed in Chapter 4, the global water cycle and accompanying phase changes of water play an important role in this planetary-scale transport of heat energy.

HYDROSPHERE

The **hydrosphere** is the water component of the climate system. Water is unique among the chemical components of the climate system in that it is the only naturally occurring substance that co-exists in all three phases (solid, liquid, and vapor) at the normal range of temperature and pressure observed near Earth's surface. Water continually cycles among reservoirs within the climate system. (We discuss the global water cycle in more detail in Chapter 5.) The ocean, by far the largest reservoir of water in the hydrosphere, covers about 70.8% of the planet's surface and has an average depth of about 3.8 km (2.4 mi). About 96.4% of the hydrosphere is ocean salt water; other saline bodies of water account for 0.6%. The next largest reservoir in the hydrosphere is glacial ice, most of which covers much of Antarctica and Greenland. Ice and snow make up 2.1% of water in the hydrosphere. Considerably smaller quantities of water occur on the land surface (lakes, rivers), in the subsurface (soil moisture, groundwater), the atmosphere (water vapor, clouds, precipitation), and biosphere (plants, animals).

The ocean and atmosphere are coupled such that the wind drives surface ocean currents. Wind-driven currents are restricted to a surface ocean layer typically about 100 m (300 ft) deep and take a few months to years to cross an ocean basin. Ocean currents at much greater depths are more sluggish and more challenging to study than surface currents because of greater difficulty in taking measurements. Movements of deep-ocean waters are caused primarily by small differences in water density (mass per unit volume) arising from small differences in water temperature and salinity (a measure of dissolved salt content). Cold sea water, being denser than warm water, tends to sink whereas warm water, being less dense, is buoyed upward by (or floats on) colder water. Likewise, saltier water is denser than less salty water and tends to sink, whereas less salty water is buoyed upward. The combination of temperature and salinity determines whether a water mass remains at its original depth or sinks to the ocean bottom. Even though deep currents are relatively slow, they keep ocean waters well mixed so that the ocean has a nearly uniform chemical composition (Table 1.2).

The densest ocean waters form in polar or nearby subpolar regions. Salty waters become even saltier where sea ice forms at high latitudes because growing ice crystals exclude dissolved salts. Chilling of this salty water near Greenland and Iceland and in the Norwegian and Labrador Seas further increases its density so that surface waters sink and form a bottom current that flows southward under equatorial surface waters and into the South Atlantic as far south as Antarctica. Here, deep water from the North Atlantic mixes with deep water around Antarctica. Branches of that cold bottom current then spread northward into the Atlantic, Indian, and Pacific basins. Eventually, the water slowly diffuses to the surface, mainly in the Pacific, and then begins its journey on the surface through the islands of Indonesia, across the Indian Ocean, around South Africa, and into the tropical Atlantic. There, intense heating and evaporation make the water hot and salty. This surface water is then transported northward in the Gulf Stream thereby completing the cycle. This *meridional overturning circulation (MOC)* and its transport of heat energy and salt is an important control of climate.

The hydrosphere is dynamic; water moves continually through different parts of Earth's land-atmosphere-ocean system and the ocean is the ultimate destination of all moving water. Water flowing in river or stream channels may take a few weeks to reach the ocean. Groundwater typically moves at a very slow pace through sediment, and the fractures and tiny openings in bedrock, and feeds into rivers, lakes, or directly into the ocean. The water of large, deep lakes moves even more slowly, in some cases taking centuries to reach the ocean via groundwater flow.

CRYOSPHERE

The frozen portion of the hydrosphere, known as the **cryosphere**, encompasses massive continental (glacial) ice sheets, much smaller ice caps and mountain glaciers, ice in permanently frozen ground (*permafrost*), and the pack ice and ice bergs floating at sea. All of these ice types except pack ice (frozen sea water) and undersea permafrost are fresh water. A **glacier** is a mass of ice that flows internally under the influence of gravity (Figure 1.10). The Greenland and Antarctic ice sheets in places are up to 3 km (1.8 mi) thick. The Antarctic

TABLE 1.2
Comparison of Composition of Ocean Water with River Water[a]

Chemical Constituent	Percentage of Total Salt Content	
	Ocean Water	River Water
Silica (SiO_2)	-	14.51
Iron (Fe)	-	0.74
Calcium (Ca)	1.19	16.62
Magnesium (Mg)	3.72	4.54
Sodium (Na)	30.53	6.98
Potassium (K)	1.11	2.55
Bicarbonate (HCO_3)	0.42	31.90
Sulfate (SO_4)	7.67	12.41
Chloride (Cl)	55.16	8.64
Nitrate (NO_3)	-	1.11
Bromide (Br)	0.20	-
Total	100.00	100.00

[a]Source: U.S. Geological Survey

FIGURE 1.10
Glaciers form in high mountain valleys where the annual snowfall is greater than annual snowmelt.

and atmospheric composition extending as far back as hundreds of thousands of years—to 800,000 years or more in Antarctica (Chapter 9).

Under the influence of gravity, glacial ice flows slowly from sources at higher latitudes and higher elevations (where some winter snow survives the summer) to lower latitudes and lower elevations, where the ice either melts or flows into the nearby ocean. Around Antarctica, streams of glacial ice flow out to the ocean. Ice, being less dense than seawater, floats, forming ice shelves (typically about 500 m or 1600 ft thick). Thick masses of ice eventually break off the shelf edge, forming flat-topped icebergs that are carried by surface ocean currents around Antarctica (Figure 1.11). Likewise, irregularly shaped icebergs break off the glacial ice streams of Greenland and flow out into the North Atlantic Ocean, posing a hazard to navigation. In 1912, the newly launched luxury liner, *RMS Titanic*, struck a Greenland iceberg southeast of Newfoundland and sank with the loss of more than 1500 lives.

Most sea ice surrounding Antarctica forms each winter through freezing of surface seawater. During summer most of the sea ice around Antarctica melts, whereas in the Arctic Ocean sea ice can persist for several years before flowing out through Fram Strait into the Greenland Sea, and eventually melting. This "multi-year" ice loses salt content with age as brine, trapped between ice crystals, melts downward, so that Eskimos can harvest this older, less salty ice for drinking water.

How long is water frozen into glaciers? Glaciers normally grow (thicken and advance) and shrink (thin and retreat) slowly in response to changes in climate. Mountain glaciers respond to climate change on time scales of a decade. Until recently, scientists had assumed that the response time for the Greenland and Antarctic ice sheets is measured in millennia; however, in 2007 scientists reported that two outlet glaciers that drain the

ice sheet contains 90% of all ice on Earth. Much smaller glaciers (tens to hundreds of meters thick) primarily occupy the highest mountain valleys on all continents. At present, glacial ice covers about 10% of the planet's land area but at times during the past 1.7 million years, glacial ice expanded over as much as 30% of the land surface, primarily in the Northern Hemisphere. At the peak of the last glacial advance, about 20,000 to 18,000 years ago, the Laurentide ice sheet covered much of the area that is now Canada and the northern states of the United States. At the same time, a smaller ice sheet buried the British Isles and portions of northwest Europe. Meanwhile, mountain glaciers worldwide thickened and expanded.

Glaciers form where annual snowfall exceeds annual snowmelt. As snow accumulates, the pressure exerted by the new snow converts underlying snow to ice. As the ice forms, it preserves traces of the original seasonal layering of snow and traps air bubbles. Chemical analysis of the ice layers and air bubbles in the ice provides clues to climatic conditions at the time the original snow fell. Ice cores extracted from the Greenland and Antarctic ice sheets yield information on changes in Earth's climate

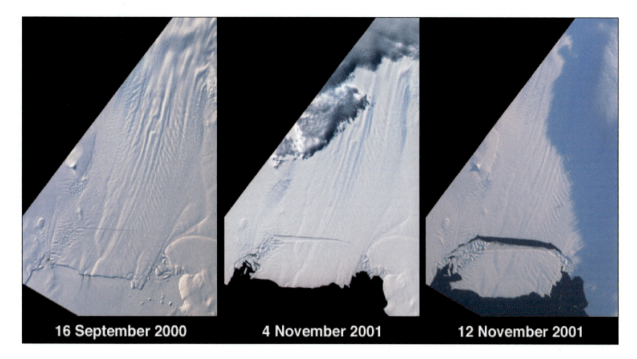

16 September 2000   4 November 2001   12 November 2001

FIGURE 1.11
A massive iceberg (42 km by 17 km or 26 mi by 10.5 mi) is shown breaking off Pine Island Glacier, West Antarctica (75 degrees S, 102 degrees W) in early November 2001 along a large fracture that formed across the glacier in mid 2000. Images of the glacier were obtained by the Multi-angle Imaging SpectroRadiometer (MISR) instrument onboard NASA's Terra spacecraft. Pine Island Glacier is the largest discharger of ice in Antarctica and the continent's fastest moving glacier. [Courtesy of NASA]

Greenland ice sheet exhibited significant changes in discharge in only a few years. This finding was confirmed by changes in ice surface elevation detected by sensors onboard NASA's Ice, Cloud, and Land Elevation Satellite (ICESat). This unexpectedly rapid discharge is likely due to the flow of large ice streams over subglacial lakes. Hence, outlet glaciers behave more like mountain glaciers, raising questions regarding the long-term stability of polar ice sheets and their response to global climate change (Chapter 12).

GEOSPHERE

The **geosphere** is the solid portion of the planet consisting of rocks, minerals, soil, and sediments. Most of Earth's interior cannot be observed directly, the deepest mines and oil wells do not penetrate the solid Earth more than a few kilometers. Most of what is known about the composition and physical properties of Earth's interior comes from analysis of seismic waves generated by earthquakes and explosions. In addition, meteorites provide valuable clues regarding the chemical composition of Earth's interior. From study of the behavior of seismic waves that penetrate the planet, geologists have determined that Earth's interior consists of four spherical shells: crust, mantle, and outer and inner cores (Figure 1.12). Earth's

interior is mostly solid and accounts for much of the mass of the planet. Earth's outermost solid skin, called the *crust*, ranges in thickness from only 8 km (5 mi) under the ocean to 70 km (45 mi) in some mountain belts. We live on the crust and it is the source of almost all rock, mineral, and fossil fuel (e.g., coal, oil, and natural gas) resources that are essential for industrial-based economies. The rigid uppermost portion of the *mantle*, plus the overlying crust, constitutes Earth's **lithosphere**, averaging 100 km (62 mi) thick. Both surface geological processes and internal geological processes continually modify the lithosphere.

Surface geological processes encompass weathering and erosion occurring at the interface between the lithosphere (mainly the crust) and the other Earth subsystems. **Weathering** entails the physical disintegration, chemical decomposition, or solution of exposed rock. Rock fragments produced by weathering are known as *sediments*. Water plays an important role in weathering by dissolving soluble rock and minerals, and participating in chemical reactions that decompose rock. Water's unusual physical property of expanding while freezing can produce sufficient pressure to fragment rock when the water saturates tiny cracks and pore spaces. More likely, however, the water is not as confined and fragmentation is due to stress caused by the growth of ice lenses within the rock.

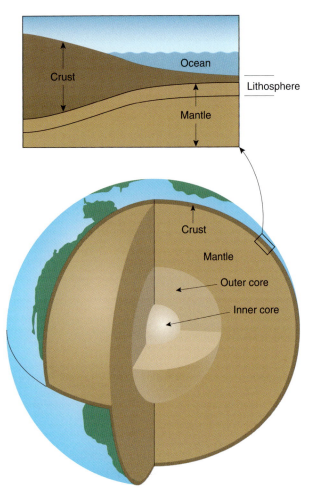

FIGURE 1.12
Earth's interior is divided into the crust, mantle, outer core, and inner core. The lithosphere is the rigid upper portion of the mantle plus the overlying crust. (Drawing is not to scale.)

The ultimate weathering product is *soil*, a mixture of organic (humus) and inorganic matter (sediment) on Earth's surface that supports plants, also supplying nutrients and water. Soils derive from the weathering of bedrock or sediment, and vary widely in texture (particle size). A typical soil is 50% open space (pores), roughly equal proportions of air and water. Plants also participate in weathering via the physical action of their growing roots and the carbon dioxide they release to the soil.

Erosion refers to the removal and transport of sediments by gravity, moving water, glaciers, and wind. Running water and glaciers are pathways in the global water cycle. Erosive agents transport sediments from source regions (usually highlands) to low-lying depositional areas (e.g., ocean, lakes). Weathering aids erosion by reducing massive rock to particles that are sufficiently small to be transported by agents of erosion. Erosion aids weathering by removing sediment and exposing fresh surfaces of rock

to the atmosphere and weathering processes. Together, weathering and erosion work to reduce the elevation of the land.

Internal geological processes counter surface geological processes by uplifting land through tectonic activity, including volcanism and mountain building. Most tectonic activity occurs at the boundaries between lithospheric plates. The lithosphere is broken into a dozen massive plates (and many smaller ones) that are slowly driven (typically less than 20 cm per year) across the face of the globe by huge convection currents in Earth's mantle. Continents are carried on the moving plates and ocean basins are formed by seafloor spreading.

Plate tectonics probably has operated on the planet for at least 3 billion years, with continents periodically assembling into supercontinents and then splitting apart. The most recent supercontinent, called *Pangaea* (Greek for "all land"), broke apart about 200 million years ago and its constituent landmasses, the continents of today, slowly moved to their present locations. Plate tectonics explains such seemingly anomalous discoveries as glacial sediments in the Sahara and fossil coral reefs, indicative of tropical climates, in northern Wisconsin (Figure 1.13). Such discoveries reflect climatic conditions hundreds of millions of years ago when the continents were at different latitudes than they are today.

Geological processes occurring at boundaries between plates produce large-scale landscape and ocean bottom features, including mountain ranges, volcanoes, deep-sea trenches, as well as the ocean basins themselves.

FIGURE 1.13
This exposure of bedrock in northeastern Wisconsin contains fossil coral that dates from nearly 400 million years ago. Based on the environmental requirements of modern coral, geoscientists conclude that 400 million years ago, Wisconsin's climate was tropical marine, a drastic difference from today's warm-summer, cold-winter climate. Plate tectonics can explain this difference between ancient and modern environmental conditions.

Enormous stresses develop at plate boundaries, bending and fracturing bedrock over broad areas. Hot molten rock material, known as **magma**, wells up from deep in the crust or upper mantle and migrates along rock fractures. Some magma pushes into the upper portion of the crust where it cools and solidifies into massive bodies of rock, forming the core of mountain ranges (e.g., Sierra Nevada). Some magma feeds volcanoes or flows through fractures in the crust and spreads over Earth's surface as lava flows (flood basalts) that cool and slowly solidify (e.g., Columbia River Plateau in the Pacific Northwest and the massive Siberian Traps). At spreading plate boundaries on the sea floor, upward flowing magma solidifies into new oceanic crust. Plate tectonics and associated volcanism are important in geochemical cycling, releasing to the atmosphere water vapor, carbon dioxide, and other gases that impact climate.

Volcanic activity is not confined to plate boundaries. Some volcanic activity occurs at great distances from plate boundaries and is due to hot spots in the mantle. A **hot spot** is a long-lived source of magma caused by rising plumes of hot material originating in the mantle (*mantle plumes*). Where a plate is situated over a hot spot, magma may break through the crust and form a volcano. The Big Island of Hawaii is volcanically active because it sits over a hot spot located within the mantle under the Pacific plate. A hot spot underlying Yellowstone National Park is the source of heat for geyser eruptions (including Old Faithful). Further complicating matters, however, both hot spots and the overlying lithospheric plate are in motion. Sometimes hot spots and spreading centers coincide, such as in Iceland.

BIOSPHERE

All living plants and animals on Earth are components of the **biosphere** (Figure 1.14). They range in size from microscopic single-celled bacteria to the largest organisms (e.g., redwood trees and blue whales). Bacteria and other single-celled organisms dominate the biosphere, both on land and in the ocean. The typical animal in the ocean is the size of a mosquito. Large, multi-cellular organisms (including humans) are relatively rare on Earth. Organisms on land or in the atmosphere live close to Earth's surface. However, marine organisms occur throughout the ocean depths and even inhabit rock fractures, volcanic vents, and the ocean floor. Certain organisms live in *extreme environments* at temperatures and pressures once considered impossible to support life. In fact, some scientists estimate that the mass of organisms living in fractured rocks on and below the ocean floor may vastly exceed the mass of organisms living on or above it.

Photosynthesis and cellular respiration are essential for life near the surface of the Earth, and exemplify how the biosphere interacts with the other subsystems of

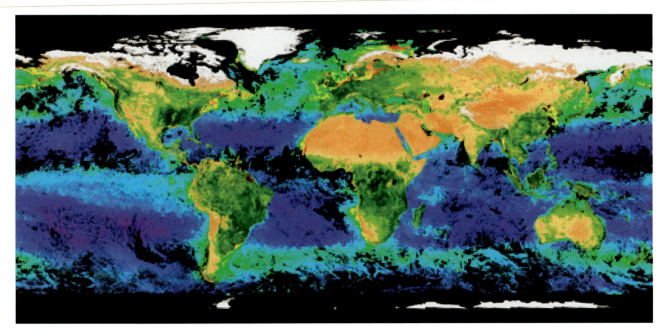

FIGURE 1.14
Earth's biosphere viewed by instruments flown onboard NASA's SeaWiFS (Sea-viewing Wide Field-of-view Sensor) on the SeaStar satellite launched in August 1997. Biological production is color-coded and highest where it is dark green and lowest where it is violet. White indicates snow and ice cover. [Provided by the SeaWiFS Project, NASA/Goddard Space Flight Center and ORBIMAGE]

the climate system. **Photosynthesis** is the process whereby green plants convert light energy from the Sun, carbon dioxide from the atmosphere, and water to sugars and oxygen (O_2). The sugars, which contain a relatively large amount of energy and oxygen, are essential for cellular respiration. Through **cellular respiration**, an organism processes food and liberates energy for maintenance, growth, and reproduction, also releasing carbon dioxide, water, and heat energy to the environment. With few exceptions, sunlight is the originating source of energy for most organisms living on land and in the ocean's surface waters.

Dependency between organisms on one another (e.g., as a source of food) and on their physical and chemical environment (e.g., for water, oxygen, carbon dioxide, and habitat) is embodied in the concept of *ecosystem*. Recall from earlier in this chapter that ecosystems consist of plants and animals that interact with one another, together with the physical conditions and chemical substances in a specific geographical area. An ecosystem is home to producers (plants), consumers (animals), and decomposers (bacteria, fungi). **Producers** (also called *autotrophs* for "self-nourishing") form the base of most ecosystems, providing energy-rich carbohydrates. **Consumers** that depend directly or indirectly on plants for their food are called *heterotrophs*; those that feed directly on plants are called *herbivores*, and those that prey on other animals are called *carnivores*. Animals that consume both plants and animals are *omnivores*. After death, the remains of organisms are broken down by *decomposers*, usually bacteria and fungi, which cycle nutrients back to the environment, for the plants to use.

Feeding relationships among organisms, called a **food chain**, can be quite simple. For example, in a land-based (terrestrial) food chain, deer (herbivores) eat plants (primary producers), and the wolves (carnivores) eat the deer. In

a food chain, each stage is called a *trophic level* (or feeding level). At most, only 10% of the energy available at one trophic level is transferred to the next. *Biomass*, the total weight or mass of organisms, is more readily measured than energy, so that scientists describe the transfer of energy in food chains in terms of so many grams or kilograms of biomass. Thus 100 g of plants are required to produce 10 g of deer, which in turn produces 1 g of wolf. Terrestrial and marine food chains are often more complex than our plants-deer-wolves example. With some notable exceptions, marine and terrestrial organisms eat many different kinds of food, and in turn, are eaten by a host of other consumers. These more realistic feeding relationships constitute a **food web**.

Climate is the principal ecological control, largely governing the location and species composition of natural ecosystems such as deserts, rain forests, and tundra. For example, the late climatologist Reid A. Bryson (1920-2008) demonstrated a close relationship between the region dominated by cold, dry arctic air and the location of Canada's coniferous boreal forest (Figure 1.15). Bryson found that the southern boundary of the

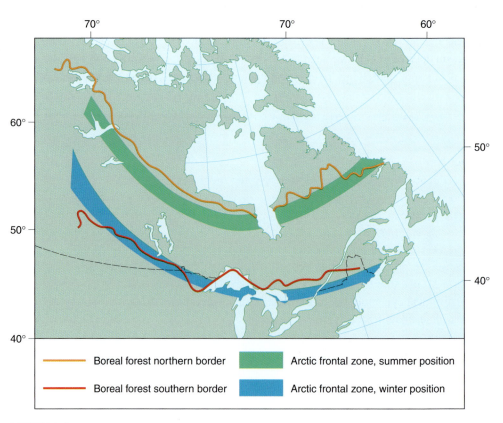

FIGURE 1.15
The northern border of Canada's coniferous boreal forest closely corresponds to the mean location of the leading edge of arctic air in summer. The southern boundary of the boreal forest nearly coincides with the mean location of arctic air in winter. The leading edge of arctic air is referred to as the arctic front. [Modified after R.A. Bryson, 1966. "Air Masses, Streamlines, and the Boreal Forest," *Geographical Bulletin* 8(3):266.]

boreal forest nearly coincides with the average winter position of the southern edge of the arctic air mass (the *arctic front*) while the boreal forest's northern border closely corresponds to the average summer position of the arctic front.

Assuming that the relationship between arctic air frequency and the boreal forest is more cause/effect than coincidence, how might a large-scale climate change affect the forest? A warmer climate would likely mean fewer days of arctic air and a northward shift of the boreal forest. What actually happens to the forest, however, could hinge on the rate of climate change. Relatively rapid warming may not only shift the ecosystem northward but also alter the ecosystem's species composition and disturb the orderly internal operation of the ecosystem. For example, rapid climate change could disrupt long-established predator/prey relationships with implications for the stability of populations of plants and animals.

Similar observations of close relationships between vegetation and climate variables on a global basis were made by the noted German climatologist Wladimir Köppen (1846-1940) in the early 20th century. This is a central aspect of his widely used climate classification system (Chapter 13). We have more to say on this topic in Chapter 12 along with the potential impact of climate change on the highly simplified agricultural ecosystems.

Subsystem Interactions: Biogeochemical Cycles

Biogeochemical cycles are the pathways along which solids, liquids, and gases move among the various reservoirs on Earth, often involving physical or chemical changes to these substances. Accompanying these flows of materials are transfers and transformations of energy. Reservoirs in these cycles are found within the subsystems of the overall planetary system (atmosphere, hydrosphere, cryosphere, geosphere, and biosphere). Examples of biogeochemical cycles are the water cycle, carbon cycle, oxygen cycle, and nitrogen cycle.

Earth is an open (or flow-through) system for energy, where *energy* is defined as the capacity for doing work. Earth receives energy from the Sun primarily and some from its own interior while emitting energy in the form of invisible infrared radiation to space. Along the way, energy is neither created nor destroyed, although it is converted from one form to another. This is the **law of energy conservation** (also known as the *first law of thermodynamics*).

The Earth system is essentially closed for matter; that is, it neither gains nor loses matter over time (except for meteorites and asteroids). All biogeochemical cycles obey the *law of conservation of matter*, which states that matter can be neither created nor destroyed, but can change in chemical or physical form. When a log burns in a fireplace, a portion of the log is converted to ash and heat energy, while the rest goes up the chimney as carbon dioxide, water vapor, creosote and heat. In terms of accountability, all losses from one reservoir in a cycle can be accounted for as gains in other reservoirs of the cycle. Stated succinctly, for any reservoir:

$$\text{Input} = \text{Output} + \text{Storage}$$

The quantity of a substance stored in a reservoir depends on the rates at which the material is cycled into and out of the reservoir. **Cycling rate** is defined as the amount of material transferred from one reservoir to another within a specified period of time. If the input rate exceeds the output rate, the amount of material stored in the reservoir increases. If the input rate is less than the output rate, the amount stored decreases. Over the long term, the cycling rates of materials among the various global reservoirs are relatively stable; that is, equilibrium tends to prevail between the rates of input and output.

Closely related to cycling rate is residence time. **Residence time** is the average length of time for a substance in a reservoir to be replaced completely, that is,

$$\text{Residence time} = \frac{\text{(amount in reservoir)}}{\text{(rate of addition or removal)}}$$

For example, the residence time of a water molecule in the various reservoirs of Earth's land-atmosphere-ocean system varies from only 10 days in the atmosphere to tens of thousands of years, or longer, in glacial ice sheets. Residence time of dissolved constituents of seawater ranges from 100 years for aluminum (Al) to 260 million years for sodium (Na). Consider the global cycling of carbon as an illustration of a biogeochemical cycle that has important implications for climate (Figure 1.16). Through photosynthesis, carbon dioxide cycles from the atmosphere to green plants where carbon is incorporated into sugar ($C_6H_{12}O_6$). Plants use sugar to manufacture other organic compounds including fats, proteins, and other carbohydrates. As a byproduct of cellular respiration, plants and animals transform a portion of the carbon in these organic compounds into CO_2 that is released to the atmosphere. In the ocean, CO_2 is cycled into and out of marine organisms through photosynthesis and

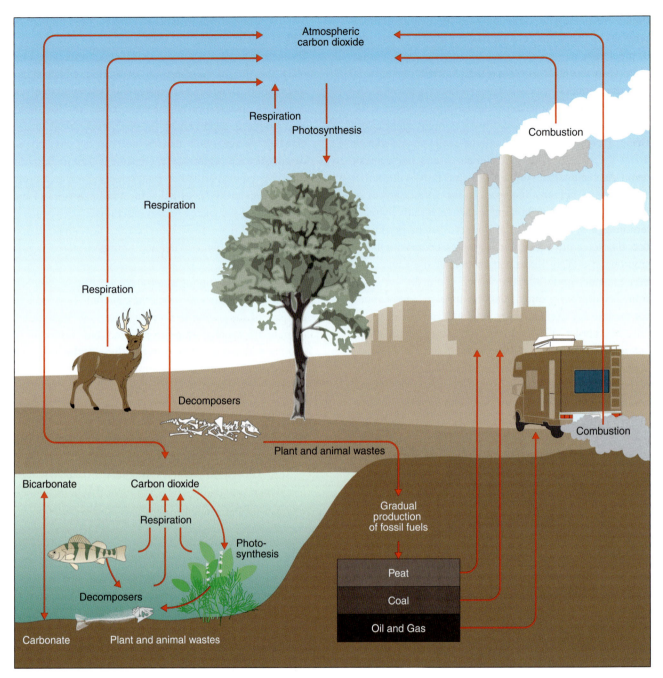

FIGURE 1.16
Schematic representation of the global carbon cycle.

respiration. In addition to the uptake of CO_2 via photosynthesis, marine organisms also use carbon for calcium carbonate ($CaCO_3$) to make hard, protective shells. Furthermore, decomposer organisms (e.g., bacteria) act on the remains of dead plants and animals, releasing CO_2 to the atmosphere and ocean through cellular respiration.

When marine organisms die, their remains (shells and skeletons) slowly settle downward through ocean waters. In time, these organic materials reach the sea floor, accumulate, are compressed by their own weight and the weight of other sediments, and gradually transform into solid, carbonate rock. Common carbonate rocks are limestone ($CaCO_3$) and dolostone ($CaMg(CO_3)_2$). Subsequently, tectonic processes uplift these marine rocks and expose them to the atmosphere and weathering processes. Rainwater contains dissolved atmospheric CO_2 producing carbonic acid (H_2CO_3) that, in turn, dissolves carbonate rock releasing CO_2. As part of the global water cycle, rivers and streams transport these weathering products to the sea where they settle out of suspension or precipitate as sediments that accumulate on the ocean floor. Over the millions of years that constitute geologic time, the formation and ultimate weathering and erosion of carbon-containing rocks have significantly altered the concentration of carbon dioxide in the atmosphere thereby changing the climate.

From about 280 to 345 million years ago, the geologic time interval known as the Carboniferous period, trillions of metric tons of organic remains (detritus) accumulated on the ocean bottom and in low-lying swampy terrain on land. The supply of detritus was so great that decomposer organisms could not keep pace. In some marine environments, plant and animal remains were converted to oil and natural gas. In swampy terrain, heat and pressure from accumulating organic debris concentrated carbon, converting the remains of luxuriant swamp forests into thick layers of coal. Today, when we burn coal, oil, and natural gas, collectively called **fossil fuels**, we are tapping energy that was originally locked in vegetation through photosynthesis hundreds of millions of years ago. During combustion, carbon from these fossil fuels combines with oxygen in the air to form carbon dioxide which escapes to the atmosphere.

Another important biogeochemical cycle operating in the Earth system is the global water cycle (Chapter 5), which is closely linked to all other biogeochemical cycles. Reservoirs in the water cycle (hydrosphere, atmosphere, geosphere, biosphere) are also reservoirs in other cycles, for which water is an essential mode of transport. In the nitrogen cycle, for example, intense heating of air caused by lightning combines atmospheric nitrogen (N_2), oxygen (O_2), and moisture to form droplets of extremely dilute nitric acid (HNO_3) that are washed by rain to the soil. In the process, nitric acid converts to nitrate (NO_3^-), an important plant nutrient that is taken up by plants via their root systems. Plants convert nitrate to ammonia (NH_3), which is incorporated into a variety of compounds, including amino acids, proteins, and DNA. On the other hand, both nitrate and ammonia readily dissolve in water so that heavy rains can deplete soil of these important nutrients and wash them into waterways.

The components of Earth's climate system co-evolved through geologic time. For more on this topic, refer to this chapter's first Essay. At times in the past, Earth's climate underwent massive changes that brought about large-scale extinctions of plants and animals. For more on this, see this chapter's second Essay.

The Climate Paradigm

The climate system determines Earth's climate as the result of mutual interactions among the atmosphere, hydrosphere, cryosphere, geosphere, and biosphere and responses to external influences from space. As the composite of prevailing weather patterns, climate's complete description includes both the average state of the atmosphere and its variations. Climate can be explained primarily in terms of the complex redistribution of heat energy and matter by Earth's coupled atmosphere/ocean system. It is governed by the interaction of many factors, causing climate to differ from one place to another and to vary on time scales from seasons to millennia. The range of climate, including extremes, places limitations on living organisms and a region's habitability.

Climate is inherently variable and now appears to be changing at rates unprecedented in relatively recent Earth history. Human activities, especially those that alter the composition of the atmosphere or characteristics of Earth's surface, play an increasingly important role in the climate system. Rapid climate changes, natural or human-caused, heighten the vulnerabilities of societies and ecosystems, impacting biological systems, water resources, food production, energy demand, human health, and national security. These vulnerabilities are global to local in scale, and call for increased understanding and surveillance of the climate system and its sensitivity to imposed changes. Scientific research focusing on key climate processes, expanded monitoring, and improved modeling capabilities are already increasing our ability to predict the future climate. Although incomplete, our current understanding of the climate system and the far-reaching risks associated with climate change call for the immediate preparation and implementation of strategies for sustainable development and long-term stewardship of Earth.[2]

Conclusions

Climate can be defined in terms of empirically derived statistical summaries based on the instrument record, specifying mean, median and extreme values of various climatic elements such as temperature and precipitation. Alternately, climate can be defined in terms of the dynamic forces that shape the climate system and its spatial and temporal variability. These two definitions

---

[2] For a timeline of key historical events in climate science, see Appendix II.

of climate (*empirical* and *dynamic*) are actually two sides of the same coin and both are utilized in our study and application of climate science (Figure 1.17). Our primary motivation for studying climate science is the link between climate and society. Society influences and is influenced by climate. By developing our basic understandings of climate science, we position ourselves to better understand the public policy and economic dimensions of climate change. In this chapter, we have seen that a central concept in this understanding is the climate system and the interaction of its component subsystems. In the next chapter, we continue building our climate science framework with a focus on spatial and temporal scales of climate, interactions of climate elements, climate models, and monitoring of the climate system.

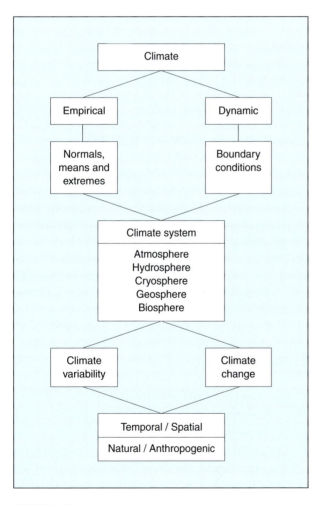

FIGURE 1.17
This flow chart identifies the major components of our framework for studying climate science.

Basic Understandings

- The study of Earth's climate began with the ancient Greek philosophers and geographers. The first climate classification, devised by Parmenides in the 5th century BCE, was based on latitudinal variations in sunshine that accompany regular changes in the angle of inclination of the Sun.

- Weather is defined as the state of the atmosphere at some place and time, described in terms of such variables as temperature, humidity, and precipitation. Meteorology is the study of the atmosphere, processes that cause weather, and the life cycle of weather systems.

- One definition of climate is empirical (based on statistical summaries) whereas another is dynamic (incorporating the governing forces). The first describes a climate state, while the second seeks to explain climate.

- With the empirical definition, climate is weather at some locality averaged over a specified time period plus extremes in weather during the same period. By international convention, normals of climatic elements are computed for a 30-year period beginning with the first year of a decade. At the close of the decade, the averaging period is moved forward 10 years. The 30-year averaging period of 1971-2000 is the reference for the first decade of the 21st century.

- With the dynamic definition, climate encompasses the boundary conditions in the planetary system (that is, the planetary system). These boundary conditions select the array of weather patterns that characterize each of the seasons. Climatology is the study of climate, its controls, and spatial and temporal variability.

- The climatic norm or normal often is equated to the average value of some climatic element such as temperature over a defined 30-year interval. More precisely, the climatic norm of some locality encompasses the total variation in the climate record, that is, both averages plus extremes. Establishing representative norms goes beyond arithmetical averages as the mean value of a climatic element may not be the same as the median value.

- In the second decade of the 19th century, the Army Medical Department was first to establish a national network of weather and climate observing stations. By the mid-1800s, Joseph Henry formed a national network of volunteer citizen observers who mailed monthly weather reports to the Smithsonian. Invention of the electric telegraph enabled Henry to obtain simultaneous weather reports and to draw the first weather maps.

- On 1 November 1870, the U.S. Army Signal Corps began operating a telegraph-linked storm warning network for the Great Lakes. Soon the network spread to other parts of the nation and encompassed networks operated by the Army Medical Department, the Smithsonian, and others. The Signal Corps was the predecessor to the U.S. Weather Bureau and today's National Weather Service (NWS).

- Derived from the old weather/climate networks operated by the Army Medical Department, the Smithsonian Institution, and the Army Signal Corps is the NWS Cooperative Observer Network. More than 11,000 volunteers record daily precipitation and maximum/minimum temperature for climatic, hydrologic, and agricultural purposes.

- Climate provides resources that can be exploited to the benefit of society as well as imposing constraints on social and economic development. It is not possible to weather- or climate-proof society to prevent damage to life and property. In the agricultural sector of the developed world, the prevailing strategy is to depend on technology to circumvent climate constraints.

- Human activity is influencing climate with significant consequences for society. In addition, some human activities are making society and ecosystems more vulnerable to climate change. Examples include the rapid growth of human population in the coastal zone and the migration of people to areas that are climatically marginal for agriculture.

- An important consideration regarding weather and climate extremes is societal resilience, that is, the ability of a society to recover from a weather- or climate-related or other natural disaster. Assessment of societal resilience must consider the regional bias of severe weather and climate extremes and the technological capabilities of a given society.

- A system is an entity whose components interact in an orderly way according to natural laws.

Earth's climate system consists of the following interacting subsystems: atmosphere, hydrosphere, cryosphere, geosphere and biosphere.

- Earth's atmosphere is a relatively thin envelope of gases and tiny suspended solid and liquid particles (aerosols) surrounding the solid planet. Based on the average vertical temperature profile, the atmosphere is divided into the troposphere, stratosphere, mesosphere, and thermosphere. The lowest layer, the troposphere, is where most weather occurs and where the atmosphere interfaces with the other subsystems of the climate system.

- Nitrogen (N_2) and oxygen (O_2), the principal atmospheric gases, are mixed in uniform proportions up to an altitude of about 80 km (50 mi). Not counting water vapor (which has a highly variable concentration), nitrogen occupies 78.08% by volume of the lower atmosphere and oxygen is 20.95% by volume.

- The significance of atmospheric gases and aerosols is not necessarily related to their concentration. Some atmospheric components that are essential for life occur in very low concentrations. Examples are water vapor (needed for the global water cycle), carbon dioxide (for photosynthesis), and stratospheric ozone (protection from solar ultraviolet radiation).

- The atmosphere is dynamic and circulates in response to temperature gradients that arise from differences in rates of radiational heating and radiational cooling within the land-atmosphere-ocean system.

- The hydrosphere consists of water in all three phases (solid, liquid, and vapor) that continually cycles among reservoirs in the planetary system. The ocean is the largest reservoir in the hydrosphere, containing 97.2% of all water on the planet and covering 70.8% of Earth's surface.

- The ocean features wind-driven surface currents and density-driven deep currents caused by small differences in temperature and salinity. An important control of climate is the meridional overturning circulation (MOC).

- The hydrosphere is dynamic, with water flowing at different rates through and between different reservoirs within the climate system. The time required for water to reach the ocean varies from days to weeks in river channels and through millennia for water locked in glacial ice sheets.

- In addition to the Antarctic and Greenland ice sheets, the frozen portion of Earth's hydrosphere, called the cryosphere, encompasses mountain glaciers, permafrost, sea ice (frozen seawater), and ice bergs.

- The geosphere is the solid portion of the planet composed of rocks, minerals, soils and sediments. The rigid uppermost portion of Earth's mantle plus the overlying crust, constitutes Earth's lithosphere. Surface geological processes (i.e., weathering and erosion) and internal geological processes (i.e., mountain building, volcanic eruptions) continually modify the lithosphere. Weathering refers to the physical and chemical breakdown of rock into sediments. Agents of erosion (i.e., rivers, glaciers, wind) remove, transport, and subsequently deposit sediments.

- Plate tectonics is responsible for the slow movement of continents across the face of the Earth, mountain building, and volcanism. These processes can explain climate change operating over hundreds of millions of years.

- The biosphere encompasses all life on Earth and is dominated on land and in the ocean by bacteria and single-celled plants and animals. Photosynthesis and cellular respiration are processes that are essential for life where sunlight is available and exemplify the interaction of the biosphere with the other subsystems of the climate system.

- The biosphere is composed of ecosystems, communities of plants and animals that interact with one another, together with the physical conditions and chemical substances in a specific geographical area. Ecosystems consist of producers (plants), consumers (animals), and decomposers (bacteria, fungi). These organisms occupy different (ascending) trophic levels in food chains.

- Climate is the principal ecological control, largely determining the location and species composition of natural ecosystems such as deserts, rain forests, and tundra.

- Biogeochemical cycles are pathways along which solids, liquids, or gases flow among the various reservoirs within subsystems of the planetary system.

- Biogeochemical cycles follow the law of energy conservation, which states that energy is neither created nor destroyed although it is converted from one form to another. Biogeochemical cycles also follow the law of conservation of matter, which states that matter can neither be created nor destroyed, but can change chemical or physical form.

- The time required for a unit mass of some substance to cycle into and out of a reservoir is the residence time of the substance in the reservoir.

Enduring Ideas

- The empirical definition of climate is based on statistical summaries of climatic elements whereas the dynamic definition incorporates the boundary conditions in the planetary system coupled with typical seasonal weather patterns.
- The climatic norm encompasses the total variability in the climate record, that is, both averages plus extremes in weather.
- Earth's climate system consists of the atmosphere, hydrosphere, cryosphere, geosphere, and biosphere that are linked by biogeochemical cycles.
- Climate imposes constraints on social and economic development by governing such essentials as fresh water supply and energy needs for space heating and cooling.

Review

1. Provide some examples of how climate operates as the principal environmental control.
2. Define weather and explain why a place and time must be specified when describing the weather.
3. How does the empirical definition of climate differ from the dynamic definition of climate?
4. Define what is meant by the climatic norm.
5. How does the operational weather observing network compare with the cooperative observer network in terms of types of data collected?
6. Identify some of the linkages between climate and society.
7. What is the significance of Earth's troposphere?
8. Under what climatic conditions would a glacier form?
9. What is the basic composition and structure of Earth's geosphere?
10. Distinguish between photosynthesis and cellular respiration. What role do these two processes play in the global carbon cycle?

Critical Thinking

1. Identify two climate controls that operate external to the land-atmosphere-ocean system.
2. In describing the climate of some locality, of what value is the record of weather extremes?
3. What are some disadvantages of computing averages of climatic elements based on a 30-year period?
4. In a study of climate change, why is it preferable to consider climate records only from stations that have not relocated?
5. Provide some examples of how the significance of an atmospheric gas is not necessarily related to its concentration.
6. Speculate on how a glacial ice sheet influences the climate.
7. What roles might plate tectonics and volcanic eruptions play in climate change?
8. How does the law of energy conservation apply to biogeochemical cycles?
9. In a biogeochemical cycle, what is the relationship between cycling rates and residence time?
10. What roles are played by water in biogeochemical cycles?

ESSAY: Evolution of Earth's Climate System

The components of Earth's climate system (atmosphere, hydrosphere, cryosphere, geosphere, and biosphere) co-evolved through the vast expanse of Earth history. According to astronomers, more than 4.5 billion years ago, Earth, the Sun, and the entire solar system evolved from an immense rotating cloud of dust, ice and gases, called a *nebula* (Figure 1). Temperature, density, and pressure were highest at the center of the nebula, gradually decreasing toward its outer limits. Extreme conditions at the nebula's center vaporized ice and light elements and drove them toward the nebula's outer reaches. Consequently, residual dry rocky masses formed the inner planets (including Earth). Farther out, meteorites and the less-dense giant planets Saturn and Jupiter formed.

FIGURE 1
The leftmost "pillar" of interstellar hydrogen gas and dust in M16, the Eagle Nebular. [Courtesy of NASA/NSSDC Photo Gallery]

Earth is known as the water planet—ocean water covers almost 71% of its surface. Yet, in view of how the solar system is believed to have formed, it is surprising that even that much water is present on Earth. Where did the hydrosphere come from? Scientists do not have a complete explanation but a popular hypothesis attributes water on Earth to the bombardment of the planet by comets and/or *planetesimals*, large meteorites a few kilometers across. While meteorites are about 10% ice by mass and the giant planets contain some ice, most of the water in the nebula condensed in comets at distances beyond Saturn and Jupiter. A *comet* is a relatively small mass composed of meteoric dust and ice that moves in a parabolic or highly elliptical orbit around the Sun.

Comets are about half ice. During the latter stages of Earth's formation, comets impacted the planet's surface forming a veneer of water. Jupiter's strengthening gravitational attraction may have drawn a multitude of ice-rich comets from the outer to the inner reaches of the solar system on a collision course with Earth. This hypothesis remained popular until scientists discovered that water on Earth and ice in comets are not chemically equivalent. Spectral analyses of three comets that approached Earth in recent years revealed that comet ice contains about twice as much deuterium as the water on Earth. *Deuterium* is an isotope of hydrogen whose nucleus is composed of one proton plus one neutron and is very rare on Earth; a normal hydrogen atom consists of a single proton. Based on this finding, some scientists suggest that comets accounted for no more than half the water on Earth and perhaps much less. The water in planetesimals, on the other hand, contains less deuterium than comet ice; they may have impacted Earth during the latter stages of the planet's formation. However, the ratio of some other chemical components of planetesimals is not the same as the ratio of those components on Earth. Another possibility is that Earth's water is indigenous; that is, the center of the nebula may have been cooler than previously assumed and some of the materials present in the inner solar system that formed Earth were water-rich.

In the beginning, Earth's atmosphere probably was mostly hydrogen (H_2) and helium (He) plus some hydrogen compounds, including methane (CH_4) and ammonia (NH_3). Because these atoms and molecules are relatively light and have high molecular speeds, Earth's weak gravitational field plus high temperatures allowed this early atmosphere to escape to space. In time, however, volcanic activity began spewing huge quantities of lava, ash, and gases. By about 4.4 billion years ago, the strength of the planet's gravitational field was sufficient to retain a thin gaseous envelope of volcanic origin, Earth's primeval atmosphere.

The principal source of Earth's atmosphere was *outgassing* from the geosphere, that is, the release of gases from rock through volcanic eruptions and the impact of meteorites on the planet's rocky surface. Perhaps as much as 85% of all outgassing took place within a million or so years of the planet's formation while outgassing continues to this day, although at a slower pace. The primeval atmosphere was mostly carbon dioxide, with some nitrogen (N_2) and water vapor (H_2O), along with trace amounts of methane, ammonia, sulfur dioxide (SO_2), and hydrochloric acid (HCl). Radioactive decay of an isotope of potassium in the planet's bedrock added argon (Ar), an inert (chemically non-reactive) gas, to the evolving mix of atmospheric gases. Dissociation of water vapor into its constituent atoms, hydrogen and oxygen, by high-energy solar ultraviolet radiation contributed a small amount of free oxygen to the primeval atmosphere. (The lighter hydrogen—with its relatively high molecular speeds—escaped to space.) Also, some oxygen combined with other elements in various chemical compounds, such as carbon dioxide.

Scientists suggest that between 4.5 and 2.5 billion years ago, the Sun was about 30% fainter than it is today. This did not mean a cooler planet, however, because the atmosphere was 10 to 20 times denser than the present one. Carbon dioxide slows the escape of Earth's heat to space, contributing to average surface temperatures that were as high as 85 °C to 110 °C (185 °F to 230 °F), levels significantly higher than currently observed (approximately 15 °C or 59 °F).

By 4 billion years ago, the planet began to cool and the Earth system underwent major changes. Cooling caused atmospheric water vapor to condense into clouds that produced rain. Precipitation plus runoff from landmasses gave rise to the ocean that eventually covered as much as 95% of the planet's surface. The global water cycle (which helped cool the Earth's surface through evaporation) and its largest reservoir (the ocean) were in place. Rains also helped bring about a substantial decline in the concentration of atmospheric CO_2. As noted elsewhere in this chapter, CO_2 dissolves in rainwater, producing weak carbonic acid that reacts chemically with bedrock. The net effect of this large-scale geochemical process was increasing amounts of carbon chemically locked in rocks and minerals with less and less CO_2 remaining in the atmosphere. The physical and chemical breakdown of rock (*weathering*) plus erosion on land delivered some carbon-containing sediment to the ocean. Also, rains washed dissolved CO_2 directly into the sea, and some atmospheric CO_2 dissolved in ocean water as sea surface temperatures fell. (Carbon dioxide is more soluble in cold water.)

Although CO_2 has been a minor component of the atmosphere for at least 3.5 billion years, its concentration has fluctuated during the geologic past, with important implications for global climate and life on Earth. All other factors being equal, more CO_2 in the atmosphere means an enhanced greenhouse effect and higher temperatures near Earth's surface. From about 5000 ppm about 550 million years ago, the concentration of atmospheric CO_2 generally declined. However, many episodes of large-scale volcanic activity were responsible for temporary upturns in CO_2 concentration and a considerably warmer global climate. These peaks in atmospheric CO_2 correspond in time with most major mass extinctions of plant and animal species on land and in the ocean (discussed in this chapter's second Essay).

During the Pleistocene Ice Age (1.7 million to 10,500 years ago), atmospheric CO_2 levels also fluctuated, decreasing during episodes of glacial expansion and increasing during episodes of glacial recession (although it is not clear whether variations in atmospheric CO_2 were the cause or effect of these global-scale climate changes).

The biosphere also played an important role in Earth's evolving atmosphere, primarily through *photosynthesis*, the process whereby green plants use sunlight, water, and CO_2 to manufacture their food. A byproduct of photosynthesis is oxygen (O_2). Although vegetation is also a sink for CO_2, photosynthesis probably was not as important as the geochemical processes described earlier in removing CO_2 from the atmosphere. Based on the fossil record, photosynthesis dates to about 2.7 billion years ago, with the first appearance of cyanobacteria in the ocean. However, it was not until 2.5 to 2.4 billion years ago that the atmosphere became oxygen-rich. Although oxygen was produced for 200-300 million years, none accumulated in the atmosphere. Why the lengthy delay?

Apparently, the ocean and land took up oxygen as fast as it was produced. In the ocean, most oxygen combined with marine sediments while very little entered the atmosphere. Eventually, oxidation of marine sediments tapered off and photosynthetic oxygen dissolved in ocean water. According to findings reported in 2007 by researchers Lee Kump of Pennsylvania State University and M. Barley of the University of Western Australia, the geologic record indicates a shift in geologic activity about 2.5 billion years ago from underwater volcanism to terrestrial volcanism. This shift was accompanied by a change in the composition of the eruptive gases, from those that react with oxygen to those that do not react with oxygen. With the subsequent build-up of atmospheric oxygen, and the concurrent decline in atmospheric CO_2, oxygen became the second most abundant atmospheric gas within the next 500 million years.

With oxygen emerging as a major component of Earth's atmosphere, the ozone shield formed. Within the *stratosphere*, incoming solar ultraviolet (UV) radiation drives reactions that convert oxygen to ozone (O_3) and ozone to oxygen. Absorption of UV radiation in these reactions prevents potentially lethal intensities of UV radiation from reaching Earth's surface. By about 440 million years ago, formation of the *stratospheric ozone shield* made it possible for organisms to live and evolve on land. UV radiation does not penetrate ocean water to any great depth, so marine life was able to exist in the ocean depths prior to the formation of the ozone shield. With the ozone shield, marine life was able to thrive in surface waters, and eventually on land.

During the past 550 million years, the concentration of oxygen in the atmosphere has fluctuated significantly. These fluctuations were linked to imbalances in the rates of the weathering of organic carbon and pyrite (FeS_2), which decreases atmospheric oxygen, and the sedimentation of these materials, which increases atmospheric oxygen. Over the 550-million-year period, the percentage of O_2 in the atmosphere has been estimated to have varied between about 13% and 31%; at present oxygen is about 21% of the air we breathe.

Nitrogen (N_2), a product of outgassing, became the most abundant atmospheric gas because it is relatively inert and its molecular speeds are too slow to readily escape Earth's gravitational pull. Furthermore, compared to other atmospheric gases, such as oxygen and carbon dioxide, nitrogen is less soluble in water. All these factors greatly limit the rate at which nitrogen cycles out of the atmosphere. While nitrogen continues to be generated as a minor component of volcanic eruptions, today the principal natural source of free nitrogen entering the atmosphere is *denitrification*, which accompanies the bacterial decay of plants and animals. This input is countered by nitrogen removed from the atmosphere by *biological fixation* (i.e., direct nitrogen uptake by leguminous plants such as clover and soybeans) and *atmospheric fixation* (i.e., the process whereby the high temperatures associated with lightning causes nitrogen to combine with oxygen to form nitrates that are washed by rains to Earth's surface).

In summary, during the more than 4.5 billion years since Earth's formation, the planet's climate system evolved gradually. Bombardment of Earth by comets and/or large meteorites delivered the water of the hydrosphere. Outgassing from the geosphere was the origin of most atmospheric gases. Geochemical processes, photosynthesis, the stratospheric ozone shield, and biogeochemical cycles explain climatically-significant fluctuations in the chemistry of the atmosphere.

ESSAY: Asteroids, Climate Change, and Mass Extinctions[a]

Geologists and other scientists have gathered evidence from the fossil record of five major mass extinctions that occurred over the past 550 million years (Table 1). Elimination of 50% or more of all species indicates drastic changes in Earth's environment, which exceeded the tolerance limits of a vast number of organisms. What caused these mass extinctions?

TABLE 1

Major Mass Extinctions of Plant and Animal Species over the past 550 Million Years

End of Ordovician period	443 million years ago
End of Devonian period	374 million years ago
End of Permian period	251 million years ago
End of Triassic period	201 million years ago
Cretaceous-Tertiary boundary	65 million years ago

Prior to 1980, the most popular explanation for mass extinctions was a gradual decrease in species number (perhaps over millions of years) due to long-term climate change coupled with ecological forces. In 1980, however, another much more dramatic explanation took center stage. The father-son team of scientists Luis (1911-1988) and Walter (1940-) Alvarez of the University of California, Berkeley, proposed that an asteroid impact on Earth was responsible for the mass extinction that took place 65 million years ago. This event was known as the *K-T mass extinction*, named for the boundary between the Cretaceous and Tertiary periods of geologic time. The Alvarez team presented convincing evidence of an asteroid impact, including the discovery of iridium (Ir) in sedimentary layers from around the world—all dating from 65 million years ago. Iridium is a silver-gray metallic element that is extremely rare in Earth's crust. Asteroids, however, contain a much higher concentration of iridium. The Alvarez hypothesis was bolstered by features found within and near the impact site.

The K-T asteroid produced the Chicxulub crater, a 180-km (112-mi) wide crater on the floor of the ancient Caribbean Sea (Figure 1). Marine sediments gradually filled the crater and geological forces later elevated a portion of the crater above

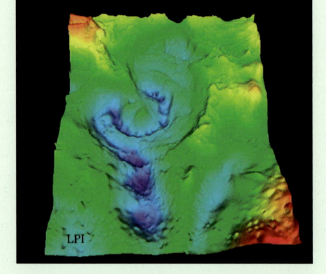

FIGURE 1
The Chicxulub Crater, centered near the town of Chicxulub on Mexico's Yucatán Peninsula, is about 180 km (112 mi) in diameter, represented here as gravity and magnetic field data. It formed about 65 million years ago when a mountain-size asteroid (at least 10 km or 6 mi across) struck Earth's surface. The effects of the impact were thought to be responsible for the extinction of the dinosaurs and about 70% of all species then living on the planet. [Courtesy of NASA, Lunar Planetary Institute, V.L. Sharpton]

---

[a]For much more on this topic, see Ward, Peter D., 2007. *Under A Green Sky*. Washington, DC: Smithsonian Books, 242 p.

sea level. Today, what remains of the Chicxulub crater forms part of Mexico's Yucatán Peninsula. Radar images obtained by the Space Shuttle *Endeavour* in 2000 revealed a 5-m (16-ft) deep, 5-km (3-mi) wide trough on the Yucatán Peninsula that may mark the outer rim of the crater. Drilling through the layers of sediment on the floor of the nearby Gulf of Mexico recovered cores of fractured and melted rock from the impact zone.

Other evidence of the asteroid impact consists of bits of tiny bead-like spherules of glassy rock, which originated as droplets of molten rock blasted into the atmosphere by the impact. These droplets cooled as they fell through the atmosphere onto the land or into the ocean. They were recovered from nearby deep-ocean sediments. Many rocks on land contain mineral grains deformed by the extreme heat and pressure produced by the impact (e.g., shocked quartz). Unusual sediment deposits were produced by enormous waves (tsunamis) generated when the asteroid (at least 10 km or 6 mi in diameter) struck the ocean surface. In addition, a layer of soot indicates considerable burning vegetation on land.

The K-T asteroid impact had a catastrophic effect on life. Best known is the extinction of the dinosaurs, which had dominated life on Earth for more than 250 million years. Dinosaurs were not the only victims, however. The asteroid impact destroyed more than 50% of the other life forms then existing on the planet and caused major extinctions among many groups of marine organisms, including plankton.

What precisely caused this ecological disaster? One widely accepted theory is that the asteroid impact vaporized large amounts of sulfur-containing deep-sea sediments. This sulfur was blown into the atmosphere, where it generated enormous clouds of tiny sulfate particles, likely augmented by meteoric and Earth materials also thrown into the atmosphere by the impact. These clouds greatly reduced the sunlight reaching Earth's surface for 8 to 13 years; most plants died because they could not photosynthesize. Furthermore, precipitation decreased by up to 90%. In this dark, cold and dry environment, dinosaurs and other animals that depended on plants for food starved and the carnivores that fed on them followed. Only small animals (as some mammals) could survive by eating the dead plants and animals until conditions improved and new food sources became available. Eventually, the aerosols settled out of the atmosphere, and photosynthesis resumed when dormant seeds sprouted. Small mammals evolved rapidly to take the place of the dinosaurs.

Another possibility is that red-hot, impact-generated particles rained down through the atmosphere making it so hot that most plants and animals were killed directly.

In the 1980s and 1990s, the Alvarez theory of asteroid impact was widely accepted as the cause of all but one of the five major mass extinctions (Table 1). However, a vocal minority of scientists took exception to the preeminent role of asteroid impact, arguing that many of the major mass extinctions were linked to volcanic activity and increased levels of atmospheric CO_2. The largest eruptions of flood basalts closely correspond in age to the times of most major mass extinctions. Flood basalts consist of many successive lava flows erupting from fissures in Earth's crust, and accompanied by toxic gases released into the atmosphere, including hydrogen sulfide (H_2S), and the greenhouse gases carbon dioxide and methane (CH_4).

Flood basalt eruptions can be enormous. The world's largest flood basalt eruptions (that produced the Siberian Traps) delivered about 4.2 million km^3 (1 million mi^3) of lava over an area of nearly 7.8 million km^2 (3 million mi^2) approximately 252 to 248 million years ago. This eruption was very near the time of the great Permian mass extinction (around 250 million years ago), when 90% of all ocean species and 70% of terrestrial vertebrates on Earth were wiped out. No evidence of an asteroid impact has been found to explain the Permian extinction. In addition, most mass extinctions took place during times when the concentration of atmospheric CO_2 was relatively high or rapidly rising.

By 2005, a new hypothesis was firmly in place that attributed most major mass extinctions to a combination of chemical and circulation changes in the ocean, coupled with global warming due to an enhanced greenhouse effect. In arriving at this alternate explanation for mass extinctions, scientists relied on analysis of biomarkers where fossils were absent. *Biomarkers* are the organic chemical residue of organisms extracted from ancient strata.

According to research conducted by Lee Kump and his colleagues at Pennsylvania State University, the late Permian ocean was stratified. The bottom water had little or no dissolved oxygen while the shallow surface layer was oxygenated. (Most of today's ocean is oxygenated from top to bottom.) With the release of greenhouse gases to the atmosphere during the eruptions that produced the Siberian Traps, the global temperature rose dramatically. This warmed the surface ocean waters, reducing the amount of oxygen absorbed from the atmosphere. A reduction in the equator to pole temperature gradient caused a weakening of wind and wind-driven surface ocean currents. Consequently, the ocean circulation changed so that great volumes of warm, nearly oxygen-free water filled the ocean bottom. In this environment, microbes were dominated by anaerobic bacteria that consumed sulfur and produced hydrogen sulfide. Biomarkers of green sulfur bacteria and photosynthetic purple sulfur bacteria were extracted from strata of this age. In time the layer of oxygen-poor, H_2S-rich water became thicker and reached the ocean surface where it escaped to the atmosphere. Highly toxic, especially at high temperatures, H_2S also reacts with and destroys stratospheric ozone, allowing lethal levels of solar ultraviolet radiation to reach Earth's surface thus causing the end of the Permian era.

MONITORING EARTH'S CLIMATE SYSTEM

Artist concept of the ice albedo feedback loop. As polar ice begins to melt, less sunlight gets reflected into space. It is instead absorbed into the oceans and land, raising the overall temperature, and fueling further melting. [Courtesy NASA]

Case-in-Point

The summer of 1816 is famous in American history for its anomalously cold weather. In fact, 1816 is often described as "the year without a summer." Unseasonable freezes and snowfalls damaged or destroyed many crops in the then mostly agrarian northeastern United States and adjacent portions of Canada. More than 90% of the corn crop, the prime food staple, was lost and serious damage was done to garden vegetables. The hardier grains like wheat and rye, however, were not as badly affected. Some superficial accounts of events of that summer give the impression that unusually cold weather persisted throughout the entire summer and affected the entire civilized world. Such was not the case.

In the Northeast, the summer of 1816 was punctuated by several cold episodes. Killing frosts struck northern and interior southern New England, as well as neighboring Quebec, from 6 to 11 June, 8 to 10 July, and again in mid and late August. The June cold snap was accompanied by moderate to heavy snowfalls in Quebec City and some highlands as far south as the Catskills of southeastern New York. No sooner had farmers replanted their crops than a killing freeze would strike again. The shortened growing season ended with a general hard freeze on 27 September.

These unusually cold episodes interrupted longer spells of seasonally warm weather. June and July mean temperatures were 1.6 to 3.3 Celsius degrees (3 to 6 Fahrenheit degrees) below average. Individually, these monthly temperature anomalies (departures from long-term averages) fall within the expected range of climate variability. Some previous and subsequent summer months (June, July, and August) have been as cold as the summer months of 1816, if not colder. What is most notable about the weather of the summer of 1816 is the persistence of a strong negative temperature anomaly through all three summer months. Nevertheless, that summer in the Northeast as a whole fell within the range of expected climate variability.

What about the weather elsewhere during the summer of 1816? We know that it was probably not the same everywhere as in the northeastern U.S., for climatic anomalies typically are geographically non-uniform in both sign (direction) and magnitude. Unfortunately, little weather information is available from the central and western United States and Canada: Few settlements existed there at the time and not many people on the frontier had weather instruments to take quantitative observations. England, France, and Germany experienced an unusually cold summer, but this is not surprising. The prevailing wave pattern in the upper-level westerly winds that was responsible for circulating unseasonably cold air into the Northeast would also cause the same anomalous weather in Western Europe. The two localities are typically one wavelength apart in the band of westerly winds that encircle the hemisphere. By contrast, available observational data indicate that east-central and eastern Europe experienced a warmer than usual summer.

This discussion is not intended to dismiss or diminish the hardships suffered by people living in the Northeast and Western Europe during the summer of 1816. Rather, it is intended to show that weather extremes, while popularly memorable, typically fall within an expected range of climate variability. Hence, the occurrence of a single exceptionally hot summer or cold winter does not necessarily signal a change in the climate. In Chapter 11, we have more to say about the cold summer of 1816 and its possible link to the April 1815 eruption of the volcano Tambora on the island of Sumbawa in the Dutch East Indies.

Driving Question:

How do climate scientists investigate the spatial and temporal characteristics of climate, climate variability and climate change?

This chapter continues the construction of a framework for our study of climate science. Climate operates over a range of spatial and temporal scales. For convenience of inquiry, climate scientists subdivide Earth's climate mosaic into macroclimates, mesoclimates, local climates, and microclimates based on spatial scale. Climate also varies over a span of timescales ranging from years to decades, to centuries, to millennia or longer, and a distinction is made between climate variability and climate change.

Subtracting the actual magnitude of a climatic element (e.g., temperature) from long-term averages yields *anomalies* that more readily portray variations and extremes. These anomalies typically are geographically non-uniform in both sign (direction) and magnitude. This same geographic non-uniformity also characterizes trends in climate with implications for the societal impacts of global- or hemispheric-scale climate change. Furthermore, feedback within the climate system can enhance or suppress climate variability.

In situ measurement and remote sensing techniques are used extensively to monitor the climate system. Because climate change knows no political boundaries, international cooperation is essential in monitoring Earth's climate system and interpreting the results of climate science research.

Numerical models in concert with monitoring of the climate system play an important role in short-term and long-term climate forecasting. Numerical climate models differ from those used in operational weather forecasting used to predict tomorrow's high temperature or precipitation totals in that the goal is to identify locations of expected positive or negative anomalies in temperature, precipitation, or other climatic elements.

This chapter closes with a summary in general terms of how climate behaves through time, that is, the lessons of the climate record. Perhaps the most obvious lesson to be gleaned is that change is an inherent characteristic of climate. The question is not *whether* climate will change in the future but *how* it will change.

Spatial Scales of Climate

Earth is a mosaic of numerous climate types. For convenience of study, climate scientists subdivide Earth's climate system based on horizontal/vertical spatial scales. From largest to smallest, scales of climate are designated macroclimate, mesoclimate, local climate, and microclimate (Table 2.1). In this hierarchical scheme, the larger scale encompasses all smaller scales. A scientist investigating the climate at one of these spatial scales typically is concerned with a specific array of processes operating within Earth's climate system (Chapter 1).

A **macroclimate** extends over large regions that can be up to continental in area. The emphasis is on describing the large-scale patterns of temperature and precipitation, together with the linkage of these patterns to the prevailing (dominant) atmospheric circulation. What happens at the macro-scale is the traditional basis for classifying the climates of the continents (Chapter 13). **Mesoclimate** characterizes smaller and physically distinctive regions such as mountains or plains and is strongly influenced by regional atmospheric circulation patterns and systems (e.g., mountain and valley breezes, chinook winds, sea breezes). An example is the moderate climate of localities situated downwind from the Great Lakes.

Local climates apply to a variety of ecosystems such as a city, cropland, marsh or forest. For example, a city tends to be somewhat warmer than the surrounding rural areas because of differences in composition (asphalt, brick, and concrete versus vegetation), water supply (more standing water in the countryside), and concentration of heat sources (motor vehicles, space heating and cooling). We have more to say on urban climatology in Chapter 4.

With **microclimates**, the focus is on the exchange of energy and matter in a small area such as an individual farm field, hillslope, or marsh. (*Phytoclimate* is the part of microclimate that deals with the climate around plants, especially in the plant canopy.) For example, the formerly glaciated landscape of northern Wisconsin includes scattered small depressions where frost often forms even in mid summer. Vegetation living in these *frost hollows* (also called *frost pockets* or *frost sags*) is mainly frost-hardy grasses and sedges, while trees and other vegetation more typical in the region surround the depression. *Cold-air drainage* accounts for the localized occurrence of sub-freezing air temperatures. On a clear night, radiational cooling chills the ground which in turn cools the overlying air. Under the influence of gravity, a shallow layer of relatively cold dense air drains into the depressions.

Climate Variability and Climate Change

The climate record clearly reveals that change over a broad range of time scales is a fundamental characteristic of climate, meaning that climate is rarely static. Climate changes over intervals ranging from years to centuries to millennia and longer. A useful distinction is made, however, between climate variability and climate change.

Climate variability refers to the variations in climate around a mean state and other statistics (e.g., standard deviation, variance, or statistics of extremes) of the climate on all time and space scales beyond that of individual weather events. (This definition is adapted from the *Intergovernmental Panel on Climate Change (IPCC)*. Climate variability is often used to describe the

TABLE 2.1
Spatial Scales of Climate

	Horizontal dimensions	*Vertical dimensions*
Macroclimate	>20,000 m (>12 mi)	>6000 m (>20,000 ft)
Mesoclimate	100 – 20,000 m (330 ft – 12 mi)	surface – 6000 m (20,000 ft)
Local climate	100 – 10,000 m (330 ft – 6 mi)	surface – 1000 m (3300 ft)
Microclimate	<1 m – 100 m (3.3 ft-330 ft)	surface – 100 m (330 ft)

deviations in climate statistics over a period of time (e.g., month, season, or year) compared to the long-term climate statistics for the same time period. For example, the average temperature of a particular year will very likely vary from the mean annual temperature over a recent 30-year period. Such variability may be due to natural processes operating within or external to the climate system (*natural climate variability*) and/or involve some anthropogenic forcing.

According to the American Meteorological Society's *Glossary of Meteorology* (2000), **climate change** is "any systematic fluctuation in the long-term statistics of climatic elements (e.g., temperature, pressure, or wind) that is sustained over several decades or longer. Such an occurrence usually has significant societal, economic, or environmental consequences. Often, climate change describes a major shift from one climatic regime to another on a global scale. During the Pleistocene Ice Age (1.7 million to 10,500 years ago), for example, the numerous shifts between glacial and interglacial climatic episodes constituted climate change. Climate change is a response to forcings that are internal and external to the climate system. Internally, the atmosphere is influenced by and interacts with other components of the climate system (e.g., ocean, sea-ice, land surface properties). External forcings include the Sun's energy output and Earth-Sun geometry. As we will see in Chapter 3, changes in internal or external forcings of the climate system can bring about change in the planet's energy balance with surrounding space as measured at the tropopause (considered to be the top of the atmosphere). Climate change ultimately is due to perturbations of the radiative forcings that are a combination of changes in incoming solar radiation, the amount of solar radiation scattered by the planetary system back to space, and alterations in the flow of infrared radiation from the planetary system to space.

At times, the term *climate trend* is used to identify a change in climate that can be described by a smooth monotonic increase or decrease in a climate variable, such as temperature, over a selected time interval. However, these trends represent only a segment of the entire period of record.

Climatic Anomalies

Climate scientists (as well as the public in general) find it useful to compare the average weather of a specific week, month, or year with the climate record. When such a comparison is carried out over a broad geographical region, departures from long-term averages, called climatic anomalies, do not occur with the same sign or magnitude everywhere. For example, as shown in Figure 2.1A, the average temperature during December 2007 was above the long-term average (*positive anomalies*) through much of the southeastern and northwestern U.S., but below the long-term average (*negative anomalies*) from the Great Lakes and Plains states westward to California and Oregon. Furthermore, the magnitude of the anomaly, positive or negative, varies from one place to another. In our example, the long-term average represents the 30-year "normals" computed for the current 1971-2000 standard interval.

As shown in Figure 2.1B, a somewhat different convention is used to represent precipitation anomalies versus temperature anomalies. Specifically, we use the percentage of normal precipitation, so that any value less than 100% means a negative precipitation anomaly. Precipitation anomalies typically form more complex patterns than temperature anomalies due to greater spatial variability in precipitation arising from shifts in storm tracks and the almost random distribution of showers and thunderstorms. For these reasons, in spring and summer in middle latitudes, even adjoining counties may experience opposite rainfall anomalies (one having above-average rainfall while the other has below-average rainfall). From an agricultural perspective, the geographic non-uniformity of climatic anomalies may be advantageous in that some compensation is implied. That is, poor growing weather and consequent reduced crop yields in one area may be compensated to some extent by better growing weather and increased crop yields elsewhere. This **agroclimatic compensation** generally applies to crops such as corn, soybeans, and other grains that are grown over broad geographical areas.

At middle and high latitudes of the Northern Hemisphere, the geographic non-uniformity of climatic anomalies is linked to the prevailing horizontal winds in the middle and upper troposphere (at altitudes between 5000 and 10,000 m or 16,000 and 33,000 ft) that flow from the west. This belt of **westerlies** circles the Northern Hemisphere in a wave-like pattern of clockwise turns (called *ridges*) and counterclockwise turns (called *troughs*). As shown in Figure 2.2, the westerly wave pattern can change in wavelength (distance from trough to trough or ridge to ridge), wave amplitude, and number of waves. The prevailing pattern ultimately governs the flow of cold and warm air masses, where storms develop, and the tracks of those storms. Hence, the pattern of the westerlies determines the location of weather extremes such as drought or very low temperatures. However, in

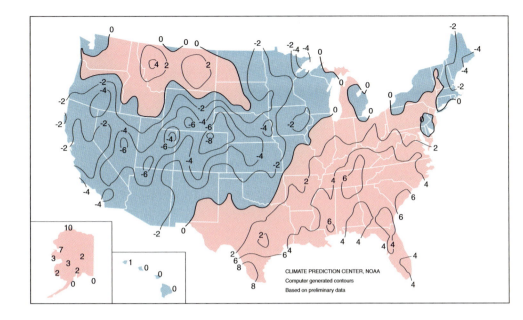

FIGURE 2.1A
Departure of average temperature from the long-term average (1971-2000) in Fahrenheit degrees for December 2007. [NOAA, Climate Prediction Center]

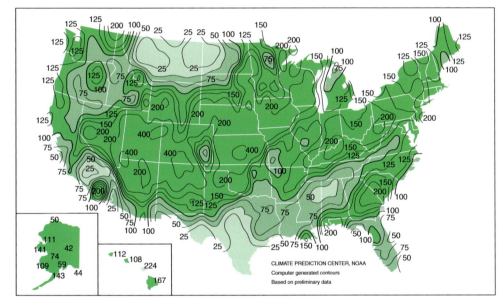

FIGURE 2.1B
Percent of long-term average (1971-2000) precipitation (rain plus melted snow) for December 2007. [NOAA, Climate Prediction Center]

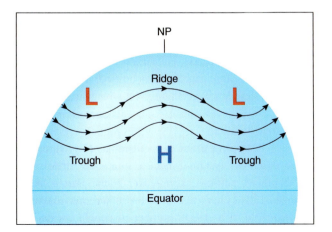

FIGURE 2.2
In the middle and upper troposphere, the belt of prevailing westerly winds circles the globe in a wave-like pattern of ridges and troughs. With time, the westerly wave pattern exhibits changes in wavelength, amplitude, and wave number.

view of the finite number of westerly waves that typically encircle the hemisphere, a single weather extreme never encompasses an area as large as the United States; that is, extreme cold or drought rarely grips the entire nation at the same time. (We have much more to say about the prevailing planetary-scale winds in Chapter 6.)

Geographic non-uniformity also applies to trends in climate. Hence, the trend in the mean annual temperature of the Northern Hemisphere is not necessarily representative of all localities within the hemisphere. During the same period, some places experience cooling whereas other places experience warming regardless of the direction of the overall hemispheric (or global) temperature trend. Not only is it misleading to assume that the direction of large-scale climatic trends applies to all localities, but also it is erroneous to assume that the magnitude of climatic trends is the same everywhere. A small change in the average hemispheric temperature typically translates into a much greater changes in some areas, but with little or no change in other areas.

FEEDBACK LOOPS

Within the climate system, many variables are linked together in complex forcing/response (or action/reaction) chains. For example, clouds have both warming and cooling effects on the lower atmosphere (troposphere). During daylight, the tops of clouds as viewed by sensors onboard Earth-orbiting satellites appear very bright indicating that sunlight is strongly reflected back to space, thereby causing cooling. On the other hand, clouds also contribute to the greenhouse effect by absorbing and emitting infrared radiation thereby causing warming at Earth's surface. For low clouds, the cooling effect dominates whereas for high clouds, the warming effect prevails.

Interactions among variables in the climate system involve feedback loops that may either amplify (*positive feedback*) or weaken (*negative feedback*) fluctuations in climate. Using terminology originally applied in electrical engineering, a feedback loop represents the processes or pathways within a system where the effects imposed by changes in the input signal result in either an amplified or damped response signal. Often these processes or pathways appear as "loops" in diagrams of the system. Figure 2.3 is a schematic diagram illustrating positive feedback. The rate of evaporation of water largely depends on temperature; as the temperature of the ocean surface rises, the rate of evaporation increases making for a more humid lower atmosphere. Water vapor is the principal greenhouse gas so that more

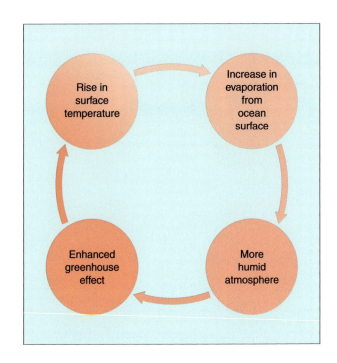

FIGURE 2.3
An example of positive feedback in Earth's climate system.

humid air means higher surface temperatures. This is an example of positive feedback.

On the other hand, another possible consequence of a warmer ocean surface and more humid air (all other factors being equal) is more low clouds (Figure 2.4). Low clouds reflect more sunlight to space and consequently the surface temperature decreases. This is an example of negative feedback.

The overall range of climate fluctuations is relatively small, suggesting that negative feedbacks are more common than positive feedbacks in Earth's climate system. Consider an example. During the glacial climatic episodes of the Pleistocene Ice Age when glaciers thickened and advanced over much of Canada, the northern tier states of what is now the United States, and northwestern Europe, the global mean annual temperature (as reconstructed from geological evidence) was only about 6 Celsius degrees (11 Fahrenheit degrees) lower than it is now. More negative than positive feedback in the climate system tends to inhibit great swings in climate.

TIPPING POINTS

A **tipping point** is a critical point or threshold in a dynamic system when a new and irreversible development, perhaps major, takes place. That is, some change in the system results in additional consequences. For example, if a critical number of people contract an

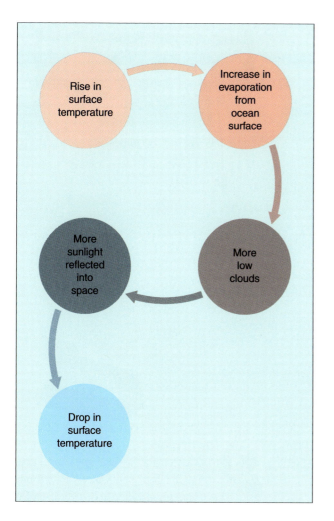

FIGURE 2.4
An example of negative feedback in Earth's climate system.

saturated with CO_2 thereby eliminating the carbon sink and dramatically increasing the amount of anthropogenic carbon dioxide that accumulates in the atmosphere. Such a turn of events would rapidly accelerate greenhouse warming. Also, once the ocean reaches equilibrium with the greenhouse-gas-warmed atmosphere, any excess heat energy will remain in the atmosphere further accelerating the warming.

Observing the Climate System

Since the 19th century, scientists have monitored the climate system not only out of curiosity and the need to know the existing climate but also to better understand the nature, causes, and implications of climate change. Achieving these goals requires observational data on the properties and processes of the climate system. Today, scientists routinely monitor the climate system using both in situ and remote sensing techniques. With **in situ measurement**, the sensor is immersed in the medium that is being monitored. **Remote sensing** refers to acquisition of data on the properties of some object without the sensor being in direct physical contact with the object.

IN SITU MEASUREMENT

In Chapter 1, we described the historical development of instrument-based weather and climate networks in the United States and elsewhere. Initially, the primary focus was on recording conditions near Earth's surface. Efforts were also underway to explore the properties of the vertical dimension of Earth's atmosphere. During the 1800s, mountaintop weather stations were established, such as the U.S. Army Signal Service's Mount Washington, NH (operating year-round 1871-1887 and in summer only through 1892) and Pikes Peak, CO (1873-1889) observatories, but the influence of the mountains meant that observations were not representative of conditions in the free atmosphere. Through the early part of the 20th century, weather instruments borne by kites, aircraft, and balloons provided data chiefly on the lowest 5000 m (16,000 ft) of the atmosphere.

On 4 August 1894 at Harvard's Blue Hill Observatory near Boston, MA, kites were used for the first time to carry aloft a self-recording thermometer (*thermograph*). This instrument provided a vertical profile of air temperature to an altitude of 427 m (1400 ft) above the ground. The next year, scientists conducted kite experiments for meteorological purposes at Mount Weather

infectious disease, local health care providers may be unable to prevent the disease from spreading more widely within the population. That critical number of people represents the tipping point.

The concept of tipping point has been applied to many disciplines including climate science. Consider an example. About half the carbon dioxide emitted to the atmosphere as a byproduct of fossil fuel combustion dissolves in the ocean. The ocean is a carbon sink as well as a major sink for heat energy. For these reasons, the ocean slows the rate of global climate change. However, with the continual rise in atmospheric CO_2 concentration and enhancement of the greenhouse effect, sea surface temperatures (SST) will rise. The saturation concentration of carbon dioxide in seawater decreases with rising SST; that is, less dissolved carbon dioxide serves to saturate warm water than cold water. At some critical point (the tipping point), surface ocean waters may become nearly

Observatory, VA, about 80 km (50 mi) west of Washington, DC. Piano wire secured box kites of various sizes and shapes to rotating steel drums. Kites were equipped with a recording instrument, called a *meteorograph*, which profiled altitude variations in air pressure, temperature, humidity, and wind speed. Wind direction was determined from the ground using a *theodolite*, an instrument that measures azimuth and elevation angles. On 3 October 1907, a box kite equipped with a meteorograph ascended to 7044 m (23,111 ft) setting a new world altitude record. On 5 May 1910, a train of 10 instrumented kites reached an altitude of 7265 m (23,835 ft).

The success of kite experiments convinced the U.S. Weather Bureau in 1907 to begin operating a network of weather kite stations at several locations, mostly in the central part of the nation (Figure 2.5). Box kites equipped with meteorgraphs profiled the atmosphere up to a maximum altitude of about 3000 m (10,000 ft). The longest and last operating station, at Ellendale, ND, was closed in 1933. Among reasons cited for discontinuing the kite network was the relatively low maximum altitude reached by kites and the interruption in observations during periods of light winds or calm air. For several years afterward, the U.S. Weather Bureau relied mostly on regular 5 a.m. (EST) aircraft observations originating at as many as 30 locations for conditions in the atmosphere up to altitudes of 4900 m (16,000 ft), but this practice proved too hazardous and costly.

A leap forward in monitoring higher altitudes came in the late 1920s with the invention of the first radiosonde. A **radiosonde** is a small instrument package equipped with a radio transmitter that is carried aloft by a helium- (or hydrogen-) filled balloon (Figure 2.6). This device transmits altitude readings, called a **sounding**, of temperature, air pressure, and dewpoint (a measure of humidity) to a ground station. Radiosonde data are received immediately; no recovery of a recording instrument is required. The first official U.S. Weather Bureau radiosonde was launched at East Boston, MA, in 1937. By World War II, meteorologists were tracking the balloon's drift from ground stations using radio direction-finding antennas, thereby monitoring variations in wind direction and speed with altitude. A radiosonde used in this way is called a **rawinsonde**.

Today, radiosondes are launched simultaneously at 12-hr intervals (just prior to 0000 UTC and 1200 UTC) from hundreds of ground stations around the world. UTC (Universel Temps Coordinné) is the time on a 24-hr clock at the prime meridian (0 degree longitude). In 2007, some 941 upper-air (RAOB) stations were operating worldwide

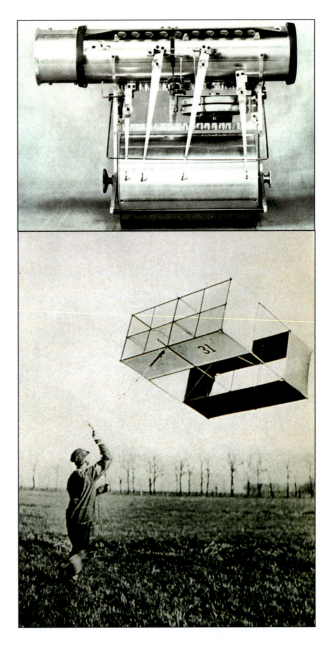

FIGURE 2.5
In the early 20th century, scientists probed the atmosphere up to a maximum altitude of about 3000 m (10,000 ft) using recording weather instruments (top) attached to a box kite (bottom). [NOAA Photo Library]

including 96 in the U.S., 35 in Canada, and 12 in Mexico (Figure 2.7). The balloon bursts at an altitude of about 30,000 m (100,000 ft), and the instrument package descends to the surface under a parachute.

As of this writing, the National Weather Service is implementing the *Radiosonde Replacement System (RRS)* to upgrade the current generation of radiosondes. The RRS consists of a global positioning system (GPS) tracking antenna, GPS radiosondes that operate at a radio

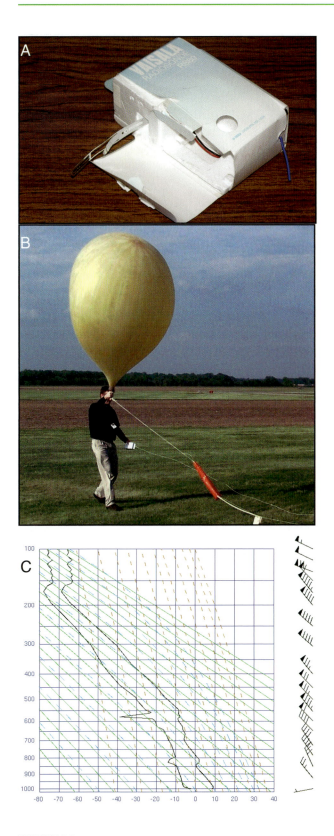

FIGURE 2.7
Locations of U.S. radiosonde observation stations. Balloons are launched simultaneously twice each day at 12-hr intervals.

frequency of 1680 MHz, plus a computer workstation. The new system will provide more detailed and accurate upper-air observations. Readings will be available at 1-second intervals, corresponding to altitude levels about 5 m (16 ft) apart.

A **dropwindsonde** is similar to a rawinsonde except that instead of being launched by a balloon from a surface station, it is dropped from an aircraft. The instrument package descends on a parachute at about 18 km per hr (11 mph) and along the way radios data back to the aircraft every few seconds. The dropwindsonde was developed at the National Center for Atmospheric Research (NCAR) in Boulder, CO, to obtain soundings over the ocean where conventional rawinsonde stations are few and far between. Dropwindsondes provide vertical profiles of air temperature, pressure, dewpoint, and wind. They are often deployed by "hurricane hunter" aircraft that fly out to investigate offshore tropical storms and hurricanes that are approaching the Gulf or East Coast.

In situ sources of oceanic data (e.g., sea-surface temperature, salinity, currents) useful in climate studies include ships at sea, instrumented buoys (Figure 2.8), floats, and gliders, piloted submersibles, autonomous instrumented platforms and vehicles, and undersea observatories. Considerable ocean research has been and is still done from ships; it is the only way to get out on the sea to investigate what cannot be detected remotely. Furthermore, satellite-borne monitoring instruments such as radar can observe directly only the ocean's surface waters because water is nearly opaque to electromagnetic radiation. Therefore, other techniques are necessary to examine the ocean below the surface. Instruments lowered from ships are used to sample ocean water at various

FIGURE 2.6
A radiosonde (A) is a small instrument package equipped with a radio transmitter and carried aloft by a hydrogen- or helium-filled balloon (B). C is a plot of radiosonde temperature and depoint at pressures shown to left. Winds are plotted to right.

FIGURE 2.8
This instrumented buoy gathers observational data from both the lower atmosphere and the surface ocean waters. [NOAA/NDBC]

FIGURE 2.9
Launch of an Argo float from a German research ship. This instrument is designed to measure profiles of ocean temperature and salinity to a depth of 2000 m. [Courtesy of the Argo Project.]

depths and instruments moored on the sea floor monitor properties of the ocean beneath the surface waters.

One in situ method uses the speed of sound waves in water to measure ocean-water temperature and its variation through time. Ocean water is highly transparent to sound, just as the atmosphere is nearly transparent to visible sunlight. A sound pulse from a transmitter on one side of the ocean can be detected by a sensitive receiver thousands of kilometers away on the other side of the ocean. (Whales, for example, take advantage of this property of seawater to communicate long distances across an ocean basin, to attract mates and to locate food.) Sound travels faster in warm water than cold water. Thus measuring the speed of sound in seawater can be used to determine the average water temperature between the transmitter and the receiver. Furthermore, variations in sound speed between many transmitters and receivers can be combined to obtain a three-dimensional representation of seawater temperature. This technique is known as **acoustic thermometry**.

A **profiling float** is an instrument package that measures vertical profiles of ocean water temperature, pressure (a measure of depth), and conductivity (a measure of salinity) (Figure 2.9). In 1998, scientists began deploying floats at 3-degree latitude/longitude intervals worldwide. The primary purpose of the float array, known as *Argo* (*Array for real-time geostrophic oceanography*), is to monitor the climate (long-term average) of the ocean's wind-driven surface layer and underlying *pycnocline* (layer of the ocean in which the density increases rapidly with depth due to changes in temperature or salinity) and beyond, typically to a depth of 2000 m (6600 ft). The initial goal to deploy 3000 floats was realized on 1 November 2007.

Research ships, commercial vessels, and low-flying aircraft drop floats into the sea. Typically, an Argo float is programmed for a 10-day cycle during which it sinks to a prescribed depth of 1000 m (3300 ft), drifts with the current for 9 days before sinking to a maximum depth of 2000 m (6600 ft), and then returns to the surface monitoring seawater properties along the way (Figure 2.10). At the surface, the float relays its collected data via satellite to computer databases. Tracking the position of the float over time also records water movements at its drifting level. The anticipated lifetime of one Argo float is 100 cycles.

While the Argo global-scale array is an international effort, various institutions in the U.S. deployed about half the floats. Among the other 22 participating nations are Australia, Canada, Japan, France, and the United Kingdom. The U.S. portion of the Argo program includes floats that are under the auspices of the University of Washington, Scripps Institution of Oceanography, and the Woods Hole Oceanographic Institution. Argo data are archived at NOAA's Atlantic Oceanographic and Meteorological Laboratory in Miami, FL.

A

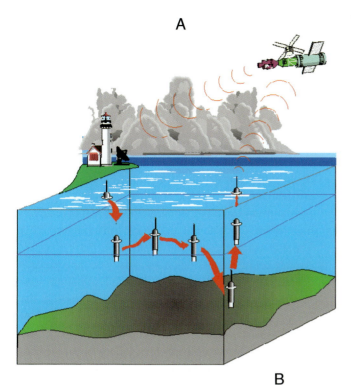

REMOTE SENSING BY SATELLITE

In March 1947, a vertically fired V2 rocket, equipped with a camera, took the first successful photographs of Earth's cloud cover from altitudes of 110-165 km (70-100 mi). This and subsequent rocket probes of the upper atmosphere convinced scientists of the value of cloud pattern photography in identifying and tracking weather systems and inspired the first serious proposals for orbiting a weather satellite. In the mid- to late-1950s, the United States' fledgling space program aimed to develop a launch vehicle (rocket) capable of putting a satellite in orbit around the planet. With the former Soviet Union's successful orbiting of Sputnik I on 4 October 1957, the age of remote sensing by satellite had begun.

Surveillance of Earth's climate system by Earth-orbiting satellites began on 13 October 1959 with the U.S. launch of *Explorer 7*. Among other instruments on board the Juno rocket was a flat-plate radiometer developed by Verner E. Suomi (1915-1995) and Robert Parent of the University of Wisconsin-Madison. The radiometer provided the first measurements of Earth's radiation budget from space and established the important role of clouds in absorbing solar radiation. *Explorer 7* transmitted data continuously until February 1961. On 1 April 1960, the U.S. orbited the world's first weather satellite, *TIROS-1* (*Television and Infrared Observation Satellite*). Beginning in 1978, satellites were launched specifically to monitor the ocean. Now instruments flown on Earth-orbiting satellites routinely provide global images of the climate system.

B

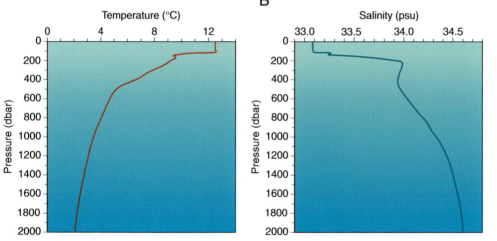

FIGURE 2.10
An Argo float (A) obtains continuous profiles of ocean temperature and salinity to a maximum depth of about 2000 m. The instrument surfaces and sends data to an orbiting satellite for downloading at a central processing facility. (B) Sample plots of float-derived temperature (left) and salinity (right) profiles were obtained at a location in the eastern North Pacific, west of Northern California, on 1 February 2008. Salinity is measured in practical salinity units (psu). [Courtesy of the U.S Global Ocean Assimilation Experiment.]

Already, the Argo float array has proven its value by increasing the accuracy of estimates of the heat storage in the ocean, an important factor in predicting the climate future and changes in sea level. These data have also made possible development of more realistic coupled atmosphere/ocean models used for seasonal climate forecasts. Finally, Argo float data provide insights on the dynamics of air-sea interactions operating in hurricanes and tropical storms.

Sensors flown on board Earth-orbiting satellites monitor the atmosphere, surface ocean waters, glacial ice sheets, vegetation, and other components of Earth's climate system. Powerful computers collect, process, and analyze enormous quantities of satellite-acquired environmental data. In only minutes, an ocean-observing satellite can collect as much data as an ocean-research vessel operating at sea continuously for a decade or longer. Sensors observing Earth from orbiting spacecraft, primarily those from NASA and NOAA, measure selected wavelengths (or frequencies) of the electromagnetic radiation reflected or emitted by the various components of Earth's climate system.

Electromagnetic radiation (or simply *radiation*) describes both a form of energy and a means of energy transfer. Forms of radiation include gamma rays, X-rays, ultraviolet (UV) radiation, visible light, infrared (IR) radiation, microwaves, and radio waves. Together, these forms of radiation make up the **electromagnetic spectrum** (Figure 2.11). All types of electromagnetic radiation travel as waves that are differentiated by wavelength or frequency. **Wavelength** is the distance between successive wave crests (or equivalently, wave troughs), as shown in Figure 2.12. **Wave frequency** is the number of crests (or troughs) that passes a given point in a specified period of time, usually one second. Passage of one complete wave is called a *cycle*, and a frequency of one cycle per second equals 1.0 hertz (Hz).

Wave frequency is inversely proportional to wavelength; that is, the higher the frequency, the shorter is the wavelength. Radio waves have frequencies in the millions of hertz with wavelengths up to hundreds of kilometers. By contrast, at the other end of the electromagnetic spectrum, gamma rays have frequencies as high as 10^{24} (a trillion trillion) Hz and wavelengths as short as 10^{-14} (a hundred trillionth) m. Furthermore, higher frequency radiation with shorter wavelengths has higher associated energy than lower frequency (longer wavelength) radiation.

Electromagnetic waves travel through space and may pass through gases, liquids, and solids. In a vacuum, all electromagnetic waves travel at maximum possible speed, 300,000 km (186,000 mi) per sec, known as the *speed of light*. All forms of electromagnetic radiation slow down when passing through materials, their speed varying with wavelength and

type of material. As electromagnetic radiation passes from one medium into another, it may be reflected or refracted (i.e., bent) at the interface. This happens, for example, where solar radiation strikes the ocean surface at an oblique angle—some radiation is reflected back to the atmosphere and some is bent (refracted) as it penetrates the water. Electromagnetic radiation is also absorbed, that is, converted to heat energy.

Satellites that monitor Earth's climate system fly in either geostationary or polar orbits. Most of us are familiar with images obtained by sensors onboard geostationary satellites, which are launched into relatively high orbits (36,000 km or 22,300 mi) (Figure 2.13). The *sub-satellite point*, the location on Earth's surface directly below the satellite, is essentially on the equator. A geostationary satellite revolves around Earth at the same rate and in the same direction as the planet rotates so that the satellite is always positioned over the same spot on Earth's surface and its sensors have a consistent field of view. For this reason, these satellites are sometimes described as *geosynchronous*.

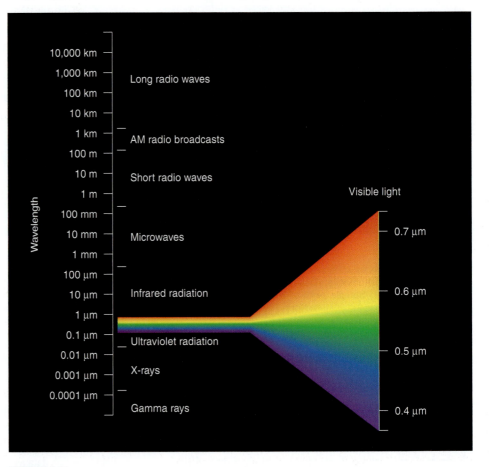

FIGURE 2.11
The electromagnetic spectrum in which the various forms of electromagnetic radiation are distinguished by wavelength in micrometers (μm), millimeters (mm), meters (m), and kilometers (km).

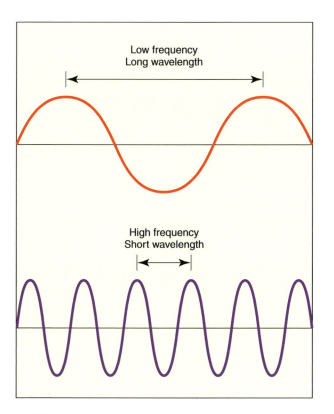

FIGURE 2.12
Wavelength is the distance between two successive wave crests or, equivalently, between two successive wave troughs. Wavelength is inversely related to wave frequency.

NOAA operates two geostationary weather satellites for the United States with one over South America (near 75 degrees W longitude) and the other over the eastern

Pacific Ocean (approximately 135 degrees W longitude). These two **Geostationary Operational Environmental Satellites (GOES-East, GOES-West)** provide a complete view of much of North America and adjacent portions of the Pacific and Atlantic Oceans. A consortium of European nations operates Meteosat-6 (63 degrees E), Meteosat-7 (57.5 degrees E, over the Indian Ocean), Meteosat-8 (3.5 degrees W), and Meteosat-9 (0 degrees, over the Atlantic Ocean). Russia's GOMS is over the equator south of Moscow; Japan's MTSAT-IR is at 140 degrees E over the mid-Pacific; and China has geostationary satellites at 105 degrees E (FY-2C) and 86.5 degrees E (FY-2D). In addition, India has launched a geostationary satellite with meteorological sensors onboard.

Each geostationary satellite views about one-third of Earth's surface and five satellites are needed to provide complete and overlapping coverage of the globe between about 60 degrees N and 60 degrees S. Considerable distortion sets in near the edge of this field of view so that polar-orbiting satellites complement geostationary satellites in monitoring Earth's climate system.

A polar-orbiting satellite travels in a relatively low-altitude (800 to 1000 km, 500 to 600 mi), nearly north/south orbit passing close to the poles (Figure 2.14). The satellite's orbit traces out a plane in space while the planet continually rotates on its axis through the plane of the satellite's orbit. With each orbit, points on Earth's surface (except near the poles) move eastward so that onboard sensors sweep out successive overlapping north/

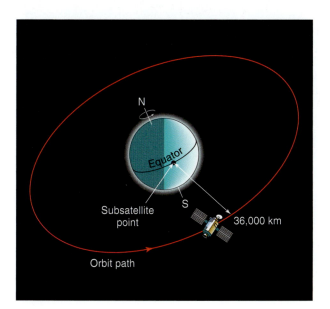

FIGURE 2.13
A satellite in geostationary orbit about the Earth.

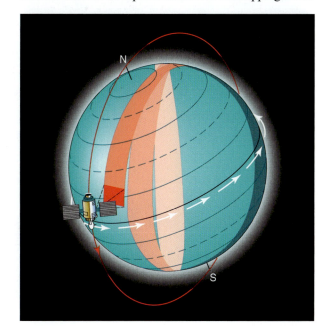

FIGURE 2.14
A satellite in polar orbit about the Earth.

south strips. A polar-orbiting satellite that follows the Sun (Sun-synchronous) passes over the same area twice during each 24-hr day. (That is, it completes one near-polar orbit about 14.1 times daily.) Other polar-orbiting satellites are positioned so that they require several days before passing over the same point on Earth's surface.

NOAA operates **Polar-orbiting Operational Environmental Satellites (POES)** that provide global coverage four times daily. They are equipped with the Advanced Very High Resolution Radiometer (AVHRR) and the Tiros Operational Vertical Sounder (TOVS). POES applications include weather analysis and forecasting, climate research and prediction, global SST measurements, atmospheric soundings of temperature and humidity, and monitoring volcanic eruptions. The clear atmosphere is essentially transparent to visible light and is selectively transparent by wavelength to other types of radiation such as infrared (Chapter 3). Satellite-borne sensors monitor these forms of radiation to gather information on atmospheric processes and properties. Ocean water is much less transparent to electromagnetic radiation than is the atmosphere so that remote sensing by satellite is essentially limited to obtaining data on surface or near-surface oceanic processes and properties.

Satellite-borne sensors are either passive or active. *Passive sensors* measure radiation coming from the Earth-atmosphere-ocean system, that is, visible solar radiation reflected by the planet and invisible infrared radiation emitted by the planet. Sunlight reflected by Earth's surface and atmosphere produces images that are essentially black and white photographs of the planet. Sun-lit, highly reflective surfaces such as cloud tops and snow-covered ground appear bright white whereas less reflective surfaces such as evergreen forests and the ocean appear much darker. Cloud patterns on a **visible satellite image** are of particular interest to atmospheric scientists (Figure 2.15). From analysis of cloud patterns displayed on the image, they can identify not only a specific type of weather system (such as a hurricane), but also the stage of its life cycle and its direction of movement when a sequence of images is animated.

A second type of passive sensor onboard a satellite detects infrared radiation (IR). IR is an invisible form of radiation emitted by all objects continually, both day and night. Hence, an **infrared satellite image** of the planet can provide useful information at any time whereas visible weather satellite images are available only during daylight hours. (Because of around-the-

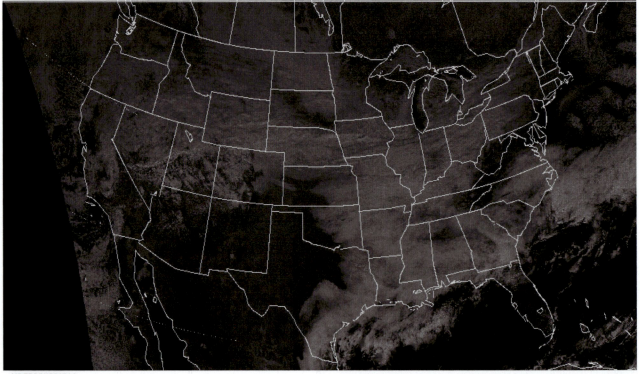

Visible Image   1815Z 25 JAN 2009

NCEP/NWS/NOAA

FIGURE 2.15
A sample visible GOES satellite image is essentially a black and white photo of the planet.

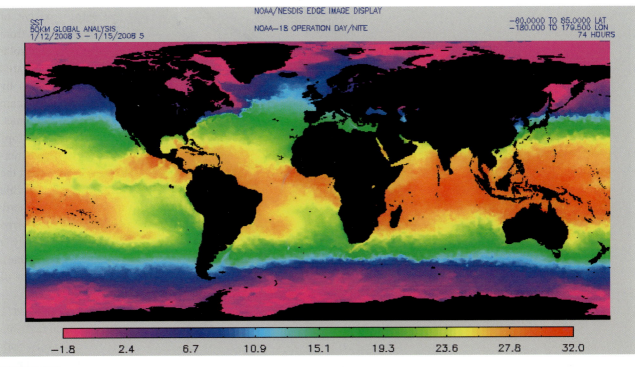

FIGURE 2.16
An infrared satellite image obtained from NOAA-18, a polar orbiting satellite, showing the sea surface temperature (SST) pattern averaged over a three-day period. Temperatures are color-coded in °C as shown in the bar at the bottom of the image. [Source: NOAA]

clock availability, infrared satellite images are usually shown on television weathercasts.) A valuable product of passive sensors for climate studies is surface temperature patterns derived by infrared (IR) sensors. The intensity of infrared radiation emitted by an object increases rapidly as the surface temperature of the object rises (Chapter 3). By calibrating temperature against IR-emission, a satellite sensor measures surface temperatures and can distinguish, for example, between warm and cold cloud tops and warm and cold ocean currents. For example, Figure 2.16 is an infrared satellite image of the global sea surface temperatures (SST) averaged over three days. SST is color coded so that oranges and reds represent the highest temperatures. In this Northern Hemisphere winter depiction, the highest SST values were in the Southern Hemisphere. Among other ocean properties measured by passive satellite sensors is water color, indicating marine production or sediment concentration.

Water vapor satellite imagery enables scientists to track the movements of moisture plumes within the atmosphere over distances of thousands of kilometers. This is a particularly valuable tool in investigating monsoon climates. Water vapor is an invisible gas and its presence is not detectable on visible or conventional infrared satellite images. But water vapor efficiently absorbs and emits certain wavelength bands of IR so that passive

sensors onboard weather satellites that are sensitive to these wavelengths can detect water vapor. Water vapor imagery displays the water vapor concentration at altitudes between about 5000 and 12,000 m (16,000 and 40,000 ft) on a gray (or color) scale (Figure 2.17). At one extreme, dark gray indicates almost no water vapor whereas at the other extreme, milky white signals a relatively high concentration of water vapor. High clouds appear milky to bright white on water vapor images, masking water vapor concentrations in the image.

Radar instruments mounted on satellites are *active sensors*; they emit pulses of microwave radiation to the ocean surface and then record the reflected signal to measure surface roughness (an indicator of surface wind speeds and wave heights) as well as ocean-surface elevation, which are used to map surface currents. Techniques to measure salinity and other ocean properties remotely are under development. Such measurements permit studies of ocean surface "weather" (week-to-week variations) as well as ocean "climate" (variability over decades to centuries).

NASA'S EARTH OBSERVING SYSTEM

The planetary system continually undergoes change (including climate change) due to both natural processes and human activity. Maintaining and improving

Water Vapor Image   0315Z 26 JAN 2009

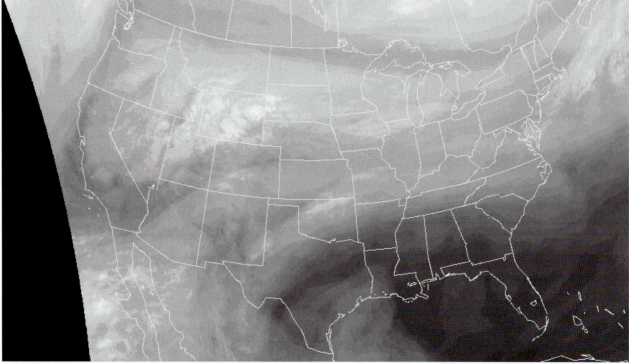

NCEP/NWS/NOAA

FIGURE 2.17
A sample GOES water vapor image displaying the water vapor concentration at altitudes between about 5000 and 12,000 m (16,000 and 40,000 ft) on a gray scale. Dark gray indicates almost no water vapor whereas milky white signals a relatively high concentration of water vapor.

the quality of life on Earth (e.g., fostering sustainable development) requires a thorough understanding of the planetary system and for that a global perspective is essential. Such a perspective provides data on the real-time state of the planetary system and the basis for predicting its future evolution. This Earth system science approach reveals information on natural variability, the forces and processes involved, and their interactions.

"How is the Earth changing and what are the consequences for life on Earth?" is a fundamental question that motivates the **National Aeronautics and Space Administration's (NASA's)** strategic goal in Earth Science. As mandated by the *National Aeronautics and Space Act of 1958*, NASA's mission in Earth Science is to ". . . conduct aeronautical and space activities so as to contribute materially to . . . the expansion of human knowledge of the Earth and of phenomena in the atmosphere and space." As will quickly become evident in this book, by conducting Earth system science from space, NASA complements the efforts of other Federal Agencies with Earth system mandates including the National Oceanic and Atmospheric Administration (NOAA), the

U.S. Geological Survey (USGS), and the Environmental Protection Agency (EPA).

In recent decades, many Earth-orbiting satellites have been launched that are equipped with a variety of sensors that monitor components of Earth's climate system. In this section, we briefly describe the objectives of selected satellites in NASA's *Earth Observing System*. In later chapters, we will refer to climate-related products of these and other satellites.

The *TOPEX/Poseidon satellite*, a joint mission of NASA and the *Centre National d'Etudes Spatiales (CNES)* in France, provided the first continuous global coverage of ocean surface topography (sea level) at 10-day intervals. Launched in 1992, it ceased operating in January 2006 after almost 62,000 orbits of the planet. Radar (microwave) altimeters on board the satellite (in a polar orbit at an altitude of 1336 km or 830 mi) bounced microwaves off the ocean surface to obtain precise measurements of the distance between the satellite and the sea surface. These data, combined with data from the Global Positioning System (GPS), generated images of sea surface height. Elevated topography (hills) indicates warmer than usual water whereas areas of low topography (valleys) indicate colder than usual water. Such images can be used to calculate ocean surface currents and identify and track the development of El Niño and La Niña (Chapter 8).

In December 2001, NASA and its French counterpart CNES, launched *Jason 1*, successor to TOPEX/Poseidon, that continues to gather data on how the ocean circulation affects Earth's climate. Among the primary objectives of Jason 1 are to improve climate forecasting, measure changes in sea level in response to global climate change, and study large-scale ocean circulation and heat transport. Sensors on Jason 1 map ocean surface topography (to an accuracy of 3.3 cm or 1.3 in.), wind speed, and wave heights over about 95% of Earth's ice-free ocean surface every 10 days. On 15 June 2008, *Jason 2*, the successor to Jason 1, was launched into the same orbit as Jason 1. This latest Ocean Surface Topography Mission (OSTM) is an international collaborative effort involving NASA, NOAA, CNES, and the *European Organisation for the Exploitation of Meteorological Satellites (EUMETSAT)*. Designed to operate for at least three years, Jason 2 will extend the continuous climate record of precise sea surface height measurements into the decade of the 2010s using the next generation of more accurate instruments.

Terra is a multi-sensor satellite that monitors Earth's land, atmosphere, ocean, and energy balance with the objective of learning more about Earth's climate system. Launched on 18 December 1999, Terra is a multi-national mission involving partnerships between NASA and the aerospace agencies of Canada and Japan. Terra carries on board five sensors. The *Advanced Spaceborne Thermal Emission and Reflection Radiometer (ASTER)* obtains high resolution images of Earth in 14 wavelengths of electromagnetic radiation ranging from visible to infrared. From this, scientists generate maps of surface temperature, reflectivity, and elevation. The *Clouds and the Earth's Radiant Energy System (CERES)* measures the planet's total radiation budget and the role of clouds. The *Multi-angle Imaging Spectro-Radiometer (MISR)* measures the amount of sunlight scattered by the planet in different directions. The *Moderate-resolution Imaging Spectroradiometer (MODUS)* sees every point on Earth's surface every 1 to 2 days in 36 spectral bands. MODUS, together with MISR and CERES, helps scientists determine the impact of clouds and aerosols on Earth's energy budget. *Measurement of Pollution in the Troposphere (MOPITT)* is an instrument that monitors the distribution, transport, sources, and sinks of carbon monoxide and methane in the troposphere.

Aqua, a NASA satellite launched on 4 May 2002, is equipped with six Earth-observing sensors intended primarily to gather data on the global water cycle. These data include measurements of evaporation from the ocean, atmospheric water vapor, clouds, precipitation, soil moisture, sea ice, land ice and snow cover, radiative energy fluxes, aerosols (Figure 2.18), vegetative cover on land, phytoplankton and dissolved organic matter in the ocean, plus surface temperatures.

CloudSat is a NASA satellite launched in 2006 and expected to operate for at least 3 years. A radar system, developed by NASA and the Canadian Space Agency, measures the altitude and properties of clouds thereby adding to our understanding of the role of clouds in climate change. For example, CloudSat data indicate that the total cloud cover over the western Arctic was about 16% less during the 2007 melt season compared to the previous year. More intense solar radiation raised the surface water temperature 2.4 Celsius degrees (4.3 Fahrenheit degrees) contributing to the record minimum summer extent of Arctic sea ice (Chapter 11).

Complementing CloudSat is the *Cloud-Aerosol Lidar and Infrared Pathfinder Satellite Observation (CALIPSO)* also launched in 2006. The CALIPSO satellite is equipped with an active lidar along with passive visible and infrared imagers. (Analogous to microwave radar, a *lidar* instrument uses a pulsed laser transmitter and optical receiver to detect various atmospheric targets.) CALIPSO investigates the vertical structure and properties of thin clouds and aerosols and, when combined with CloudSat, provides a three-dimensional perspective of how clouds form, evolve, and influence weather and climate. CALIPSO is a joint U.S. (NASA) and French (CNES) mission with a life expectancy of 3 years.

NASA's *Ice, Cloud, and Land Elevation Satellite (ICESat)* launched in 2003, is equipped with the *Geoscience Laser Altimeter System (GLAS)*. ICESat's primary objectives are to monitor the mass balance of the Antarctic and Greenland ice sheets, determine how they contribute to sea level change, and provide data essential for predicting future changes in ice volume and sea level. GLAS measures the distance from the spacecraft to the ice sheet with an accuracy near 10 cm (3.9 in.) and is the basis for determining ice sheet topography. ICESat also measures altitudes of clouds and aerosols, land topography, and vegetation characteristics. Its expected life is 3 to 5 years.

Aqua, CloudSat, and CALIPSO are part of a group of U.S. and international satellites flying in

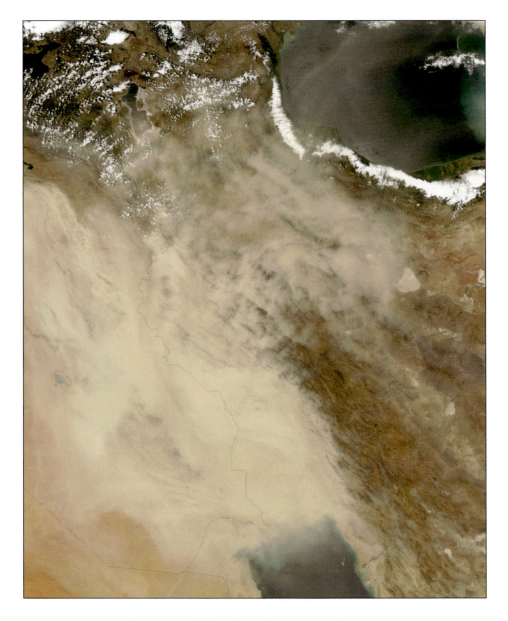

FIGURE 2.18
A severe dust storm swept over Iraq during the first week of July 2009 causing respiratory distress for many people and disrupting travel by land and air for almost a week. This photo-like image was obtained by the Moderate Resolution Imaging Spectroradiometer (MODIS) on NASA's Aqua satellite at the height of the storm on 5 July. [Source: NASA]

formation around the globe. Each member of this Afternoon Constellation of satellites, sometimes called the A-Train, has complementary measurement capabilities that together provide new insight on the role of clouds in weather and climate prediction.

The flood of observational data from satellite-based sensors is analyzed and stored by computers. Via technical advances, computer speed doubles about every 18 months. Ordinary personal computers, now inexpensive and widely available, far exceed the capabilities of the most powerful, room-sized supercomputers in use only a few decades ago. Internet Websites now make available

in near-real time Earth's climate system data that were impossible to obtain only a decade ago.

Studies of Earth's climate system also require satellite-based communications systems. These communications capabilities are now routinely combined as one system for observing both the ocean and atmosphere. Hence, in addition to serving as a platform for remote sensing instruments that make measurements of the planet, some of the satellites serve as relay stations where data obtained, say from ARGO floats, are uploaded to the satellite and then transmitted (or relayed) to a central processing location.

International Cooperation in Understanding Earth's Climate System

Earth's climate system knows no political boundaries and climate change can have far-reaching global ramifications. A number of efforts currently are underway to standardize international monitoring of the climate system and to promote and synthesize contributions to the understanding of climate science.

GLOBAL EARTH OBSERVATION SYSTEM OF SYSTEMS (GEOSS)

Of great potential value in monitoring the climate system is the international cooperative effort currently underway to build the Global Earth Observation System of Systems (GEOSS). In July 2003, representatives of 33 nations plus many international organizations assembled at the Earth Observation Summit in Washington, D.C., and adopted a resolution pledging a political commitment to establish an Earth Observation System, that is, a worldwide public infrastructure that integrates existing and new observation systems, making available environmental data and analyses for a variety of users. In April 2004, representatives of 47 nations met in Tokyo, Japan at the second Earth Observation Summit and adopted a framework document for a 10-year implementation plan (2005-2015) for GEOSS. As of this writing, some 74 nations and 51 international organizations are participating in building GEOSS.

GEOSS will monitor and measure continually the state of the environment at Earth's surface, in the atmosphere, and from space resulting in improved understanding of dynamic Earth processes and enhanced predictive capabilities. Through GEOSS, decision-makers and stakeholders will have access to an extraordinary range of timely, high quality, and long-term data and models for the benefit of society. GEOSS will link together existing and new observation systems, hardware, and analytical software packages in compatible formats. Through GEOSS, nations will participate in the free and open exchange of information and coherent data sets at no cost via the Internet and a network of telecommunications satellites. According to NOAA, GEOSS aims "to provide the right information, in the right format, to the right people, at the right time, to make the right decisions." The U.S. and other developed nations will assume most costs, assemble and maintain the system, gather environmental data, enhance data distribution, and make available predictive models. Adaptation to climate variability and climate change is among nine areas of critical importance to society being addressed by GEOSS. Others include natural and human-induced disasters, managing energy resources, and promoting sustainable agriculture.

In embarking on the task of building GEOSS, participating nations have to confront many major obstacles. These include agreeing on common data formats for the systems and deciding which data will be shared, especially in view of possible implications for national security.

INTERGOVERNMENTAL PANEL ON CLIMATE CHANGE (IPCC)

In addition to a coordinated international effort to monitor the climate system, the global nature of climate change calls for international cooperation in the climate science enterprise. To this end, the **Intergovernmental Panel on Climate Change (IPCC)** was formed in 1988 by the World Meteorological Organization (WMO) and the United Nations Environmental Programme (UNEP). The IPCC is charged with evaluating the state of climate science as the basis for policy action and serving the interests of scientists, public policy-makers, and through them the public at large. Specifically, the IPCC assesses the "scientific, technical and socio-economic information relevant to understanding the scientific basis of risk of human-induced climate change, its potential impacts and options for adaptation and mitigation." The IPCC does not plan or conduct research or monitor climate data. In fact, for the IPCC to be objective, transparent, and rigorous in its evaluation of information, assessment must be independent of research. More than 2000 scientists representing 154 nations have participated in the IPCC process.

The IPCC has three working groups and a task force. *Working Group I* assesses the scientific aspects of the climate system and climate change. *Working Group II* assesses the vulnerability of socio-economic and natural systems to climate change, negative and positive consequences of climate change, and options for adapting to it. *Working Group III* assesses options for limiting greenhouse gas emissions and otherwise mitigating climate change. *The Task Force on National Greenhouse Gas Inventories* is responsible for the IPCC National Greenhouse Gas Inventories Programme. Assessment Reports were completed in 1990, 1995, 2001, and 2007.

In 2007, the IPCC concluded that global warming since the mid-20th century *very likely* (estimated probability of greater than 90%) was caused mostly by human activities. In a report issued six years earlier, the

IPCC described the human role in global warming as *likely* (estimated probability of higher than 66%). The *Synthesis Report of the IPCC Fourth Assessment Report*, issued in November 2007, concluded that: "Warming of the climate system is unequivocal as is now evident from observations of increases in global average air and ocean temperatures, widespread melting of snow and ice, and rising global average sea level." The Report goes on to state that: "Most of the increase in globally-averaged temperatures since the mid-20th century is *very likely* due to the observed increase in anthropogenic greenhouse gas concentrations."

Modeling Earth's Climate System

A **model** is an approximate representation or simulation of a real system, incorporating only the essential features (or variables) of a system while omitting details considered non-essential or non-predictable. Models are useful in conveying the basic understandings of climate science. Models are widely used to investigate Earth's climate system, the interactions of its components, and predicting the climate future. Essential in the design and validation of a model are reliable and representative observational data. Models are conceptual, graphical, physical, or numerical.

A *conceptual model* is a statement of a fundamental law or relationship. This type of model organizes data or describes the interactions among components of a system. For example, conceptual models are used to explore linkages among physical and biological subsystems in the ocean and they enable us to understand how and why the atmosphere and ocean circulate. We use conceptual models throughout this book. A *graphical model* compiles and displays data in a format that readily conveys meaning. For example, climate scientists make extensive use of maps on which are plotted mean values of climatic elements such as temperature, precipitation, and snowfall to represent spatial patterns of these climatic elements. A *physical model* is a small-scale (miniaturized) portrayal of a system. For example, before powerful computers were widely available, physical models were used to simulate the time-averaged planetary-scale circulation of the atmosphere (dishpan models). Today, because powerful computers are readily available, numerical models have essentially replaced physical models in investigating the climate system.

The physical evolution of some systems (e.g., the orbit of the Moon) can be determined precisely by solving a specific mathematical equation. Other systems such as the fluid atmosphere or ocean are too complex to be described by a simple single mathematical formula. In these cases, statistics or mathematics approximates the behavior of the system on a computer. A *numerical model* consists of many mathematical equations that simulate the processes under study. Weather forecasts are probably the most familiar products of numerical models; in fact, numerical models of the Earth-atmosphere system have been used to forecast weather since the 1950s. Observational data are used as initial and boundary conditions as well as to guide and verify model predictions. Usually, a numerical model for weather prediction is initialized using current observational data. With the current state of the atmosphere/ocean as a starting point, a prediction is made for some subsequent time interval, say 10 minutes hence. Repetition of this process eventually generates a prediction of conditions for the subsequent 12, 24, 36 hrs or longer. Through this iterative process, a numerical model predicts the future state of a complex system. Very powerful computers are needed to handle the complex mathematics and huge data sets required for ocean, atmosphere, or coupled ocean-atmosphere numerical models. Equations may be altered or new data provided to simulate different situations.

Computer-based numerical climate models are either empirical or dynamic. Empirical climate models, based as they are on data sets of actual weather observations, are most appropriately used to predict climate variability. Dynamic climate models, based on interacting forcing mechanisms internal and external to the climate system, are used to predict climate change. These same two types of numerical models also are employed to predict the onset and life cycle of El Niño and La Niña (Chapter 8).

In the U.S., numerical weather and climate forecasting is the purview of NOAA's **National Centers for Environmental Prediction (NCEP)**. Listed in Table 2.2 are the nine components of NCEP and their specific locations and responsibilities. According to its mission statement, NCEP aims to deliver "analyses, guidance, forecasts and warnings for weather, ocean, climate, water, land surface and space weather to the Nation and the world." Furthermore, NCEP "provides science-based products and services through collaboration with partners and users to protect life and property, enhance the Nation's economy and support the Nation's growing need for environmental information." NCEP produces a "seamless suite" of National Weather Service (NWS) forecast products and services using sophisticated

TABLE 2.2
Components of NOAA's National Centers for Environmental Prediction (NCEP)[a]

- *Hydrometeorological Prediction Center*, College Park, MD, provides analysis and forecast products, specializing in quantitative precipitation forecasts to five days, weather forecast guidance to seven days, real-time weather model diagnostics discussions and surface pressure and frontal analyses.

- *Climate Prediction Center*, College Park, MD, serves the public by assessing and forecasting the impacts of short-term climate variability, and emphasizing enhanced risks of weather-related extreme events for use in mitigating losses and maximizing economic gain.

- *Aviation Weather Center*, Kansas City, MO, provides aviation warnings and forecasts of hazardous flight conditions at all levels within domestic and international airspace.

- *Ocean Prediction Center*, College Park, MD, issues weather warnings and forecasts out to five days, in graphical, text and voice formats for coastal areas and the Atlantic and Pacific Oceans, north of 30 degrees North.

- *Space Weather Prediction Center*, Boulder, CO, provides space weather alerts and warnings for disturbances that can affect people and equipment working in space and on Earth.

- *Storm Prediction Center (SPC)*, Norman, OK, provides accurate tornado and severe weather forecasts and watches for the contiguous United States along with a suite of hazardous weather and mesoscale products. The SPC continually monitors mesoscale atmospheric processes related to severe weather and tornado outbreaks, extreme winter weather events and critical fire weather conditions.

- *Tropical Prediction Center (TPC)*, Miami, FL, provides official NWS forecasts of the movement and strength of tropical weather systems and issues the appropriate watches and warnings for the U.S. and surrounding areas. The TPC also issues a suite of marine products for the tropical Atlantic and Pacific.

- *Environmental Modeling Center*, College Park, MD, develops and improves numerical weather, climate, hydrological and ocean prediction through a broad program of applied research in data analysis modeling and product development.

- *NCEP Central Operations*, College Park, MD, sustains and executes the operational suite of the numerical analyses and forecast models and prepares NCEP products for dissemination.

[a]From NOAA's National Centers for Environmental Prediction (NCEP)

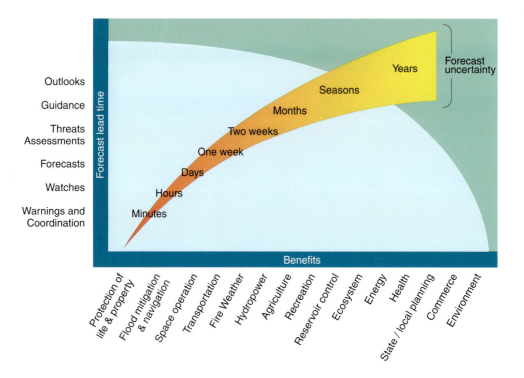

FIGURE 2.19
NOAA's National Centers for Environmental Prediction produces a seamless suite of forecast products and services using sophisticated numerical models and supercomputers. [From NOAA, NCEP]

numerical models and supercomputers, and is informed by scientific and technological advances in the research community (Figure 2.19).

SHORT-TERM CLIMATE FORECASTING

NCEP's **Climate Prediction Center** "assesses and forecasts the impacts of short-term climate variability, emphasizing enhanced risks of weather-related extreme events for use in mitigating losses and maximizing economic gain." Forecasters prepare 30-day (monthly), 90-day (seasonal), and multi-seasonal generalized climate outlooks that map areas where the probability of temperature and total precipitation is expected to depart from long-term averages ("normals") (Figure 2.20). Outlooks are issued from two weeks to 12.5 months in advance for the coterminous U.S., Hawaii, and other Pacific islands.

Successfully forecasting the prevailing circulation pattern at the 700-mb level of the atmosphere (i.e., where the air pressure is everywhere 700 millibars, at an average altitude of approximately 3000 m or 10,000 ft above sea level) is the first step in predicting the probability of monthly temperature and precipitation anomalies. The present circulation pattern is extrapolated into the future, although an effort is also made, based on historical data,

to identify features of the present pattern that are most likely to persist. The prevailing westerly flow at the 700-mb level is a steering wind that permits identification of the flow of warm and cold air masses as well as principal storm tracks. Predictions of surface temperature and precipitation anomalies derive from this analysis.

A somewhat different approach is taken for 90-day outlooks (issued once each month near mid-month). Forecasters rely more on long-term trends and recurring events, attempting to isolate persistent circulation features from prior months and seasons. A suite of 40 NCEP *Climate Forecasting System (CFS)* runs yields a mean prediction. Model runs depend heavily on tropical Pacific sea-surface temperatures (SST) specified by a coupled ocean-atmosphere model. The model-generated outlook accounts for the influence of not only El Niño and La Niña, but also low frequency fluctuations such as the North Atlantic Oscillation (Chapter 8).

In mid-December 1994, climatologists at NOAA's Climate Analysis Center (now the Climate Prediction Center) began issuing multi-seasonal outlooks for temperature and precipitation anomalies in 102 mega-climate divisions of the coterminous U.S. Forecasts were issued for a dozen 3-month periods each overlapping the prior period by 1 month. Hence, lead times successively increased by 1

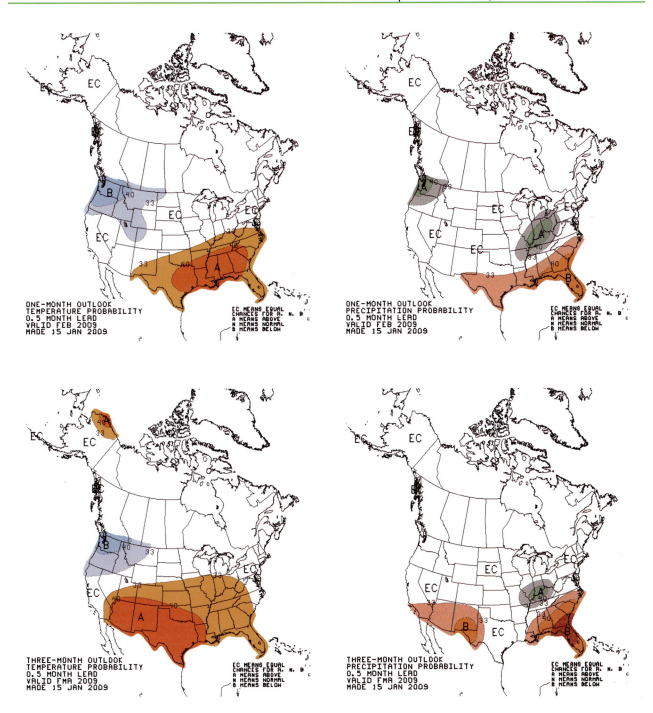

FIGURE 2.20
Sample monthly and seasonal outlook maps show the probability of total precipitation and temperature departing from normal, i.e., long-term average. *A* indicates a probability of higher than normal values whereas *B* indicates a probability of lower than normal values. *EC* stands for equal chances of above or below normal. [NCEP's Climate Prediction Center]

month out to 12.5 months. For example, in mid November 2004, a 12.5 month lead forecast was issued for December 2005 through February 2006. Temperature and precipitation anomalies are designated as above (A), near (N), or below (B) "normal" or equal chances (EC).

Following the first decade of operation, CPC climatologists R.E. Livezey and M.M. Timofeyeva conducted an assessment of multi-seasonal forecasting. Skill was rated on a scale of 0% to 100% where 0% represents a mere chance occurrence whereas 100% indicates perfection. Forecasting showed a modest increase in skill through the 10-year period. Overall, temperature forecast skill was 13%, up from 8% during the prior decade. However, during El Niño or La Niña, winter temperature forecasts with a lead time of up to more than 8 months had a skill score greater than 85% over much of the eastern U.S. The overall skill level for precipitation forecasts was only 3%. But during El Niño or La Niña, precipitation skill scores increased to 50% to more than 85% along the southern tier states and the West Coast with a lead time of about 6 months. We have more on the role of El Niño and La Niña in inter-annual climate variability in Chapter 8.

LONG-TERM CLIMATE FORECASTING

Other types of numerical models are used to predict the potential long-term impact on the global climate of rising levels of atmospheric carbon dioxide and other greenhouse gases. As discussed in more detail in Chapters 3 and 12, the atmospheric carbon dioxide concentration has been rising, especially since the beginning of the Industrial Revolution, due to our increased reliance on fossil fuels. (As discussed in this chapter's first Essay, the actual onset of the human impact on climate may predate the Industrial Revolution by thousands of years.) The likely consequence is higher air temperatures because, as noted in Chapter 1, carbon dioxide slows the escape of Earth's heat to space.

A **global climate model (GCM)** simulates Earth's climate system. One type of global climate model consists of mathematical equations that describe the physical interactions between the atmosphere and the other components of the climate system, that is, the ocean, land, ice-cover, and biosphere. The focus of global climate models differs from that of numerical weather prediction (NWP) models in that climate forecasting is a *boundary value problem* whereas weather forecasting is an *initial value problem*. NWP models begin with conditions representing the state of the atmosphere at the present time and apply the laws of physics to work

forward iteratively. Climate models are designed so that certain known boundary conditions (e.g., solar radiational forcing, concentrations of greenhouse gases, or surface characteristics) can be changed to determine how the climate adjusts to these new conditions. A climate model predicts broad regions of expected positive and negative temperature and precipitation anomalies (departures from long-term averages) and the mean location of circulation features such as jet streams and principal storm tracks over much longer time scales.

Climate scientists employ numerical global climate models in experiments to compute the magnitude of warming and other impacts that might accompany a continued increase in atmospheric carbon dioxide concentration. A global climate model is designed to incorporate the various boundary conditions imposed by the climate system and accurately depict the long-term average air temperature worldwide. Then, holding constant all other variables in the model, the concentration of atmospheric CO_2 (or other greenhouse gas) is elevated. Two different approaches are taken in adding CO_2: In a *transient run*, CO_2 is slowly added to the model and the effects are evaluated from moment to moment whereas with an *equilibrium run*, CO_2 is added all at once and the model is run until it achieves a new equilibrium. By comparing the new climate state with the present climate, scientists deduce the impact of an enhanced greenhouse effect on patterns of temperature and precipitation. Using boundary conditions derived from proxy climate data sources, global climate models have also been used to predict climates that prevailed in the geologic past such as the last glacial maximum (20,000 to 18,000 years ago).

Monitoring and numerical modeling of the climate system are essential partners in generating realistic predictions of the climate future. Monitoring the climate system is essential for determining the *actual* trends in climatic parameters. Global and regional climate models are used to predict how parameters such as mean annual temperature and sea level are likely to respond to anthropogenic forcings of the climate system. As noted, a model is an approximation of a real system and there may be advantages in comparing different models and model runs in an effort to understand the sensitivity of the models as well as the climate. Climate change studies, for example, routinely compare the output of multiple numerical models. Interacting atmospheric and oceanic processes are simulated using coupled models of the ocean, atmosphere, and biosphere for predicting responses to global climate change.

Lessons of the Climate Record

What does the climate record tell us about the behavior of climate through time? The following lessons of the climate record are useful in assessing prospects for the climate future and the possible impacts of climate change. We will return to them throughout this book.

- *Climate is inherently variable over a broad spectrum of time scales ranging from years to decades, to centuries, through millennia.*

- *Climate variability and climate change are geographically non-uniform in both sign (e.g., warmer or cooler, wetter or drier) and magnitude.*

- *Global- and hemispheric-scale trends in climate are not necessarily duplicated at particular locations although the magnitude of temperature change tends to amplify with increasing latitude (known as polar amplification).*

- *Climate change may consist of a long-term trend in various elements of climate (e.g., mean temperature or average precipitation) and/or a change in the frequency of extreme weather events (e.g., drought, excessive cold).*

- *Climate change tends to be more abrupt than gradual.*

- *Only a few cyclical variations can be discerned from the long-term climate record, which are associated with several known periodic cycles in climate forcing.*

- *Climate change impacts society.* For a discussion of climate change and the fate of the first Norse settlements in Greenland, refer to this chapter's second Essay.

Conclusions

To this point in our discussion, we have explored many of the basic spatial and temporal characteristics of climate with the objective of building a framework for identifying and applying the basic understandings of climate science. With this background, we take a closer look at Earth's climate system. We begin in the next chapter by applying the laws of radiation in examining the flow of energy into and out of the climate system, the interactions of solar radiation with the atmosphere and Earth's surface, the flow of terrestrial infrared radiation to space, and the greenhouse effect.

Basic Understandings

- From largest to smallest, scales of climate are designated macroclimate, mesoclimate, local climate, and microclimate.
- Change through time is a fundamental characteristic of climate. A distinction is made between climate variability and climate change. Climate variability refers to temporal variations of the climate around a mean state and typically applies to timescales of months to millennia or longer. Climate change is any systematic fluctuation in the climate state usually detected through deviations in long-term statistics of climate elements such as temperature.

- Climate scientists compare the average weather of a specific week, month, or year with the climate record. When these comparisons are carried out over a broad geographical area, departures from long-term averages, called anomalies, do not appear with the same sign or magnitude everywhere. Precipitation anomalies typically form more complex patterns than temperature anomalies.

- At middle and high latitudes, the geographical non-uniformity of climatic anomalies is linked to the prevailing winds blowing from the west in the middle and upper troposphere. In view of the finite number of waves in these westerly winds that typically encircle the hemisphere, a single weather extreme such as drought never encompasses an area as large as the coterminous United States or Canada.

- Geographic non-uniformity also applies to trends in climate. Hence, the trend in average annual temperature of the Northern Hemisphere is not necessarily representative of all localities within the hemisphere. Geographic non-uniformity also applies to the magnitude of a climate trend.

- Interactions among variables in the climate system involve feedback loops that may either amplify (positive feedback) or weaken (negative feedback) fluctuations in climate.

- A tipping point is a critical point or threshold in a dynamic system when a new and irreversible development, perhaps major, takes place. That is, some change in the system results in additional consequences.

- Using in situ and remote sensing techniques, climate scientists monitor Earth's climate system to better understand the nature, causes, and implications of climate variability and climate change. With in situ measurement, the sensor is immersed in the medium that is under surveillance. Remote sensing refers to acquisition of data on the properties of some object without the sensor being in direct physical contact with the object.

- A radiosonde is a small instrument package equipped with a radio transmitter that is carried aloft by a helium- or hydrogen-filled balloon and transmits altitude readings (soundings) of air temperature, pressure, and humidity to a ground station. Tracking the balloon's drift from ground stations (a rawinsonde observation) gives the variation in wind direction and speed with altitude.

- In situ sources of oceanic data, such as sea surface temperature, salinity and currents, useful in climate studies include ships at sea, instrumented buoys, floats, and gliders, piloted submersibles, autonomous instrumented platforms and vehicles, and undersea observatories.

- A profiling float (e.g., Argo) is an instrument package that measures vertical profiles of ocean water temperature, pressure (a measure of depth), and conductivity (a measure of salinity) to a maximum depth of about 2000 m (6600

ft). In 1998, scientists began deploying floats at 3-degree latitude/longitude intervals worldwide.

- Sensors on board Earth-orbiting satellites monitor the atmosphere, surface ocean waters, glacial ice sheets, vegetation, and other components of Earth's climate system. These sensors detect selected wavelengths (or frequencies) of electromagnetic radiation reflected or emitted by the various components of Earth's climate system.

- Electromagnetic radiation is a form of energy, a means of energy transfer, and a means for transmitting collected data. The various types of electromagnetic radiation make up the electromagnetic spectrum, travel as waves, and are differentiated by wavelength or frequency.

- Satellites that monitor Earth's climate system are in either geostationary or polar orbits. A geostationary satellite revolves around Earth at the same rate and in the same direction as the planet rotates so that satellite sensors have a consistent field of view. A polar-orbiting satellite travels in a relatively low altitude nearly north/south orbit passing close to the poles.

- Sensors on board Earth-orbiting satellites are either passive (measuring reflected or emitted radiation) or active (emit and receive radiation such as microwaves). Widely used products of passive satellite sensors include visible, infrared, and water vapor images.

- Earth's climate system knows no political boundaries and climate change can have global consequences. Hence, of great potential value in monitoring the climate system is the international cooperative effort currently underway to build the Global Earth Observation System of Systems (GEOSS).

- A model is an approximate representation or simulation of a real system, incorporating only the essential features (or variables) of the system while omitting details considered non-essential. Models used in climate science are conceptual, graphical, physical, or numerical. A numerical model consists of one or more mathematical equations that simulate the processes under study.

- Numerical global or regional climate models are important tools in predicting the short-term and long-term climate future. NOAA's National Centers for Environmental Prediction (NCEP) produces a "seamless suite" of National Weather

Service (NWS) forecast products and services using sophisticated numerical models and supercomputers, and is informed by scientific and technological advances by the research community.

- NCEP's Climate Prediction Center "assesses and forecasts the impacts of short-term climate variability, emphasizing enhanced risks of weather-related extreme events for use in mitigating losses and maximizing economic gain." Forecasters prepare 30-day (monthly), 90-day (seasonal), and multi-seasonal generalized climate outlooks that map areas where the temperature or total precipitation are expected to depart from long-term averages. Outlooks are issued from two weeks to 13 months in advance for the coterminous U.S., Hawaii, and other Pacific islands.

- Climate scientists employ numerical climate models in experiments to compute the magnitude of warming and other impacts that might accompany a continued increase in the atmospheric concentration of carbon dioxide and other greenhouse gases.

- Monitoring and modeling the climate system are essential partners in developing realistic predictions of the climate future.

- According to the climate record: Climate is inherently variable over a broad range of timescales; climate change is geographically non-uniform in both sign and magnitude; climate change may consist of a trend in mean values of climatic elements and/or a change in frequency of extremes; climate change tends to be more abrupt than gradual; few cyclical variations can be discerned from the long-term climate record; and climate change impacts society (Figure 2.21).

FIGURE 2.21
Drought apparently forced the Indians of Mesa Verde, in southwestern Colorado, to abandon their cliff dwellings around CE 1300. History recounts numerous instances when climate change significantly impacted society. Modern societies may be more capable of dealing with climate change than early peoples, but a rapid and significant change in climate would seriously impact all sectors of modern society. [Photo courtesy of Randolph Femmer/National Biological Information Infrastructure/USGS]

Enduring Ideas

- Change is an inherent characteristic of climate.
- Climate change is geographically non-uniform in both sign (direction) and magnitude.
- Interactions among variables in Earth's climate system involve feedback loops that may either amplify (positive feedback) or weaken (negative feedback) fluctuations in climate.
- A global climate model is an approximate simulation of Earth's climate system that is used to predict anomaly patterns in temperature and precipitation resulting from changes in boundary conditions.

Review

1.   Distinguish among macroclimate, mesoclimate, local climate, and microclimate.
2.   How does climate variability differ from climate change?
3.   Why are precipitation anomaly patterns typically more complex than temperature anomaly patterns?
4.   How does geographic non-uniformity apply to global or hemispheric trends in mean annual temperature?
5.   Distinguish between in situ measurement and remote sensing of Earth's climate system.
6.   What is the purpose of a radiosonde observation?
7.   What is the principal goal of the Argo float array?
8.   What is a major advantage of an infrared satellite image versus a visible satellite image?
9.   Distinguish between an active and passive sensor flown on board a satellite.
10. How is a global climate model used to predict the effect of fossil fuel combustion on global mean air temperature?

Critical Thinking

1.   How are climatic anomalies computed?
2.   Why does agroclimatic compensation generally apply to crops such as grains that are cultivated over broad geographical areas?
3.   Speculate on whether positive or negative feedback prevails in Earth's climate system. Justify your response.
4.   Explain how measuring the speed of sound in seawater can be used to determine the average water temperature between the sound transmitter and receiver.
5.   How do Argo float data add to our understanding of Earth's climate system?
6.   What advantage does a polar-orbiting satellite offer over a geostationary satellite in monitoring Earth's surface?
7.   What is GEOSS and what is its potential value in climate studies?
8.   Explain why monitoring and numerical modeling of the climate system are essential partners in generating realistic predictions of the climate future.
9.   Climate change may be expressed as a change in frequency of extreme weather events. Provide some possible examples.
10. What might be the implication for societal resilience if climate change is abrupt rather than gradual?

ESSAY: Beginnings of the Human Impact on Climate

It is popularly assumed that the anthropogenic contribution to the build-up of greenhouse gases in the atmosphere began at the onset of the Industrial Revolution, near the end of the 18th century. Humankind's increasing reliance on fossil fuels (coal at first, then oil and natural gas) released to the atmosphere increasing amounts of carbon dioxide (CO_2) with global climate change as a consequence. Recently, William F. Ruddiman, a marine geologist at the University of Virginia, took issue with this assumption arguing that humans' role in global climate change may have pre-dated the Industrial Revolution by thousands of years. Ruddiman proposes that agricultural activity was responsible for increasing the concentrations of CO_2 and methane (CH_4).

Ruddiman's hypothesis stemmed from his study of the 3-km (1.9-mi) length Vostok glacial ice core from Antarctica spanning the past 400,000 years. Glacial ice cores contain air bubbles trapped as snow accumulated and gradually transformed into layers of ice, so that the core serves as a proxy record of past climate. Chemical analysis of air bubbles enables climate scientists to develop a chronology of fluctuations in the concentrations of atmospheric gases including carbon dioxide and methane. These analyses show that until the beginnings of agriculture, the atmospheric concentrations of CO_2 and CH_4 faithfully followed a cycle with higher concentrations during the relatively mild interglacial climatic episodes and lower concentrations during the relatively cold glacial climatic episodes. To understand the reasons for these natural variations in CO_2 and CH_4, we need to briefly examine the cause of the interglacial/glacial climatic fluctuations of the Pleistocene Ice Age (1.7 million to 10,500 years ago).

Regular changes in Earth-Sun geometry are responsible for the large-scale variations in Earth's glacial ice cover. As discussed in more detail in Chapter 11, the three Milankovitch cycles (operating at periodicities of 22,000, 41,000, and 100,000 years) result in regular changes in the amount of solar radiation received on Earth by latitude and season. When the cycles favor less than the usual input of solar radiation in summer at high latitudes (e.g., northern Canada), temperatures are lower, and some of the winter snows persist through summer. Repetition of such summers eventually culminates in the formation of a glacier. At other times, the Milankovitch cycles favor more than the usual summer solar radiation, higher temperatures, and melting glacial ice or no ice at all.

Precession of Earth's rotational axis (similar to the wobble of a top) is the Milankovitch cycle that sets the tempo for the natural variability of methane (Figure 1). Earth's rotational axis completes one revolution every 22,000 years. The primary natural source of methane is anaerobic (without oxygen) decay of organic matter in wetlands. CH_4 concentration is highest during interglacial climatic episodes when the Northern Hemisphere is closest to the Sun in summer, temperatures are highest, and wetlands are most extensive. About 11,000 years into the cycle, the Northern Hemisphere is farthest from the Sun in summer, temperatures are lower, wetlands are less extensive, and the methane concentration dips to a minimum. About 5000 years ago, when Earth's precession cycle would normally favor a downturn in methane concentration, the concentration actually began to increase. That upward trend in CH_4 continued until the beginning of the Industrial Revolution with a total increase of 250 ppb (parts per billion).

The three Milankovitch cycles together drive the natural variability of atmospheric CO_2. As with methane, carbon dioxide concentration peaked during interglacial climatic episodes and bottomed out during glacial climatic episodes. After peaking about 10,500 years ago, the CO_2 concentration declined but then about 8000 years ago the trend reversed direction—rising about 40 ppm higher than expected by the beginning of the Industrial Revolution.

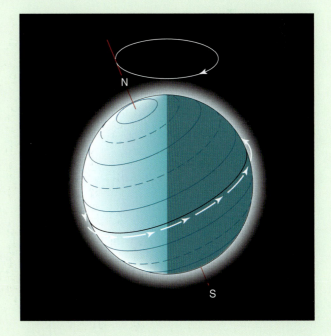

FIGURE 1
Earth's rotational axis completes one revolution every 22,000 years. This precession sets the tempo for the natural variability of methane concentration in the atmosphere.

Ruddiman attributes the unexpected upturns in methane and carbon dioxide concentrations to agricultural innovation and expansion. He notes that the increase in CH_4 corresponded in time to the onset and subsequent spread of irrigation for growing rice in southern Asia. Rice paddies are a major source of methane. Furthermore, Ruddiman claims that deforestation of land for agriculture was responsible for the unexpected upturn in carbon dioxide. CO_2 is released when felled trees are burned or left to decay. Deforestation for agriculture began in Europe and southern Asia about 8000 years ago.

Global climate models predict that a rise in methane by 250 ppb and carbon dioxide by 40 ppm in combination would elevate the global mean temperature by almost 0.8 Celsius degrees (1.4 Fahrenheit degrees) just prior to the Industrial Revolution. This warming trend appears to have partially offset a natural cooling trend. According to Ruddiman, uncompensated by anthropogenic greenhouse warming, that cooling trend would have resulted in temperatures averaging perhaps 2 Celsius degrees (3.6 Fahrenheit degrees) lower than at present.

At the December 2008 meeting of the American Geophysical Union (AGU), researchers S. Vavrus, J.E. Kutzbach, and G. Philippon of the University of Wisconsin-Madison reported that without Ruddiman's proposed early beginning to anthropogenic warming, glaciers around the world would now be growing instead of shrinking in a cooling climate. Regular changes in Earth-Sun geometry (the Milankovitch cycles) would drive a new ice age (Chapter 11). The researchers based their findings on computer models that compared a world unaffected by human activity with one in which anthropogenic warming began 8000 to 5000 years ago.

ESSAY: Climate Change and the Norse Settlements in Greenland

The late 9th century marked the beginning of a lengthy episode of unusually mild conditions throughout much of the Northern Hemisphere. During the *Medieval Warm Period*, a relatively warm climatic regime enabled Viking explorers to probe the northerly reaches of the Atlantic Ocean. Previously, severe cold and extensive drift ice were insurmountable obstacles to European navigators. By CE 930, the Vikings established the first permanent settlement in Iceland, some 970 km (600 mi) west of Norway and just south of the Arctic Circle.

Among Iceland's early inhabitants was Erik the Red (950-*ca.*1003), a troublesome individual whose exploits eventually caused his banishment from the island settlement in 982. He sailed west and discovered a new land, which he named Greenland. In spite of its name, then as now, much of Greenland was buried under a massive glacial ice sheet. The only habitable lands were small patches scattered along the coast, hemmed in by the sea and ice sheet, and separated by treacherous mountain ridges and deep fjords. Some historians speculate that Erik gave the land its name to entice others to follow him. At the time, however, the climate was so mild that some sheltered valleys were greener than they are today.

In such a place on Greenland's southwest coast in the year 986, Erik founded the first of two Norse settlements (the Eastern Settlement) (Figure 1). The second settlement (the Western Settlement) was located about 400 km (248 mi) to the north. Although never prosperous, Norse colonization of Greenland persevered for nearly five centuries. The Norse subsistence economy was primarily agrarian and depended on raising cattle, sheep, and goats. In addition, the Norse hunted migratory harp seals in spring and caribou in autumn. They even embarked on long, hazardous hunting expeditions northward along Greenland's west coast in search of walrus and polar bear, whose valuable tusks and hides they traded to Europeans for durable goods.

Initially, the Greenland colonists faired relatively well; the population of the Eastern Settlement climbed to an estimated 4000-5000, and that of the Western Settlement peaked at about 1000-1500. By about CE 1360, however, the Western Settlement was vacant, apparently the victim of some sudden calamity. The larger Eastern Settlement also succumbed, but more gradually, so that by 1500, Norse society no longer existed in Greenland. What happened to the Norse in Greenland can be inferred only from their graves, the ruins of their homes and barns, and a few chronicles. There were no survivors. In 1921, an expedition from Denmark examined the Norse remains and found evidence that the inhabitants had suffered a painful annihilation. Some grazing land was buried under advancing lobes of glacial ice while most farmland was made useless by permafrost. Near the end, the descendants of a robust and hardy people were ravaged by famine; they were crippled, dwarflike, and diseased.

In subsequent years, many explanations have been proposed for the extinction of Norse society in Greenland, and many factors probably contributed. Climate change and the inability (or unwillingness) of the Norse to adapt to an increasingly stressful climate and unreliable food sources, however, appear to have been the major causes of the disaster. Based on analysis of ice-core borings, by the 1300s, the climate was cooling rapidly, heralding the *Little Ice Age*. Sea ice expanded over the North Atlantic, hampering and eventually halting, navigation between Greenland and Iceland. All contact between the Greenland Norse and the outside world ended shortly after 1400.

FIGURE 1
General locations of the Eastern Settlement and Western Settlement in Norse Greenland.

In a novel approach to reconstructing the magnitude of cooling endured by the Norse in Greenland, Henry C. Fricke of the University of Michigan and his colleagues applied the oxygen-isotope technique of temperature reconstruction to human teeth. The teeth were recovered from skeletons of Norse and Thule Culture Eskimos from three Norse archeological sites in Greenland. Oxygen occurs as two isotopes, the more common ^{16}O and the less common ^{18}O. The ratio of ^{18}O to ^{16}O in the atmosphere depends on variables such as temperature and humidity, giving scientists a way of determining the average air temperature from the oxygen isotope ratio. The ratio of ^{18}O to ^{16}O in the dental enamel of modern animals is the same as that of the precipitation and spring water ingested by them during their formative years. Based on their analysis of teeth dated from CE 1100 to 1450, Fricke and colleagues reconstructed a drop in mean annual temperature of about 1.5 Celsius degrees (2.7 Fahrenheit degrees) during the 350-year period.

Using the same technique, other scientists detected a 3% decline in ^{18}O to ^{16}O ratio extracted from samples at various Greenland sites over the period 1400 to 1700, equivalent to cooling of about 6 Celsius degrees (11 Fahrenheit degrees). This 300-year period of cooling was followed by a 250-year warming trend of about 3 Celsius degrees (5.4 Fahrenheit degrees).

Deteriorating climate in Greenland meant wetter summers with poor haying conditions that caused major livestock losses. Longer, snowier winters probably claimed an even greater toll of livestock—especially among newborn animals. Evidence points to a dietary shift from terrestrial sources of food to marine sources. Climate change also disrupted the harp seal migration and decimated caribou herds, further reducing food sources. About the same time, the Norse had to compete for shrinking food sources with the Inuit people who had migrated southward along Greenland's west coast. Competition for food evidently led to hostilities between the two groups.

Unable to survive the stress of the colder climatic regime, the Norse succumbed to famine and yet the neighboring Inuit people survived. The Inuit probably succeeded because their hunting skills and techniques were better suited to the hostile climatic conditions. On the other hand, the Norse probably refused to adopt the hunting practices of the Inuit and clung tenaciously to their traditional subsistence methods. Unfortunately, those methods were no longer suited to the new climatic regime.

The Norse tragedy in Greenland may be one of only a very few historical examples of the extinction of a European society in North America. What lesson does it teach contemporary society? The lesson is three-fold: Climate changes and it can change rapidly—sometimes with serious, even disastrous, consequences. Secondly, nowhere are people more vulnerable to climate change than in regions where the climate is just marginal for their survival. These are regions where barely enough rain falls to sustain crops and livestock, or where mean temperatures are so low and the growing season so short that only a few hardy crops can be cultivated successfully. In such regions, even a small change in climate can make agriculture impossible. Thirdly, societies that can adapt survive changes in climate, whereas those that do not will fail. If people are unable to locate and utilize new food sources, or if they cannot or will not migrate to more hospitable lands, their fate may be similar to that of the Greenland Norse.

CHAPTER 3

PLANETARY ENERGY BUDGET IN EARTH'S CLIMATE SYSTEM

Solar power drives Earth's climate. The setting sun, photographed from the International Space Station. [Courtesy NASA/JSC]

Case-in-Point

Ancient people were well aware of the annual solar cycle and the march of the seasons. Knowing when the seasons begin and end is critical to the timing of planting and harvesting of crops (e.g., to reduce the likelihood of being exposed to a killing frost). In monsoon climates, the timing of the rainy and dry seasons is closely tied to the solar cycle. While these people did not possess a calendar in the modern sense, their knowledge of the path of the Sun through the sky and other regular astronomical events

inspired them to construct astronomical calculators in the form of elaborate megaliths.

Probably the best known of the ancient astronomical calculators is Stonehenge located in southern England, the earliest portion of which dates to about BCE 2950. As early as the 18th century, scientists noticed that the horseshoe arrangement of great stones opened in the direction of sunrise on the summer solstice. Stones also point in the direction of the mid-winter sunset. In more recent years, scientists discovered that the arrangement of stones and other features at Stonehenge could be used to predict solar and lunar eclipses. People living at Cahokia (a community just east of modern St. Louis, MO, that may have had as many as 40,000 residents) erected similar solar calendars consisting of wooden posts arranged in circles. These so-called Woodhenge calendars date from about CE 900 to 1100.

Predating Stonehenge by some two thousand years, the oldest astronomical calculator discovered so far consists of megaliths and a stone circle, located near Nabta in the Nubian Desert of southern Egypt. A present-day global positioning satellite confirms the stone's alignment with the position of the rising Sun on the summer solstice (as it would have been 6000 years ago). For people of that time and place, this was a significant date because monsoon rains typically begin shortly after the summer solstice.

Chankillo is a 2300 year-old Peruvian ruin located in a coastal desert about 400 km (250 mi) north of Lima. Until recently, archaeologists were unsure of Chankillo's function. Three thick concentric walls and hilltop location suggested fortress but there was no water supply. Relicts suggested rituals and possibly a temple. In 2007, researchers Iván Ghezzi of the Pontificia Universidad Católica del Peru and Clive Ruggles of the University of Leicester, UK, presented convincing evidence that at least part of Chankillo served as a solar observatory, the oldest one in the Americas.

While visiting Chankillo in 2001, Ghezzi observed a string of 13 nearly evenly spaced towers on a nearby ridge, oriented in a roughly north-south direction, and having a total length of about 300 m (985 ft). The squat towers are about 5 m (16 ft) apart and have a height of 2-6 m (7-20 ft), with flat tops measuring 11-13 m (36-43 ft) by 6-9 m (20-30 ft). Ghezzi suspected that the array of towers was an astronomical tool, pre-

FIGURE 3.1
Sun rise on the June solstice at the ancient solar observatory at Chankillo, Peru, erected 2300 years ago. [Photo © Iván Ghezzi]

dating by 1800 years the solar calendars that reputedly were used by the Incas of Peru. Returning to the site in 2004, Ghezzi discovered two observation sites, one located about 200 m (660 ft) east of the towers and the other about 200 m west of the towers. Ghezzi proposed that ancients followed the progress of the seasons from day to day by viewing the towers and noting the location of the rising Sun (from the western observation site) and setting Sun (from the eastern observation site). Over the course of a solar year, the location of sunrise and sunset shifted from the June solstice at the northern end of the tower array to the December solstice at the southern end, and then back again to complete the annual solar cycle (Figure 3.1). The Chankillo solar calendar had an accuracy of 2 to 3 days per year. Furthermore, the dual observation sites offered the ancients an opportunity for a second daily observation in the event that a thick cloud cover obscured their view of the Sun during the other observation.

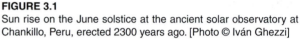

Driving Question:
How does solar and infrared radiation interact with Earth's climate system?

The Sun drives Earth's climate system; that is, the Sun is the ultimate source of energy that drives the circulation of the atmosphere and ocean, and powers winds and storms. **Energy** is defined as the capacity for doing work and occurs in many different forms such as

radiation and heat. The circulation of the atmosphere and ocean ultimately is responsible for the temporal and spatial variability of weather and is the major control of climate.

The Sun ceaselessly emits energy to space in the form of *electromagnetic radiation*. A very small portion of that energy is intercepted by the Earth-atmosphere-land-ocean system and converted to other forms of energy including, for example, heat and the kinetic energy of the winds and ocean currents. Energy can neither be created nor destroyed although it can be converted from one form to another; as noted in Chapter 1, this is the **law of conservation of energy** (also known as the *first law of thermodynamics*).

In this chapter, we examine the basic forms of electromagnetic radiation, laws governing electromagnetic radiation, how solar radiation interacts with the components of the Earth-atmosphere-land-ocean system, and the conversion of solar radiation to heat. The Earth-atmosphere-land-ocean system continually emits infrared radiation, some of which is absorbed by certain atmospheric gases (e.g., water vapor, carbon dioxide). In turn, those gases emit infrared radiation. Downward directed radiation elevates the temperature of Earth's surface and lower atmosphere to levels that make life possible (the *greenhouse effect*). We begin with a description of the various forms of electromagnetic radiation.

Forms of Electromagnetic Radiation

Essentially all objects absorb and emit electromagnetic radiation. Forms of electromagnetic radiation include radio waves, microwaves, infrared radiation, visible light, ultraviolet radiation, X-rays, and gamma radiation. Together, these forms of radiation make up the **electromagnetic spectrum**, illustrated in Figure 2.11. Although the electromagnetic spectrum is continuous, different names are assigned to different segments because we detect, measure, generate, and use those segments in different ways. Furthermore, the various types of electromagnetic radiation do not begin or end at precise points along the spectrum. For example, red light shades into invisible infrared radiation (infrared, meaning *below red*). At the other end of the visible portion of the electromagnetic spectrum, violet light shades into invisible ultraviolet radiation (ultraviolet, meaning *beyond violet*).

Beyond visible light on the electromagnetic spectrum, and in order of increasing energy level, increasing frequency, and decreasing wavelength, are ultraviolet radiation (UV), X-rays, and gamma radiation. All three types of radiation occur naturally and can be produced artificially. All have medical uses: ultraviolet radiation is a potent germicide; X-rays are used as a powerful diagnostic tool; and both X-rays and gamma

radiation are used to treat cancer patients. These three highly energetic types of radiation are dangerous as well as useful. Ultraviolet radiation can cause irreparable damage to the light-sensitive cells of the eye. Staring at the Sun (for instance, during a partial solar eclipse) can permanently blind a person unless a filter is used to block ultraviolet radiation. Also, overexposure to UV, X-rays, or gamma radiation can cause sterilization, cancer, mutations, or damage to a fetus. Fortunately, Earth's atmosphere blocks most incoming ultraviolet radiation and virtually all X-rays and gamma radiation through absorption by the rarefied gases of the upper atmosphere. Without this protective atmospheric shield, life as we know it would not exist. (Refer to the first Essay in Chapter 1.)

At lower frequencies and longer wavelengths, UV radiation shades into **visible radiation**, that is, radiation that is perceptible by the human eye. White light, as we call visible sunlight, is *polychromatic*, composed of multiple colors associated with individual wavelength bands. The wavelength of visible light ranges from about 0.40 micrometer at the violet end to approximately 0.70 micrometer at the red end. (One *micrometer* is a millionth of a meter, about one-tenth the thickness of a human hair.) Visible light is essential for many activities of plants and animals. In plants, light provides the energy needed for photosynthesis; it also regulates the opening of buds and flowers in spring and the dropping of leaves in autumn. For animals, light governs the timing of reproduction, hibernation, and migration and, for many species, makes vision possible thereby ensuring survival.

Between red light and microwave radiation on the electromagnetic spectrum is **infrared radiation (IR)**. IR is not visible, but we can feel the heat it generates when it is intense, as it is, for example, when emitted by a hot stove. Actually, every known object emits some amount of infrared radiation. As discussed later in this chapter, absorption and emission of IR by certain atmospheric gases are responsible for significant warming of the lower atmosphere (the *greenhouse effect*), resulting in a habitable planet.

At longer wavelengths is the microwave segment of the electromagnetic spectrum, which spans wavelengths of about 0.1 to 100 millimeters. Some microwave frequencies are used for radio communication, in microwave ovens, and in radar. At the low energy, low frequency, long wavelength end of the electromagnetic spectrum are radio waves. Wavelengths range from a fraction of a centimeter up to hundreds of kilometers, and frequencies can extend to a billion Hz. FM (frequency modulation) radio waves, for example, span 88 million to

108 million Hz; hence, the familiar 88 and 108 at opposite ends of the FM radio dial.

Radiation Laws

Several physical laws describe the properties of electromagnetic radiation emitted by a perfect radiator, a so-called blackbody. By definition, a **blackbody** at a constant temperature absorbs all radiation that is incident on it and emits all the radiant energy it absorbs. A blackbody is both a perfect absorber and perfect emitter of radiation. The wavelengths of emitted radiation are related to the temperature of the blackbody. Surfaces of real objects may approximate blackbodies for certain wavelengths of radiation but not for others. Freshly fallen snow, for example, is very nearly a blackbody for infrared radiation but not for visible light. (Note that *blackbody* does not refer to color.) Although neither the Sun nor Earth is a precise blackbody, their absorption and emission of radiation is sufficiently close to that of a blackbody so that we can apply blackbody radiation laws to them with some very useful results. Here we apply two blackbody radiation laws: Wien's displacement law and the Stefan-Boltzmann law.

All known objects emit and absorb all forms of electromagnetic radiation. However, the wavelength of most intense radiation (λ_{max}) emitted by a blackbody is inversely proportional to the absolute temperature (T) of the object. That is,

$$\lambda_{max} = C/T$$

where C, a constant of proportionality, has the value of 2897 if λ_{max} is expressed in micrometers, and T is in kelvins. *Absolute temperature* is the number of kelvins above absolute zero (-273.15 °C or -459.67 °F). This is a statement of **Wien's displacement law**. As shown in Figure 3.2, a significant portion of the radiation emitted (or absorbed) by an object is found around this peak maximum wavelength.

According to Wien's displacement law, hot objects (such as the Sun) emit radiation that peaks at relatively short wavelengths, whereas relatively cold objects (such as the Earth-atmosphere-land-ocean system) emit peak radiation at longer wavelengths (and lower frequencies). The top heating elements of an electric kitchen range provide an illustration of Wien's displacement law. After switching on a heating element, its metal coils warm and we readily feel the heat. We are actually feeling invisible

infrared radiation. As the coil temperature continues to rise, the coils emit more intense infrared radiation with peak emission at progressively shorter wavelengths. That is, as the coil temperature rises, the coil's peak radiation is displaced from the infrared toward the near-infrared portion of the electromagnetic spectrum. The highest setting on the stove will cause the heating element to glow dull red as some of the radiation is in the longer wave portion of the visible light spectrum.

Radiation emitted by the Sun is similar to that emitted by a blackbody at a temperature of about 6100 K (11,000 °F). Figure 3.3 shows the flux (energy transfer per unit time per unit area) of solar radiation received at the top of the atmosphere as a function of wavelength. The Sun emits a band of radiation (at wavelengths mostly between 0.25 and 2.5 micrometers) that is most intense at a wavelength of about 0.5 micrometer (in the green of visible light). The flux of radiation emitted by Earth's surface is similar to that emitted by a blackbody at a temperature of

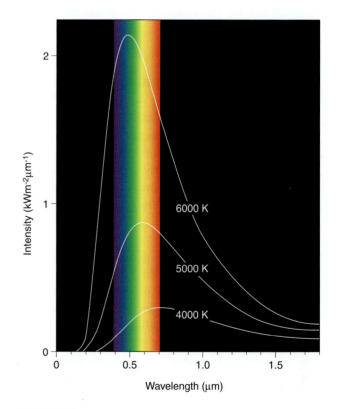

FIGURE 3.2
Family of radiation curves for ideal radiators or absorbers (black bodies) at different temperatures. With increasing absolute temperature (K), the wavelength of maximum radiation emitted or absorbed by an object decreases. This is known as Wien's displacement law. The total amount of energy emitted or absorbed by the object, as indicated by the area under the individual curves, increases by the fourth power of the temperature, as specified by the Stefan-Boltzmann law.

Spectral distribution of extra-atmospheric irradiance
from SOLAR2000

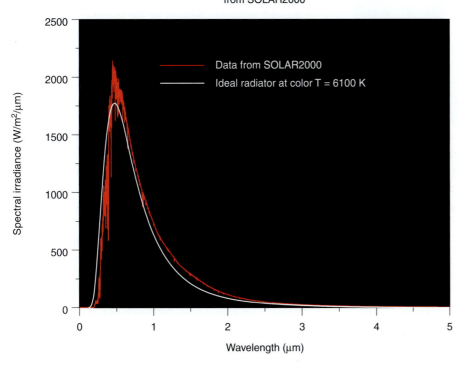

FIGURE 3.3
The flux of solar radiation incident at the top of the atmosphere as a function of wavelength. Radiation emitted by the Sun is similar to that emitted by a blackbody at a temperature of about 6000 °C (about 6100 K). [NOAA, Space Environment Center and Space Environment Technologies/SpaceWx]

about 15 °C (59 °F) peaking in the infrared (Figure 3.4). Earth's surface emits a broad band of infrared radiation (at wavelengths mostly between 4 and 24 micrometers) with peak intensity at a wavelength of about 10 micrometers.

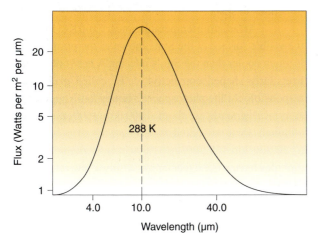

FIGURE 3.4
Flux of radiation as a function of wavelength emitted by a blackbody radiating at about the same average temperature as Earth's surface, that is, 15 °C (288 K).

The area under the blackbody curves in Figure 3.3 (for the Sun) and Figure 3.4 (for Earth's surface) represent the total radiational energy emitted per unit time per unit surface area at all wavelengths. Note that the vertical scales in the two figures are much different because the Sun emits immensely more total radiational energy than does Earth's surface. According to the **Stefan-Boltzmann law**, the total energy flux emitted by a blackbody across all wavelengths (E) is proportional to the fourth power of the absolute temperature (T^4) of the object; that is,

$$E = \sigma T^4$$

where σ is the Stefan-Boltzmann constant of proportionality. This relationship implies that a small change in the temperature of a blackbody results in a much greater change in the total amount of radiational energy emitted by the blackbody during a given time interval. The Sun radiates at a much higher temperature than does Earth's surface, so that the Stefan-Boltzmann law predicts that the rate of the Sun's energy output per square meter is almost 190,000 times that of the Earth-atmosphere-ocean system to space.

In addition, the graphics show that electromagnetic radiation at shorter wavelengths is more energetic than radiation at longer wavelengths. Hence, X-rays are described as high energy radiation whereas radio waves are considered relatively low energy.

As solar radiation travels away from the Sun and spreads outward into space in all directions, its intensity (energy per unit area) diminishes rapidly, as the inverse square of the distance traveled. The radiation emanating in all directions from a nearly point source like the Sun spreads out as ever larger spheres, with the intensity being reduced as a function of the surface area of the sphere (where the area is a function of the square of the sphere's radius.) According to this **inverse square law**, doubling the distance traversed by radiation reduces its intensity to $(1/2)^2$ or 1/4 of its initial value. Earth orbits the Sun at a

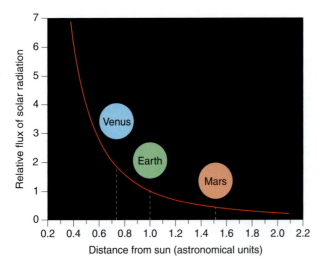

FIGURE 3.5
As radiation flows from the Sun into space, its intensity decreases as the inverse square of the distance traveled. Venus, closer to the Sun than Earth, receives about twice as much solar radiation as Earth. Mars, on the other hand, is farther from the Sun than Earth and receives roughly half as much solar radiation as Earth. [Courtesy of E.J. Hopkins]

distance such that it receives an intensity of solar radiation resulting in a temperature range that supports life as we know it (Figure 3.5). Venus is closer to the Sun, receives nearly twice as much solar radiation per unit intercepted area as Earth, and is too hot for life, with surface temperatures averaging about 460 °C (860 °F). Mars is farther away from the Sun, receives solar radiation less than one-half as intense as Earth, and is currently too cold for life, with surface temperatures averaging about −53 °C (−63 °F).

Because Earth rotates on its axis relatively rapidly, making one rotation with respect to the Sun on a daily basis, the amount of incident solar radiation is effectively spread out around the planet's surface. This more even distribution of sunlight makes the planet more habitable than if only one side of the planet were exposed to sunlight for a longer time. Our Moon rotates on its axis on a slower basis (one rotation period is approximately the same as one lunar revolution of Earth), resulting in a large temperature variation on the lunar surface facing Earth between new Moon (no incident sunlight) and full Moon (maximum sunlight intensity) over a 14-day period. According to NASA's *World Book*, the temperature on the lunar equator varies between -173 °C (-280 °F) at night to 127 °C (260 °F) in the daytime. The absence of an atmosphere also contributes to this considerable temperature range.

Several decades of satellite measurements as well as a relatively constant planetary temperature indicates that the total energy (in the form of solar radiation) absorbed by planet Earth is equal to the total energy (in the form of infrared radiation) emitted by the Earth-atmosphere-land-ocean system to space. This balance between energy input and energy output is known as **global radiative equilibrium** and is an example of the *law of conservation of energy*.

Incoming Solar Radiation

The Sun, the closest star to Earth, is a huge gaseous body composed almost entirely of hydrogen (about 80% by mass) and helium with internal temperatures that may exceed 20 million °C. The ultimate source of solar energy is a continuous nuclear fusion reaction in the Sun's interior. Simply put, in this reaction, four hydrogen nuclei (protons) fuse to form one helium nucleus (an alpha particle). However, the mass of one helium nucleus is about 0.7% less than the mass of the four hydrogen nuclei. This mass lost in the fusion of hydrogen to helium is converted to energy as described by Albert Einstein's mass-energy equivalence principle:

$$E = mc^2$$

where mass, m (in kg), is related to energy, E (in joules), with c being the speed of light (300,000 km per sec or 186,000 mi per sec). Note that c^2 is a huge number so that even a very small mass is converted to an enormous quantity of energy. Some of the energy produced by nuclear fusion in the Sun is used to bind the helium nucleus together; the rest of the energy is radiated and transferred by convection to the Sun's surface, and then radiated in all directions to space.

At radiating temperatures near 6100 K (11,000 °F), the visible surface of the Sun, known as the **photosphere**, is much cooler than the Sun's interior. A network of huge, irregularly shaped convective cells, called *granules*, is responsible for the photosphere's honeycomb appearance. A typical granule is about 1000 km (600 mi) across, although some so-called *supergranules* may be 30,000 to 50,000 km (18,500 to 31,000 mi) in diameter. Most granules have a life expectancy of only a few minutes and consist of a broad central area of rising hot gas surrounded by a thin layer of cooler gas sinking back toward the center of the Sun. This zone of convective activity encompasses the outer 200,000 km (125,000 mi) of the Sun.

Relatively dark, cool areas, called **sunspots**, usually occurring in pairs, dot the surface of the photosphere. Sunspot temperatures may be 400 to 1800

Celsius degrees (720 to 3240 Fahrenheit degrees) lower than the photosphere's average temperature. Bright areas, known as *faculae*, usually occur near sunspots. Changes in the number of sunspots and faculae accompany changes in solar energy output and may influence Earth's climate (Chapter 11).

Outward from the photosphere is the **chromosphere**, consisting of ions of hydrogen and helium at 4000 °C to 40,000 °C (7200 °F to 72,000 °F). Beyond this zone is the outermost portion of the Sun's atmosphere, the **solar corona**, a region of extremely hot (1 to 4 million °C) and highly rarefied ionized gases (predominantly hydrogen and helium) that extends millions of kilometers into space. The solar wind originates in the corona and is intensified by solar flares that erupt from the photosphere into the corona. In its orbit about the Sun, Earth intercepts only about one two-billionth of the enormous quantity of energy continually radiated by the Sun to space.

SOLAR ALTITUDE

At middle latitudes, the intensity of solar radiation striking Earth's surface varies significantly with the time of day and the season. At local solar noon, the summer Sun is higher in the sky than the winter Sun, and solar rays striking Earth's surface are more concentrated in summer than in winter. Even over the course of a single day, regular changes occur in incoming solar radiation striking Earth's surface; that is, the solar beam is more concentrated at noon than at sunrise or sunset. Hence, the angle of the Sun above the horizon, called the **solar altitude**, influences the intensity of solar radiation received at Earth's surface. At the place on Earth where the Sun is directly overhead, the local solar altitude has its maximum value of 90 degrees and solar rays are most concentrated (Figure 3.6). Wherever the Sun is positioned lower in the sky (that is, as the solar altitude is lower), solar radiation spreads over a larger area of Earth's horizontal surface and thus is less intense. *Local solar noon* is the time of day at essentially any location on Earth when the Sun appears at its highest point in its daily path across the local sky. At that time, the Sun would be on your local meridian (due south in the Northern Hemisphere or due north in the Southern Hemisphere).

Earth is so distant from the Sun (a mean distance of about 150 million km or 93 million mi) that solar radiation reaches the planet essentially as parallel beams of uniform intensity. But the nearly spherical Earth presents a curved surface to incoming solar radiation so that the noon solar altitude always varies with latitude (Figure 3.7). The intensity of solar radiation actually striking Earth's atmosphere is greatest at the latitude where the noon Sun

is in the zenith (solar altitude is 90 degrees) and decreases with distance north and south of that latitude.

Solar altitude also influences the interaction between solar radiation and the atmosphere. Decreasing solar altitude lengthens the path of the Sun's rays through the atmosphere (Figure 3.8). As the path lengthens, the greater interaction of incoming solar radiation with clouds, gases and aerosols reduces its intensity. Even if the sky were cloud-free, the longer the path of solar radiation through the atmosphere, the less intense is the radiation striking Earth's surface. The interactions between incoming solar radiation and atmosphere (scattering, reflection, and absorption) are described later in this chapter.

While solar altitude influences the intensity of solar radiation striking Earth's horizontal surface per unit area, the length of daylight affects the total amount

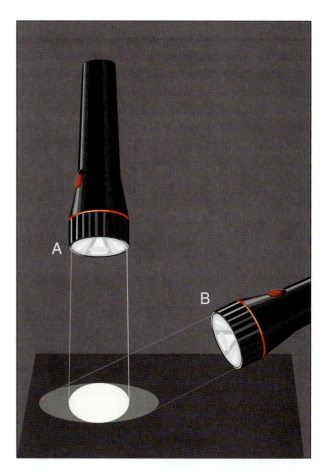

FIGURE 3.6
The intensity of solar radiation striking Earth's surface per unit area varies with the solar altitude. Consider this analogous situation: (A) A flashlight beam shines on a horizontal surface most intensely when the flashlight shines from directly overhead (analogous to a solar altitude of 90 degrees). (B) At an angle decreasing from 90 degrees, the flashlight beam spreads over an increasing area of the horizontal surface so that the light is less concentrated (less radiational energy received per unit area).

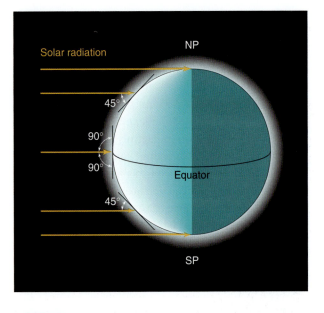

FIGURE 3.7
On any day of the year, the noon solar altitude always varies with latitude because Earth presents a curved surface to the incoming solar beam. In this example for an equinox, the solar altitude is 90 degrees at the equator and decreases with latitude (toward the poles). Hence, solar radiation striking horizontal surfaces per unit area is most intense at the equator and least intense at the poles.

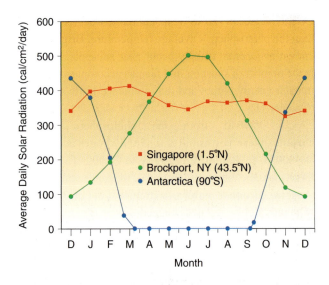

FIGURE 3.9
Average daily solar radiation (in calories per cm² per day) by month at Singapore, Brockport, NY, and Antarctica. During part of the year, the polar (Antarctic) and mid-latitude (Brockport) locations receive more solar radiation daily than the near-equator location (Singapore).

of solar radiational energy that is received each day. For example, in summer the altitude of the noon Sun is lower at high latitudes than at middle latitudes. Nonetheless, the greater length of daylight at high latitudes may translate into more total radiation striking Earth's surface during a 24-hr day (Figure 3.9). In the tropics, nearly uniform length of daylight plus little variation in the maximum daily solar altitude means that the total solar radiation received varies little through the course of a year. This may be an important factor in explaining the shrinkage of the glacial ice at the summit of Mount Kilimanjaro, located near the equator in East Africa. For details on this, refer to this chapter's first Essay.

Variations in both solar altitude and length of daylight accompany the annual march of the seasons. Before examining these relationships, first consider the fundamental motions of Earth in space: rotation of the planet on its spin axis and orbiting of the planet about the Sun.

EARTH'S MOTIONS IN SPACE AND THE SEASONS

Rotation of Earth on its axis accounts for day and night. Once every 24 hrs, Earth completes one rotation on its axis. At any instant, half the planet is illuminated by solar radiation (day) while the other half is in darkness (night).

Over the course of one year, which is actually 365.2422 days, Earth makes one complete revolution about the Sun in a slightly elliptical orbit (Figure 3.10). Earth's *orbital eccentricity*, that is, its departure from a circular orbit, is so slight that the Earth-to-Sun distance varies by only about 3.3% through the year. Earth is

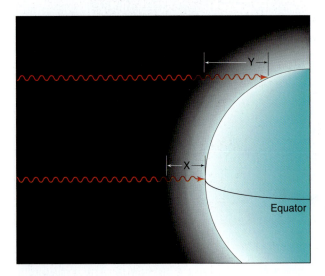

FIGURE 3.8
Solar radiation's path through the atmosphere lengthens with decreasing solar altitude, that is, as the Sun moves lower in the sky. *X* is the path length at high solar altitude and *Y* is the path length at low solar altitude. Path *Y* may be 30 times longer than *X*, especially when the Sun is near the local horizon.

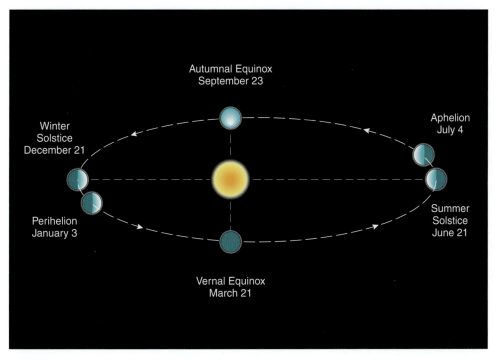

FIGURE 3.10
Earth's orbit is an ellipse with the Sun located at one focus. Earth is closest to the Sun at *perihelion* (about 3 January) and farthest from the sun at *aphelion* (about 4 July). Note that the eccentricity of Earth's orbit is greatly exaggerated in this drawing.

closest to the Sun (147 million km or 91 million mi) on about 3 January and farthest from the Sun (152 million km or 94 million mi) on about 4 July. These are the current dates of **perihelion** and **aphelion**, respectively. In the Northern Hemisphere, Earth is closest to the Sun in winter and farthest from the Sun in summer. Because of the spreading of radiation described by the inverse square law, Earth intercepts about 6.7% more solar radiation at perihelion than at aphelion. The eccentricity of Earth's orbit about the Sun cannot explain the seasons. What does account for seasons?

The answer is found in the tilt of Earth's spin axis. Because of this tilt, Earth's equatorial plane at the present time is inclined 23 degrees 27

minutes to the plane defined by the planet's annual orbit about the Sun (Figure 3.11). Equivalently, Earth's spin axis is tilted 23 degrees 27 minutes from a perpendicular drawn to Earth's orbital plane. During Earth's annual revolution about the Sun, its spin axis remains in the same alignment with respect to the background stars (the North Pole always points toward *Polaris*, the North Star, during the current millennium) while its orientation to the Sun changes continually. Accompanying these changes are regular variations in solar altitude and length of daylight, which in turn affect the intensity and total amount of solar radiation received at different latitudes on Earth's surface. If Earth's rotational axis were perpendicular to

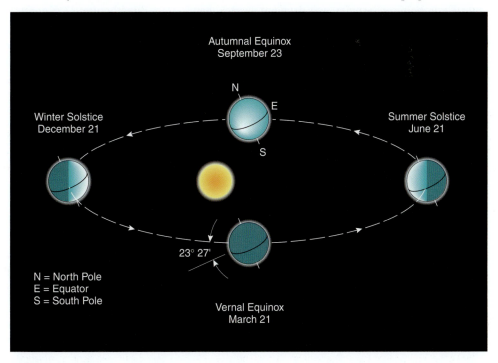

FIGURE 3.11
The seasons change because Earth's equatorial plane is inclined (at 23 degrees, 27 minutes) to its orbital plane. The seasons given are for the Northern Hemisphere. Note that the eccentricity of Earth's orbit is greatly exaggerated in this drawing.

its orbital plane (no tilt), Earth's axis would always have the same orientation to the Sun. Without an axial tilt, only changes in Earth-Sun distance between aphelion and perihelion would produce a seasonal contrast and it would be slight.

How does Earth's orientation to the Sun change over the course of a year? Viewed from Earth's surface, the latitude where the Sun's rays are most intense (overhead with a solar altitude of 90 degrees) shifts from 23 degrees 27 minutes south of the equator to 23 degrees 27 minutes north of the equator, and then back to 23 degrees 27 minutes south. On about 21 March and again on about 23 September, the Sun's noon position is directly over the equator. At these times, day and night are approximately equal in length (12 hrs) everywhere, except at the poles (Figure 3.12). For this reason, these dates are the **equinoxes** (from the Latin for "equal nights").

Following the equinoxes, the Sun continues its apparent journey toward its maximum poleward locations. On or about 21 June, the Sun's noon rays are vertical at 23 degrees 27 minutes N, the latitude circle known as the **Tropic of Cancer**. As shown in Figure 3.13, daylight is continuous north of the **Arctic Circle** (66 degrees 33 minutes N) and absent south of the **Antarctic Circle** (66 degrees 33 minutes S). Elsewhere, days are longer than nights in the Northern Hemisphere, where it is the first day of astronomical summer, and days are shorter than nights in the Southern Hemisphere, where it is the first day of astronomical winter. Hence, this is a **solstice** date—the summer solstice in the Northern Hemisphere and winter

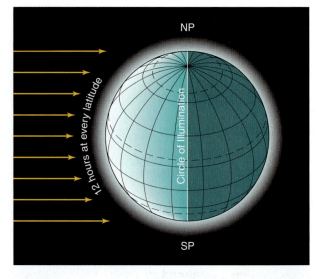

FIGURE 3.12
At the autumnal and spring equinoxes, the noon solar altitude is greatest (90 degrees) at the equator. Day and night are about equal in length except at the poles, where the Sun appears to travel along the local horizon for the entire day.

solstice in the Southern Hemisphere. Solstice is from the Latin, *solstitium*, referring to the Sun standing still as its latitudinal position changes very little from day to day at this time of year.

On or about 21 December, the noon Sun is directly over 23 degrees 27 minutes S, the latitude circle known as the **Tropic of Capricorn**, and the situation is reversed (Figure 3.14). Daylight is continuous south of the Antarctic Circle and absent north of the Arctic Circle.

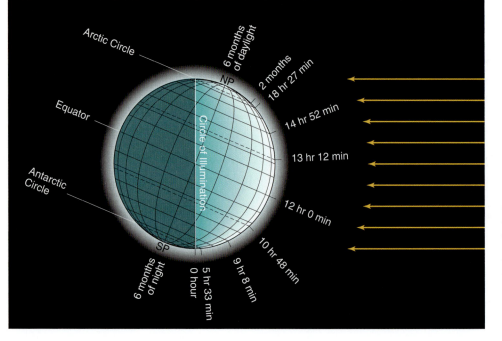

FIGURE 3.13
On the Northern Hemisphere summer solstice (about 21 June), the noon solar altitude is greatest (90 degrees) at 23 degrees 27 minutes N, and days are longer than nights everywhere north of the equator. Duration of daylight is given for every 20 degrees of latitude.

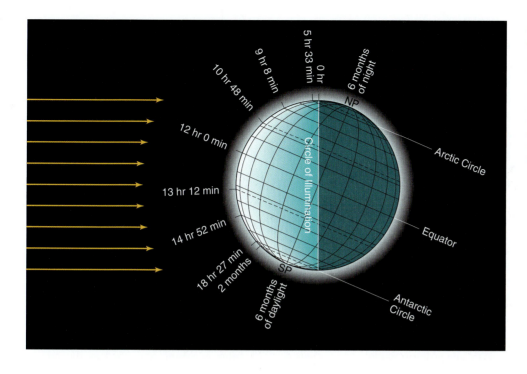

FIGURE 3.14
On the Northern Hemisphere winter solstice (about 21 December), the noon solar altitude is greatest (90 degrees) at 23 degrees 27 minutes S, and days are shorter than nights everywhere north of the equator. Duration of daylight is given for every 20 degrees of latitude.

Elsewhere, nights are longer than days in the Northern Hemisphere, where it is the first day of astronomical winter, and days are longer than nights in the Southern Hemisphere, where it is the first day of summer. Thus this is the date of the winter solstice in the Northern Hemisphere and the summer solstice in the Southern Hemisphere.

As Earth's orientation to the Sun changes through the course of a year, so does the daily path of the Sun through the local sky. Figure 3.15 portrays the path of the Sun through the sky from sunrise to sunset for solstices and equinoxes at the equator, a middle latitude Northern Hemisphere location, and the North Pole. The high point of each path is the Sun's position at local solar noon. (Because of one's location within a civil time zone and the time of the year, along with daylight saving time, local solar noon may differ by as much as one hour from noon on a clock.) At the middle latitude location and North Pole, the altitude of the noon Sun is greatest on the summer solstice, but at the equator, the noon solar altitude is greatest on the equinoxes (when the Sun is directly overhead). Note that on the summer solstice at the North Pole, the Sun circles the sky at a constant solar altitude (about 23.5 degrees) over the 24-hr day.

Ignoring atmospheric effects, solar radiation incident on a horizontal Earth surface is at maximum intensity where the noon Sun is directly overhead. North and south of that latitude, the intensity of noontime solar radiation diminishes because the solar altitude decreases. At the equinoxes, solar rays are most intense at the equator at noon and decrease with latitude toward the poles. On the Northern Hemisphere summer solstice, solar rays are most intense at local noon along the Tropic of Cancer and decrease to zero at the Antarctic Circle. On that day, the noon Sun has the same altitude at 47 degrees N as at the equator, although the Sun appears 23.5 degrees on opposite sides of each latitude's respective zenith. On the Northern Hemisphere winter solstice, solar rays are most intense along the Tropic of Capricorn and decrease to zero at the Arctic Circle.

As noted earlier, days and nights are approximately equal in length (12 hrs) everywhere on Earth (except at the poles) on only two days of the year, the spring and autumnal equinoxes. The length of daylight on the equinoxes is not precisely 12 hrs because of the optical effects of the atmosphere on the solar beam. Times of sunrise and sunset refer to the hour and minute when the outer edge of the Sun (not its center) is on the horizon. The atmosphere refracts (bends) the incoming solar beam downward so that the Sun appears to be higher than it actually is, lengthening the period of daylight. For example, at Washington, DC, the day length is 12 hrs and 6 minutes on the autumnal equinox (on or about 23 September).

Between the March and September equinoxes, days are longer than nights in the Northern Hemisphere and days are shorter than nights in the Southern Hemisphere. Between the September and March equinoxes, days

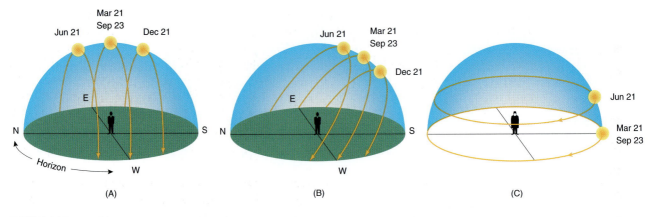

FIGURE 3.15
Path of the Sun through the sky on the solstices and equinoxes at (A) the equator, (B) middle latitudes of the Northern Hemisphere, and (C) the North Pole.

are shorter than nights in the Northern Hemisphere and days are longer than nights in the Southern Hemisphere. Furthermore, the seasonal (winter-to-summer) contrast in length of daylight increases with increasing latitude (Figure 3.16). At the equator, the period of daylight is essentially the same (slightly more than 12 hrs) year round. At 60 degrees N, the length of daylight varies from 5 hrs 53 minutes on 22 December to 18 hrs 52 minutes on 21 June, while approaching the North Pole, daylight length varies from continuous winter darkness to 24-hr summer daylight.

THE SOLAR CONSTANT

For convenience of study, the solar energy input into the planetary system is often described as the solar constant. The **solar constant** is defined as the rate at which solar radiation falls on a unit area of a flat surface located at the outer edge of the atmosphere and oriented perpendicular to the incoming solar beam when Earth is at its mean distance from the Sun. The *constant* designation is misleading because solar energy output actually fluctuates by a very small fraction of a percent over a year and exhibits longer-term climatically significant variations (Chapter 11). The solar constant averages about 1.97 calories per square centimeter per minute (cal/cm^2/min), or 1368 watts per square meter (W/m^2). Figure 3.17 summarizes the distribution of solar radiation received at the top of the atmosphere by latitude and day of the year in watts per square m.

Suppose that the incoming solar radiation were spread evenly over the rapidly rotating planet. What would be the intensity of solar radiation per unit area of Earth's surface (neglecting atmospheric effects)? Visualize Earth as intercepting a continuous stream of disks of solar energy (Figure 3.18). The

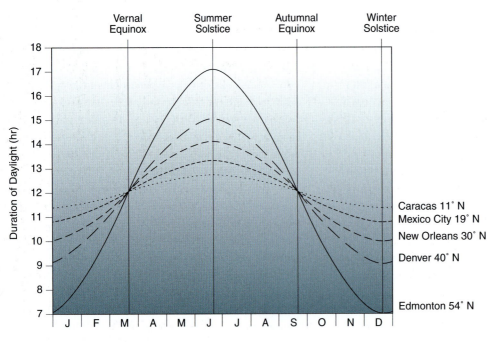

FIGURE 3.16
Variation in the length of daylight through the year increases with increasing latitude.

Extra-atmospheric irradiance in W/m² as a function of season and latitude for a solar constant of 1370 W/m²

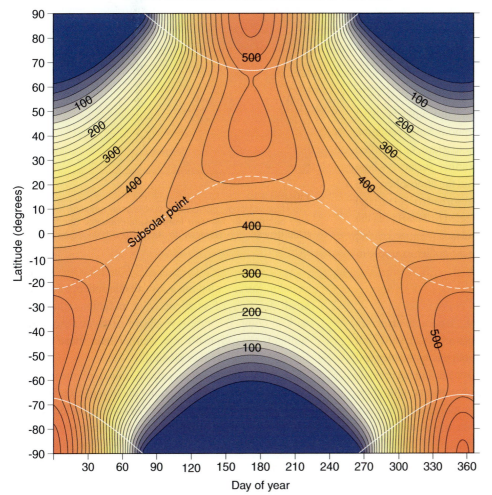

FIGURE 3.17

Distribution of solar radiation received at the top of the atmosphere by latitude and day of the year in watts per square m. Solid white lines in polar latitudes mark the equatorward limits of the region experiencing 24 hours of uninterrupted daylight. [Courtesy of E.J. Hopkins]

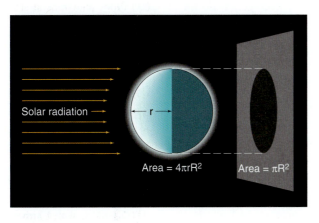

FIGURE 3.18

The area of a disk is ¼ of the surface area of a sphere. Hence, if incoming solar radiation is spread uniformly over Earth, the flux of solar radiation is reduced to ¼ of the solar constant.

radius, R, of each disk is the same as that of the planet and the surface area of a disk is given by

$$\pi R^2$$

where π is about 3.1416. At any point in time, a disk of energy is spread uniformly over the nearly spherical Earth having a surface area given by

$$4\pi R^2$$

Hence, the solar constant is reduced to $(\pi R^2/4\pi R^2)$ or ¼ of its original magnitude, that is, about 0.5 cal/cm²/min or 342 W/m².

The rate of total solar energy input for the planet varies through the course of a year from a maximum when Earth is closest to the Sun (perihelion) to a minimum when Earth is farthest from the Sun (aphelion). At perihelion, Earth is about 3.3% closer to the Sun than at aphelion. Applying the inverse square law, the planet intercepts about 6.7% more radiation at perihelion (2.04 cal/cm²/min or 1417 W/m²) than at aphelion (1.91 cal/cm²/min or 1326 W/m²).

The perihelion/aphelion contrast in solar energy input coupled with the seasonal variation in radiation has implications for global climate. The Southern Hemisphere receives more radiation during summer and less radiation during winter than does the Northern Hemisphere. Consequently, all other factors being equal, we might expect a greater winter-to-summer temperature contrast in the Southern Hemisphere. However, the ocean is more dominant in the Southern Hemisphere than the Northern Hemisphere. The Southern Hemisphere is 19.1% land and 80.9% ocean whereas the Northern Hemisphere is 39.3% land and 60.7% ocean. In the Southern Hemisphere the larger percentage of ocean surface area coupled with the relatively great thermal inertia of ocean water moderates seasonal temperature differences and largely offsets the greater seasonal contrast in incoming solar radiation (Chapter 4).

The Atmosphere and Solar Radiation

Solar radiation interacts with gases and aerosols as it travels through the atmosphere. These interactions consist of scattering, reflection, and absorption. Solar radiation that is not scattered or reflected back to space, or absorbed by gases and aerosols, reaches Earth's surface, where further interactions take place. Within the atmosphere, the percentage of solar radiation that is absorbed (*absorptivity*) plus the percentage scattered or reflected (*albedo*) plus the percentage transmitted to Earth's surface (*transmissivity*) must equal 100%. This relationship is another example of the *law of conservation of energy*.

SCATTERING AND REFLECTION

In the process of **scattering**, a particle disperses solar radiation in all directions—forward, backward, and sideways. Within the atmosphere, both gas molecules and aerosols (including the tiny water droplets and ice crystals composing clouds) scatter solar radiation but with some important differences. Scattering by molecules and other particles that are less than one-tenth the wavelength of light is wavelength dependent with the preferential scattering of blue-violet light by nitrogen and oxygen molecules being the principal reason for the color of the daytime clear sky. On the other hand, the tiny water droplets and ice crystals that compose clouds (but are much larger than molecules) scatter visible solar radiation equally at all wavelengths so that clouds appear white. Airborne particles (aerosols) suspended in the atmosphere with diameters similar to the wavelength of light also scatter the sunlight giving a milky appearance to the sky.

Reflection, a special case of scattering, takes place at the interface between two different media, such as air and cloud, when some of the radiation striking that interface is redirected (backscattered). The fraction of incident radiation that is backscattered by airborne particles or reflected by a surface (or interface) is the **albedo** of that surface, that is,

albedo = [(reflected radiation)/(incident radiation)]

where albedo is expressed either as a percentage or a fraction. Surfaces having a high albedo reflect a relatively large fraction of incident solar radiation and appear light

FIGURE 3.19
Cloud tops strongly backscatter visible solar radiation and appear very bright as in this view from the International Space Station. [Courtesy of *The Gateway to Astronaut Photography of Earth*, NASA]

in color. Surfaces having a low albedo reflect a relatively small fraction of incident solar radiation and appear dark in color. Because low-albedo surfaces absorb more sunlight, they typically are warmer than those that are lighter (having a higher albedo).

Within the atmosphere, much of the incident sunlight is backscattered making the tops of clouds appear as the most important reflectors of solar radiation (Figure 3.19). Cloud top albedo depends primarily on cloud thickness and varies from under 40% for thin clouds (less than 50 m, or 165 ft, thick) to 80% or more for thick clouds (more than 5000 m, or 16,500 ft, thick). For this reason, during daytime, a high thin veil of cirrus clouds with relatively few ice crystals appears much brighter than the underside of a much thicker thunderstorm (cumulonimbus) cloud with numerous cloud droplets. The average albedo for all cloud types and thickness is about 55%, and at any point in time, clouds cover about 60% of the planet. All other factors being constant, solar radiation reaching Earth's surface is more intense and daytime surface temperatures are higher when the sky is clear rather than cloudy.

ABSORPTION

Scattering and reflection within the atmosphere alter only the direction of incoming solar radiation. **Absorption**, however, is an energy conversion process whereby some of the radiation striking an object is converted to heat energy. Oxygen, ozone, water vapor, and various aerosols (including cloud particles) absorb a portion of the incoming solar radiation. Absorption by atmospheric gases varies by wavelength; that is, a specific gas may absorb strongly in some wavelengths, but weakly or not at all in other wavelengths. As a consequence of absorption, radiation in that wavelength band is removed before the Sun's rays reach Earth's surface.

Within the stratosphere, oxygen (O_2) and ozone (O_3) strongly absorb solar ultraviolet radiation at wavelengths shorter than 0.3 micrometer. Oxygen absorbs UV at very short wavelengths (less than 0.2 micrometer), and ozone absorbs longer UV (0.22 to 0.29 micrometer). The net effect of this absorption is twofold: (1) a significant reduction in the intensity of UV that reaches Earth's surface and (2) a marked warming of the upper stratosphere (Figure 1.8). The clear atmosphere is essentially transparent to solar radiation in the wavelength range between about 0.3 and 0.8 micrometer (mostly visible radiation). Water vapor absorbs solar infrared radiation in certain wavelength bands greater than 0.8 micrometer. Clouds are relatively poor absorbers of solar radiation, typically absorbing less than 10% of the solar radiation that strikes the cloud top, although exceptionally thick clouds such as thunderclouds absorb somewhat more.

Stratospheric Ozone Shield

Ozone (O_3) is a relatively unstable molecule made up of three atoms of oxygen, but has certain physical and chemical properties that differentiate it from free oxygen (O_2). As shown in Figure 3.20, the greatest concentration of ozone occurs naturally in the atmosphere at altitudes

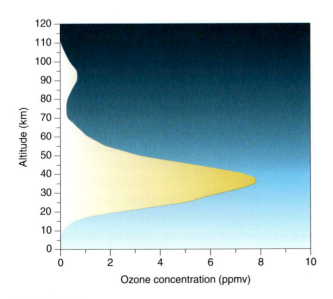

FIGURE 3.20
The concentration of ozone (O_3) peaks in the stratosphere. [Source: U.S. Standard Atmosphere, 1976, NASA, NOAA, and US Air Force]

near 40 km (25 mi). Depending on its location in the atmosphere, ozone has either a positive or negative impact on life. On the negative side, ozone in the lower troposphere is a serious air pollutant and one of the chief constituents of photochemical smog. On the positive side, chemical reactions involved in the formation and dissociation of ozone in the stratosphere shields organisms at Earth's surface from exposure to potentially lethal intensities of solar ultraviolet radiation. Without this **stratospheric ozone shield**, life as we know it could not exist on Earth.

Within the stratosphere, two sets of competing chemical reactions, both powered by solar ultraviolet radiation, continually generate and destroy ozone (Figure 3.21). During ozone production, UV strikes an oxygen molecule (O_2) causing it to split into two free oxygen atoms (O). Free oxygen atoms then collide with molecules of oxygen to form ozone molecules (O_3). At the same time, ozone is destroyed. Ozone absorbs ultraviolet radiation, splitting the molecule into one free oxygen atom (O) and one molecule of oxygen (O_2). The free oxygen atom then collides with an ozone molecule to form two molecules of oxygen. The net effect of these opposing sets of chemical reactions is a minute reservoir of ozone that peaks at only about 10 parts per million (ppm) in the middle stratosphere.

Ultraviolet radiation (at different wavelengths) powers each set of chemical reactions so that much, but not all, UV radiation is prevented from reaching Earth's surface. Prolonged exposure of the skin to the UV that does penetrate the ozone shield can cause serious health problems. A thinner ozone shield would likely mean more intense UV radiation received at Earth's surface and, for humans, a greater risk of skin cancer, cataracts of the eye, and immune deficiencies. As a general rule, every 1% decline in stratospheric ozone concentration is accompanied by a 2% increase in the intensity of UV that passes through the ozone shield. Various studies suggest that a 2.5% thinning of the ozone shield could boost the rate of human skin cancer by 10%. The most dangerous portion of UV radiation that reaches Earth's surface, designated *UVB*, spans the wavelength band from 0.29 to 0.32 micrometer. The actual increase in UVB that reaches Earth's surface hinges on the cloudiness and dustiness of the atmosphere.

Chemicals that threaten to destroy stratospheric ozone are of natural and industrial origin. Since the 1970s, a group of chemicals known as *chlorofluorocarbons (CFCs)* has received considerable attention for the potential impact on the ozone shield. First synthesized in 1928, CFCs were widely used as chilling (heat-transfer)

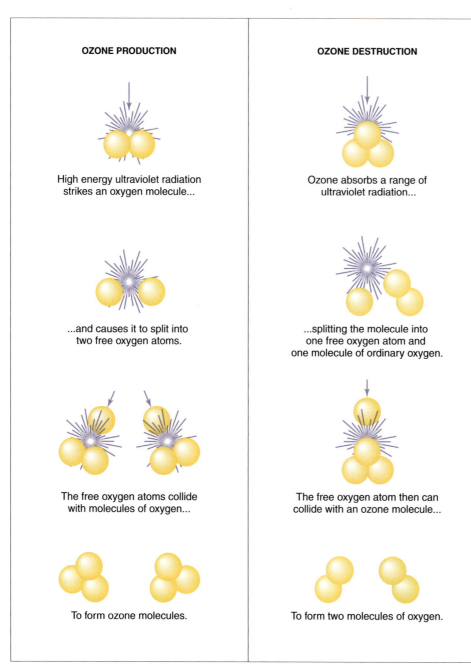

OZONE PRODUCTION

High energy ultraviolet radiation strikes an oxygen molecule...

...and causes it to split into two free oxygen atoms.

The free oxygen atoms collide with molecules of oxygen...

To form ozone molecules.

OZONE DESTRUCTION

Ozone absorbs a range of ultraviolet radiation...

...splitting the molecule into one free oxygen atom and one molecule of ordinary oxygen.

The free oxygen atom then can collide with an ozone molecule...

To form two molecules of oxygen.

FIGURE 3.21
Within the stratosphere, two sets of competing chemical reactions continually generate and destroy ozone (O_3). [Adapted from "Ozone: What is it and why do we care about it?" *NASA Facts*, NASA Goddard Space Flight Center, Greenbelt, MD, 1993.]

chlorine (Cl), a gas that readily reacts with and destroys ozone. Products of this reaction are chlorine monoxide (ClO) and molecular oxygen (O_2). Chlorine (Cl) is a catalyst in chemical reactions that convert ozone to oxygen. In this way, each chlorine atom destroys perhaps tens of thousands of ozone molecules.

F.S. Rowland and M.J. Molina of the University of California at Irvine first warned of the threat of CFCs to the stratospheric ozone shield in 1974. Consequently, use of CFCs as propellants in common household aerosol sprays such as deodorants, hairsprays, and furniture polish was banned in the United States, Canada, Norway, and Sweden in 1979. For their pioneering research on the depletion of stratospheric ozone, Rowland, Molina (now at the Massachusetts Institute of Technology), and P.J. Crutzen (of the Max Planck Institute for Chemistry, Germany) were awarded the 1995 Nobel Prize in chemistry.

The first sign of thinning of the stratospheric ozone shield came from Antarctica. For about six weeks during the Southern Hemisphere spring (mainly in September and October), the ozone layer in the Antarctic stratosphere (mostly at altitudes from 11 to 23 km, or 7 to 14 mi) thins drastically and then recovers during November (Figure 3.22). The **Antarctic ozone hole** is defined as the thinning of the ozone layer over the continent to significantly below pre-1979 levels. (Prior to formation of the ozone hole, ozone levels in

agents in refrigerators and air conditioners, for cleaning electronic circuit boards, and in the manufacture of foams used for insulation. Certain CFCs are inert (chemically non-reactive) in the troposphere, where they have accumulated for decades. Atmospheric circulation transports CFCs into the stratosphere where, at altitudes above about 25 km (15 mi), intense UV radiation breaks down CFCs, releasing

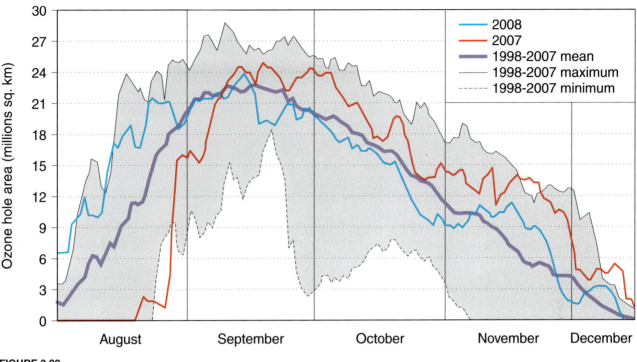

FIGURE 3.22

The Antarctic ozone hole is a widespread area of stratospheric ozone depletion that develops over Antarctica during the Southern Hemisphere spring. The mean area encompassed by the Antarctic ozone hole during 1998-2007 increased from near zero at the beginning of August, peaked in September, and declined to zero in December. [NOAA and NASA Goddard Space Flight Center]

Antarctica were normally lowest in autumn.) Research conducted during the National Ozone Expeditions to the U.S. McMurdo Station in 1986-87 plus NASA aircraft flights into the Antarctic stratosphere in 1987 revealed relatively high concentrations of chlorine monoxide (ClO) in the Antarctic stratosphere. This discovery established a convincing link between the Antarctic ozone hole and CFCs.

World-wide acceptance of the threat of CFCs and other ozone-depleting substances (ODSs) to the stratospheric ozone shield prompted the United Nations Environmental Programme (UNEP) in 1987 to draft the *Montreal Protocol on Substances That Deplete the Ozone Layer*, which has been ratified subsequently by all 192 UN member states. The original goal of the Montreal Protocol was to cut CFC production in half by 1992 (compared to 1986 levels). However, the seriousness of the problem led to seven subsequent amendments expanding the list of regulated substances (e.g., to include bromine-containing halons) and requiring the worldwide phase out of the manufacture and use of CFCs beginning in January 1996.

CFCs and halons have long atmospheric lifetimes so that these substances will continue to threaten the stratospheric ozone shield for some time to come in spite of the phase out. The 2006 UNEP *Scientific Assessment of Ozone Depletion Concentrations* reported that ozone-depleting chlorine levels in the stratosphere are on the decline, but recovery of stratospheric ozone to pre-1980 levels is not likely until the mid 21st century in middle latitudes and a decade or two later in polar regions. Humankind's successful intervention in this global environmental problem encourages scientists and policymakers that similarly effective action can also be taken to curb anthropogenic climate change (Chapters 14 and 15).

Earth's Surface and Solar Radiation

Incoming solar radiation has both direct and diffuse components. Solar radiation that passes directly through the atmosphere to Earth's surface is the direct component that provides bright sunshine and casts shadows. Solar radiation that is scattered and/or reflected to Earth's surface is the diffuse component, such as found during daytime under an overcast sky. Direct plus diffuse solar radiation that strikes Earth's surface is either reflected or absorbed depending on the surface albedo. The fraction that is not reflected at the surface is absorbed (that is, converted to heat).

As noted earlier, light surfaces reflect more solar radiation than dark surfaces. Skiers and snow boarders who have been sunburned on the slopes on a sunny day are well aware of the high reflectivity of a snow cover. The albedo of fresh-fallen snow typically ranges between 75% and 95%; that is, 75% to 95% of the solar radiation striking a fresh snow cover is reflected, while the rest (5% to 25%) is absorbed (converted to heat). A fresh snow cover is highly reflective because the snow surface consists of a multitude of randomly oriented crystals each having many reflecting surfaces. As snow ages, the surface albedo declines as snow crystals convert to rounded ice particles having fewer reflecting surfaces. The albedo of old snow typically ranges between 40% and 60%. At the other extreme, the albedo of a dark surface, such as a black topped road or a spruce forest, may be as low as 5%. Differences in albedo explain why light-colored clothing is usually a more comfortable choice than dark-colored clothing on a sunny, hot day. Albedo values of some common surface types are listed in Table 3.1.

FIGURE 3.23
In this MODIS satellite image of the Great Lakes region, water surfaces appear appear dark because of their low albedo. Note the appearance of Lakes Superior, Michigan and Huron. Light green along shorelines displays sediment in the water while land in the center is generally brown at this time of year. Snow covered land surfaces across the top of the image are light gray while convective clouds over the lower right are bright white. [NASA]

TABLE 3.1

Average Albedo (Reflectivity) of some Common Surface Types for Visible Solar Radiation

Surface	Albedo (% reflected)
Deciduous forest	15-18
Coniferous forest	9-15
Tropical rainforest	7-15
Tundra	15-35
Grasslands	18-25
Desert	25-30
Sand	30-35
Soil	5-30
Green crops	15-25
Sea ice	30-40
Fresh snow	75-95
Old snow	40-60
Glacial ice	20-40
Water body (high solar altitude)	3-10
Water body (low solar altitude)	10-100
Asphalt road	5-10
Urban area	14-18
Cumulonimbus cloud	90
Stratocumulus cloud	60
Cirrus cloud	40-50

In the visible satellite image in Figure 3.23, the ocean surface appears dark because of its low albedo and strong absorption of solar radiation. The albedo of the ocean (or lake) surface varies with the angle of the Sun above the horizon (*solar altitude*). Under clear skies, the albedo of a flat, tranquil water surface decreases with increasing solar altitude (Figure 3.24). The albedo approaches a mirror-like 100% near sunrise and sunset (when the solar altitude is near 0 degrees) but declines sharply as the solar altitude increases and approaches 20 degrees. With overcast skies, only diffuse solar radiation strikes the water surface and the albedo varies little with solar altitude and is uniformly less than 10%. On a global basis, the albedo of the ocean surface averages only about 8%; that is, the ocean absorbs on average 92% of incident solar radiation.

Whereas the clear atmosphere is relatively transparent to solar radiation, the ocean absorbs most solar radiation within relatively shallow depths. As shown in Figure 3.25, the ocean's absorption of the visible portion of solar radiation is selective by wavelength. Water absorbs the longer wavelengths of visible light

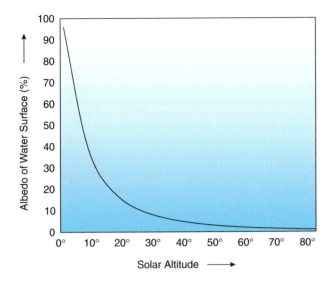

FIGURE 3.24
The albedo of a flat and undisturbed water surface under clear skies decreases with increasing solar altitude. A wave-covered water surface has a slightly higher albedo at high solar altitudes and a slightly lower albedo at low solar altitudes.

(i.e., reds and yellows) more efficiently than the shorter wavelengths (i.e., greens and blues) so that green and blue penetrate to greater depths. Within clear, clean water, red light is completely absorbed within about 15 m (50 ft) of the surface, whereas green and blue-violet light may penetrate to depths approaching 250 m (800 ft). More green and blue light is scattered to our eyes, explaining the blue/green color of the open ocean. Particles suspended in the water significantly boost absorption so that sunlight often is completely absorbed at shallower depths. In fact, some near-shore waters are so turbid (murky) that little if

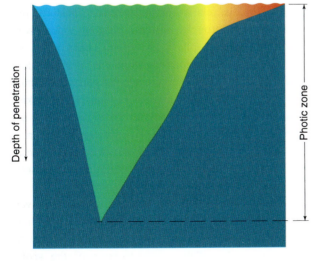

FIGURE 3.25
Visible solar radiation is selectively absorbed by wavelength as it penetrates the ocean's surface waters.

any sunlight reaches much below 10 m (35 ft). Suspended particles preferentially scatter yellow and green light giving these waters their characteristic color.

Significant changes in surface albedo occur seasonally and affect the fraction of incident solar radiation that is converted to heat. In autumn in forested areas, loss of leaves from deciduous trees raises the surface albedo. Forests tend to have low albedo values because a large fraction of the incident sunlight penetrates into the canopy and is not scattered back to space. At middle and high latitudes, significant increases in surface albedo accompany the winter freeze-over of lakes, the formation of sea ice, and development of a persistent snow cover.

Global Solar Radiation Budget

Measurements by sensors on board Earth-orbiting satellites indicate that the Earth-atmosphere-land-ocean system scatters or reflects back to space on average about 31% of the solar radiation intercepted by the planet. This is Earth's **planetary albedo**. By contrast, the albedo of the Moon is only about 7%, primarily because of the absence of clouds in the highly rarefied lunar atmosphere. Hence, Earth viewed from the Moon (by astronauts) is more than four times brighter than the full Moon viewed from Earth on a clear night.

The atmosphere (i.e., gases, aerosols, clouds) absorbs only about 20% of the total solar radiation intercepted by Earth. In other words, the atmosphere is relatively transparent to solar radiation. The remaining 49% of solar radiation (31% was scattered or reflected to space) is absorbed by Earth's surface, chiefly because of the low average albedo of ocean water covering about 71% of the surface of the globe. The global annual solar radiation budget is summarized in Table 3.2.

Earth's surface is the principal recipient of solar heating, and heat is transferred from Earth's surface to the atmosphere, which eventually radiates this energy to space. Earth's surface is thus the main source of heat for

TABLE 3.2	
Earth's Solar Radiation Budget	
Reflected by the Earth-atmosphere system	31%
Absorbed by the atmosphere	20%
Absorbed by Earth's surface	49%
Total	100%

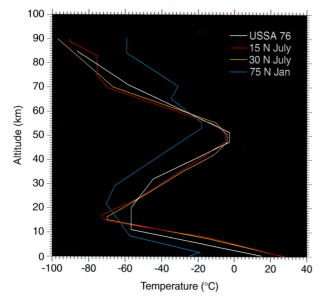

FIGURE 3.26
Average temperature profile for the atmosphere showing seasonal and latitudinal variations. Except at polar latitudes in winter, the average temperature decreases with altitude within the troposphere. [US Standard Atmosphere, 1976, NOAA, NASA, and U.S. Air Force]

the atmosphere; that is, the atmosphere is heated from below. This is evident in the average vertical temperature profile of the troposphere (Figure 3.26). Normally, air is warmest close to Earth's surface, and air temperature drops with increasing altitude within the troposphere, that is, away from the main source of heat. An exception is high latitudes in winter where the air temperature initially increases with increasing altitude above the surface (known as a *temperature inversion*). A secondary heat source is found at the top of the stratosphere, where absorption of ultraviolet radiation by oxygen and ozone molecules at altitudes below 50 km (31 mi) results in the relatively warm *stratopause*.

Outgoing Infrared Radiation

If solar radiation were continually absorbed by the Earth-atmosphere-land-ocean system without any compensating flow of heat out of the system to space, Earth's surface temperature would rise steadily. Eventually, life would be extinguished and the ocean would boil away. Actually, the global air temperature changes very little from one year to the next. **Global radiative equilibrium** keeps the planet's temperature in check; that is, the outgoing emission of heat to space in the form of infrared radiation balances the incoming solar radiational heating of the Earth-

atmosphere-land-ocean system. Although solar radiation is supplied only to the illuminated half of the planet at any instant, infrared radiation is emitted to space ceaselessly, day and night, from over the entire planet. This explains why nights are usually colder than days and why air temperatures typically drop throughout the night.

While the clear atmosphere is relatively transparent to solar radiation, certain gases in the atmosphere impede the escape of infrared radiation to space thereby elevating the temperature of the lower atmosphere. This important climate forcing mechanism is known as the *greenhouse effect*.

GREENHOUSE EFFECT

The **greenhouse effect** refers to the heating of Earth's surface and lower atmosphere caused by strong absorption and emission of infrared radiation by certain atmospheric gases, known as **greenhouse gases**. As noted earlier, solar radiation and terrestrial infrared radiation peak in different portions of the electromagnetic spectrum, their properties differ, and they interact differently with the atmosphere. The atmosphere directly absorbs only about 20% of the solar radiation intercepted by the planet. The atmosphere absorbs a greater percentage of the infrared radiation emitted by Earth's surface, and the atmosphere, in turn, radiates some IR to space and some back to Earth's surface. Hence, Earth's surface is heated by absorption of both solar radiation and atmosphere-emitted infrared radiation.

The similarity in radiational properties between infrared-absorbing atmospheric gases and the glass or plastic glazing of a greenhouse is the origin of the term greenhouse effect. Greenhouse glazing, like the atmosphere, is relatively transparent to visible solar radiation but strongly absorbs infrared radiation. A greenhouse, where plants are grown, takes advantage of the radiational properties of glazing (Figure 3.27). Sunlight readily penetrates greenhouse glazing and much of it is absorbed (converted to heat) within the greenhouse. Objects in the greenhouse emit infrared radiation that is strongly absorbed by glazing. The glazing, in turn, emits IR to both the atmosphere and the greenhouse interior, thereby raising the temperature within the greenhouse. The analogy between the atmosphere and a greenhouse is not strictly correct, however. A greenhouse also functions as a shelter from the wind and this is the principal reason for the elevated temperature observed within most greenhouses. Nonetheless, *greenhouse effect* is such a commonly used term (especially by the media) that we use it in this book.

FIGURE 3.27
The glazing of a greenhouse behaves similarly to certain gases (e.g., water vapor, carbon dioxide) in the atmosphere that are transparent to sunlight but strongly absorb and emit infrared radiation. [Courtesy of Oak Ridge National Laboratory, U.S. Department of Energy]

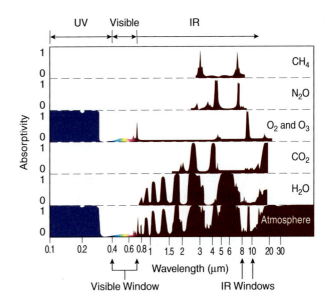

FIGURE 3.28
Absorption of radiation by selected gaseous components of the atmosphere as a function of wavelength. *Absorptivity* is the fraction of radiation absorbed and ranges from 0 to 1 (0% to 100% absorption). Absorptivity is very low or near zero in *atmospheric windows.* Note the visible window between 0.3 and 0.7 micrometers and the infrared windows near 8 and 10 micrometers.

The greenhouse effect is responsible for considerable warming of Earth's surface and lower atmosphere. Viewed from space, the planet radiates at about −18 °C (0 °F), whereas the average temperature at the Earth's surface is about 15 °C (59 °F). The temperature difference is due to the greenhouse effect and amounts to

$$[15 \text{ °C} - (-18 \text{ °C})] = 33 \text{ Celsius degrees}$$
or
$$[59 \text{ °F} - 0 \text{ °F}] = 59 \text{ Fahrenheit degrees.}$$

Without the greenhouse effect, Earth would be too cold to support most forms of plant and animal life.

GREENHOUSE GASES

Water vapor is the principal greenhouse gas having a clear-sky contribution of 60% to the greenhouse effect. Other gases contributing to the greenhouse effect are carbon dioxide (26%), ozone (8%), and methane plus nitrous oxide (6%). As shown in Figure 3.28, the percentage of infrared radiation absorbed by each gas varies with wavelength. An **atmospheric window** is a range of wavelengths over which little or no radiation is absorbed. A *visible atmospheric window* extends from about 0.3 to 0.7 micrometers and the major *infrared atmospheric window* is from about 8 to 13 micrometers. Significantly, this latter window includes the wavelength of the planet's peak infrared emission (about 10 micrometers). Through this window, most heat from Earth's surface and atmosphere

escapes to space as infrared radiation. IR sensors on board Earth-orbiting satellites monitor this upwelling radiation which is calibrated in terms of the surface temperature of the radiating object: the higher the temperature, the more intense is the emission of IR radiation.

Warming caused by atmospheric water vapor is evident even at the scale of mesoclimates or local climates. Consider an example. Locations in the Desert Southwest and along the Gulf Coast are at about the same latitude and receive essentially the same input of solar radiation. In both places, summer afternoon high temperatures commonly top 32 ºC (90 ºF). At night, however, air temperatures often differ markedly. Air is relatively dry (low humidity) in the Southwest so that infrared radiation readily escapes to space and air temperatures near Earth's surface may drop well under 15 ºC (59 ºF) by dawn. People who camp in the desert are aware of the dramatic fluctuations in temperature between day and night. Infrared radiation does not escape to space as readily through the Gulf Coast atmosphere where the air is more humid. Water vapor strongly absorbs outgoing IR and emits IR back towards Earth's surface so that early morning low temperatures may dip no lower than the 20s Celsius (70s Fahrenheit). The smaller diurnal (day to night) temperature contrast along the Gulf Coast is due to more water vapor and a stronger greenhouse effect.

Clouds are composed of IR-absorbing water droplets and/or ice crystals and also contribute to the greenhouse effect. All other factors being equal, nights usually are warmer when the sky is cloud-covered than when the sky is clear. Even high, thin cirrus clouds through which the Moon is visible can reduce the nighttime temperature drop at Earth's surface by several Celsius degrees. Clouds thus affect climate in two opposing ways: By absorbing and emitting IR, clouds warm Earth's surface and by reflecting solar radiation, clouds cool Earth's surface. On a global scale, which one of these two opposing effects is more important? Analysis of satellite measurements of incoming and outgoing radiation indicates that clouds have a net cooling effect on global climate. That is, all other factors being equal, a more extensive cloud cover would cool the planet whereas less extensive cloud cover would warm the planet.

Although water vapor is the principal greenhouse gas, it does not instigate warming or cooling trends in climate. The amount of water vapor in the atmosphere (the humidity) varies in response to changes in temperature (brought on, for example, by the buildup of other greenhouse gases such as CO_2). Water vapor's role in climate change is to amplify rather than to trigger temperature trends. In this regard, based on analysis of satellite data from 2003 to 2008, researchers reported in 2008 that global climate models do a relatively good job at simulating the relationship between temperature and humidity.

THE CALLENDAR EFFECT

The **Callendar effect** is the theory that global climate change can be brought about by enhancement of Earth's natural greenhouse effect by increased levels of atmospheric CO_2 from anthropogenic sources, principally the burning of fossil fuels. The theory is named for the British engineer Guy Stewart Callendar (1898-1964) who investigated the link between global warming and fossil fuel combustion beginning in the late 1930s.

In 1957, systematic monitoring of atmospheric carbon dioxide levels began at NOAA's Mauna Loa Observatory in Hawaii under the direction of Charles D. Keeling (1928-2005) of Scripps Institution of Oceanography. The observatory is situated on the northern slope of Earth's largest volcano 3397 m (11,140 ft) above sea level in the middle of the Pacific Ocean, sufficiently distant from major sources of air pollution such that carbon dioxide levels are considered representative of at least the Northern Hemisphere. Also since 1957, atmospheric CO_2 has been monitored at the South Pole station of the U.S. Antarctic Program and that record closely parallels the one at Mauna Loa. The Mauna Loa record (the *Keeling curve*) shows a sustained increase in average annual atmospheric carbon dioxide concentration from about 316 ppmv (parts per million by volume) in 1959 to 388 ppmv near the end of 2009 (Figure 3.29). During the ten year period ending in 2007, the growth rate of atmospheric CO_2 was faster than any other 10-year period since the Mauna Loa record began. Superimposed on this upward trend is an annual carbon dioxide cycle caused by seasonal changes in

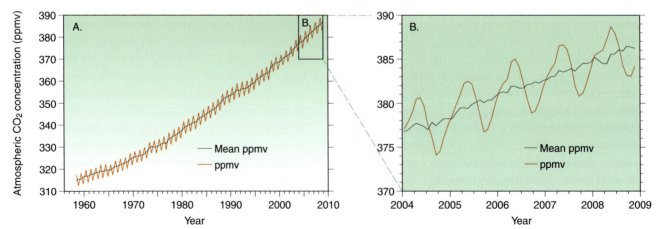

FIGURE 3.29
Concentration of atmospheric carbon dioxide (CO_2) as measured at the Mauna Loa Observatory, Hawaii (the *Keeling curve*). (A) The entire Mauna Loa record to date. (B) Recent monthly mean values centered on the middle of each month. Red curves include the influence of annual cycles in photosynthesis and cellular respiration. Black curves have been corrected for the average seasonal cycle. [Source: C.D. Keeling *et al.*, Scripps Institution of Oceanography, University of California, La Jolla, CA; Dr. Pieter Tans, NOAA/ESRL]

photosynthesis by Northern Hemisphere vegetation. The level of carbon dioxide falls during the growing season to a minimum in October and recovers over winter to a maximum in May. Northern Hemisphere vegetation dominates because there is more of it; only 19.1% of the Southern Hemisphere is land.

The upward trend in atmospheric carbon dioxide was underway long before Keeling's monitoring and appears likely to continue well into the future. As pointed out in the first Essay of Chapter 2, the anthropogenic contribution to the buildup of atmospheric CO_2 may have begun thousands of years ago with land clearing for agriculture and settlement. Land clearing contributes CO_2 to the atmosphere via burning of vegetation, decay of wood residue, and reduced photosynthetic removal of carbon dioxide from the atmosphere. By the middle of the 19th century, growing dependency on coal burning associated with the beginnings of the Industrial Revolution spurred a more rapid rise in CO_2 concentration. Carbon dioxide is a by-product of the burning of coal and other fossil fuels. Based on analysis of ancient air bubbles trapped in glacial ice cores, the concentration of atmospheric CO_2 is now about 35% higher than it was in the pre-industrial era. Today, fossil fuel combustion accounts for roughly 75% of the increase in atmospheric carbon dioxide while deforestation (and other land clearing) is likely responsible for the balance. With continued growth in fossil fuel combustion, the atmospheric carbon dioxide concentration could top 550 ppmv (double the pre-industrial level) by the end of the present century.

Besides carbon dioxide, rising levels of other infrared-absorbing gases (e.g., methane, nitrous oxide, and chlorofluorocarbons) enhance the greenhouse effect. Based on analysis of the chemical composition of air bubbles trapped in cores extracted from glacial ice sheets, the concentration of methane (CH_4) in the atmosphere is now greater than at any time in the past 400,000 years. In the period from 1750 through the 1970s, the global average atmospheric concentration of methane increased by about 220%, up from an estimated 700 ppbv (parts per billion by volume) to 1560 ppbv. Since then the growth rate slowed, leveled off to a steady state of about 1770 ppbv during 1999-2006, and then began to increase again (Figure 3.30A). Methane, the principal component of natural gas, is a product of the decay of organic matter in the absence of oxygen (*anaerobic decay*); its concentration is rising because of more rice cultivation, cattle, landfills, and/or termites, all sources of methane. Additional human-influenced sources include coal mining, wastewater treatment, and petroleum systems.

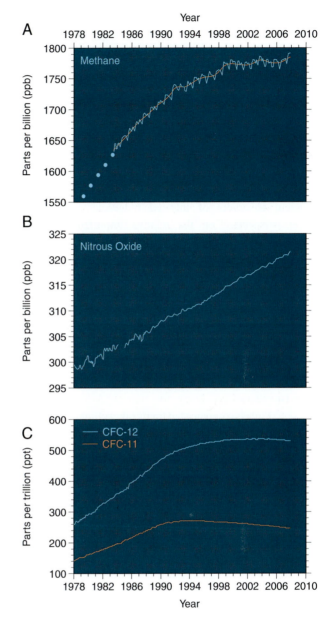

FIGURE 3.30
Average atmospheric concentration of the greenhouse gases (A) methane, (B) nitrous oxide, and (C) CFC-11 and CFC-12 in ppbv beginning in 1978. [NOAA Earth System Research Laboratory, Global Monitoring Division]

The chief sink for methane is oxidation by chemical reaction with hydroxyl radicals (OH) in the troposphere producing CH_3 and water. The atmospheric lifetime of methane is about 12 years.

The global average atmospheric concentration of nitrous oxide (N_2O) has increased from about 270 ppbv in 1750 to recent levels near 322 ppbv (Figure 3.30B). The principal human-related sources of N_2O in the U.S. include fertilizer application, fossil fuel combustion, and nitric acid production. Natural sources account for more than 60% of total N_2O emissions. *Photolysis* (breakdown by sunlight) in the stratosphere is the main N_2O removal mechanism. The atmospheric lifetime of nitrous oxide is about 110 years. Spurred by international efforts to protect the stratospheric ozone shield (discussed earlier in this chapter), the atmospheric concentrations of the chlorofluorocarbons CFC-11 and CFC-12 began leveling off in the early 1990s (Figure 3.30C).

Although occurring in extremely low concentrations, methane and nitrous oxide are very efficient absorbers of infrared radiation. To provide some perspective on the relative importance of the various infrared-absorbing gases, the **global warming potential (GWP)** was developed. According to the U.S. Environmental Protection Agency, the "GWP for a particular greenhouse gas is the ratio of heat trapped by one unit mass of the greenhouse gas to that of one unit mass of CO_2 over a specified time period." Methane is about 21 times more effective at trapping heat in the atmosphere when compared to CO_2 over 100 years and nitrous oxide is about 310 times more effective than CO_2 over 100 years.

Unless compensated for by other climate forcings, enhancement of Earth's greenhouse effect is likely to cause further global warming. The prospect of a continued rise in global temperature and other climate changes has helped to spur interest in reducing our dependency on fossil fuels in favor of alternative energy sources such as solar power. For information on solar power, see this chapter's second Essay. We have much more on the potential implications of an enhanced greenhouse effect in Chapter 12.

Global Radiative Equilibrium and Climate Change

Over the long-term history of the planet, equilibrium has prevailed between solar energy absorbed by the Earth-atmosphere-land-ocean system and energy emitted by Earth to space in the form of infrared (IR) radiation. That is, *energy in equals energy out*, following the law of conservation of energy. As noted earlier in this chapter, this is known as *global radiative equilibrium.*

Earth's mean temperature at global radiative equilibrium depends on the solar constant, planetary albedo, and how closely the planet approximates a blackbody (the *emissivity*). A change in any of these parameters could shift Earth to a new radiative equilibrium, altering the radiative equilibrium temperature and the global climate.

Suppose, for example, that for some reason the solar constant were to increase. Initially, this would create a radiative imbalance on Earth with more energy input than energy output. But more intense incoming solar radiation that is absorbed would raise the temperature at Earth's surface and in the lower atmosphere. The now warmer Earth-atmosphere-land-ocean system would respond by emitting more intense IR until eventually a new radiative equilibrium were established—but at a higher radiative equilibrium temperature. A change in the radiative equilibrium temperature would translate into a global climate change.

In another example, suppose that Earth's planetary albedo were to increase (perhaps because of greater cloud cover). Again, initially this would cause a radiative imbalance with less available solar energy as input and therefore greater initial output of energy. Less solar radiation absorbed by the system would lower the temperature of Earth's surface and the lower atmosphere. The Earth-atmosphere-land-ocean system would respond by emitting less intense IR radiation to space until a new radiative equilibrium were established—but at a lower radiative equilibrium temperature, with global climate change as a consequence.

In both of these examples, an interval of time elapsed before the new equilibrium was established. This implies that although global radiative equilibrium is the prevailing condition on Earth, there may be episodes of imbalance while the climate undergoes change. In fact, there is some indication that such an imbalance is occurring at present with global warming (associated with an enhanced greenhouse effect) partially offset by an increased storage of heat energy in the ocean. We have more on this topic in Chapters 11 and 12.

Conclusions

Earth intercepts only a tiny fraction of the total energy output of the Sun. Movements of the planet in space (rotation on its spin axis and revolution about the Sun) distribute this energy unequally within the Earth-atmosphere-land-ocean system. As solar radiation passes through the atmosphere, a portion is scattered and reflected back to space, a portion is absorbed (that is, converted to heat), while the rest is transmitted to Earth's surface either as direct or diffuse radiation. Overall, however, the atmosphere is relatively transparent to solar radiation and that which reaches Earth's surface is either reflected or absorbed. The Earth-atmosphere-land-ocean system responds to solar heating by emitting infrared radiation to space. Greenhouse gases and clouds absorb and emit IR in all directions including toward Earth's surface, thereby significantly elevating the average temperature of the lower troposphere (the greenhouse effect).

Net incoming solar radiation is balanced by the infrared radiation emitted to space by the Earth-atmosphere-land-ocean system. Absorption of solar radiation causes system warming whereas emission of infrared radiation to space causes system cooling. Within the Earth-atmosphere system, however, the rates of radiational heating and radiational cooling are not the same everywhere. In the next chapter, we distinguish between heat and temperature, describe heat transfer processes, examine the significance of thermal response and heat capacity for climate, and identify the various controls of air temperature.

Basic Understandings

- All objects both emit and absorb energy in the form of electromagnetic radiation. The many forms of electromagnetic radiation make up the electromagnetic spectrum and are distinguished on the basis of wavelength, frequency, and energy level.

- Earth and the Sun closely approximate perfect radiators (blackbodies) so that blackbody radiation laws may be applied to them to describe and predict their radiational characteristics.

- Wien's displacement law predicts that the wavelength of most intense radiation is inversely proportional to the absolute temperature of a radiating object. Hence, solar radiation peaks in the visible whereas Earth (terrestrial) radiation peaks in the infrared (IR).

- According to the Stefan-Boltzmann law, the total energy radiated by an object at all wavelengths is directly proportional to the fourth power of the object's absolute temperature. Hence, the much hotter Sun emits considerably more radiational energy per unit area per unit time than does the much cooler Earth-atmosphere-land-ocean system.

- The intensity of radiation emitted by the Sun, assumed to be nearly a point source as viewed from planetary distances, decreases as the inverse square of the distance traversed by the radiation. Thus, the intensity of the solar radiation reaching

Earth's orbit, equivalent to the solar constant, is sufficient to help maintain Earth as a habitable planet.

- The total energy (in the form of solar radiation) that is absorbed by the Earth-atmosphere-land-ocean system equals the total energy (in the form of infrared radiation) emitted by Earth to space. This balance between energy input and energy output is known as global radiative equilibrium and is the prevailing condition on planet Earth.

- Solar altitude, the angle of the Sun above the local horizon, influences the intensity of solar radiation that strikes Earth's surface at that location. All other factors being equal, as the solar altitude increases, the intensity of solar radiation received at Earth's surface (energy per unit horizontal surface area) also increases.

- As a consequence of Earth's elliptical orbit about the Sun, the planet's nearly spherical shape, rotation, and tilted spin axis, solar radiation is distributed unevenly over Earth's surface and changes throughout the course of a year.

- In the summer hemisphere, maximum local solar altitudes are higher and daylight is longer, resulting in the receipt of more solar radiation, and ultimately higher air temperatures. In the winter hemisphere, maximum local solar altitudes are lower, daylight is shorter, there is less solar radiation, and air temperatures are lower.

- Solar radiation that is not absorbed by the atmosphere (converted to heat) or scattered or reflected to space reaches Earth's surface. With scattering, a particle disperses solar radiation in all directions. Reflection, a special case of scattering, takes place at the interface between two media when some of the radiation striking the interface is backscattered.

- Absorption of ultraviolet radiation during the natural formation and destruction of ozone within the stratosphere shields organisms from exposure to potentially lethal levels of UV radiation.

- Solar radiation that reaches Earth's surface is either reflected or absorbed depending on the surface albedo. Surfaces that appear light-colored have a relatively high albedo for visible radiation, whereas surfaces that appear dark-colored have a relatively low albedo and typically have a higher surface temperature when exposed to sunlight.

- The atmosphere is heated primarily from below; that is, on average heat flows from Earth's surface to the overlying air. This is evident in the average temperature profile of the troposphere; that is, air temperature drops with increasing altitude.

- Water vapor, carbon dioxide, and several other atmospheric trace gases absorb outgoing IR and emit infrared radiation in all directions including downward, thereby significantly elevating the average temperature of Earth's surface and the lower atmosphere. Clouds also contribute to this greenhouse effect. Water vapor is the principal greenhouse gas.

- Clouds affect climate in two opposing ways: By absorbing and emitting IR, clouds warm Earth's surface and by reflecting incoming solar radiation, clouds cool Earth's surface. On a global scale, clouds have a net cooling effect on climate.

- The Callendar effect is the theory that global climate change can be brought about by enhancement of Earth's natural greenhouse effect by increased levels of atmospheric CO_2 from anthropogenic sources, principally the burning of fossil fuels.

- Rising levels of atmospheric CO_2 and other IR-absorbing gases enhance the natural greenhouse effect and contribute to global climate change. Combustion of fossil fuels and, to a lesser extent, deforestation is responsible for an upward trend in atmospheric carbon dioxide.

- Earth's mean temperature at global radiative equilibrium depends on the solar constant, planetary albedo, and how closely the planet approximates a blackbody. A change in any of these parameters could shift Earth to a new state of radiative equilibrium, changing the radiative equilibrium temperature and the global climate.

Enduring Ideas

- The much hotter Sun emits considerably more radiational energy per unit area per unit time and mostly at shorter wavelengths than does the much cooler Earth-atmosphere-land-ocean system.
- Global radiative equilibrium is the prevailing condition on Earth; that is, the total energy (in the form of solar radiation) absorbed by the planetary system ultimately equals the total energy (in the form of infrared radiation) emitted by Earth to space.
- Solar radiation that is not absorbed by the atmosphere or scattered or reflected to space reaches Earth's surface where it is either reflected or absorbed.
- Water vapor, carbon dioxide, and several other infrared-absorbing atmospheric gases are responsible for the greenhouse effect that significantly elevates Earth's average surface temperature so that the planet is hospitable for life as we know it.

Review

1. Describe the properties of a blackbody. Can we apply blackbody radiation laws to both Earth and the Sun?
2. How does the temperature of a radiating object affect the wavelength of maximum radiation emission?
3. Describe how the intensity of solar radiation changes as it moves away from a point source and into space.
4. How does the intensity of solar radiation striking a horizontal surface on Earth vary with changes in the solar altitude?
5. All other factors being equal, when exposed to the same flux of solar radiation, a low albedo opaque surface typically warms more than a high albedo surface. Explain why.
6. Distinguish between scattering and absorption of solar radiation.
7. The troposphere is heated from below. Explain the significance of this statement.
8. Describe the greenhouse effect and identify the principal greenhouse gas.
9. What is responsible for the current upward trend in atmospheric carbon dioxide concentration?
10. Explain why global radiative equilibrium is an example of the law of conservation of energy.

Critical Thinking

1. How would the magnitude of the solar constant change if the distance between the Sun and Earth were three times its present value?
2. In terms of astronomical seasons, what is the significance of the Tropic of Cancer and the Tropic of Capricorn?
3. Why is the length of daylight on the equinoxes not precisely 12 hours?
4. All other factors being equal, mid-afternoon temperatures tend to be lower when the ground is snow-covered rather than bare. Explain why.
5. What is the climatic significance of the ocean's relatively low average albedo for solar radiation?
6. Why is the albedo of the Moon so much less than Earth's planetary albedo?
7. On a global scale, do clouds have a net cooling or net warming effect on Earth's surface? Explain your response.
8. Comment on the validity of the greenhouse analogy to the atmosphere.
9. What role does the ocean play in regulating the atmospheric concentration of carbon dioxide?
10. Speculate on some possible impacts of global warming in the short-term and long-term.

ESSAY: Shrinking Glaciers of Mount Kilimanjaro

One of the most widely reported impact of global climate change is shrinkage of mountain glaciers, that is, glaciers occupying the summits and valleys of mountain ranges. Rising temperatures due to an enhanced greenhouse effect apparently are causing mountain glaciers to melt. According to Roger G. Barry, director of NOAA's *National Snow and Ice Data Center* at the University of Colorado at Boulder, the rate of melting of most of the world's mountain glaciers accelerated after the mid-1900s and especially since the mid-1970s. Some mountain glaciers have disappeared entirely. For example, today only 26 glaciers remain of the 150 glaciers that existed in Montana's Glacier National Park a century ago. According to scientists with the U.S. Geological Survey, climate models predict that Glacier National Park will have lost all its glaciers by 2030.

Alpine glaciers are also shrinking at accelerating rates. According to Frank Paul at the University of Zurich-Irchel in Zurich, Switzerland, the combined area covered by almost 940 Alpine glaciers diminished by about 18% from 1973 to 1999. Paul based his findings on analyses of satellite images, aerial photographs, and land surveys. Swiss glaciers are now shrinking at an annual rate that is more than six times the rate that prevailed from 1850 to 1973. At the present rate of shrinkage, Alpine glaciers at elevations below 2000 m (6500 ft) likely will disappear by the year 2070. While most of the glaciers around the globe are waning, some in Alaska appear to be expanding due to increases in snowfall.

A warming trend is probably the chief reason for glacial recession at temperate latitudes, but the causative mechanism may be different for shrinking tropical glaciers such as those occupying the upper reaches of Mount Kilimanjaro at 3.1 degrees S, 37.4 degrees E in northeastern Tanzania (Figure 1). Research conducted by Philip W. Mote of the University of Washington and Georg Kaser of the University of Innsbruck in Austria indicates that shrinkage of the permanent ice cover on Kilimanjaro is the consequence of solar radiation and drier conditions, not directly because of higher temperatures.

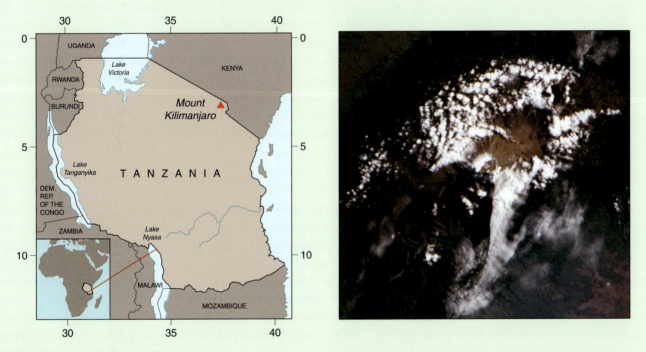

FIGURE 1
Location of Mount Kilimanjaro in Tanzania in eastern Africa (left), and Kilimanjaro as photographed by the crew of Space Shuttle mission STS-97 on December 2, 2000. [Courtesy NASA]

Kilimanjaro consists of three inactive volcanic cones. It is the highest mountain in Africa, with its highest peak at 5895 m (19,341 ft) above sea level. A plateau-shaped ice cap up to 40 m (130 ft) thick with nearly vertical edges occupies the 2.4 km (1.5 mi) wide flat crater at the summit of Kibo, the mountain's highest peak. Draped on the slopes of Kilimanjaro are small tongues of glacial ice, one of which extends down to an elevation of 4800 m (15,750 ft) above sea level. Based on ground surveys and aerial photographs, records of the surface area of glacial ice on Kilimanjaro date back to 1880 when the area was estimated at 20 km^2. Coverage plunged to 12.1 km^2 by 1912 and 2.5 km^2 in 2003, a total shrinkage of about 90%. In 2009, Lonnie G. Thompson, a glaciologist at Ohio State University, reported that more than 25% of the mountain's ice cover in 2000 had disappeared by late 2007.

As summarized in Table 1, growth or decay of a typical glacier depends on the local energy balance (input versus output of heat) and mass balance (input of snow versus output of meltwater and ice). Although Kilimanjaro is located near the equator, temperatures at and near the summit are always well below freezing (0 °C) so that all precipitation falls as snow and very little ice mass is lost by melting. (Little if any runoff is observed from the glaciers.) Also because of its tropical location, the flux of incoming solar radiation varies little throughout the course of a year and hence, the air temperature is also relatively stable. In fact, the day-night difference in temperature is greater than the annual range in mean monthly temperature.

TABLE 1

Processes involved in the Energy Balance and Mass Balance of a Glacier.

Energy Balance

Losses	*Gains*
Net emitted infrared radiation	Net absorbed solar radiation
Conduction of heat to air	Conduction of heat to ice
Latent heat for sublimation	Geothermal heat to ice
Latent heat for melting	Latent heat from deposition

Mass Balance

Losses	*Gains*
Melting plus runoff	Snowfall
Sublimation	Avalanching
Removal by wind	Addition by wind
	Freezing rain
	Deposition (*rime*)

Mote and Kaser propose that much of the shrinkage of the glaciers on Kilimanjaro is the result of sublimation plus an extended episode of relatively dry conditions responsible for reduced snowfall. *Sublimation* is the process whereby ice and snow become vapor without first becoming liquid and can operate at temperatures well below freezing. About eight times more heat energy (supplied by solar radiation) is required to sublimate versus melt an equivalent mass of ice. Scientists have not detected a significant warming trend at the summit of Kilimanjaro since 1958 and any warming dating back to 1880 could not account for the amount of shrinkage of ice during that period. A possible relationship between glacier shrinkage and global warming (due to an enhanced greenhouse effect) is the proposed link between sea-surface temperatures (SST) in the Indian Ocean and the strength of the East African monsoon.

How long will it be before the snows and ice of Kilimanjaro are gone? There is considerable disagreement among scientists. A great unknown is the future role of the monsoon circulation; a particularly intense wet monsoon can deliver very heavy accumulations of snow that could temporarily reverse the shrinking trend. In 2009, Thompson, estimated that based on current rates of shrinkage, permanent ice fields on Kilimanjaro could be gone by 2022. In 2007, a team of scientists from the University of Innsbruck estimated that there would be nothing left of the plateau ice cap by 2040.

ESSAY: Climatology of Solar Power

Solar power is a renewable energy source that should last as long as the Sun, billions of years. According to N.S. Lewis of the California Institute of Technology, "more energy from the Sun hits the Earth in 1 hour than all of the energy consumed by humans in an entire year." Solar power is an environmentally attractive alternative to conventional energy sources, especially coal, oil, and nuclear fuels. Furthermore, the cost of solar power is becoming more competitive with the price of conventional fuels in a number of applications and environments.

If all the solar radiation that is transmitted through the day-lit atmosphere were distributed uniformly over Earth's surface, about 180 watts would illuminate every square meter. (By comparison, the *solar constant* is 1368 watts per m^2.) This surface flux amounts to about 43 kilowatt-hours (kWh) per square meter per day, sufficient to meet nearly half the heating and cooling requirements of a typical American home. The total solar energy that falls yearly on America's land surface is about 600 times greater than the nation's total annual energy demand.

All technologies based on the collection of solar radiation are limited by the variability of the energy source. There is no solar radiation at night and its intensity varies with time of day, day of the year, latitude and cloud cover. Seasonal fluctuations in incident solar radiation are especially pronounced at middle and high latitudes. Figure 1 shows the average daily solar radiation per month (in kWh per m^2 per day) for the United States derived from the 1961-1990 National Solar Radiation Data Base (NSRDB) for January and July. Figure 2 is a global map of the annual average daily solar radiation for the entire globe based on the period July 1983 through June 2005.

In relatively sunny eastern Washington State, average annual solar radiation striking Earth's surface is 4.6 kWh per m^2 per day, but monthly mean values range from a low of 1.2 to a high of 8.0 kWh per m^2 per day, almost a sevenfold difference through the course of the year. In the U.S., the Desert Southwest is the most promising area for solar power because of minimal cloud cover and less seasonal variability in solar radiation. Fluctuations in consumer demand for energy further complicate matters so that matching energy supply with demand requires use of energy storage systems. Household solar collectors may produce more heat during the day than can be used at that time so that excess heat is usually stored in insulated water tanks or compartments filled with rocks. At middle and high latitudes, a conventional heating system is usually needed as a backup to solar collectors, particularly during extended spells of cloudy or very cold weather.

A traditional flat plate *solar collector* is basically a framed panel of glass designed to tap solar power. Sunlight passes through two layers of glass before being absorbed by a blackened (low albedo) metal plate. Heat is then conducted from the absorbing plate to either air or a liquid, which is conveyed by a fan or pump to wherever the heat is needed. Solar collectors typically capture 30%

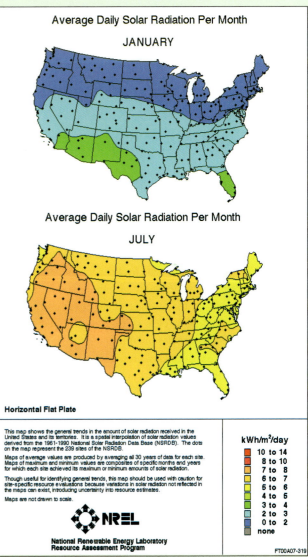

FIGURE 1
The average daily solar radiation received on a flat-plate collector facing south on a horizontal surface (in kWh per m^2 per day) for January and July. Data from measurements by pyranometers and from sunshine data using regression equations. [National Renewable Energy Laboratory, Resource Assessment Program]

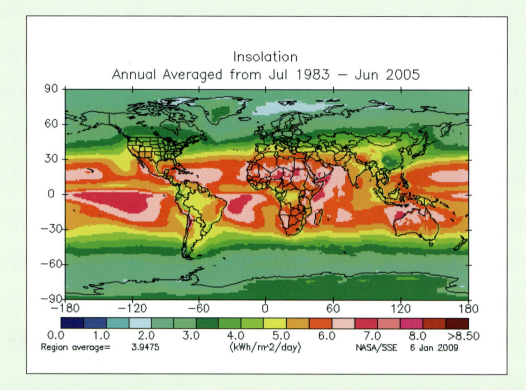

FIGURE 2
Annual average incoming solar radiation (in kWh per m² per day) based on the period July 1983 through June 2005. Data from satellite sensors. [Source: NASA Surface Meteorology and Solar Energy]

to 50% of the solar energy that strikes them and are most commonly used for space or water heating in small buildings such as homes, schools, and apartment houses. Flat plate solar collectors do not produce temperatures high enough to turn water into steam, so these devices are not suitable for most industrial purposes.

To partially compensate for the variability of incoming solar radiation due to changes in solar altitude, solar collectors are tilted toward the equator (to the south in the Northern Hemisphere). The advantage of tilted collectors over ones that are horizontal depends on the average cloud cover and latitude of the site. An optimal situation occurs in winter at a middle latitude locality favored by clear skies, where a tilted solar collector can double the amount of solar radiation that is absorbed. Some solar collectors are designed to track the sun so that they are always oriented perpendicular to the solar beam (Figure 3). However, the cost of tracking devices usually exceeds the value of the energy obtained.

Scientists and engineers are developing ways to more efficiently convert solar energy to electricity on a large scale. In one conversion method, a *power tower system*, an array of computer-controlled flat mirrors, called *heliostats*, track the Sun and focus its radiation on a single heat collection point at the top of a tower. Concentrated sunlight in these systems can produce temperatures up to 480 ºC (900 ºF), high enough to convert water to high-pressure steam for driving turbines that generate electricity. Other designs utilize a large parabolic dish or trough as a collection device.

An alternative to solar-driven turbines for generating electricity is the *photovoltaic cell* (*PV cell*), also known as a *solar cell*. Solar cells convert solar radiation directly into electricity and routinely power many devices, including handheld calculators, parking lot lights, and space vehicles. Solar cells use sunlight to create a *voltage*, that is, a difference in electrical potential, in a *diode*. When sunlight strikes the diode, an electric current flows through the circuit to which it is connected. Only special materials develop the necessary voltage to produce a direct current when they are exposed to sunlight. These materials, called *semiconductors*, are composed of highly purified silicon to which tiny amounts of certain impurities have been added. Semiconductors are manufactured in the form of tiny wafers or sheets. Electricity travels through metal contact wires on the front and back of the wafer. Groups of wafers are wired together to form photovoltaic modules, and these are interconnected to form a photovoltaic panel. About 40 wafers must be linked to match the power output of a single automobile battery.

A serious drawback to today's solar cells is relatively low efficiency. In this context, *efficiency* is defined as the percentage of solar radiation striking a photovoltaic cell that is converted to electrical energy. The efficiency of a handheld

FIGURE 3
A panel of interconnected photovoltaic cells, computer-controlled and designed to follow the Sun and always present a horizontal surface to the incoming solar beam.

solar-powered calculator, for example, is only 3% or less. At the other extreme, solar cells that power Earth-orbiting satellites have the highest efficiencies—approaching 50%. Crystalline silicon cells typically are 22% to 23% efficient. Mass-produced photovoltaic panels have conversion efficiencies of 15% to 18%. Among the many factors that contribute to low solar-cell efficiency are cell reflectivity, conversion of radiation to heat, and the sensitivity of cells to only a portion of the solar spectrum.

The main obstacle to greater use of solar cells for generating electricity is cost, although future prospects are encouraging. The cost of solar-cell generated electricity has dropped substantially in recent decades, but the cost per kilowatt-hour of solar-derived electricity is still more than 7 times that of fossil fuel-derived electricity. Future cost reductions are unlikely to be achieved by boosting the efficiency of traditional solar cells because, as a general rule, production costs soar with increasing solar cell efficiency. A more promising alternative is further development of solar cells which are less efficient but also much less expensive to manufacture compared to traditional solar cells. An example is a *thin-film solar cell* consisting of a film of silicon or other light-sensitive substance deposited on a base material (whereas traditional solar cells consist of individual crystals).

In the future, multi-megawatt solar-cell power plants are expected to routinely feed electricity into regional grids at costs that are competitive with conventional coal-fired power plants. This outlook is based on current technological trends in developing more efficient thin-film solar cells, plus declining manufacturing costs made possible by mass production and economy of scale. In addition, international efforts to offset anthropogenic contributions to global climate change and the inevitable decline in supplies of fossil fuels should spur greater demand for solar power in the future.

CHAPTER 4

THERMAL RESPONSE OF THE CLIMATE SYSTEM

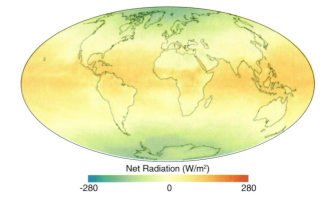

Net Radiation (W/m²)
-280   0   280

This map of net radiation (incoming sunlight minus reflected light and outgoing heat) shows global energy imbalances in September 2008, the month of an equinox. Areas around the equator absorbed about 200 watts per square meter more on average (orange and red) than they reflected or radiated. Areas near the poles reflected and/or radiated about 200 more watts per square meter (green and blue) than they absorbed. Mid-latitudes were roughly in balance. (NASA map by Robert Simmon, based on CERES data.)

Case-in-Point

Death Valley has the hottest and driest climate of any place in North America. Designated a National Monument in 1933 and a National Park in 1994, it is located in eastern central California near the Nevada border. The first permanent weather station in Death Valley was established in 1911 at Greenland Ranch, now known as Furnace Creek Ranch. Based on observations from 1911-2001, the average daily high temperature ranges from 18 °C (65 °F) in December and January to 46 °C (115 °F) in July. The average daily low temperature for the same period ranges from 4 °C (39 °F) in December and January to 31 °C (88 °F) in July. Death Valley holds many North American high temperature records. On 10 July 1913, the temperature topped out at 56.7 °C (134 °F), second only to the 58 °C (136 °F) reading at El'Azizia, Libya on 13 September 1922, recognized as the highest temperature ever recorded on Earth. The temperature at Death Valley reached or exceeded 37.8 °C (100 °F) for a record 154 consecutive days during the summer of 2001. During the summer of 1996, the hottest summer on record at Death Valley, the temperature topped 48.9 °C (120 °F) on 40 successive days and 43.3 °C (110 °F) for 105 days in a row.

Air temperatures are recorded by standard thermometers mounted within instrument shelters (Figure 4.1). Shelters are located about 1.5 m (4.9 ft) above the surface and shield instruments from direct sunshine and precipitation. Temperatures at ground level are often considerably higher than shelter temperatures. For example, on 15 July 1972 the ground temperature at Furnace Creek reached a record high 93.9 °C (201 °F) while the shelter temperature was 53.3 °C (128 °F).

A combination of topographic setting, atmospheric circulation, and intense solar radiation is

FIGURE 4.2
A combination of many environmental factors is responsible for Death Valley's extreme climate, the hottest and driest place in North America.

responsible for the extreme heat and desert conditions of Death Valley (Figure 4.2). The valley is a narrow basin—about 210 km (130 mi) long and 10 to 23 km (6 to 14 mi) wide—surrounded by high steep mountains. At its lowest point, the valley is 86 m (282 ft) below sea level. The prevailing west to east moist air flow inland from the Pacific Ocean encounters four major mountain ranges prior to entering Death Valley. For reasons discussed in detail in Chapter 5, clouds and precipitation develop where winds blow upslope while warm and dry conditions prevail where winds blow down slope. By the time the winds reach Death Valley, they are exceptionally dry so that clear skies prevail. Average annual precipitation is only 4.9 cm (1.92 in.). The greatest total rainfall in a calendar year was 11.53 cm (4.54 in.) in 1913 and again in 1983; the greatest rainfall in any 12-month period totaled 15.47 cm (6.09 in.) in 1997-98. During the driest stretch on record, 1.63 cm (0.64 in.) of rain accumulated over a 40-month period, from 1931 to 1934. The Greenland Ranch site had a run of 385 days without measurable precipitation (30 December 1952 to 18 January 1954). The Furnace Creek site also recorded 385 days without measurable precipitation (29 December 1988 through 17 January 1990). High temperatures and little moisture mean exceptionally stressful conditions for plants and animals living in Death Valley.

Solar radiation heats the dry desert floor consisting of rock, soil, and sparse vegetation. Scant moisture means that most of the heat energy is used to raise the temperature of the ground surface rather than evaporate water. The hot surface heats the overlying air via conduction and convection (confined to the valley by the surrounding mountain rim). The air is so dry to begin

FIGURE 4.1
The official thermometer for Death Valley, CA, is housed in this instrument shelter at Furnace Creek Ranch, the site of the highest officially recorded temperature in North America.

with that the upward moving branch of convection currents usually does not produce clouds. The air in the downward moving branch of convection currents is compressed and heated.

On the rare occasions when significant rain does fall in Death Valley, a spectacular bloom of colorful wild flowers spreads over the valley floor. However, if rainfall were locally heavy, dangerous flash flooding is likely (Figure 4.3). With little protective vegetative cover to anchor the soil, even a brief downpour can cause considerable erosion and property damage. For example, on 15 August 2004, rainfall estimated in the range of 2.5 to 5 cm (1 to 2 in.) washed out large sections of roadways and eroded trails in Death Valley. Two fatalities were reported.

FIGURE 4.3
Road sign alerts travelers to the possibility of flash flooding due to a brief period of heavy rainfall in Death Valley National Park, CA.

Driving Question:
What are the causes and consequences of heat transfer within Earth's climate system?

Temperature is an important and widely used climatic parameter; it is our indicator of energy flow within the climate system. This chapter focuses on the various boundary conditions in the climate system that govern air temperature. In response to a temperature gradient, heat energy is transferred from warmer locales to colder locales via radiation, conduction, convection, and phase changes of water. The specific heat of a substance is a measure of the magnitude of temperature change of that substance associated with an input or output of a quantity of heat energy. An unusually high specific heat is one of the reasons why water plays an important role in Earth's climate system. Large bodies of water have a relatively great capacity for storing heat energy, helping to dampen changes in temperature (high *thermal inertia*). For one, this helps to explain the difference between maritime and continental climates.

Operating at a global scale, differences in rates of radiational heating and radiational cooling produce temperature gradients between Earth's surface and atmosphere, as well as between the tropics and higher latitudes. In response to these temperature gradients, heat is transported from (1) Earth's surface to atmosphere via phase changes of water (latent heating) and conduction and convection (sensible heating), and (2) from the tropics

to higher latitudes via air mass exchange, storms, and ocean circulation.

The combination of the local radiation budget plus air mass advection controls the annual and diurnal variations in air temperature near Earth's surface. Climatic boundary conditions include, for example, incoming solar radiation, proximity to large bodies of water, Earth's surface characteristics, and cloud cover. This chapter opens with the distinction between temperature and heat energy and how temperature is measured.

Distinguishing Temperature and Heat

From everyday experience, we know that temperature and heat are closely related. Heating a pan of soup on the stove raises the temperature of the soup, whereas dropping an ice cube into a warm beverage lowers the temperature of the beverage. Although sometimes used interchangeably, temperature and heat are different concepts.

All matter is composed of atoms or molecules that are in continual vibrational, rotational, and/or translational motion. The energy represented by this motion is called kinetic molecular energy or just *kinetic energy*, the energy of motion. In any substance, atoms or molecules actually exhibit a range of kinetic energy. **Temperature** is directly proportional to the *average* kinetic energy of the atoms or molecules composing a substance. At the same temperature, the average kinetic molecular energy of water is the same regardless of the volume of the water.

Internal energy encompasses all the energy in a substance, that is, the kinetic energy of atoms and molecules plus the potential energy arising from forces between atoms or molecules. If two objects have different temperatures (different average kinetic molecular energies) and are brought into contact, molecular collisions will transfer energy between the objects; we call this energy in transit **heat**. Heat transferred from an object reduces the internal energy of that object, whereas heat absorbed by an object increases its internal energy and hence, will affect its average kinetic energy (temperature).

Differences in temperature rather than differences in internal energy govern the direction of heat transfer. For example, a cup of water at 20 °C has more internal energy than a teaspoon of water at 20 °C, but no heat is transferred between them. Heat energy is always transferred from a warmer object to a colder object. Heat is not necessarily transferred from an object having greater internal energy to an object with less internal energy. Consider, for example, a hot marble (at 40 °C) that is dropped into 5 liters of cold water (at 5 °C). The water has much more internal energy than the marble; nonetheless, heat is transferred from the warmer marble to the cooler water.

The following clarifies the distinction between temperature and heat. A cup of water at 60 °C (140 °F) is much hotter than a bathtub of water at 30 °C (86 °F); that is, the average kinetic energy of water molecules is greater at 60 °C than at 30 °C. Although lower in temperature, the much greater volume of water in the bathtub means that it contains more total kinetic molecular energy than does the cup of water. If in both cases, the water is warmer than its environment, heat is transferred from water to its surroundings. However, much more heat energy must be removed from the bathtub water than from the cup of water for both to cool to the same temperature.

From the above discussion, we might assume that any substance, including water, always changes temperature whenever it gains or loses heat. However, this is not necessarily the case. Water is a component of air and can occur in all three phases (ice crystals, droplets, and vapor), and, as we will see later in this chapter, heat must be either added from, or is released to the environment, for water to change phase. Furthermore, air

is a compressible mixture of gases; that is, an air sample can change volume. Heat energy is required for the work of expansion or compression of air. Hence, as a sample of air gains or loses heat, that heat may be involved in some combination of temperature change, phase change of water, or volume change.

TEMPERATURE SCALES AND HEAT UNITS

When measuring temperature, a temperature scale is needed that is based upon recognizable and reproducible reference points. For most scientific purposes, temperature is measured on the Celsius scale. First popularized in 1742 by the Swedish astronomer Anders Celsius (1701-1744), the Celsius temperature scale has the numerical convenience of a 100-degree interval between the readily reproducible freezing point and boiling point of pure water at sea level. (These reference points are crucial because the temperature remains fixed during the calibration process at the phase change.) A German physicist, Gabriel Daniel Fahrenheit (1686-1736), introduced another scale in 1714 that was also based on the phase changes of water. Today, the United States is one of only a few nations still making public reports of surface weather and climate conditions using the numerically less convenient Fahrenheit temperature scale. If a thermometer graduated in both scales were immersed in a beaker containing an equilibrium mixture of ice and water, the temperature will reach 0 degrees on the Celsius scale (0 °C) and 32 degrees on the Fahrenheit scale (32 °F). In boiling water at average sea level air pressure, an equilibrium between water and water vapor will have temperature readings of 100 °C and 212 °F.

Average kinetic molecular energy decreases with falling temperature. The theoretical temperature at which all molecular motion ceases, called **absolute zero**, corresponds to −273.15 °C (−459.67 °F). Actually, some atomic-level activity likely occurs at absolute zero, but an object at that temperature emits no electromagnetic radiation. On the Kelvin scale, temperature is the number of *kelvins* above absolute zero; hence, the Kelvin scale is a more direct measure of average kinetic molecular activity than either the Celsius or Fahrenheit temperature scales. The Kelvin scale is named for Lord Kelvin (William Thomson, 1824-1907) who developed it in the mid 19th century. Whereas units of temperature are expressed in degrees Celsius (°C) on the Celsius scale and degrees Fahrenheit (°F) on the Fahrenheit scale, temperature is expressed simply in kelvins (K) on the Kelvin scale. Nothing can be colder than absolute zero so the Kelvin scale has no negative values and a one-

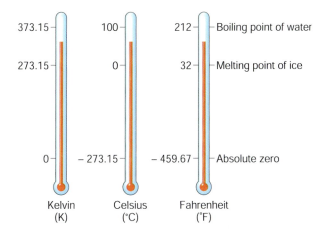

FIGURE 4.4
Comparison of three temperature scales: Kelvin, Celsius, and Fahrenheit.

TABLE 4.1
Temperature Conversion Formulas

$$^\circ F = 9/5 \; ^\circ C + 32^\circ$$
$$^\circ C = 5/9 \; (^\circ F - 32^\circ)$$
$$K = 5/9 \; (^\circ F + 459.67)$$
$$K = {}^\circ C + 273.15$$

---

kelvin increment corresponds precisely to a one-degree increment on the Celsius scale. The three temperature scales are contrasted in Figure 4.4, and conversion formulas are presented in Table 4.1.

Temperature is a convenient and useful way of describing the degree of hotness or coldness of an object. At the same time, heat energy can be directly quantified. Until recently, atmospheric scientists commonly measured heat energy in units called calories, where one **calorie (cal)** is defined as the quantity of heat equivalent to that needed to raise the temperature of 1 gram of water 1 Celsius degree (technically, from 14.5 °C to 15.5 °C). (The term *calorie* has two definitions, which can cause confusion. The calorie unit is also used to measure the energy content of food and is actually 1000 heat calories or 1.0 kilocalorie. Here we refer to the "small" calorie.) Today, the more acceptable unit for energy of any form, including heat, is the *joule (J)*. The joule is energy in units of kg m^2 per sec^2 and one calorie equals 4.1868 J. In the English system, heat is quantified as British thermal units. One **British thermal unit (Btu)** is defined as the amount of heat equivalent to that required to raise the temperature of 1 pound (lb) of water 1 Fahrenheit degree (technically, from 62 °F to 63 °F). One Btu is equivalent to 252 cal and to 1055 J.

MEASURING AIR TEMPERATURE

A **thermometer** is the usual instrument for measuring air temperature, an important element of climate. Through the years, thermometers became more sophisticated and more reliable with important implications for the integrity of the long-term instrument-based climate record (Chapter 10). Galileo Galilei (1564-1642) is credited with inventing the thermometer in 1592. (Actually, Galileo invented the *gas thermoscope*, a simple instrument that provided relative temperature measurements without a scale. Air in a glass bulb expands or contracts as the air temperature changes, thus changing the level of colored water in the neck of the instrument.) A common type of thermometer consists of a liquid-in-glass tube attached to a graduated scale (Figure 4.5). Typically, the liquid is either alcohol (which freezes at −117 °C or −179 °F) or mercury (which freezes at −39 °C or −38 °F). Both the glass and the liquid (alcohol or mercury) expand when heated and contract when cooled but the liquid much more so than the glass. As air warms, heat is transferred to the thermometer, and the expanding liquid rises in the glass tube. The temperature can be read from the scale etched on the glass or printed on the surface of the plate on which the glass tube is mounted. As air cools, heat is transferred from the thermometer, and the liquid contracts, dropping in the tube. While this type of thermometer is relatively inexpensive and accurate, it must be read directly and not remotely.

During the 19th century, self-recording thermometers, a variation of the common liquid-in-glass thermometers, were perfected to record the daily maximum and minimum temperatures that often did not occur at regularly scheduled observation times. A maximum recording thermometer, similar to a clinical thermometer, contains a constriction in the bore of the glass tube between the bulb and the etched portion of the stem, designed to break the mercury thread as the temperature begins to fall from the daily maximum reading. A minimum recording thermometer uses

FIGURE 4.5
Liquid-in-glass thermometer. In this instrument, the liquid is alcohol.

FIGURE 4.6
Air temperature recorded over a 24-hr period indicating the passage of a cold front, as measured by an electronic thermometer.

alcohol and contains a small glass index located inside the alcohol column in the bore of the thermometer. Surface tension of the liquid keeps the index located just below the meniscus of the liquid column. This index remains at the daily minimum temperature until reset.

A second type of thermometer employs an electrical conductor whose resistance changes with fluctuations in the temperature, permitting a calibration between electrical resistance and temperature. Figure 4.6 shows the change of temperature during passage of a cold front as determined by an electronic thermometer. A radiosonde is equipped with this type of thermometer. Digital read-out thermometers widely available in a variety of consumer products are also of this type. Some electronic thermometers give remote temperature readings by using wireless transmission or mounting the sensor at the end of a long cable joined to the instrument sensor. The latter system has replaced standard liquid-in-glass thermometers at National Weather Service facilities nationwide. The liquid-in-glass maximum/minimum thermometers used in the *NWS Cooperative Observing Network* for more than a century are now being replaced by electronic thermometers and a remote digital readout display, called the *Maximum-Minimum Temperature System (MMTS)*.

Another common type of thermometer uses a bimetallic sensing element to take advantage of the expansion and contraction that accompany the heating and cooling of metals. A bimetallic sensing element consists of strips of two different metals welded together back to back. The two metals have different rates of thermal expansion; that is, one metal expands more than the other in response to the same amount of heating. Because the two metals are bonded together, heating causes the bimetallic strip to bend; the greater the heating, the greater the bending. For example, the rate of thermal expansion of brass is about

twice that of iron so that a bimetallic strip composed of those two metals will bend in the direction of the iron when heated. Gears or levers translate the response of the bimetallic strip to a pointer and a dial calibrated to read in °C or °F. Alternatively, this device may be connected to a pen and a clock-driven drum to produce a continuous trace of temperature with time; this instrument is called a **thermograph**. Also, as described in Chapter 3, satellite sensors (radiometers) monitor surface temperatures remotely by measuring the intensity of emitted infrared radiation.

Regardless of the type of thermometer used, two important properties of the instrument are accuracy and response time. For most climate purposes, a thermometer that is accurate to within 0.3 Celsius degree (0.5 Fahrenheit degree) will suffice. *Response time* refers to the rapidity at which an instrument resolves changes in temperature. Electrical resistance thermometers have rapid response times, liquid-in-glass somewhat less, whereas bimetallic thermometers tend to be more sluggish. Because of differences in response time, a switch from liquid-in-glass thermometers to electronic thermometers may result in changes in the long-term climate record of a station.

For in situ measurements of air temperature to be representative of the atmospheric environment, ideally a thermometer should be adequately ventilated, in thermal equilibrium with its surroundings, and shielded from precipitation, direct sunlight (to reduce unwanted heating of the instrument), and the night sky (preventing excessive radiative heat loss). Shielding prevents heat exchanges with the thermometer (other than air) due to solar energy, radiation loss, or phase changes. Enclosing temperature sensors (and other weather instruments) in a white, louvered wooden shelter had been standard practice for official temperature measurements for more than a century (Figure 4.1). The sensor for the National Weather Service electronic thermometer is mounted inside a ventilated shield made of white plastic and the digital read-out box is located indoors (Figure 4.7). So that temperature readings acquired at different locations are comparable, an instrument shelter should be located in an open grassy area well away from trees, buildings, or other obstacles, and at a standard height (1.5 m or about 5 ft) above the ground. As a general rule, the shelter should be no closer than four times the height of the nearest obstacle. If a shelter is not available, mounting a thermometer outside a window on the shady north side of a building is usually sufficient for general purposes.

FIGURE 4.7
Enclosure for the National Weather Service electronic temperature sensor shields the instrument from direct sunshine and precipitation. This shield also protects a humidity sensor.

Heat Transfer Processes

Air temperature varies from one place to another in large part because of imbalances in rates of radiational heating and radiational cooling within the planetary system. A change in temperature over a distance is known as a **temperature gradient**. A familiar temperature gradient prevails between the hot equator and the cold poles (a horizontal temperature gradient). Another is the vertical temperature gradient between Earth's surface and the relatively cold tropopause (Figure 1.8).

Heat flows in response to a temperature gradient, according to the **second law of thermodynamics**. Simply put, this law states that all systems tend toward a state of disorder. You probably have personal experience with some implications of the second law. If you avoid cleaning up your home, for example, it rapidly becomes more and more disorganized. The presence of a gradient of any kind within a system signals order within that system. Hence, as a system tends toward disorder, gradients decrease. The second law predicts that where a temperature gradient exists, heat is transferred in a direction so as to eliminate the gradient; that is, heat flows from where the temperature is

higher toward where the temperature is lower. In addition, the greater the temperature difference (i.e., the steeper the temperature gradient), the more rapid the rate of heat flow. Within the planetary system, heat is transferred via radiation, conduction, convection, and phase changes of water in an attempt to eliminate the temperature gradients. Only the input of solar energy with its seasonal and diurnal variations maintains these gradients.

RADIATION

Radiation is both a form of energy and a means of energy transfer (Chapter 2). Electromagnetic radiation consists of a spectrum of waves traveling at the speed of light. Unlike conduction and convection, radiation requires no intervening physical medium; that is, it can travel through a vacuum. Although not actually a vacuum, interplanetary space is so highly rarefied that conduction and convection play essentially no role in transporting energy from the Sun to Earth. Rather, radiation is the principal means whereby the planetary system gains energy from the Sun. Radiation is also the principal means whereby energy escapes from the planet to space maintaining a habitable environment on Earth (Chapter 3).

Absorption of radiation consists of the conversion of electromagnetic energy to heat. By contrast, emission of electromagnetic energy is a loss of heat from the radiating object to the environment. All objects both absorb and emit electromagnetic radiation. If an object absorbs radiation at a greater rate than it emits radiation, the internal energy increases and the temperature of the object will rise. This type of imbalance in the flux of radiation is known as **radiational heating**. If an object emits radiation at a greater rate than it absorbs radiation, the internal energy decreases and the temperature of the object will fall. This type of imbalance in the flux of radiation is called **radiational cooling**. At radiative equilibrium, when absorption and emission of radiation are equal, the object's internal energy and temperature remain constant.

Radiative equilibrium does not necessarily mean that the temperature stays constant among all components of a system. Over the long-term, the entire planetary system is in radiative equilibrium with surrounding space so that the effective planetary temperature remains relatively constant from year to year. Nonetheless, heat may be redistributed among the various components of the climate system (for example, among the ocean, land, and glaciers) so that air temperature at a specified location may undergo significant short- and long-term variations. Hence, global radiative equilibrium does not preclude climate variability or climate change.

CONDUCTION AND CONVECTION

Heat can be conducted within a substance or between substances that are in direct physical contact. **Conduction** (of heat) refers to the transfer of kinetic energy of atoms or molecules via collisions between neighboring atoms or molecules. This is why the temperature of a metal spoon rises when placed in a steaming hot cup of coffee. As the more energetic molecules of the hot coffee collide with the less energetic atoms of the cooler spoon, some kinetic energy is transferred to the atoms of the spoon. These atoms then transmit some of their heat energy, via collisions, to neighboring atoms, so that heat is conducted up the handle of the spoon and eventually the handle becomes warm to the touch.

Some substances are better conductors of heat than others. **Heat conductivity** is defined as the ratio of the rate of heat transport across an area to the temperature gradient. Hence, in response to a specified temperature gradient, substances having higher heat conductivities have greater rates of heat transport. As a rule, solids are better conductors of heat than are liquids, and liquids are better heat conductors than gases. At one extreme, metals are excellent conductors of heat whereas at the other extreme still air is a very poor conductor of heat. Heat conductivities of some common substances are listed in Table 4.2, ranked from most to least conductive.

Differences in heat conductivity can be the reason one object feels colder than another, even though both objects have the same temperature. For example, at the same room temperature, a metallic object feels colder than a wooden object. The heat conductivity of metals is much greater than that of wood so that when you grasp the two objects, your hand more rapidly loses heat by conduction to the metallic object than to the wooden object; thus you have the sensation that the metal is colder than the wood.

The relatively low heat conductivity of air makes it a good heat insulator. Heat conductivity is lower for still air than for air in motion, so to avoid additional heat loss by convection, air must be confined. For example, in a thick fiberglass blanket used as attic insulation, the motionless air trapped between individual fiberglass filaments is primarily responsible for inhibiting heat loss. Similarly, the heat conductivity of a fresh snow cover is low because of the air trapped between individual snowflakes. A thick snow cover (20 to 30 cm or 8 to 12 in.) can thus inhibit or prevent freezing of the underlying soil, even though the temperature of the overlying air drops well below freezing. In time, however, a snow cover loses some of its insulating property as the snow settles. As snow settles, air escapes, snow density increases, and heat conductivity increases (Figure 4.8).

TABLE 4.2
Heat Conductivity of Some Familiar Substances[a]

Copper	0.92
Aluminum	0.50
Iron	0.16
Ice (at 0 °C)	0.0054
Limestone	0.0048
Concrete	0.0022
Water (at 10 °C)	0.0014
Dry sand	0.0013
Air (at 20 °C)	0.000061
Air (at 0 °C)	0.000058

[a]*Heat conductivity* is defined as the quantity of heat (in calories) that would flow through a unit area of a substance (cm²) in one second in response to a temperature gradient of one Celsius degree per centimeter. Hence, heat conductivity has units of calories per cm² per sec per C° per cm.

During the day, heat is conducted from Sun-warmed ground to cooler overlying air, but because air is a poor conductor of heat, conduction is significant only in a very thin layer of air in immediate contact with Earth's surface. Much more important than conduction in transporting heat vertically within the troposphere is convection. **Convection** is the transport of heat within a fluid via motions of the fluid itself. Although conduction takes place in solids, liquids, or gases, convection generally occurs only in fluids (liquids or gases).

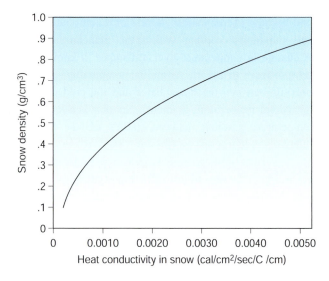

FIGURE 4.8
A thick layer of fresh snow (low density) is a good heat insulator primarily because of air trapped between the individual snowflakes. (Air is a poor conductor of heat.) But in time, the snow settles, air escapes, snow density increases, and the snow cover's insulating property diminishes.

Convection in the atmosphere is the consequence of differences in air density. At the same pressure, cold air is denser than warm air. As heat is conducted from the relatively warm ground to cooler overlying air, that air is heated, becoming less dense and thus more buoyant compared to the surrounding air. Cool, denser air from above sinks and replaces the warmer, less dense air at the ground forcing it to rise (buoyancy). (This also happens when cold tap water flows into a tub of hot water; that is, the denser cold water sinks and forces the less dense hot water to rise.) Ascending warm air expands, cools, and eventually sinks back to the ground. Meanwhile, air now in contact with the warm ground is heated and rises—displaced by the sinking cooler denser air. In this way, as illustrated in Figure 4.9, convection currents transport heat upward from Earth's surface sometimes reaching thousands of meters into the troposphere.

We can readily observe convection currents in a pan of water on a hot stove. By adding a drop or two of food coloring to the water, we actually see the circulating water redistributing heat that is conducted from the bottom of the pan into the water. In this example, as well as in the troposphere, conduction and convection work together in transferring heat. The combination of conduction and convection is known as **sensible heating**, as we can *sense* a change in temperature in response to such heating.

PHASE CHANGES OF WATER

Another important heat transfer process operating within the climate system involves phase changes of water. Water is one of the very few substances that can occur naturally in all three phases within the temperature and pressure ranges found at and near Earth's surface. Water occurs as a crystalline solid (ice or snow), liquid, and gas (water vapor), and is continually changing phase as environmental conditions vary (Figure 4.10). Depending on the type of phase change, water either absorbs heat from its environment or releases heat to its environment. The quantity of heat that is involved in phase changes of water is known as **latent heat**, where the term "latent" refers to heat that is "hidden" until released. Heat is absorbed or released during phase changes because of differences in molecular activity represented by the three physical phases of water, which for water, is of unusually great magnitude. For the reason behind this and other unusual thermal properties of water, refer to this chapter's first Essay.

In the solid phase (ice), water molecules are relatively inactive and vibrate about a fixed location in an ice crystal lattice. Hence, an ice cube or any other piece of ice tends to retain its shape. In the liquid phase, molecules are less strongly bonded and move about with greater freedom, so that liquid water takes the shape of its container. In the vapor phase, water molecules are

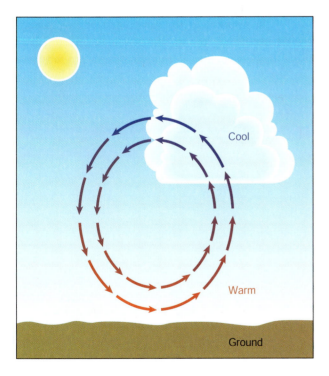

FIGURE 4.9
Convection currents transport heat conducted from Earth's surface into the troposphere.

FIGURE 4.10
At and near Earth's surface, water occurs in all three phases: solid (ice and snow), liquid, and vapor. Clouds are composed of tiny water droplets and/or ice crystals.

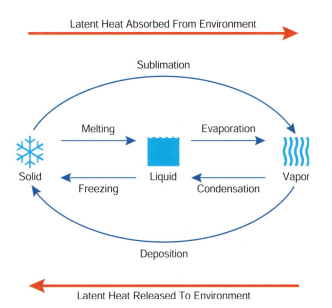

Latent Heat Absorbed From Environment

Latent Heat Released To Environment

FIGURE 4.11
When water changes phase, heat energy is either absorbed from or released to the environment.

essentially unbonded and exhibit maximum activity, diffusing readily throughout the entire volume of its container. A change in phase is thus linked to a change in level of molecular activity, which is brought about by either an addition or loss of heat (Figure 4.11). Heat is absorbed from the environment during those changes to higher energy states: *melting* (phase change from solid to liquid), *evaporation* (phase change from liquid to vapor), and *sublimation* (phase change directly from solid to vapor). Heat is released to the environment during those changes to lower energy states: *freezing* (phase change from liquid to solid), *condensation* (phase change from vapor to liquid), and *deposition* (phase change directly from vapor to solid). During any phase change, heat is exchanged between water and its environment. Although the temperature of the environment changes in response, the temperature of the water undergoing the phase change remains constant until the phase change is complete. That is, the available heat (latent heat) is involved exclusively in changing the phase of water and not its temperature.

Latent heating refers to the transport of heat from one location to another as a consequence of changes in the phase of water. The latent heat that is supplied to change liquid water to vapor (evaporation) in one place (e.g., evaporation at the ocean surface) is released to the environment when that water vapor condenses at some other place (e.g., cloud formation in the atmosphere). Prior to condensation, winds can transport water vapor thousands of kilometers horizontally and to altitudes of thousands of meters so that latent heating can involve long-range transport of heat energy, helping to make the planet habitable.

Thermal Response and Specific Heat

Whether by radiation, conduction, convection, or latent heating, transfer of heat from one place to another within the climate system is accompanied by changes in temperature. A heat gain causes a rise in temperature whereas a heat loss causes a drop in temperature.

The temperature change associated with an input (or output) of a specified quantity of heat varies from one substance to another. The amount of heat that will raise the temperature of 1 gram of a substance by 1 Celsius degree is defined as the **specific heat** of that substance. Joseph Black (1728-1799), a Scottish chemist, first proposed the concept of specific heat in 1760. The specific heat of all substances is measured relative to that of liquid water, which is defined as 1 calorie per gram per Celsius degree (at 15 °C). The specific heat of ice is about 0.5 calorie per gram per Celsius degree (near 0 °C). Specific heats of some other familiar substances are listed in Table 4.3, ranked from high to low values. The variation in specific heat from one substance to another implies that different materials have different capacities for storing internal energy.

In response to the same input of heat energy, a substance with a low specific heat undergoes a greater rise in temperature than an equivalent mass of a substance

TABLE 4.3 Specific Heat of Some Familiar Substances[a]	
Water	1.000
Wet mud	0.600
Ice (at 0 °C)	0.478
Wood	0.420
Aluminum	0.214
Brick	0.200
Granite	0.192
Sand	0.188
Dry air[b]	0.171
Copper	0.093
Silver	0.056
Gold	0.031

[a]Calories per gram per Celsius degree.
[b]At constant volume.

with a high specific heat. Water has the greatest specific heat of any naturally occurring substance. From Table 4.3, water's specific heat is about five times that of dry sand. One calorie of heat will raise the temperature of 1 gram of water 1 Celsius degree, whereas 1 calorie of heat will raise the temperature of 1 gram of dry sand by 5 Celsius degrees. This contrast in specific heat helps explain why at the beach in summer the sand feels considerably hotter to bare feet than the water. The specific heat contrast plus evaporative cooling explains why wet sand feels cooler than dry sand.

THERMAL INERTIA

Water's exceptional capacity to store heat has important implications for climate. A large body of water (such as the ocean or Great Lakes) can significantly influence the climate of downwind localities. The most persistent influence is on air temperature. Compared to an adjacent landmass, a body of water does not warm as much during the day (or in summer) and does not cool as much at night (or in winter). In other words, a large body of water exhibits a greater resistance to temperature change, called **thermal inertia**, than does a landmass. Whereas the specific heat of water versus that of land is a major reason for this contrast in thermal inertia, differences in heat transport also contribute. Sunlight penetrates water to some depth and is absorbed (converted to heat) through a significant volume of water (Chapter 3). But sunlight cannot penetrate the opaque land surface and is therefore absorbed only at the surface. Furthermore, circulation of ocean and lake waters transports heat through great volumes of water, whereas heat is conducted only very slowly into soil. The input (or output) of equal amounts of heat energy causes a land surface to warm (or cool) more than the equivalent surface area of a body of water over a given span of time.

MARITIME AND CONTINENTAL CLIMATES

Air temperature is regulated to a considerable extent by the temperature of the surface over which the air resides or travels via heat exchanges. With sufficient residence time over a large body of water, an air mass tends to take on similar temperature characteristics as the surface water. Hence, places immediately downwind of the ocean experience much less contrast between average winter and summer temperature; such places have a **maritime climate**. Localities at the same latitude (especially at mid and high latitudes), but well inland, experience a much greater contrast between winter and summer temperature; such places have a **continental climate**. That is, at the same latitude, summers are cooler and winters are milder in maritime climates than in continental climates.

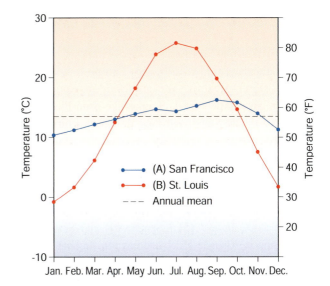

FIGURE 4.12
Variation in the march of monthly mean temperature for (A) maritime San Francisco, CA, and (B) continental St. Louis, MO. The two cities are located at about the same latitude.

Consider an example of the contrast in temperature regime between continental and maritime climates (Figure 4.12). The latitude of San Francisco, CA (37.8 degrees N) is almost the same as that of St. Louis, MO (38.8 degrees N) so that the seasonal variation in the amount of incoming solar radiation (due to astronomical factors) is similar at both places. (San Francisco has an average incoming radiation flux of approximately 196 W/m² while St. Louis has about 176 W/m².) St. Louis is situated far from the moderating influence of the ocean and the climate is continental. St. Louis' average summer (June, July, and August) temperature is 25.6 °C (78.0 °F) and its average winter (December, January, and February) temperature is 0.6 °C (33.0 °F), giving an average summer-to-winter seasonal temperature contrast of 25 Celsius degrees (45 Fahrenheit degrees). San Francisco, on the other hand, is located on the West Coast, immediately downwind of the Pacific Ocean; the climate is maritime. The average summer temperature at San Francisco is 17.0 °C (62.6 °F), whereas the average winter temperature is 10.2 °C (50.4 °F), giving an average seasonal temperature contrast of only 6.8 Celsius degrees (12.2 Fahrenheit degrees).

In Western Europe, the air temperature contrast between summer and winter is less than it is over most of North America (outside of the Pacific coastal region) because of prevailing onshore winds from the North Atlantic Ocean. Sea-surface temperatures (SST) change relatively little through the course of a year and this stable SST regime dampens the summer-to-winter temperature contrast of air flowing over the ocean to downwind Western Europe. Whereas summer average air temperatures are somewhat lower, winter average air temperatures are milder in Western Europe compared to upwind North America. The northward-moving warm Gulf Stream parallels the U.S. coastline from Florida to the Mid-Atlantic States and then the current turns east and northeast across the North Atlantic. In winter, the relatively warm ocean surface moderates cold air masses as they surge from polar areas southeastward toward the British Isles and Western Europe.

Compare, for example, January and July temperatures at Cork, Ireland (51.9 degrees N) and Saskatoon, Saskatchewan, Canada (52.1 degrees N). At both places on average, January and July are the coldest and warmest months of the year respectively. Although located at about the same latitude, the two cities have markedly different climates. The average temperature contrast between July and January is about 36.5 Celsius degrees (65.7 Fahrenheit degrees) at continental Saskatoon but only about 11 Celsius degrees (20 Fahrenheit degrees) at maritime Cork. The reduced seasonal contrast at Cork is mostly due to much higher winter temperatures. January average temperature is 4.5 °C (40.1 °F) at Cork but –18.5 °C (–1.3 °F) at Saskatoon. July average temperatures are not much different at the two locations with 18 °C (64.4 °F) at Saskatoon and 15.5 °C (59.9 °F) at Cork.

Along Canada's east coast, some maritime influence is noted but with prevailing winds directed offshore, the average temperature contrast between summer and winter is greater than in western Ireland where prevailing winds are onshore. For Cartwright along the southeastern Labrador coast at 53.7 degrees N, the average temperature contrast between July and January is 26.9 Celsius degrees (48.4 Fahrenheit degrees), somewhat less than the seasonal temperature range at Saskatoon but significantly greater than the range at Cork.

Heat Imbalance: Atmosphere versus Earth's Surface

Sensors on board Earth-orbiting satellites detect imbalances in rates of radiational heating and radiational cooling. One important aspect of this heating imbalance involves Earth's surface versus the atmosphere and results in a net transfer of heat from Earth's surface to the troposphere.

Figure 4.13 shows how solar radiation intercepted by Earth interacts with the atmosphere and Earth's surface. Numbers are global annual averages, normalized so that the units represent a percentage of the average solar radiation incident at the top of the atmosphere. For every 100 units of solar radiation that enter the upper atmosphere, the planetary system scatters or reflects 31 units (31%) back to space, the atmosphere absorbs 20 units (20%), and Earth's surface (principally the ocean) absorbs 49 units (49%). In response to radiational heating, Earth's surface emits 114 units of infrared radiation. Atmospheric gases and clouds absorb 105 units of infrared radiation and emit 95 units to Earth's surface (the *greenhouse effect*). A total of 69 units of IR radiation are emitted out the top of the atmosphere to space, equal to the amount of solar radiation absorbed by the planetary system. Note that the 69 units of IR plus

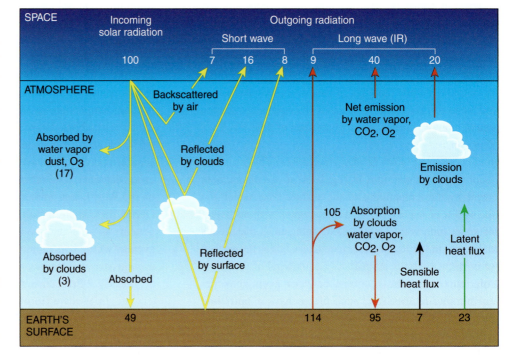

FIGURE 4.13
Globally and annually averaged disposition of 100 units of solar radiation entering the top of the atmosphere. Solar radiation fluxes are depicted at the left, infrared radiation fluxes in the middle, and latent and sensible heat fluxes to the right.

TABLE 4.4
Global Radiation Balance

Solar radiation intercepted by Earth	100 units
Solar radiation budget	
Scattered and reflected to space (7 + 16 + 8)	31
Absorbed by the atmosphere (17 + 3)	20
Absorbed at the Earth's surface	49
Total	100 units
Radiation budget at the Earth's surface	
Infrared cooling (95 − 114)	−19
Solar heating	+49
Net heating	+30 units
Radiation budget of the atmosphere	
Infrared cooling (− 40 − 20 + 105 − 95)	−50
Solar heating	+20
Net cooling	−30 units
Non-radiative heat transfer: Earth's surface to atmosphere	
Sensible heating (conduction plus convection)	7
Latent heating (phase changes of water)	23
Net transfer	30 units

the 31 units of incoming radiation scattered or reflected equals the 100 units of solar radiation entering the upper atmosphere.

The global annual average distribution of incoming solar radiation and outgoing infrared radiation implies net warming of Earth's surface and net cooling of the atmosphere (Table 4.4). At Earth's surface, absorption of solar radiation is greater than emission of infrared radiation. In the atmosphere, on the other hand, emission of infrared radiation to space is greater than absorption of solar radiation. That is, on a global annual average basis, Earth's surface undergoes net radiational heating and the atmosphere undergoes net radiational cooling.

The atmosphere is not actually cooling relative to Earth's surface because radiation is not the only heat transfer mechanism at work here. In response to the radiationally induced temperature gradient between Earth's surface and atmosphere, heat is transferred from Earth's surface to the atmosphere by non-radiative heat transfer. A combination of latent heating (phase changes of water) and sensible heating (conduction and convection) is responsible for this transfer of heat. As shown in Figure 4.13, on a global annual average basis, 30 units of heat energy are transferred from Earth's surface to the atmosphere: 23 units (about 77% of the total) by latent heating and 7 units (about 23%) by sensible heating.

LATENT HEATING

As noted earlier in this chapter, latent heating refers to the transfer of heat energy from one place to another as a consequence of phase changes of water. When water changes phase, heat energy is either absorbed from the environment (i.e., melting, evaporation, sublimation) or released to the environment (i.e., freezing, condensation, deposition). As part of the global water cycle, latent heat that is used to vaporize water at the Earth's surface is transferred to the atmosphere when clouds form. Significantly for Earth's climate, ocean water covers a large portion of Earth's surface and is the principal source of water vapor that eventually condenses and returns to Earth's surface as precipitation. In general, only well inland does most precipitation originate as evaporation from the continents (Figure 4.14).

As Earth's surface absorbs radiation (both solar and infrared), some of the heat energy is used to vaporize water from the ocean, glaciers, lakes, rivers, soil, and vegetation (*transpiration*). The latent heat required for vaporization (evaporation or sublimation) is supplied at the Earth's surface, and latent heat subsequently is released to the atmosphere during the formation of clouds. Within the troposphere, clouds develop as some of the water vapor condenses into liquid water droplets or deposits as ice crystals, releasing latent heat to the atmosphere. Through latent heating, then, heat is transferred from Earth's surface to the troposphere. In fact, latent heat transfer is more important than either radiational cooling or sensible heat transfer in cooling Earth's surface (Figure 4.15).

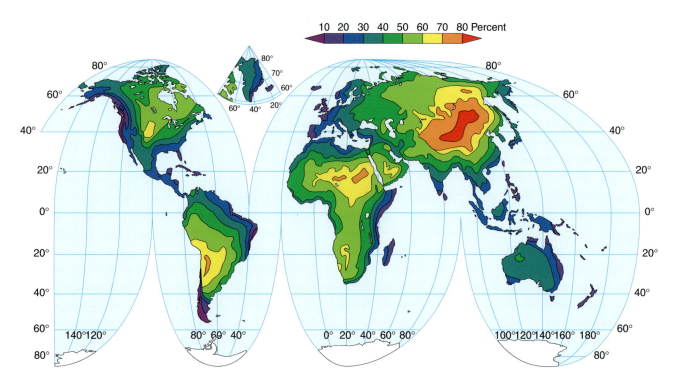

FIGURE 4.14
The percentage of annual precipitation that falls on land, originating as water that vaporized from the continents and averaged over 15 years. In many land areas, the principal source of water for precipitation is evaporation from the ocean. [World Climate Research Programme, Global Energy and Water Cycle Experiment]

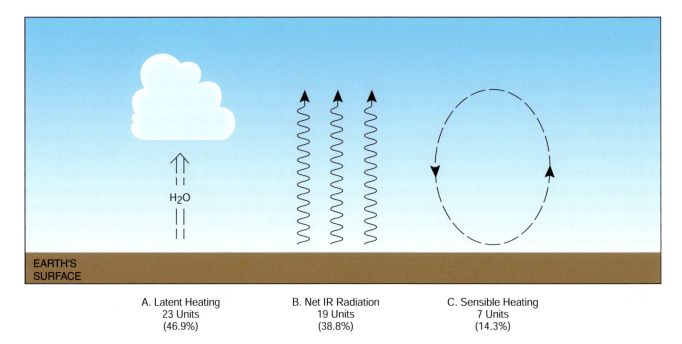

FIGURE 4.15
Earth's surface is cooled via (A) vaporization of water, (B) net emission of infrared radiation, and (C) conduction plus convection. Numbers are global annual averages.

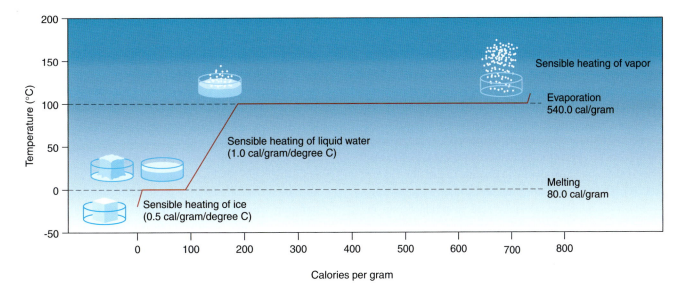

FIGURE 4.16
Heating one gram of ice causes a rise in temperature plus phase changes.

Unusually large quantities of heat are required to bring about phase changes of water as compared to phase changes of other naturally occurring substances. Consider, for example, the quantity of heat involved in changing the phase and temperature of one gram of ice at sea level as it is heated from an initial temperature of −20 °C (−4 °F) to water vapor at a final temperature of 100 °C (212 °F) (Figure 4.16). The specific heat of ice is about 0.5 cal per gram per Celsius degree, which means that 0.5 cal of heat must be supplied for every 1 Celsius degree rise in temperature. Hence, warming our ice from −20 °C to 0 °C (−4 °F to 32 °F) requires an input of 10 cal of heat. Once the melting (or freezing) point is reached, an additional 80 cal of heat, called the **latent heat of fusion**, must be supplied per gram to break the forces that bind water molecules in the ice phase. The temperature of the water and ice mixture remains at 0 °C until all the ice melts.

Once the ice melts completely, additional heating of the water causes the temperature of the liquid water to rise. The specific heat of liquid water is 1 cal per gram per Celsius degree so that 1 cal is needed for every 1 Celsius degree rise in water temperature. While evaporation can occur at any temperature, suppose that no evaporation takes place until the temperature of the system reaches the normal boiling point of 100 °C (212 °F). This 100 Celsius degree rise in temperature requires the addition of 100 calories. When the water boils at sea-level, the temperature of the water will remain at 100 °C (212 °F) until all the liquid is vaporized. The **latent heat of vaporization** is 540 cal per gram at 100 °C (212 °F). Any additional heating after evaporation will cause the temperature of the vapor

to increase. If the one gram had evaporated at 0 °C after melting, the latent heat of vaporization at that temperature would be 600 cal per gram as more energy is required to break the hydrogen bonds at lower temperatures when the water molecules are less active than at higher temperatures. For our 1 gram of ice to vaporize directly without melting first (*sublimation*), the latent heat of fusion plus the latent heat of vaporization must be supplied to the ice. This amounts to 680 cal per gram at 0 °C.

If the processes just described are reversed, that is, if the water vapor were cooled until it becomes liquid (*condensation*) and then ice (*freezing*), the water temperature drops and phase changes take place as equivalent amounts of latent heat are released into the environment. When water vapor becomes liquid, the latent heat of vaporization is released to the environment, and when water freezes, the latent heat of fusion is released. If water vapor changes to ice without first becoming liquid (*deposition*), the latent heats of vaporization plus fusion are released to the environment. Regardless of when phase and temperature changes occur, energy is always conserved.

SENSIBLE HEATING

Heat transfer via conduction and convection can be monitored (sensed) by temperature changes; hence, sensible heating encompasses both of these processes. Heat is conducted from the relatively warm surface of the Earth to the cooler overlying air. Heating reduces the density of that air, which is forced to rise by cooler denser air replacing it at the surface. In this way, convection transports heat from Earth's surface into the troposphere. Because air is a relatively poor conductor of heat, convection is much more important than conduction as a heat transfer mechanism within the troposphere.

Often sensible heating combines with latent heating to channel heat from Earth's surface into the troposphere. This happens during thunderstorm development. Updrafts (ascending branches) of vapor-laden air in convection currents often produce *cumulus clouds*, which resemble puffs of cotton floating in the sky (Figure 4.17A). These clouds are sometimes referred to as *fair-weather cumulus* because they seldom produce rain or snow. However, if atmospheric conditions are favorable, these convective currents can surge to great altitudes, and cumulus clouds merge and billow upward to form

FIGURE 4.17
Cumulus clouds form in the ascending branch of a convective circulation (A). A cumulonimbus (thunderstorm) cloud forms when atmospheric conditions favor deep convection (B).

cumulonimbus clouds, also known as thunderstorm clouds (Figure 4.17B). In retrospect, two important heat transfer processes (latent heating and sensible heating) took place last summer when a thunderstorm sent you scurrying for shelter at the beach.

At some times and places, heat transfer is directed from the troposphere to Earth's surface, the reverse of the global annual average situation. This reversal in direction of heat transport occurs, for example, when mild winds blow over cold, snow-covered ground or when warm air moves over a relatively cool ocean surface. And heat transport from the lower atmosphere to Earth's surface is the usual situation at night (especially when the sky is clear) as radiational cooling causes Earth's land surface to become colder than the overlying air.

An enlightening concept is the comparison of how the heat energy received at Earth's surface (by absorption of solar and infrared radiation) is partitioned between sensible heating and latent heating. That comparison frequently is shown as a ratio, the **Bowen ratio**, named for Ira Bowen (1898-1973), an American astronomer who studied heat exchanges at surfaces. That is,

Bowen ratio = [(sensible heating)/(latent heating)]

We have already seen that at the global scale,

Bowen ratio = [(7 units)/(23 units)] = 0.3.

As shown in Table 4.5, the average Bowen ratio varies from one place to another depending on the

TABLE 4.5
Bowen Ratio[a] of Various Geographical Areas

All Oceans	0.11
Atlantic Ocean	0.11
Pacific Ocean	0.10
Indian Ocean	0.09
All Land	0.96
North America	0.74
South America	0.56
Europe	0.62
Asia	1.14
Africa	1.61
Australia	2.18
Globe	0.30

[a]Ratio of sensible heating to latent heating is a non-dimensional number.

amount of surface moisture. The moister the surface, the more dominant latent heating is over sensible heating. The Bowen ratio ranges from about 0.1 (one-tenth as much sensible as latent heating) for the ocean to as high as about 20 (twenty times as much sensible as latent heating) in deserts. Ocean waters cover much of Earth's surface and evaporation is continuous from the ocean surface so it is not surprising that the global Bowen ratio is relatively low (0.3).

Partitioning of net radiation absorbed at Earth's surface into sensible heating and latent heating varies through the course of a year. Consider two examples.

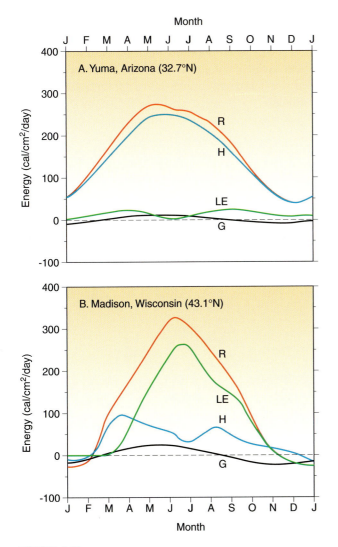

FIGURE 4.18
Surface energy budget through the course of a year at (A) Yuma, AZ, and (B) Madison, WI, where *R* is net radiation absorbed, *H* is sensible heating (conduction plus convection), *LE* is latent heating (phase changes of water), and *G* is storage. [Modified after Sellers, W.D., 1965. *Physical Climatology*. Chicago: The University of Chicago Press, p. 106.]

Figure 4.18A shows the average annual variation of the surface energy budget for Yuma, AZ. *R* is the net radiation absorbed (solar plus downwelling infrared), *H* is the energy used for conduction and convection, *LE* is the energy used for latent heating, and *G* is the energy stored in the ground. In this desert locality, note that sensible heating exceeds latent heating throughout the year and the Bowen ratio peaks in mid-summer. Figure 4.18B shows the average annual variation of the surface energy budget for Madison, WI. In this much more humid locality, latent heating is greater than sensible heating from April until November and peaks in early summer.

Heating Imbalance: Tropics versus Middle and High Latitudes

On a global scale, imbalances in radiational heating and cooling occur not only between Earth's surface and atmosphere but also between the tropics and higher latitudes. Because the planet is nearly a sphere, parallel beams of incoming solar radiation strike the tropics more directly than higher latitudes. (That is, noon solar altitudes are higher in the tropics and lower at higher latitudes.) At higher latitudes, solar radiation spreads over a greater area and is less intense per unit horizontal surface area than in the tropics. Furthermore, solar radiation reaching the surface in polar latitudes is depleted more than in tropical latitudes because of a greater path length through the atmosphere, as indicated by Figure 3.8.

Emission of infrared radiation by the planetary system also varies with latitude but less than solar radiation. Because air temperatures are generally lower at higher latitudes, IR emission also declines with increasing latitude. (Recall from Chapters 2 and 3 that radiation emission is temperature dependent.) Consequently, over the period of a year at higher latitudes, the rate of infrared cooling to space exceeds the rate of warming caused by absorption of solar radiation. At lower latitudes the reverse is true; that is, over the course of a year, the rate of solar radiational heating is greater than the rate of infrared radiational cooling (Figure 4.19). Averaged over the entire globe, incoming energy (absorbed solar radiation) must equal outgoing energy (IR emitted to space). That is, the areas under the two curves in Figure 4.19 are equal. Recall that the balance between energy entering and leaving the planetary system (*global radiative equilibrium*) is the prevailing condition.

Measurements by sensors on board Earth-orbiting satellites indicate that the division between regions of net

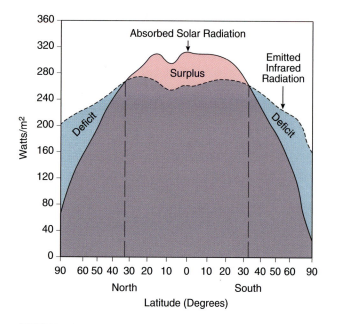

FIGURE 4.19
Variation by latitude of absorbed solar radiation and outgoing infrared radiation based on measurements by sensors flown on board Earth-orbiting satellites. [NOAA/NESDIS]

source region) or travels (Figure 4.20). Air masses that form at high latitudes over cold, often snow- or ice-covered surfaces are relatively cold. Air masses that form over warm surfaces at low latitudes are relatively warm. Air masses that develop over the ocean are humid and those that form over land are relatively dry. Hence, there are four basic types of air masses: cold and humid, cold and dry, warm and humid, and warm and dry. (Further discussion of air masses appears in Chapter 7.)

Warm air masses that form in lower latitudes flow toward the poles and are replaced by cold air masses flowing toward the equator from source regions at high latitudes. Air masses modify (become cooler or warmer, drier or more humid) to some extent as they move away from their source region, gaining or losing heat energy and/or moisture in the process. In this north-south exchange of air masses, a net transport of mostly sensible heat energy occurs directed from lower to higher latitudes.

radiational cooling and regions of net radiational warming is near the 35-degree latitude circle in both hemispheres. By implication, latitudes poleward of about 35 degrees N and 35 degrees S should experience net cooling over the course of a year, while tropical latitudes are sites of net warming. In fact, lower latitudes do not become progressively warmer nor do higher latitudes become colder because heat is transported from the tropics poleward into middle and high latitudes. **Poleward heat transport** is brought about by (1) air mass exchange, (2) storm systems, and (3) ocean circulation.

HEAT TRANSPORT BY AIR MASS EXCHANGE

North-south exchange of air masses transports mainly sensible heat from the tropics into middle and high latitudes. An **air mass** is a huge volume of air covering thousands of square kilometers that is relatively uniform horizontally in temperature and humidity. The properties of an air mass largely depend on the characteristics of the surface over which the air mass forms (its

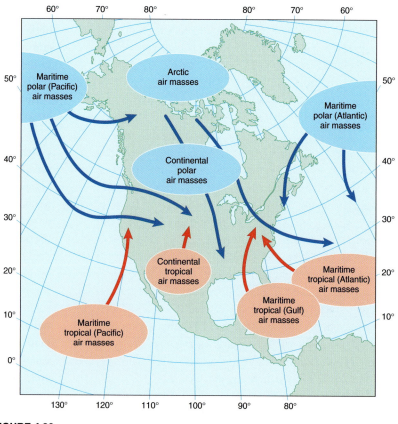

FIGURE 4.20
Source regions of air masses that regularly invade North America. Warm air masses flow poleward while cold air masses flow equatorward so that the net transport of sensible heat is from lower to higher latitudes.

HEAT TRANSPORT BY STORMS

Acquisition and subsequent release of latent heat in migratory storm systems (*cyclones* or *lows*) also plays an important role in the poleward transport of heat. At low latitudes, water that evaporates from the warm ocean surface is drawn into the circulation of a developing storm system. As the storm travels poleward, some of that water vapor condenses into clouds, thereby releasing latent heat to the troposphere. Latent heat of vaporization from low latitudes is thereby delivered to middle and high latitudes. Because they convey much more water vapor and latent heat, tropical storms and hurricanes are greater contributors to poleward heat transport than ordinary middle latitude (*extratropical*) storms.

HEAT TRANSPORT BY OCEAN CIRCULATION

The ocean too contributes to poleward heat transport via wind-driven surface currents and the deeper thermohaline circulation. Surface water that is warmer than the overlying air is a *heat source* for the atmosphere; that is, heat is transferred from sea to air by conduction and convection. Surface water that is cooler than the overlying air is a *heat sink* for the atmosphere; that is, heat is conducted from air to sea. Warm surface currents, such as the Gulf Stream, flow from the tropics into middle latitudes, supplying sensible and latent heat to the cooler middle latitude troposphere (Figure 4.21). At the same time, cold surface currents, such as the California Current, flow from high to low latitudes, absorbing heat from the relatively warm troposphere and greater solar radiation in the tropics.

The ocean's **thermohaline circulation** is the density-driven movement of water masses, traversing the lengths of the ocean basins. As noted in Chapter 1, the density of seawater increases with decreasing temperature and increasing salinity. More dense water tends to sink while less dense water rises. The thermohaline circulation transports heat energy, salt, and dissolved gases (e.g., the greenhouse gas carbon dioxide) over great distances and to great depths in the world ocean, and plays an important role in Earth's climate system. In the North Atlantic, for example, a warm surface ocean current flows north and eastward from the Florida Strait. At high latitudes, surface waters cool, sink, and flow southward as cold bottom water. This heat transporting circulation is also known as the **meridional overturning circulation (MOC)**.

In recent years, based on the output of coupled ocean/atmosphere global climate models, some scientists predicted that global warming could cause the MOC to gradually weaken, having implications for the climate of northern latitudes, particularly Europe. To test this

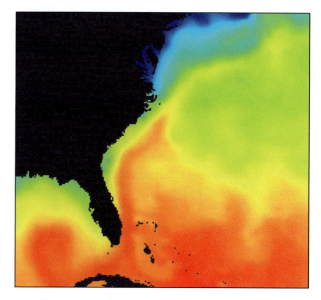

FIGURE 4.21
A composite infrared satellite image showing sea surface temperatures (SST) color-coded in ºC. The Gulf Stream is clearly discernible as a ribbon of relatively warm water flowing along the East Coast from Florida north to off the Delaware coast. [NOAA, National Environmental Satellite, Data, and Information Service]

hypothesis, a team of U.S. and British scientists from the *Rapid Climate Change Program* (begun in 2001) strung an array of moored instruments across the North Atlantic Ocean along the 26.5 degree N latitude circle. They reported their findings in *Science* magazine in August 2007. Measurements made over a one-year period indicated considerable intra-annual variability in the MOC. The MOC varied by a factor of 8, from a low of 4.0 Sv to a high of 34.9 Sv with an average of 18.7 ± 5.6 Sv. (1.0 Sv = 1.0 Sverdrup = 1.0 million cubic m per sec.) The challenge is to separate out any long-term trend from the substantial natural variability of the MOC but this would require a much more lengthy observational record. For now, there is no indication that the MOC is weakening.

Controls of Air Temperature

Air temperature is variable, fluctuating from hour to hour, from one day to the next, with the seasons, and from one place to another. Our discussion of heat transfer processes, Earth's radiation budget, and poleward heat transfer provides some insight as to why air temperature is so variable. The radiation budget plus movements of air masses regulate air temperature locally. Although these two controls actually work in concert, for purpose of study, we initially consider them separately.

LOCAL RADIATION BUDGET

Many factors govern the local radiation budget and air temperature, including the following: (1) latitude along with time of day and day of the year, which determine the solar altitude and the intensity and duration of solar radiation striking Earth's surface; (2) cloud cover, because cloudiness affects the flux of both incoming solar and outgoing terrestrial radiation; and (3) surface characteristics, which determine the albedo and the percentage of absorbed radiation (heat) used for sensible heating and latent heating. Hence, air temperature is generally higher in July than in January (in the Northern Hemisphere), during the day than at night, under clear rather than overcast daytime skies, when the ground is bare instead of snow-covered, and when the ground is dry rather than wet.

The annual temperature cycle (also called the march of mean monthly temperature) reflects the systematic variation in incoming solar radiation over the course of a year at selected latitudes ranging from the tropics to subpolar latitudes (Figure 4.22). In the latitude belt between the Tropics of Cancer and Capricorn, incoming solar radiation varies little through the course of a year so that average monthly air temperatures exhibit minimal seasonal contrast. In fact, in the tropics, the average temperature difference between summer and winter often is less than the average day-to-night temperature contrast. At middle latitudes, solar radiation features a pronounced annual maximum and minimum. At high latitudes, poleward of the Arctic and Antarctic circles, the seasonal difference in solar radiation is extreme, varying from near or at zero in fall and winter to a maximum in spring and summer. This marked periodicity of solar radiation outside of the tropics is the primary boundary condition for the

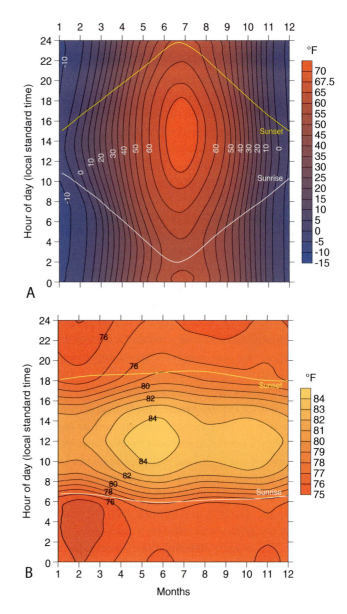

FIGURE 4.23
Average hourly temperatures through the course of a year (1961-1990) in °F for (A) Fairbanks, Alaska (64.8 degrees N, 148.7 degrees W), and (B) Agana, Guam (13.5 degrees N, 144.8 degrees E). [Courtesy of Edward J. Hopkins]

distinct winter-to-summer temperature contrasts observed in middle and high latitudes. Note the contrast in average hourly temperatures through the course of a year at a high latitude location (Figure 4.23A) versus a tropical location (Figure 4.23B).

At middle and high latitudes, the march of mean monthly temperature lags behind the monthly variation in solar radiation so that the warmest and coldest months of the year typically do not coincide with the times of maximum and minimum incoming solar radiation, respectively. The troposphere's temperature profile takes time to adjust to

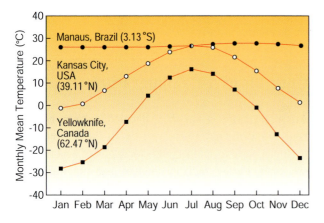

FIGURE 4.22
March of monthly mean temperature at Manaus, Brazil, Kansas City, MO, and Yellowknife, Canada. [Source: NOAA/NCDC]

seasonal changes in solar energy input. Typically, the warmest portion of the year is about a month after the summer solstice, and the coldest part of the year usually occurs about a month after the winter solstice. In the United States, the temperature cycle lags the solar cycle by an average of 27 days. However, in coastal localities with a strong maritime influence (e.g., Florida, the shoreline of New England, and coastal California), the average lag time is up to 36 days. In addition, as we saw earlier in this chapter, the maritime influence reduces the amplitude of the annual march of mean monthly temperature; that is, the winter-to-summer temperature contrast is less in maritime climates.

The lag between mean monthly temperature and the monthly variation in solar radiation means that the astronomical method of delineating the seasons (based on dates of solstices and equinoxes as presented in Chapter 3) is not very satisfactory for climatic purposes. For example, the word summer conjures thoughts of long days and short nights. Following the astronomical definition of seasons, however, summer commences when the daylight length begins to wane following the summer solstice. Also, summer-like weather episodes often occur well before the summer solstice. Hence, to better match climatology, by international convention, atmospheric scientists define **meteorological seasons** as successive three-month intervals centered on the typical occurrence of the warmest and coldest months of the year. Meteorological spring consists of the months of March, April, and May; summer encompasses June, July, and August; autumn is September, October, and November; and meteorological winter consists of December, January, and February. Length is more uniform for meteorological seasons than for astronomical seasons. The maximum difference in length among astronomical seasons is about 4.7 days, as compared to 2 days for the meteorological seasons.

Over the course of a 24-hour day, surface air temperature responds to regular variations in the flux of radiation. With clear skies and light winds or calm air, the day's lowest (minimum) temperature typically occurs shortly after sunrise as solar radiation just begins to heat the ground (Figure 4.24). The day's highest (maximum) temperature is usually recorded in early or mid-afternoon, even though solar radiation peaks around local solar noon. Air temperature depends on the relative magnitudes of the incoming solar radiation and net outgoing infrared radiation. Beginning shortly after sunrise, solar radiation exceeds outgoing IR and the air temperature rises. By early to mid-afternoon, down welling of IR radiation from the

atmosphere (the *greenhouse effect*) coupled with incoming solar radiation causes the air temperature to reach its daily maximum. Within an hour or two prior to sunset, outgoing IR exceeds incoming solar radiation (declining solar altitude) and the air temperature falls. Overnight, the only radiational flux is outgoing IR radiation, continually cooling the surface and air above.

The diurnal lag between solar radiation and air temperature explains why in summer the greatest risk of sunburn is during the several hours centered on local solar noon (the time of peak solar altitude) and not during the warmest time of day. Incoming solar ultraviolet radiation, the cause of sunburn, is most intense at local noon, but the air temperature does not reach a maximum until several hours later.

As a further illustration of local radiational controls of air temperature, consider the influence of ground characteristics. **Aspect** refers to the direction faced by a sloping surface and is important in the study of microclimates in that it affects the intensity of solar radiation incident on Earth's surface. For example, in middle latitudes of the Northern Hemisphere, snow tends to persist longer and the growing season is shorter on the shaded north-facing hill slopes (facing away from the Sun) than on south-facing slopes (facing the Sun). In portions of the Rocky Mountains, the sunnier, drier, and warmer south-facing slopes are sparsely vegetated by grasses, ponderosa pine, and juniper. Meanwhile, dense stands of fir and spruce grow on the shaded, moist, and cooler north-facing slopes.

All other factors being equal, in response to the same intensity of solar radiation striking Earth's surface, air over a dry surface (e.g., bare soil) warms more than air over a moist or vegetated surface. When the surface is dry, absorbed radiation is used primarily for sensible

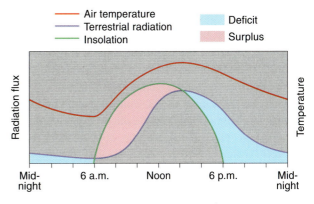

FIGURE 4.24
In the absence of cold or warm air advection, the variation in air temperature through the course of a 24-hour day depends upon the relative fluxes of incoming solar radiation and outgoing terrestrial infrared radiation.

heating of the air (mainly by conduction and convection). Hence, the air temperature is higher. On the other hand, when the surface is moist, much of the absorbed radiation is used to evaporate water, so the sensible heating and the air temperature are lower.

Dry soil helps explain why unusually high air temperatures often accompany *drought*, a lengthy period of moisture deficit. Soils dry out, crops wither and die, and lakes and other reservoirs shrink. Because less surface moisture is available for vaporization, more of the available heat is channeled into raising the air temperature through conduction and convection. Consider as an example the severe drought that gripped a ten-state area of the southeastern United States between December 1985 and July 1986. In most places, rainfall was less than 70% of the long-term average, and in the hardest hit areas, portions of the Carolinas, it was less than 40%. By July, many weather stations in the drought-stricken region were setting new high temperature records. Columbia, SC, Savannah, GA, and Raleigh-Durham, NC, reported the warmest July on record. Also contributing to record heat was the more intense solar radiation that reached the ground, a consequence of less than the usual daytime cloud cover. The same association between exceptionally dry surface conditions and unusually high air temperatures was observed during the severe drought that afflicted the Midwest and Great Plains during the summer of 1988. At many long-term weather stations, the summer of 1988 was one of the driest and hottest on record.

Snow has a high albedo and substantially reduces the amount of solar radiation that is absorbed at the surface and converted to heat. Furthermore, snow-covered ground reduces sensible heating of the overlying air because some of the available heat is used to vaporize or melt snow. Consequently, a snow cover lowers the day's maximum air temperature. Because snow is also an excellent emitter of infrared radiation, nocturnal radiational cooling is extreme where the ground is snow-covered, especially when skies are clear (minimum *greenhouse effect*). Cooling near Earth's surface is further enhanced if winds are very light or the air is calm. Light winds or calm conditions reduce vertical mixing of air allowing maximum cooling by radiation. On such nights, the air temperature near the surface may be 10 Celsius degrees (18 Fahrenheit degrees), or more, lower than if the ground were bare of snow. By reducing both the maximum and minimum daily air temperatures, a snow cover significantly lowers the 24-hr mean temperature. For these reasons, the extent and average duration of a snow-cover has an important influence on the mean winter temperature.

COLD AND WARM AIR ADVECTION

Air mass advection refers to the movement of an air mass from one locality to another. With advection, one air mass replaces another air mass having different temperature (and/or humidity) characteristics. **Cold air advection** occurs when the wind transports colder air into a previously warmer area. On a weather map cold air advection is indicated by winds blowing across regional isotherms from a colder area to a warmer area (arrow A in Figure 4.25). Cold air advection occurs behind a cold front. **Warm air advection** takes place when the wind blows across regional isotherms from a warmer area to a colder area (arrow B in Figure 4.25). Warm air advection occurs behind a warm front and ahead of a cold front. **Isotherms** are lines drawn on a map through localities having the same air temperature. Recall from earlier in this chapter that air mass exchange is a major contributor to poleward heat transport.

The significance of air mass advection for local temperature variations depends on (1) the initial temperature of the new air mass, and (2) the degree of modification the air mass undergoes as it travels over Earth's surface. For example, a surge of bitterly cold arctic air loses much of its punch when it travels over ground that is not snow covered, because the arctic air is

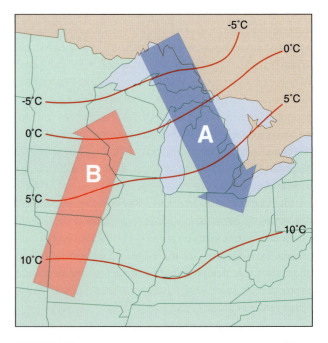

FIGURE 4.25
Cold air advection occurs when (A) horizontal winds blow across regional isotherms from colder areas toward warmer areas, and warm air advection occurs when (B) horizontal winds blow across regional isotherms from warmer areas toward colder areas. Solid lines are isotherms in ºC.

warmed from below by sensible heating (conduction and convection). In contrast, modification of an arctic air mass by sensible heating is minimized when the air mass travels over a cold, snow-covered surface.

So far, we have been describing how horizontal movement of air (advection) might influence air temperature at some locality. However, as we saw in our discussion of convection currents, air also moves vertically. As air ascends and descends, its temperature changes; air cools as it rises but warms as it descends. The reasons for these temperature changes are given in Chapter 5.

Although we have considered the local radiation budget and air mass advection separately, the two actually combine in regulating air temperature. Sometimes air mass advection acts with or against, or even overwhelms, local radiational influences on air temperature. As noted earlier, in response to the local radiation budget, the air temperature usually climbs from a minimum near sunrise to a maximum in early or mid afternoon. This typical pattern can change, however, if an influx of cold air occurs at the same time. Depending on how cold the incoming air is, air temperatures may climb more slowly than usual, remain steady, or even fall during daylight hours. If cold air advection were extreme, air temperatures may drop precipitously throughout the day, in spite of bright, sunny skies. In another example, air temperatures may climb through the evening hours as a consequence of strong warm air advection, so the day's high temperature would occur at night.

ANTHROPOGENIC INFLUENCE

A major focus of this book is the human influence on macroscale climates particularly as related to enhancement of the greenhouse effect primarily by the combustion of fossil fuels. In addition, anthropogenic activity has already demonstrably changed local climate by altering the local radiation budget. In some cases, this action is intentional as in efforts to prevent freezes in Florida citrus groves. (For information on freeze-prevention strategies, see this chapter's second Essay.) In other cases, the change in local climate is an unintentional byproduct of some other activity. The development of urban heat islands is an example.

A city is an island of warmth surrounded by cooler air, a so-called **urban heat island**. Scientists first discovered this phenomenon in London in the early 19th century. Snow melts faster and flowers bloom earlier in a city than in the surrounding countryside. Reporting in 2004, Xiaoyang Zhang and colleagues at Boston University, investigated urban heat islands using land surface temperature and vegetation data derived by the *Moderate Resolution Imaging Spectroradiometer (MODIS)* flown on board NASA's Terra satellite (Chapter 2). Researchers found that in 70 eastern North American cities (each covering an area of more than 10 km² or 4 mi²), springtime land surface temperatures were on average 2.3 Celsius degrees (4.1 Fahrenheit degrees) higher than surrounding rural areas. In late autumn to winter, urban temperatures were 1.5 Celsius degrees (2.7 Fahrenheit degrees) higher than in the surrounding areas. In cities, nonagricultural vegetation began to bud about 7 days earlier in spring and retained foliage about 8 days longer in autumn when compared to vegetation in non-urban areas.

The relative lack of moisture in cities is one of the reasons why the average annual surface temperature is a degree or two higher in a city than in nearby rural areas (Figure 4.26). City surfaces (e.g., facades, roofs, pavement, and sidewalks) are made of mostly impervious materials; hence, to prevent flooding, sewer systems are designed to carry off most runoff from rain and snowmelt. On the other hand, the countryside typically has considerable standing water (e.g., lakes, rivers, moist soils) and much more vegetative cover for transpiration (emission of water vapor to the air by plants). In a city, more of the available heat from absorbed radiation is used to raise the temperature of surfaces (sensible heating) and less for evaporation of water (latent heating). In the moister countryside, heat from absorption of radiation is used less for sensible heating and more for latent heating so that the average Bowen ratio is about four times greater in a city than in the countryside.

Other factors that contribute to an urban heat island include the greater concentration of heat sources in a city (e.g., motor vehicles, space heaters, air conditioners). On a cold winter day in New York City, heat from urban sources may approach 100 W/m², equal to about 7% of the solar constant. Calculations made by David J. Sailor, a mechanical engineer at Portland State University, OR, indicate that a motor vehicle with a fuel efficiency of 10 km per liter (24 mpg) releases sufficient heat to melt about 4.5 kg (10 lb) of ice for each kilometer traveled.

City surfaces generally have a lower albedo than the vegetative cover of rural areas. A city's canyon-like terrain of narrow streets and tall buildings causes multiple reflections of sunlight, thus increasing the amount of solar radiation that is absorbed. Urban building materials (e.g., concrete, asphalt, and brick) conduct heat more readily than the soil and vegetation of rural areas so that release

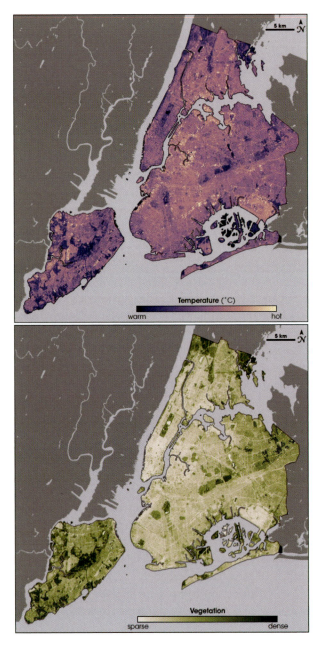

FIGURE 4.26
Thermal infrared satellite data of New York City measured by NASA's Landsat Enhanced Thematic Mapper Plus on 14 August 2002. Top image is the pattern of surface temperatures ranging from blue (warm) to yellow (hot). Bottom image displays density of vegetation from beige (sparse) to deep green (dense). Note that the temperature is lower where the vegetation is denser. [Courtesy of NASA Earth Observatory, posted 2 August 2006.]

of heat to the urban atmosphere from streets and the interior of buildings partially counters radiational cooling, especially at night. In the downtown core of a city, the vertical facades of skyscrapers radiate heat to the surfaces of neighboring buildings rather than the cold sky, further contributing to warming.

An urban heat island is best developed at night when the air is calm and the sky is clear. Under those conditions, the nighttime temperature contrast between a city and its surroundings can be as great as 10 Celsius degrees (18 Fahrenheit degrees). When winds are strong, however, the temperature contrast is greatly diminished as city and country air is mixed.

With continued population growth and global warming, more people will be exposed to urban-accentuated heat. By 2030, the global human population is projected to increase by about 25% (1.65 billion). Much of this population growth is expected to occur in cities currently with populations of less than 500,000 and with greater urbanization will come more intense urban heat islands. Even with higher energy efficiency, a rise in total energy consumption (e.g., for air conditioning) and greater emission of greenhouse gases appear likely in cities. This positive feedback on urban temperatures means more stressful environmental conditions for urban dwellers (Chapter 14).

Among the strategies proposed to ameliorate the urban heat island effect is to cover all or a portion of the upper surfaces of a building with a layer of soil and vegetation. A *green roof* is more reflective, reduces sensible heating, and offers better insulation. A simpler approach is to replace a low albedo roof with a high albedo (e.g., white) roof. Research recently conducted in the American Southwest demonstrates that green roofs and white roofs significantly reduce the demand for air conditioning, thereby decreasing energy use and emission of greenhouse gases.

Conclusions

Heat and temperature are distinct yet closely related quantities. Unequal rates of radiational heating and cooling give rise to temperature gradients within the planetary system. In response to these temperature gradients, heat is transferred from warmer to colder localities via radiation, conduction, convection, and latent heating. The temperature response of a substance to a gain or loss of heat depends primarily on its specific heat. The contrast in specific heat is one of the primary reasons why the temperature of the surface of a body of water is much less variable than that of land surfaces. This difference in thermal inertia influences the temperature of the overlying air mass so that regions downwind of the ocean (or large lakes) exhibit less temperature contrast between summer and winter than locations far from large water bodies.

Imbalances in radiational heating and radiational cooling within the planetary system are ultimately responsible for the circulation of the atmosphere and ocean. These imbalances occur both vertically (between Earth's surface and the atmosphere) and horizontally (between tropical and higher latitudes). Through atmospheric circulation, heat is redistributed within the planetary system. All this explains how air temperature is regulated by a combination of the local radiation budget and air mass advection.

As we have seen in this chapter, water and its phase changes are an important heat transfer process. Also operating within Earth's climate system is the global water cycle and the formation of clouds and precipitation, key elements of climate. The next chapter takes a closer look at water in the climate system.

Basic Understandings

- Temperature is directly proportional to the average kinetic energy of the atoms or molecules composing a substance. Heat is the name given to energy transferred from a warmer object to a colder object.

- As a volume of air gains or loses heat, that heat may be used for some combination of changes in temperature, phase of water, or volume.

- The Fahrenheit and Celsius temperature scales are based upon the reproducible phase changes in water. While the Fahrenheit temperature scale is still commonly used in the United States, the Celsius temperature scale is more convenient in that a 100-degree interval separates the freezing and boiling points of pure water at sea level. The Kelvin scale is based on absolute zero and is a more direct measure of average kinetic molecular activity than either the Fahrenheit or Celsius scale. An object at absolute zero (0 kelvins) would emit no electromagnetic radiation. Heat energy is quantified as calories, joules, or British thermal units.

- Common types of thermometers are liquid-in-glass, electronic, and bimetallic strips. Another type of thermal sensor is a radiometer that measures IR radiation emitted by an object. Thermometers should be mounted where they are well ventilated but sheltered from precipitation, direct sunshine, and the night sky.

- In response to a temperature gradient, heat always flows from locations of higher temperature to locations of lower temperature. This is a consequence of the second law of thermodynamics. Heat transfer occurs via radiation, conduction, convection, as well as by phase changes of water (latent heating).

- All objects absorb and emit radiation, an energy transport mechanism that does not require a physical medium. If absorption exceeds emission (radiational heating), the temperature of an object rises, but if emission exceeds absorption (radiational cooling), the temperature of an object falls. At radiative equilibrium, absorption and emission balance and the temperature of the object is constant.

- As a rule, solids (especially metals) are better conductors of heat than are liquids, and liquids are better conductors than gases. Motionless air is a very poor conductor of heat. Convection is the transport of heat within a fluid via motion of the fluid itself and is much more important than conduction in transporting heat within the troposphere.

- When water changes phase, heat is either absorbed from or released to the environment. Latent heat is absorbed during melting, evaporation, and sublimation. Latent heat is released during freezing, condensation, and deposition.

- The temperature response to an input or output of heat differs from one substance to another depending primarily on the specific heat of each substance. Water bodies, such as the ocean or large lakes, exhibit less temperature variability from day to night, and from summer to winter, than do landmasses because the specific heat of water is higher than that of land. Solar radiation penetrates water but soil is opaque to solar radiation, and water circulates. The winter-to-summer seasonal temperature contrast is greater in continental climates than in maritime climates. In addition, compared to continental locales, maritime locales experience a greater lag in the temperature response to seasonal variations in solar radiation.

- On a global annual average basis, radiational cooling is greater than radiational heating of the atmosphere. On the other hand, radiational heating is greater than radiational cooling of Earth's surface. In response, heat is transported from the warmer Earth's surface to the cooler troposphere via latent heating (vaporization of water followed by cloud development) and sensible heating (conduction plus convection).

- The ratio of sensible heating to latent heating, the Bowen ratio, depends on the amount of moisture at the Earth's surface. Sensible heating is greater than latent heating for dry surfaces (higher Bowen ratio) and latent heating plays an increasingly important role relative to sensible heating for wet surfaces (lower Bowen ratio). The global Bowen ratio is 0.30.

- Poleward of about 35 degrees latitude, over the course of a year, the rate of cooling due to infrared emission to space is greater than the rate of warming due to absorption of incoming solar radiation. In tropical latitudes, on the other hand, the rate of warming due to absorption of solar radiation is greater than the rate of cooling due to emission of infrared radiation. Poleward heat transport within the Earth system is the consequence.

- Poleward heat transport is brought about by north-south exchange of air masses, release of latent heat in storm systems, and ocean circulation (wind-driven surface currents and the thermohaline circulation).

- The radiation budget plus air mass advection govern variations in local air temperature. The latitude of a locale has a major influence on the radiation budget as it is related to the amount of incident solar radiation. The local radiation budget varies with time of day, time of the year, cloud cover, and properties of Earth's surface. Air mass advection occurs wherever warm air replaces cold or cold air replaces warm. Cold or warm air advection reinforces, compensates for, or even overwhelms the influence of the local radiation budget on air temperature.

- At middle and high latitudes, the march of mean monthly temperature lags behind the monthly variation in solar radiation so that the warmest and coldest months of the year do not coincide with the times of maximum and minimum incoming solar radiation.

- With clear skies, light winds or calm air, the 24-hr daily minimum air temperature occurs near sunrise and the maximum air temperature is recorded in the early to mid afternoon, a consequence of the diurnal variation in solar radiation.

- Record high temperatures often accompany a prolonged and intense drought, as more of the absorbed solar radiation is used for sensible heating due to the lack of soil moisture.

- By reducing both the maximum and minimum air temperatures, a snow cover significantly lowers the 24-hr mean temperature. For this reason, the average seasonal duration of snow-covered ground influences the climate.

- Mean annual air temperature is a degree or two higher in a city than in the surrounding countryside. Contributing to the formation of an urban heat island are several factors. Compared to rural areas, cities have (1) less standing water and moist surfaces (higher Bowen ratio), (2) a greater concentration of heat sources (e.g., motor vehicles), (3) a lower surface albedo, (4) multiple reflections of sunlight within the cityscape, and (5) component materials that more readily conduct heat (i.e., brick, asphalt, concrete) and store heat for release at a later time.

Enduring Ideas

- Temperature and heat are closely related concepts. Temperature is directly proportional to the average kinetic energy of the atoms or molecules composing a substance. Heat is the name we give to energy transferred from a warmer object to a colder object.

- The winter-to-summer seasonal temperature contrast is greater in continental climates than in maritime climates. Also, compared to continental localities, maritime localities experience a greater lag in temperature response to the regular seasonal variation in incoming solar radiation.

- In response to differences in rates of radiational heating and cooling, heat is transported from Earth's surface to atmosphere (via latent and sensible heating) and from tropical latitudes to higher latitudes (via air mass exchange, storm systems, and ocean circulation).

- The radiation budget plus air mass advection govern variations in local air temperature.

Review

1. What is the distinction between temperature and heat?
2. Compare the roles of conduction and convection in transferring heat from Earth's surface to the troposphere.
3. How is latent heat involved in the transport of heat energy from one place to another?
4. On a global annual average basis, what is the net direction of heat transport between Earth's surface and atmosphere?
5. Why is latent heating more important than sensible heating in transporting heat from Earth's surface to the atmosphere?
6. What determines the basic properties (temperature and humidity) of an air mass?
7. Identify and describe the processes involved in poleward heat transport.
8. All other factors being equal, would you expect the day's maximum temperature to be lower with a dry or moist surface? Explain your choice.
9. How does cloud cover affect the daily range in air temperature?
10. Distinguish between cold and warm air advection.

Critical Thinking

1. Explain why a thick snow cover can prevent freezing of the underlying soil even if the air temperature drops well below freezing.
2. Would you expect a West Coast locality to have a more continental climate than an East Coast locality at the same latitude? Justify your answer.
3. Identify the heat transfer process that is most significant in cooling Earth's surface.
4. As a drought begins and intensifies, describe what happens to the ratio of sensible heating to latent heating.
5. How do hurricanes contribute to poleward heat transport?
6. What drives the thermohaline circulation of the ocean?
7. How are imbalances in radiational heating and cooling at the global scale related to atmospheric circulation?
8. Identify the various factors in the local radiation budget that influence variations in air temperature.
9. How is it possible for a 24-hr day's maximum air temperature to occur just prior to midnight?
10. Speculate on what might happen to the pole-to-equator temperature gradient and the strength of the westerly winds in the absence of poleward heat transport.

ESSAY: Unique Thermal Properties of Water

Compared to other naturally occurring substances, water's thermal properties are unique. For example, based on water's molecular weight as well as the freezing and boiling temperatures of chemically related substances, fresh water should freeze at about –90 °C (–130 °F) and boil at about –70 °C (–94 °F). Actually, fresh water's freezing point is 0 °C (32 °F) and its boiling point is 100 °C (212 °F) at average sea level air pressure. Water's unusual properties arise from the physical structure of the water molecule (H_2O) and the bonding that occurs between water molecules. Without this intermolecular force of attraction, known as *hydrogen bonding*, water would exist only as a gas within the range of surface temperature and pressure on Earth. If such a condition were to prevail, the planet would have no water cycle, no ocean, no glaciers, and probably no life as we know it.

A water molecule consists of two hydrogen (H) atoms bonded to an oxygen (O) atom (Figure 1). Within the water molecule, bonding between hydrogen and oxygen atoms involves sharing of electrons, one from each hydrogen atom and two from the oxygen atom. (An *electron* is a negatively charged subatomic particle.) In this bonding, the electrons spend more time near the oxygen atom so that the oxygen acquires a small negative charge while the hydrogen is left with a small positive charge. Because of the strength of this *covalent bonding*, a water molecule resists dissociation into its constituent hydrogen and oxygen atoms. The 105-degree angle formed by the arrangement of the hydrogen-oxygen-hydrogen atoms produces a charge separation in the water molecule. Molecules having a separation of positive and negative charges are described as *polar*.

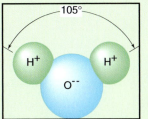

FIGURE 1
The water molecule consists of two hydrogen atoms bonded to one oxygen atom.

Opposite electrical charges attract so that, like tiny magnets, neighboring water molecules link together. The positively charged (hydrogen) pole of one water molecule attracts the negatively charged (oxygen) pole of another water molecule; this attractive force constitutes *hydrogen bonding*. Each water molecule can form as many as four hydrogen bonds with surrounding water molecules. An inter-molecular hydrogen bond is only about 5% to 10% as strong as the intra-molecular covalent bond between hydrogen and oxygen atoms in individual water molecules. Nonetheless, hydrogen bonding is sufficiently strong to significantly influence the physical and chemical properties of water. Hydrogen bonding inhibits changes in water's internal energy so that it absorbs or releases unusually great quantities of heat energy (latent heat) when changing phase. Because of hydrogen bonding, greater additions or losses of heat are required to change the temperature of water as compared to other chemically related substances.

Like all crystalline solids, ice has a regular internal three-dimensional framework consisting of a repeating pattern of molecules. A physical model of the *crystal lattice* of ice is shown in Figure 2. Each water molecule is bound tightly to its neighbors but intermolecular bonds are elastic (acting like springs) so that molecules vibrate about fixed locations in the lattice. Hydrogen bonding is responsible for the ordered arrangement of water molecules in the crystal lattice and the hexagonal (six-sided) structure of ice crystals. (Many of us are familiar with the six-sided symmetry of snow flakes.) Because ice's internal framework is an open network of water molecules, the molecules in ice crystals are not as closely packed as a similar number of molecules in liquid water. At 0 °C (32 °F), ice has a density of about 0.92 g per cubic cm whereas pure liquid water at the same temperature has a density of nearly 1.0 g per cubic cm. This density difference explains why ice floats on the surface of liquid water. Most common solids would sink if placed in their liquid phase. Formation of ice in a confined area causes sufficient expansional pressure to burst water pipes or fracture rocks.

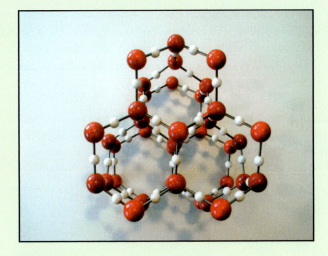

FIGURE 2
A physical model of the crystal lattice of ice.

When ice melts, it becomes liquid water. After observing ice crystals disappearing during melting, we might expect that all hydrogen bonds between water molecules break during the transition from ice to liquid water. This is not the case. Instead, many water molecules remain linked by hydrogen bonding as transient clusters of molecules surrounded by non-bonded (free) water molecules (Figure 3). Although molecular clusters persist into the liquid phase, water molecules exhibit much greater activity in the liquid than solid phase. In the liquid phase, water molecules undergo vibrational, rotational, and translational (straight-line) motions.

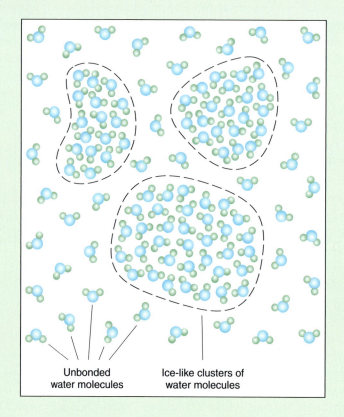

Unbonded water molecules   Ice-like clusters of water molecules

FIGURE 3
When water changes phase from ice to liquid, many water molecules remain linked by hydrogen bonds as transient clusters of molecules surrounded by non-bonded water molecules.

Why is the latent heat of vaporization so much greater than the latent heat of fusion? As noted earlier, during the melting of ice not all hydrogen bonds are broken. But when liquid water changes to vapor, energy is absorbed until essentially all hydrogen bonds are broken. Hence, individual molecules move about with even greater freedom in the vapor phase than in the liquid phase, diffusing rapidly to fill the entire volume of its container.

ESSAY: Freeze Prevention Strategies

During the 1980s, three exceptional cold waves invaded the Florida peninsula with devastating impacts on the state's citrus industry. For three days, beginning on Christmas Day 1983, an arctic air mass gripped Florida, dropping temperatures well below freezing throughout much of the state. Citrus trees covering 93,000 hectares (230,000 acres) were either damaged or killed for a total loss of $1 billion. Thirteen months later, on 21 January 1985, a cold wave of even greater severity further damaged surviving citrus trees. The next day, subfreezing temperatures were reported as far south as Miami. This double blow reduced citrus-producing acreage by almost 90% in central Florida's Lake County, formerly the state's second largest citrus producing county. Some growers replanted and the groves were recovering when yet another deep freeze over Christmas weekend 1989 ruined much of central Florida's citrus crop. By 1992, Lake County had dropped to 16th place among Florida's orange-producing counties, accounting for less than 1% of the state's total production.

The threat of freeze damage to crops is one of many weather extremes that farmers must cope with almost everywhere (Figure 1). As with the 1980s freezes in Florida, crop losses to excessive cold can cause considerable economic loss. Two principal types of regional weather patterns are responsible for freezes. A *radiational freeze* occurs when the regional weather pattern features a large high pressure system accompanied by clear skies and light winds or calm air, conditions conducive to extreme nocturnal radiational cooling. An *advective freeze* occurs when the large-scale circulation pattern transports a cold air mass into a region, often on strong and gusty winds. Whereas radiational freezes develop only at night, advective freezes can occur day or night. With a radiational freeze, the lowest temperatures typically occur in lowlands such as river valleys, marshes, or hollows. Hence, crop damage tends to be localized. With an advective freeze (such as described for Florida above), strong winds ensure considerable mixing of air so that minimum air temperatures tend to be more uniform over a broad geographical area so that crop damage may be more widespread.

FIGURE 1
Checking for freeze damage to the citrus crop. [Courtesy USDA]

The vulnerability of crops to subfreezing temperatures varies with crop species and the stage of the crop's life cycle. For example, tomatoes and rice are more sensitive to low air temperatures than are oats or barley. Young and rapidly growing seedlings (such as flowers) are more sensitive than mature, slowly growing plants (such as ripened fruit). Based on crop vulnerability, a variety of strategies have been developed for protecting crops from the potentially damaging effects of cold weather. Consider some strategies directed at minimizing the impact of a radiational freeze on crops.

Site selection is a key consideration in protecting crops from radiational freezes. Growers should avoid cultivating sites that are routinely on the receiving end of *cold-air drainage*, the down-slope flow of a shallow layer of cold dense air. In freeze-prone areas such as valley bottoms, the growing season may be several weeks shorter than on surrounding highlands. For this reason, orchards and vineyards are often situated on hill slopes rather than valley floors. But even on hill slopes, site selection must be done with care. Obstructions such as hedgerows, roads, or railroad embankments can impede cold-air drainage so that crops immediately upslope from the obstruction may be damaged by pooling of cold air behind the barrier. Growers solve this problem by cutting channels through the barrier so that cold air can continue flowing down slope.

Based on knowledge of the conditions that favor extreme nocturnal radiational cooling, scientists and growers have developed strategies that attempt to reduce the intensity and duration of radiational freezes. One factor that contributes to such freezes is clear night skies. To protect crops on such nights, growers create their own clouds using a fine water spray. Such clouds provide some protection for crops by enhancing the local greenhouse effect. That is, cloud droplets absorb infrared radiation emitted by the ground and vegetation and emit IR to the crop, thereby slowing the nighttime drop in air temperature.

In another strategy, growers spray crops with a fine water mist when the temperature of plant tissues drops to 0 °C (32 °F). The mist freezes on contact with the plant surfaces. One might question how a coating of ice could possible help plants survive subfreezing temperatures. Although water on the plant surface is 0 °C, the latent heat released during the phase change from liquid to ice helps stabilize plant temperatures. Stabilization of the temperature near 0 °C often prevents plant damage because the actively growing tissues of most plants are not injured until their temperature drops to −5 °C to −1 °C (23 °F to 30 °F). Nonetheless, this strategy requires careful monitoring. As long as sprinkling and freezing continues, the temperature of the ice remains at about 0 °C. However, if sprinkling were discontinued before the ambient air temperature rises high enough to melt the ice, then heat is conducted from the plant to the ice and the leaf temperature drops to potentially lethal levels. In addition, care must be taken that the ice burden does not become so great that the plants are damaged by excessive weight. For this reason, the sprinkling method of freeze control is most suited to low-lying vegetable crops such as cucumbers and strawberries.

The need for a ready supply of water means that cranberries are often grown in sandy lowland sites (cranberry bogs) that are also prone to cold-air drainage. Prior to harvest in late September and early October, chilly nights promote development of the berries' rich red color. Potentially killing freezes, however, may also occur during this ripening period. When subfreezing temperatures are forecast, growers turn on a sprinkler system that sprays the low-growing cranberry vines with a water mist that freezes on contact. Latent heat released when the water freezes is often sufficient to keep the vine temperatures above about −2 °C (28 °F), the threshold for vine injury. Sprinkling is continued until the air temperature rises above 0 °C (32 °F) or the ice begins to melt.

For small crops, radiation screens can be effective. For example, plastic "hot caps" placed over individual plants create a protective microclimate. During the day, the Sun warms the soil and plants. In the late afternoon, hot caps are placed over the individual plants so that the heat gained during the day is better conserved at night, reducing the chance that the plants will be exposed to subfreezing air. For somewhat larger plants, other radiation screens (such as cold frames, wooden slates or cheesecloth) restrict exposure to the clear night sky without significantly blocking the daytime solar radiation essential for photosynthesis and crop growth.

A water spray reduces the threat of freeze damage by enhancing the local greenhouse effect or by adding latent heat. Sensible heat can also be added through fuel combustion in oil heaters, also known as *smudge pots*. At one time, it was believed that smoke emanating from smudge pots inhibited freezing temperatures by enhancing the local greenhouse effect. However, smoke particles are actually nearly transparent to infrared radiation, so that their influence on the local greenhouse effect is negligible. Furthermore, smoky smudge pots are banned in many localities because of their adverse effect on air quality. Smudge pots primarily benefit plants that are directly exposed to the warm plume of air, so that an array of many small heaters is considerably more effective than a few large ones.

In response to extreme radiational cooling and calm air, a temperature inversion often develops within the air layer adjacent to the ground. The ground is a better infrared radiator than the overlying air so that the ground surface cools more rapidly and heat is conducted from the warmer air to the cooler ground. Consequently, the air temperature is lowest next to the ground and increases with altitude (a *temperature inversion*). While the air temperature at ground level may drop below freezing, air temperatures at the top of the inversion (perhaps 15 m or 50 ft above the ground) may be several degrees above freezing. The extreme stability of the inversion layer prevents the warmer air aloft from mixing with the colder air at ground level. Heat generated by smudge pots, however, spurs convection (vertical mixing) that may break up the inversion locally and cause air temperatures at low levels to rise. In citrus groves and some orchards it is standard practice to accomplish the same mixing using large motor-driven fans or propellers mounted on towers.

Crop protection strategies that work for radiational freezes (e.g. site selection, radiation screens, water spray, smudge pots, fans) typically have little success with advective freezes. An air mass advecting southward on strong gusty winds is well mixed and uniformly cold over both highlands and lowlands. A low-level temperature inversion does not develop (until winds weaken) and winds quickly disperse the heat generated by smudge pots.

CHAPTER 5

WATER IN EARTH'S CLIMATE SYSTEM

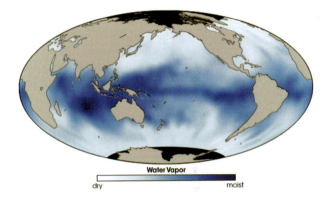

Global Water Vapor map. While carbon dioxide is commonly touted as a strong greenhouse gas because of its ability to trap heat near the surface of Earth, moisture in the atmosphere (water vapor) is, in fact, a more powerful greenhouse gas than carbon dioxide. [NASA Earth Observatory]

Case-in-Point

Today, most discussions regarding global climate change tend to focus on trends in air temperature. We should not, however, overlook trends in precipitation. Water is a key component of Earth's climate system, and climate change resulting in a significant increases or decreases in fresh water can have serious consequences. In fact, a long-term decline in rainfall was responsible for one of the most dramatic environmental changes of the past 10,000 years, that is, the transition of North Africa from a green savanna to the world's largest warm desert, the desolate Sahara.

In 2008, Stefan Kröpelin of the University of Cologne, Germany and his colleagues reported in *Science* magazine on their reconstruction of environmental changes in the Sahara Desert over the past six millennia. Their primary source of information was analysis of sediments extracted from the bottom of Lake Yoa, located in northern Chad at 19.03 degrees N, 20.31 degrees E. Lake Yoa is one of the very few permanent lakes in the Sahara that has a continuous record of sedimentation extending from the present back to the middle of the Holocene Epoch (the past 10,500 years). Today the climate is exceptionally hot and dry and the lake is maintained by inflow of groundwater from an aquifer that has not been recharged in almost 10,000 years. From the lake sediments, researchers were able to reconstruct vegetation (from fossil pollen), prevailing winds (from dust), and precipitation (from the salinity of organic remains). Radiocarbon dating provided the chronology of events.

The *African Humid Period* began in the early Holocene and featured a strong monsoon circulation and rainfall sufficient to maintain a broad savanna (grassland with scattered tropical trees) where hyperarid desert is found today. About 5600 years before present (BP), a sudden influx of wind-borne dust, discovered in deep-sea sediments extracted from the floor of the equatorial Atlantic (downwind from the Sahara), signaled the beginning of a long-term gradual drying of North Africa. The monsoon weakened and estimated annual rainfall declined from about 250 mm 6000 years BP, to less than 150 mm by 4300 years BP, to less than 50 mm by 2700 years BP. Today, the climate of northern Chad is characterized by high daytime temperatures and scant rainfall. In some years, no rain falls. An influx of wind-blown dust into Lake Yoa, beginning about 4300 years BP followed by sand 3700 years BP indicated the loss of vegetative cover and soil. By 2700 years BP, true desert plants appeared and today's barren desert landscape was in place. The monsoon belt had shifted southward and was replaced by dry trade winds blowing from the northeast year round.

With the gradual transition from the "green" Sahara to today's desert, the region became less and less habitable except for scattered settlements near groundwater-fed oases. A striking example of human adaptation to the changing climate comes from a location well to the northwest of Lake Yoa in the Ubari Sand Sea of western Libya. There the Garamantes people adapted to the gradually drying climate by developing an irrigation system that relied on withdrawal of groundwater. For about 1000 years (from about BCE 500 to CE 500), agriculture thrived, as they raised animals and cultivated crops, and impressive cities were built in the desert. Eventually, however, the groundwater resource diminished to the point that the Garamantian civilization could no longer adapt to the hostile conditions, and it collapsed.

Driving Question:
How does the global hydrologic cycle contribute to making Earth's climate habitable?

The significance of water in Earth's climate system stems from water's unique physical properties (Chapter 4). Within the range of air temperature and pressure on and near Earth's surface, water can co-exist in all three phases, and as water changes phase unusually large amounts of heat energy (latent heat) are absorbed from or released to the environment. The continuous flow of water and energy within and among the reservoirs of the global water cycle is a central focus of this chapter. We describe how the water vapor component of air is quantified, how air becomes saturated, what processes form clouds and precipitation, and how rainfall and snowfall are measured.

This discussion introduces scientific concepts underlying the link between climate change and the freshwater supply. Climate change is one of the major reasons why climate scientists anticipate that precipitation patterns will change and freshwater scarcity will become more common in many parts of the world.

Global Water Cycle

The total amount of water on Earth is neither increasing nor decreasing, although natural processes continually generate and break down water at essentially equal rates.

Water vapor accounts for perhaps half of all gases emitted during a volcanic eruption; at least some of this water originally was sequestered in magma and solid rock. Volcanic activity is more or less continuous on Earth and adds to the supply of water. Also, a minute amount of water is delivered to Earth by meteorites and other extraterrestrial debris that continually bombard the upper atmosphere. At the same time, intense solar radiation entering the thermosphere, converts (*photo-dissociates*) a small amount of water vapor into its constituent hydrogen and oxygen atoms, which may escape to space. Also, water chemically reacts with other substances and thereby is locked up in various compounds. Some of the water that is removed may be released eventually through other chemical reactions involving the new compounds. Annually, additions of water from volcanic eruptions and extraterrestrial sources roughly equal losses of water through photo-dissociation of water vapor and chemical reactions. This balance of give and take has prevailed on Earth for perhaps hundreds of millions of years.

The essentially fixed quantity of water on Earth is distributed in three phases among various reservoirs (Table 5.1). The ocean is the largest of these reservoirs by far, accounting for 97.2% of all water on the planet; most of the rest is tied up as ice sheets up to 3 km (1.8 mi) thick that cover most of Antarctica and Greenland. Relatively small amounts of water occur in living organisms (plants and animals), rivers and lakes, and occupy the tiny pore spaces and fractures within soil, sediment, and bedrock (soil moisture and groundwater). Even smaller amounts of water occur in the atmosphere as clouds, precipitation, and invisible water vapor.

The ceaseless movement of water among the various reservoirs on a planetary scale is known as the global water cycle (Figure 5.1). In brief, water vaporizes from ocean and land surfaces to the atmosphere where winds transport water vapor to other locations, sometimes thousands of kilometers away. Clouds form and rain, snow and other forms of precipitation fall from clouds to Earth's surface, recharging the oceanic and terrestrial (land-based) reservoirs of water. From terrestrial reservoirs, water flows into the ocean basins. The residence time of water

TABLE 5.1
Water Stored in Reservoirs of the Global Water Cycle

Reservoir	Percent of total water
Ocean	97.20
Ice sheets and glaciers	2.15
Groundwater	0.62
Lakes (freshwater)	0.009
Inland seas, saline lakes	0.008
Soil moisture	0.005
Atmosphere	0.001
Rivers and streams	0.0001

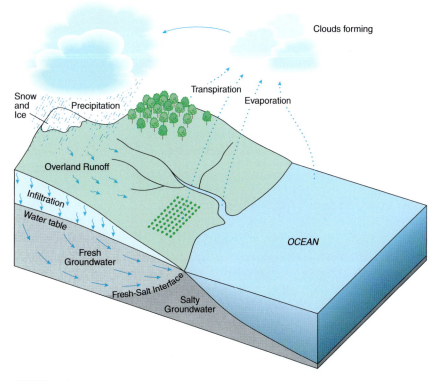

FIGURE 5.1
The global water cycle is a continuous flow of water and energy among oceanic, terrestrial (land-based), and atmospheric reservoirs.

molecules in the atmosphere is on time scales associated with weather systems (8-10 days), whereas the residence time of water in the ocean and glacial ice sheets has time scales of the order of climate (millennia).

The Sun drives the global water cycle. As we saw in Chapter 4, some of the radiation that strikes Earth's surface is absorbed, that is, converted to heat, and some of this heat evaporates water or sublimates ice and snow. If water did not vaporize, there would be no clouds, no precipitation, and no global water cycle. While solar radiation powers the global water cycle, gravity is important in keeping water molecules from escaping to space, as well as causing water to fall from clouds as precipitation and to flow from the continents to the ocean. This section focuses on the links among the atmospheric, oceanic, and terrestrial reservoirs of water.

TRANSFER PROCESSES

As part of the global water cycle, water is transferred between Earth's surface and the atmosphere via phase changes (evaporation, condensation, transpiration, sublimation, and deposition) and by precipitation. At the interface between liquid water and air (e.g., lake or sea surface), water molecules continually change phase: some crossing the interface from water to air and others from air to water. If more water molecules enter the atmosphere as vapor than return as liquid, a net loss occurs in liquid water mass, the process known as **evaporation**. About 85% of the total annual evaporation on Earth takes place at the seawater/air interface making the ocean the principal source of water in the atmosphere (Figure 5.2). On the other hand, at the interface between liquid water and air, if more water molecules return to the water surface as liquid than enter the atmosphere as vapor, then a net gain of liquid water mass results, the process called **condensation**.

FIGURE 5.2
The ocean is the principal source of water in the atmosphere.

Transpiration is the process whereby water that is taken up from the soil by plant roots eventually escapes as vapor through tiny pores on the surface of green leaves. On land during the growing season, transpiration often is more important than direct evaporation of water in supplying water vapor to the atmosphere. For example, a single hectare (2.5 acres) of corn typically transpires about 34,000 liters (L) (8985 gal) of water per day. Measurements of direct evaporation from Earth's surface plus transpiration are often combined as **evapotranspiration**.

At the interface between ice and air (e.g., the surface of a snow cover), water molecules also continually change phase: from ice directly to vapor and from vapor directly to ice. If more water molecules enter the atmosphere as vapor than transition to ice, a net loss of ice mass occurs. **Sublimation** is the process whereby ice or snow becomes vapor without first becoming a liquid. Sublimation explains the gradual disappearance of snow and ice on sidewalks even though the air temperature remains well below freezing. (Recall from the first Essay in Chapter 3 that sublimation is responsible for shrinkage of the snow and ice on Mount Kilimanjaro.) On the other hand, if more atmospheric water molecules transition to ice than move from ice to vapor, a net gain of ice mass results. **Deposition** is the process whereby water vapor becomes ice without first becoming a liquid. During a winter night, formation of frost on automobile windows is an example of deposition. Condensation or deposition within the atmosphere produces clouds. **Precipitation** is water in liquid, frozen or freezing form (i.e., rain, drizzle, snow, ice pellets, hail, and freezing rain) that falls from clouds under the influence of gravity to Earth's surface.

Evaporation (or sublimation) followed by condensation (or deposition) purifies water. As water vaporizes from the Earth's surface, all suspended and dissolved substances such as sea salts and other contaminants are left behind. Through this natural cleansing mechanism, called **distillation**, salty ocean water is the source of much of what eventually falls to Earth's surface as freshwater precipitation.

GLOBAL WATER BUDGET

Return of water from the atmosphere to the land and ocean via condensation, deposition, and precipitation completes an essential subcycle of the global water cycle. (Figure 4.14 shows the recycling of water over the continents.) To learn more about this subcycle, compare the movement of water between the continents and the atmosphere with that between the ocean and the

TABLE 5.2
Global Liquid Water Budget

Source	Cubic meters per year	Gallons per year
Precipitation on the ocean	$+3.24 \times 10^{14}$	$+85.5 \times 10^{15}$
Evaporation from the ocean	-3.60×10^{14}	-95.2×10^{15}
Net loss from the ocean	-0.36×10^{14}	-9.7×10^{15}
Precipitation on land	$+0.98 \times 10^{14}$	$+26.1 \times 10^{15}$
Evapotranspiration from land	-0.62×10^{14}	-16.4×10^{15}
Net gain on land	$+0.36 \times 10^{14}$	$+9.7 \times 10^{15}$

TABLE 5.3
Average Annual Depth of Liquid Water

	Precipitation (mm)	Evaporation (mm)	Net (mm)
Land	663	419	+ 244
Ocean	895	944	− 49
Global	828	828	0

atmosphere. The balance sheet for inputs and outputs of water to and from the various global reservoirs is called the **global water budget** and may be expressed in terms of water volume (Table 5.2) or depth of water (Table 5.3).

Over the course of a year, the volume of precipitation (rain plus melted snow) that falls on land exceeds the total volume of water that vaporizes from land by about one-third. This imbalance occurs because landmasses favor certain precipitation-forming mechanisms. Over the same period, the volume of precipitation falling on the ocean is less than the volume of water that evaporates from the ocean. Evaporation is greater because the ocean surface is essentially a limitless source of water vapor. The global water budget indicates an annual net gain of water mass on the continents and an annual net loss of water mass from the ocean. The annual excess of water on the continents equals the deficit from the ocean. The excess water on land drips, seeps and flows by gravity back to the sea thereby completing the global water cycle. The net flow of water from land to sea implies a return net flow of water within the atmosphere directed from sea to land.

Precipitation strikes the ground directly or it may be intercepted by vegetation and then evaporates or drips to the ground (Figure 5.3). Also, some trees collect moisture from drifting fog or low clouds and that water drips to the ground (known as *fog drip*). This allows specialized vegetation to grow in foggy desert climates where there is little rainfall. Once water reaches Earth's land surface, it follows various pathways. Some water vaporizes directly back into the atmosphere while some is temporarily stored in lakes, snow and ice fields, or glaciers. The remainder either flows on the surface as rivers or streams (*runoff component*) or seeps into the ground as soil moisture or groundwater (*infiltration component*). About one-third of the precipitation that falls on land runs off to the ocean. The ratio of the portion of water that infiltrates the ground to the portion that runs off depends on rainfall intensity, vegetation, topography, and physical properties of the intercepting land surface. For example, rain falling

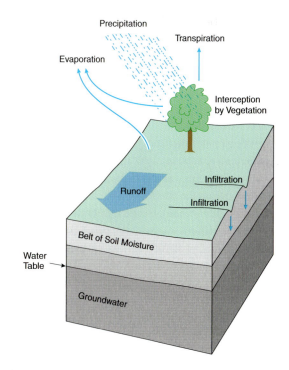

FIGURE 5.3
Various pathways taken by precipitation falling on land.

on mountainous terrain usually runs off quickly. Likewise, rain falling on frozen ground or city streets mostly runs off whereas rain falling on unfrozen sandy soil readily soaks into the ground.

Rivers and streams plus their tributaries drain a fixed geographical area known as a *drainage basin* (or *watershed*). The quantity and quality of water flowing in a river depends on the climate, vegetation, topography, geology, and land use in its drainage basin. For example, in places where the climate features distinct rainy and dry seasons, stream flow can vary considerably through the year. A drainage basin may also include lakes, wetlands, glaciers, and other temporary impoundments of surface water.

Water Vapor in the Atmosphere

It's not the heat, it's the humidity is a popular statement that attributes the discomfort a person feels on a hot, muggy day to the water vapor component of air. **Humidity** is a general term referring to any one of many ways of describing the amount of water vapor in the air. Experience tells us that humidity varies with the season, from one day to the next, within a single day, and from one place to another. In most places summer days feel more humid than a typical winter day. In cold climates, dry winter air also causes some discomfort. This section covers some quantitative measures of humidity. Annual issues of the *Local Climatological Data* for major U.S. cities include mean values of dewpoint and wet-bulb temperatures, along with "normal" values of relative humidity for 00, 06, 12, 18, and 2400, Local Standard Time.

VAPOR PRESSURE

Gas molecules composing air (including water molecules) are always in rapid, random motion, and each molecule exerts a force as it collides with other molecules, including those on the surface of a solid (e.g., the ground) or liquid (e.g., the ocean). In one millionth of one second billions upon billions of gas molecules bombard every square centimeter of Earth's surface. *Pressure* is defined as a force per unit area. The total air pressure is the cumulative force of a multitude of molecules colliding with a unit surface area of any object in contact with air.

Each gaseous component of air contributes to the total air pressure as described by **Dalton's law of partial pressures**, named for its discoverer the British scientist John Dalton (1766-1844). According to Dalton's law, the total pressure exerted by a mixture of gases equals the sum of the pressures produced by each constituent gas; that is, each gas species in the mixture acts independently of all the other molecules. Stated another way, each gas exerts a pressure as if it were the only gas present.

The pressure produced by the gas molecules composing air depends on (1) the mass of the molecules, and (2) the kinetic molecular activity. In the larger sense, we can think of **air pressure** at a given location on the Earth's surface as the weight per unit area of the column of air above that location. The pressure at any altitude within the atmosphere is equal to the weight per unit area of the atmosphere above that altitude. *Weight* is the force exerted by gravity on a mass, that is,

weight = (mass) × (acceleration of gravity)

The average air pressure at sea level is equivalent to the weight exerted by Earth's gravity on an area of one square cm by a mass of approximately 1.0 kg (equivalent to 14.7 lb per square in.).

When water enters the atmosphere as vapor, water molecules disperse and mix with the other gases composing air (mostly N_2 and O_2), thereby contributing to the total pressure exerted by the atmosphere. The amount of pressure produced by water vapor molecules is a measure of the humidity; the more water vapor, the greater the pressure exerted by water molecules. Water vapor's contribution to the total air pressure is known as **vapor pressure**.

Water vapor is a highly variable component of air but composes at most no more than about 4% by mass of the atmosphere's lowest kilometer—even in the sultry air over the tropical ocean and rainforests. The total pressure exerted by all atmospheric gases at sea level averages about 1000 millibars (mb). This means that the vapor pressure is very unlikely anywhere to exceed 40 mb (4% × 1000 mb = 40 mb) at sea level, and in most places the vapor pressure is much less than 40 mb.

SATURATED AIR

The amount of water vapor in air at a specified temperature has an upper limit. At its maximum humidity, air is described as *saturated* with water vapor. Earlier in this chapter, we described a two-way exchange of water molecules at the interface between water and air (or between ice and air). Water molecules are in a continual state of flux between the liquid (or ice) and vapor phases. During evaporation, more water molecules become vapor than return to the liquid phase, and during condensation, more water molecules return to the liquid phase than enter the vapor phase. Eventually, a dynamic equilibrium may develop such that the flux of water molecules is the same in both directions; that is, liquid water becomes vapor at the same rate that water vapor becomes liquid. At equilibrium, above a plane surface of liquid water or ice, the air is saturated with water vapor. The vapor pressure at this equilibrium condition is called the **saturation vapor pressure**.

Altering the temperature disturbs this dynamic equilibrium at least temporarily. Heating water to a higher temperature causes the average kinetic energy of individual water molecules to increase so that they more readily escape the water surface as vapor. Initially, evaporation prevails. If the supply of water were sufficient, and as long as water vapor is not continually carried away by the wind, eventually a new dynamic equilibrium would be established. That is, the flux of water molecules becoming liquid again balances the flux of water molecules becoming vapor. This new equilibrium is achieved with more water vapor in the air at the higher temperature. Hence, raising the air temperature increases the saturation vapor pressure.

Conversely, with a drop in water temperature, the average kinetic energy of individual water molecules decreases and molecules less readily escape the water surface as vapor. Initially, at the lower temperature, condensation prevails but eventually a new equilibrium is established; that is, the flux of water molecules becoming vapor again balances the flux of water molecules becoming liquid. This new equilibrium is achieved with less water vapor in the air at the lower temperature. Decreasing the air temperature reduces the saturation vapor pressure.

The response of the saturation vapor pressure to a change in temperature is an example of Le Chatelier's Principle, named for the French chemist and engineer Henri-Louis Le Chatelier (1850-1936) who first proposed it in 1884. According to **Le Chatelier's Principle**, if a system in dynamic equilibrium experiences a change or perturbation (e.g., in temperature, concentration, volume, or stress), then the equilibrium shifts in such a way as to counteract or compensate for the change or perturbation.

Ultimately, the water vapor component of air depends on the rate of vaporization of water, which is regulated chiefly by temperature, and on the presence of a water supply. The dependence of the saturation vapor pressure on temperature, shown in Figure 5.4 and Table 5.4, is non-linear. That is, the saturation vapor pressure approximately doubles in value for every 10 Celsius

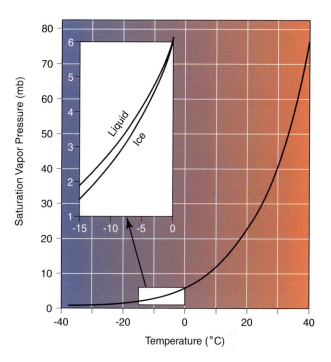

FIGURE 5.4
Variation in saturation vapor pressure with changing air temperature. Note that at subfreezing temperatures, the saturation vapor pressure is greater over supercooled water than over ice.

degree (20 Fahrenheit degree) rise in temperature. At 100 °C (212 °F) at sea level, the saturation vapor pressure is 1013.25 mb, the same as standard sea level air pressure. Water boils when the saturation vapor pressure and ambient (surrounding) air pressure are equal. Because air pressure decreases with altitude, the boiling point of water correspondingly decreases, such that at Denver, CO (the "mile-high city"), where the average air pressure is 830 mb, the boiling point is approximately 95 °C (203 °F).

The relationship between temperature and saturation vapor pressure is popularly interpreted to mean that warm air can "hold" more water vapor than can cold air. That is, air is likened to a sponge that can soak up only so much water depending on the temperature. While intuitively appealing, this analogy can be misleading. Air does not literally hold water vapor like a sponge; rather, water vapor coexists with the other gases that form the mixture known as air. Recall from Dalton's law that each gas in a mixture of gases exerts a pressure as though it were the only gas present. Water vapor is just one of the many gases that compose air. If water vapor were added to air at constant pressure and temperature, it follows that water vapor displaces some of the other gaseous components of air. Because temperature largely governs the rate of vaporization of water, the saturation vapor pressure is actually a measure of water's vaporization

TABLE 5.4
Variation of Saturation Vapor Pressure with Temperature[a]

Temperature °C (°F)	Saturation Vapor Pressure (mb)	
	Over water	Over ice
50 (122)	123.40	
45 (113)	95.86	
40 (104)	73.78	
35 (95)	56.24	
30 (86)	42.43	
25 (77)	31.67	
20 (68)	23.37	
15 (59)	17.04	
10 (50)	12.27	
5 (41)	8.72	
0 (32)	6.11	6.11
−5 (23)	4.21[a]	4.02[a]
−10 (14)	2.86	2.60
−15 (5)	1.91	1.65
−20 (−4)	1.25	1.03
−25 (−13)	0.80	0.63
−30 (−22)	0.51	0.38
−35 (−31)	0.31	0.22
−40 (−40)	0.19	0.13
−45 (−49)	0.11	0.07

[a]Note that for temperatures below freezing, two different values are given: one over supercooled water and the other over ice. Supercooled water remains liquid at subfreezing temperatures.

rate. At any specified temperature, the saturation vapor pressure would have the same value even if water vapor were the only gas in the atmosphere. For information on measuring the rate of evaporation, refer to this chapter's first Essay.

RELATIVE HUMIDITY

Relative humidity is the water vapor measure frequently reported by television and radio weathercasters and is probably familiar to most of us. **Relative humidity** compares the actual amount of water vapor in the air with the amount of water vapor that would be present if that same air were saturated. Relative humidity (RH) is expressed as a percentage and can be computed from the vapor pressure. That is,

$$RH = \frac{\text{vapor pressure}}{\text{saturation vapor pressure}} \times 100\%$$

When the actual concentration of water vapor in air equals the water vapor concentration at saturation, the relative humidity is 100%; that is, the air is saturated with respect to water vapor.

Consider an example of how relative humidity is computed. Suppose that the air temperature is 10 °C (50 °F) and the vapor pressure is 6.1 mb. From Table 5.4, we determine that the saturation vapor pressure of air at 10 °C is 12.27 mb. Using the formula above, we compute a relative humidity of about 50%, that is,

$$RH = \frac{6.1 \text{ mb}}{12.27 \text{ mb}} \times 100\% = 49.7\%$$

At constant temperature and pressure, the relative humidity varies directly with the vapor pressure; that is, the relative humidity increases as water vapor is added to air as long as the air temperature and pressure do not change. But because the saturation vapor pressure also varies <u>directly</u> with temperature, the relative humidity varies <u>inversely</u> with temperature. If no water vapor were added to or removed from unsaturated air, the relative humidity would increase as the temperature drops, but decrease as the temperature rises. Consider a common example.

With a clear sky and calm air, the air temperature usually rises from a minimum near sunrise to a maximum during early to mid-afternoon, and then falls through the evening hours and overnight (Chapter 4). If the amount of water vapor in air remains essentially constant throughout the day, then the relative humidity will vary inversely with air temperature. As shown in Figure 5.5,

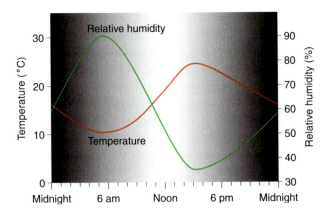

FIGURE 5.5
Variation in relative humidity and air temperature on a day when the air is calm and there is essentially no variation in vapor pressure. The relative humidity is highest when the temperature is lowest (near sunrise) and the relative humidity is lowest when the temperature is highest (early to mid afternoon).

the relative humidity is highest when the air temperature is lowest and the relative humidity is lowest when the temperature is highest. After sunrise, as the air warms, the relative humidity drops because the saturation vapor pressure increases as the air temperature rises.

DEWPOINT

Dewpoint, cited on some television and radio weathercasts, is another useful measure of humidity. **Dewpoint** is the temperature to which air must be cooled at constant pressure to achieve saturation of air relative to liquid water without the addition or removal of water vapor. The higher the dewpoint, the greater is the concentration of water vapor in air. Cooling unsaturated air at constant pressure increases its relative humidity. When the relative humidity reaches 100%, the air is saturated and the air temperature is the same as the dewpoint. Warming the air without adding water vapor via evaporation lowers the relative humidity and increases the difference between the actual air temperature and the dewpoint. Usually, the dewpoint is less than or equal to the air temperature.

Dew consists of tiny droplets of water formed when water vapor condenses on a cold surface such as blades of grass on a clear, calm night. Dew is not a form of precipitation because it does not fall from clouds to Earth's surface. Dew forms as a consequence of radiational cooling of the surface of an object. For water vapor to condense as dew on the surface of an object, that surface must cool to a temperature below the dewpoint. When cooling at constant pressure produces saturation at an air temperature below freezing, water vapor deposits as **frost**, ice crystals that form on exposed surfaces such as vegetation. The air temperature at which such cooling initially causes frost to form is known as the **frost point**.

From the above discussion, dewpoint is an ideal measure of atmospheric humidity, with wide climatological applications. Average daily, monthly, or annual dewpoints can be computed from hourly dewpoint readings made at automated weather stations. On average, within the United States, summer dewpoints are highest at localities bordering the Gulf of Mexico where the July mean dewpoint typically ranges between 21 °C and 24 °C (70 °F and 75 °F). During the oppressive heat waves that sometimes sweep over the continent east of the Rocky Mountains, the dewpoint may top the low 20s Celsius (low 70s Fahrenheit) even in the northern states. On 30 July 1999, during a particularly humid heat wave, the dewpoint at Milwaukee, WI, tied a record high of 27.8 °C (82 °F). Summer dewpoints are lowest in the Rocky Mountain States and the American Southwest. From New Mexico northward into western Montana, the July mean dewpoint generally ranges between −1 °C and 7 °C (30 °F and 45 °F). The dewpoint (actually frost point) is exceptionally low in polar and arctic air masses that invade broad regions of the United States in winter.

PRECIPITABLE WATER

Precipitable water is another way of describing the amount of water vapor in the atmosphere. Unlike the other humidity measures that represent the amount of water vapor at a specific location within the atmosphere, **precipitable water** is the depth of water that would be produced if all the water vapor in a vertical column of air were condensed into liquid water. The air column is usually taken to extend from Earth's surface to the top of the troposphere, the portion of the atmosphere where most water vapor occurs. A reasonably good measure of precipitable water is obtained from in situ measurements by radiosondes or remotely by Earth-orbiting satellites operating in the sounding mode. Precipitable water is often used in numerical climate models to simulate radiative transfer and determine the attenuation of solar radiation in the atmosphere due to water vapor.

Condensing all the water vapor in the global atmosphere would produce a layer of water that would cover the entire Earth's surface to a depth of about 2.5 cm (1.0 in.). The global pattern of monthly average precipitable water in cm is plotted in Figure 5.6A for January and in Figure 5.6B for July. These data were supplied by NASA/Surface meteorology and Solar Energy (SSE) and obtained via satellite. The average precipitable water depth decreases with latitude in response to the poleward decline in mean air temperature; that is, evaporation and precipitable water are lower in cold regions. Precipitable water varies from more than 4.0 cm (1.6 in.) in the humid tropics to less than 0.5 cm (0.2 in.) in the polar regions. Seasonal variations in average precipitable water also occur at middle and high latitudes, with highest values generally occurring in summer when the troposphere is warmest.

Precipitable water is not always indicative of the amount of precipitation that might fall at a particular location. Numerous other factors, including horizontal advection of water vapor into or from the air column along with recycling of water locally, ultimately determine the amount of precipitation.

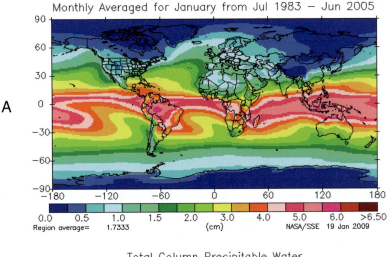

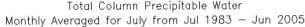

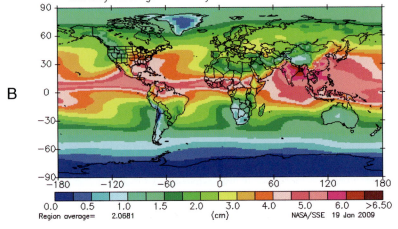

FIGURE 5.6
Global map of long-term average precipitable water in cm for (A) January and (B) July. [From NASA/Surface meteorology and Solar Energy (SSE)]

Monitoring Water Vapor

The **hygrometer** is an instrument that measures the water vapor concentration of air; several different designs are available. NOAA's National Weather Service employs a **dewpoint hygrometer** as a component of its Automated Surface Observing System (ASOS) (Figure 5.7). With this instrument, air passes over the surface of a metallic mirror that is cooled electronically. An electronic sensor continually monitors the temperature of the mirror at the same time that an infrared beam is pointed at the mirror. With sufficient cooling, a thin film of water condenses on the mirror changing its reflectivity and altering the reflection of the infrared beam. The mirror temperature is automatically recorded as the dewpoint. Then the mirror

is warmed electronically to evaporate the dew to prepare for the next measurement.

Some hygrometers take advantage of the sensitivity of organic materials to changes in humidity. One common design, called a **hair hygrometer**, uses human hair as the sensing element. Cells in the hair *adsorb* (collect on the surface) water and swell, causing the hair to lengthen slightly as the relative humidity increases. Typically, hair changes length by about 2.5% over the full range of relative humidity from 0% to 100%. Usually, a sheaf of blond hair is linked mechanically to a pointer on a dial that is calibrated to read in percent relative humidity. Unfortunately, hair hygrometers do not measure extremes in relative humidity accurately nor do they respond quickly to rapid changes in humidity.

An **electronic hygrometer** is based on changes in the electrical resistance of certain chemicals as they adsorb water vapor from the air. The adsorbing element may be a thin carbon coating on a glass or plastic strip. The more humid the air, the more water adsorbed, and the lower is the resistance to an electric current passing through the sensing element. Variations in electrical resistance are calibrated in terms of percent relative humidity or dewpoint. An electronic hygrometer is flown on board radiosondes (Chapter 2).

For more than a century, most meteorological observations of water

FIGURE 5.7
The temperature/dewpoint sensor (hygrothermometer) that is a component of NOAA's National Weather Service Automated Surface Observing System (ASOS).

sounding indicates that the temperature of the ambient air were dropping more rapidly with altitude than the dry adiabatic lapse rate (that is, more than 9.8 Celsius degrees per 1000 m), then the ambient air would be unstable for both saturated and unsaturated air parcels. This situation is called **absolute instability**. If the sounding lies between the dry adiabatic and moist adiabatic lapse rates, **conditional stability** prevails; that is, the air layer is stable for unsaturated air parcels, but unstable for saturated air parcels. With this relatively common situation, unsaturated air must be forced upward in order to reach saturation. But once saturation is achieved, the now cloudy air cools at the moist adiabatic lapse rate.

An air layer is stable for both saturated and unsaturated air parcels when the sounding indicates any of the following conditions: (1) the temperature of the ambient air drops more slowly with altitude than the moist adiabatic lapse rate; (2) the temperature does not change with altitude (*isothermal*); (3) the temperature increases with altitude (*temperature inversion*). Any one of these three types of temperature profiles indicates **absolute stability**.

What happens when a sounding coincides with either the dry or moist adiabatic lapse rate? A sounding that equals the dry adiabatic lapse rate is neutral for unsaturated air parcels and unstable for saturated air parcels. A sounding that is the same as the moist adiabatic lapse rate is neutral for saturated air parcels but stable for unsaturated air parcels. Within a **neutral air layer**, a rising or descending air parcel always has the same temperature (and density) as its surroundings. Hence, a neutral air layer neither impedes nor spurs upward or downward motion of air parcels.

In summary, atmospheric stability influences weather and, in turn, climate by affecting vertical motion of air. Stable air suppresses vertical motion whereas unstable air enhances vertical motion, convection, expansional cooling, and cloud development.

LIFTING PROCESSES

Ascending unsaturated air cools and its relative humidity increases. With sufficient ascent and expansional cooling, the relative humidity nears 100% and condensation or deposition begins, thus forming clouds. What causes air to rise? Air rises (1) as the ascending branch of a convection current, (2) along the surface of a front, (3) up the slopes of a hill or mountain, or (4) where surface winds converge.

As we saw in Chapter 4, sensible heating (conduction and convection) is a means of heat transfer between Earth's surface and atmosphere. Cumulus clouds may form where convection currents ascend, and the sky is generally cloud-free where convection currents descend (Figure 5.14). The higher the altitude reached by ascending convection currents, the greater the amount of expansional cooling, and the more likely it is that clouds (and precipitation) will form.

Clouds and precipitation are often triggered by **frontal uplift**, which occurs where contrasting air masses meet. A **front** is a narrow zone of transition between two

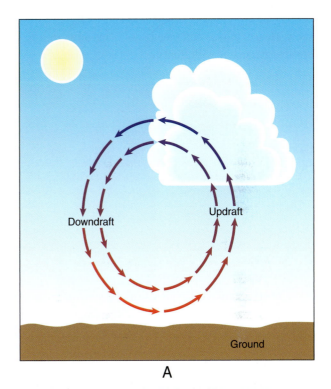

A

B

FIGURE 5.14
A convection current consists of (A) an updraft and a downdraft; (B) cumulus clouds form where air ascends while the surrounding sky is cloud-free where the air descends.

air masses that differ in temperature and/or humidity. A warm and humid air mass is less dense than a cold and dry air mass and hence, as a cold air mass retreats, the warm air advances by riding up and over the cold air (Figure 5.15A). The leading edge of the advancing warm air at Earth's surface is known as a *warm front*. In contrast, cold and dry air displaces warm and humid air by sliding under it and forcing the warm air upward (Figure 15.15B). The leading edge of advancing cold air at Earth's surface is known as a *cold front*. The net effect of the replacement of one air mass by another air mass is uplift and expansional cooling of air, cloud development, and perhaps rain or snow. Hence, clouds and precipitation are often (but not always) associated with fronts.

Orographic lifting occurs where air is forced upward by topography, the physical relief of the land. Horizontal winds sweeping across the landscape alternately ascend hills and descend into valleys. With sufficient topographical relief, the resulting expansional cooling and compressional warming of air affects cloud and precipitation development.

A mountain range that intercepts the prevailing winds forms a natural barrier that is responsible for a cloudier, wetter climate on one side of the range than on the other side (Figure 5.16). Air that is forced to ascend the *windward slopes* (facing the oncoming wind) expands and cools, which increases its relative humidity. With sufficient cooling, saturation is achieved and clouds

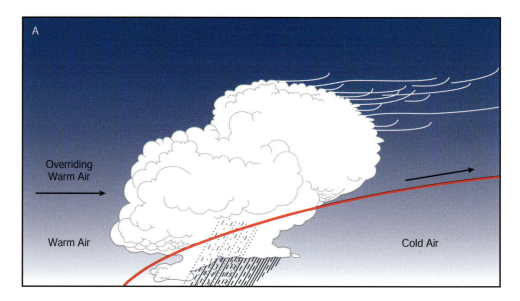

FIGURE 5.15
As shown in these vertical cross-sections of a (A) warm front and (B) cold front, less dense air ascending along a front expands and cools so that clouds and precipitation may develop. The vertical scale is greatly exaggerated.

FIGURE 5.16
Winds ascend the windward slopes of a mountain range and descend along the leeward slopes. The climate is wetter along the windward slopes than the leeward slopes.

and precipitation develop. Meanwhile, air descending the *leeward slopes* (downwind side) is compressed and warms, raising its saturation vapor pressure while the remaining cloud particles evaporate or sublimate. The relative humidity of descending air decreases and existing clouds vaporize so that precipitation is less likely. In this way, mountain ranges induce contrasting climates: moist climates on the windward slopes and dry climates on the leeward slopes. Dry conditions often extend many hundreds of kilometers downwind of a prominent mountain range; this region is known as a **rain shadow**. The Rocky Mountain rain shadow, for example, extends eastward from the Continental Divide to about the 100th meridian (100 degrees W longitude), traditionally considered to be the boundary between dry land with irrigated agriculture to the west and rain-fed agriculture to the east.

An orographically induced contrast in precipitation is apparent in the Pacific Northwest, where the north-south trending Coastal and Cascade Mountain Ranges intercept the prevailing west-to-east flow of humid air from the Pacific Ocean. Exceptionally rainy conditions prevail in western Washington and Oregon, whereas semiarid conditions characterize much of the eastern portions of those states, some 300 km (200 mi) inland. For example, the average annual precipitation (rain plus melted snow) at Astoria, OR, on the Pacific coast is 172 cm (67.7 in.) but only 20.8 cm (8.2 in.) at Yakima, WA, in the rain shadow of the Cascade Mountains. This climate contrast affects the indigenous plant and animal communities, domestic water supply, demand for irrigation water, types of crops that can be grown, and requirements for human shelter.

The influence of topography on precipitation patterns is also impressive on the mountainous islands of Hawaii, where a few volcanic peaks top 4000 m (13,000 ft). Over the ocean waters surrounding the islands, estimated mean annual rainfall is uniformly between about 560 mm (22 in.) and 700 mm (28 in.). On the islands, however, mean annual precipitation is much more variable due to orographic effects, ranging from only 190 mm (7.5 in.) leeward of the Kohala Mountains on the Big Island of Hawaii to 11,990 mm (39.3 ft) on the windward slopes of Mount Waialeale on the island of Kauai, arguably one of the rainiest spots on Earth.

Another mechanism responsible for uplift and cloud formation is convergence of surface winds. Associated upward motion means expansional cooling, increasing relative humidity, and eventually cloud and perhaps precipitation formation. In Chapter 6, we describe weather systems in which convergence of surface winds plays an important role in triggering stormy weather. For example, converging surface winds are largely responsible for cloudiness and precipitation in a low-pressure system (cyclone). Also, converging sea breezes is a major factor in the relatively high frequency of thunderstorms in central Florida.

Clouds

A **cloud** is the product of condensation or deposition of water vapor within the atmosphere; it consists of a sufficiently large and visible aggregate of minute water droplets and/or ice crystals. Clouds are important players in Earth's climate system, affecting the incoming and outgoing flux of radiation. Some cirrus clouds (high thin clouds composed of mostly ice crystals) originate as contrails produced in the exhaust of jet aircraft engines and may contribute to climate change. For details, see this chapter's second Essay.

At any given time, clouds shroud about 60% of the planet (Figure 5.17). How do cloud droplets (or ice crystals) form? Suspended in the atmosphere are tiny solid and liquid particles known as **nuclei** that provide surfaces on which condensation or deposition initially takes place. Nuclei are abundant, continually cycling into the atmosphere from Earth's surface. Sources of nuclei include volcanic eruptions, wind erosion of soil, forest fires, and ocean spray. When sea waves break, drops of salt water enter the atmosphere and the water evaporates leaving behind tiny sea-salt crystals that function as nuclei. Emissions from domestic and industrial chimneys also contribute nuclei to the atmosphere.

FIGURE 5.17
Fraction of an area that was cloudy on average for November 2009. Color scale ranges from blue (no clouds) to white (totally cloudy) representing the portion of each pixel that was covered by clouds. Measurements obtained by the Moderate Resolution Imaging Spectroradiometer (MODIS) on NASA's Terra satellite. [NASA Earth Observatory]

Depending on the product (liquid water droplets or ice crystals), a distinction is made between cloud condensation nuclei and ice-forming nuclei. **Cloud condensation nuclei (CCN)** promote condensation of water vapor at temperatures both above and below the freezing point of water. Within the atmosphere, water vapor can condense into cloud droplets that remain liquid even at temperatures well below 0 °C (32 °F). Droplets at such temperatures are described as *supercooled*. **Ice-forming nuclei (IN)** are much less common than CCN and promote formation of ice crystals only at temperatures well below freezing.

Hygroscopic nuclei are CCN that possess a chemical attraction for water molecules. Condensation begins on hygroscopic nuclei at a relative humidity well under 100%. Magnesium chloride ($MgCl_2$), a salt in sea-spray, can promote condensation at a relative humidity as low as 70%. Clouds form more efficiently where hygroscopic nuclei are abundant. Many sources of hygroscopic nuclei exist in urban-industrial areas and this helps explain why localities downwind of large cities tend to be somewhat cloudier and rainier than upwind localities. The *Metropolitan Meteorological Experiment (METROMEX)* conducted in the 1970s demonstrated that the average summer rainfall was 5% to 25% greater within and up to 50 to 75 km (31 to 47 mi) downwind of

St. Louis, MO, than upwind of the city. In more recent years, similar urban effects on precipitation were detected in Tokyo, Phoenix, and other cities. Besides being a source of hygroscopic nuclei, cities also spur cloud and precipitation development by contributing water vapor (raising the relative humidity) and heat (adding to the buoyancy of air). Furthermore, the relative roughness of a city surface induces convergence of horizontal winds, uplift of air, and expansional cooling.

CLOUD CLASSIFICATION

Clouds come in a variety of sizes and shapes, something that is obvious from even a cursory glance at the sky. Clouds are classified on the basis of (1) general appearance (texture) as viewed from the ground, (2) altitude of cloud base, (3) temperature, or (4) composition (Table 5.6). Based on a cloud's general appearance, the simplest distinction is among cirriform, stratiform, and cumuliform clouds (Figure 5.18). A **cirriform cloud** is wispy or fibrous, a **stratiform cloud** is layered, and a **cumuliform cloud** is heaped or puffy. On the basis of altitude of cloud base, a distinction is made among high, middle, and low clouds, and clouds having significant vertical development. The altitude of a cloud affects its temperature and ultimately its composition. High, middle, and low clouds generally occur in layers and are produced

TABLE 5.6
Summary of Cloud Classification Schemes

Temperature	Composition	Altitude	
		Stratiform	Cumuliform
Cold	ice crystals	high	clouds with
	supercooled droplets	middle	vertical
Warm	water droplets	low	development

by gentle uplift of air (typically less than 5 cm per sec or 0.1 mph) over a broad geographic area. These layered clouds often form along and ahead of a warm front. Clouds having significant vertical development usually cover smaller areas but are associated with much more vigorous uplift (sometimes in excess of 30 m per sec or 70 mph). These vertically developed clouds may form along or ahead of a cold front and develop into thunderstorms.

The temperature in a **warm cloud** is above 0 °C (32 °F) whereas the temperature in a **cold cloud** is at or below 0 °C (32 °F). Because average air temperature decreases from sea level to the tropopause and ascending cloudy air expands and cools, high clouds are colder than low clouds. Vertically developed clouds that surge to great altitudes, such as thunderstorm clouds (cumulonimbus), may be warm near their base but cold aloft. To a large extent, temperature of the atmosphere at cloud level dictates the composition and appearance of clouds. Cold clouds are made up of ice crystals and/or supercooled water droplets; ice crystal clouds have wispy edges. Warm clouds are made up of water droplets and have sharply defined boundaries.

Scientists have identified ten fundamental cloud types, organized by altitude of cloud base in middle latitudes (Table 5.7). The actual altitude of cloud base varies seasonally and with latitude. For example, the base of high clouds may be as low as 3000 m (10,000 ft) in polar latitudes and as high as 18,000 m (60,000 ft) in the tropics.

Usually only drizzle falls from stratus clouds, but significant amounts of rain or snow may fall from the much thicker **nimbostratus (Ns) clouds**. Nimbostratus resemble stratus except that they are darker gray and have a less uniform, more ragged base. Precipitation from nimbostratus tends to be light to moderate and may be continuous for 12 hrs or longer. By contrast, relatively brief but heavy showery-type precipitation is often associated with a cumulonimbus (thunderstorm) cloud.

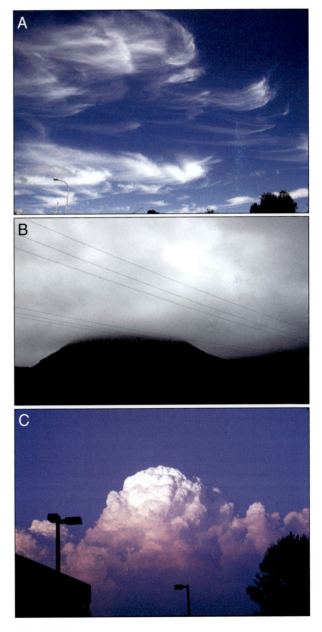

FIGURE 5.18
Examples of (A) cirriform, (B) stratiform, and (C) cumuliform clouds.

TABLE 5.7
Cloud Classification

Cloud group	Altitude of base (m)[a]
High clouds	
Cirrus (Ci)	5000-13,000
Cirrostratus (Cs)	5000-13,000
Cirrocumulus (Cc)	5000-13,000
Middle clouds	
Altostratus (As)	2000-5000
Altocumulus (Ac)	2000-5000
Low clouds	
Stratus (St)	surface-2000
Stratocumulus (Sc)	surface-2000
Nimbostratus (Ns)	surface-2000
Clouds with vertical development	
Cumulus (Cu)	to 3000
Cumulonimbus (Cb)	to 3000

[a]Average for middle latitudes.

Gentle rainfall from nimbostratus more readily infiltrates the soil, whereas the intense downpour of a thunderstorm quickly saturates the upper soil and then mostly runs off, perhaps causing flash flooding.

Clouds having significant vertical development form in the updrafts of convection currents. The altitude at which condensation begins to occur through convection is typically between 1000 and 2000 m (3600 and 6600 ft). Once cumulus clouds form, the stability profile of the troposphere determines the extent of vertical cloud development. Cumulonimbus clouds have tops that sometimes reach altitudes of 20,000 m (60,000 ft) or higher.

FOG

Fog is a visibility-restricting suspension of tiny water droplets or ice crystals (called *ice fog*) in an air layer next to Earth's surface. Simply put, fog is a cloud in contact with the ground. By international convention, fog is defined as restricting visibility to 1000 m (3250 ft) or less. *Dense fog* reduces visibility to 100 m (330 ft) or less. Fog may develop when air becomes saturated through radiational cooling, advective cooling, addition of water vapor, or expansional cooling.

By restricting visibility, fog adversely affects travel by motor vehicles, ships, and aircraft. Fog tends to be a major problem in some coastal areas. The world's foggiest place is Argentia, Newfoundland, which reports roughly 205 days of fog per year. One of the foggiest places in the United States is Cape Disappointment, WA, at the mouth of the Columbia River, where dense fog occurs on average about 2556 hrs annually (30% of the time). Moose Peak Lighthouse on Maine's Mistake Island has about 1560 hrs of dense fog per year.

With a clear night sky, light winds, and an air mass that is humid near the ground and relatively dry aloft, radiational cooling may cause the air near the ground to approach saturation. When this condition occurs, a ground-level cloud forms and is called **radiation fog** (Figure 5.19). High humidity at low levels within the air mass is usually due to evaporation of water from a moist surface. Hence, radiation fog is most common over marshy areas or where the soil has been saturated by recent rainfall or snowmelt. Air that is chilled by radiational cooling is relatively dense and, in hilly terrain, drains downslope and settles in low-lying areas such as river valleys and wetlands. Because of this air movement, known as *cold-air drainage*, hilltops may be clear of fog while fog is thick and persistent in deep valleys.

Often radiation fog lasts for only a few hours after sunrise. Fog gradually thins and disperses as saturated air at low altitudes mixes with drier air above the fog bank; the relative humidity drops and fog droplets vaporize. Mixing may be caused by convection triggered

FIGURE 5.19
Radiation fog seen here in the distance develops as a consequence of radiational cooling on a night when there is some air movement and the atmosphere is humid at low levels.

by solar heating of the ground or by strengthening of regional winds. In winter, however, the upper surface of a fog layer readily reflects the weak rays of the Sun and radiation fog may linger. For example, in the valleys of the Great Basin and in California's San Joaquin Valley, winter radiation fogs may persist for many days to weeks at a time.

Fog sometimes accompanies air mass advection. The temperature and the water vapor concentration of an air mass depend on the nature of the surface over which the air mass forms and travels. As an air mass moves from one place to another, termed *air mass advection*, those characteristics change, partly because of the modifying influence of the surfaces over which the air mass travels. When the advecting air passes over a relatively cold surface, the air mass may be chilled to saturation in its lowest layers. This type of cooling occurs, for example, in early spring in the northern states when mild, humid air flows over relatively cold, snow-covered ground. Snow on the ground may chill the overlying air to the dewpoint and fog develops. Fog formed by advective cooling is known as **advection fog**.

Advection fog also develops when warm and humid air streams over the relatively cold surface of a lake or the ocean. Persistent, dense fogs form in this way over the Great Lakes in summer (Figure 5.20). Thick sea fog develops when mild maritime air from the warm Gulf Stream flows northward and encounters the frigid waters of the Grand Banks off Newfoundland.

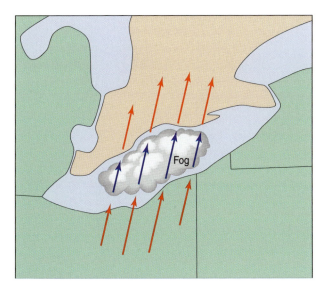

FIGURE 5.20
In summer, fog forms over the Great Lakes when relatively warm and humid air passes over the cold water surface and is chilled to its dewpoint.

Nocturnal radiational cooling sometimes combines with air mass advection to produce fog. An initially dry air mass becomes more humid in its lower levels following a long trajectory over the open water of a lake or the ocean. Over land areas, downwind from the water body, if the night sky is clear, the modified air mass is subjected to extreme radiational cooling and fog develops. San Francisco's famous fog forms in this way. Onshore winds transport cool, humid air into the city and nocturnal radiational cooling chills the air to saturation.

Steam fog (sometimes called *Arctic sea smoke*) is fog that develops in late fall or winter when extremely cold and dry air flows over a large unfrozen body of water. Evaporation and sensible heating cause the lower portion of the air mass to become more humid and warmer than the air above. Heating from below destabilizes the air and the consequent mixing of mild, humid air with cold, dry air brings the air to saturation and fog forms. Because the air is destabilized, fog appears as rising streamers that resemble smoke or steam. Steam fog also develops on a cold day over a heated outdoor swimming pool or hot tub and sometimes over a wet highway or field when the Sun comes out after a rain shower.

Fog may develop on hillsides or mountain slopes as a consequence of the upslope movement of humid air. Ascending humid air undergoes expansional cooling and eventually reaches saturation. Any further ascent of the saturated air produces **upslope fog**. In the coastal mountain ranges of California, upslope fog sometimes overtops the range crest and spreads as stratus clouds over the leeward valleys. A cloud layer formed in this way is known as *high fog*.

Precipitation

Clouds are no guarantee that rain or snow will reach the ground. Nimbostratus and cumulonimbus clouds produce the bulk of precipitation, but most clouds do not yield any significant rain or snow. This is because a special combination of circumstances is required for precipitation to develop. **Precipitation** is water in solid or liquid form that falls from clouds to Earth's surface under the influence of gravity. This section summarizes the two mechanisms of precipitation formation: the collision-coalescence process and the Bergeron process.

Terminal velocity, the speed at which a particle falls through a fluid, helps explain why the tiny particles that compose clouds (water droplets and ice crystals) will not fall to Earth's surface as precipitation unless they undergo considerable growth. Generally, terminal velocity increases with the size of the particle. For a particle to descend from a cloud to Earth's surface as precipitation, the particle's terminal velocity must be greater than the rising motion of air within the cloud (updrafts). Cloud water droplets and ice crystals are so small (typically 10 to 20 micrometers in diameter) that their terminal velocities are very low (usually only 0.3 to 1.2 cm per sec). Hence, even weak updrafts will keep them suspended in clouds. If droplets or ice crystals fall through the base of a cloud, their terminal velocity is so slow that it would take 24 hrs or longer for the cloud particles to reach Earth's surface. Long before that, the particles would vaporize in the unsaturated air under the cloud.

For clouds to precipitate, cloud particles must grow large enough that their terminal velocities overwhelm the updrafts. This is no minor task! It takes the water content of about 1 million cloud droplets (having diameters of 10 to 20 micrometers) to form a single raindrop (about 2 mm in diameter). Condensation or deposition alone cannot account for the formation of raindrop-sized particles. Cloud droplets grow into raindrops through collision-coalescence of cloud droplets in warm clouds whereas a combination of the Bergeron process and collision-coalescence causes ice crystals in cold clouds to grow into snowflakes.

WARM-CLOUD PRECIPITATION

In a warm cloud (at temperatures above 0 °C), droplets may grow by colliding and coalescing (merging) with one another, the so-called **collision-coalescence process**. This process takes place in a cloud composed of a mixture of droplets of different sizes, ideally with some droplets having diameters of at least 20 micrometers. Cloud droplets of unequal diameters have different terminal velocities, greatly increasing the likelihood that faster falling (larger) droplets will overtake, and then collide and coalesce with slower falling (smaller) droplets in their paths. Through repeated collisions and coalescence, droplets grow larger and some become sufficiently large that they survive a fall from clouds to Earth's surface as raindrops (Figure 5.21).

COLD-CLOUD PRECIPITATION

Although precipitation formation through the collision-coalescence process in warm clouds is important, especially in the tropics, most precipitation in middle and high latitudes originates in cold clouds (clouds or portions of clouds at temperatures below 0 °C) that have some ice crystals. At middle and high latitudes, clouds routinely

reach altitudes where the temperature is below freezing regardless of season. Precipitation is most likely to fall from a cloud initially composed of a mixture of supercooled water droplets and ice crystals. Around 1930, the Swedish meteorologist Tor Bergeron (1891-1977) contributed to an understanding of how precipitation forms in cold clouds and for this reason, this is called the **Bergeron process**, also known as the *ice-crystal process*.

As noted above, cloud condensation nuclei are much more abundant than ice-forming nuclei so that a mixed cloud (at least initially) consists of far more supercooled water droplets than ice crystals. Supercooled water droplets quickly vaporize as ice crystals grow. At subfreezing temperatures, water molecules more readily vaporize from a liquid (supercooled) water surface than from an ice surface because water molecules are more

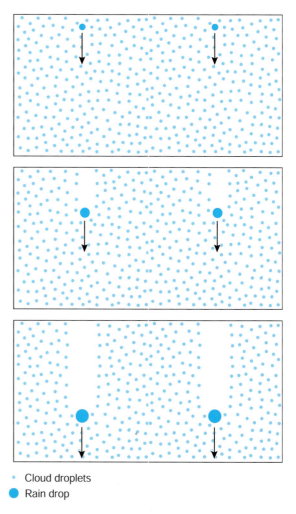

• Cloud droplets
● Rain drop

FIGURE 5.21
Within a warm cloud, a relatively large droplet falls through a cloud of much smaller droplets. The larger droplet falls faster and collides with the smaller droplets in its path and grows by coalescence. This is the collision-coalescence process of precipitation formation.

strongly bonded in the solid phase than the liquid phase. At the same subfreezing temperature, the saturation vapor pressure is greater over supercooled water than over ice (Table 5.4).

Within a cloud composed of a mixture of ice crystals and supercooled water droplets, a vapor pressure that is saturated for water droplets is actually supersaturated for ice crystals at the same temperature. Suppose, for example, that a cloud of supercooled water droplets at −15 °C (5 °F) is saturated relative to water (relative humidity = 100%). According to Table 5.4, the saturation vapor pressure is 1.91 mb. Imagine the sudden appearance of an ice crystal in the same region of the cloud. Table 5.4 indicates that the saturation vapor pressure over ice at −15 °C (5 °F) is 1.65 mb so that air surrounding the ice crystals is supersaturated (relative humidity = 116%). Water vapor deposits on the ice crystals and the crystals grow. Deposition lowers the vapor pressure surrounding the water droplets to unsaturated conditions, causing water droplets to evaporate. Hence, via the Bergeron process, ice crystals grow at the expense of supercooled water droplets.

Larger, faster-falling ice crystals overtake, collide and agglomerate with smaller, slower falling ice crystals and supercooled water droplets (Figure 5.22). Ice crystals grow still larger and may fall out of the cloud base. If air temperatures are below freezing at least much of the way to the ground, ice crystals reach Earth's surface as snowflakes. If air temperatures are above freezing in the lower reaches of the cloud or between the cloud and Earth's surface, snowflakes melt to raindrops.

Regardless of the mechanism of formation, once a falling raindrop or a snowflake leaves the base of a cloud, it enters unsaturated air and begins to vaporize. The longer the distance to Earth's surface, (i.e., the higher the cloud base) and the lower the relative humidity of air beneath the clouds, the greater is the quantity of rain or snow that vaporizes. This is one of the reasons why highlands, being closer to the base of clouds, usually receive more precipitation than lowlands.

FORMS OF PRECIPITATION

Depending upon environmental conditions, precipitation occurs in different forms. Included are liquid precipitation (rain, drizzle), freezing precipitation (freezing rain, freezing drizzle), and frozen precipitation (snow, ice pellets, hail).

Rain consists of liquid water drops with diameters generally in the range of 0.5 to 6 mm (0.02 to 0.2 in.) that fall mostly from nimbostratus and cumulonimbus clouds. At middle and high latitudes, raindrops usually begin as

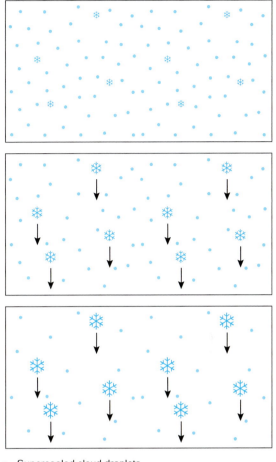

- Supercooled cloud droplets
- ❄ Ice crystals

FIGURE 5.22
Within a cold cloud, ice crystals grow at the expense of supercooled water droplets. As they grow larger, ice crystals fall faster and collide with droplets and other ice crystals in their paths. Eventually, they may grow large enough to fall out of the cloud as snowflakes. This is the Bergeron process of precipitation formation.

snowflakes (or sometimes would-be hailstones) that melt and coalesce as they descend through air at temperatures above 0 °C (32 °F).

Drizzle consists of liquid water drops having diameters between 0.2 and 0.5 mm (0.01 and 0.02 in.) that drift very slowly toward Earth's surface. Drizzle originates mostly in stratus clouds that are so low and thin that droplets undergo only limited growth by collision-coalescence.

Snow is an agglomeration of ice crystals in the form of flakes. Although constituent ice crystals are hexagonal (six-sided), snowflakes vary in shape and size depending on water vapor concentration and temperature in the portion of the cloud where they form and grow. **Ice pellets**, commonly called **sleet**, are spherical or irregularly shaped, transparent or translucent particles of ice that are

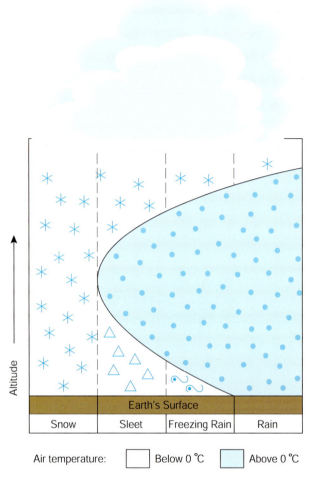

FIGURE 5.23
Schematic cross-section of the lower atmosphere showing temperature conditions required for formation of frozen, freezing, and liquid forms of precipitation.

5 mm (0.2 in.) or less in diameter. Ice pellets form when snowflakes partially or completely melt as they fall through above-freezing air beneath the cloud base (Figure 5.23). These raindrops or partially melted snowflakes then fall into a relatively thick layer of subfreezing air where they refreeze into ice particles prior to striking the ground.

Freezing rain (or *freezing drizzle*) consists of rain (or drizzle) drops that become supercooled and at least partially freeze on contact with cold surfaces (at subfreezing temperatures), forming a coating of ice (*glaze*) on roads, tree branches, and other exposed surfaces. Even a thin coating of ice can create hazardous driving and walking conditions. Sometimes glaze grows so thick that its weight brings down tree limbs and snaps power lines. Freezing rain develops in much the same way as sleet except that the layer of subfreezing air at Earth's surface is shallower (Figure 5.24).

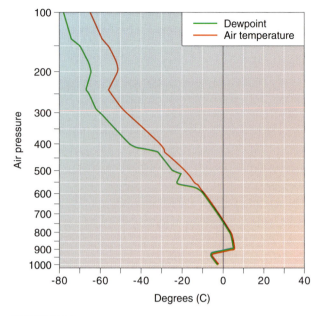

FIGURE 5.24
Radiosonde soundings of temperature and dewpoint indicating conditions favorable for freezing rain.

Freezing rain can be highly localized and persistent where the terrain is hilly or mountainous such as the Appalachians of Pennsylvania and West Virginia. Memorable freezing rain and ice storm events also can occur in the Columbia Gorge near Portland, OR. Cold, dense air drains into valley bottoms. Rain that falls from warm air aloft into this shallow layer of subfreezing air becomes supercooled and forms a glaze on cold surfaces. This situation often persists until the wind shifts and flushes the subfreezing air out of the valley bottom.

Hail consists of balls or jagged lumps of ice, often characterized by concentric internal layering resembling the structure of an onion (Figure 5.25). Layering is

FIGURE 5.25
A hailstone consists of internal concentric layers of clear and opaque ice as shown in this cross section photographed in polarized light. [National Center for Atmospheric Research/University Corporation for Atmospheric Research/National Science Foundation]

related to the various moisture collection environments that the hail encounters as it travels through the cloud. Hail forms within intense thunderstorms (cumulonimbus clouds) characterized by vigorous updrafts, an abundant supply of supercooled water droplets, and great vertical cloud development. Updrafts transport ice pellets into the middle and upper portions of the cumulonimbus cloud. Along the way, ice pellets grow by collecting supercooled water droplets, and eventually become too heavy to be supported by updrafts. Ice pellets then descend through the cloud, exit the cloud base, and enter air that is typically above the freezing point. Pellets begin melting, but if large enough initially, some ice will survive the journey to Earth's surface as *hailstones*. Most hailstones are harmless granules of ice less than 1 cm in diameter, but violent thunderstorms may spawn destructive hailstones the size of golf balls or larger. Hail is usually a spring or summer phenomenon that is particularly devastating to crops.

Measuring Precipitation

Rainfall and snowfall are routinely measured in terms of depth of accumulation over a specified interval of time, usually hourly, every 6 hrs, or 24 hrs. Measurements are made directly by collecting a volume of rain or snow in a rain or snow gauge or estimated remotely by weather radar or satellite sensors. Land-based precipitation data are collected, verified, and tabulated before becoming part of the official climate record of the station.

RAIN AND SNOW GAUGES

Today precipitation is collected and measured in essentially the same way as it was more than 2000 years ago; by using a container open to the sky. The first reference to a rain gauge appears in the Indian manuscript *Arthasastra*, authored by Kautila (also known as Vishnugupta Chanakya) sometime in the 4th century BCE. The gauge consisted of a 45-cm (18-in.) diameter bowl. Kautila, India's Chancellor of Exchequor, formulated a plan for taxing lands based on local rainfall. Rain gauges were reported in Palestine in the 1st century CE. The first snow gauges appeared in China in 1287 and by the middle of the 15th century, rain gauges were introduced to Korea. The first reported use of a rain gauge in Europe was in 1639 by the Italian Benedetto Castelli (1577-1644).

A standard modern non-recording **rain gauge** design consists of a cylinder equipped with a cone-shaped funnel at the top (Figure 5.26). The funnel directs rainwater into a narrower cylinder seated inside the larger outer cylinder. The funnel and narrow cylinder magnify the scale of measurement so that the instrument can resolve rainfall into increments as small as 0.01 in. (0.25 mm). A simple graduated stick is used to measure the depth of water that accumulates in the inner cylinder. (The stick is graduated in inches in the U.S. but in millimeters in Canada and most other nations.) Rainfall of less than 0.005 in. (0.1 mm) is recorded as a *trace*. Rainfall is usually measured at some fixed time once every 24 hrs and then the gauge is emptied. This type of rain gauge must be read manually and many of the NWS Cooperative Observer Stations across the United States still rely upon these gauges.

Continuous monitoring of rainfall is often useful, especially in flood-prone areas. Accordingly, some rain gauges are designed to provide a cumulative record of rainfall through time, thereby yielding rainfall rates, an indicator of precipitation intensity. Two common designs are the weighing-bucket and tipping-bucket rain gauges. These recording rain gauges can be used in remote locations. A **weighing-bucket rain gauge** calibrates the

FIGURE 5.26
A standard National Weather Service non-recording rain gauge.

weight of accumulating rainwater in terms of water depth. This instrument has a device that marks a chart on a clock-driven drum or sends an electronic signal to a computer for processing. At subfreezing temperatures, antifreeze in the collection bucket or a heater melts snow or ice pellets producing a cumulative record of melt water.

A **tipping-bucket rain gauge** consists of a free-swinging container partitioned into two compartments, each of which can collect the equivalent of 0.01 in. of rainfall. Each compartment alternately fills with water, tips and spills its contents, and trips an electric switch that either marks a chart on a clock-driven drum or sends an electrical pulse to a computer for recording. Compared to a weighing-bucket rain gauge, a tipping-bucket rain gauge does not perform as well at subfreezing air temperatures or at very high rainfall rates. A heated tipping-bucket gauge has been a standard component of the NWS Automated Surface Observing System (ASOS) but is currently being replaced by a weighing-bucket precipitation gauge (Figure 5.27).

Across the U.S., 6154 stations, including both Cooperative Observer Stations and NOAA National Weather Service sites, are part of the Hourly Precipitation Data network. In addition, nearly 3400 stations are capable of measuring precipitation at 15-minute intervals.

For snow, scientists are interested in (1) the depth of snow that falls during the period between observations, (2) the melt water equivalent of that snowfall, and (3) the depth of snow on the ground at observation time. New snowfall accumulates on a simple wooden board placed on top of the old snow cover. Using a ruler graduated in tenths of an inch (or a meter stick graduated in centimeters), snow depth is measured to the board. The board is then swept clean and moved to a new location. Snowfall measurements, reported to the nearest 0.1 in., are made at regularly scheduled times, typically once daily.

The melt water equivalent of new snowfall can be determined by measurements taken by a weighing-bucket gauge or by melting the snow collected in a non-recording gauge (with the funnel and inner cylinder removed). The average density of fresh snow is about 0.1 gram per cubic centimeter so that as a very general rule, 10 cm of fresh snow melts to 1 cm of water. However, the actual snow/melt-water ratio varies considerably depending on the crystalline form of the snowflakes, and the temperature of the air through which the snow falls. Snow consisting of columnar crystals can be relatively dense whereas snow made up of dendritic (star-like) crystals has very low density, with the ratio of snowfall depth to melt-water depth as great as 50 to 1. Wet snow falling at surface air

FIGURE 5.27
Weighing-bucket rain gauge that is a component of NOAA's National Weather Service Automated Surface Observing System (ASOS). The array of slats surrounding the mouth of the gauge is the windshield that helps reduce wind-induced errors in the catch of precipitation.

temperatures at or above 0 °C (32 °F) has much greater water content than dry snow falling at very low surface air temperatures. The ratio of snowfall depth to melt-water depth may vary from 3 to 1 for very wet snow to 30 to 1 or higher for dry fluffy snow. As a snow cover ages, snow converts to tiny granules of ice, air-filled space decreases, snow settles, and snow density increases.

Snow depth represents an accumulation of any fresh snow plus the remnants of past snowfalls. The depth of snow on the ground is usually determined using a graduated ruler inserted vertically into the snow at several representative locations. These measurements are averaged, rounded to the nearest inch, and reported daily. Care must be taken to avoid areas where the wind has whipped the snow into drifts.

Rainfall and snowfall can vary considerably from one place to another, especially when produced by convective showers and thunderstorms. Precipitation gauges are sited so that measurements are as accurate and representative as possible. Instruments must be sheltered from strong winds which cut the collection efficiency of rain gauges. Winds tend to accelerate over the open top of a gauge, reducing the amount of rain that enters the instrument. One study found that winds blowing at 16 km per hr (10 mph) reduced the amount of rainwater collected by 10% compared to the amount that would enter the gauge if the air were calm. Many rain gauges are equipped with wind shields, protective attachments of freely hanging slats placed around the orifice of these gauges to eliminate the influence of wind eddies on the catch (Figure 5.27). Gauges should not be placed in windswept locations, but must be sited well away from buildings and tall vegetation that might shield the instrument from precipitation. As a general rule, obstacles should be no closer to the gauge than four times the height of the obstacle.

Traditionally, atmospheric scientists derive spatial patterns of atmospheric variables from measurements made by instruments at discrete points, that is, at weather stations. This approach works reasonably well for variables such as air temperature and pressure that vary continuously from one place to another. Hence, the temperature is likely to be near 65 °F at a location half way between two weather stations, one reporting 60 °F and the other reporting 70 °F. Basing spatial patterns on measurements at discrete points does not work as well for discontinuous variables such as precipitation. Rainfall and snowfall exhibit considerable spatial variability. Hence, it does not always follow that moderate amounts of rain fell at a location halfway between two weather stations, one reporting heavy rainfall and the other reporting light rainfall. Furthermore, regular precipitation measurements are missing from vast stretches of the ocean where significant amounts of precipitation fall. To help remedy this situation,

scientists developed radar technologies and satellite-borne instruments that provide more continuous coverage of precipitation over broad areas.

REMOTE SENSING OF PRECIPITATION

Weather radar sends out pulses of microwave energy that are intercepted and scattered back as echoes. In this way, a radar system locates areas of precipitation and displays them as blotches on a television-type screen. The strength of the echo is an indication of precipitation intensity that is displayed according to a color scale. Weather radar also provides estimates of the total amount of rainfall. Whereas networks of precipitation gauges measure rainfall at discrete locations often more than 100 km (60 mi) apart, radar measures rainfall within continuous volume scans of the atmosphere centered on the unit. A special computer algorithm uses radar monitoring of rainfall rate to generate color-coded maps of rainfall totals over a specific period of time such as the last hour or since the start of the storm (Figure 5.28).

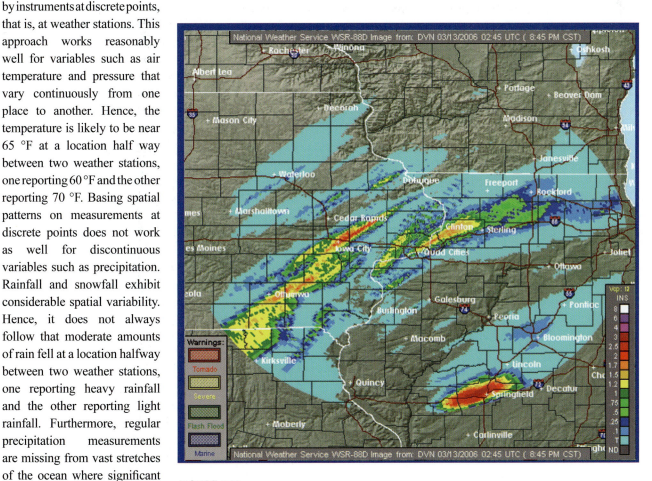

FIGURE 5.28
Cumulative rainfall as determined by computer analysis of weather radar echoes. Color bar to the right provides range of radar-estimated precipitation totals.

Satellite sensors are also valuable remote sensing tools for monitoring precipitation. The *Tropical Rainfall Measuring Mission (TRMM)* utilizes satellite technology primarily to measure rainfall over the area between 40 degrees N and 40 degrees S. Knowledge of tropical rainfall and the associated flow of heat energy promises to improve our understanding of Earth's climate system and make possible the development of more realistic global climate models. About two-thirds of all precipitation falls in the tropics and much of the planetary-scale heat transport originates in the tropics. This satellite mission is also helping scientists detect the onset and follow the evolution of El Niño and La Niña (Chapter 8). Launched in November 1997, TRMM is a joint venture of NASA and Japan's National Space Development Agency.

The TRMM satellite, in orbit about 350 km (215 mi) above Earth's surface, utilizes three sensors to measure rainfall remotely: radar, a microwave imager, and a visible/IR scanner. Satellite-based precipitation radar (PR) is similar to ground-based radar except that the signal has a much shorter wavelength. PR obtains vertical profiles of rain and snow from Earth's surface to an altitude of about 20,000 m (65,500 ft) and has a horizontal resolution on Earth's surface of 4 km (2.5 mi). The TRMM Microwave Imager (TMI) measures the minute flux of microwave energy emitted by the planetary system. The instrument can "see through" clouds and measure sea surface temperatures (SST). Variations in the amount of energy received at different wavelengths are interpreted to measure water vapor, cloud water, and rainfall intensity. Similar sensors have been flown on board polar-orbiting Defense Meteorological Satellites since 1987. The TRMM Visible and Infrared Scanner (VIRS) system monitors five spectral channels from visible to infrared (between 0.63 and 12 micrometers). The height of convective clouds (determined from IR-derived cloud-top temperature) is the basis for rainfall estimates; that is, the higher the cloud, the greater is the rainfall. The same type of precipitation-measuring scanner is used routinely on NOAA's Polar Orbiting Environmental Satellites (POES) and Geostationary Operational Environmental Satellites (GOES) for weather analysis and forecasting.

Scientists also measure snow cover remotely, information that is valuable for water resource managers and flood forecasters. In mountainous areas where snowfall is considerable, the melt water equivalent of snow can be determined remotely using snow pillows. A *snow pillow* is a device that is filled with antifreeze solution and fitted with a manometer (an instrument that measures pressure changes) that is calibrated to give the water equivalent from the weight of the overlying snow cover. Data are radioed to a satellite and from there to a data center for downloading and analysis. The *Airborne Gamma Radiation Snow Survey Program* also provides remote sensing of the water content of snow packs. At designated times, low-flying aircraft measure terrestrial gamma ray emission along selected flight lines. Gamma radiation emitted by Earth materials attenuates as the snow pack thickens. Comparison of gamma emission over the same flight line with and without snow cover is calibrated in terms of the melt water equivalent of snow (or soil moisture).

Conclusions

As air nears saturation, water vapor begins condensing or depositing on airborne nuclei. With continued cooling, usually due to expansion of rising air, clouds form. Clouds are distinguished on the basis of appearance, altitude of their base, temperature, and composition. Through the collision-coalescence and Bergeron processes, cloud droplets and ice crystals grow large enough to survive the fall to Earth's surface as precipitation. Precipitation occurs as rain, drizzle, snow, ice pellets (sleet), freezing rain (or freezing drizzle), and hail. Precipitation type depends upon cloud conditions and the temperature of the air column through which the precipitation falls to Earth's surface.

A variety of direct and remote sensing techniques are employed by scientists to monitor and measure precipitation. Weather radar sends out pulses of microwave energy that locate areas of precipitation and determine the intensity of rainfall. Instruments onboard Earth-orbiting satellites are also used to estimate rainfall over tropical latitudes.

Atmospheric circulation plays a key role in bringing air to saturation and triggering cloud and precipitation development, an important aspect of climate. The next chapter covers the time-averaged planetary-scale circulation and its relationship to mesoclimates. We begin with a discussion of the forces that drive and shape atmospheric circulation.

Basic Understandings

- The global water cycle is the ceaseless circulation of a fixed quantity of water among Earth's oceanic, atmospheric, and terrestrial reservoirs. Powered by solar radiation and influenced by gravity, water and heat energy move between Earth's surface and the atmosphere via evaporation, condensation, transpiration, sublimation, deposition, and precipitation.

- The global water budget indicates conservation of the mass of water, with globally and annually averaged precipitation equal to the global and annual evaporation average. However, an annual surplus of precipitation over evapotranspiration occurs on the land and more evaporation than precipitation over the ocean. Thus, a net flow of liquid water is directed from land to sea while water vapor flows from ocean to land. Precipitation falling on land vaporizes, infiltrates the ground, is taken up by plant roots and transpired to the atmosphere, is temporarily stored in lakes, ponds, ice fields, and glaciers, and/or runs off into rivers or streams.

- Vapor pressure is a direct measure of the water vapor component of air. At a specific temperature, relative humidity compares the actual amount of water vapor in air to the amount of water vapor if the air were saturated.

- Saturation vapor pressure indicates equilibrium in the flux of water molecules entering and leaving the vapor phase. The saturation vapor pressure increases with rising temperature because temperature largely regulates the rate of vaporization of water.

- When the water vapor concentration of air equals the water vapor concentration at saturation, the relative humidity is 100%. For unsaturated air on a calm day, the relative humidity varies inversely with air temperature; that is, the relative humidity is highest when the air temperature is lowest (near sunrise), and the relative humidity is lowest when the air temperature is highest (early to mid-afternoon).

- Dewpoint is the temperature to which air is cooled at constant pressure to achieve saturation. Dew consists of tiny droplets of water that condense onto cold surfaces from air that has been chilled to its dewpoint and below. When cooling at constant pressure produces saturation at air temperatures below freezing, water vapor deposits on cold surfaces as frost. The air temperature at which frost begins forming is called the frost point. Tropical air masses typically have higher dewpoints than polar air masses.

- Precipitable water is the depth of water that would be produced if all the water vapor in a vertical column of air, extending from Earth's surface to the tropopause, were condensed into liquid water. Generally, the amount of precipitable water decreases with increasing latitude in response to lower average temperatures and reduced evaporation.

- Hygrometers are instruments that directly measure the water vapor concentration of air and include dewpoint hygrometers, hair hygrometers, and electronic hygrometers. A less direct measure of humidity is provided by a psychrometer consisting of dry-bulb and wet-bulb thermometers mounted side-by-side. The greater the wet-bulb temperature depression, the lower is the relative humidity. When the wet-bulb and dry-bulb temperatures are the same, the relative humidity is 100%.

- The relative humidity of unsaturated air increases when the air is cooled (lowering the saturation vapor pressure) or when water vapor is added to the air (increasing the vapor pressure) at constant temperature. As the relative humidity nears 100%, clouds are increasingly likely to form.

- Expansional cooling is the principal means whereby clouds form in the atmosphere. As ascending currents of unsaturated (clear) air cool, the relative humidity increases and approaches saturation.

- Atmospheric stability is determined by comparing the temperature (or density) of an air parcel moving vertically (up or down) within the atmosphere with the temperature (or density) of the surrounding (ambient) air.

- Stable air inhibits vertical motion, convection, and cloud formation. Unstable air enhances vertical motion, convection, and cloud formation. Local radiational heating or cooling, air mass advection, or large-scale ascent or subsidence can change the stability of an air mass.

- Expansional cooling and cloud development occur through uplift of air in convection currents, along fronts or mountain slopes, or where surface winds converge. Clouds form in the updraft of convection currents but dissipate in the downdraft. The ascent of warmer, less dense air along a front may give rise to clouds and precipitation.

- Winds ascend along the windward slopes of a mountain range and descend along the leeward slopes. For this reason, the climates of the windward slopes tend to be cloudier and wetter than the climates of the leeward slopes. A rain shadow with semiarid to arid conditions may

- extend hundreds of kilometers downwind of a prominent mountain range.
- Clouds, the visible product of condensation or deposition within the atmosphere, are composed of large numbers of tiny water droplets, ice crystals, or both.
- As the relative humidity nears 100%, condensation and deposition occur on nuclei, tiny solid and liquid particles suspended in the atmosphere. Cloud condensation nuclei are much more abundant than ice-forming nuclei. Most ice-forming nuclei are active at temperatures well below the freezing point.
- Many condensation nuclei are hygroscopic; that is, they have a chemical affinity for water and induce condensation at a relative humidity less than 100%.
- Clouds are classified by general appearance (cirriform, stratiform, cumuliform), altitude of base (high, middle, low, significant vertical development), temperature (cold, warm), and composition (water droplets, ice crystals).
- High, middle, and low clouds are produced by relatively gentle uplift of air over a broad geographic area. These clouds are layered, that is, stratiform. Clouds having significant vertical development are the consequence of more vigorous uplift in more restricted geographic areas and are heaped or puffy in appearance, that is, cumuliform.
- Nimbostratus and cumulonimbus are the principal precipitation-producing clouds. Precipitation from nimbostratus clouds is typically lighter and lasts longer than showery precipitation from cumulonimbus clouds.

- Fog is a visibility-restricting suspension of tiny water droplets or ice crystals in an air layer next to Earth's surface. By international convention, fog is defined as restricting visibility to 1000 m (3250 ft) or less.
- Two mechanisms, the collision-coalescence process and Bergeron process, describe how cloud particles grow large enough to counter updrafts and fall to Earth's surface as precipitation under the influence of gravity.
- The collision-coalescence process occurs in warm clouds and requires the presence of relatively large cloud droplets that grow through collision and coalescence with smaller cloud droplets.
- The Bergeron process requires the coexistence of ice crystals and supercooled water droplets in cold clouds. At the same subfreezing temperature, the saturation vapor pressure surrounding a supercooled water droplet is higher than the saturation vapor pressure surrounding an ice crystal. Hence, air that is saturated for supercooled droplets is supersaturated for ice crystals. Ice crystals grow by deposition while water droplets vaporize.
- Principal forms of precipitation are rain, drizzle, snow, ice pellets (sleet), freezing rain, and hail. The form of precipitation depends on the source cloud and the vertical temperature profile of the air beneath the cloud.
- Precipitation is measured directly by collecting a volume of rain or snow in a rain or snow gauge or estimated remotely using weather radar or sensors on board Earth-orbiting satellites.

Enduring Ideas

- Powered by solar radiation and influenced by gravity, the global water cycle transports water and heat energy within Earth's climate system.
- On a global basis, a net flow of water is directed from the continents to the ocean while a net flow of water vapor occurs from over the ocean to land.
- Where unsaturated air ascends in the atmosphere, the air expands and cools, the relative humidity approaches 100%, and clouds form. Where air descends in the atmosphere, the air is compressed and warms, the relative humidity decreases, and clouds vaporize or fail to form.
- Stable air inhibits vertical motion, convection, and cloud formation. Unstable air enhances vertical motion, convection, and cloud formation.

Review

1. Distinguish between sublimation and deposition.
2. How does the saturation vapor pressure vary with air temperature?
3. Under what condition does the actual air temperature equal the dewpoint?
4. Frost not considered a form of precipitation. Why?
5. How does atmospheric stability influence the formation of clouds?
6. Why do cumulus clouds develop in the updraft of a convective current but not in the downdraft?
7. Contrast the climate on the windward slopes of a mountain range with the climate on the leeward slopes.
8. What is the significance of hygroscopic nuclei in cloud formation?
9. Describe the collision-coalescence precipitation process in warm clouds.
10. Explain how freezing rain forms.

Critical Thinking

1. If most of the water in the atmosphere originates in the ocean, how is it that precipitation consists of fresh water?
2. If the difference between the air temperature and dewpoint increases, how does the relative humidity change?
3. Explain the difference between the dry adiabatic lapse rate and the moist adiabatic lapse rate.
4. How does the stability of a cold air mass change as it moves from snow-covered ground to bare ground?
5. What is meant by "supercooled" water droplets?
6. What is the significance of the relatively low terminal velocities of cloud droplets?
7. Why does the Bergeron process of precipitation formation take place only in cold clouds?
8. Why is heavy rain unlikely to fall from stratus clouds?
9. How is sleet distinguished from hail?
10. In the effort to document evidence of climate change, why is it important to know how precipitation data were collected?

ESSAY: Climate Change and Evaporation Rates

The simplest and least expensive instrument for measuring evaporation is a *pan evaporimeter*, a large cylindrical metal pan filled with fresh water and equipped with a water height gauge, rain gauge, and floating thermometer (for monitoring temperature at the air/water interface). Use of these devices for measuring evaporation dates back to the 19th century but the instrument was not standardized until 1951. The standard NOAA National Weather Service evaporation pan (often found at agricultural experiment stations) measures 121.9 cm (48 in.) in diameter and is 25.4 cm (10 in.) deep (Figure 1). Changes in water level are recorded as long as temperatures remain high enough for the water surface to remain ice-free. The drop in water level from one day to the next (plus any rainfall) is recorded as evaporation.

FIGURE 1
Standard evaporation pan used by the National Weather Service.

A more sophisticated instrument for estimating direct evaporation plus transpiration (*evapotranspiration*) is the lysimeter. A *lysimeter* consists of an enclosed block of soil with a natural vegetation cover (usually grass) that is seated on a scale. Continuous measurements are made of rainfall and water percolating through the soil. Day-to-day changes in the weight of the soil block that are not accounted for by rainfall, runoff, or percolation are attributed to evapotranspiration.

A lysimeter can also be used to measure potential evapotranspiration, a useful parameter in arid and semiarid climates. *Potential evapotranspiration* refers to the maximum possible water loss (to vapor) given the available heat energy (required for latent heating). Potential evapotranspiration is measured in the same way as evapotranspiration except that the soil/vegetation surface of the lysimeter is continually irrigated. Under these conditions, evapotranspiration proceeds at the maximum possible rate. Lysimeters are even sunk into a snow pack to monitor sublimation and evaporation (of meltwater).

Numerical models are used to compute evapotranspiration based on direct measurements or estimates of selected environmental factors that influence the rate of vaporization of water. Evapotranspiration increases with rising temperature, increasing wind speed, and as the difference in vapor pressure between a water surface and the overlying air increases. Transpiration also varies with dewpoint, intensity of incident solar radiation, length of daylight, type of vegetation, stage in a plant's life cycle, and soil moisture.

We expect that increasing evaporation rates would accompany higher air temperatures. However, in 1995, scientists reported that in spite of a general warming trend, pan evaporimeter measurements in the United States and the former Soviet Union steadily decreased from 1950 to 1990. Several possible explanations have been proposed for this counter-intuitive discovery. For one, more extensive cloud cover would translate into cooler days, warmer nights, and less evaporation. Pan evaporimeter measurements may not accurately represent the rate of evaporation from the variety of terrestrial surfaces. If higher air temperatures resulted in increased terrestrial evaporation, the air over a pan evaporimeter would become more humid thus inhibiting evaporation. A third possibility is that a decrease in sunlight received at Earth's surface from 1957-58 to 1990 due to an increase in atmospheric aerosols may explain the reduction in pan evaporation. (This dimming trend reversed beginning in 1990.)

ESSAY: Contrails and Climate Change

A familiar sight in the daytime sky almost anywhere is *contrails*, bright white streamers of ice crystals that form in the exhaust of jet aircraft (Figure 1). Contrails are modifying the cloud cover along heavily traveled air corridors between major urban areas with possible implications for climate (Figure 2). According to some studies, contrails now cover about 5% of the sky in some portions of the eastern United States. A contrail, short for *con*densation *trail*, develops when the hot humid air in jet engine exhaust mixes with the cold, drier air at high altitudes. A similar process is the reason you can *see your breath* when exhaling on a cold day. You are actually seeing a small cloud formed by the mixing of your warm, humid breath with the colder, drier ambient air.

FIGURE 1
Contrails form when the hot humid exhaust from jet aircraft engines mixes with cold drier ambient air.

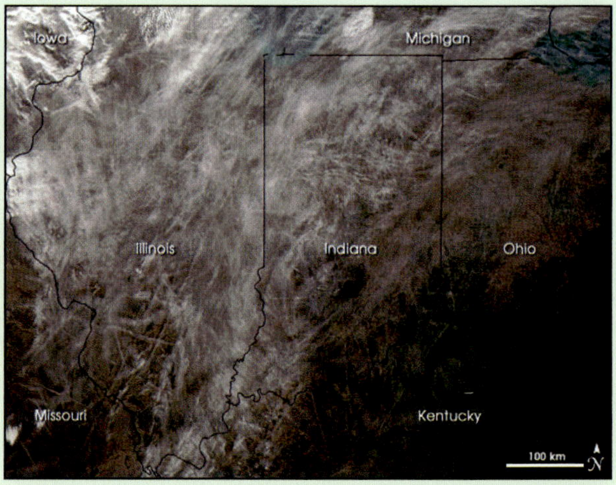

FIGURE 2
MODIS image of contrails over the Midwest on 25 November 2006, obtained by NASA's Terra satellite. [NASA Earth Observatory]

Cloud formation by mixing of air masses that differ in temperature and vapor pressure provides insight as to the cause of jet aircraft contrails. Surprisingly, a cloud may form even though the two air masses initially are unsaturated. As noted elsewhere in this chapter, the saturation vapor pressure increases rapidly with rising temperature. This relationship is depicted schematically in Figure 3 where temperature is plotted on the horizontal axis and vapor pressure is plotted on the vertical axis. The average temperature and average vapor pressure of a specific air mass plot as a single point on the diagram. Air of temperature and vapor pressure conditions given by a point plotted above the curve is supersaturated, on the curve is saturated, and below the curve is unsaturated.

FIGURE 3
Variation of saturation vapor pressure with temperature. Air of temperature and pressure (humidity) conditions given by a point above the curve is supersaturated, on the curve is saturated, and under the curve is unsaturated.

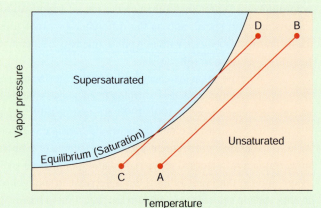

Suppose that two different unsaturated air masses plot as points *A* and *B* on the diagram. It is reasonable to assume that mixing the two air masses would produce a new air mass having properties (average temperature and vapor pressure) that would plot somewhere along a straight line connecting the points *A* and *B*. The precise location of that point (the mixture) along the line depends on the relative volumes of the two air masses. In any case, the mixture of the two air masses is unsaturated so no cloud forms. Consider, however, two other unsaturated air masses, *C* and *D*, one cold and dry, the other warm and humid. In this case the straight line linking *C* and *D* intersects the saturation vapor pressure curve. The new air mass resulting from the mixing of equal volumes of air masses *C* and *D* is saturated and a cloud forms.

In a similar manner, the exhaust from a jet engine mixes with the ambient air and may form a contrail behind the aircraft. Heat and water are among the combustion products of jet engines so that the exhaust is hot and humid. Turbulence in the wake of a jet engine promotes the mixing of the exhaust with ambient air. If the ambient air has the appropriate combination of vapor pressure and temperature, the mixture will be saturated and a contrail forms. Such conditions are most likely in the upper troposphere where commercial jetliners travel.

Depending on atmospheric conditions, contrails may dissipate (sublimate) within minutes or hours or they may spread laterally forming wispy cirrus clouds that persist for a day or so. Scientists hypothesize that increased cloud cover caused by contrails would affect the local radiation budget in two ways. During daylight hours, contrails and their cirrus byproducts reflect incoming solar radiation to space thereby cooling Earth's surface. At night, contrails and associated cirrus clouds absorb and emit infrared radiation welling up from below thereby enhancing the greenhouse effect and elevating Earth's surface temperature. Cooler days and warmer nights reduce the *diurnal temperature range (DTR)*, the difference between the day's maximum and minimum temperatures. Furthermore, contrails may stimulate precipitation locally by supplying ice crystal nuclei for lower clouds.

Scientists had an opportunity to test their hypothesis regarding the impact of contrails on the radiation budget immediately after the terrorist attacks of 11 September 2001 when the Federal Aviation Administration (FAA) ordered a three-day shutdown of all commercial air traffic in the United States. Almost immediately, contrails began to dissipate. David J. Travis, a climate scientist at the University of Wisconsin-Whitewater, and colleagues at Pennsylvania State University analyzed the effects of a contrail-free sky on surface temperatures at 4000 U.S. weather stations. They found that the DTR during the three-day period was about 1.1 Celsius degree (2 Fahrenheit degrees) greater than the long-term (1971-2001) average. Furthermore, the DTR for the three-day periods preceding and following the aviation shutdown was below the long-term average. These findings prompted Travis and colleagues to attribute the increase in DTR during the grounding of aircraft to the absence of contrails. With global aircraft traffic expected to increase by 2% to 5% per year through 2050, some scientists predict that the contribution of contrails to climate change will become more significant.

In 2008, however, researchers at Texas A&M University and NASA's Langley Research Center took another look at whether the absence of contrails was responsible for the increase in DTR during 11-14 September 2001. They examined cloud cover, humidity, winds, and surface temperatures over the coterminous U.S. for the period 1971-2001. They concluded that the increase in DTR during the grounding of aircraft was likely the result of anomalies in low cloud cover rather than the absence of contrails. Low clouds are thicker and warmer than high clouds or contrails and reflect more sunlight to space and emit more infrared radiation to Earth's surface. Furthermore, the researchers pointed out that the DTR during 11-14 September was within the range of natural variability.

CHAPTER 6

GLOBAL ATMOSPHERIC CIRCULATION

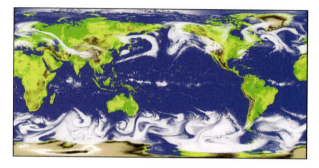

Global Mesoscale Circulation model at NOAA's Geophysical Fluid Dynamics Laboratory. Cloud distributions are of great importance in weather and climate. The distribution of latent heat produced by the moist convextion feeds directly into the dynamics that genetate weather systems. [Courtesy NOAA]

Case-in-Point

The devastating 1930s drought and Dust Bowl of the southern Great Plains were products of both anthropogenic and natural climatic factors. People living in the Dust Bowl and elsewhere on the Great Plains endured wrenching physical and economic hardship that ultimately forced many of them to abandon their homes and farms. Between 1931 and 1935, for example, almost 1 million people fled their farms across the Great Plains.

In middle latitudes, prevailing surface winds blow from the west and southwest. East of the Rocky Mountains, descending air produces a broad rain shadow over the Great Plains with a steady increase in average annual rainfall from west to east. West of the 100th degree meridian (longitude), average annual rainfall is 50 cm (20 in.) or less, considered the threshold for cultivation of non-irrigated crops. In fact, conditions were so dry in the western plains that in 1820, U.S. Army Major Stephen Long described this semi-arid region as the *Great American Desert*. (Long agreed to make weather observations for the American Philosophical Society during his 1819 expedition to the Rocky Mountains.) The name stuck until just after the American Civil War when the region was more commonly referred to as the Great Plains. Following Long's expedition and prior to the droughts of the 1850s and 1860s, conditions across the Great Plains were relatively wet, peaking in 1829.

The chain of events that culminated in the Dust Bowl began during World War I and the strong demand for wheat that sent prices for the grain soaring. Propped up by government subsidies, high wheat prices persisted through the post-war recovery period. The promise of riches spurred numerous homesteaders to settle in the previously agriculturally undeveloped western and southern Great Plains. Developers tried to lure skeptics to the semi-arid region by the now discredited claim that *rain follows the plow*, that is, cultivating the soil in marginally arable areas would bring additional rain. Encouraged by relatively wet conditions from 1926-29 and the introduction of more efficient mechanized farming equipment, sod busters ripped out the drought-resistant native short Buffalo grass, roots and all, and planted wheat. So much wheat was harvested, however, that huge surpluses accumulated and wheat prices plunged. By early 1930, the price of wheat had dropped to about one-eighth of what it had been some ten years earlier. By early summer 1931, the nationwide wheat harvest set a record, but with supply greatly outstripping demand, wheat prices dropped to a level that was about half of what it cost farmers to grow the crop. A year later, with mounting debts, roughly one-third of farmers on the Great Plains faced foreclosure.

Then came a prolonged drought (1932-38) when the practice of removing the native grasses proved to be the making of an environmental disaster. With no root system to anchor the soil, winds sweeping across the essentially tree-less plains readily lifted the fine-textured topsoil thousands of meters into the atmosphere during a series of dust storms, also known as *dusters* or *black blizzards* (Figure 6.1). The first dust storm appeared on

FIGURE 6.1
A dust storm approaching Strafford, TX on 18 April 1935. [NOAA's National Weather Service Collection, George E. Marsh Album]

14 September 1930. The greatest number of dusters in any one year was 134 in 1937. In these storms, fine dust particles filled the air restricting visibility and in some cases blocking out the mid-day Sun. Dust caused respiratory distress and illness among people and livestock and readily penetrated even the tiniest cracks in the walls of a house. On some occasions, winds transported dust as far as the Eastern Seaboard and to the ocean beyond. On one day alone, 14 April 1935, it is estimated that winds eroded more than 300,000 tons of topsoil from the Great Plains. The most severely impacted area, named the *Dust Bowl* in 1935, included parts of Colorado, Kansas, Nebraska, New Mexico, Texas, and Oklahoma (Figure 6.2).

In investigating the circumstances leading up to the Dust Bowl, a special committee appointed by President Franklin D. Roosevelt (1882-1945) concluded that climate change was not responsible for the calamity. Rather, drought is to be expected across the Great Plains as part of the natural climate variability of the region. The real culprit was the failure to implement an agricultural system that was appropriate for the climate of that region.

FIGURE 6.2
During the Dust Bowl, fallout from dust storms buried farm equipment on the Great Plains. [NOAA's National Weather Service Collection]

Driving Question:

What are the principal features of the planetary-scale atmospheric circulation and how does that circulation influence climate?

As discussed in Chapter 1, the seasonally varying boundary conditions of the planetary system select for an array of possible weather patterns that is characteristic of each season. These boundary conditions (or forcings) are at the core of our dynamic definition of climate. The time-averaged planetary-scale circulation is an imposing response to these boundary conditions, influencing climates around the globe. For example, the planetary-scale circulation determines the locations of warm desert climates (e.g., the Sahara) and the monsoon climates that feature a dry season and rainy season (e.g., Southeast Asia).

Prior to describing the planetary-scale circulation, we first examine the various forces that initiate and shape the wind (air in motion). We consider each force separately and then show how they work together to control the speed and direction of the wind in weather systems such as anticyclones (highs) and cyclones (lows). The continuity and variability of the wind imply a linkage between the horizontal and vertical (up and down) motions of air and, as we saw in the previous chapter, vertical motion is key to cloud and precipitation development as well as expanses of clear, fair-weather skies.

We describe the various components of the planetary-scale circulation on the rotating Earth including the wind belts, semi-permanent pressure systems, and the intertropical convergence zone (ITCZ). The locations of these components shift north and south with the seasons, following the Sun, as they impose their distinctive signatures on regional climates. A special emphasis on the westerlies of middle latitudes focuses on the relationship between prevailing flow patterns and the occurrence of weather extremes such as drought and excessive heat or cold.

Wind: The Forces

Wind is the local motion of air measured relative to the rotating Earth; the atmosphere is an integral and coupled component of the rotating planet. Once every 24 hrs, all points on Earth's surface and in the atmosphere (except right at the poles) complete a circular path in space. In addition to this rotation, the atmosphere circulates in response to temperature gradients that develop within the planetary system. As described in Chapter 4, these temperature gradients are due to differences in rates of radiational heating and radiational cooling between (1) Earth's surface and atmosphere, and (2) the tropics and high latitudes. Circulation of the atmosphere and ocean ultimately transports heat from warmer locations to colder locations. In this section, we describe the principal forces operating in atmospheric circulation: the pressure gradient force, centripetal force, Coriolis Effect, friction, and gravity.

PRESSURE GRADIENT FORCE

Just as water pushes in on a submerged object from all directions, air exerts a force on the surfaces of all objects immersed in it. (A *force* is a push or pull on an object, and is computed as mass times acceleration. A force is a *vector* quantity; that is, it has both magnitude and direction.) As noted in Chapter 5, we can think of **air pressure** at a particular location in the atmosphere or at Earth's surface as the weight per unit area of the column of air above that location. Air is highly compressible. The weight (a force) of the air above compresses the atmosphere so that the maximum air density and pressure are at Earth's surface, while air density and pressure decrease rapidly with increasing altitude (Figure 6.3). The average air pressure at sea level is about 1013.25 millibars (mb). At an altitude of only 5500 m (18,000 ft), air pressure is about half of its average value at sea level. The rapid drop in air pressure with altitude means that significant changes in air pressure accompany relatively minor changes in land elevation. For example, the average air pressure at Denver, CO, the Mile-High City, is about 83% of the average air pressure at sea level.

Air pressure varies with both space and time, and these variations shape the circulation of the atmosphere. Variations in air pressure are not always due to topography. In fact, we are most interested in air pressure variations that arise from factors other than land elevation. Hence, weather observers determine an equivalent sea-level air pressure value; that is, for weather stations located above sea level, they adjust air pressure readings upward (using documented equations that describe the pressure change

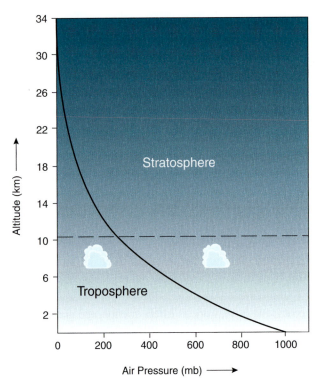

FIGURE 6.3
Average variation in air pressure in millibars (mb) with altitude (km). At sea level, the average air pressure is 1013.25 mb. The horizontal dashed line represents the tropopause. [Source: *U.S. Standard Atmosphere, 1976*, produced by NASA, NOAA, and U.S. Air Force]

with altitude) to approximate what the pressure would be if the station were actually located at sea level. When this adjustment to sea level is carried out everywhere, air pressure is observed to vary from one place to another and fluctuate from day to day; even from one hour to the next. Spatial and temporal changes in air pressure adjusted to sea level (or to any specific constant elevation) arise from variations in air temperature (principally), humidity, and atmospheric circulation.

In the free atmosphere, air density varies inversely with both temperature and humidity. That is, air density increases with falling temperature and decreasing humidity. Cold, dry air masses are denser and usually produce higher surface pressures than warm, humid air masses. Warm, dry air masses, in turn, often exert higher surface pressures than equally warm, but more humid air masses. As one air mass replaces another at a specific location, the air pressure at that location typically changes. Falling air pressure often signals a turn to stormy weather, whereas rising air pressure indicates clearing skies or continued fair weather.

A change in air pressure over some distance is known as an **air pressure gradient**. Air pressure gradients occur both vertically and horizontally within the atmo-

sphere. A vertical air pressure gradient is a permanent feature of the atmosphere because air pressure always decreases with increasing altitude (at a rate dependent on the density of the air column). A horizontal air pressure gradient refers to pressure changes along a surface of constant altitude (e.g., mean sea level). Horizontal air pressure gradients can be determined on weather maps from patterns of **isobars**, lines joining places having the same air pressure (adjusted to sea level). Usually isobars are drawn on weather maps at 4-millibar intervals (Figure 6.4).

On a weather map, a *HIGH* or *H* symbol designates places where sea-level air pressure is relatively high compared to the air pressure in surrounding areas. A high is also known as an **anticyclone** and is usually a fair weather system. A *LOW* or *L* symbol signifies regions where sea-level air pressure is relatively low compared to the air pressure in surrounding areas. A low is also known as a **cyclone** and often brings stormy weather.

In response to horizontal air pressure gradients, the wind blows from where the pressure is relatively high toward where the pressure is relatively low. The force that causes air to move in this way as the consequence of an air pressure gradient, known as the **pressure gradient force**, is always directed perpendicular to isobars and toward lowest pressure. The magnitude of the pressure gradient force is inversely related to the spacing of isobars. The wind is relatively strong where the pressure gradient is steep (closely spaced isobars), and light where the pressure gradient is weak (widely spaced isobars).

CENTRIPETAL FORCE

Isobars plotted on a surface weather map are almost always curved, indicating that the pressure gradient force changes direction from one place to another. Consequently, the horizontal wind blows in curved paths. Curved motion indicates the influence of the centripetal force.

A simple demonstration illustrates the centripetal force. A rock is tied to a string and then whirled about so that the tethered rock describes a circular orbit of constant radius (Figure 6.5). The string exerts a net force on the rock by confining it to a curved (circular) path. At any instant, the net force is directed inward, perpendicular to the direction of motion, and toward the center of the circular orbit. For this reason, the net force is known as the **centripetal** (*center-seeking*) **force**. If we cut the string, the centripetal force no longer operates; that is, a net force no longer confines the rock to a curved path. The rock flies off in a straight line as described by **Newton's first law of motion**; that is, an object in straight-line, un-accelerated motion remains that way unless acted upon by an unbalanced force.

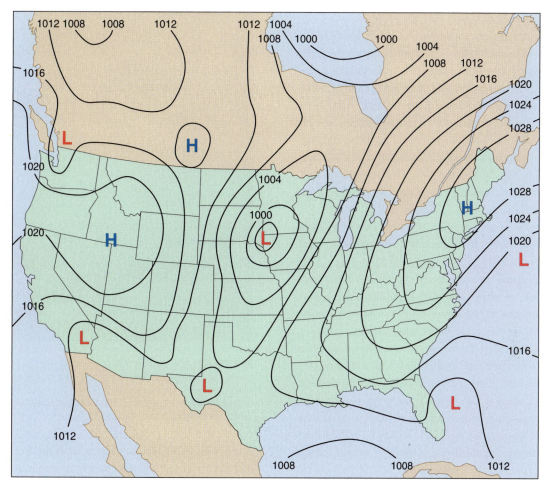

FIGURE 6.4
A surface weather map shows spatial variations in air pressure (reduced to sea level). Dark lines are isobars (in mb) passing through localities having the same air pressure. *L* is plotted where the air pressure is relatively low and *H* is plotted where the air pressure is relatively high.

A net (i.e., unbalanced) force causes acceleration. We usually think of acceleration as a change in speed, as

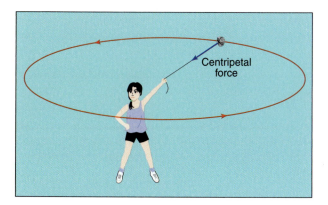

FIGURE 6.5
A rock tied to a string follows a circular path as it is whirled about. Centripetal (*center-seeking*) is the name of the force that confines an object to a curved path. If the string is cut, the centripetal force is eliminated, and the rock travels in a straight line (tangent to the circular path).

when an automobile speeds up. But velocity is a *vector* quantity; that is, it has both magnitude and direction. Acceleration is also a vector quantity and consists of a change in either speed or direction, or both. In our rock-on-a-string example, the centripetal force is responsible only for a continual change in the direction of the rock (curved rather than straight-line path); the rock neither speeds up nor slows down. The acceleration imparted to a unit mass by the centripetal force is directed toward the center of curvature, with a magnitude directly proportional to the square of the speed and inversely proportional to the radius of curvature of the path.

The centripetal force is not an independent force; rather, it arises from the action of other forces and may be the consequence of imbalances in other forces. In our rock-on-a-string example, the tension of the string is responsible for the centripetal force. Within the atmosphere, a centripetal force operates whenever air parcels follow a curved path.

CORIOLIS EFFECT

If Earth did not rotate, surface winds would blow directly from the cold poles (where surface air pressure is relatively high) to the hot equator (where surface air pressure is relatively low) in response to horizontal pressure gradient forces directed toward the equator. And these winds would push ocean surface currents directly toward the equator. But because Earth rotates, anything moving freely over the planet's surface, including air and water, is deflected to the right in the Northern Hemisphere and to the left in the Southern Hemisphere (Figure 6.6). This deflection is known as the **Coriolis Effect**, named for Gaspard-Gustave de Coriolis (1792-1843), the French mathematician who first described the phenomenon quantitatively in 1835.

As noted above, according to *Newton's first law of motion*, an object in constant, straight-line motion remains that way unless acted upon by an unbalanced force. Winds in the atmosphere (and currents moving in the ocean) exhibit this behavior. But these motions occur on a rotating Earth so that as air (or water) moves in a straight line, Earth rotates beneath the moving air (or water). Except at the equator, the wind is displaced from a straight-line path when its motion is measured with respect to the rotating Earth (Figure 6.7). While no force is causing this turning motion, this inconsistency can be incorporated in Newton's first law of motion by explaining the deflection to be the result of an imaginary

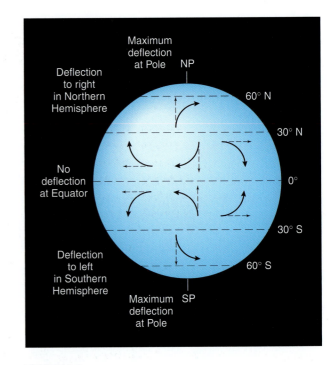

FIGURE 6.6
Large-scale winds are deflected to the right of their initial direction in the Northern Hemisphere and to the left of their initial direction in the Southern Hemisphere. This deflection, known as the Coriolis Effect, varies from zero at the equator to a maximum at the poles.

force. This deflection is referred to as the Coriolis Effect and the apparent force invented to describe its magnitude and direction is called the *Coriolis Force*.

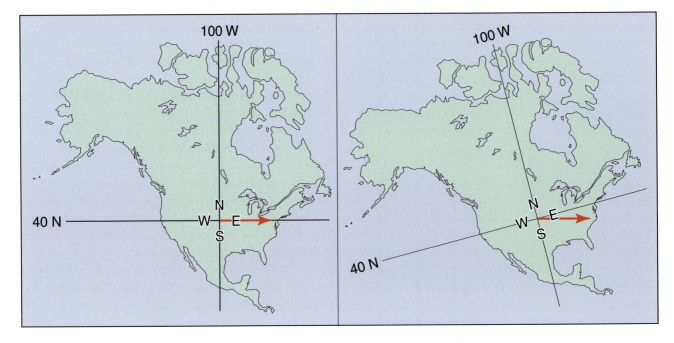

FIGURE 6.7
Viewed from a fixed point in space, the familiar north-south, east-west frame of reference rotates eastward in space as Earth rotates on its axis. Rotation of this coordinate system gives rise to the Coriolis Effect.

Reversal in the direction of the Coriolis Effect between the Northern and Southern Hemispheres is related to differences in our perspective of Earth's rotation direction in the two hemispheres. To an observer looking down from high above the North Pole, the planet rotates counterclockwise, whereas to an observer looking down from high above the South Pole, the planet rotates clockwise.

For an Earth-bound observer, this reversal in the sense of Earth's rotation between the two hemispheres translates into a reversal in Coriolis Effect. That is, moving air (and water) is deflected to the right in the Northern Hemisphere, but to the left in the Southern Hemisphere. Because of these hemispheric differences, we refine the definition of the Coriolis Effect as always acting at 90 degrees and to the right of the direction of motion in the Northern Hemisphere, and at 90 degrees and to the left in the Southern Hemisphere.

Although the Coriolis Effect influences the wind blowing in any direction, the amount of deflection varies significantly with latitude; that is, the magnitude of the Coriolis Effect varies from zero at the equator to a maximum value at the poles. This variation with latitude can be understood by visualizing the daily rotation of towers about their vertical axes when located at different latitudes. In a 24-hr day, Earth completes one rotation, as would towers located at the North Pole and South Pole. In the same period, a tower at the equator would not rotate at all about its vertical axis because of its orientation perpendicular to Earth's axis of rotation. At any latitude in between, some rotation of a tower occurs but not as much as at the poles.

The magnitude of the Coriolis Effect also varies with wind speed and spatial scale of atmospheric circulation. The Coriolis Effect increases as the wind strengthens because, in the same time interval, faster moving air parcels cover greater distances than slower moving air parcels. The longer the trajectory, the greater is the rotation of the underlying Earth. The Coriolis Effect significantly influences the wind only in large-scale weather systems, that is, systems larger than ordinary thunderstorms. Large-scale weather systems also have longer life expectancies than small-scale systems so that air parcels cover greater distances over longer periods of time, allowing the impact of Earth's rotation to manifest itself.

FRICTION

Friction is the resistance that an object or medium encounters as it moves in contact with another object or medium. We are familiar with the friction or resistance to

motion associated with solid objects, as when we attempt to slide a heavy appliance across the floor. But friction also affects fluids, both liquids and gases. The friction of fluid flow, known as **viscosity**, is of two types: molecular viscosity and eddy viscosity. One source of fluid friction is the random motion of molecules composing a liquid or gas; this type of fluid friction is called **molecular viscosity**. Considerably more important, however, is fluid friction that arises from much larger irregular motions, called *eddies*, which develop within fluids; this type of fluid friction is known as **eddy viscosity**.

A swiftly flowing stream illustrates the effects of eddy viscosity. Rocks in the streambed obstruct the flow of water causing the current to break into eddies immediately downstream of the rocks (Figure 6.8). Eddies, visible as swirls of water, tap some of the stream's kinetic energy so that the stream slows. In an analogous manner, obstacles on Earth's surface such as trees and houses break the wind into eddies, of various sizes, to the lee of each obstacle. Consequently, the near surface wind slows.

The rougher the surface of the Earth, the greater is the eddy viscosity of the wind. A forest thus offers more frictional resistance to the wind than does the smoother surface of a freshly mowed lawn. Eddy viscosity diminishes rapidly with altitude above Earth's surface, away from obstacles mainly responsible for frictional resistance. Hence, horizontal wind speed increases with altitude. Above an average altitude of about 1000 m (3300 ft), friction is a minor force that has little impact on the smooth flow of air. The atmospheric zone to which frictional resistance (eddy viscosity) is essentially confined is called the **atmospheric boundary layer**.

FIGURE 6.8
Rocks in a riverbed are obstacles that break the current into turbulent eddies downstream from the rocks, slowing the stream flow. A similar frictional interaction takes place as wind encounters obstacles on Earth's surface.

GRAVITY

Air parcels, like all other objects with mass are subject to the force of **gravity**, and are pulled towards Earth. Gravity is the net result of two forces working together: gravitation and centripetal force. *Gravitation* is the force of attraction between Earth and some object; its magnitude is directly proportional to the product of the masses of Earth and the object, but inversely proportional to the square of the distance between their centers of mass. The much weaker *centripetal force* is imparted to all objects because of their rotation with Earth on its axis. Combined as gravity, the two forces accelerate a unit mass of any object toward Earth's surface at the rate of 9.8 m per sec each second.

Gravity always acts directly downward. For this reason, gravity, unlike the Coriolis Effect and friction, does not modify the horizontal wind. Gravity influences air that is ascending or descending, such as the updrafts and downdrafts in convection currents, and gravity is responsible for the downhill drainage of cold, dense air.

SUMMARY

Having examined individually the various forces that influence the wind, we can draw the following conclusions:

1. The horizontal pressure gradient force, which is responsible for initiating essentially all air motion, accelerates air parcels perpendicular to isobars away from regions of high air pressure and toward regions of low air pressure. The magnitude of the force is directly proportional to the pressure gradient; that is, the closer the spacing of isobars, the greater is the magnitude of the pressure gradient force.

2. A centripetal force is an imbalance of actual forces and exists whenever the wind describes a curved path. It is responsible for a change in wind direction, but not wind speed.

3. The Coriolis Effect arises from the rotation of Earth on its axis and deflects large-scale winds to the right of their initial direction in the Northern Hemisphere, but to the left of their initial direction in the Southern Hemisphere. The Coriolis Effect increases with latitude from zero at the equator to a maximum at the poles and is directly proportional to wind speed.

4. Friction always opposes motion, acting opposite to the wind direction and increasing with increasing surface roughness. Friction slows horizontal winds blowing within about 1000 m (3300 ft) of Earth's surface.

5. Gravity pulls air downward toward Earth's surface and because of its vertical direction does not modify the horizontal wind.

Wind: Joining Forces

To this point in our discussion, we have examined forces operating in the atmosphere as if each force acted independently of all the others. In reality, these forces act together in governing both wind speed and direction. In some cases, two or more forces achieve a balance or equilibrium. From *Newton's first law of motion*, when the forces acting on an air parcel are balanced, no net force operates and the parcel either remains stationary or continues to move along a straight path at constant speed. When forces are balanced, the net acceleration is zero.

In this section, we build an increasingly sophisticated model of atmospheric motion by examining how forces interact in the atmosphere to control the vertical and horizontal motions of air, that is, the wind. These interactions are responsible for (1) the geostrophic wind, (2) the gradient wind, and (3) surface winds (horizontal winds within the atmospheric boundary layer).

GEOSTROPHIC WIND

The existence of air pressure gradients in a horizontal direction would result in a wind blowing from high toward low pressure. Inspection of upper-air weather maps shows that horizontal winds above the atmospheric boundary layer tend to blow parallel to isobars with low pressure to the left in the Northern Hemisphere. In an attempt to explain this observation, we make a set of assumptions contained in the geostrophic wind model. The term *geostrophic* means *Earth turning*. This frictionless model assumes that isobars are straight and parallel.

The **geostrophic wind** is an un-accelerated, horizontal movement of air that follows a straight path at altitudes above the atmospheric boundary layer. It results from a balance between the horizontal pressure gradient force and the Coriolis Effect. The Coriolis Effect is significant only in broad-scale circulations so that the geostrophic wind develops only in large-scale weather systems.

Consider the traditional description of the evolution of geostrophic equilibrium: An air parcel is placed in a horizontal pressure field where isobars are straight and parallel (Figure 6.9). In response to the horizontal pressure gradient force (P_H), the air parcel initially accelerates directly across isobars from high pressure toward low pressure. As the air parcel accelerates, however, the Coriolis Effect (C) comes into play, strengthens and causes the air parcel to turn gradually to the right of its initial flow direction (in the Northern Hemisphere). The Coriolis Effect changes direction as the parcel turns, always remaining at

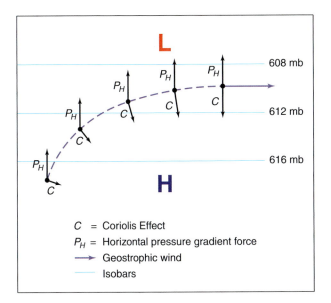

FIGURE 6.9
This diagram approximates the evolution of geostrophic equilibrium, a balance between the horizontal pressure gradient force and the Coriolis Effect. The geostrophic wind blows parallel to straight isobars at altitudes above the atmospheric boundary layer.

right angles to the parcel's direction of motion. The parcel continues turning until the two forces are acting in opposite directions and balancing one another, so that the geostrophic wind blows at a constant speed in a straight path parallel to isobars with the lowest air pressure to the left of the direction of air motion. In the Southern Hemisphere, the Coriolis Effect causes the air parcel to turn to the left until the flow is parallel to isobars and the lowest air pressure is to the right.

GRADIENT WIND, HIGHS, AND LOWS

A more common model of atmospheric flow is the **gradient wind** that shares many of the same characteristics as the geostrophic wind. It is also large-scale, horizontal, and frictionless, and blows parallel to isobars. However, the important distinction between the two models is that the geostrophic wind blows in a straight path, whereas the path of the gradient wind is curved. Forces are not balanced in the gradient wind because a net centripetal force constrains air parcels to a curved trajectory. Recall from our earlier discussion that the centripetal force changes only the direction and not the speed of an air parcel. The horizontal pressure gradient force, the Coriolis Effect, and the centripetal force interact in the gradient wind.

A gradient wind develops at altitudes above the atmospheric boundary layer around a dome of high air pressure, called an anticyclone (or *High*), or around a

center of low air pressure, called a cyclone (or *Low*). In an idealized anticyclone, isobars form a series of concentric circles about the location of highest air pressure, as shown in Figure 6.10. Under gradient wind conditions, the horizontal pressure gradient force (P_H) is directed radially outward, away from the center of the high. The Coriolis Effect (C) is directed inward. The Coriolis Effect is slightly greater than the pressure gradient force, with the difference between the two forces giving rise to the inward-directed centripetal force (C_E). (This is what was meant earlier when we indicated that a centripetal force results from an imbalance of other forces.) Viewed from above in the Northern Hemisphere, the gradient wind in an anticyclone blows clockwise and parallel to the curved isobars above the atmospheric boundary layer.

In an idealized cyclone, isobars form a series of concentric circles about the location of lowest air pressure. As indicated in Figure 6.11, the horizontal pressure gradient force (P_H) is directed inward toward the cyclone center, and the Coriolis Effect (C) is directed radially outward from the center of the low. The pressure gradient force is slightly greater than the Coriolis Effect,

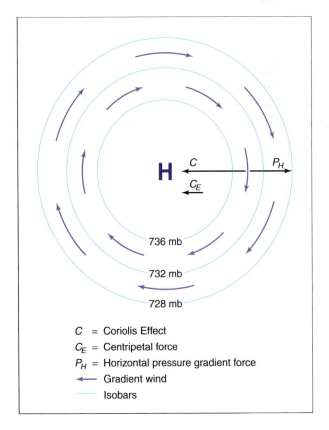

FIGURE 6.10
Viewed from above in the Northern Hemisphere, the gradient wind blows clockwise and parallel to isobars in an anticyclone. In this idealized case, isobars form a pattern of concentric circles.

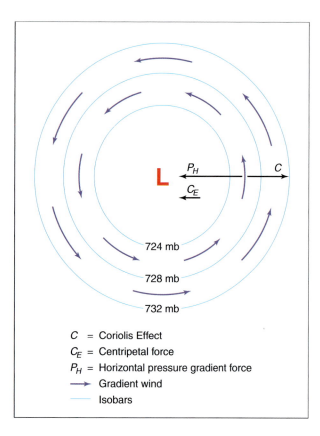

FIGURE 6.11
Viewed from above in the Northern Hemisphere, the gradient wind blows counterclockwise and parallel to isobars in a cyclone. In this idealized case, isobars form a pattern of concentric circles.

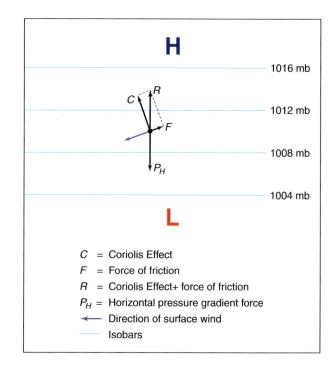

FIGURE 6.12
Within the atmospheric boundary layer, friction slows the wind and shifts the wind across isobars toward low pressure.

with the difference being equal to the net inward-directed centripetal force (C_E). Viewed from above, in the Northern Hemisphere, the gradient wind in a cyclone blows counterclockwise and parallel to the curved isobars above the atmospheric boundary layer.

The geostrophic and gradient wind models only approximate the actual behavior of horizontal winds above the atmospheric boundary layer. These approximations are nonetheless quite useful, and atmospheric scientists routinely rely on such approximations in analyzing weather maps.

SURFACE WINDS IN HIGHS AND LOWS

Geostrophic and gradient winds are frictionless; that is, they occur at altitudes where frictional resistance is insignificant. How does friction affect horizontal winds within the atmospheric boundary layer? Intuitively, we know that friction should slow the wind, but in addition, friction interacts with the other horizontal forces and alters the wind direction.

As shown in Figure 6.12, for large-scale air motion along a straight path, the frictional force (F)

combines with the Coriolis Effect (C) to balance the horizontal pressure gradient force (P_H). Friction always acts directly opposite (180 degrees to) the wind direction whereas the Coriolis Effect is always at a right angle (90 degrees) to the wind direction. Friction slows the wind and thereby weakens the Coriolis Effect so that the Coriolis no longer balances the horizontal pressure gradient force. The horizontal pressure gradient force (P_H) is balanced by the resultant (R) of the Coriolis (C) plus friction (F). Friction (due to the roughness of Earth's surface) slows the horizontal wind and shifts the wind direction obliquely across isobars and toward lower pressure.

The angle between near-surface wind direction and isobars depends on the roughness of Earth's surface. That angle varies from 10 degrees or less over relatively smooth surfaces, such as over the ocean, where friction is minimal, to almost 45 degrees over rough terrain, where friction is greater, such as over a forest. As noted earlier, friction's influence on the horizontal wind diminishes with altitude and where it becomes negligibly small marks the top of the atmospheric boundary layer. Thus horizontal winds strengthen with altitude through the atmospheric boundary layer. Furthermore, the angle between wind direction and isobars is greatest near Earth's surface, decreases with altitude, and is essentially zero at the top of the atmospheric boundary layer (Figure 6.13). Above

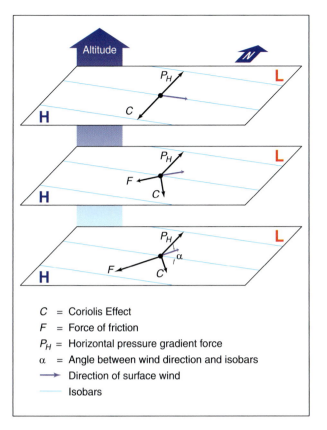

FIGURE 6.13
For the same horizontal air pressure gradient, the angle between the wind direction and isobars decreases with altitude within the atmospheric boundary layer.

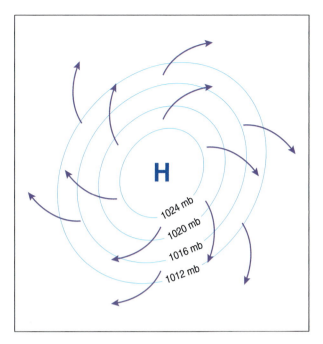

FIGURE 6.14
Viewed from above, in the Northern Hemisphere surface winds blow clockwise and outward in an anticyclone.

blow counterclockwise and spiral outward. Above the atmospheric boundary layer, Southern Hemisphere cyclonic winds blow clockwise and parallel to isobars, whereas Southern Hemisphere anticyclonic winds blow counterclockwise and parallel to isobars.

the atmospheric boundary layer, the horizontal wind is either geostrophic (where isobars are straight) or gradient (where isobars are curved).

How does surface roughness affect horizontal surface winds blowing in an anticyclone and cyclone? As with surface winds in a pressure field in which isobars are straight and parallel, friction slows anticyclonic and cyclonic winds and combines with the Coriolis Effect to shift horizontal winds so that they blow across isobars and toward low pressure. Viewed from above, surface winds in a Northern Hemisphere anticyclone blow clockwise and spiral outward, as shown in Figure 6.14, and surface winds in a Northern Hemisphere cyclone blow counterclockwise and spiral inward, as shown in Figure 6.15.

In the Southern Hemisphere, anticyclonic and cyclonic winds blow opposite their Northern Hemisphere counterparts. The difference is due to the Coriolis Effect acting to the left of the direction of motion in the Southern Hemisphere, opposite to its effect in the Northern Hemisphere. Viewed from above in the Southern Hemisphere, surface winds in a cyclone blow clockwise and spiral inward, whereas surface winds in an anticyclone

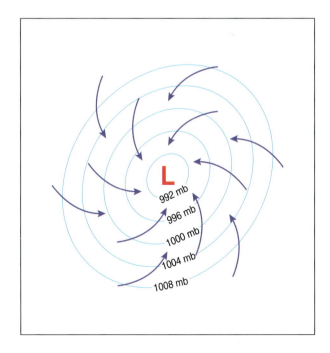

FIGURE 6.15
Viewed from above, in the Northern Hemisphere surface winds blow counterclockwise and inward in a cyclone.

FIGURE 6.16
On a typical surface weather map, isobars (lines of equal air pressure) exhibit clockwise curvature (*ridges*) and counterclockwise curvature (*troughs*).

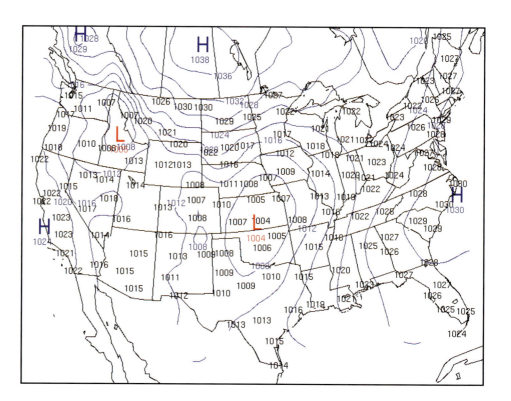

A glance at almost any national weather map reveals that isobars seldom describe lengthy straight segments or circular patterns (Figure 6.16). Isobars often form more complicated patterns of *ridges* (anticyclonic curves) and *troughs* (cyclonic curves). In ridges and troughs, winds tend to parallel isobars above the atmospheric boundary layer and cross isobars toward low pressure near Earth's surface. An additional consideration in analyzing isobaric patterns for wind is the spacing of isobars. The greater the horizontal air pressure gradient, the faster is the wind. Where isobars are closely spaced, the geostrophic and gradient winds are relatively strong. Where isobars are widely spaced, these winds are weak. The same rule applies to surface winds.

Continuity of Wind

Like all fluids, air is continuous (no voids) and this *continuity* implies a link between the horizontal and vertical components of the wind. For example, surface winds are forced to follow Earth's undulating topography, ascending hills and descending into valleys. In addition, uplift occurs along frontal surfaces as one air mass advances and either overrides or pushes under another retreating air mass (Chapter 5). Having examined the horizontal circulation in anticyclones and cyclones, we can identify other important connections between the horizontal and vertical components of the wind.

As noted earlier in this chapter, surface winds in a Northern Hemisphere anticyclone spiral clockwise and outward from its high pressure center. Consequently, horizontal surface winds diverge away from the center of the high. A vacuum does not develop at the center, however, because air descends towards Earth's surface and replaces the air that is diverging. Aloft, horizontal winds converge above the center of the surface high to replace the descending air (Figure 6.17). Recall from Chapter 5 that adiabatic compression raises the temperature and saturation vapor pressure of descending air, lowering the relative humidity and causing clouds to vaporize.

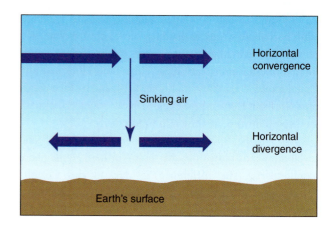

FIGURE 6.17
In this idealized vertical cross-section of an anticyclone, horizontal winds converge aloft, air descends, and surface winds diverge.

Skies therefore tend to be clear within anticyclones, and anticyclones are appropriately described as fair weather systems. Furthermore, within an anticyclone, the horizontal air pressure gradient is typically very weak over a broad area around the center of the system. Light winds or calm air coupled with clear skies and low humidity favor intense nocturnal radiational cooling. Air adjacent to the ground may be chilled to saturation so that nighttime dew, frost, or fog may develop. As discussed later, air masses develop under large, slow moving high pressure systems because the air is relatively uniform in temperature and humidity over a broad area.

Surface winds in a Northern Hemisphere cyclone spiral counterclockwise and inward. Surface winds therefore converge toward the center of a low. Air does not simply pile up at the center of the system; rather, air ascends in response to converging surface winds and diverging winds aloft (Figure 6.18). Recall from Chapter 5 that adiabatic expansion of ascending air lowers the temperature and saturation vapor pressure, thereby increasing the relative humidity of unsaturated air. Clouds and precipitation may eventually develop, so that cyclones are typically stormy weather systems. Because air flows into a low pressure system from all directions, at middle and high latitudes, these weather systems tend to bring together different air masses forming fronts.

Continuity of the wind also means that vertical motion can accompany downwind changes in surface roughness. The rougher the Earth's surface, the greater the resistance it offers to horizontal winds. When the

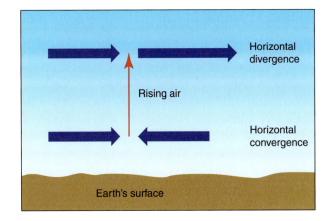

FIGURE 6.18
In this idealized vertical cross-section of a cyclone, surface winds converge, air ascends, and winds aloft diverge.

horizontal wind blows from a rough surface to a relatively smooth surface, as when it blows from land to sea, the wind accelerates. As shown in Figure 6.19, this acceleration causes the wind to diverge (stretch), thereby inducing downward motion of air. In contrast, when the horizontal wind blows from a smooth to a rough surface, the wind slows and converges (piles up), thereby inducing upward air motion. This is one reason why, along a coastline, cumuliform clouds (e.g., cumulus) tend to develop with an onshore wind (directed from sea to land) and tend to dissipate with an offshore wind (directed from land to sea). Frictionally-induced convergence of surface winds also plays an important role in the development of lake-effect snow (Chapter 7).

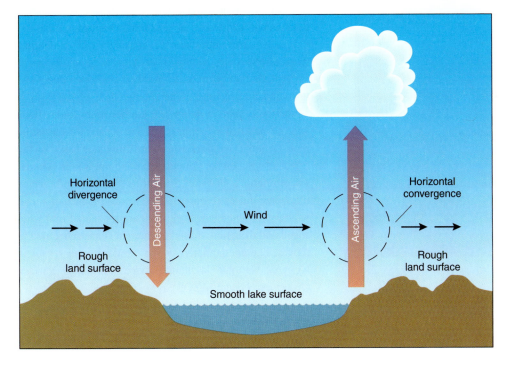

FIGURE 6.19
Surface winds accelerate and undergo horizontal divergence when blowing from a rough to a smooth surface (e.g., from land to water). Surface winds slow and undergo horizontal convergence when blowing from a smooth to a rough surface (e.g., from water to land). Divergence of surface winds causes air to descend whereas convergence of surface winds causes air to ascend.

Wind Measurement

A distinction is usually made between the *horizontal* and *vertical* components of the wind. Except in small, intense weather systems such as thunderstorms, the magnitude of vertical air motion is typically only 1% to 10% of the horizontal wind speed. Nonetheless, as demonstrated in Chapter 5, the vertical component of the wind plays the key role in cloud formation with ascending air that promotes expansional cooling and in creating broad expanses of fair weather by the compressional heating of sinking air. Furthermore, as described above, vertical and horizontal components of the wind are linked so that a change in one may be accompanied by a change in the other.

The most common wind-monitoring instruments are designed to measure only the horizontal component of the wind. For some specialized research purposes, very sensitive instruments are available that measure vertical wind speeds or a combination of the vertical and horizontal components of moving air. An ordinary **wind vane** consists of a free-swinging horizontal shaft with a vertical plate at one end and a counterweight (arrowhead) at the other end (Figure 6.20). The counterweight always points directly into the wind. Wind direction is always

FIGURE 6.21
A cup anemometer measures wind speed; the greater the wind speed, the faster the cups spin.

designated as the direction *from which* the wind blows. For example, a wind blowing from the east toward the west is described as an east wind. A wind vane may be linked electronically or mechanically to a dial that is calibrated to read and/or record in points of the compass or in degrees measured clockwise from true north. Hence, an east wind is specified as 90 degrees, a south wind as 180 degrees, a west wind as 270 degrees, and a north wind as 360 degrees. Wind direction is reported as 0 degrees only during calm conditions (when the wind speed is zero).

A **cup anemometer** consists of 3 or 4 open hemispheric cups mounted to spin horizontally on a vertical shaft (Figure 6.21). At least one open cup faces the wind at any time. The rotation rate of the cups is calibrated to read in m per sec, km per hr, or knots. (One *knot* = 1 nautical mi per hr = 0.515 m per sec = 1.15 statute mph.) Another wind-measuring instrument is based on the effect of wind on the propagation of sound waves. A **sonic anemometer** consists of three arms that send and receive ultrasonic pulses (Figure 6.22). The travel times of sound waves with and against the wind are translated into wind speed and direction. Sonic anemometers are scheduled to replace cup anemometers as the standard wind sensor in NOAA's National Weather Service Automated Surface Observing System (ASOS).

Ideally, a wind vane and anemometer system should be mounted on a tower so that the instruments monitor horizontal winds 10 m (33 ft) above the ground.

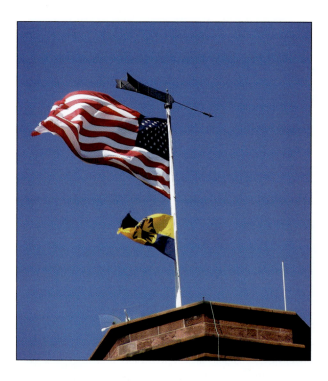

FIGURE 6.20
Wind vane atop the Smithsonian Building in Washington, DC. The arrow of the wind vane points in the direction from which the wind is blowing, as confirmed by the flags that are stretched out in the downwind direction. [Photo by Robert S. Weinbeck]

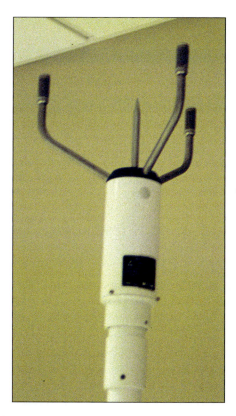

FIGURE 6.22
A sonic anemometer is based on the effect of wind on the propagation of sound waves. This instrument is scheduled to replace cup anemometers at NWS Automated Surface Observing Systems (ASOS).

This is the standard height for National Weather Service anemometers. Rooftop locations should be avoided because the wind tends to accelerate as it flows over buildings. In addition, the system should be sited well away from (1) structures that might shelter the instruments, and (2) any obstacles that might channel (and thus accelerate) the wind or alter its direction.

Wind vanes and anemometers monitor winds near Earth's surface but primarily over land. In the past, information on winds at sea came from infrequent and sometimes unreliable measurements by ships and instrumented buoys. Today, sensors known as scatterometers are flown on satellites (or aircraft) to monitor near-surface ocean wind speeds and directions accurately and continually. An example is NASA's QuikSCAT mission which launched SeaWinds in June 1999 (Figure 6.23). A scatterometer is a radar system that emits pulses of microwave energy to the sea surface where waves backscatter some of the energy to the instrument (an echo). The stronger the wind, the higher the waves, and the greater is the backscattering. Sea-surface wind vectors are determined from the strength of the echoes.

FIGURE 6.23
NASA's Quick Scatterometer, or "QuikScat," mission launched SeaWinds in June 1999 to measure remotely, near ocean-surface wind vectors using a scatterometer sensor. [NASA, Jet Propulsion Laboratory]

How are winds measured aloft? In Chapter 2, we saw that winds in the troposphere are measured by tracking the movements of a balloon-borne radiosonde (a *rawinsonde observation*). In the 1970s, sensors onboard geostationary satellites began to indirectly measure winds aloft over the ocean and other regions where weather observations are sparse. A sequence of satellite images of the same Earth view formatted as an animated loop is used to determine the speed of individual recognizable cloud elements that are assumed to be carried by the wind at a given level in the atmosphere. Since the early 1990s, the *Doppler Effect* (the same principle used to measure the speed of traffic or a pitched baseball) has been used in *wind profilers* to monitor winds up to an altitude of about 16,000 m (52,500 ft). This chapter's first Essay, Climatology of Wind Power, describes how modern technology is tapping the energy of air in motion.

Scales of Atmospheric Circulation

Although the atmosphere is a continuous fluid, for convenience of study we subdivide atmospheric circulation into discrete weather systems that operate at different spatial and temporal scales (Table 6.1). The large-scale wind belts encircling the planet (polar easterlies, westerlies of middle latitudes, and trade winds) are global or **planetary-scale systems**. **Synoptic-scale systems** are continental or oceanic in scale; extratropical cyclones, hurricanes, and air masses are examples. **Mesoscale systems** include thunderstorms and sea and lake breezes, circulation systems that are so small that they may influence the weather in only a portion of a large city or county. A weather system covering a very small area such as several city blocks or a small town represents the smallest spatial subdivision of atmospheric motion, **microscale systems**. A weak tornado is an example of a microscale system.

Circulation systems not only differ in spatial scale, they also contrast in life expectancy. Essentially the same pattern of planetary-scale circulation may persist for weeks or even months. Synoptic-scale systems typically last for several days to a week or so as they travel over distances of thousands of kilometers. Mesoscale systems usually complete their life cycles in a matter of hours to perhaps a day, whereas microscale systems might persist for hours/minutes or less. Other differences exist among the various scales of atmospheric circulation systems. At the micro- and meso-scale, vertical wind speeds at times may be comparable in magnitude to horizontal wind speeds. At the synoptic and planetary scales, however, horizontal winds are considerably stronger than vertical flow. Furthermore, at the micro- and meso-scale, the Coriolis Effect is usually negligibly small. By contrast, the Coriolis Effect is very important in synoptic- and planetary-scale circulation systems.

Each smaller scale weather system is part of, and dependent on, the larger scale atmospheric circulation; that is, the various scales of atmospheric motion form a kind of hierarchy. For example, extreme nocturnal radiational cooling requires a synoptic weather pattern that favors clear skies and light winds or calm air. At the micro-scale, such weather conditions may lead to formation of frost or fog in a river valley. But regardless of size, all scales of atmospheric circulation are ultimately governed by the boundary conditions of the planetary system.

Planetary-Scale Circulation

This section begins our description of atmospheric circulation, focusing initially on the planetary-scale and its relationship to Earth's mosaic of climate types. The atmosphere's planetary-scale circulation is shaped by boundary conditions that involve the forces discussed above. These boundary conditions include the decrease in solar radiation with increasing latitude, Earth's rotation on its axis (the Coriolis Effect), and the physical properties of Earth's surface (e.g., land versus ocean). The atmosphere's planetary-scale circulation plays an important role in the transport of heat poleward from energy surplus regions to energy deficit regions (Chapter 4) and the atmospheric transport of water as part of the global water cycle (Chapter 5).

BOUNDARY CONDITIONS

To understand how the planetary-scale circulation is shaped, we begin with an idealized model of planet Earth. Imagine Earth as a non-rotating sphere with a uniform solid surface. Also assume that the Sun heats the equatorial regions more intensely than the poles as it does on the real Earth. A temperature gradient develops between the equator and poles on the side of the Earth

---

TABLE 6.1
Scales of Atmospheric Circulation

Circulation	*Space scale*	*Time scale*
Planetary scale	10,000 to 40,000 km	weeks to months
Synoptic scale	100 to 10,000 km	days to a week
Mesoscale	1 to 100 km	hours to a day
Microscale	1 m to 1 km	seconds to hours

facing the Sun. In response, two huge convection cells form, one in each hemisphere (Figure 6.24A). Cold, dense air sinks at the poles and flows at the surface toward the equator, where it forces warm, less dense air to rise. Aloft, the equatorial air flows toward the poles, completing the convective circulation.

If our idealized planet begins to rotate, the Coriolis Effect comes into play (Figure 6.24B). In the Northern Hemisphere, surface winds observed relative to Earth's surface shift to the right and blow toward the southwest, and in the Southern Hemisphere, surface winds shift to the left and blow toward the northwest. Hence, on our hypothetical planet, surface winds blow counter to the planet's direction of rotation, which is from west to east. Circulation is maintained in the atmosphere of our idealized Earth because the planetary-scale winds split into three belts in each hemisphere, so

that some winds blow with and some winds blow against the planet's rotational direction (Figure 6.24C). In the Northern Hemisphere, average surface winds are from the northeast between the equator and 30 degrees latitude, from the southwest between 30 and 60 degrees, and from the northeast between 60 degrees and the North Pole. In the Southern Hemisphere, surface winds blow from the southeast between the equator and 30 degrees, from the northwest between 30 and 60 degrees, and from the southeast between 60 degrees and the South Pole.

Surface winds converge along the equator and along the 60-degree latitude circles. Convergence leads to ascending air, expansional cooling, cloud development, and precipitation. These convergence zones are belts of relatively low surface air pressure (Figure 6.24D). On the other hand, surface winds diverge at the poles and along the 30-degree latitude circles. In these regions, air descends,

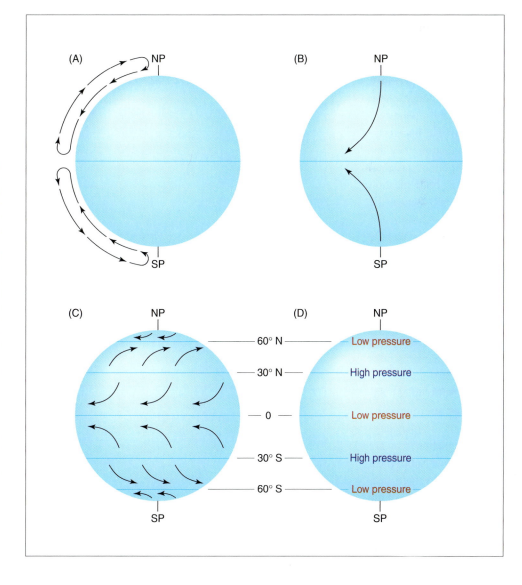

FIGURE 6.24
Planetary-scale atmospheric circulation on an idealized spherical model of Earth featuring a uniform solid surface. (A) If the sphere initially is non-rotating, huge convection currents develop on the sunlit portion of the planet's Northern and Southern Hemispheres so that air circulates between the hot equator and cold poles. (B) With a rotating Earth, surface winds blow from the northeast in the Northern Hemisphere and from the southeast in the Southern Hemisphere owing to the Coriolis Effect. (C) In reality, surface winds divide into three belts in each hemisphere of the rotating planet. (D) Zones of converging and diverging surface winds give rise to east-west belts of low pressure and high pressure. Adding the continents and ocean basins produces a more realistic model of the time-averaged planetary-scale circulation.

is compressed and warms, and the weather is generally fair. These divergence zones are belts of relatively high surface air pressure.

If continents and ocean basins are added to our idealized Earth, the temperature characteristics of Earth's surface become more complicated, and so do the planetary-scale air pressure pattern and winds. Some of the pressure belts break into separate cells, and important contrasts in air pressure develop over land versus sea. Now our idealized model of planetary-scale atmospheric circulation more closely approximates the actual time-averaged circulation pattern.

PRESSURE SYSTEMS AND WIND BELTS

Maps of global average air pressure at sea level for January and July reveal several areas of relatively high and low air pressure (Figure 6.25). These are **semipermanent pressure systems**. Although these systems are persistent features of the planetary-scale circulation, they undergo important seasonal changes in both location and strength, hence the modifier *semi-permanent*. Pressure systems include subtropical anticyclones, the intertropical convergence zone (ITCZ), subpolar lows, and polar highs. These pressure systems are in turn, linked by planetary-scale wind belts (Figure 6.26).

Subtropical anticyclones are imposing features of the planetary-scale circulation that are centered over subtropical latitudes (on average, near 30 degrees N and S) of the North and South Atlantic, the North and South Pacific, and the Indian

Ocean. These highs extend from the ocean surface up to the tropopause and exert major influences on climate over vast areas of the ocean and continents.

Stretching from the center of each subtropical high, outward over its eastern flank, are extensive areas of subsiding stable air. Subsiding air undergoes compressional warming, which produces low relative humidity and generally fair skies. The world's major subtropical deserts, including the Sahara of North Africa and the Sonora of Mexico and southwest United States, are located under the eastern flanks of subtropical anticyclones. On the far western portions of the subtropical highs,

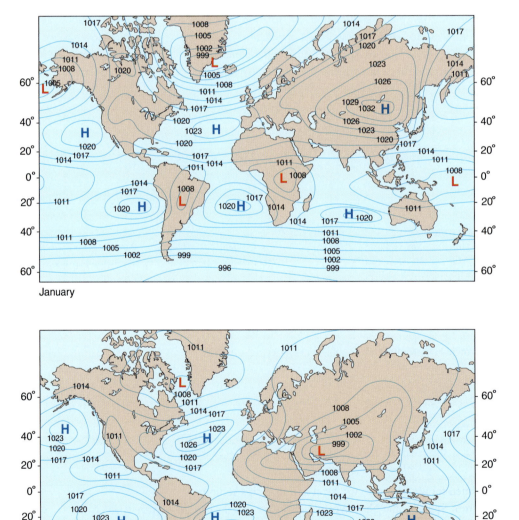

January

July

FIGURE 6.25
Mean sea-level air pressure during January and July. Contour lines are isobars in millibars (mb).

FIGURE 6.26
Schematic representation of the components of the time-averaged planetary-scale surface circulation of the atmosphere.

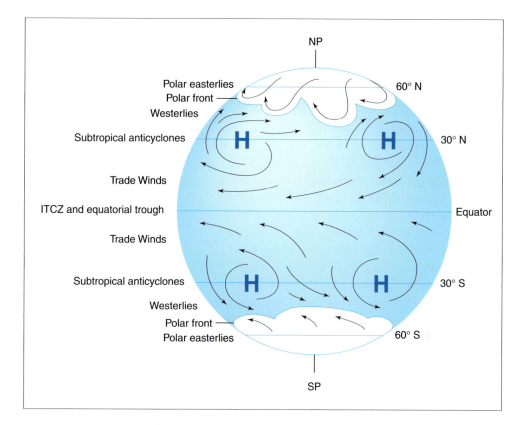

however, subsidence is less, the air is not as stable, and episodes of cloudy, stormy weather are more frequent. The contrast in climate between the eastern and western flanks of a subtropical high is apparent across southern North America. The climate of the American Southwest (on the eastern side of the North Pacific high, also called the *Hawaiian high*) is considerably drier than the climate of the American Southeast (on the western side of the North Atlantic high, also called the *Bermuda-Azores high*).

As is typical of anticyclones, a subtropical high features a weak horizontal air pressure gradient over a broad area surrounding the system's center. Hence, surface winds are very light or the air is calm over extensive areas of the subtropical ocean. This situation played havoc with ancient sailing ships, which were becalmed for days or even weeks at a time. Ships setting sail from Spain to the New World were often caught in this predicament, and crews were forced to jettison their cargo of horses when supplies of water and food ran low. For this reason, early mariners referred to this region of calm air as the **horse latitudes**, a name now applied to all latitudes between about 30 and 35 degrees N and S under subtropical highs.

In the Northern Hemisphere, viewed from above, surface winds blow clockwise and outward,

away from the centers of the subtropical highs, forming the westerlies and trade winds. Surface winds north of the horse latitudes constitute the highly variable **midlatitude westerlies** (which on average actually blow from the southwest). Surface winds blowing from the northeast out of the southern flanks of the anticyclones are known as the **trade winds**. The trades are the most persistent winds on the planet, in some regions blowing from the same direction more than 80% of the time. Analogous winds develop in the Southern Hemisphere, but recall that the Coriolis Effect is to the left in the Southern Hemisphere. A counterclockwise and outward surface airflow thus causes southeast trade winds on the northern flanks of the Southern Hemisphere subtropical highs and a belt of northwesterly winds on the southern flanks.

Mariners were aware of the westerlies and trade winds of the North Atlantic at least as early as the 15th century. On his venture to find a route to India, Christopher Columbus (1451-1506) took advantage of what we now know as the circulation about the Bermuda-Azores subtropical high. In his westward voyage, Columbus first sailed southward from Spain along the northwest African coast into the northeast trade winds that eventually took his ships westward to the Caribbean. On his return trip to Spain, he sailed north into the westerlies.

During the formative stages of their life cycles, tropical cyclones (such as tropical storms and hurricanes) are carried along by the trade winds. The position of the subtropical high pressure systems helps dictate the path of these cyclones, initially to the west and then poleward.

Trade winds of the two hemispheres converge into a broad east-west equatorial belt of light and variable winds, called the **doldrums**. In that belt, the converging air produces ascending air flow that induces cloudiness and rainfall. The most active weather develops along the **intertropical convergence zone (ITCZ)**, a discontinuous low-pressure belt with thunderstorms paralleling the equator. The average location of the ITCZ corresponds approximately to the latitude where Earth's mean annual surface temperature is highest, the so-called **heat equator**. Primarily because more land is in the Northern Hemisphere than in the Southern Hemisphere, the world-wide average location of the heat equator is near 10 degrees N latitude.

On the poleward side of the subtropical highs, surface westerlies flow into regions of low pressure, found primarily over subpolar ocean basins. In the Northern Hemisphere, there are two **subpolar lows**: the *Aleutian low* over the North Pacific Ocean and the *Icelandic low* over the North Atlantic Ocean. These pressure cells mark the convergence of the midlatitude southwesterlies with the polar northeasterly winds. By contrast, in the Southern Hemisphere, the midlatitude northwesterlies and the polar southeasterlies converge along a nearly continuous belt of low pressure surrounding the Antarctic continent (the *Antarctic circumpolar vortex*).

Surface westerlies meet and override the polar easterlies along the **polar front**. A *front* is a narrow zone of transition between air masses that differ in temperature, humidity, or both. In this case, dense, cold air masses flowing toward the equator meet milder, less dense air masses moving toward the pole. The polar front is not continuous around the globe; rather, it is well-defined in some areas and not in others, depending on the temperature contrast across the front. Where the temperature gradient across the boundary between air masses is great, such as in winter, the front is well defined and is a potential site for development of extratropical cyclones. On the other hand, where the air temperature contrast is minimal, as is often found in summer, the polar front is poorly defined and inactive or non-existent. The polar front is usually apparent on a surface weather map of North America, dividing colder air to the north from warmer air to the south. At any time, segments of

the polar front may be stationary or moving northward as a warm front or southward as a cold front.

At high latitudes, air subsides and flows away at the surface from the centers of shallow, cold anticyclones. In the Northern Hemisphere, **polar highs** are well developed only in winter over the continental interiors. In the Southern Hemisphere, cold highs persist over the glacier-bound Antarctic continent year-round.

WINDS ALOFT

What is the pattern of the planetary-scale winds aloft, that is, in the middle and upper troposphere? As noted earlier, air subsides in subtropical anticyclones, sweeps toward the equator as the surface trade winds, and then ascends in the ITCZ. Aloft, in the middle and upper troposphere, winds blow poleward, away from the doldrums and into the subtropical highs. The Coriolis Effect shifts these upper-level winds toward the right in the Northern Hemisphere (southwest winds), and toward the left in the Southern Hemisphere (northwest winds). In the tropics, therefore, winds aloft blow in a direction opposite that of the underlying surface trade winds.

A north-south vertical profile of this low-latitude circulation resembles a huge convection current (Figure 6.27). This circulation is known as the **Hadley cell**, named for the English meteorologist George Hadley (1685-1768), who first proposed its existence in 1735 as part of his explanation of the trade winds. Hadley cells are situated on either side of the ITCZ and extend poleward to the subtropical highs. It was once proposed that separate cells, similar to Hadley cells, occurred in both middle and polar latitudes. But upper-air monitoring does not provide definitive evidence of such well-defined cells.

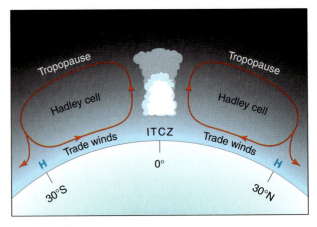

FIGURE 6.27
Idealized north-south vertical cross-section of the Hadley cell circulation in tropical latitudes of the Northern and Southern Hemispheres. Air rises in the Intertropical Convergence Zone (ITCZ) and sinks in the subtropical anticyclones.

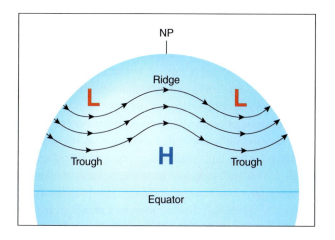

FIGURE 6.28
In the middle and upper troposphere, the Northern Hemisphere westerlies blow from west to east in a wave-like pattern of ridges (clockwise turns) and troughs (counterclockwise turns).

Aloft in middle latitudes, winds blow from west to east in a wavelike pattern of ridges and troughs, as shown in Figure 6.28. These winds are responsible for the development and movement of the synoptic-scale weather systems (highs, lows, air masses, and fronts) discussed in Chapter 7. Also, their north/south components contribute to poleward heat transport described in Chapter 4. Because the upper-air westerlies are so important for the weather and climate of middle latitudes, we examine them more extensively in a separate section of this chapter.

Figure 6.29 is a vertical cross-sectional profile of prevailing winds in the troposphere from the North

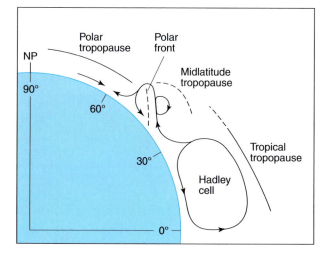

FIGURE 6.29
Vertical cross-section of the north-south (meridional) component of the atmospheric circulation of the Northern Hemisphere. Note that the tropopause occurs in three segments, with the highest altitudes in the tropics and lowest altitudes in the polar regions. The vertical scale is greatly exaggerated.

Pole to the equator. In this perspective, we are viewing only the north-south and vertical (up-down) components of the wind and are neglecting the west-east component. The altitude of the tropopause is directly related to the mean air temperature of the troposphere; that is, the lower the mean temperature, the lower is the altitude of the tropopause. Note that the altitude of the tropopause does not steadily increase from pole to equator but occurs in discrete steps. The tropical tropopause is at a higher altitude than the midlatitude tropopause, which, in turn, is at a higher altitude than the polar tropopause.

TRADE WIND INVERSION

The circulation on the eastern flank of subtropical anticyclones gives rise to the **trade wind inversion**, a persistent and climatically significant feature of the planetary-scale circulation over the eastern portions of tropical ocean basins. Key to formation of the trade wind inversion is the descending branch of the Hadley cell. As air subsides in the subtropical highs as part of the Hadley circulation, it is warmed by compression and its relative humidity decreases. Descending air encounters the *marine air layer*, a shallow layer of air that overlies the ocean surface. Where sea-surface temperatures (SST) are relatively low, the marine air layer is cool, humid, and stable. Where sea-surface temperatures are relatively high, the marine air layer is warm, more humid, less stable, and well mixed by convection.

Air subsiding from above is warmer (and much drier) than the upper portion of the marine air layer. The trade wind inversion forms at the altitude where air subsiding from above meets the top of the marine air layer. Within a temperature inversion, the air temperature increases with increasing altitude and an air layer characterized by a temperature inversion is extremely stable, strongly inhibiting vertical motion of air. The trade wind inversion is an elevated temperature inversion that essentially acts as a lid over the lower atmosphere.

As noted above, most subsidence occurs on the eastern flank of a subtropical high. A trade wind inversion develops to the east and southeast of the center of a subtropical high, in the region dominated by the trade winds. The inversion slopes downward from the center of the high toward the western coasts of the continents. For example, the trade wind inversion slopes downward from an average altitude of about 2000 m (6560 ft) over Hawaii to about 800 m (2400 ft) over coastal southern California.

Because of its extreme stability, the trade wind inversion limits the vertical development of convective clouds and rainfall, and inhibits formation of tropical storms and hurricanes. The inversion also limits orographic precipitation (Chapter 5). For example, some high volcanic peaks on the Hawaiian Islands protrude through the trade wind inversion into the dry air above. Because of orographic lifting, rainfall generally increases with elevation along windward slopes of the volcanic mountains but only up to the trade wind inversion. Above the inversion, conditions are dry. Traveling west of the high pressure center, the trade wind inversion continues to become elevated and weaker, essentially disappearing near Oceanus in the western Pacific. Convective clouds develop to higher altitudes, and thunderstorm activity is more frequent.

Seasonal Shifts and Climates

Between winter and summer, important changes take place in the planetary-scale circulation with implications for climates in various parts of the world. Pressure systems, the polar front, the planetary wind belts, and the ITCZ follow the Sun, shifting toward the poles in spring and toward the equator in autumn. Because the seasons are reversed in the two hemispheres, the planetary-scale systems of both hemispheres move north and south in tandem. In addition, the strength of pressure cells varies seasonally. Subtropical anticyclones, such as the Bermuda-Azores high, exert higher surface pressures in summer than in winter as they are displaced poleward. The Icelandic low deepens in winter, but greatly weakens in summer, and, though well developed in winter, the Aleutian low disappears in summer.

Seasonal reversals in surface air pressure also occur over the continents. These pressure changes stem from the contrast in solar heating of land versus sea. For reasons presented in Chapter 4, the ocean surface exhibits smaller temperature variations over the course of a year than does Earth's land surface. Hence, continents at middle and high latitudes are dominated by relatively high pressure in winter and relatively low pressure in summer. In winter, in response to extreme radiational cooling, cold anticyclones develop over northwestern North America and over the interior of Asia, the most prominent of which is the massive Siberian high. In summer, in response to intense solar heating, a belt of low pressure forms across North Africa and stretches from the Arabian Peninsula eastward into Southeast Asia. Warm low-pressure cells (called *thermal lows*) also develop in summer over arid and semi-arid regions of Mexico and the southwestern United States.

Seasonal shifts in the planetary-scale wind belts, pressure systems, and the ITCZ leave their mark on the world's climates. Northward migration of the ITCZ triggers summer monsoon rains in Central America, North Africa, India, and Southeast Asia. The *monsoon* is a large-scale circulation feature that has distinctive seasonal variations and is discussed in detail in Chapter 7. As shown in Figure 6.30, north-south movements of the ITCZ tend to be greater over continents than over the ocean. Over the ocean, the mean latitude of the ITCZ varies by only about 4 degrees through the year (between 4 degrees N in April and 8 degrees N in September). Anchoring of the ITCZ over the ocean is a consequence of the ocean's greater thermal inertia.

The influence of the subtropical anticyclones on precipitation regimes is illustrated by the contrast in the march of mean monthly precipitation between Charleston, SC, and San Diego, CA, (Figure 6.31). Although the two cities have about the same latitude, San Diego has a distinct rainy season (winter) and dry season (summer),

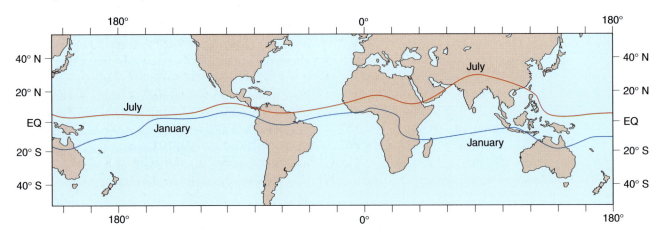

FIGURE 6.30
The Intertropical Convergence Zone (ITCZ) follows the Sun, reaching its most northerly latitudes in July and its most southerly latitudes in January.

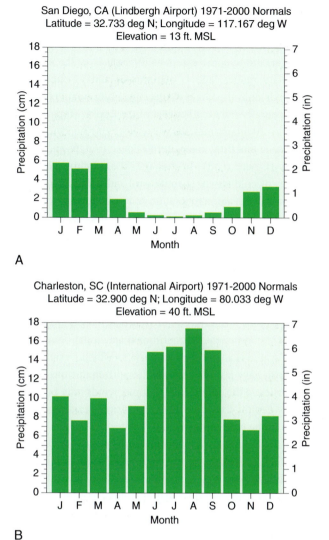

San Diego, CA (Lindbergh Airport) 1971-2000 Normals
Latitude = 32.733 deg N; Longitude = 117.167 deg W
Elevation = 13 ft. MSL

A

Charleston, SC (International Airport) 1971-2000 Normals
Latitude = 32.900 deg N; Longitude = 80.033 deg W
Elevation = 40 ft. MSL

B

FIGURE 6.31
Mean monthly precipitation (rain plus melted snow) in centimeters and inches for (A) San Diego, CA, and (B) Charleston, SC, for the period 1971-2000. San Diego has a distinct dry season (summer) when its weather is dominated by subsiding air on the eastern flank of the Hawaiian subtropical high. [NOAA, National Climatic Data Center]

whereas Charleston has a subtropical humid climate and receives abundant precipitation throughout the year. In summer, San Diego is under the dry eastern flank of the Hawaiian high, while Charleston is on the receiving end of the humid air flow on the western flank of the Bermuda-Azores high. Consequently, San Diego is dry while Charleston has frequent episodes of convective showers. After the subtropical highs shift toward the equator in autumn, the climate of both cities is influenced by widespread precipitation associated with west-to-east moving extratropical cyclones so that winters are wet in both places.

The climate of San Diego is similar to that which rims the Mediterranean Sea and is called a **Mediterranean climate**. This type of climate occurs on the western side of continents between about 30 and 45 degrees latitude. In North America, mountain ranges confine the Mediterranean climate to a narrow coastal strip of California. Although precipitation exhibits a regular seasonality (dry summers and wet winters), the temperature regime is quite variable. Along the coast, cool onshore breezes prevail, lowering the mean annual temperature and reducing seasonal temperature contrasts. Well inland, however, away from the ocean's moderating influence, summers are considerably warmer; hence, inland mean annual temperatures are higher and seasonal temperature contrasts are greater than in coastal areas. The climatic contrast between coastal San Francisco and inland Sacramento, CA, illustrates the variability of the temperature regime within Mediterranean climates. Although the two cities are only about 145 km (90 mi) apart, the climate of Sacramento is much more continental (much warmer summers and somewhat cooler winters) than the climate of San Francisco.

Westerlies of Middle Latitudes

The middle latitude westerlies of the Northern Hemisphere merit special attention here because they govern the weather and climate over much of North America. As noted earlier, in the middle and upper troposphere, the westerlies flow from west to east about the hemisphere in wavelike patterns of ridges and troughs. Winds exhibit a clockwise (anticyclonic) curvature in ridges, and a counterclockwise (cyclonic) curvature in troughs. Between two and five waves typically encircle the hemisphere at any one time. These long waves are called **Rossby waves**, after Carl-Gustav Rossby (1898-1957), the Swedish-American meteorologist who discovered them in the late 1930s. Rossby waves characterize the westerlies above the 500-mb level, that is, above the altitude where the atmospheric pressure is 500 mb. At lower levels, waves are distorted somewhat by friction and topographic irregularities of Earth's surface.

Atmospheric scientists describe the upper-air westerlies in terms of: (1) wavelength (distance between successive troughs or, equivalently, successive ridges), (2) amplitude (north-south extent), and (3) number of waves encircling the hemisphere. The westerlies exhibit changes in all three characteristics, as well as their positions relative to Earth's surface and, as a direct consequence, the weather changes. The westerlies are more vigorous in winter than in summer. In winter, they strengthen and exhibit fewer waves of longer length and greater amplitude. These seasonal changes stem from variations in the north-south air pressure gradient, which is steeper in winter because of the greater temperature contrast between north and south at that time of year. Furthermore, the location of the largest north-south temperature contrast follows the Sun. In summer, north-south temperature differences are less and shifted toward polar latitudes, horizontal air pressure gradients are weaker and displaced poleward, and so are the westerlies.

ZONAL AND MERIDIONAL FLOW PATTERNS

The weaving westerlies have two components of motion: a north-south wind superimposed on a west-to-east wind. The north-south airflow is the westerlies' *meridional component* and the west-to-east airflow is its *zonal component*. The meridional component of Rossby waves brings about a north-south exchange of air masses and poleward transport of heat (Chapter 4). In the Northern Hemisphere, winds blowing from the southwest carry warm air masses toward the northeast, and winds blowing from the northwest transport cold air masses southeastward. Cold air is exchanged for warm air and heat is transported poleward. However, as Rossby waves change length, amplitude, and number, concurrent changes take place in the advection of air masses. Consider some examples.

Occasionally, the westerlies blow almost directly from west to east, nearly parallel to latitude circles, with only a weak meridional component (Figure 6.32). This is a **zonal flow pattern** in which north-south exchange of air masses is minimal. Cold air masses stay to the north

while warm air masses remain in the south. At the same time, air that originated over the Pacific Ocean floods the contiguous United States and southern Canada. **Pacific air** dries out to some extent as it passes over the western mountain ranges, and is compressed and warmed as it descends onto the Great Plains, spreading relatively mild and generally fair weather east of the Rocky Mountains.

At other times, the westerlies exhibit considerable amplitude and flow in a pattern of deep troughs and sharp ridges (Figure 6.33). In this **meridional flow pattern**, masses of cold air surge southward and warm air streams northward. Greater temperature contrasts develop over the United States and southern Canada. Where contrasting air masses collide, warm air overrides cold air, and the stage is set for the development of extratropical cyclones that are then swept along by the westerlies.

These two illustrations of Rossby wave configurations are opposite extremes of a wide range of possible westerly wind patterns, each featuring different components of meridional and zonal flow. More complicated is a **split flow pattern**, in which westerlies to the north have a wave configuration different from that of westerlies to the south. For example, winds may be zonal across central Canada while winds are meridional over much of the coterminous United States.

The westerly wind pattern typically shifts back and forth between dominantly zonal and dominantly meridional flow. For example, zonal flow might persist for a week and then give way to a more meridional flow that lasts for a few weeks, before returning to zonal flow. The transition from one wave pattern to another is usually abrupt and sometimes takes place within a single day. This abruptness poses a challenge to weather forecasters because a sudden shift in the upper-air winds may divert

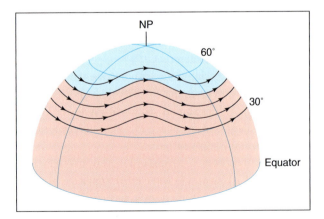

FIGURE 6.32
Midlatitude westerlies exhibit a zonal flow pattern when winds blow almost directly from west to east, with only a small meridional (north-south) component.

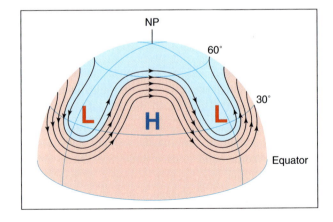

FIGURE 6.33
Midlatitude westerlies exhibit a meridional flow pattern when west to east winds have a strong meridional (north-south) component.

a cyclone toward or away from a locality or cause an unanticipated influx of colder air that could, for example, change rain to snow.

Unfortunately for the long-range weather forecaster, shifts in westerly wave patterns appear to have little regularity; that is, no predictable zonal/meridional cycle is readily apparent. The only observation useful to forecasters is that meridional patterns tend to persist for longer periods than zonal patterns. During the winter of 1976-77, for example, a strong meridional wave pattern persisted over North America from late October through mid-February. Northwesterly winds brought surge after surge of bitterly cold arctic air into the midsection of the United States resulting in one of the coldest winters of the 20th century for that region. During the same period, however, southwesterly winds brought unseasonably mild air to far-western North America, including Alaska.

BLOCKING SYSTEMS AND WEATHER EXTREMES

For the continent as a whole, North American weather is more dramatic when the westerlies are strongly meridional. Sometimes undulations of the westerlies become so great that huge whirling masses of air actually separate from the main westerly air flow. This situation, shown schematically in Figure 6.34, is analogous to whirlpools that form in rapidly flowing rivers. In the atmosphere, cutoff masses of air whirl in either a cyclonic or an anticyclonic direction (viewed from above). A *cutoff low* or a *cutoff high* that prevents the usual west-to-east movement of weather systems is referred to as a **blocking**

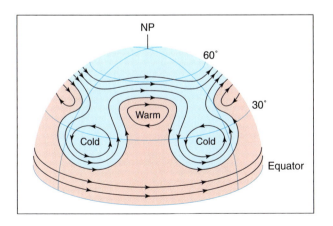

FIGURE 6.34
Blocking pattern in the middle latitude westerlies in which huge pools of rotating air are cut off from the main west to east air flow. The pool of relatively cool air rotating in a counterclockwise direction is a cutoff low, and the pool of relatively warm air rotating in a clockwise direction is a cutoff high. The latter system is sometimes referred to as an *omega block* because of its resemblance to the Greek letter.

system. Because a blocking circulation pattern tends to persist for extended periods (often several weeks or longer), extremes of weather such as drought or flooding rains or excessive heat or cold can result. Consider a few examples.

The drought that affected the Midwest and Great Plains during the spring and summer of 1988 is an example of the linkage between a weather extreme and a blocking weather pattern. (For a discussion of what constitutes drought, refer to this chapter's second Essay.) In Figure 6.35, the major upper-air circulation features that dominated the summer of 1988 are contrasted with long-term average conditions. From early May through mid-August, the prevailing westerlies were more meridional than usual and featured a huge stationary, warm high-pressure system over the nation's midsection and troughs over the West Coast and East Coast. The belt of strongest westerlies was displaced north of its usual location so that moisture-bearing weather systems were diverted into central Canada, well north of their normal paths. In the Corn Belt, the May through June period was the driest since 1895. By late July, drought was categorized as either severe or extreme over 43% of the land area of the coterminous United States. By the end of the growing season, the impact on the nation's grain harvest was severe: Corn production was down by 33%, soybeans by 20%, and spring wheat by more than 50%. The National Climatic Data Center (NCDC) estimated that the 1988 drought was one of the most costly weather-related disasters to hit the nation in the previous twenty-five years, resulting in total damage of $71.2 billion (in 2007 dollars).

A blocking circulation pattern during the summer of 1993 was responsible for record flooding in the Midwest and drought over the Southeast (Figure 6.36). A cold upper-air trough stalled over the Pacific Northwest and northern Rocky Mountains, bringing unseasonably cool weather to those regions. Meanwhile, the Bermuda-Azores high shifted west of its usual location over the subtropical Atlantic, causing the worst drought since 1986 over the Carolinas and Virginia. Between the northwestern trough and the subtropical high to the southeast, the principal storm track weaved over the Midwest. This circulation pattern persisted through June, July, and part of August, bringing a nearly continual procession of clusters of thunderstorms to that region. Across Iowa, Illinois, and Wisconsin, the June-July period was the wettest on record. Meanwhile over the Southeast, subsiding, stable air inhibited thunderstorm development so that hot and dry conditions prevailed through most of the summer.

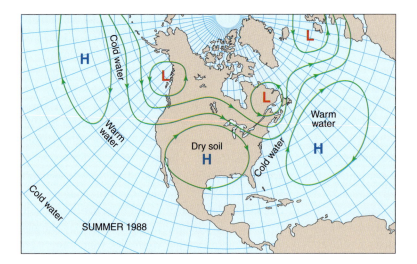

A

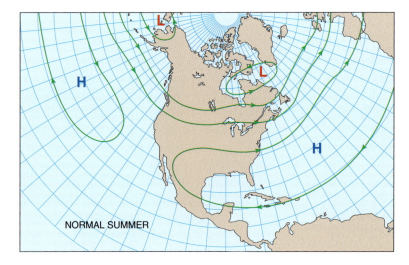

B

FIGURE 6.35
Prevailing circulation pattern in the mid to upper troposphere (A) during the summer of 1988 as compared to (B) the long-term average circulation pattern. The blocking warm anticyclone over the central U.S. contributed to severe drought.

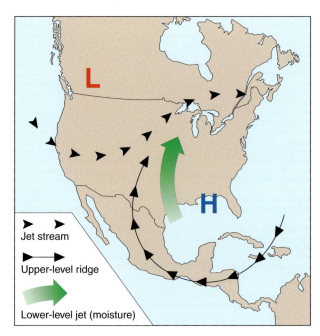

FIGURE 6.36
Principal features of the prevailing atmospheric circulation pattern during the summer of 1993.

Heavy rains falling on the drainage basins of the Missouri and Upper Mississippi River valleys saturated soils and triggered excessive runoff, all-time record river crests, and flooding that impacted all or part of nine states. Setting the stage for flooding in the upper Mississippi River valley was the wet autumn of 1992, heavy winter snowfall, and abundant spring snowmelt. The worst flooding occurred between Minneapolis, MN and Cairo, IL on the Mississippi River, and between Omaha, NE and St. Louis, MO on the Missouri River. Flooding was unprecedented in magnitude and persistence with some places reporting more than one record river crest. At St. Louis, the Mississippi River remained above the previous record crest for three weeks. Societal and economic impacts of the 1993 flood were devastating. In many cities (such as Des Moines, IA and St. Joseph, MO), the freshwater supply was cut off. Barge traffic was halted on the Upper Mississippi River and a portion of the Missouri River from late June until early August. At least 50,000 homes were damaged or destroyed; and more than 4 million hectares (10 million acres) of cropland were inundated. All told, property damage totaled $30.2 billion (in 2007 dollars), about one-third in crop losses. The death toll from flooding was 48.

In the summer of 1993, some parts of the Midwest received more than twice the long-term average seasonal rainfall while some localities in the Southeast received less than half of their long-term average seasonal rainfall (Figure 6.37). As a whole, the Southeast experienced its second driest July on record, and from Alabama and Georgia north to Tennessee and Virginia, July 1993 was the hottest on record. The combination of drought and heat stress caused severe crop damage, especially in South Carolina where over 95% of the corn crop and 70% of the soybean crop were lost. Total crop losses were estimated at $1.4 billion (in 2007 dollars).

As illustrated by weather events of the summers of 1988 and 1993, a persistent meridional flow pattern in the westerlies can block the usual movement of weather systems and cause weather extremes such as drought

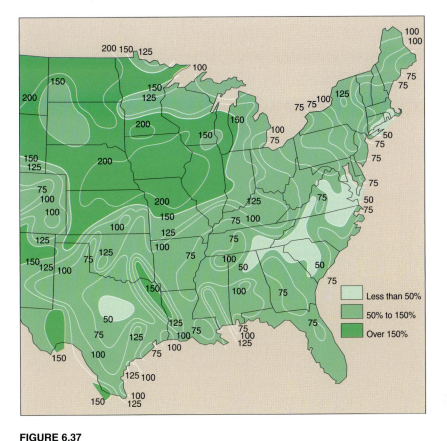

FIGURE 6.37
Total rainfall for the period June-August 1993, expressed as a percentage of the long-term average. [From NOAA, Climate Analysis Center]

or flooding rains. In fact, any westerly wave pattern, meridional or zonal, can cause extremes in weather if the pattern persists for a sufficiently long period of time. A persistent westerly wave pattern means the same type of air mass advection, the same storm tracks, and basically the same weather type.

Wind-Driven Ocean Gyres

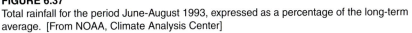

Surface ocean currents are wind-driven. Hence, the horizontal movement of ocean surface waters to a large extent mirrors the long-term average planetary-scale atmospheric circulation. Early on, mariners took advantage of surface ocean currents to hasten their voyage. The long-term average pattern of ocean currents is plotted in Figure 6.38.

The trade winds and the westerlies associated with the semi-permanent subtropical highs drive the **subtropical gyres**. The subtropical gyres are centered near 30 degrees latitude in the North and South Atlantic, North and South Pacific, and the Indian Oceans. Subtropical gyres are similar in the Northern and Southern Hemisphere

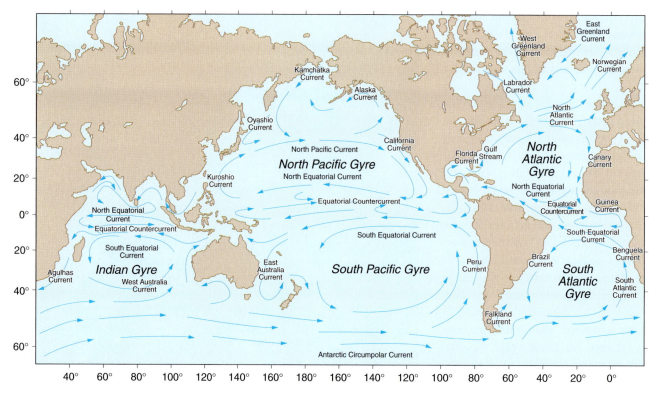

FIGURE 6.38
Long-term average pattern of wind-driven ocean-surface currents. Gyres in the ocean basin are driven by the planetary-scale atmospheric circulation.

except that they rotate in opposite directions because the Coriolis Effect acts in opposite directions in the two hemispheres. Viewed from above, subtropical gyres rotate in a clockwise direction in the Northern Hemisphere and in a counterclockwise direction in the Southern Hemisphere. As part of the subtropical gyre in the North Atlantic Ocean, the Gulf Stream is a warm current that moves northward along the East Coast of North America before turning east toward northwest Europe, while the cold southward flowing Canary Current moves along the African coast. Surface currents flow westward across the equatorial Atlantic. In the North Pacific Ocean, the Kuroshio Current moves northward off Asia, then travels eastward across the Pacific before turning south as the now cold California Current.

Sub-polar gyres, smaller than their subtropical counterparts, occur at high latitudes of the Northern Hemisphere; they are the Alaska gyre in the far North Pacific and the gyre south of Greenland in the far North Atlantic. The counterclockwise surface winds in the Aleutian and Icelandic sub-polar low pressure systems drive the sub-polar gyres. Hence, viewed from above, the rotation in these sub-polar gyres is opposite that of the Northern Hemisphere subtropical gyres.

Conclusions

The pressure gradient force, centripetal force, Coriolis Effect, friction, and gravity interact in initiating and shaping atmospheric circulation. Horizontal and vertical air motions in anticyclones and cyclones largely determine the type of weather associated with these pressure systems. With this as background, we examined the time-averaged planetary-scale atmospheric circulation. The principal features of that circulation are the ITCZ, trade winds, subtropical highs, the westerlies, polar front, subpolar lows, polar easterlies, and polar highs. We saw that aloft (in the middle and upper troposphere), winds blow counter to the surface trade winds in tropical latitudes and in a west-to-east wave pattern at middle and high latitudes. Through the course of a year, these components of the planetary-scale circulation shift in tandem north and south with the Sun and influence regional climates.

The next chapter focuses primarily on synoptic climatology, emphasizing the linkages between the westerly winds aloft and the weather and climate of middle latitudes. Specifically, the westerlies (1) produce horizontal divergence aloft for the development of cyclones, (2) steer storms, and (3) control air mass advection. We also explore monsoon climates and weather regimes that characterize certain mesoclimates.

Basic Understandings

- Wind is the movement of air measured relative to Earth's surface. Wind has both direction and magnitude and is usually divided into horizontal and vertical components, although atmospheric scientists often use the term "wind' when referring to the horizontal movement of air.

- The horizontal wind is governed by interactions of the pressure gradient force, centripetal force, Coriolis Effect, friction, and gravity.

- The pressure gradient force initiates air motion and arises in part from spatial variations in air temperature and, to a lesser extent, water vapor concentration. In response to air pressure gradients, air accelerates away from areas of relatively high air pressure, perpendicular to isobars, and toward areas of relatively low air pressure.

- The centripetal force operates whenever the wind follows a curved path; it is responsible for changes in wind direction but not changes in wind speed.

- The Coriolis Effect arises from Earth's rotation on its axis. Wind is deflected to the right of its initial direction of motion in the Northern Hemisphere, but to the left in the Southern Hemisphere. The force invented to describe and quantify this deflection is zero at the equator and increases with latitude to a maximum at the poles. The Coriolis Effect also increases as wind speed increases and is most important in large-scale (planetary- and synoptic-scale) circulation systems.

- Friction affects horizontal winds blowing within about 1000 m (3300 ft) of Earth's surface (the atmospheric boundary layer). Obstacles on Earth's surface slow the wind by breaking it into turbulent eddies.

- Gravity always pulls objects directly downward and is important in the vertical motion of air (e.g., cold air drainage).

- The geostrophic wind is an un-accelerated, horizontal wind that blows in a straight path parallel to isobars at altitudes above the atmospheric boundary layer. The geostrophic wind results from a balance between the horizontal pressure gradient force and the Coriolis Effect.

- The gradient wind is a horizontal wind that parallels curved isobars at altitudes above the atmospheric boundary layer. Viewed from above in the Northern Hemisphere, the gradient wind blows clockwise in anticyclones (*Highs*) and counterclockwise in cyclones (*Lows*).

- In large-scale (synoptic- and planetary-scale) circulation systems, friction slows the near-surface wind and interacts with the Coriolis Effect to shift the wind direction obliquely across isobars and toward lower pressure.

- Within the atmospheric boundary layer, viewed from above, horizontal winds blow clockwise and spiral outward in Northern Hemisphere anticyclones but counterclockwise and inward in Northern Hemisphere cyclones.

- In an anticyclone, horizontal divergence of surface winds causes air above to descend and warm by compression. Hence, an anticyclone is generally a fair-weather system. In a cyclone, horizontal convergence of surface winds causes ascending air, which expands and cools. Hence, a cyclone is frequently a stormy weather system.

- Along a coastline, surface winds blowing offshore speed up and undergo horizontal divergence (inducing descending air), whereas onshore winds are slowed by surface roughness and undergo horizontal convergence (inducing ascending air).

- A wind vane and anemometer are the usual instruments for monitoring surface wind direction and speed. Winds can be determined remotely by tracking rawinsondes, precipitation particles (by Doppler radar), cloud elements, and wind-driven ocean wave patterns (from sensors on Earth-orbiting satellites).

- Atmospheric circulation is typically divided into four spatial/temporal scales: planetary, synoptic, meso-scale, and micro-scale.

- Earth's planetary circulation regime is driven by the unequal distribution of solar radiation on a very nearly spherical planet, leading to higher air temperatures in tropical latitudes and lower air temperatures in polar latitudes. The resulting circulation is deflected by the relatively rapidly rotating planet (the Coriolis Effect).

- The principal features of the atmosphere's planetary-scale circulation are the intertropical convergence zone (ITCZ), trade winds, subtropical anticyclones, westerlies of middle latitudes, subpolar lows, polar front, polar easterlies, and polar highs. The ITCZ, wind belts, subtropical highs, and polar front lag the Sun's annual north and south shift relative to the equator.

- Trade winds blow out of the equatorward flanks of the subtropical anticyclones, while the westerlies blow out of their poleward flanks. The east side of a subtropical anticyclone features subsiding air and dry climates, whereas the west side is more humid and receives more rainfall.

- Aloft, in the middle and high tropical troposphere, winds reverse direction, completing the Hadley cell circulation. At higher latitudes, upper-air winds meander from west to east as Rossby long waves, each consisting of a ridge (clockwise turn) and trough (counterclockwise turn).

- Contrasts in Earth's surface temperatures in winter favor relatively high air pressure over continents and low air pressure over the ocean.

In summer, this pattern reverses, with relatively low pressure prevailing over continents and high pressure over the ocean.

- The wave pattern exhibited by the upper-air westerlies varies in length, amplitude, and number. At one extreme, westerlies can be strongly zonal, that is, they blow west-to-east with little north/south amplitude. At the other extreme, westerlies can be strongly meridional, that is, they blow west-to-east with considerable north/south amplitude. Shifts between zonal and meridional flow patterns alter north-south air mass exchange, poleward heat transport, and the paths of cyclones.

- Cutoff cyclones and cutoff anticyclones block the usual west to east progression of weather systems and may lead to extended periods of extreme weather conditions such as drought, excessive rainfall, or periods of unusually high or low temperature.

- Horizontal movements of ocean surface waters to a large extent mirror the long-term average planetary-scale atmospheric circulation.

Enduring Ideas

- The speed and direction of the large-scale surface winds are governed by interactions among the pressure gradient force, centripetal force, Coriolis Effect, friction, and gravity.
- Surface winds blow clockwise and spiral outward in Northern Hemisphere anticyclones but counterclockwise and inward in Northern Hemisphere cyclones. In response to downward vertical motion, an anticyclone is a fair-weather system whereas in response to upward vertical motion, a cyclone is a stormy weather system.
- The principal features of the planetary-scale circulation consists of the ITCZ, trade winds, subtropical anticyclones, westerlies, subpolar lows, polar front, polar easterlies, and polar highs.
- Shifts between zonal and meridional flow patterns in the westerlies alter the tracks of cyclones and air masses and can include blocking systems that are responsible for weather extremes.

Review

1. Describe the general relationship between isobar spacing and horizontal wind speed.
2. How does the Coriolis Effect influence the direction of winds that initially are blowing from southwest to northeast in the Northern Hemisphere?
3. What role is played by gravity in cold-air drainage?
4. Compare and contrast the geostrophic wind with the gradient wind.
5. How does Earth's surface roughness affect the speed and direction of surface winds in a Northern Hemisphere cyclone?
6. Describe the circulation of surface winds in a Southern Hemisphere anticyclone (as viewed from above).
7. Explain why anticyclones tend to be fair-weather systems and cyclones tend to be stormy weather systems.
8. Describe the movement of the intertropical convergence zone (ITCZ) in the Northern Hemisphere spring.
9. Compare the climate on the western side of the Bermuda-Azores subtropical high with the climate on the eastern side of that semi-permanent system.
10. What is the relationship between blocking systems and weather extremes?

Critical Thinking

1. Explain why the Coriolis deflection reverses between the Northern Hemisphere and Southern Hemisphere.
2. Why is there no horizontal component of the Coriolis Effect at the equator?
3. Compare the pattern of horizontal surface winds in a Northern Hemisphere cyclone with winds in a Southern Hemisphere cyclone. Why the difference?
4. Is the gradient wind a consequence of balanced forces? Explain your response.
5. What is the relationship between the trade wind inversion and the semi-permanent subtropical anticyclones?
6. In terms of the planetary-scale atmospheric circulation, explain why summer is the dry season in Southern California.
7. Are Pacific air masses more common in the Midwestern U.S. during a zonal flow pattern or a meridional flow pattern?
8. Distinguish between a zonal flow pattern and a meridional flow pattern in the midlatitude westerlies.
9. Why is the tropopause at a higher altitude in the tropics than at midlatitudes?
10. With a blocking high in place, why does a drought tend to intensify with time?

ESSAY: Climatology of Wind Power

Harnessing the kinetic energy of the wind is a technology that was established as early as the 12th century in portions of the Middle East where water power was not available. In North America, the energy crisis of the 1970s spurred renewed interest in this ancient technology. More recent efforts to reduce our nation's dependency on fossil fuels in view of concerns over global climate change have rekindled interest in *wind power*. Today, scientists and engineers employ modern aerodynamic principles and space-age materials in designing and constructing modern wind-driven turbines that convert some of the wind's kinetic energy into electricity.

In Chapter 4, we saw how the Sun drives the atmosphere. Only about 2% of the solar energy that reaches Earth is ultimately converted to the kinetic energy of wind, but that is still a tremendous amount of energy. Theoretically, windmill blades can convert a maximum of about 60% of the wind's energy into mechanical energy. In actual practice, however, wind generators extract at best only about 25% of the wind's energy. Furthermore, minimum wind speeds must be at least 15 km per hr (9 mph) for most wind-powered electricity-generating systems to operate economically. The amount of power that a wind turbine can extract from air in motion varies directly with: (1) the cube of the instantaneous wind speed at hub height (V^3), (2) the area swept out by the windmill blade, and (3) air density.

Wind speed is by far the most important consideration in evaluating a location's wind energy potential. Even small changes in wind speed translate into great variations in energy harvested. For example, doubling the wind speed (a common occurrence) multiplies the available wind power by a factor of eight ($2 \times 2 \times 2$). Wind speed usually increases with altitude above Earth's surface, especially within the lowest meters of the atmosphere, so that a wind turbine is mounted on the top of a tower to take advantage of the stronger winds at that elevation. As a general rule, a minimum of several years of detailed wind monitoring is needed for a preliminary evaluation of the wind power potential of any site. Also, the long-term climate record should be checked for the frequency of potentially destructive winds. Wind data from nearby weather stations can be very useful as long as care is taken in extrapolating wind data between localities and from one elevation to another.

Recall that standard National Weather Service anemometers are located 10 m or 33 ft above the ground. Usually, routine surface wind speeds have to be extrapolated upward from anemometer level to 60 m (200 ft), which is the typical hub height of wind energy conversion systems. This extrapolation depends upon surface roughness and atmospheric stability (Chapter 5).

At ordinary speeds, wind is a relatively diffuse energy source, comparable in magnitude to incoming solar radiation. Hence, a wind turbine's power generation potential also depends on the area swept out by the windmill blades; larger windmill blades can harvest more energy. Until recent decades, design and strength-of-materials problems limited the size of wind turbines. Today, availability of stronger and lighter materials for blades is making possible larger wind turbines that can generate as much as 0.5 megawatt of electricity. One wind turbine design uses three 23-m (75-ft) blades (Figure 1).

Tapping the wind's energy for electric power generation faces several challenges. Both wind speed and direction vary with exposure of the site, roughness of the terrain, and season of the year. The most formidable obstacle to the development

FIGURE 1
This windmill farm in northeastern Wisconsin is designed to meet the electrical energy needs of about 3600 households.

of wind power potential stems from the inherent variability of the wind. The electrical output of a wind turbine varies as a consequence. Hence, a wind power system must include a means of storing the energy generated during gusty periods for use when the wind is light or the air is calm. Banks of batteries may serve this purpose. Wind power has its greatest immediate potential in regions where average winds are relatively strong and consistent in direction. Favorable regions include the western High Plains, the Pacific Northwest coast, portions of coastal California, the Great Lakes region, the south coast of Texas, and exposed summits and passes in the Rockies and Appalachians.

Economy of scale favors centralized arrays of many wind turbines, called *windmill farms*, over individual household wind turbines. Windmill farms consist of a dozen or more super wind turbines, each capable of producing as much as 0.5 megawatt of electricity. Windmill farms currently operating in California supplement conventional sources by feeding electricity directly to power grids. Since 1981, more than 16,000 turbines have been installed in windy mountain passes in California. According to the California Energy Commission, in 2007, windmill farms in California generated 6802 gigawatt-hours of electricity, about 2.3% of the state's gross system power.

ESSAY: Drought

Drought is an extended period of moisture deficit that is of particular concern for agricultural interests and water managers (Figure 1). Drought also affects the supply of water for domestic and industrial use and hydroelectric power generation. Soils dry out; crops wither and die; lakes and other reservoirs shrink; and river and stream flow slackens, thereby impeding navigation. Unusually high temperatures often accompany summer drought further stressing crops.

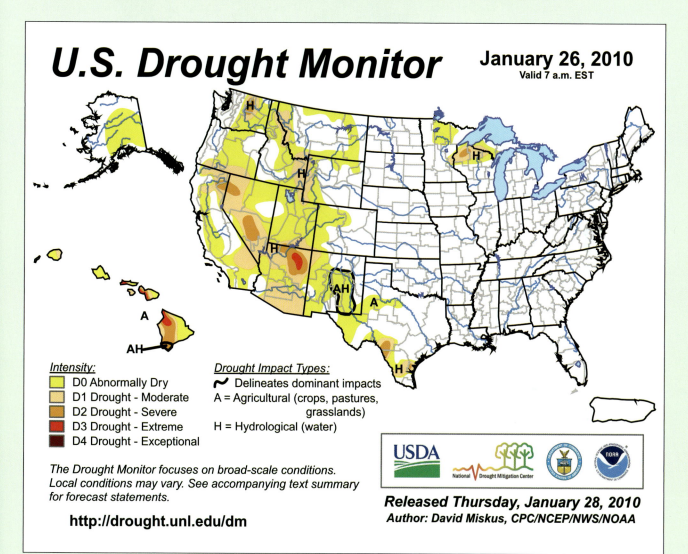

FIGURE 1
A sample U.S. Drought Monitor indicating the intensity of drought and the impacts.

A drought usually begins gradually and without warning. In fact, it is difficult if not impossible to tell whether a spell of dry weather actually signals the onset of drought. Similarly, the end of a drought is always uncertain because one rain event, even if substantial, does not necessarily break a drought. Furthermore, whether a dry spell is a drought depends on its impact, so that a distinction is made among hydrologic drought, agricultural drought, and meteorological drought.

For water resource interests, a *hydrologic drought* is a period of moisture deficit that reduces stream or river discharge and groundwater supply to levels that seriously impede water-based activities such as irrigation, barge traffic, or hydroelectric power generation. Hydrologic drought develops when the water supply is inadequate during one or more successive water

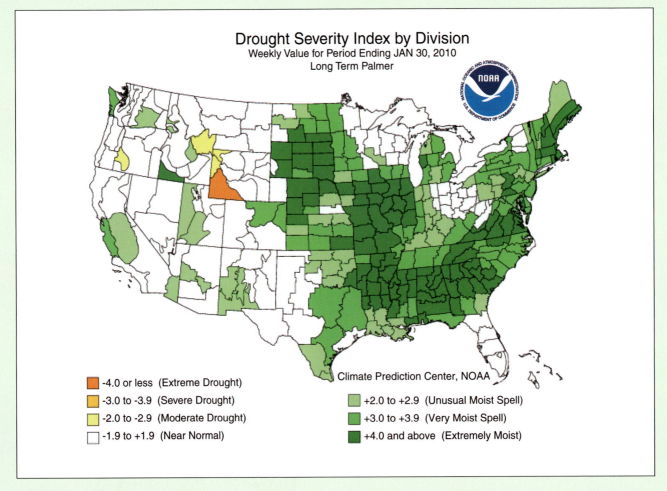

Drought Severity Index by Division
Weekly Value for Period Ending JAN 30, 2010
Long Term Palmer

Climate Prediction Center, NOAA

- ■ -4.0 or less (Extreme Drought)
- ■ -3.0 to -3.9 (Severe Drought)
- ■ -2.0 to -2.9 (Moderate Drought)
- □ -1.9 to +1.9 (Near Normal)
- ■ +2.0 to +2.9 (Unusual Moist Spell)
- ■ +3.0 to +3.9 (Very Moist Spell)
- ■ +4.0 and above (Extremely Moist)

FIGURE 2
A sample Palmer Drought Severity Index Map by climate divisions.

years. A *water year* is defined to extend from 1 October of one year through 30 September of the next year. *Agricultural drought* depends on the shorter-term supply of rainfall and soil moisture for crops during the growing season. Complicating the criterion for agricultural drought is the fact that different crops have different water requirements, and the water needs of a specific crop species change as the crop progresses through its life cycle. Inadequate moisture at a critical stage of crop growth and maturation, especially over successive growing seasons, may constitute agricultural drought. Hydrologic and agricultural droughts do not always occur at the same time.

Varying criteria have been used to define *meteorological drought*. Some atmospheric scientists define drought as a period when the seasonal or annual precipitation falls below a certain threshold percentage (e.g., 85%) of the long-term (e.g., 30-year) average. But basing the criterion for drought on precipitation alone ignores the influence of temperature and wind on the evaporation rate. Higher temperatures and/or stronger winds increase the rate of evaporation, exacerbating the severity of drought.

One of the most popular drought indicators is incorporated in the *Palmer Drought Severity Index* (Figure 2). The Palmer Index uses temperature and rainfall data in a formula that gauges unusual dryness or wetness over extended intervals from months to years. NOAA's National Weather Service and the U.S. Department of Agriculture jointly compute the Palmer Index every week for each of the 344 climatic divisions across the U.S. A Palmer Index map portrays those divisions experiencing drought with negative values, while those regions receiving excess precipitation have positive index values. Index values range from greater than +4.00 for extremely wet conditions to under –4.00 for extreme drought; zero indicates long-term average moisture levels.

CHAPTER 7

ATMOSPHERIC CIRCULATION AND REGIONAL CLIMATES

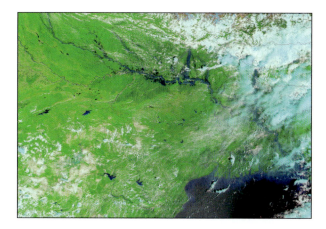

Monsoon floods swell rivers in northeastern India on 2 September 2008, when the Moderate Resolution Imaging Spectroradiometer (MODIS) on NASA's Terra satellite acquired this image. [Courtesy NASA]

Case-in-Point

Monsoon climates are found in regions of the world subject to a seasonally reversing monsoon circulation and typically characterized by dry winters and wet summers. The amount of rainfall during the wet season can vary considerably from one year to the next, and over longer periods with important implications for agriculture and food supply. One key to a better understanding of the monsoon circulation and its variability is interpretation of instrument-based and reconstructed climate records in monsoon regions. An example of a long-term reconstructed record of rainfall comes from a cave located at the northern fringe of the Asian monsoon in northern China.

In the 7 November 2008 issue of *Science*, Pingzhong Zhang of Lanzhou University, Lanzhou, China, and his Chinese and American colleagues reported on their reconstruction of a continuous rainfall record spanning the period from CE 190 to 2003. Researchers analyzed a 118-mm-long section of a *speleothem* (stalagmite) extracted from the floor of Wanxiang Cave located at 33.3 degrees N, 105 degrees E, and 1200 m (3900 ft) above sea level. In that area, today's average annual precipitation is 480 mm, with about 80% falling from May to September.

A speleothem consists of layers of calcium carbonate ($CaCO_3$) precipitated from groundwater that drips onto the cave floor. Groundwater flow into the cave varies with local rainfall so the thickness of a speleothem layer can serve as a proxy for rainfall. Little growth (a relatively thin layer) may indicate drought whereas rapid growth (a relatively thick layer) may indicate heavy rainfall. Scientists also extract rainfall information by analyzing the *oxygen isotope ratio* of speleothem layers. Water is composed of two isotopes of oxygen: heavy oxygen (^{18}O) and light oxygen (^{16}O). The ratio of these oxygen isotopes in water varies with air temperature, the amount of water locked up in glaciers worldwide, and the local precipitation. More light oxygen compared to heavy oxygen indicates heavy rainfall.

The age of these speleothem layers is determined via the uranium/thorium radiometric dating technique. This method differs from other radiometric dating techniques in that it is not based on the amount of stable decay product that accumulates as an unstable isotope decays. Rather, the age of the sample is determined from the degree to which equilibrium has been restored between thorium-230, a radioactive isotope, and its radioactive parent uranium-234. Uranium-234 decays to thorium-230 (half-life of 245,000 years). Thorium-230 is also radioactive (half-life of 75,000 years) and approaches equilibrium with its parent isotope. At equilibrium, the number of thorium decays per year equals the number of uranium-234 decays per year. For the speleothem from Wanxiang Cave, this radiometric dating technique produced an average resolution of 2.5 years.

Comparing the reconstructed 1800-year rainfall record from Wanxiang Cave with Chinese historical records, Zhang et al. noted that three multi-century dynasties ended following several decades of weak monsoon circulation and low rainfall. Specifically, weak monsoons during the periods CE 850-940, 1350-1380, and 1580-1640 encompass the decline and demise of the Tang, Yuan, and Ming Dynasties respectively. Perhaps drought led to food shortages and social turmoil. Conversely, the Northern Song Dynasty thrived during CE 960-1020, a period when the Asian monsoon circulation was vigorous and rainfall was heavy. At this time, harvests were bountiful and the population exploded. The apparent correspondence between weakening of the Asian monsoon and collapse of multi-century dynasties in China is intriguing. However, while climate may have played a role in these events, it was likely not the only factor involved.

Zhang et al. also noted that the demise of the Tang Dynasty occurred at about the same time that Alpine glaciers advanced and the Maya Classic Period collapsed (about CE 900) due to severe drought. This suggests that climate change was planetary in scale.

Driving Question:
What role is played by atmospheric circulation systems in shaping the climate of specific regions of the world?

In profiling the climate of a particular location or region, scientists measure, record and analyze a variety of quantitative elements, including temperature, precipitation, humidity, and wind. The mean value and variability of these climatic elements largely depend on the influence of atmospheric circulation operating on scales ranging from the micro- to planetary-scale. In the previous chapter, the primary focus was on the components of the planetary-scale circulation. In this chapter, the emphasis is chiefly on **synoptic climatology**, a subfield of climatology that relates regional and local climates to prevailing atmospheric circulation patterns. Synoptic-scale analysis reveals patterns over distances of 100 to 10,000 km (60 to 6000 mi) and synoptic-scale weather systems (air masses, fronts, cyclones, and anticyclones) having periods ranging from days to weeks.

Subjective and objective schemes are used to identify atmospheric circulation patterns and their relationship to various weather types. Subjective schemes often rely on analysis of a sequence of daily weather maps to determine the synoptic pattern associated with a specific climatic episode. For example, such an analysis can distinguish seasonal weather patterns. Objective schemes depend on statistical techniques to interpret digital observational data.

In all cases, boundary conditions within Earth's climate system determine the array of atmospheric circulation patterns and systems plus their seasonal variations that characterize the climate of a particular locale. For example, the mean summer daily maximum temperature tends to be lower where sea breezes are frequent. The mean winter temperature (from December through February) in Minneapolis, MN, depends to a large extent on the number of days with arctic air from December through February; that is, winters are colder when there are more days with arctic air and milder when there are fewer days with arctic air. The day-to-day variability of winter and spring temperatures tends to be greater where Chinook winds are common, such as Rapid City, located on the eastern slopes of the Black Hills in western South Dakota. Although the latitude of Rapid City (44.08 degrees N) is close to that of Minneapolis, MN (45.08 degrees N), the warming influence of Chinook winds is evident in higher long-term average January temperatures at Rapid City. The January average daily high temperature is about 1.1 °C (34 °F) at Rapid City and −5.6 °C (22 °F) at Minneapolis while the January average daily low temperature is approximately −11.7 °C (11 °F) at Rapid City and −15.6 °C (4 °F) at Minneapolis.

Air Masses

An **air mass**, a huge expanse of air covering thousands of square kilometers, is relatively uniform horizontally in temperature and humidity, both typically decreasing with altitude. The properties of an air mass depend upon the type of surface over which it develops, its source region. An air mass *source region* features nearly homogeneous surface characteristics over a broad area with little topographic relief, such as a great expanse of snow-covered ground or a vast stretch of ocean water. To be conditioned by the temperature and moisture characteristics of the underlying surface, an air mass must reside in its source region for several days to weeks.

Often, large semi-permanent anticyclones such as the *Siberian high* dominate air mass source regions (Chapter 6). Subsidence of air and divergence of surface winds in an anticyclone cause the air to develop relatively uniform physical properties. Light winds that inhibit mixing coupled with the long residence time that is typical of high pressure systems transfer the thermal properties of Earth's surface to the air mass. Air mass source regions are typically not found where cyclones develop or along the principal storm tracks.

Air masses are classified as either cold (polar, abbreviated as *P*) or warm (tropical or *T*), and either dry (continental or *c*) or humid (maritime or *m*). Air masses that form over the cold snow-covered surfaces of high latitudes are relatively cold, whereas those that develop over the warm surfaces of low latitudes are relatively warm. Air masses that form over land tend to be relatively dry, whereas those that develop over the ocean are relatively humid. Based on combinations of temperature and humidity, the four basic types of air masses are: cold and dry, *continental polar* (*cP*); cold and humid, *maritime polar* (*mP*); warm and dry, *continental tropical* (*cT*); and warm and humid, *maritime tropical* (*mT*). A fifth air mass type, *arctic* (*A*) air, is dry like continental polar air but colder.

NORTH AMERICAN TYPES AND SOURCE REGIONS

All the air mass types listed above occur over North America, forming over either the landmass or surrounding ocean water. Source regions for these air masses are plotted in Figure 7.1.

Continental tropical air (cT) develops primarily in summer over the subtropical deserts of Mexico and the southwestern United States primarily in summer and is hot and dry. **Maritime tropical air (mT)** is very warm and humid because its source regions are tropical and

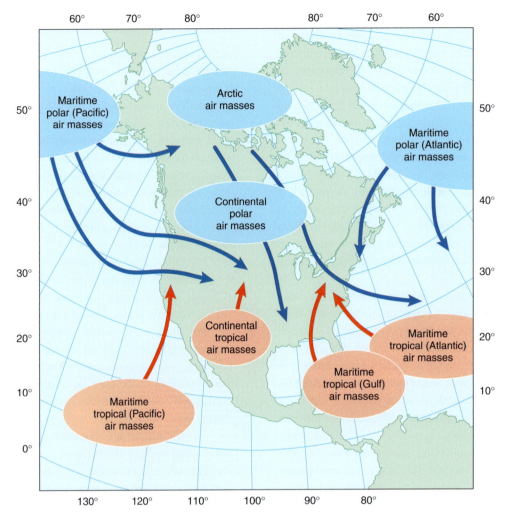

FIGURE 7.1
Source regions of North American air masses. The temperature and humidity of an air mass depend on the properties of its source region.

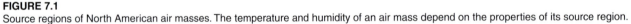

subtropical seas (e.g., Gulf of Mexico). This air mass retains these properties year-round and is responsible for oppressive summer heat and humidity east of the Rocky Mountains. The source regions for **maritime polar air (mP)** are the cold ocean waters of the North Pacific Ocean and North Atlantic Ocean, especially north of 40 degrees N. Along the West Coast, *mP* air brings heavy winter rains, snows in the mountains, and persistent coastal fogs in summer. Dry **continental polar air (cP)** develops over the northern interior of North America. In winter, *cP* air is typically very cold because the ground in its source region is often snow covered, daylight is short, solar radiation is weak, and radiational cooling is extreme. In summer, when the snow-free source region warms in response to extended hours of bright sunshine, *cP* air is quite mild and pleasant.

Arctic air (A) forms over the snow or ice covered regions of Siberia, the Arctic Basin, Greenland, and North America, north of about 60 degrees N, in much the same

way as continental polar air, but in a region that receives very little solar radiation in winter, although it still radiates strongly to space in the infrared. These exceptionally cold and dry air masses are responsible for the bone-numbing winter cold waves that sweep across the Great Plains, at times penetrating as far south as the Gulf of Mexico and central Florida. For example, the *Siberian Express* is the name given to an arctic air mass that forms over Siberia, crosses over the North Pole into Canada, plunges onto the Great Plains, and then moves south and eastward.

Air masses differ not only in temperature and humidity but also in stability. As noted in Chapter 5, stability is an important property of air because it influences vertical motion and the consequent development of clouds and precipitation as well as the persistence of the air mass. Table 7.1 is a list of the usual stability, temperature, and humidity characteristics of North American air masses within their source regions.

TABLE 7.1
Stability, Temperature, and Humidity Characteristics of North American Air Masses

Air mass type	Source region stability		Characteristics	
	Winter	Summer	Winter	Summer
A	Stable		Bitter cold, dry	
cP	Stable	Stable	Very cold, dry	Cool, dry
cT	Unstable	Unstable	Warm, dry	Hot, dry
mP (Pacific)	Unstable	Unstable	Mild, humid	Mild, humid
mP (Atlantic)	Unstable	Stable	Cold, humid	Cool, humid
mT (Atlantic)	Unstable	Unstable	Warm, humid	Warm, humid

AIR MASS MODIFICATION

As air masses move out of their source regions, their properties modify, some air masses changing more than others. Changes may occur in temperature, humidity, and/or stability. **Air mass modification** occurs primarily by (1) exchange of heat or moisture, or both, with the surface over which the air mass travels; (2) radiational heating or cooling; and (3) adiabatic heating or cooling associated with large-scale vertical motion.

In winter, as a *cP* air mass travels southeastward from Canada into the coterminous United States, its temperature can modify rapidly. Although daily minimum temperatures across the Northern Plains might dip well below −18 °C (0 °F), after the polar air arrives in the southern states, minimum air temperatures may not drop much below the freezing point. Rapid air mass modification occurs because, outside of its source region, polar air is usually colder than the ground over which it travels. The Sun warms the snow-free ground, and the warmer ground heats the bottom of the air mass, destabilizing it and triggering convective currents that distribute heat vertically within the air mass.

When *cP* air travels over snow-covered ground, however, modification is more limited because much of the incoming solar radiation is reflected rather than being converted to heat (via absorption). The relatively cold snow surface chills the air, increasing stability and weakening convection. Extreme nocturnal radiational cooling plays an important role in inhibiting modification of the *cP* air mass, especially when nights are long.

A tropical air mass does not modify as readily as a polar air mass because, outside of its source region, tropical air is often warmer than the ground over which it travels. The bottom of the air mass cools by contact with the ground; this cooling stabilizes the air mass by increasing the density of the near-surface air and suppresses convective currents. Hence, cooling is restricted to the lowest portion of the air mass. By contrast, if a tropical air mass moves over a warmer surface, the air mass can become even warmer. Hence, a cold wave loses much of its punch as it pushes southward, but a summer heat wave can retain its warmth as it journeys from the Gulf of Mexico northward into southern Canada.

Air masses undergo significant changes in temperature and humidity through orographic lifting (Chapter 5). When cool and humid *mP* air sweeps inland off the Pacific Ocean, the air is forced up the windward slopes of coastal mountain ranges, such as the Cascades and the Sierras, and expands and cools adiabatically. Cooling brings air to saturation and triggers condensation or deposition (cloud development) and precipitation, primarily on the windward slopes. Latent heat released during condensation (and deposition) partially offsets adiabatic cooling. Then, as the air mass descends the leeward slopes into the Great Basin, it warms adiabatically and clouds dissipate. Some evaporative cooling is associated with cloud dissipation, but net heating (and drying) prevails as compressional heating is greater. Because water precipitated out on the windward side of the mountains, descending leeward air loses its saturation at a higher altitude than the altitude at which saturation was achieved by air ascending the windward slopes. Descending unsaturated air, subjected to compressional heating only, warms at the greater dry adiabatic rate.

The same modification processes are repeated as the air mass is forced to flow up and over the Rockies. Eventually, the air mass emerges on the Great Plains considerably milder and drier than the original *mP* air mass. East of the Rockies, such an air mass is described as modified **Pacific air**. When the westerly wave pattern aloft is dominantly zonal (Chapter 6), Pacific air floods the eastern two-thirds of the United States and southern Canada. Much of eastern North America experiences lengthy episodes of mild and generally dry weather because polar air masses stay far to the north, while tropical air masses stay to the south.

In 1966, the late climatologist Reid A. Bryson published a map showing the regions of North Amer-ica dominated by various air mass types. As shown in Figure 7.2, shaded areas are occupied more than 50% of the time by the specified air mass. Mild Pacific air dominates the region that closely corresponds to the wedge-shaped area of the U.S. and Canada where most grain is grown, that is, from Alberta south to Colora-do and eastward to western New York. Prior to settle-ment by Europeans, the indigenous biome of the area was prairie, consisting of grasslands, herbs, and shrubs, with only scattered trees (Figure 7.3). If the correspon-dence between the area of mild Pacific air dominance and the location of the prairie wedge is more than just coincidental, we might reasonably expect that shifts in the prevailing atmospheric circulation patterns accom-panying climate change would also alter the frequency of occurrence of air masses with important implications for agriculture.

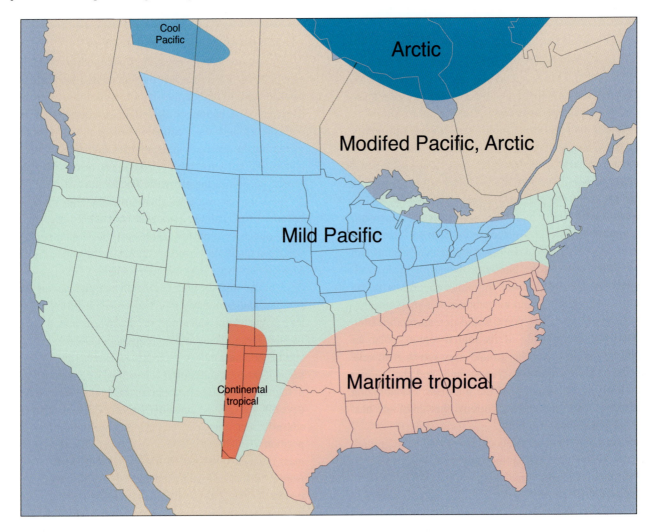

FIGURE 7.2
Regions of North America occupied by various air mass types. Blue, pink, and red shaded regions are occupied by the indicated air mass more than 50% of the time. [Modified after Bryson, R.A., 1966. "Air Masses, Streamlines, and the Boreal Forest." *Geographical Bulletin* 8(3): 249.]

FIGURE 7.3
The presence of mild Pacific air apparently plays an important role in determining the location of the prairie and grain belt. Tallgrass Prairie National Preserve, Flint Hills, Chase County, eastern Kansas, is intended to help conserve what little remains of the original tall grass ecosystem of North America. Cultivation and development reduced the original indigenous vegetation to less than 4% of the original area of about 1,000,000 km² (400,000 mi²).

Fronts

A **front** is a narrow zone of transition between air masses that differ in density. Density differences are usually due to temperature contrasts; for this reason we use the nomenclature *cold* and *warm* fronts. However, density differences may also arise from contrasts in humidity. Although the transition zone associated with an actual front may be up to a hundred or more kilometers wide, traditionally a line representing a front is drawn on a weather map along the warm (less dense) edge of the transition zone. Air temperatures are nearly constant on the warm side of the front, but fall with distance from the front to a region of nearly uniform temperature in the cold air mass.

A front is also associated with a trough in the sea-level pressure pattern, a corresponding wind shift, and converging winds. Where contrasting air masses meet, the colder (denser) air forces the warmer (less dense) air to ascend. Often the ascending air cools sufficiently (via expansion) that clouds and precipitation develop. Depending on the slope of the front in the vertical plane and the motion of air relative to the front, frontal weather may be confined to a narrow band (tens of kilometers wide), or it may extend over a broad region (hundreds of kilometers wide). In addition, the slope of the front influences the types of clouds and precipitation forming along the front.

Properties that define a front (differences in temperature and/or humidity, wind shift, convergence, and trough) change with time similar to the way air masses modify. If processes in the atmosphere, such as a region

of converging surface winds, increase the density contrast between air masses, then a front forms or grows stronger; this process is called **frontogenesis**. On the other hand, if the density contrast between air masses decreases, perhaps because of diverging surface winds, the front weakens; this process is known as **frontolysis**. This section covers the basic types of synoptic-scale fronts: stationary, warm, cold, and occluded.

STATIONARY FRONT

A **stationary front** exhibits essentially no forward movement. For example, a stationary front develops along the Front Range of the Rocky Mountains when a shallow pool of cold air (typically about 1000 m or 3300 ft thick) surges south and southwestward out of the Prairie Provinces of Canada. The vertical extent of the air mass is below the level of mountain passes, so the shallow cold dry air mass abuts the mountain range and can push no further westward; therefore, its leading edge is marked by a stationary front paralleling the Continental Divide, the crest of the mountain range. Milder air remains in the Great Basin to the west of the Rockies. On occasion, the cold air mass may be sufficiently deep that cold air pours westward through mountain passes into the Great Basin. Similarly, a stationary front can form when any type of preexisting front becomes parallel to the upper-level wind pattern. Under the proper conditions, a stationary front can also form along a boundary in the surface temperature pattern, such as a coastline or the edge of a regional snow cover.

Many features of a stationary front are common to all fronts. As shown in vertical cross section in Figure 7.4, a front slopes from Earth's surface toward colder air (or more precisely, denser air). A front lies in a trough in the pressure pattern on any horizontal surface intersecting the front; this is especially evident in the sea-level isobars drawn on a surface weather map. Recall that the wind blows approximately parallel to the isobars, with friction turning the wind somewhat toward lower pressure. Winds on the two sides of a front exhibit different directions and often different speeds. The differences in wind direction and speed across a front are usually associated with convergence, which leads to upward motion, clouds, and perhaps precipitation.

A stationary front does not always have a broad region of associated clouds and precipitation as depicted in Figure 7.4. Frontal weather can vary considerably from case to case, depending on the supply of water vapor and the specifics of air motion relative to the front. In cases that do produce precipitation, the rain or snow falls mostly on the cold side of the stationary front. Warm humid air flows

FIGURE 7.4
A stationary front; the vertical scale is greatly exaggerated.

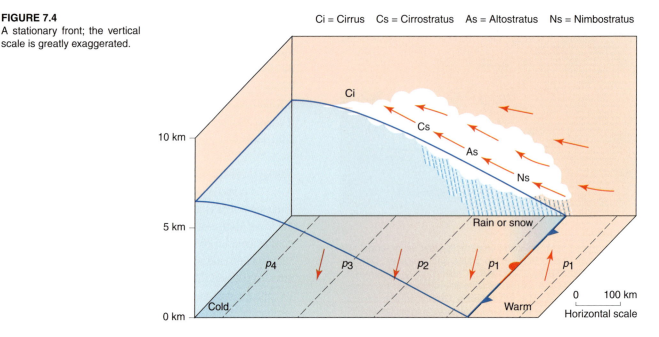

up and over the cooler air mass, more or less along the sloping frontal surface. Ascending air cools by expansion and becomes saturated, which triggers condensation and perhaps precipitation. This situation is often referred to as **overrunning** and can result in an extended period of relatively widespread cloudiness, drizzle, light rain, or light snow.

WARM FRONT

If a stationary front begins to move such that the warm (less dense) air advances while the cold (more dense) air retreats, the front becomes a **warm front**. The overall characteristics of a warm front are very similar to those of the stationary front, as shown in Figure 7.5.

Differences between a warm front and a stationary front are evident in a comparison of Figures 7.4 and 7.5. The slope of the warm frontal surface (ratio of vertical rise to horizontal distance) is less near Earth's surface because surface roughness (friction) has slowed the warm front. Winds on the warm side of the front are quite similar in both instances, but air on the cold side of the warm front is retreating. Thus, the warm air advances relative to Earth's surface, rather than just gliding up and over the cold air as in the case of a stationary front.

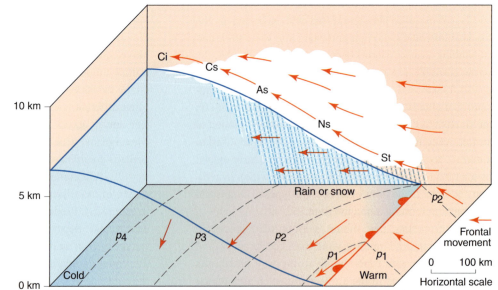

FIGURE 7.5
A warm front is the leading edge of a warm air mass that advances by overrunning denser colder and/or drier air; the vertical scale is greatly exaggerated.

FIGURE 7.6
Thin wispy cirrus clouds may appear more than 1000 km (620 mi) in advance of a surface warm front.

As a warm front approaches, clouds develop, gradually lowering and thickening in the following general sequence: cirrus, cirrostratus, altostratus, nimbostratus, and stratus. This sequence of stratiform clouds reflects gentle uplift associated with overrunning in a stable environment. The initial wispy cirrus clouds may appear more than 1000 km (620 mi) in advance of the surface warm front (Figure 7.6). Slowly, clouds spread laterally as thin sheets of cirrostratus, turning the sky a bright milky white.

In time, cirrostratus clouds give way to altostratus. Soon after altostratus clouds thicken enough to block out the Sun (or Moon), additional lowering of the cloud deck often occurs, accompanied by light rain or snow. Steady precipitation falls from low, gray nimbostratus clouds and persists until the warm front finally passes, a period that may exceed 24 hrs. Copious amounts of rain may fall ahead of the surface warm front, and, because precipitation intensity is usually only light to moderate, much of the water infiltrates the soil (as long as the ground is unfrozen and not already saturated with water). If the air is cold enough for the precipitation to fall in the form of snow, accumulations may be substantial.

Just ahead of the surface warm front, steady precipitation associated with nimbostratus clouds usually gives way to drizzle falling from low stratus clouds and sometimes fog. **Frontal fog** develops when rain falling through the shallow layer of cool air at the ground evaporates and increases the water vapor concentration to saturation. After the warm front finally passes, frontal fog dissipates and skies clear, at least partially, because the zone of overrunning has also passed. The weather tends to become warmer and more humid.

The cloud and precipitation sequence just described for a warm front applies when the advancing warm air is stable. If the warm air were unstable, uplift would be more vigorous and often would give rise to cumulonimbus (thunderstorm) clouds embedded within the zone of overrunning ahead of the surface warm front. Lightning, thunder, and brief periods of heavy rainfall, or perhaps snowfall, may punctuate the otherwise steady fall of light-to-moderate precipitation.

COLD FRONT

An air mass boundary becomes a **cold front** if it begins to move in such a way that colder (more dense) air displaces warmer (less dense) air. Over North America in winter, the temperature contrast is typically greater across a cold front than across stationary or warm fronts. In summer, however, maximum air temperatures on either side of a cold front are sometimes essentially the same. When this occurs, the density contrast between the two air masses arises from differences in water vapor content rather than differences in temperature. At the same temperature and pressure, drier air is denser than more humid air (Chapter 5). Following passage of a cold front, both temperature and humidity usually drop. Dramatic declines in temperature are less likely in summer but humidity changes can be significant.

Friction associated with the roughness of Earth's surface slows the advancing colder air and is responsible for the nose-shaped profile that characterizes a cold front (Figure 7.7). The slope of a cold frontal surface is steeper than the slope of a warm frontal surface. Because of the steep frontal slope and the typical flow aloft, across the front, from the cold to the warm side, uplift is restricted to a narrow zone at or near the cold front's leading edge. Low-level air motion also differs from warm and stationary fronts; the low-level air motion in the cold air is, at least in part, toward the front and forces the warm air aloft.

If the cold front advances at a moderate but steady pace, around 30 km per hr (20 mph), the type of frontal weather will depend on the stability of the warmer air. Brief showers are likely as uplift along the steep cold front is sufficiently vigorous to produce a narrow band of cumuliform clouds and convective showers at or just ahead of the front. If the warm air ahead of the front is unstable, uplift is more vigorous, giving rise to towering cumulus congestus that build into cumulonimbus clouds with cirrus clouds blown downstream by high-altitude winds. These thunderstorms may be accompanied by strong and gusty surface winds, hail, or other severe weather. If a well-defined cold front progresses at a relatively rapid pace, say 45 km per hr (28 mph), then a **squall line**, a band of intense thunderstorms, may

FIGURE 7.7
A cold front is the leading edge of a colder and/or drier air mass that advances by displacing warmer and/or more humid air; the vertical scale is greatly exaggerated.

Ci = Cirrus   Cb = Cumulonimbus

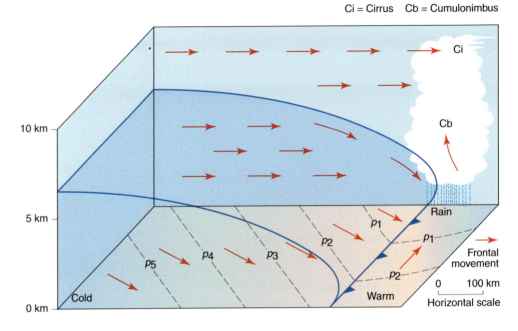

develop either along the front or as much as 300 km (180 mi) ahead of the front. On the other hand, if the warm air is relatively stable, nimbostratus and altostratus clouds may form. Following the passage of a cold front, colder air sweeping across the warmer surface often becomes unstable, leading to gusty westerly winds along with stratocumulus clouds and intermittent light rain or snow showers (depending on the temperature).

A typical Northern Hemisphere cold front trails south or southwestward from the center of an extratropical cyclone, as discussed later in this chapter, and progresses from west to east. In winter, cold fronts often drop southward out of Canada onto the Great Plains, even traveling as far south as the Gulf of Mexico. Southward or southwestward moving cold fronts occur east of the Appalachians in New England, but in that region, they are seen most frequently in summer and fall and are referred to as **back-door cold fronts**. These fronts often usher in welcome relief from hot weather, in the form of cP air from Canada or mP air from the Atlantic Ocean. The Appalachian Mountains block the westward progression of the cooler air, which then moves farther south on the eastern side of the Appalachians along the Piedmont and coastal plains (Figure 7.8).

OCCLUDED FRONT

Late in its life cycle, as a middle latitude cyclone moves into colder air, a front forms, known as an **occluded front**, or simply an *occlusion*. There are three types of occlusions, distinguished by the temperature contrast between the air behind the cold front and the air

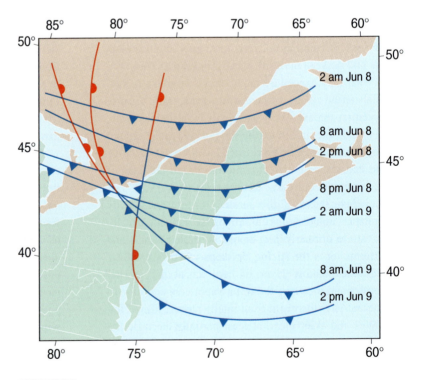

FIGURE 7.8
A back-door cold front moves across New England toward the south and southwest.

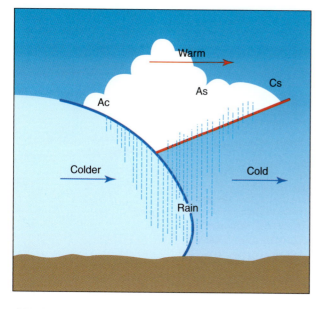

FIGURE 7.9
Schematic vertical cross-section of a cold-type occlusion with the vertical scale greatly exaggerated.

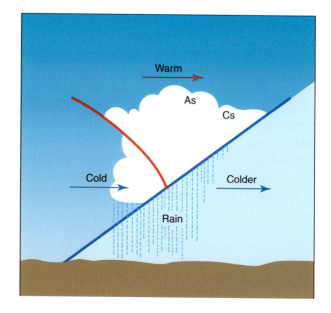

FIGURE 7.10
Schematic vertical cross-section of a warm-type occlusion with the vertical scale greatly exaggerated.

ahead of the warm front: cold occlusion, warm occlusion, and neutral occlusion. If air behind the advancing cold front (*cP*) is colder than the cool air (*mP*) ahead of the warm front, the cold air slides under and lifts the air in the warm sector, the cool air, and the warm front (Figure 7.9). This motion produces a *cold occlusion* that has the characteristics of a cold front at the surface, but the temperature contrast between the cold and cool air masses is less than across a typical winter cold front. Weather ahead of the occlusion is similar to that in advance of a warm front, but the actual frontal passage may be marked by more showery conditions, similar to a cold front. This type of occlusion is most common over the eastern half of the North American continent, where the coldest air follows behind the front on northwest winds.

A *warm occlusion* develops when air behind the advancing cold front is not as cold as the air ahead of the warm front (Figure 7.10). This type of occlusion often occurs in the northerly portions of western coasts, such as in Europe or in the Pacific Northwest. In this case, the air behind the cold front is relatively mild (*mP*), having traversed ocean waters, whereas the air ahead of the warm front is relatively cold (*cP*), having traveled over land. With this type of occlusion, the air behind the cold front slides under the warm air but rides over the colder air. Weather ahead of a warm occlusion is similar to that ahead of a warm front, with the surface front behaving as a warm front.

Extratropical Cyclones

Frontal weather occurs in combination with an **extratropical cyclone**, a major weather maker of middle and high latitudes. Viewed from above in the Northern Hemisphere, surface winds blow counterclockwise and inward toward the low-pressure center of a cyclone (Chapter 6). Surface winds converge, air ascends, expands, and cools, resulting in clouds and precipitation. This section explores the life cycle and characteristics of these synoptic-scale weather systems.

During World War I and continuing into the 1920s, scientists at the Bergen School of Meteorology in Norway conducted ground-breaking studies of synoptic-scale weather systems. Led by Vilhelm Bjerknes (1862-1951) and including his son Jacob Bjerknes (1897-1975) and Swedish meteorologist Tor Bergeron (1891-1977), researchers investigated air masses, fronts, and cyclones. They were the first to describe the basic stages in the life cycle of an extratropical cyclone and for this reason this conceptual model is referred to as the **Norwegian cyclone model**. The model was derived primarily from surface weather observations; at the time researchers were unable to monitor winds aloft routinely. Subsequent advances in atmospheric monitoring techniques—especially remote sensing by satellite—have verified the main features of the Norwegian cyclone model. Amazingly, the model remains a close approximation of our current understanding of middle latitude cyclones, even though individual cyclones may not follow the model precisely.

UPPER-AIR SUPPORT

With adequate *upper-air support*, an extratropical cyclone can form and intensify. **Cyclogenesis**, the birth of a cyclone, usually occurs along the polar front, directly under an area of strong horizontal divergence in the upper troposphere. Strong horizontal divergence aloft occurs (1) to the east of an upper-level *trough* and (2) under the left-front quadrant of a *jet streak*, a particularly high-speed segment of a jet stream.

The upper-level westerly winds of middle and high latitudes feature both long waves (Chapter 6) and short waves. **Short waves** are ripples in Rossby long waves. Although Rossby waves (and associated troughs and ridges) usually drift very slowly eastward, short waves propagate rapidly through the Rossby waves. Whereas five or fewer Rossby waves encircle the hemisphere, the number of short waves may be a dozen or more.

Both short waves and long waves in the westerlies can provide upper-air support for the development of extratropical cyclones. For the same isobar spacing (air pressure gradient), gradient and geostrophic wind speeds are not equal; that is, anticyclonic gradient winds are stronger than geostrophic winds, and cyclonic gradient winds are weaker than geostrophic winds. Hence, as shown in Figure 7.11, westerly winds tend to speed up downwind in a ridge and weaken downwind in a trough. The result is horizontal divergence of mid- to upper-tropospheric winds to the east of a trough (and west of a ridge). Short and long waves thus favor cyclone development by inducing horizontal divergence aloft. Figure 7.12 portrays the relationship between an upper-air wave and surface cyclone and anticyclone.

Within the atmosphere are relatively narrow corridors of very strong winds, known as *jet streams*. In middle latitudes, the most prominent jet stream is located above the

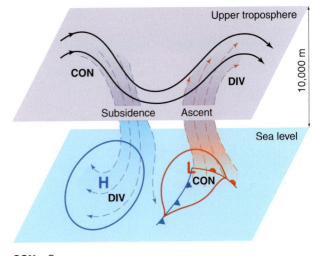

CON = Convergence
DIV = Divergence

FIGURE 7.12
Schematic representation of the relationship between westerly waves aloft and a surface high and low.

polar front in the upper troposphere between the midlatitude tropopause and the polar tropopause. Because of the close association with the polar front, it is known as the **polar front jet stream**. This jet follows the meandering path of the planetary westerly waves and attains wind speeds that frequently top 160 km per hr (100 mph).

Why is a jet stream associated with the polar front? The polar front is a narrow zone of transition between relatively cold and warm air masses. The link between a jet stream near the tropopause and a horizontal air temperature gradient at the Earth's surface hinges on the influence of air temperature on air density. Cold air is denser than warm air, so that air pressure drops more rapidly with altitude within a column of cold air than it does in a column of warm air (Figure 7.13). So, even if the air pressure at the Earth's surface is nearly the same everywhere, with increasing altitude a horizontal pressure gradient develops between adjacent cold and warm air masses. At any specified altitude within the troposphere, the pressure is higher in the warm air mass than in the cold air mass. In response to this horizontal air pressure gradient, air accelerates away from high pressure (the warm air column) and toward low pressure (the cold air column). Simultaneously, the Coriolis Effect comes into play and eventually balances the horizontal pressure gradient force. Consequently, the wind blows parallel to isotherms (that is, parallel to the underlying surface front) with the cold air to the left of the direction of motion in the Northern Hemisphere.

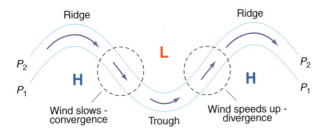

FIGURE 7.11
Westerlies speed up in ridges and slow down in troughs, inducing horizontal convergence aloft ahead of ridges and horizontal divergence aloft ahead of troughs. Solid lines are isobars; P_1 is greater than P_2.

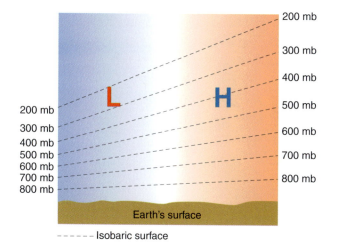

--- Isobaric surface

FIGURE 7.13
Air pressure drops more rapidly with altitude in cold air (left) than in warm air (right) giving rise to a horizontal pressure gradient directed from warm air toward cold air. Vertical scale is exaggerated.

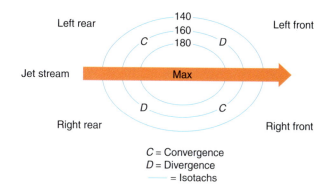

C = Convergence
D = Divergence
— = Isotachs

FIGURE 7.14
Map view of a jet streak, a segment of accelerated winds within the polar front jet stream with associated regions of horizontal divergence (D) and horizontal convergence (C) aloft. In this view from above, the jet streak is outlined by *isotachs*, lines of equal wind speed (in km per hr). In this case of a straight jet streak, the strongest horizontal divergence is in the left-front quadrant, supplying upper-air support for development of an extratropical cyclone.

With the horizontal pressure gradient strengthening with increasing altitude, the horizontal wind speed also increases. Wind speed is highest near the tropopause because, at altitudes above the tropopause (in the stratosphere), the horizontal temperature gradient reverses direction. Absorption of sunlight by ozone (O_3) heats the polar stratopause more continuously than in the tropics. In the troposphere, the coldest air is at higher latitudes, but in the stratosphere, the coldest air is at lower latitudes. In the stratosphere, the horizontal pressure gradient weakens with increasing altitude so that the horizontal wind speed also weakens with altitude above the tropopause. Hence, the maximum wind speed (the polar front jet stream) is located near the tropopause and above the polar front.

Similar to the front with which it is associated, the polar front jet stream is not uniformly defined around the globe. The polar front is well defined where surface horizontal temperature gradients are particularly steep, and so the jet stream winds are stronger. Such a segment, in which the wind may strengthen by as much as an additional 100 km per hr (62 mph), is known as a **jet streak**. The strongest jet streaks develop in winter along the east coasts of North America and Asia where the contrast in temperature between snow-covered land and ice-free sea surface is particularly great. Over those areas, jet streak wind speeds have on rare occasions exceeded 350 km per hr (217 mph). A typical jet streak might be about 160 km (100 mi) wide, 2 to 3 km (1 to 2 mi) thick, and several hundred kilometers in length.

Air flowing through a jet streak changes speed and direction; these changes induce an associated pattern of horizontal divergence and horizontal convergence. Consider Figure 7.14, a schematic representation of a straight jet streak viewed from above. Blue lines are *isotachs*, lines of equal wind speed (in km per hr). Air accelerates as it enters the jet streak and decelerates as it exits the jet streak. Associated with these changes in wind speed is diverging and converging air. Viewed from above, a jet streak can be divided into four quadrants: left-rear, right-rear, left-front, and right-front. Horizontal divergence occurs in both the left-front and right-rear quadrants, and horizontal convergence takes place in both the right-front and the left-rear quadrants.

A jet streak provides upper-air support for a cyclone by contributing horizontal divergence aloft. For an essentially straight jet streak, the strongest horizontal divergence is in the left-front quadrant, so it is under this sector of the jet streak that a cyclone typically has the best chance of developing. This rule also applies to a cyclonically curved jet streak; for an anticyclonically curved jet streak, the strongest divergence is in the right-rear quadrant. Horizontal divergence aloft is strongest when a jet streak is situated on the east side of a trough.

Also similar to the polar front, the jet stream undergoes seasonal changes, strengthening in winter (when the north-south air temperature contrast is greatest), and weakening in summer (when temperature contrasts are less). As shown in Figure 7.15, the average summer location of the polar front jet stream is across southern Canada, whereas the average winter position is across the southern United States. These locations represent long-term averages; the jet stream actually weaves over

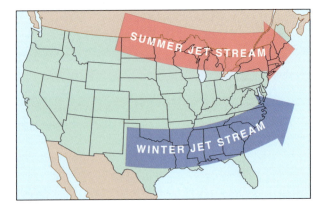

FIGURE 7.15
Approximate average location of the polar front jet stream in winter (December through March) and summer (June through October).

a considerable range of latitude from week to week, and even from one day to the next. As a general rule, when the polar front jet stream is south of your location, the weather tends to be relatively cold, and when the polar front jet stream is north of your location, the weather tends to be relatively warm.

LIFE CYCLE

If horizontal divergence aloft removes more mass from a column of air than is supplied by converging surface winds, the air pressure at the bottom of the column falls. Consequently, a horizontal air pressure gradient develops in the lower reaches of the atmosphere and a cyclonic circulation begins; that is, a storm is born. Westerlies aloft then steer the cyclone as it progresses through its life cycle (Figure 7.16).

Just prior to the formation of a typical cyclone, the polar front is often stationary with surface winds blowing parallel to the front. As the surface air pressure drops, surface winds respond and turn toward the front and converge as segments of the front begin to move (Figure 7.16A). West of the low center (where the sea-level air pressure is lowest), the polar front pushes toward the southeast as a *cold front*. East of the low center, the polar front advances northward as a *warm front*. The minimum pressure for the incipient low might be near 1000 mb, with a single closed isobar on a standard surface weather map. Satellite imagery shows that the narrow cloud band associated with the stationary front develops a bulge at the low center and extends along the developing warm front. The upper-level circulation pattern depicted in Figure 7.16A shows a trough to the west of the surface low, a position that favors further development (*deepening*) of the system. With more horizontal divergence aloft than

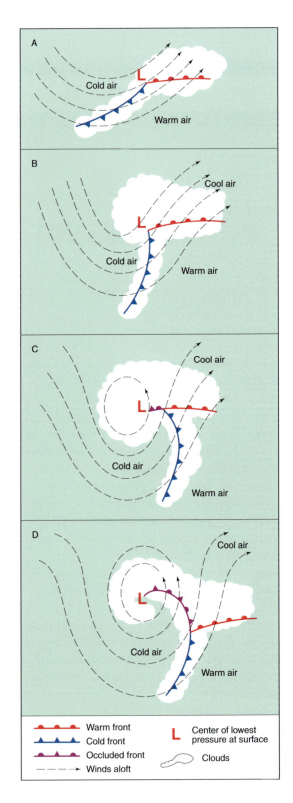

FIGURE 7.16
The life cycle of an extratropical cyclone (low-pressure system): (A) incipient cyclone, (B) wave cyclone, (C) beginning of occlusion, and (D) bent-back occlusion. As a wave cyclone, the system's center is located east of the upper-level trough; at occlusion, the system's center is under the upper-level trough. Air temperatures refer to conditions at Earth's surface.

convergence near the surface, the surface pressure at the cyclone center continues to fall.

If the circulation pattern in the upper troposphere does not favor further development of the cyclone, the low center typically ripples along the stationary front without deepening, producing cloudiness and light precipitation along the way. Such cyclones affect only a small area and typically travel along the front at 50 to 70 km per hr (30 to 45 mph). On the other hand, if atmospheric conditions support development of the incipient cyclone, the central pressure continues to drop, the associated horizontal air pressure gradient becomes steeper, and counterclockwise winds strengthen. The upper-level trough also frequently deepens while remaining to the west of the low center, as cold air is brought into the system from the northwest.

The *warm sector* of the cyclone (the area between the warm and cold fronts, occupied by warm air at the surface) becomes better defined during the early stage of storm development. At this stage of the cyclone's life cycle, fronts form a pronounced wave pattern (Figure 7.16B), hence the descriptive name *wave cyclone*. The intensifying cyclone may now have a central pressure approaching 990 mb, identified by perhaps three closed isobars on a weather map, and typically is moving eastward or northeastward at 40 to 55 km per hr (25 to 35 mph). A large-scale comma-shaped cloud pattern, known as a **comma cloud**, typically appears in satellite images at this stage, and reflects the strengthening of the system's circulation (Figure 7.17). The head of the comma extends from the low center to the northwest, with its tail trailing southward or southwestward along the cold front. Extensive stratiform cloudiness caused by overrunning appears north of the warm front.

The cold front generally moves faster than the warm front so that the angle between the two fronts gradually closes; that is, the area occupied by the surface warm sector diminishes. As the cold front closes in on the warm front, an *occluded front* begins forming near the low center, forcing the warm air aloft and causing the warm sector at the surface to occupy a smaller area away from the surface low center. Figure 7.16C represents the beginning of the occlusion stage of the cyclone's life cycle. The upper-level pattern now features a closed circulation almost above the surface cyclone. Dry air descending behind the cold front is drawn nearly into the center of the cyclone as a **dry slot** that separates the cloud band along the cold front from the comma head, now more west than northwest of the low center. The central pressure of the cyclone has dropped significantly to perhaps 985 mb in the time that elapsed between Figures 7.16B and 7.16C.

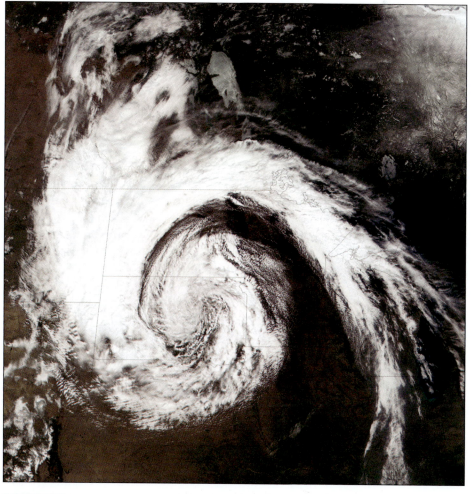

FIGURE 7.17
The Moderate Resolution Imaging Spectroradiometer (MODIS) on NASA's Terra satellite captured this image of a storm over the Dakotas on the morning of 19 April 2006. The comma-shaped pattern of clouds in this satellite image is characteristic of a well-developed extratropical cyclone. Note the *dry slot* to the east and north-east of the system's center.

Cold, warm, and occluded fronts intersect at the point of occlusion, or **triple point**, where conditions can favor development of a new cyclone, sometimes called a *secondary cyclone*. (Note the similar appearance of the triple point with its fronts in Figure 7.16D and the incipient low with its fronts in Figure 7.16A). At this stage, the cloud pattern typically becomes a spiraling swirl with enhanced bands associated with the fronts. The spiraling cloud bands can circle the center of an intense low (central pressure of 960 mb or lower) several times.

Eventually, the cyclone weakens as its central pressure rises, the horizontal pressure gradient weakens, and winds diminish; this process is called **cyclolysis** or *filling*. The cyclone can weaken during any of the stages described above if its upper-air support decreases to the point that horizontal divergence aloft becomes less than horizontal convergence near the surface. Then the surface pressure at the low center rises, and the surrounding horizontal pressure gradient weakens. As the central pressure rises, the cyclone loses its identity in the sea-level pressure field and is marked only by an area of cloudy skies and drizzle.

The cyclone life cycle described above may occur over many days, or it may be completed in a day or so. If upper-air support is less favorable, the storm may spend a longer time in any one of the early stages, even weaken temporarily, and still become fully occluded. Sometimes cyclones develop with meager upper-air support (weak divergence aloft) and are poorly defined. At other times, widespread cloudiness and precipitation are linked to an upper-air or surface trough with no closed cyclonic circulation at the surface. When upper-level conditions are ideal, the entire life cycle from incipient cyclone to occlusion can occur in less than a day and a half.

CYCLONE WEATHER

As an illustration of typical extratropical cyclone weather, consider a winter wave cyclone that developed a well-defined warm sector as it moved into the Upper Midwest. Although the storm is still intensifying at map time, its circulation, clouds, and precipitation already affect weather over a broad region. The typical diameter of such a cyclone is between 1000 and 2000 km (about 600 to 1200 mi). Figure 7.18 is a schematic representation of the surface winds, air temperatures, cloud shield, and precipitation pattern associated with this mature low.

Ideally, based on surface weather, a mature wave cyclone can be divided into four sectors about the low center. The lowest air temperatures occur to the northwest of the storm center, where continental polar or arctic air flow southward and eastward. Because of the relatively steep air pressure gradient typically found on the west side of the surface low, winds are strong. Stratiform clouds and non-convective precipitation in this northwest sector are associated with the head of the comma cloud; precipitation may be either rain or snow depending upon air temperature. The cold front is south of the low center; it is accompanied by a relatively narrow band of ascending air, cumuliform clouds, and convective showers and thunderstorms.

Sinking air and generally clear skies characterize much of the southwest sector of the storm system. The mildest air is in the southeast (warm) sector of the cyclone, where south and southeast winds advect maritime tropical air northward from the Gulf of Mexico. Skies are generally partly cloudy, dewpoints are high, and scattered convective showers are possible, triggered by afternoon solar heating. To the north and northeast of the storm center is an extensive zone of overrunning as maritime tropical air surges over a wedge of cold air maintained by east and northeast winds at the surface. Skies in the

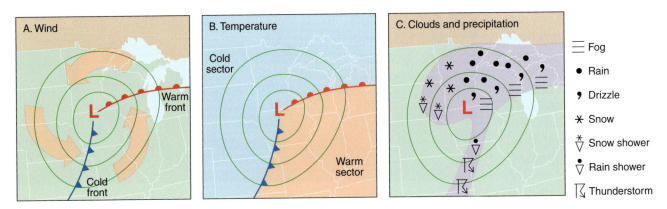

FIGURE 7.18
A mature extratropical cyclone centered on the border between Minnesota and Iowa, showing typical patterns of (A) surface winds, (B) surface air temperatures, and (C) areas of cloudiness and precipitation.

northeast sector are cloudy with an extensive shield of stratiform clouds, and the non-convective precipitation is steady and substantial.

PREFERRED REGIONS OF CYCLOGENESIS

Extratropical cyclones tend to originate and develop in certain geographical regions. A combination of atmospheric conditions and topography is key to the climatology of cyclogenesis (Figure 7.19). Extratropical cyclones tend to form where atmospheric stability is relatively weak thereby favoring ascending motion of air. A nearby source of warm, humid air favors condensation of water vapor at low levels, release of latent heat, and increased buoyancy. In addition, great horizontal temperature gradients in the lower troposphere give rise to fronts associated with the developing low and jet streaks that provide the low with upper-air support (i.e., divergence aloft).

Regions of preferred cyclogenesis occur to the lee of mountain chains and along the east coasts of middle and high latitude continents. Prevailing westerly winds are distorted as they flow over the Rocky Mountains, forming a *lee-of-the-mountain trough* over eastern Alberta and Colorado. These troughs provide upper-air support for cyclogenesis. Also, migrating lows are observed to reform on the leeward slopes of the Rockies.

An anticyclone that sweeps ashore along the west coast of North America seems to disappear from the sea-level pressure pattern as it tracks inland over the mountainous West. A few days later, a low develops to the east of the Rocky Mountain Front Range, typically on the plains of Alberta or eastern Colorado. What actually happened to the storm as it crossed the mountains?

Visualize an extratropical cyclone tracking eastward onshore from the Pacific Ocean as a huge cylinder of air spinning about a vertical axis in a counterclockwise direction as viewed from above (Figure 7.20). The cylinder has Earth's surface as its lower boundary and the tropopause as its upper boundary. Because the tropopause remains at essentially the same altitude over the region,

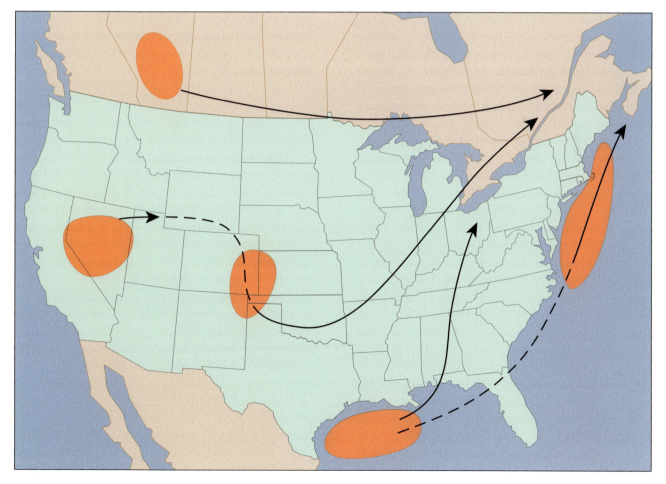

FIGURE 7.19
Principal areas of cyclogenesis and associated cyclone tracks.

FIGURE 7.20
The cyclonic spin of a cylinder of air weakens as it ascends the windward slopes of a mountain range and strengthens as it descends the leeward slopes of the mountain range.

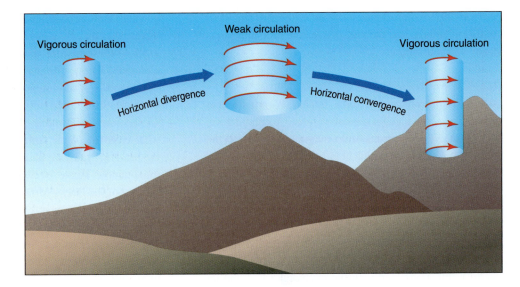

the cylinder is forced to shrink in height as it moves up the windward slopes of a mountain range. The cylinder widens as it shrinks and the spin of the cylinder slows; that is, the storm's circulation weakens. As the cylinder of air then descends the leeward slopes of the mountain range, it stretches vertically and contracts horizontally, and the cyclonic spin strengthens. Weakening cyclonic circulation upslope followed by strengthening cyclonic circulation downslope account for the seeming disappearance and reappearance of the storm as it traverses mountainous terrain.

The changes in the storm's circulation are analogous to what happens to an ice skater performing a spin. The skater changes her spin rate by extending or drawing in her arms. When she extends her arms (analogous to horizontal divergence and the widening of our cylinder), her spin rate slows. When the skater brings her arms close to her body (analogous to horizontal convergence and the contracting of our cylinder), she spins faster. Both the spinning skater and the cyclone passing through the mountains conserve angular momentum. Simply put, *conservation of angular momentum* means that a change in the radius of a rotating mass is balanced by a change in its rotational speed. An increase in radius (horizontal divergence) due to shrinking of the vertical column is accompanied by a reduction in rotational rate whereas a decrease in radius (horizontal convergence) due to stretching of the column is accompanied by an increase in rotational rate.

Along the east coasts of middle and high latitude continents, cyclogenesis is associated with the strong horizontal temperature gradient between a cold continent and warm ocean (especially well-developed in winter

and enhanced by a warm ocean surface current such as the Gulf Stream). Contributing to cyclogenesis is weak atmospheric stability caused by cold air masses from the continent flowing over warmer ocean water. Also, a strong upper tropospheric jet stream is located over such regions in winter because of strong horizontal temperature gradients, again providing upper-air support for development of the storm system.

PRINCIPAL CYCLONE TRACKS

The specific track taken by an extratropical cyclone depends on the pattern of upper-level westerlies in which the storm is embedded. The storm center tends to move in the direction of the wind that blows directly above the surface low pressure system in the mid-troposphere (i.e., at or near the 500-mb level). As a general rule, the cyclone center moves forward at about one-half the speed of the 500-mb winds. Keep in mind, however, that the upper-level circulation pattern also changes with time, and the cyclone's path shifts accordingly.

Principal storm tracks across the coterminous United States are plotted in Figure 7.21. All storms tend to converge toward the northeast; their ultimate destination is usually the semi-permanent Icelandic low of the North Atlantic Ocean or Western Europe. Although many storm tracks appear to originate just east of the Rocky Mountains, in reality they form over the Pacific Ocean, near the Aleutians or in the Gulf of Alaska. For reasons presented above, as a cyclone travels through mountainous terrain, it may temporarily lose its identity, but redevelops on the Great Plains just east of the Front Range of the Rockies.

Cyclones that consistently follow some of the principal tracks have nicknames. The *Panhandle hook*

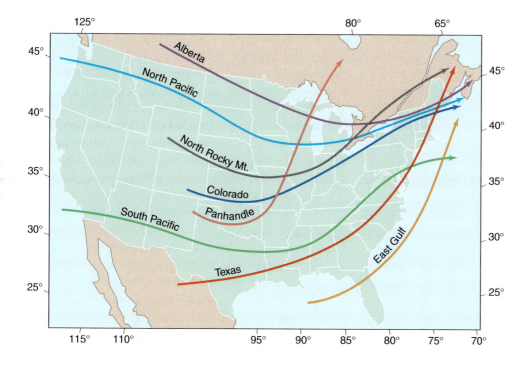

FIGURE 7.21
Principal extratropical cyclone tracks across the lower 48-states of the United States.

forms to the lee of the central Rockies, moves southeastward to the Oklahoma and Texas Panhandles, before turning almost due north over the southern Plains in a hook-like path, hence the name. The *Alberta clipper* develops to the lee of the Canadian Rockies in Alberta and travels rapidly from west to east across southern Canada or the northern tier states. *Nor'easters* track toward the northeast along the East Coast. (Typical tracks of nor'easters are plotted as "Texas" and "East Gulf" in Figure 7.21.) It may be confusing to think of a storm as moving toward the northeast, while the winds in the northeast sector of the storm blow from the northeast. Actually, nor'easters usually intensify off the North Carolina coast and then track in a northeasterly direction up the coast. In effect, two motions occur simultaneously: (1) forward movement of the storm center along the coast and (2) the counterclockwise circulation of winds about the storm center. The circulation of a storm relative to its center is for the most part independent of the storm's path, much as the spin of a Frisbee® is independent of its trajectory.

Generally, cyclones that form in the south yield more precipitation than those that develop in the north. The southerly flow in the warm sector of southern cyclones is better positioned to draw moisture-rich maritime tropical air from source regions such as the Gulf of Mexico. For this reason, Alberta cyclones typically yield only light amounts of rain or snow, whereas Colorado and Gulf-track storms often produce heavy accumulations of rain or snow. The forward speed of the cyclone also affects the total precipitation received at a particular location. For places in the path of a cyclone, a fast-moving system may yield rain or snow for only a few hours whereas a slower moving storm may precipitate for 12 hrs or longer.

Just as the planetary-scale circulation undergoes seasonal changes, so too do cyclones. In summer, when the mean position of the polar front and jet stream is across southern Canada, few well-organized cyclones occur in the United States, and the Alberta storm track shifts northward across Canada. In winter, however, when the mean position of the polar front and jet stream shifts southward, cyclogenesis is more frequent in the coterminous United States. Alberta-track storms are the most common because they occur year-round, whereas storms with more southerly tracks develop primarily in winter. On longer time scales, cyclone tracks would be displaced equatorward during extended cold intervals, but move poleward during warm episodes.

COLD SIDE/WARM SIDE

An extratropical cyclone has a cold side and a warm side. In Figure 7.18, the coldest air is northwest of the low center, where surface winds are blowing from the northwest, while the warmest air is southeast of the low center, where surface winds are blowing from the south. Hence, as an extratropical cyclone travels eastward across the continent, the weather to the north or left (cold side) of the storm track is quite different from the weather to the south or right (warm side) of the storm track.

Consider, for example, the two storm tracks plotted in Figure 7.22. A winter cyclone develops over eastern Colorado. As the storm matures, it moves northeastward toward the Great Lakes region, following either track A or track B, depending on the direction of the steering winds aloft. In both cases, the center of the cyclone passes within 150 km (95 mi) of Chicago. With track A, the storm moves west and then north of Chicago whereas track B takes the storm to the south and then east of the city. The storm's influence on Chicago's weather depends on its track.

If the storm follows track A, Chicago residents experience the warm side of the storm. As the storm approaches, stratiform clouds thicken and lower, with steady rain—or perhaps snow briefly at the onset—giving way to drizzle and fog after 12 to 24 hrs, signaling the arrival of the warm front. As the warm front passes over the city, skies partially clear and winds shift abruptly from east to southeast and then to the south, advecting

warm and humid (mT) air at the surface. At this time, the warm sector is over Chicago as the low center passes to the north. Clearing is short-lived, however, as convective clouds develop and are accompanied by scattered showers and thunderstorms that herald the arrival of colder air behind the cold front. The surface cold front sweeps through the city and then toward the east and northeast. Winds *veer* (turn clockwise with time), blowing initially from the southwest, then west, and finally northwest. Stratocumulus clouds follow the frontal passage and then skies clear again as the air temperature and dewpoint fall.

In contrast, if the storm takes track B, Chicago residents experience the cold side of the storm and no frontal passages. With the approach of the storm, lowering and thickening stratiform clouds accompany gusty east and northeast winds that drive steady snow or rain for 12 hrs or longer. As the storm passes to the south of the city, winds gradually *back* (turn counterclockwise with time) to a northerly direction, precipitation tapers off to

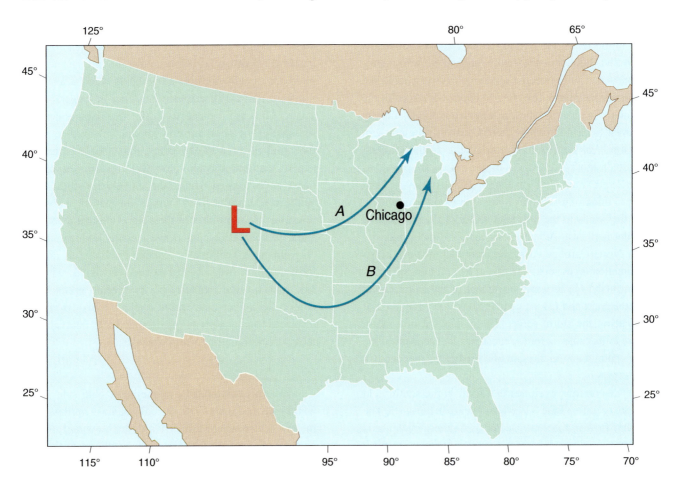

FIGURE 7.22
A winter cyclone develops over eastern Colorado and tracks northeastward toward the western Great Lakes. Track *A* takes the storm to the west and north of Chicago so that the city is on the warm side of the system. Track *B* takes the storm to the south and east of Chicago so that the city is on the cold side of the system.

snow flurries or showers, and air temperatures drop. Finally, winds shift to northwest, the sky begins to clear, and air temperatures continue to fall in response to strong cold air advection as the low moves northeast and away from Chicago.

In summary, if you are located on the warm side of an extratropical cyclone's path, the wind direction veers with time and a cold front follows a warm front. However, if you are on the cold side of a cyclone's path, the wind direction backs with time without the passage of fronts. In winter in a northern location, substantial snowfall is much more likely on the cold side than on the warm side of a cyclone's track, with the axis of heaviest snowfall running parallel and about 240 km (150 mi) to the northwest of the storm track.

COLD- AND WARM-CORE SYSTEMS

An occluded extratropical cyclone is a **cold-core cyclone**; that is, the lowest temperatures occur within the column of air above the low-pressure center. Furthermore, a cyclonic circulation prevails throughout the troposphere and is most intense at high altitudes. On the other hand, the lowest air temperatures in a non-occluded wave cyclone occur northwest of the system's center and the highest temperatures are to the southeast. Until occlusion, the low center aloft is not located above the low center near the surface, but rather is displaced to the cold side of the storm implying that the system tilts with altitude. This structure is consistent with Figure 7.16B, which shows the upper-level trough lagging behind the surface cyclone.

A different type of low that sometimes is plotted on a surface weather map has characteristics markedly different from those of cold-core or tilted lows. Cyclones of this type are stationary, have no fronts, and generally are associated with fair weather. In response to intense solar heating of the ground, they form over a broad expanse of arid or semiarid land, including the interior of Mexico and the southwestern United States.

The hot surface heats the overlying air, lowering the density of the air column over an area broad enough for a synoptic-scale low to appear on a surface weather map. This **warm-core cyclone** (or *thermal low*) is very shallow, and its circulation weakens rapidly with altitude, away from the heat source (the ground). The surface counterclockwise circulation frequently reverses at some altitude, and an anticyclone overlies the thermal low at middle and upper levels of the troposphere. Few clouds and little precipitation develop near the center of the thermal low that overlies the lower Colorado River Valley. To the east, the flow of humid air on southeast

winds produces afternoon showers and thunderstorms over the mountains of Arizona and New Mexico, part of the Southwest Monsoon (discussed later in this chapter). In addition, tropical cyclones (e.g., tropical storms, hurricanes) are warm-core lows that have no associated fronts (Chapter 10).

Anticyclones

In the daily march of weather at middle and high latitudes, migratory anticyclones follow cyclones. As noted earlier, the circulation in an extratropical cyclone favors convergence of contrasting air masses and development of fronts, clouds, and precipitation. In an **anticyclone**, by contrast, subsiding air and diverging surface winds favor formation of a uniform air mass, no fronts, and generally fair skies (Figure 7.23). Air modifies as it moves away from a center of high pressure so that a semi-permanent anticyclone can be the source of different types of air masses. Like cyclones, anticyclones have either cold or warm cores.

ARCTIC AND POLAR HIGHS

A **cold-core anticyclone** is actually a dome of continental polar (*cP*) or arctic (*A*) air and, depending on the specific type of air mass, is either a **polar high** or an **arctic high**. Cold anticyclones are products of extreme radiational cooling over the often snow-covered continental interior of North America well north of the polar front. They are shallow systems in which the clockwise circulation weakens with altitude and frequently reverses direction aloft. Hence, a cold upper tropospheric trough typically overlies a cold surface anticyclone.

FIGURE 7.23
A slow-moving anticyclone is responsible for bright sunny skies with just a few scattered clouds and light winds.

Cold-core anticyclones exert the highest surface pressures in winter, when the associated air mass is coldest and air is most dense. The air is extremely stable in these systems; soundings indicate a temperature inversion in the lowest kilometer or two, associated with strong subsidence and adiabatic heating above the inversion. Massive arctic or polar anticyclones with very high central pressures tend to remain stationary over their source region. The *Siberian high*, for example, is centered near Lake Baikal and from late November to early March has an average sea level air pressure greater than 1030 mb. Lobes of cold air (smaller cold highs) often break away from stationary arctic or polar highs and drift equatorward. In North America, cold air masses move out of source regions in Alaska and northwest Canada and push southeasterly across the Prairie Provinces of southern Canada and into the United States east of the Rockies. Cold anticyclones interact with the circulation of extratropical cyclones, helping to maintain and strengthen the temperature contrast across the cyclone's cold front. Hence, clearing skies and sharply lower temperatures usually follow winter storms as the arctic high pushes into the region.

On some occasions in winter, a particularly strong arctic high brings a surge of bitterly cold air that sweeps as far south as Florida, and rarely can even traverse the Gulf of Mexico into Central America. The resulting subfreezing temperatures can spell disaster for citrus growers and other agricultural interests.

WARM HIGHS

A **warm-core anticyclone** forms south of the polar front and consists of extensive areas of subsiding warm, dry air. Like cold-core cyclones, warm-core anticyclones strengthen with altitude. They are massive systems with a circulation extending from Earth's surface up to the tropical tropopause. The semipermanent subtropical anticyclones, such as the *Bermuda-Azores high*, are examples of warm-core highs (Chapter 6). While extensions of these subtropical highs may stretch across North America, other warm-core anticyclones may develop over the interior of North America, especially in summer.

A cold high coincides with a shallow mass of cold, dense air and produces high surface air pressures compared to surrounding areas. How does a warm anticyclone produce high surface pressures? Whereas the central surface air pressure may be the same for both cold and warm anticyclones, an important difference is found in their vertical structures and the density difference between warm and cold air. The column of cold dense air in a cold-core anticyclone is much shallower than the column of less dense warm air in a warm-core anticyclone. The total mass of air over the center of a warm-core anticyclone, compared to the surrounding atmosphere, is responsible for the warm anticyclone's high surface pressure.

Cold-core anticyclones modify as they travel and may eventually become warm-core systems. As noted earlier, a cold-core anticyclone is actually a dome of relatively cold air, and as that air mass traverses land that is snow free, it is heated from below and moderates considerably, its pressure decreasing significantly. As a cold-core high drifts southeastward over the coterminous United States, air mass modification may be sufficiently great that the pressure system eventually merges with the warm-core subtropical high over the Atlantic Ocean.

ANTICYCLONE WEATHER

An anticyclone is a fair-weather system because surface winds diverging outward induce subsidence of air over a broad area. Because subsiding air is compressionally warmed, the relative humidity drops and clouds usually dissipate or fail to develop. Although anticyclones are fair-weather systems, they can produce weather and climate extremes. The lowest air temperatures of the winter are usually associated with an arctic high, whereas drought and excessive heat in summer are associated with a warm anticyclone that stalls. Sinking air also becomes more stable and a temperature inversion forms as we saw with the trade wind inversion (Chapter 6).

The horizontal air pressure gradient is weak over a broad region about the center of an anticyclone so that prevailing winds are light or the air is calm. Clear skies and light winds favor intense radiational cooling, perhaps resulting in dew, frost, or fog as nighttime air temperatures fall to the dewpoint or frost point. Away from the broad central region of an anticyclone, the horizontal air pressure gradient strengthens and so does the wind. With stronger winds, significant advection can occur. Typically, well to the east of the center of a Northern Hemisphere high, northwest winds advect cold air southeastward, whereas to the west of the high center, southeast winds advect warm air northwestward. Air mass advection helps to increase the temperature contrast across the low pressure trough that separates highs, thereby favoring *frontogenesis*.

An understanding of the circulation in an anticyclone enables us to anticipate the weather sequence as a high moves into and out of a middle latitude location. Consider what happens in winter as a cold anticyclone slides southeastward out of southern Canada and into the northeastern United States. Depending on the anticyclone's forward speed, it may take several days to a week to play out.

Ahead of the anticyclone, strong northwest winds bring a surge of cold continental polar or arctic air. Strong winds and falling temperatures produce low wind-chill temperatures. To the lee of the Great Lakes, heavy lake-effect snow showers break out. Even hundreds of kilometers downwind of the lakes, instability showers bring light accumulations of snow (e.g., over the hills of West Virginia and western Pennsylvania). However, as the anticyclone drifts closer, winds slacken, skies clear, and radiational cooling produces very low surface temperatures over night. Under these conditions, air temperatures dip to their lowest readings, especially where the ground is snow-covered. Then, as the anticyclone drifts away toward the southeast, winds again strengthen, now blowing from the south, and warm air advection begins. The first sign of warm air advection is the appearance of high, thin cirrus clouds in the western sky.

In summer, a Canadian high-pressure system causes the same advection patterns as in winter, except the temperature contrast between air masses is considerably less. Air advected ahead of the high on northwesterly winds may be not much cooler than the air advected behind the high on southerly winds because of the extensive sunlit daytime hours. Often, the most noticeable difference between northerly and southerly winds at middle latitudes is a contrast in humidity. Air advected ahead of the high is often less humid (lower dewpoint), and therefore is more comfortable than the air advected behind the high.

At times in summer, however, an anticyclone becomes established east of the Rocky Mountains and may stall there for weeks. Such anticyclones are warm-core systems with a very deep circulation that is not readily displaced. Aloft, over the high is a warm ridge that may become cut-off from the prevailing westerly flow. With a *block* in place (Chapter 6), subsiding air associated with the anticyclone suppresses cloudiness and precipitation. Persistence of this pattern over many weeks or months causes unusually high temperatures and drought-producing conditions.

The pattern of air mass advection in an anticyclone also applies to an upper-level ridge. Cold air advection usually occurs ahead (to the east) of a ridge, while warm air advection occurs behind (to the west of) a ridge. The circulation in an anticyclone (or ridge) does not occur in isolation from the circulation in a cyclone (or trough). The atmosphere is a continuous fluid, with anticyclones following cyclones and cyclones following anticyclones. Northwest winds develop ahead of the high and on the back (west) side of a retreating low. Winds are caused by horizontal air pressure gradients that develop between migrating anticyclones and cyclones.

Monsoon Climates

Monsoon is derived from the Arabic word *mausim* for season. A **monsoon climate** characterizes regions where seasonal reversals in prevailing winds typically cause wet summers and dry winters. Monsoon was first applied to winds over the Arabian Sea, which blow from the northeast for about six months and then from the southwest for another six months. In this section, we describe the vigorous monsoon circulation over portions of Africa and Asia, where more than 2 billion people depend on monsoon rains for their drinking water and agriculture. Over much of India, monsoon rains, falling between June and September, account for 80% or more of total annual precipitation. We then discuss the weaker monsoon that affects the American Southwest.

ASIAN AND AFRICAN MONSOON

What causes the Asian and African monsoons? As first proposed in 1686 by the English astronomer Edmund Halley (1656-1742), monsoons depend on seasonal contrasts in the heating of land and ocean surfaces. The ocean has a greater *thermal inertia* than does the land (Chapter 4). Beginning in spring, relatively cool air over the ocean and relatively warm air over the land give rise to a horizontal air pressure gradient directed from sea to land, generating an onshore flow of humid air (Figure 7.24A). Over the land, intense solar heating triggers convection. Due to stability considerations, hot, humid air rises, and consequent expansional cooling leads to condensation, cloud development, and rain. Release of latent heat during condensation intensifies the buoyant uplift, triggering even more clouds and rainfall. Aloft, the air spreads seaward and subsides over the relatively cool ocean surface, thus completing the monsoon circulation. The trajectory of monsoon winds is sufficiently long and persistent to be influenced by the Coriolis Effect. Surface monsoon winds are deflected to the right in the Northern Hemisphere and to the left in the Southern Hemisphere.

By early autumn, radiational cooling chills the land more than the adjacent ocean surface, resulting in higher surface air pressures over land and setting up a horizontal air pressure gradient directed from land toward sea. Air subsides over the land, and dry surface winds sweep seaward (Figure 7.24B). Air rises over the relatively warm ocean surface and aloft drifts landward, completing the winter monsoon circulation. Over land, therefore, the summer monsoon is wet whereas the winter monsoon is dry.

FIGURE 7.24
Streamlines showing surface wind patterns over Asia during (A) wet monsoon (summer) and (B) dry monsoon (winter).

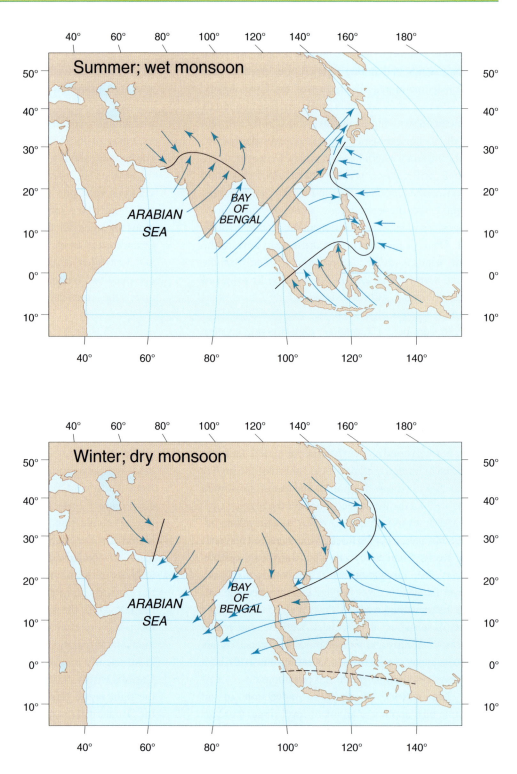

Topography complicates the monsoon circulation and the geographical distribution of rainfall. For example, the massive Tibetan plateau, with elevations topping 4000 m (13,000 ft) over a broad area, strongly influences the Asian monsoon. In winter, the westerly jet stream splits into two branches, one to the south and the other to the north of the plateau. The southerly branch steers cyclones that originate over the Mediterranean Sea across northern India, bringing significant precipitation to that region. Meanwhile, the rest of India experiences the dry monsoon flow. In spring, the southern branch weakens and by late May shifts northward over the plateau. It is not until this happens that the moist monsoon flow begins over India.

Monsoon rainfall is neither uniform nor continual. On the contrary, the rainy season typically consists of a sequence of active and dormant phases. During a *monsoon active phase*, the weather is mostly cloudy with frequent deluges of rain, but during a *monsoon dormant phase*, the weather is sunny and hot. The monsoon shifts from active to dormant phases as bands of heavy rainfall surge inland. Heavy rains first strike coastal areas and soak the ground. As the soil becomes saturated with water, more solar radiation is used for evaporation and less is available for sensible heating. Coastal areas cool, uplift weakens, and skies partially clear (monsoon dormant phase). Meanwhile, the area of maximum heating, most vigorous uplift and heavy rains shifts inland. Back in the coastal areas, however, the hot sun eventually dries the soil, sensible heating intensifies, uplift strengthens, and rains resume (monsoon active phase). This sequence of active and dormant phases repeats about every 15 to 20 days during the wet monsoon.

Solar radiation, land and water distribution, and topography impose some regularity on the monsoon circulation and monsoon climates; that is, summers are wet and winters are dry. The planetary-scale circulation (especially shifts of the ITCZ) and the strength and distribution of convective activity, however, vary from year to year. Consequently, the intensity and duration of monsoon rains are not the same from one year to the next, an example of climate variability. Monsoon failure and drought is always possible in monsoon climates. This chapter's first Essay describes a particularly tragic example of long-term monsoon failure and drought.

Recently, researchers discovered some previously unexpected aspects of the Asian monsoon that may have implications for the impact of future climate change. At the northern edge of the Asian monsoon, in northern China, are fields of sand dunes and to the north is the windswept Gobi desert. During wet periods, vegetation anchors the dunes in place whereas during dry periods, plants die and dunes become active. Researchers are able to reconstruct past dune activity (and rainfall) using *optically stimulated luminescence (OSL)*, a technique that dates the last time the sand was exposed to sunlight and thereby determines times in the past when dunes were active or stable. Joseph Mason and Zhengyu Liu of the University of Wisconsin-Madison and their Chinese and American colleagues used *OSL* to map dune activity and reconstruct rainfall across northern China.

In the October 2009 issue of the journal *Geology*, the researchers reported that between 11,500 and 8000 years ago, climate modeling indicates a strong monsoon with heavy summer rainfall over central and southern China. However, in northern China, during the same period, sand dunes were active indicating dry conditions. The contrast in monsoon rainfall between central and northern China may relate to the prevailing summer monsoon circulation pattern that produced strong upward airflow and heavy rain over central China and a compensating sinking airflow to the north and west that suppressed cloud development and rainfall.

This finding is another example of the geographical nonuniformity of climate change. That is, a strengthening of the southern Asian monsoon (predicted by some climate scientists for the present century) may not result in wetter conditions throughout China. In northern China, in fact, drier conditions may occur, reducing water supply and available grazing lands.

MONSOON OF THE AMERICAN SOUTHWEST

The **North American Monsoon System (NAMS)**, also called the **Southwest Monsoon**, is a prominent feature of the climate of the American Southwest including Arizona, New Mexico, southern Nevada, and parts of southern Colorado. The monsoon brings a dramatic increase in rainfall to this region mainly during July and August. Many places in New Mexico, for example, receive 40% to 50% of their total annual precipitation during these two months.

The Southwest Monsoon originates over northern Mexico during May and June. In fact, the Southwest Monsoon is actually the northern extension of the Mexican monsoon which is responsible for up to 70% of annual precipitation in portions of Mexico. Intense solar radiational heating of the ground raises the temperature of the overlying air, reducing the air pressure at low levels of the troposphere over a broad area (a *thermal low*). Meanwhile a warm high pressure system (the *monsoon high*) develops aloft in the middle and upper troposphere. A horizontal air pressure gradient develops at low levels, which is directed from higher pressure over the waters of the Gulf of California and the Gulf of Mexico toward lower pressure over interior Mexico. In response to this air pressure gradient, warm humid air moves inland giving rise to considerable cloudiness and rainfall. Southerly winds on the western flank of the monsoon high transport moisture northward between the Sierra Madre Mountains to the east and smaller mountains in Baja California to the west.

The monsoon high shifts northward and by July is centered over New Mexico. Winds across the American Southwest shift from the dry westerlies to a humid flow from the south and southeast with some of the flow originating over the Gulf of California and some from the Gulf of Mexico. With this wind shift, the Southwest Monsoon is underway. As a rough guide, the monsoon is considered to be underway in Phoenix, AZ, when the dewpoint reaches 13 °C (55 °F) or higher on three consecutive days. The beginning date on average is 7 July, which corresponds closely with the onset of more frequent thunderstorm activity across central Arizona.

During the monsoon, rainfall across the American Southwest is not continuous but consists of isolated showers and thunderstorms, with the heaviest rains in the mountains. Subsiding dry air associated with the monsoon high is responsible for frequent dry episodes. Nonetheless, the Southwest Monsoon, which is sometimes augmented by rainfall from the remnants of tropical cyclones tracking inland from the eastern Pacific Ocean, is responsible for 70% to 80% of total annual precipitation across Arizona and the Sonora Desert (Figure 7.25). Monsoon rains usually end in September with the return of the dry westerlies.

The Southwest Monsoon is the source of water for dry land farming, ranching, and wildfire control but the timing and strength of the monsoon vary each year. A weak monsoon leads to water shortages and greater susceptibility to forest and brush fires. Also, heavy rains cause flash flooding of drainage ways that are usually dry most of the year. In New Mexico, for example, flash floods are most likely in July and August.

Although sharing some of the fundamental seasonal wind flow characteristics, the Southwest Monsoon is not as strong or as persistent as its south Asian counterpart. This difference can be explained in part by the topography. The Mexican Plateau is not as high as the Himalayan Massif, meaning that the source of heating is not as elevated in the Southwest compared to south Asia. Furthermore, the

general north-south orientation of the mountain ranges in Mexico and the Southwest, along with the narrowness of the land mass to the south does not help develop significant land-sea contrasts as are found in Asia.

SOUTHWEST DRYING TREND

Climate change during the present century is expected to significantly reduce the supply of fresh water in America's Southwest, a region where the supply is already limited and the demand for water is increasing rapidly because of the growing population. In May 2007, Richard Seager of Columbia University's Lamont Doherty Earth Observatory and colleagues reported that an ensemble of 19 numerical climate models (cited in the *IPCC Fourth Assessment Report*) predicted a decline in precipitation over the southwest United States and parts of northern Mexico (25 degrees N to 40 degrees N, 125 degrees W to 95 degrees W) that would persist throughout this century. This prediction assumes that CO_2 emissions would continue to increase until about 2050 followed by a modest decline to about 720 ppmv in 2100.

The decline in precipitation is expected to persist, while the climate models predict that evaporation rates will decrease in summer and be unchanged or slightly higher in winter. The magnitude of this drying trend is expected to be somewhat greater than the Dust Bowl drought (1932-1938) but not quite as extreme as the 1948-1957 southwest drought. Similar drying during this century is predicted elsewhere in the subtropics, across the southern Europe-Mediterranean-Middle East region. According to model projections, the descending branch of the Hadley cell (responsible for aridity) and the belt of middle latitude westerlies will shift poleward during the 21st century, causing the subtropical dry zone to expand poleward (Chapter 6).

Interests that depend on Colorado River water are particularly vulnerable to the Southwest drying trend (Chapter 1). Allocation of Colorado River water to the seven user states and Mexico is regulated by the Colorado River Compact (signed in 1922 and amended in 1944 to include Mexico). According to the Compact, each year up to 7.5 million acre-feet (about 9.3 billion m³) of Colorado River water is allocated to the upper Colorado basin states (Utah, Wyoming, Colorado, and New Mexico), the same amount to the lower basin states (California, Nevada, and Arizona), and at least 1.5 million acre-feet to Mexico (Figure 1.6). The total annual allocation is 16.5 million acre-feet. (One *acre-foot* of water would cover one acre to a depth of one foot, that is, approximately 326,000 gallons, or roughly the volume of water used by two to three households in one year.)

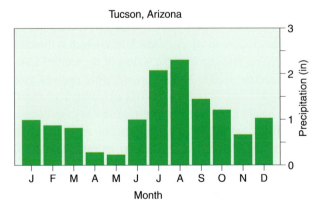

FIGURE 7.25
Mean monthly precipitation in inches (based on 1971-2000) at Tucson International Airport, AZ, showing the abrupt increase in rainfall from June to July signaling the onset of the Southwest Monsoon. [NOAA data]

It turns out that the river flow data that were the basis for water allocations specified by the Colorado River Compact came from a 26-year period when the climate was relatively wet. Figure 7.26 is a graph of the annual flow in million acre-feet (MAF) of the Colorado River at Lee's Ferry, located near the division between the upper and lower basin states. Note the considerable interannual variability and the 10-year running mean (useful for determining trends). Prior to 1922, there was no river gauge at Lee's Ferry so that flow data were interpolated from gauges up and down the river but only as far back as 1896. From 1906 to 1921, the annual flow averaged 18.5 MAF but from 1930 to 1977, the annual flow averaged only 13.6 MAF. The average flow of the Colorado River during the 20th century was about 15 MAF. Hence, if all the Colorado water allocations were actually used, the water demand would exceed the long-term average flow in the river by about 10%.

During the first eight years of the 21st century, drought prevailed in the Colorado River basins but at temperatures higher than the prior droughts of the early 20th century, 1930s, and the mid-century. Consequently, the Colorado River was flowing at 60% of its long-term average and the levels of Lakes Powell and Mead dropped considerably. Melting snow supplies more than 75% of the water in the Colorado River. Higher temperatures will mean earlier melting of the mountain snowpack, more sublimation of snow, greater evapotranspiration, and drier soils that will absorb more snowmelt.

Tim P. Barnett and David W. Pierce of Scripps Institution of Oceanography in La Jolla, CA, point to the likely impact of the drying trend on reservoirs along the Colorado River. In an average year about 18.5 km³ of water flows into Lake Mead behind Hoover Dam. Of this, about 2.1 km³ either evaporates or infiltrates the ground. In 2008, more than 16.6 km³ of water was diverted to cities such as San Diego and Los Angeles. Based on the expected drying trend and current water allocations along the Colorado River, Barnett and Pierce predict a 50% chance that by 2023, Lake Mead will not provide water without pumping. (Already the recent severe drought the reservoir water volume at half capacity.) Furthermore, Barnett and Pierce project a 50% chance that by 2017, Hoover Dam will no longer be a source of hydroelectric power because of insufficient water flow.

To the west, in California, the concern is also the diminishing mountain snowpack due to higher

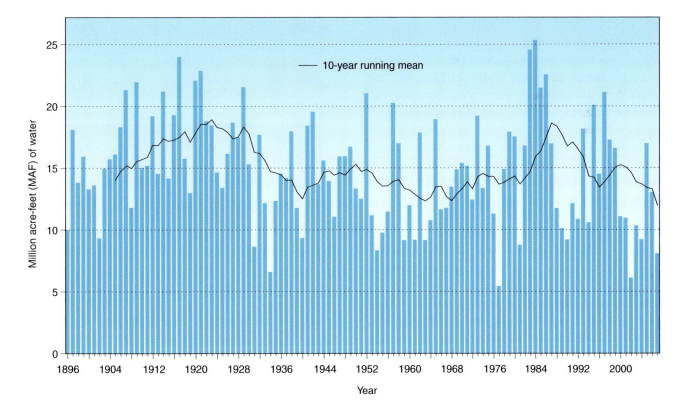

FIGURE 7.26
Annual flow of the Colorado River at Lee's Ferry in millions of acre-feet (MAF). Lee's Ferry is located near the Arizona/Utah border just downstream of the Glen Canyon Dam. Black line indicates the 10-year running mean. [U.S. Bureau of Reclamation]

temperatures. The West relies on a late-melting thick snowpack to supply reservoirs in late spring and beyond. According to Barnett, the warming trend, probably anthropogenically induced, is shrinking mountain snowpacks because more precipitation is falling as rain and the snowmelt is occurring earlier in spring.

Local and Regional Circulation Systems

Air masses, fronts, cyclones, and anticyclones are synoptic-scale systems that dominate the weather and climates of middle and high latitudes. Many other circulation systems operating at smaller spatial and temporal scales also contribute to the weather and climate worldwide. In some cases, larger scale systems set the boundary conditions that make smaller scale circulation systems possible. In this section, we describe sea and land breezes, mountain and valley breezes, Chinook winds, katabatic winds, desert winds, and lake-effect snow. In this chapter's second Essay, we examine the atmospheric circulation that is responsible for tropical and subtropical cloud forest climates.

SEA (OR LAKE) AND LAND BREEZES

For people lucky enough to live near the ocean or a large lake, a sea or lake breeze can bring welcome respite from the oppressive heat of a summer afternoon. On warm days, if the synoptic-scale air pressure gradient is relatively weak, a cool wind sweeps inland from over the sea or large lake. Depending on the source, this refreshing wind is called either a **sea breeze** or a **lake breeze**. Both breezes owe their existence to differential heating of land and water surfaces.

When land and water are exposed to the same intensity of solar radiation, the land surface warms more than the water surface (Chapter 4). The relatively warm land heats the overlying air, thereby lowering air density. Compared to the land, the water surface is relatively cool, as is the air overlying the water. A local horizontal air pressure gradient develops between land and water, with higher air pressure over the water surface and lower air pressure over the land (Figure 7.27A). In response to this horizontal air pressure gradient, cool air sweeps inland. Aloft, continuity requires a return flow of air directed from land to water, with air rising over the relatively warm land and sinking over the relatively cool water.

Sea (or lake) breezes are shallow circulation systems, generally confined to the lowest 1000 m (3300 ft) of the troposphere. Typically, the breeze begins near the shoreline several hours after sunrise and gradually expands both inland and out over the body of water, attaining maximum strength by middle afternoon. The inland extent of the breeze varies from only a few hundred meters to tens of kilometers, depending in part on local topography.

Near sunset, the sea (or lake) breeze dies down. By late evening, surface winds begin to blow offshore as a **land breeze**, due to the reversal in heat differential between land and water. At night, radiational cooling chills the land surface more than the water surface. Air

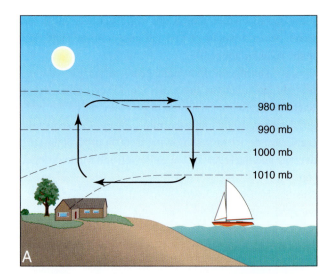

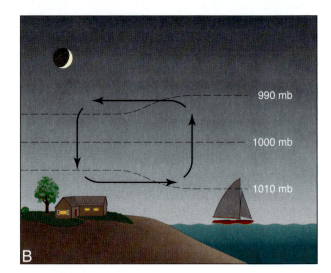

FIGURE 7.27
Vertical cross-sections showing isobaric (constant pressure) surfaces and circulation in (Top) a sea (or lake) breeze and (Bottom) a land breeze. A sea (or lake) breeze blows onshore during the day whereas a land breeze blows offshore at night.

over the land surface thus becomes cooler and denser than air over the water surface. A horizontal gradient in air density gives rise to a horizontal air pressure gradient with higher pressure over the land and lower pressure over the sea (or lake). A cool offshore breeze develops, along with a return airflow aloft; air sinks over the relatively cool land and rises over the relatively warm water (Figure 7.27B). A land breeze attains maximum strength around sunrise but is generally weaker than a sea (or lake) breeze.

Earth's rotation typically does not significantly influence the direction of sea (or lake) and land breezes because the duration of these wind regimes is too brief and distances too small. In some places, however, the Coriolis Effect is responsible for a gradual shift in the direction of a sea breeze through the course of a day. Furthermore, in some localities, uplift produced along a sea-breeze front spurs the development of convective clouds and perhaps showers or thunderstorms.

MOUNTAIN BREEZE AND VALLEY BREEZE

In summer, another localized diurnal circulation may develop in wide, deep mountain valleys that face the Sun (Figure 7.28). After the winter snows have melted, bare valley walls strongly absorb solar radiation and sensible heating raises the temperature of air in contact with the valley walls. Air adjacent to the valley walls becomes warmer and less dense than air at the same altitude out over the valley floor. The cooler, denser air over the valley sinks as air adjacent to the valley walls blows upslope as a **valley breeze**. The ascending valley breeze expands and cools and may trigger development of cumulus clouds near the summit.

A valley breeze is best developed between late morning and sunset. By midnight, the circulation reverses direction and persists until about sunrise. Under clear skies, nocturnal radiational cooling chills the valley walls and the air in contact with the walls also cools. Now air adjacent to the valley walls is colder and denser than the air at the same altitude above the valley floor. Air over the valley is forced to ascend as the cold, gusty **mountain breeze** blows downslope. Cold, dense air accumulates in the valley bottom where additional radiational cooling may lead to formation of fog or low stratus clouds.

Mountain and valley breezes are most common during fair weather and when synoptic-scale winds are light. Hence, these localized winds typically occur when mountainous regions are under the influence of a slow-moving anticyclone.

Valley breeze

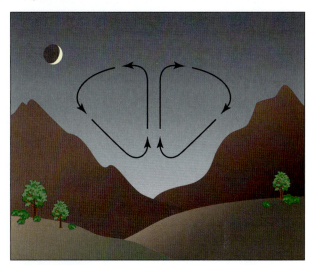

Mountain breeze

FIGURE 7.28
Schematic representation of the circulation in a valley breeze and a mountain breeze.

CHINOOK WIND

A **chinook wind** is a relatively warm and dry wind, which develops when air aloft is adiabatically compressed as it descends the leeward slopes of a mountain range. (The name comes from the Chinook Indians who lived along the lower Columbia River valley and coastal Washington and Oregon.) For every 1000 m of descent, the air temperature rises about 9.8 Celsius degrees (the *dry adiabatic lapse rate*). Air that flows down the slopes of high mountain ranges such as the eastern side of the Rocky Mountains in the U.S. and Canada can undergo considerable warming. As this unsaturated air warms, its relative humidity decreases. During a chinook, a band of clouds, known as the *chinook arch*, overhangs the Rocky Mountain crest marking the location where clouds that formed on the windward slopes vaporize.

Typically, a chinook develops when strong winds force a layer of stable air in the lower troposphere to ascend the windward slopes of a mountain range. When the air reaches the leeward slopes, its stability causes it to descend to its original altitude. The larger-scale circulation causes further descent of the air. For example, chinook winds may be directed down the leeward slopes of the Rocky Mountain Front Range by strong west winds associated with cyclones and anticyclones centered over the Great Plains well east of the mountains. In another case, downslope winds may be associated with the circulation produced by a large high pressure system centered over the Great Basin (in Utah) and a low pressure system tracking across the Canadian Rockies.

At the onset of a chinook, surface air temperatures often abruptly climb tens of degrees in response to the arrival of compressionally warmed air. On 6 January 1966, at Pincher Creek, Alberta, a chinook sent the temperature soaring 21 Celsius degrees (38 Fahrenheit degrees) in only 4 minutes. The sudden spring-like warmth may just as quickly give way to bitter cold as chinook winds abate. Along the foot of the Rockies, a shift of synoptic-scale winds from a westerly direction to a northerly direction brings an abrupt end to the chinook, the return of polar or arctic air, and plunging temperatures.

According to Native American tradition, chinook means *snow eater*, referring to the catastrophic effect of the warm, dry wind on a snow cover. As noted in Chapter 5, air ascending the windward slopes of a mountain range loses much of its water vapor to condensation and deposition (cloud formation) and precipitation. As air descends the leeward slopes and is compressionally warmed, the relative humidity drops dramatically. Because the chinook is both warm and very dry, a snow cover sublimates and melts rapidly. It is not unusual for a half meter of snow to disappear in this way in only a few hours.

Chinook winds tend to be strong and gusty and, locally, may reach destructive speeds, especially along the foothills of the Front Range of the Rocky Mountains. At Boulder, CO, in the foothills just northwest of Denver (Figure 7.29), violent downslope winds, sometimes gusting to 160 km per hr (100 mph) or higher, unroof buildings and topple power poles. On average, the community sustains about $1 million in property damage each year because of these destructive winds.

In southern California, the **Santa Ana wind** is a chinook-type wind that typically occurs in autumn and winter. A strong high pressure system centered over the Great Basin sends northeast winds over the southwestern United States, driving air downslope from the desert plateaus of Utah and Nevada, around the Sierra Nevada, and as far west as coastal southern California (Figure 7.30). Adiabatic compression produces hot, dry winds that desiccate vegetation and contribute to outbreaks of forest and brush fires (Chapter 13). Santa Ana winds sometimes gust to 130 to 145 km per hr (80 to 90 mph) and almost always cause property damage.

Chinook-like winds are not restricted to North America. A similar wind, called the **foehn** (or *föhn*) blows into the Alpine valleys of Austria, Germany, and Switzerland. The same type of warm, dry wind is drawn down the leeward slopes of the Andes in Argentina, where it is known as the **zonda**.

FIGURE 7.29
Boulder, CO, situated in the foothills of the Rockies, experiences some particularly strong and destructive downslope winds.

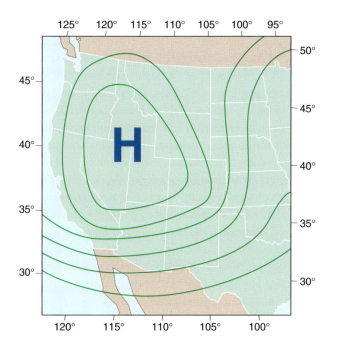

FIGURE 7.30
Schematic representation of the surface weather pattern that favors development of hot and dry Santa Ana winds over Southern California. Solid lines are isobars.

KATABATIC WIND

A shallow layer of cold, dense air flows downhill under the influence of gravity. This **katabatic wind** usually originates in winter over an extensive snow-covered plateau or other highlands. Although the descending air warms to some extent by adiabatic compression, the air is so cold to begin with that these winds are still quite cold when they reach the lowlands. Among the best known katabatic winds are the mistral and bora. The **mistral** descends from the snow-capped Alps down the Rhone River Valley of France and into the Gulf of Lyons along the Mediterranean coast. The **bora** originates in the high plateau region of Croatia and cascades onto the narrow Dalmatian coastal plain along the Adriatic Sea. The mistral and bora are winter phenomena.

Katabatic winds typically are weak with speeds averaging less than 10 km per hr (6 mph). But in some places, such as inlets of the coastal mountain ranges of British Columbia and Alaska, katabatic winds are forced to blow through narrow valleys, and this constricted flow sometimes accelerates the wind to potentially destructive speeds. Steep slopes also accelerate katabatic flow. Along the edge of the Greenland and Antarctic ice sheets, for example, katabatic winds frequently top 100 km per hr (62 mph). The bora also may produce gusts of 50 to 100 km per hr (31 to 62 mph).

DESERT WINDS

Deserts are windy places primarily because of intense solar heating of the ground. In deserts, most of the absorbed radiation goes into sensible heating because water is scarce, vegetation is sparse, and relatively little heat is used for evapotranspiration. In some spots, the summer midday temperature of the ground surface can exceed 55 °C (131 °F). Such a hot surface generates a *superadiabatic lapse rate* within the lowest air layer (that is, a lapse rate greater than 9.8 Celsius degrees per 1000 m). A superadiabatic lapse rate means great instability, vigorous convection, and gusty surface winds. But with negligible water vapor, few clouds form. The strength and gustiness of the wind vary with the intensity of solar radiation so that wind speeds and gustiness usually peak in the early afternoon and during the warmest months.

A **dust devil**, a whirling mass of dust-laden air, is a common sight over flat, dry terrain (not exclusively deserts). Dust devils develop on sunny days in response to local variations in surface characteristics (i.e., albedo, moisture supply, topography) that give rise to localized hot spots. Air over a hot spot is heated and forced to rise by cooler surface winds that converge toward the hot spot. Changes in wind speed and/or direction, known as *wind shear*, cause the column of rising hot air to spin about a nearly vertical axis. The source of wind shear may be nearby obstacles that disturb the horizontal wind, or the overturning of air induced by extreme instability. In the process, dust is lifted off the ground and whirled about, making the system visible. Unlike a tornado, a dust devil is not linked to a cloud.

Most dust devils are small whirls less than 1.0 m (about 3 ft) across, which typically last less than a minute. On rare occasions, a dust devil exceeds 100 m (330 ft) in diameter and whirls for 20 minutes or longer. Such intense dust devils may be visible to altitudes topping 900 m (3000 ft), but the invisible portion of the rising air column may reach 4500 m (15,000 ft).

Most dust devils are too weak to cause serious property damage. The larger ones, however, are known to produce winds in excess of 75 km per hr (45 mph), and may cause some damage. According to NOAA's National Weather Service, every year in New Mexico several large dust devils cause substantial damage to mobile homes, travel trailers, and buildings under construction. In the spring of 1991, a powerful dust devil that fortuitously passed over anemometers at the NWS Forecast Office at Albuquerque produced a wind gust of 113 km per hr (70 mph).

Thunderstorms or migrating cyclones produce larger scale winds in deserts. Surface winds associated with these weather systems can give rise to dust storms or sandstorms, the difference between the two hinging on the size range of the loose surface sediments lifted by the wind. Dust consists of very small particles (diameters less than 0.06 mm or 0.002 in.) that can be carried by winds to great altitudes. Sand, typically covering only a small fraction of desert terrain, consists of larger particles (diameters of 0.06 to 2.0 mm or 0.002 to 0.08 in.) that are transported by the wind within about a meter of the ground. Dust storms and sandstorms can be hazardous, abruptly reducing visibility to perhaps only a few meters, contributing to vehicular accidents on highways.

The strong, gusty downdraft of a thunderstorm generates one of the most spectacular dust storms, known as a **haboob**. (The name derives from the Arabic word *habb*, meaning wind.) In a desert, rain falling from thunderstorm clouds often evaporates completely in the dry air beneath cloud base and does not reach the ground. A thunderstorm downdraft, however, exits the cloud base and strikes the ground as a surge of cool, gusty air, which lifts dust off the ground. The dusty mass rolls along the ground as a huge ominous black cloud, severely restricting visibility. A haboob may be more than 100 km (60 mi) wide and may reach altitudes of several thousand meters. These dust storms are most common in the northern and central Sudan, especially near Khartoum, between May and September. They also occur in the American southwest deserts.

LAKE- EFFECT SNOW

A highly localized fall of snow immediately downwind of an open lake at middle latitudes of the Northern Hemisphere is known as **lake-effect snow**. Typically, such snows extend inland only a few tens of kilometers. In fact, the snowfall is often confined to such a small area that weather stations do not detect the event except by radar or local observer reports. Residents of the affected area, however, may be swamped by snow. On 21 January 1994, Adams, NY at the eastern end of Lake Ontario, received 152 cm (60 in.) of road-clogging snow in only 18 hrs. Such an extreme lake-effect snowfall is sometimes called a *snowburst*.

Lake-effect snow is most common in autumn and early winter when lake surface temperatures are still relatively mild. As an early season outbreak of cold (arctic) air streams over the lake, water readily evaporates and raises the vapor pressure of the lowest portion of the advecting air mass. In addition, the warmer lake water heats the cold air from below, reducing its stability and enhancing convection and cloud development. Often this is all that is needed to trigger snowfall over the lake.

As the modified (milder, more humid, and less stable) air flows toward the lake's lee (downwind) shore, the contrast in surface roughness between the lake and land becomes important. The rougher land surface slows onshore winds, and the consequent horizontal convergence induces ascent of air, further development of clouds, and lake-effect snow. The topography of the downwind shore also affects the amount of precipitation. Hilly terrain forces greater uplift and enhances snowfall.

Ultimately, the frequency and intensity of lake-effect snows hinge on the degree of air mass modification, which in turn, depends on the temperature contrast between the relatively warm lake surface and overlying cold air, and the distance the wind travels over open water (*fetch*). Field studies conducted in the Great Lakes region indicate that lake-effect snow is most likely when the temperature difference between the lake-surface and an altitude of about 1500 m (5000 ft) is greater than 13 Celsius degrees (23 Fahrenheit degrees), and the temperature contrast between the lake-surface and the adjacent land surface exceeds 10 Celsius degrees (18 Fahrenheit degrees). Ideally, the fetch of the wind must be at least 160 km (100 mi) and the horizontal wind direction must exhibit very little shear with altitude (varying by less than 30 degrees between the lake-surface and an altitude of about 3000 m or 10,000 ft).

To the lee of the Great Lakes, the bulk of lake-effect snows falls between mid-November and mid-January, the period when the temperature contrast between lake-surface and overlying air is greatest. Cold air usually sweeps into the Great Lakes region on west and northwest winds. Considering the maximum possible fetch, the greatest potential for substantial lake-effect snows is along the downwind southern and eastern shores of the Great Lakes. In these *snowbelts*, lake-effect snow accounts for a substantial portion of total seasonal snowfall (Figure 7.31).

Occasionally, lake-effect snows develop on the western shores of the lakes which are normally upwind. For example, an early winter extratropical cyclone tracking through the lower Great Lakes region may produce strong northeast winds which blow onshore on the western shores. In this case, the lake's snow-generating mechanism adds to the snowfall produced by the cyclone. This is sometimes distinguished as **lake-enhanced snow** and can mean paralyzing accumulations of snow for places such as the Milwaukee-Chicago metropolitan areas.

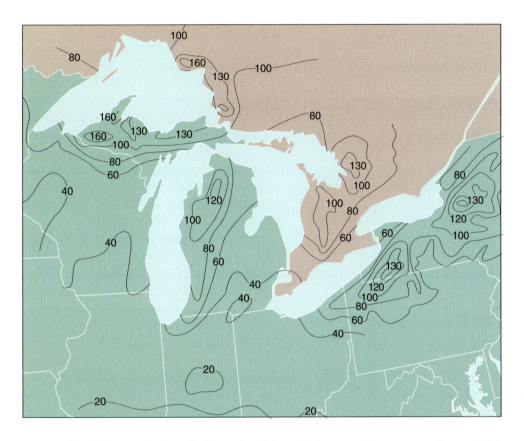

FIGURE 7.31
Average annual snowfall in the Great Lakes region (in inches). The greatest snowfall totals occur downwind of the lakes and are largely due to lake-effect snows.

In addition to the Great Lakes region, lake-effect snows also affect the valleys southeast of Utah's Great Salt Lake. Each year perhaps six or so significant lake-effect snowfalls occur in the Tooele and Salt Lake valleys, bowl-shaped depressions that slope up and away from Great Salt Lake. Both valleys parallel the northwest-southeast-trending long axis of the lake. Which of the two valleys receives the heavier snowfall depends on wind direction. Cold air on northwest winds is moderated by contact with the relatively warm surface of the lake and downwind, the mountain valley topography forces horizontal convergence and uplift. Clouds billow upward locally releasing bursts of heavy snow.

Great Salt Lake snows are generally less intense than their Great Lakes counterparts. Major reasons are the smaller area of the lake and the shorter fetch of winds blowing over the lake, so that air masses can not modify as much. The waters of Great Salt Lake cover an area that is only about 13% of that of Lake Ontario, the smallest of the five Great Lakes. Maximum fetch on Great Salt Lake is only about 120 km (75 mi), whereas maximum fetches on the Great Lakes are measured in the hundreds of kilometers.

Similar localized snowfalls sometimes accompany onshore winds along Atlantic coastal localities. Cape Cod, MA, for example, may experience this *ocean-effect snow* with northerly winds. Rarely, north winds following the passage of a coastal low pressure system have even produced *bay-effect snows* over eastern Virginia off the Chesapeake Bay.

Conclusions

The prevailing atmospheric circulation is a major player in Earth's climatic regimes at all spatial scales. At middle and high latitudes, the climate is strongly influenced by the regular march of air masses, fronts, cyclones, and anticyclones. The track and types of weather associated with these systems depend on the atmospheric circulation operating at the planetary- and synoptic-scale. The monsoon circulation is linked to the planetary-scale circulation (particularly shifts of the ITCZ) and is responsible for monsoon climates featuring wet summers and dry winters. In certain locales, boundary conditions such as topography, Earth's surface properties, and proximity to large bodies of water give rise to local or regional weather systems such as Chinook winds, sea breezes, and lake-effect snows that leave their signature on the climate.

The next chapter covers the role of air/sea interactions in Earth's climate system. Special emphasis is on El Niño and La Niña, important contributors to inter-annual climate variability.

Basic Understandings

- Air masses are classified on the basis of temperature and humidity, characteristics that are acquired in their source region. The four basic air mass types are cold and dry, cold and humid, warm and dry, and warm and humid.

- Air masses form over land (continental) and ocean (maritime); at high latitudes (polar) and low latitudes (tropical). Arctic air is distinguished from polar air by its bitter cold.

- Air masses modify as they travel from one place to another with the degree of modification depending on air mass stability and the nature of the surface over which the air travels.

- A front is a narrow zone of transition between air masses that contrast in temperature or humidity. The four types of fronts are: stationary, warm, cold, and occluded. A stationary front exhibits essentially no horizontal movement. A stationary front becomes a cold front if it begins to move in such a way that colder (more dense) air displaces warmer (less dense) air. If a stationary front begins to move in such a way that the warm (less dense) air advances while the cold (more dense) air retreats, the front changes in character and becomes a warm front.

- An occluded front forms late in the life cycle of a cyclone as the system moves into colder air. The cold front sweeps around the low to overtake and merge with the warm front, forcing the warm air aloft.

- Weather associated with stationary or warm fronts typically consists of a broad stratiform cloud and precipitation shield that may extend hundreds of kilometers on the cold side of the surface front. Weather along or ahead of a cold front usually consists of a narrow band of cumuliform clouds and brief rain or snow showers, or thunderstorms.

- With adequate upper-air support, an extratropical cyclone can form and intensify. Cyclogenesis, the birth of a cyclone, usually takes place along the polar front directly under an area of strong horizontal divergence in the upper troposphere. Strong horizontal divergence aloft occurs to the east of an upper-level trough and under the left-front quadrant of a jet streak, a particularly strong segment of a jet stream.

- As an extratropical cyclone progresses through its life cycle, it is supported and steered by the upper-level circulation toward the east and northeast. An extratropical cyclone typically begins as a wave along the polar front and deepens as the surface air pressure continues to drop. Winds strengthen and frontal weather develops. The cyclone finally occludes as the faster-moving cold front approaches the slower-moving warm front and the upper-level trough becomes vertically stacked over the surface cyclone.

- The track followed by a cyclone is key to the type of weather experienced at a given locality. On the cold side of the storm track, winds back with time (turn counterclockwise) with no frontal passages. On the warm side of the storm track, winds veer (turn clockwise) with time and a cold front follows a warm front.

- Across eastern North America, Alberta-track cyclones are most frequent, but Colorado and coastal storm tracks are responsible for heavier precipitation, especially in winter.

- A warm-core cyclone (thermal low) is stationary, has no fronts, and is associated with hot dry weather. These relatively shallow systems develop in the American southwest desert.

- A cold-core anticyclone is a shallow system that coincides with a dome of continental polar or arctic air. Cold air advection occurs ahead of a high and warm air advection occurs behind a high.

- A warm-core anticyclone, such as a semipermanent subtropical high, extends from Earth's surface to the tropopause, and produces a broad area of subsiding warm, dry air.

- A monsoon circulation depends on seasonal contrasts in the heating of land and ocean surfaces. The ocean has a greater thermal inertia than does the land. In summer, prevailing onshore surface winds produce monsoon rains whereas in winter prevailing offshore surface winds are responsible for the dry season.

- The North American Monsoon System (NAMS), also called the Southwest Monsoon, is a prominent feature of the climate of the American Southwest including Arizona, New Mexico, and parts of southern Colorado. The monsoon brings a dramatic increase in rainfall to this region mainly during July and August.

- Climate change during this century is expected to significantly reduce the supply of fresh water in the American Southwest, a region where supply is already limited and the demand for water is growing rapidly.
- When synoptic-scale winds are weak, localized horizontal air pressure gradients can develop between land and sea (or lake). In response, winds blow onshore during the day (as a sea or lake breeze) and offshore at night (as a land breeze).
- Mountain breezes and valley breezes develop in summer in deep, wide mountain valleys that face the Sun. A valley breeze is an upslope wind that forms during the day, and a mountain breeze is a downslope wind that forms at night.
- A chinook wind consists of compressionally warmed, stable air that is forced down the leeward slopes of a mountain range by the circulation about cyclones or anticyclones.
- Usually a katabatic wind originates in winter over an extensive snow-covered plateau or other highland. Under the influence of gravity, a shallow layer of cold, dense air flows downhill, creating a katabatic wind.
- Intense solar heating of desert terrain produces a steep temperature lapse rate in the lowest air layer. Dust devils may develop in such unstable air. In addition, strong winds associated with thunderstorms or migrating cyclones may cause dust storms or sandstorms.
- Highly localized snow downwind of an open lake at middle latitudes of the Northern Hemisphere is known as lake-effect snow. Typically, such snowfall extends inland only a few tens of kilometers.

Enduring Ideas

- Air masses, fronts, cyclones, and anticyclones are the principal weather makers of middle and high latitude climates.
- With adequate upper-air support (horizontal divergence), an extratropical cyclone develops and passes through its life cycle steered by winds aloft toward the east and northeast.
- A monsoon circulation depends on seasonal contrasts in the heating of land and ocean surfaces such that summers are wet and winters are dry.
- Topography, Earth's surface properties, and proximity to large bodies of water give rise to local or regional weather systems that impose their signature on the climate.

Review

1. What controls the temperature and humidity of an air mass?
2. Compare and contrast the characteristics of a warm front with those of a cold front.
3. What are the sources of upper-air support for development of an extratropical cyclone?
4. Describe the weather in the southeast sector of a mature extratropical cyclone.
5. Why do Alberta-track cyclones tend to produce less precipitation than Colorado-track cyclones?
6. Why is a thermal low (warm-core cyclone) stationary?
7. Compare and contrast a cold high with a warm high.
8. How does a monsoon climate differ from Mediterranean climates (described in Chapter 6)?
9. Why do sea breezes develop during daylight but not at night?
10. How does a chinook wind differ from a katabatic wind?

Critical Thinking

1. Why does a polar air mass modify more rapidly than a tropical air mass?
2. What factors govern the strength of the polar front jet stream?
3. Why do the principal cyclone tracks across the coterminous United States tend to converge toward the northeast?
4. Distinguish between a cold-core cyclone and a warm-core cyclone.
5. What causes the shift from a monsoon active phase to a monsoon dormant phase?
6. How are monsoon climates linked to seasonal shifts in the planetary-scale circulation?
7. What is the effect of frequent sea breezes on the climate of a coastal locality?
8. How do Santa Ana winds contribute to the wildfire hazard in Southern California?
9. What is the effect of frequent chinook winds on the winter climate of localities downwind from a mountain range?
10. Why are deserts frequently windy places?

ESSAY: Monsoon Failure and Drought in Sub-Saharan Africa

Perhaps nowhere in the world has prolonged drought caused more human misery than sub-Saharan Africa, much of which is known as the Sahel. The Sahel, an Arabic word meaning "shore" or "border," is the transition zone in West Africa between the Sahara Desert to the north and the rainforest of the Guinea coast to the south. As shown on the map in Figure 1, the Sahel includes all or part of Mauritania, Senegal, Mali, Burkina Faso, Niger, and Chad. These are among the poorest nations on Earth. Low average annual rainfall, considerable year-to-year variability in rainfall, and prolonged droughts has brought considerable hardship to the people of the Sahel. They are particularly vulnerable to the effects of drought because most of them depend on agriculture for their livelihood. Droughts forced them off their lands that, even in the best of times, are marginal for survival of crops and livestock. They migrated to cities in search of food and work, and many ended up in refugee camps. In the worst of times, horrible scenes of starving children, emaciated livestock, and withered crops were televised to a worldwide audience.

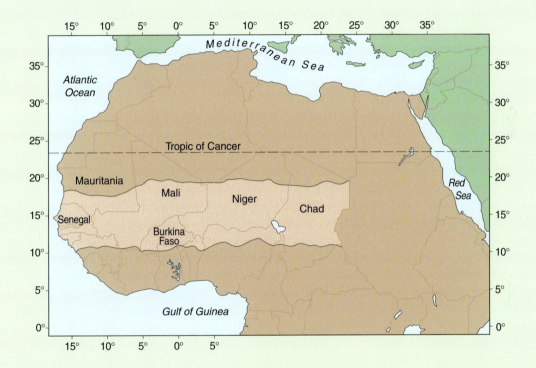

FIGURE 1
The Sahel of Africa, transitional between the Sahara Desert to the north and tropical rainforest to the south, is subject to monsoon failure and prolonged drought.

Drought in the Sahel is the product of interactions involving the atmosphere, ocean, land, and people. The climate of the Sahel is semiarid, featuring a rainy "high Sun" season (Northern Hemisphere summer) and a dry "low Sun" season (Northern Hemisphere winter). The average length of the rainy season and mean annual rainfall increases from north to south across the Sahel. At the northern edge of the Sahel (about 18 degrees N), the rainy season usually does not begin until June and may last only a month or two; mean annual rainfall typically is less than 100 mm (about 4 in.). In the extreme southern Sahel (about 10 degrees N), the rainy season is underway as early as April and persists for up to 5 or 6 months; mean annual rainfall exceeds 500 mm (about 20 in.).

Shifts between the dry season and rainy season are linked to changes in prevailing winds associated with the location of the intertropical convergence zone (ITCZ). As the ITCZ follows the Sun, its northward surge in spring triggers rainfall, whereas a southward shift in fall brings the rainy season to an end. During the rainy season, surface winds blow from the south and southwest, transporting humid air from both the Atlantic and Indian Oceans. During the dry season, the Sahel is under the eastern flank of the Bermuda-Azores subtropical high and surface winds blow from the dry north and northeast.

In a summer when the ITCZ fails to move as far north as usual, or arrives late, or shifts southward early, rainfall is below average. A succession of such summers means drought. In the 20th century, the people of the Sahel endured three major long-term droughts: 1910-1914, 1940-1944, and 1970-1990s. Rainfall during the 1961-1990 period was 20% to 40% lower than it was during the prior three decades, constituting the greatest 30-year anomaly in precipitation recorded anywhere in the world. Rainfall increased in the late 1990s but drought returned in 2005. Drought and famine claimed over 600,000 lives in 1972-1975 and about the same number again in 1984-1985. The reconstructed long-term climate record (based on lake-level fluctuations and historical accounts of landscape changes) indicates that droughts lasting 1 or 2 decades (*megadroughts*) are the norm in the Sahel.

Some scientists argue that the principal reason for drought in the Sahel is changes in atmospheric circulation patterns, that is, seasonal shifts of the moisture-bearing ITCZ that are linked to anomalies in sea-surface temperature (SST). Initial studies showed that when SST is higher than normal in the eastern tropical Atlantic, south of the equator and southwest of West Africa (e.g., the Gulf of Guinea), rainfall is below the long-term average in the Sahel. In 2003 scientists at the International Research Institute for Climate Prediction at Palisades, NY, announced the results of a NASA global climate model study of rainfall in the Sahel covering the period 1930 to 2000. They found that rising sea-surface temperature in the Indian Ocean was more important than changes in Atlantic SST in simulating long-term trends in Sahel rainfall. They also found that El Niño and La Niña accounted for much of the year-to-year variability in Sahel rainfall (Chapter 8). More recently, runs of another global climate model developed at the Max Planck Institute for Meteorology in Germany found a similar dependence of decadal trends in Sahel rainfall on Indian Ocean SST.

But that is not the final word on the matter. In 2005, James Hurrell of the National Center for Atmospheric Research (NCAR) and Martin Hoerling of NOAA reported on new findings regarding drought in the Sahel. In experiments utilizing 5 different numerical models to generate 80 simulations, Hurrell and Hoerling discovered that drought in southern Africa is linked to warming of the Indian Ocean. However, their model simulations indicated that moisture conditions in the Sahel depends on SST in the Atlantic Ocean rather than the Indian Ocean. Apparently, monsoon failure and drought in the Sahel were linked to late 20th century cooling of the North Atlantic Ocean. A shift to wetter conditions in the Sahel during the 1990s corresponded to the post-1990 warming of the North Atlantic.

Human activity (e.g., overgrazing, conversion of woodland to agriculture) may also contribute to persistent drought in sub-Saharan Africa by reinforcing dry conditions. Overgrazing and deforestation alter the regional radiation budget by denuding the soil's vegetative cover and degrading the quality and moisture holding capacity of soils. Soil albedo increases, moisture supply to the atmosphere decreases, convection weakens, and rainfall is less likely. What has happened in parts of the Sahel may be *desertification*, the conversion of arable land to desert due to a combination of climate change and human land management practices.

What does the future hold for the Sahel? Today's climate models disagree on future trends in rainfall in the Sahel. For example, Kerry Cook and Edward Vizy of Cornell University selected three models from the *IPCC Fourth Assessment Report* that best reproduced the climate of the 20th century. One model predicted relatively wet conditions in the Sahel for the entire 21st century, another predicted severe drying late in the century, and the third called for modest drying through the period.

To upgrade forecasting of the West African monsoon, a 10-year field experiment began in 2001, known as the *African Monsoon Multidisciplinary Analysis (AMMA)*. This consortium of more than 140 European, American, and African institutions monitors almost all aspects of the monsoon including rainfall, cloud properties, dust transport, and SST in the Gulf of Guinea. A key priority of AMMA involves increasing the density of observation stations. As of 2006, the entire African continent had only 1152 climate stations, about one-eighth the coverage recommended by the World Meteorological Organization (WMO). A greater density of surface and upper-air (radiosonde) stations promises to improve the performance of climate models by providing more data for initializing them and verifying their predictions.

Better understanding of the West African monsoon will also benefit the performance of global climate models. The Sahara-Sahel is a major source of heat energy for the atmosphere. The region is also a major source of airborne dust that may influence the global radiation budget and climate as well as the development of tropical cyclones (tropical storms and hurricanes) over the Atlantic Ocean basin.

ESSAY: Cloud Forests and Climate Change

About 2.5% of the world's tropical forests are perpetually shrouded in clouds and mist. These *cloud forests* are an important source of clean fresh water for millions of people living in the surrounding lowlands and are home to thousands of rare plants and animals, most of which have yet to be studied by scientists. However, pressure from land developers (i.e., clearing forests for cattle grazing, coca plantations, logging, and mining), perhaps exacerbated by climate change, make cloud forests one of the world's most threatened ecosystems. So far efforts by conservationists to halt the loss of cloud forests by establishing nature reserves have had little success.

Cloud forests are confined to mountainous areas of the tropics and subtropics generally at elevations from 2000 to 3000 m (6500 to 9800 ft) above sea level. In 2004, the World Conservation Monitoring Centre, a component of the UN Environmental Programme (UNEP), reported that about 60% of cloud forests are located in Asia (mostly in Indonesia and Papua New Guinea), 25% in Central and South America (from Honduras, Panama, and Costa Rica to as far south as northern Argentina), and the remaining 15% in Africa. They are also found at lower elevations (as low as 500 m or 1600 ft above mean sea level) on humid portions of Caribbean and Hawaiian Islands (Figure 1). Worldwide, based on satellite monitoring, cloud forests cover about 380,000 square km (94 million acres).

FIGURE 1
Tropical montane cloud forest on East Maui, Hawaii at 1220 m elevation, on the south-facing slope of Haleakala volcano. [Courtesy USGS]

The prevailing onshore and upslope flow of warm, humid air is responsible for low orographic clouds, fog, and mist that characterize cloud forests. The air is initially unsaturated but in flowing up the mountain slopes, it expands and cools. Expansional cooling raises the relative humidity to saturation (100%) and water vapor condenses into low clouds and fog. By trapping tiny cloud droplets and promoting their coalescence into larger drops, the tree canopy strips moisture from the windblown clouds and this water drips to the forest floor. The volume of water provided in this way may amount to 20% to 60% or more of the local rainfall and is especially important where there is a dry season. Hence, the cloud forest is a steady source of water for mountain streams that, in turn, supply water to downstream cities and towns. Persistent cloud cover also reduces incoming solar radiation and suppresses evapotranspiration. Cloud forests are biologically diverse ecosystems that provide unique habitats for lichens, tree ferns, orchids, and numerous species of birds and mammals.

Since the 1970s, dry season moisture from orographic clouds has decreased at Monteverde, Costa Rica. This has been attributed to higher orographic cloud basis arising from climate change and may be responsible for population declines of species in the area at the same time. Using numerical modeling and NASA satellite imagery, Udaysankar Nair of the University of Alabama in Huntsville determined that clearing forests in the lowlands upwind of Monteverde reduced the amount of moisture and raised the air temperature so that orographic clouds formed less readily and at higher elevations. NASA's Moderate Resolution Imaging Spectroradiometer (MODIS) mapped the geographical distribution of cloud forests from cloud immersion measurements, land use, and vegetation characteristics.

Higher sea surface temperatures (SST) in the tropics associated with global climate change could affect the cloud forests. Higher SST would have two opposing impacts on the properties of the air flowing onshore and upslope. For one, the air would be warmer so that greater ascent (i.e., more expansional cooling) would be required to produce clouds. Clouds would form at higher elevations and perhaps lift off some mountaintops entirely. The original cloud forest would become drier, dramatically altering plant and animal habitats.

Higher SST would also increase the rate of evaporation of ocean water resulting in a more humid onshore and upslope airflow. More humid air would condense into clouds at lower elevations. Using a numerical climate model to simulate the effects of higher SST on tropical airflow, Stephen Schneider and his colleagues at Stanford University concluded that higher air temperatures would be the more important factor in controlling the elevation of cloud forests. Cloud forests appear to be particularly sensitive to climate and may be among the early indicators of the environmental effects of global climate change.

CHAPTER 8

CLIMATE AND AIR/SEA INTERACTIONS

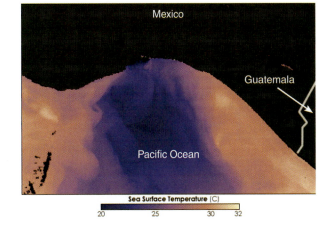

During the winter in southern Mexico, powerful winds called the Tehuano winds roar out toward the Pacific Ocean through breaks in the western coastal mountains. This image from the Moderate Resolution Imaging Spectroradiometer (MODIS) on NASA's Aqua satellite shows sea surface temperature, with deep purple indicating the coolest temperatures and pinkish yellow indicating the warmest. A swath of intense dark purple—indicating cold water—stretches southward away from the coast. This strip of cold water shows where the Tehuano winds pushed surface water away from shore, and allowed cold, deep water to well up and replace it. [Courtesy NASA]

Case-in-Point

During the winter of 1997-98, the San Francisco Bay area experienced its worst flooding in more than 40 years due to unusually high sea levels and wind-driven waves. Property damage from storms and flooding amounted to hundreds of millions of dollars. On 3 February 1998, waters from the Bay inundated U.S. Highway 101 up to a depth of 1.2 m (4 ft) north of the Golden Gate Bridge. Several factors control sea level in San Francisco Bay with the most important being the regular astronomical tides (lunar and solar tides). Each day San Francisco Bay experiences two high tides and two low tides, with a maximum tidal range of about 2 m (6.5 ft) based on long-term measurements at the Fort Point tide gauge in the Presidio at the south end of the Golden Gate Bridge. On 3 February, the morning high tide was about 0.6 m (2 ft) higher than the tide anticipated based on astronomical forcing alone.

A second factor is the slow long-term trend of rising sea level due to global warming that causes expansion of sea water and melting of land-based glaciers. This sea-level rise only amounts to about 0.1 to 0.2 cm (0.04 to 0.08 in.) per year. The coupling of surface winds and water, causing sea level along the California coast to drop in spring and rise in winter also affects the San Francisco Bay's sea level. In spring and summer, prevailing surface winds blow from the north along the coast. As discussed later in this chapter, the frictional coupling of wind and water, plus the Coriolis Effect, transports surface water westward away from the coast thus inducing upwelling of cold water. In autumn and winter, coastal winds relax somewhat and sea level rises slightly.

El Niño is also a major factor affecting sea level. El Niño refers to an episode of anomalous warming of surface waters in the eastern tropical Pacific Ocean. During the 1997-98 El Niño, one of the most intense of the 20th century, winter winds blew strongly from the south along the California coast, driving surface water toward the shore and raising sea level. Reinforcing these south winds was the circulation in strong storms that swept inland from the Pacific. Higher sea level and landward winds increased flooding by Bay waters.

Driving Question:
How do interactions between the ocean and atmosphere impact worldwide climate and short-term climate variability?

Some weather extremes, such as drought and episodes of unusually heavy rains, are linked to changes in the coupled circulation of the atmosphere and ocean. The principal focus of this chapter is seasonal to inter-annual climate variability involving interactions between the ocean and atmosphere. El Niño/La Niña, one of the most extensively studied of these interactions, occurs in the tropical Pacific Ocean. During El Niño, trade winds weaken, upwelling diminishes off the South American coast and along the equatorial Pacific, sea-surface temperatures (SST) rise well above long-term averages over the central and eastern tropical Pacific, and areas of heavy rainfall shift from the western into the central equatorial Pacific.

Occasionally, La Niña follows El Niño, and is a period of exceptionally strong trade winds in the tropical Pacific, vigorous coastal and equatorial upwelling in the eastern Pacific, and lower than usual SST in the central and eastern tropical Pacific. Based on changes in SST in the eastern tropical Pacific, some scientists refer to El Niño as the *warm phase* and La Niña as the *cold phase* of El Niño/Southern Oscillation (**ENSO**) variability.

Broad-scale changes in SST patterns over the tropical Pacific that accompany El Niño and La Niña influence the prevailing circulation of the atmosphere in some portions of the middle latitudes, especially in winter. Weather extremes that most often accompany El Niño are essentially opposite those that usually occur during La Niña. Although we devote much of this chapter to El Niño and La Niña, we also consider other examples of short-term climate variability stemming from air/sea interactions including the North Atlantic Oscillation, the Arctic Oscillation, and the Pacific Decadal Oscillation. We set the stage for this discussion by first describing the basics of air/sea interactions within Earth's climate system.

Air/Sea Interactions

The ocean is a major player in Earth's climate system operating on temporal scales from days to millennia and spatial scales from local to global. The ocean influences radiational heating and cooling of the planet. The ocean, covering about 71% of Earth's surface, is the primary control of the amount of solar radiation absorbed (converted to heat) at Earth's surface. (Recall from Chapter 3 that the average albedo of the ocean surface is only 8%.) Also, the ocean is the main source of the most important *greenhouse gas* (water vapor) and a major regulator of the concentration of atmospheric carbon dioxide (CO_2), another significant greenhouse gas.

On average, the ocean absorbs about 92% of the solar radiation striking its surface and reflects the balance to space. Most of this absorption takes place within 100 to 200 m (330 to 650 ft) of the ocean surface. The amount of suspended particles and discoloration caused by dissolved substances limits the depth of sunlight's penetration. Alternatively, at polar latitudes highly reflective multi-year pack ice and low Sun angles greatly reduce the amount of solar radiation absorbed by the ocean. The snow-covered surface of sea ice absorbs only about 15% of incident solar radiation and reflects away the rest. At present, multi-year pack ice covers about 7% of the ocean surface, with greater coverage in the Arctic Ocean than the Southern Ocean (mostly in Antarctica's Weddell Sea). (The Arctic is an ocean surrounded by continents whereas the Antarctic is a continent roughly centered on the pole and surrounded by ocean.)

The atmosphere is nearly transparent to incoming solar radiation but much less transparent to outgoing infrared radiation. This is the basis of the *greenhouse effect* (Chapter 3). Most water vapor, the principal greenhouse gas, enters the atmosphere via evaporation of seawater (Figure 8.1). Carbon dioxide, a lesser greenhouse gas, cycles into and out of the ocean depending on the sea-surface temperature, circulation patterns, salinity, and biological activity in surface waters. CO_2 is more soluble in cold water than warm water. Chilled surface waters at higher latitudes absorb CO_2 while upwelling, which brings cold water to the surface at warmer lower latitudes, releases CO_2 to the atmosphere. Organisms take up carbon dioxide via photosynthesis and release carbon dioxide via cellular respiration.

The ocean influences the planetary energy budget not only by affecting radiational heating and radiational cooling, but also by contributing to non-radiative latent heat and sensible heat fluxes at the air/sea interface. Recall from Chapter 4 that heat is transferred from Earth's surface to the atmosphere via latent heating (vaporization of water at the surface followed by cloud formation in the atmosphere) and sensible heating (conduction plus convection). On average, about ten times more heat is

FIGURE 8.1
The ocean interacts with the atmosphere exchanging heat energy, water, and gases. The ocean is the primary source of atmospheric water vapor, the principal greenhouse gas, and exchanges carbon dioxide, another greenhouse gas, with the atmosphere.

transferred from the ocean surface to the atmosphere via latent heating than sensible heating.

The ocean and atmosphere are closely coupled. This coupling is most apparent near the ocean's surface where variations in atmospheric conditions can impact temperatures and wind-driven currents within hours to days. On the other hand, the deeper basin-scale *thermohaline circulation* responds more sluggishly to changes in atmospheric conditions, taking decades to centuries, or longer, to fully adjust. In turn, ocean surface currents strongly influence climate. Cold surface currents, such as the California Current, are heat sinks; they chill and stabilize the overlying air, thereby increasing the frequency of sea fogs and reducing the likelihood of thunderstorms. Relatively warm surface currents, such as the Gulf Stream, are heat sources; they supply latent and sensible heat and moisture to the overlying air, destabilizing the air, thereby energizing storm systems. As noted in Chapter 4, ocean surface currents and the thermohaline circulation transport heat from the tropics to higher latitudes.

Although the importance of the ocean in Earth's climate system is apparent, it is important to remember that the ocean and atmosphere work together in governing climate. Simple observations underscore the ocean-atmosphere climate connection in somewhat surprising ways. At the same latitude, winters are significantly milder in Western Europe than in Eastern North America. Since at least the mid-1800s, scientists have attributed this climate contrast primarily to the moderating influence of the warm Gulf Stream and North Atlantic Current on Western Europe. However, recent research results question this view. Some scientists argue that while the Gulf Stream continually replenishes the relatively warm waters of the North Atlantic, it may not be as important as the global-scale atmospheric circulation in explaining Western Europe's relatively mild winters.

In 2002, Richard Seager of Columbia University's Lamont-Doherty Earth Observatory and David Battisti of the University of Washington in Seattle, WA, reported that the key to Western Europe's mild winters is the pattern exhibited by the prevailing westerlies. Recall from Chapter 6 that the westerlies in the mid and upper troposphere blow from west to east in a wave-like pattern of ridges and troughs (Figure 6.28). In winter, a cold pool of air (a *trough*) commonly prevails over eastern North America resulting in cold northwesterly winds on the western flank of the trough, but an intrusion of warm air over the North Atlantic (a *ridge*) is associated with milder southwest winds over Western Europe. That is, in winter, the westerlies tend to blow from the colder northwest over Eastern North America but from the milder southwest over Western Europe. In addition, the relatively great thermal inertia of ocean water means that summer heat persists in North Atlantic surface waters long after the North American continent has cooled in fall. Hence, southwest winds blowing toward Western Europe cross relatively warm waters and are heated from below. The direction of the prevailing winds over the North Atlantic Ocean and Western Europe is primarily responsible for delivering relatively mild air masses in winter. This finding also implies that shifts in the direction of the prevailing winds could alter the winter climate of Western Europe.

Broad scale patterns of sea-surface temperature (SST) strongly influence the location of major features of the atmosphere's planetary scale circulation. When SST patterns change, so too do the locations of planetary-scale circulation features. For example, changes in the location of the highest sea-surface temperatures in the tropical Atlantic affect the north-south shifts of the *intertropical convergence zone (ITCZ)*. Recall that the ITCZ is a discontinuous belt of showers and thunderstorms paralleling the equator, marking the convergence of the trade winds of the Northern and Southern Hemispheres (Figure 6.26). The ITCZ encircles the globe and shifts north and south with the seasonal excursions of the Sun—more so over land and less so over the ocean—reaching its most northerly location in July and most southerly location in January. Displacement of the ITCZ over the tropical Atlantic Ocean affects the timing and amount of rainfall along the east coast of South America from Brazil northward into the Caribbean Sea from March to May and in the western part of sub-Saharan Africa in August and September.

Mean State of the Ocean Circulation

Once the wind sets surface waters in motion as a current, the Coriolis Effect, Ekman transport, and the configuration of the ocean basin modify the speed and direction of the current. This section begins with a discussion of the forces involved in the coupling of wind and ocean surface waters.

EKMAN TRANSPORT

Wind blowing over the ocean exerts a frictional drag that moves surface waters. Ripples or waves produce the surface roughness necessary for the wind to couple with surface waters. A wind blowing steadily over deep water for 12 hrs at an average speed of about 100 cm per sec (2.2 mph) produces a 2 cm per sec current (about 2% of the wind speed).

If Earth did not rotate, frictional coupling between moving air and the ocean surface would push a thin layer of water in the same direction as the wind. This surface layer in turn would drag the layer beneath it, putting it into motion. This interaction would propagate downward through successive ocean layers, like cards in a deck, each moving forward at a slower speed than the layer above. However, because Earth rotates, the shallow layer of surface water set in motion by the wind is deflected to the right of the wind direction in the Northern Hemisphere and to the left of the wind direction in the Southern Hemisphere. (We discussed the *Coriolis Effect* in Chapter 6.) Except at the equator, where the Coriolis Effect is zero, each layer of water put into motion by the layer above shifts direction because of Earth's rotation.

Using arrows to represent the direction and speed of layers of water at successive depths, a simplified model of the three-dimensional current pattern caused by a steady horizontal surface wind emerges (Figure 8.2A). This model is known as the **Ekman spiral**, named for the Swedish physicist V. Walfrid Ekman (1874-1954) who first described it mathematically in 1905. Ekman based his model on observations made by the Norwegian explorer Fridtjof Nansen (1861-1930). Nansen was interested in learning about the currents of polar seas. In 1893, he allowed his 39-m (128-ft) wooden ship, the *Fram*, to freeze into the Arctic pack ice about 1100 km (685 mi) south of the North Pole. His goal was to drift with the ice and cross the North Pole thereby determining how ocean currents affect the movement of pack ice. The *Fram* remained locked in pack ice for 35 months but came no closer than about 394 km (245 mi) to the North Pole. As the *Fram* slowly drifted with the ice, Nansen noticed that the direction of ice and ship movement was consistently 20 to 40 degrees to the right of the prevailing wind direction.

The Ekman spiral indicates that the direction of water movement changes with increasing depth. In an ideal case, a steady wind blowing across an ocean of unlimited depth and extent causes surface waters to move at an angle of 45 degrees to the right of the wind in the Northern Hemisphere (45 degrees to the left of the wind in the Southern Hemisphere). Each successively lower layer moves more toward the right and at a slower speed. At a depth of about 100 to 150 m (330 to 500 ft), the Ekman spiral has gone through less than half a turn. Yet water moves so slowly (about 4% of the surface current) that

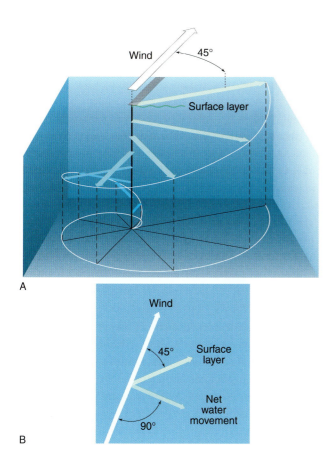

FIGURE 8.2
The Ekman spiral describes how the horizontal wind sets surface waters into motion. (A) As represented by horizontal arrows, the speed and direction of water motion change with increasing depth. (B) Viewed from above in the Northern Hemisphere, the surface layer of water moves at 45 degrees to the right of the wind. The net transport of water through the entire wind-driven column (Ekman transport) is 90 degrees to the right of the wind.

cient for a full spiral to develop so that the angle between the horizontal wind direction and surface-water movement can be as little as 15 degrees. As the water deepens, the angle increases and approaches 45 degrees. Also, tides can have an important influence on Ekman transport.

Ekman transport piles up surface waters in some areas of the ocean and removes surface waters from other areas, producing variations in the height of the sea surface and causing it to slope gradually. One consequence of a sloping ocean surface is the generation of horizontal differences (gradients) in water pressure. These pressure gradients, in turn, give rise to geostrophic flow.

GEOSTROPHIC FLOW AND OCEAN GYRES

The horizontal movement of surface water arising from a balance between the pressure gradient force and the Coriolis Effect is known as **geostrophic flow**. Geostrophic flow characterizes **gyres**, large-scale roughly circular surface current systems in the ocean basins. Recall from Chapter 6 that a distinction is made between subtropical and sub-polar gyres. The trade winds, on the equatorward flank of a subtropical high pressure system, and the westerlies, on the poleward flank of a subtropical high pressure system, drive **subtropical gyres**. The subtropical gyres are centered near 30 degrees latitude in the North and South Atlantic, the North and South Pacific, and the Indian Ocean. Subtropical gyres in the Northern and Southern Hemispheres are similar except that they rotate in opposite directions because the Coriolis Effect reverses direction in the two hemispheres. Viewed from above, subtropical gyres rotate in a clockwise direction in the Northern Hemisphere but in a counterclockwise direction in the Southern Hemisphere.

Driven by the long-term average (prevailing) winds in the semi-permanent subtropical high pressure systems, Ekman transport causes waters to converge from all sides toward the central region of a subtropical gyre. This transport produces a broad mound of water as high as 1 m (3 ft) above mean sea level near the center of the gyre (Figure 8.3). As more water is transported toward the gyre center, the slope of the mound becomes steeper, and the horizontal water pressure gradient increases. In response to the horizontal gradient in water pressure, water parcels move from where the pressure is higher toward where the pressure is lower, that is, downhill from the center of the gyre. Simultaneously, the Coriolis Effect arises because of the water's motion and shifts the direction of the parcels to the right in the Northern Hemisphere (to the left in the Southern Hemisphere). The water parcels speed up until the Coriolis Effect balances the *pressure gradient force*.

this depth is considered to be the lower limit of the wind's influence on the movement of ocean waters.

In the Northern Hemisphere, the Ekman spiral predicts net water movement through a depth of about 100 to 150 m (330 to 500 ft) at 90 degrees to the right of the wind direction (Figure 8.2B). That is, if one adds up all the arrows in Figure 8.2A, the resulting flow is at 90 degrees to the right of the surface wind direction. In the Southern Hemisphere, the net water movement is 90 degrees to the left of the surface wind direction. This net transport of water due to the coupling between the surface wind and water is known as **Ekman transport** and is a dominant type of flow in the surface layer of the ocean (the *mixed layer*).

The real ocean departs from the idealized conditions of the Ekman spiral; that is, wind-induced water movements often differ appreciably from theoretical predictions. In shallow water, for example, the water depth is insuffi-

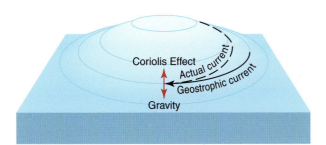

FIGURE 8.3
Ekman transport causes surface waters to converge toward the central region of a subtropical gyre from all sides, producing a broad mound of water. A balance develops between the Coriolis Effect and the force arising from the horizontal water pressure gradient such that surface currents flow parallel to the contours of elevation of sea level. This current is known as geostrophic flow.

With the balancing, the water parcels flow around the gyre and parallel to contours of elevation of sea level as geostrophic flow.

Sub-polar gyres, smaller than their subtropical counterparts, occur at higher latitudes of the Northern Hemisphere; they are the Alaska gyre in the far North Pacific and the gyre south of Greenland in the far North Atlantic. The counterclockwise surface winds in the Aleutian and Icelandic low pressure systems drive the sub-polar gyres. (Recall from Chapter 6 that the *Aleutian low* and *Icelandic low* are persistent features of the atmosphere's planetary-scale circulation.) Hence, viewed from above, the rotation in the sub-polar gyres is opposite that of the Northern Hemisphere subtropical gyres. Ekman transport causes surface waters to diverge away from the central region of the sub-polar gyres. The thinner surface layer permits more nutrient-rich waters from deeper in the ocean to move upward into the Sun-lit photic zone, thereby increasing biological productivity in these regions, as evidenced by the major fisheries in the North Pacific and North Atlantic Oceans.

The long-term average pattern of ocean surface currents is plotted in Figure 6.38. Some currents are relatively warm (red) whereas others are cold (blue). Some (i.e., western boundary currents such as the Gulf Stream) are faster moving than others. Winds associated with passing storm systems disturb the ocean surface and can cause the local flow of ocean currents to deviate temporarily from the long-term average pattern.

Most wind-driven surface currents transport water within a specific ocean basin. One of the few areas of the world ocean where inter-basin transport takes place is the *Indonesian Through-flow*. The islands of Indonesia mark the boundary between the Indian and Pacific Oceans

but only partially block the flow of seawater between the two ocean basins. Although lacking some details, it is clear that warm, low-salinity water from the Pacific is transported into the Indian Ocean around the thousands of Indonesian islands. These waters replenish the large amounts of water removed by evaporation from the northern Indian Ocean. The summer Asian monsoon circulation transports this water vapor from sea to land over India and Southeast Asia, where it falls as torrential monsoon rains (Chapter 7). After flowing westward across the Indian Ocean, waters enter the South Atlantic via the Agulhas Current flowing around southern Africa. In addition, partial blocking by the Indonesian Islands causes warm surface waters to accumulate in the western equatorial Pacific Ocean, an important aspect of ENSO.

UPWELLING AND DOWNWELLING

In some coastal areas of the ocean (and large lakes, such as the Great Lakes of North America), the combination of persistent winds, Earth's rotation (Coriolis Effect), and restrictions on lateral movements of water caused by shorelines and shallow bottoms induces upward (upwelling) and downward (downwelling) movements of water.

As explained above, the Coriolis Effect plus the frictional coupling of wind and water (Ekman transport) cause net movement of surface water at about 90 degrees to the right of the wind direction in the Northern Hemisphere and to the left of the wind direction in the Southern Hemisphere. **Coastal upwelling** occurs where Ekman transport moves surface waters away from the coast; surface waters are replaced by water that wells up from below (Figure 8.4). Where Ekman transport moves surface waters toward the coast, the water piles up and sinks in the process known as **coastal downwelling** (Figure 8.5).

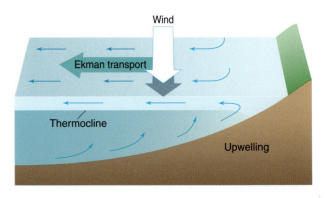

FIGURE 8.4
Where Ekman transport moves surface waters away from the coast, surface waters are replaced by water that wells up from below in the process known as coastal upwelling.

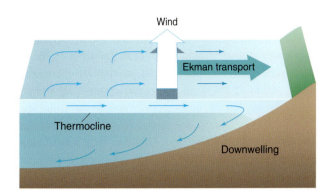

FIGURE 8.5
Where Ekman transport moves surface waters toward the coast, the water piles up and sinks in the process known as coastal downwelling.

Upwelling is most common along the west coast of continents (eastern sides of ocean basins) bordering the subtropical gyres. In the Northern Hemisphere, upwelling occurs along west coasts (e.g., coasts of California, Northwest Africa) when winds blow equatorward (causing Ekman transport of surface water away from the shore). Figure 8.6 shows the influence of upwelling on sea surface temperatures along the central and northern California coast. Winds blowing poleward cause upwelling along continents' eastern coasts, although not as noticeable because of the western boundary currents. Upwelling also occurs along the west coasts in the Southern Hemisphere (e.g., coasts of Chile, Peru, and southwest Africa) when the wind direction is equatorward because the net transport of surface water is westward, away from the shoreline. In the Southern Hemisphere, as in the north, winds blowing poleward cause upwelling along the continents' eastern coasts.

Upwelling and downwelling also occur in the open ocean where winds cause surface waters to diverge from a region (causing upwelling) or to converge toward some region (causing downwelling). For example, upwelling takes place along much of the equator (Figure 8.7). Recall that the deflection due to the Coriolis Effect reverses direction on either side of the equator. Hence, westward-flowing, wind-driven surface currents near the equator turn northward on the north side of the equator but southward on the south side. Surface waters are moved away from the equator and replaced by upwelling waters; this is known as **equatorial upwelling**.

Upwelling and downwelling influence sea-surface temperatures and biological productivity. Upwelling waters originate below the pycnocline and are usually colder than the surface waters they replace.

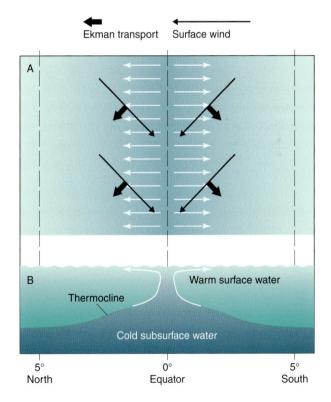

FIGURE 8.7
Equatorial upwelling. (A) In this plan view of the ocean from 5 degrees S to 5 degrees N, the trade winds of the two hemispheres are shown to converge near the equator. The consequent Ekman transport away from the equator gives rise to upwelling as shown in (B) a vertical cross section from 5 degrees S to 5 degrees N.

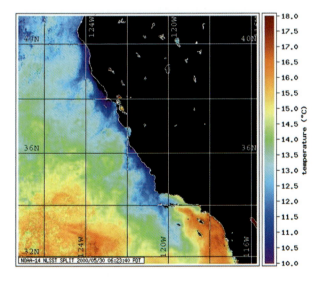

FIGURE 8.6
An infrared satellite image shows the relatively cold upwelling waters along the coast of central and northern California in response to the Ekman transport caused by winds blowing from the north. The lowest sea-surface temperatures are plotted as blue and purple. Note how in places, strong upwelling extends up to 100 km (62 mi) or so offshore from the coast. [NOAA Ocean Explorer]

The *pycnocline* is the layer of ocean water in which density changes rapidly with depth (due to changes in temperature and/or salinity), situated between the mixed layer and the deep layer. If temperature is the primary reason for the change in density, the layer is called the *thermocline*, but if salinity is the primary reason for the change in density, the layer is known as the *halocline*. Coastal upwelling also transports waters rich in dissolved nutrients (nitrogen and phosphorus compounds) from the ocean depths into the photic zone where sunlight penetrating the water supports the growth of phytoplankton populations. The world's most productive fisheries are in areas of coastal upwelling (especially in the eastern boundary regions of the subtropical gyres); about half the world's total fish catch comes from upwelling zones. On the other hand, in zones of coastal downwelling, the surface layer of warm, nutrient-deficient water thickens as water sinks. Downwelling reduces biological productivity and transports heat, dissolved materials, and surface waters rich in dissolved oxygen to greater depths. This occurs, for example, along the west coast of Alaska in the eastern boundary region of the Gulf of Alaska sub-polar gyre.

Alternate weakening and strengthening of upwelling off the coast of Ecuador and Peru are associated with ENSO variability in the tropical Pacific. During an El Niño event, upwelling wanes in the central and eastern tropical Pacific. Cold nutrient-rich water remains so deep that weak upwelling brings only warm, nutrient-poor water into the photic zone. In extreme cases, nutrient-deficient waters coupled with over-fishing cause fisheries to collapse with severe economic impacts.

Coastal upwelling and downwelling also influence weather and climate. Along the central and northern California coast, upwelling lowers sea surface temperatures. Relatively cold surface waters chill the overlying humid marine air to saturation so that dense fog develops, especially in summer. Also, seasonal upwelling and downwelling reduce the annual temperature range along the coast of California, Oregon, and Washington. During El Niño and La Niña, changes in sea-surface temperature patterns associated with weakening and strengthening of upwelling off the northwest coast of South America and along the equator in the tropical Pacific affect the distribution of precipitation in the tropics and elsewhere. Upwelling cold water also inhibits formation of tropical cyclones (i.e., hurricanes and tropical storms). Tropical cyclones derive their energy in the form of latent heat from warm surface waters.

THERMOHALINE CIRCULATION

As noted above, the upper portion of the ocean is put into motion by wind forcing at the surface. Below the pycnocline, the deep ocean is shielded from the direct action of the wind and ocean currents are driven primarily by density differences between water masses. These density contrasts are caused by variations in water temperature and salinity, with cold and salty water being the densest combination. The deep-ocean circulation driven by variations in density is known as the **thermohaline circulation**, the name coming from *thermo* meaning heat and *haline* referring to salinity.

The thermohaline circulation is a significant player in the Earth system for several reasons. Although deep-ocean currents are relatively weak, the volume of deep waters is much greater than the volume of surface waters so the magnitude of water transport is about the same. The temperature difference between the warm surface waters and the cold deep waters controls the basic stratification of the ocean, except at high latitudes where the key factor is salinity. The deep circulation contributes to poleward heat transport (Chapter 4) and its circulation varies over a broad range of temporal scales from years to millennia, modulating climate.

Density gradients in the deep ocean are established by dense waters sinking in a few select locations. These deep-water formation regions are characterized by very low temperatures and, in some cases, high salinity. In the Northern Hemisphere, deep water forms in the Greenland-Norwegian Sea and Labrador Sea in late winter when the surface ocean water reaches its lowest temperature and hence its greatest density. The primary mechanism of deep water formation in the Northern Hemisphere is **open ocean convection**. Cold winds cool the surface water to the extent that its density becomes greater than that of the water beneath it, creating an unstable water column and driving overturning. Deep waters formed in the Greenland-Norwegian Sea are separated from the main body of the Atlantic Ocean by a submarine ridge running east from Greenland, through Iceland, and on to Europe. Dense waters overflow through gaps in this ridge, cascading down-slope into the Atlantic Ocean and entraining overlying waters along the way. In contrast, convection in the Labrador Sea directly injects water into the mid-depth Atlantic. Unlike other sources of deep water in the Atlantic, deep water formation in the Labrador Sea undergoes considerable inter-annual variability. Some winters lack any deep convection whereas other winters are characterized by vigorous overturning.

A third source of deep waters in the North Atlantic Ocean consists of very salty outflows from the Mediterranean Sea. These waters originate in the northwestern Mediterranean in winter as cold, dry *mistral winds* cool the surface waters and enhance evaporation. This dense salty water fills the deep Mediterranean basin to a depth of about 400 m (1300 ft) and spills over the sill at the Strait of Gibraltar into the Atlantic basin.

In the Southern Hemisphere, deep waters form at several locations around the Antarctic continent, but primarily in the Weddell Sea. Unlike the Northern Hemisphere, much of the deep water formation occurs underneath floating sea ice and details are not as well known. The water is cooled by cold winds acting on openings (leads) in the sea ice cover. In addition, water becomes denser during formation of sea ice so salt left behind during the freezing of seawater increases the salinity of water immediately below the freezing interface by a process called **brine rejection**. Water made denser by cooling and brine rejection sinks along the continental slope of Antarctica into the deep ocean. As this cold, dense water sinks, it entrains additional water thereby increasing the total flow. In addition to the Weddell Sea, cold dense bottom waters are created in the Ross Sea as well as scattered locations along the continental shelf of Antarctica.

For an example of how external processes may have influenced the thermohaline circulation in the past, refer to this chapter's first Essay.

El Niño, La Niña, and the Southern Oscillation

For more than a century, some scientists have been aware of short-term (inter-annual) variations in climate at many locations worldwide. One of the regions where these inter-annual variations in Earth's climate system are readily apparent is the tropical Pacific Ocean. There, ocean conditions varying on a quasi-periodic basis are referred to as El Niño and La Niña. Within the last several decades, the scientific community has realized that these episodes can cause weather extremes not only in the tropical Pacific but also in many other parts of the world.

HISTORICAL PERSPECTIVE

Originally, El Niño was the name given by fishermen to a local wind-driven warm ocean current that was accompanied by poor fishing off the coast of Peru and Ecuador. This phenomenon occurs every year, often coinciding with the Christmas season. (*El Niño* is the Spanish reference to the Christ child.) Typically, these warm water episodes are relatively brief, lasting perhaps a month or two, before sea-surface temperatures and the fisheries return to normal. About every three to seven years, however, El Niño persists for 12 to 18 months, or even longer, and is accompanied by significant changes in SST over vast stretches of the tropical Pacific, major shifts in planetary-scale atmospheric and oceanic circulations, and collapse of important South American fisheries. Today, atmospheric and oceanic scientists reserve the term **El Niño** for these long-lasting ocean/atmosphere anomalies.

An important step in understanding El Niño came in 1924 with the discovery of the Southern Oscillation by the Englishman Sir Gilbert Walker (1868-1958). Monsoon failure in 1899-1900 caused terrible famines in India, with the loss of more than a million lives. In 1904, Walker was appointed director general of observatories in India and charged with developing a method to predict the Indian monsoon. He set out on an extensive search for any possible relationship between monsoon rains and weather conditions in various parts of the world. One of his discoveries was a seesaw variation in air pressure across the tropical Indian and Pacific Oceans that he named the **Southern Oscillation**. When air pressure was low over the Indian Ocean and the western tropical Pacific, it was high east of the international dateline (180 degrees longitude) in the eastern tropical Pacific. During these conditions, monsoon rains were plentiful over India. With the opposite pressure pattern (high pressure west of the dateline and low pressure east of the dateline), monsoon rains were lighter than usual.

Today, the *Southern Oscillation Index (SOI)* is based on the difference in the anomalous air pressure between Darwin (on the north coast of Australia at 12 degrees S, 130 degrees E) and Tahiti (an island in the central south Pacific at about 18 degrees S, 149 degrees W). When air pressure is anomalously low at Darwin, air pressure tends to be anomalously high at Tahiti. Conversely, when air pressure is high at Darwin, it is low at Tahiti (Figure 8.8). (Some scientists question the consistency and quality of air pressure readings at Tahiti prior to 1935.)

More than four decades would pass before the broader significance of Walker's discovery of the Southern Oscillation was fully recognized. In 1966, the Norwegian-American meteorologist Jacob Bjerknes (1897-1975), while at the University of California in Los Angeles, demonstrated a relationship between El Niño

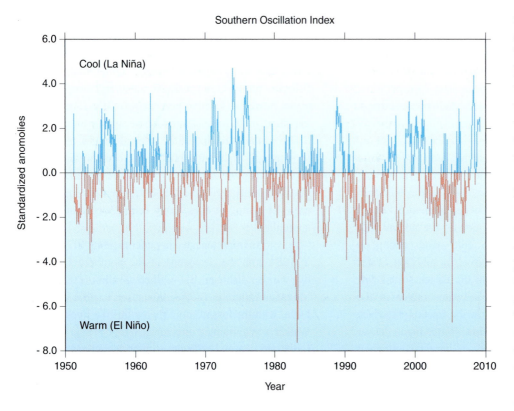

FIGURE 8.8
Variation in the Southern Oscillation Index (SOI) based on monthly mean sea level pressure anomalies at Darwin, Australia, and Tahiti. Strongly positive values of the index indicate La Niña conditions and strongly negative values of the index indicate El Niño conditions. [NOAA, Earth System Research Laboratory, Physical Sciences Division]

and the Southern Oscillation, proposed a mechanism whereby they interacted, and identified some of the effects on North American weather. The air pressure difference (*air pressure gradient*) across the tropical Pacific changes as air pressure to the west rises and air pressure to the east falls (and vice versa). These changes in air pressure, in turn, may be linked to changes in SST. Bjerknes analyzed oceanic/atmospheric observations gathered from the tropical Pacific during 1957-58, the *International Geophysical Year*, which happened to coincide with a strong El Niño. He found that an El Niño episode begins when the air pressure gradient across the tropical Pacific starts to weaken, heralding the slackening of the trade winds. Scientists refer to this relationship between El Niño and the Southern Oscillation by the acronym **ENSO**.

ENSO is a *coupled* phenomenon, its variability in Earth's climate system cannot be explained as exclusively an oceanic or atmospheric event. *Coupled* refers to more than merely something that occurs in both the ocean and atmosphere; in this case, the phenomenon depends on feedbacks between the ocean and atmosphere. Changes in ocean conditions (primarily SST) can and do drive changes

in atmospheric circulation and precipitation patterns. What is unique about ENSO is the strength of the coupling: changes in the ocean drive changes in the atmosphere which then feedback and further alter the ocean.

Not until the El Niño event of 1982-83 (one of the two most intense of the 20th century) and satellites revealing its planetary scale, did the scientific community and general public fully realize the potential worldwide impact of ENSO. That event spurred development of numerical models to simulate ENSO as well as deployment of a network of instrumented buoys and satellites to provide advance warning of a developing El Niño. Also, the last two decades have seen increasing interest in **La Niña**, the name coined in the mid 1980s for an atmosphere/ocean interaction that is essentially opposite El Niño. Some scientists refer to the warm El Niño and cold La Niña as opposite extremes of the *ENSO cycle*.

One of several possible triggers for ENSO is the **Madden-Julian Oscillation (MJO)**. The MJO is a large-scale (1000 km) disturbance of the near-equatorial troposphere that slowly propagates from the Indian Ocean into the Western Pacific over the course of about 30 to 50 days. Although coupled with the ocean, the depth of the MJO's influence in the ocean is unknown. Associated winds heat or cool the ocean's mixed layer by as much as 1 Celsius degree, while associated cloudiness affects the flux of solar radiation incident on the ocean surface. These changes in ocean temperature and solar radiation flux likely affect the rate of evaporation.

So important are El Niño and La Niña in year-to-year climate variability that signs of a developing El Niño or La Niña are now routinely incorporated into long-range seasonal climate outlooks worldwide. Such outlooks identify areas of expected anomalies (departures from long-term averages) in temperature and precipitation, and guide development of regional agricultural and water management strategies. Adoption of these strategies helps lessen the impact of weather extremes on water supply and food production in various parts of the world.

NEUTRAL CONDITIONS IN THE TROPICAL PACIFIC

El Niño and La Niña represent departures from the long-term average or *neutral* oceanic conditions in the tropical Pacific. Understanding El Niño and La Niña initially requires a look at neutral conditions. The prevailing trade winds impact ocean currents, sea-surface temperatures, and rainfall across the tropical Pacific.

Prevailing winds blow from the south or southwest along the west coast of South America so that most of the time Ekman transport drives warm surface waters to the left (westward), away from the coast. This causes cold, nutrient-rich waters to well up from below the thermocline, which is normally only 50 to 100 m (165 to 325 ft) deep along the coast, replacing the warm, nutrient-poor surface waters that are transported offshore. Although this zone of *coastal upwelling* is narrow (typically less than 15 km or 10 mi wide), the abundance of nutrients conveyed into the photic zone spurs an explosive growth of phytoplankton populations for an additional 15 km offshore. Those populations in turn, support a diverse marine ecosystem and highly productive fishery. In addition, *equatorial upwelling* produces a strip of relatively low sea-surface temperatures along the equator from the coast of South America westward to near the international dateline.

Meanwhile, the trade winds drive a pool of relatively warm surface waters westward toward Indonesia and northern Australia. The wedge of warm water increases the depth of the thermocline and raises sea level in the western tropical Pacific. The thermocline is at a depth of about 150 m (490 ft) in the western tropical Pacific but slopes upward to a depth of about 50 m (165 ft) in the eastern tropical Pacific. Water expands when heated (as well as being piled up by trans-Pacific trade winds) so that sea level is about 60 cm (2 ft) higher in the west than in the east. The contrast in sea-surface temperature between the western and eastern tropical Pacific (averaging a difference of about 8 Celsius degrees or 14.4 Fahrenheit degrees) has important implications for precipitation across the tropical Pacific (Figure 8.9). Relatively cool surface waters in the central and eastern tropical Pacific chill the overlying air and suppress convection so that rainfall is light in that region and along the adjacent western coastal plain of South America. Meanwhile, over the western tropical Pacific, warm surface waters heat the overlying air, strengthening convection currents that produce heavy rainfall.

The contrast between relatively high air pressure over the central and eastern tropical Pacific and relatively low air pressure over the western tropical Pacific ultimately drives the trade winds. As discussed in Chapter 6, winds initially blow from regions where air pressure is relatively high toward regions where air pressure is relatively low. The greater the air pressure contrast (i.e., horizontal air pressure gradient), the stronger are the winds. Once in motion, the Coriolis Effect deflects winds to the right in the Northern Hemisphere and to the left in the Southern Hemisphere. Trade winds blow from the northeast in the Northern Hemisphere and from the southeast in the Southern Hemisphere. As long as high pressure and low SST persist over the eastern basin and low pressure and

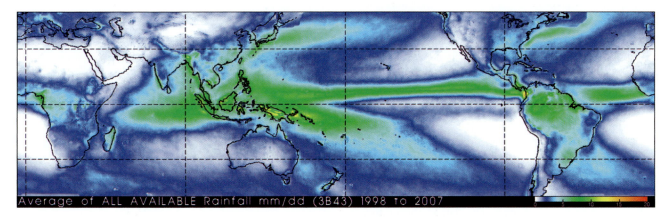

FIGURE 8.9
Benchmark average rainfall in millimeters per day (mm/d) across the tropical Pacific Ocean for the ten-year period 1998 through 2007. The heaviest rainfall is in the western tropical Pacific where sea-surface temperatures are highest. These data were obtained from the TRMM Microwave Imager and IR sensors on board geosynchronous satellites supplemented by conventional rain gauge measurements. [Source: NASA, Tropical Rainfall Measuring Mission (TRMM)]

Normal Conditions

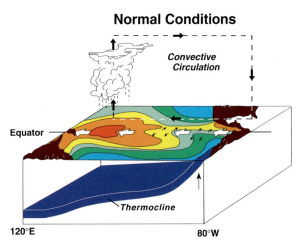

FIGURE 8.10
Schematic block diagram showing ocean/atmosphere conditions in the tropical Pacific during normal (neutral) episodes. Red indicates areas of highest sea-surface temperatures. [NOAA, Pacific Marine Environmental Laboratory (PMEL), Tropical Atmosphere Ocean Project]

El Niño Conditions

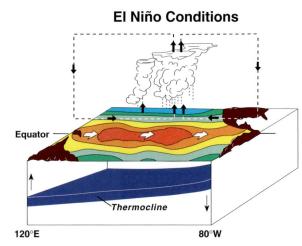

FIGURE 8.11
Schematic block diagram showing ocean/atmosphere conditions in the tropical Pacific during El Niño conditions. Red indicates areas of highest sea-surface temperatures. [NOAA, Pacific Marine Environmental Laboratory (PMEL), Tropical Atmosphere Ocean Project]

high SST persist over the western basin, the trade winds are strong and upwelling remains vigorous along the coast of Ecuador and Peru, and the equator east of the international dateline.

High SST in the western tropical Pacific lower the surface air pressure, whereas low SST in the eastern tropical Pacific raise the surface air pressure. Hence, during neutral conditions the east-west SST gradient reinforces the trade winds by strengthening the east-west air pressure gradient. In flowing over the ocean surface, the trade winds become warmer and more humid so in the western tropical Pacific warm humid air rises, expands, and cools (Figure 8.10). Water vapor condenses into towering cumulonimbus (thunderstorm) clouds that produce heavy rainfall. Aloft this air flows back eastward and sinks over the cooler waters of the eastern tropical Pacific. Sinking air is compressed and warmed so that clouds vaporize or fail to develop. This completes the large convective-type circulation known as the **Walker Circulation** (named for Sir Gilbert Walker), an east-west oriented atmospheric circulation across the equatorial Pacific and shown by the dashed lines in Figure 8.10.

WARM PHASE

With the onset of El Niño, air pressure falls over the eastern tropical Pacific and rises over the western tropical Pacific as part of the Southern Oscillation. The air pressure gradient across the tropical Pacific weakens and the trade winds slacken in the western and central equatorial Pacific. During a particularly intense El Niño, trade winds west of the international dateline may reverse direction.

In response to these shifts in atmospheric circulation over the tropical Pacific, changes take place in oceanic conditions (Figure 8.11). In the western tropical Pacific, SST drops, sea-level falls, and the thermocline rises whereas in the eastern tropical Pacific, SST rises, sea-level climbs, and the thermocline deepens (Figure 8.12). With relaxation of the trade winds, the westward flow of the equatorial currents weakens and at times reverses direction. Hence, the thick layer of warm surface water normally in the west drifts eastward across the tropical Pacific, until deflected toward the north and south by the continental landmasses. Eastward drift of warm water is

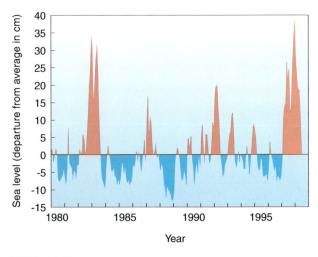

FIGURE 8.12
Sea level variation at Galápagos in the eastern tropical Pacific based on tide gauge records and expressed in cm as departure from the long-term average. Relatively high sea levels correspond to El Niño episodes. [NOAA PMEL]

so slow that it may take several months for higher SST to reach the west coasts of North and South America sometimes spreading as far north as the coast of western Canada, and as far south as central Chile.

The impacts of these environmental changes can be severe on marine ecosystems. Arrival of warm surface waters in the eastern tropical Pacific reduces upwelling of nutrient-rich waters along the coast of Ecuador and Peru. Deprived of nutrients, phytoplankton populations decline and the commercial fish harvest plummets. By the 1970s, Peru's fishing industry was one of the largest in the world, accounting for about 20% of the total global catch of anchovies and generating almost one-third of the nation's foreign exchange income. However, over-fishing combined with the effects of the 1972-73 El Niño sent the Peruvian fishery into a tailspin from which it has not recovered.

Warmer surface waters can also severely stress coral reefs living in shallow tropical waters. In response to unusually high sea-surface temperatures, coral expels zooxanthallae, the symbiotic microscopic algae that through photosynthesis supplies coral with oxygen and organic compounds. Without zooxanthallae, coral polyps have little pigmentation and appear nearly transparent on the coral's white skeleton, a condition referred to as **coral bleaching**. Excessive bleaching can kill coral polyps, destroying habitats for a wide variety of marine organisms. Extensive coral bleaching was reported during the 1997-98 El Niño. For more on coral bleaching and global climate change see Chapter 12.

During El Niño, changes in the trade wind circulation give rise to anomalous weather patterns in the tropics and subtropics. Long-term average winds blow onshore over Indonesia so that rainfall is normally abundant, but during El Niño, prevailing winds are directed offshore and the weather of Indonesia becomes anomalously dry. El Niño-related droughts may also grip India, eastern Australia, northeastern Brazil, and southern Africa. Meanwhile, warmer than usual surface waters off the west coast of South America spur convection and heavy rainfall along the normally arid coastal plain, causing flash flooding. Wetter conditions tend to occur in southern Brazil, Uruguay, and equatorial East Africa.

Martin Hoerling, a NOAA meteorologist, and colleagues reported that during the period 1871 to 2002, central India endured 10 severe droughts during the monsoon rainy season (June through August). All 10 droughts occurred during an El Niño episode. However, the converse did not hold; that is, not all El Niño events during the 132-year period were accompanied by drought in central India. In fact, in 13 cases rainy season precipitation was at or slightly greater than the long-term average. In examining the various cases of drought, Hoerling and colleagues found that the greatest SST anomalies during El Niño occurred either in the central or eastern equatorial Pacific, and drought occurred in central India when the highest SST were in the central equatorial Pacific. Numerical models predicted that these exceptionally warm waters produced warm, humid air that ascended high into the tropical atmosphere and then flowed over central India, where it subsided. The subsiding air inhibited development of clouds and precipitation and soon a drought would be underway. This is one of many examples that demonstrate the fact that no two El Niño events are the same.

Apparently, El Niño influences the intensity, frequency, and spatial distribution of tropical cyclones (e.g., tropical storms, hurricanes). Stronger than usual winds aloft inhibit development of tropical cyclones over the Atlantic Basin. The few tropical cyclones that do form usually are weaker and shorter lived than usual. In the Pacific and Indian Oceans, changes in SST that accompany El Niño appear to alter the intensity and spatial distribution of tropical cyclones rather than their frequency. Because of the extensive area of warmer water over the eastern tropical Pacific, hurricanes that form there can travel farther north and west than usual.

El Niño typically brings dry weather to the Hawaiian Islands. Almost all of Hawaii's major droughts during the 20th century coincided with an El Niño event. As part of the atmospheric circulation changes that take place during El Niño, the North Pacific subtropical anticyclone shifts so that the associated region of descending air moves closer to the islands. The greater than usual frequency of sinking air over Hawaii is responsible for a persistent dry weather pattern.

El Niño also has a ripple effect on the weather and climate of middle latitudes, especially during winter. A linkage between changes in atmospheric circulation occurring in widely separated regions of the globe, often over distances of thousands of kilometers, is known as a **teleconnection**. What causes teleconnections? Latent heat released to the atmosphere during the buildup of thunderstorms in the tropical troposphere is one of the major controls of the planetary-scale circulation. Changes in the location of these heat sources alter wind and weather patterns worldwide. Higher than usual SST over the central and eastern tropical Pacific during El Niño heats and destabilizes the troposphere. Deep convection generates towering thunderstorms that help drive atmospheric circulation, governing the course of jet streams, storm tracks, and moisture transport by winds at higher latitudes.

Events taking place in the ocean and atmosphere are somewhat analogous to the effect of large boulders on a swiftly flowing stream. Boulders in the streambed induce a train of turbulent eddies that extends downstream. Moving the boulders will displace the train of eddies. Thunderstorm clouds building high into the tropical troposphere deflect the upper air winds in much the same manner as boulders in a stream; that is, a shift in location of the principal area of convection eastward over the tropical Pacific redirects the atmospheric circulation.

During typical El Niño winters, prevailing storm tracks bring abundant rainfall and cooler than usual conditions to the Gulf Coast states, from Texas to Florida. Over the northern U.S. and Canada, prevailing winds tend to blow from west to east, so extremely cold air masses move eastward across the Arctic and northern Canada. Persistence of this circulation pattern keeps exceptionally cold air masses in the far north, so that mild weather prevails over much of Canada, Alaska, and parts of the northern U.S. West-to-east flow in the westerlies also diminishes the usual spring contrast between warm, humid air masses moving northeastward from the Gulf of Mexico and cold, dry air masses sweeping southeastward from Canada. Consequently, severe thunderstorms and tornadoes may be less frequent than usual in the Ohio and Tennessee River Valleys.

Although some weather extremes almost always accompany El Niño, no two events are the same because El Niño is only one of many factors that influence inter-annual climate variability. In Southern California, for example, heavy winter rains (snows at higher elevations) have occurred during some but not all El Niño events. Record heavy rainfall in Southern California during January 1995 was linked to a shift of the jet stream (and storm track) south of its usual position over the eastern Pacific. In that case, a change in atmospheric circulation associated with El Niño was the culprit. Whereas the 1982-83 El Niño brought severe drought to eastern Australia, dry conditions in Australia during the 1997-98 El Niño were far less severe.

COLD PHASE

Occasionally La Niña follows El Niño. **La Niña** is a period of unusually strong trade winds and exceptionally vigorous upwelling in the eastern tropical Pacific (Figure 8.13). Like its warm counterpart, La Niña tends to persist for 12 to 18 months. During La Niña, SST anomalies are

essentially opposite those observed during El Niño; that is, surface waters are colder than usual over the central and eastern tropical Pacific and somewhat warmer than usual over the western tropical Pacific. SST anomalies over the eastern tropical Pacific typically are greater in magnitude during El Niño than during La Niña. SST usually rise about 5 to 6 Celsius degrees (9 to 11 Fahrenheit degrees) above the long-term average during an intense El Niño but drop only 2 to 3 Celsius degrees (3.6 to 5.4 Fahrenheit degrees) below the long-term average during a strong La Niña.

Accompanying La Niña are worldwide weather extremes that are often opposite those observed during El Niño. As with El Niño, the most consistent middle latitude teleconnections appear in boreal winter. In the tropical Pacific, lower than usual SST in the east inhibit rainfall while higher than usual SST in the west enhance rainfall in Indonesia, Malaysia, and northern Australia during the Northern Hemisphere winter, and the Philippines during the Northern Hemisphere summer. Elsewhere around the globe, the summer Indian monsoon rainfall tends to be heavier than average, especially in northwest India, and during the Northern Hemisphere winter, wet conditions prevail over southeastern Africa and northern Brazil. From southern Brazil to central Argentina, the winter is relatively dry. In addition, weak winds aloft during La Niña are conducive to tropical cyclone formation over the Atlantic Basin.

Across middle latitudes of the Northern Hemisphere, winds tend to be more *meridional* during La Niña; that is, prevailing winds encircle the globe

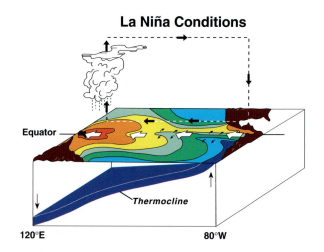

La Niña Conditions

FIGURE 8.13

Schematic block diagram showing ocean/atmosphere conditions in the tropical Pacific during La Niña conditions. Red indicates areas of highest sea-surface temperatures. [NOAA, Pacific Marine Environmental Laboratory (PMEL), Tropical Atmosphere Ocean Project]

from west to east in great north-south loops. These winds steer cold air masses toward the southeast and warm air masses toward the northeast. Occasionally, a meridional flow pattern becomes so extreme that a broad pool of rotating air essentially separates from the main current and persists over the same geographical area for weeks to months. Recall from Chapter 6 that a pool of air rotating in a clockwise direction (viewed from above) is known as a *cutoff anticyclone* (or *high*). For as long as a cutoff high persists over the same area, the weather remains dry and the probability of drought increases. This caused the severe summer drought that afflicted the central U.S. during the La Niña year of 1988.

In spring, a more meridional flow pattern in the winds may increase the likelihood of severe thunderstorms and tornadoes across the central U.S. by bringing together air masses with great contrasts in temperature and humidity. Also in the U.S., La Niña tends to be accompanied by below average winter precipitation and mild temperatures in a band from the Southwest, through the central and southern Rockies, and eastward to the Gulf Coast. Winter in the Pacific Northwest tends to be cool and wet. Lower than usual winter temperatures also occur in the northern Intermountain West and north-central states.

Changes in the atmospheric circulation patterns associated with a La Niña episode in 1998-2001 were likely responsible for a persistent cut-off high that brought drought to southwestern Asia, from Iran eastward to western Pakistan. Near the center of the dry belt, Afghanistan experienced its most severe drought in 50 years, compounding the

human misery brought on by decades of political instability and civil strife. Heidi M. Cullen, at the time a climatologist with the National Center for Atmospheric Research (NCAR) in Boulder, CO, proposed that changes in the atmospheric circulation pattern were linked to exceptionally low SST in the central tropical Pacific and unusually high SST in the western tropical Pacific. These La Niña conditions brought heavy winter rains to the eastern Indian Ocean (e.g., Malaysia), a shift in the location of the subtropical jet stream, and development of a cut-off high over southwestern Asia. The cut-off high blocked the usual west to east movement of storm systems and was responsible for an extended period of dry weather in parts of southwestern Asia. Cullen cautions, however, that not all La Niña events are accompanied by drought in the Iran-Afghanistan-Pakistan region.

FREQUENCY

In September 2003, NOAA scientists announced that they had reached a consensus for an index that could form the basis for operational definitions of El Niño and La Niña (Figure 8.14). The Index is based on six variables measured in the tropical Pacific: sea-level air pressure, zonal (east-west) component of surface wind, meridional (north-south) component of surface wind, surface air temperature, sky cloud cover, and sea-surface temperature. Sea-surface temperatures are drawn from an area of the tropical Pacific that includes the equatorial cold tongue and is bounded by longitude 120 degrees W and 170 degrees W, latitude 5 degrees N and 5 degrees S. By this index, El Niño is characterized by a *positive*

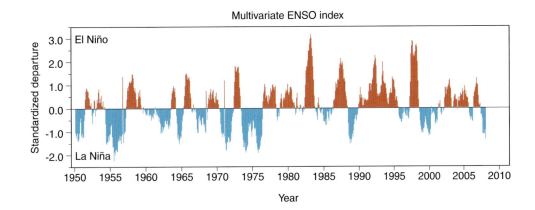

FIGURE 8.14
Variations in the Multivariate ENSO Index showing the sequence of El Niño and La Niña events since 1950. The Index is based on six variables measured in the tropical Pacific: sea-level air pressure, zonal (east-west) component of surface wind, meridional (north-south) component of surface wind, surface air temperature, sky cloud cover, and sea-surface temperature. Sea-surface temperature anomalies (departures from 1971-2000 averages) are measured for the area in the tropical Pacific Ocean between 5 degrees N and 5 degrees S latitude and from 120 degrees W to 170 degrees W longitude. Warm anomalies (greater than about 0.5 Celsius degree) generally indicate El Niño whereas cold anomalies (less than about −0.5 Celsius degree) generally indicate La Niña. [NOAA/ESRL/Physical Science Division-University of Colorado at Boulder/CIRES/CDC]

sea-surface temperature departure from normal (based on the 1971-2000 period) greater than or equal to 0.5 Celsius degree, averaged over three consecutive months. La Niña is characterized similarly, but by a *negative* sea-surface temperature departure from normal.

NOAA's Climate Prediction Center launched its new **ENSO Alert System** in February 2009. An El Niño or La Niña *watch* is issued when conditions in the equatorial Pacific are favorable for the development of El Niño or La Niña conditions within the next three months. An El Niño or La Niña *advisory* is issued when El Niño or La Niña conditions have developed and are expected to continue.

Although El Niño does not always give way to La Niña, La Niña appears more likely to follow an intense El Niño rather than a weak one. This may be explained by the fact that during an intense El Niño, the global water cycle conveys a great amount of heat out of the eastern tropical Pacific and into higher latitudes. Unusually cold water situated just below the warm surface water is poised to well up to the surface as soon as the trade winds strengthen. For example, a significant La Niña followed the intense 1997-98 El Niño.

During the second half of the 20th century, El Niño conditions prevailed 31% of the time and La Niña occurred 23% of the time. The balance of the time neutral or near-neutral conditions prevailed. As of this writing, the most recent El Niño was a significant episode beginning in 2009 (Table 8.1). During the 1980s and 1990s, La Niña was less frequent than El Niño. The most recent La Niña was in 2008-09. While all El Niño and La Niña episodes share common characteristics described earlier in this chapter, they differ from one another in duration, intensity, and development.

TABLE 8.1
El Niño and La Niña Events since 1950

El Niño

2009- , 2006, 2004-05, 2002-03, 1997-98, 1991-95, 1986-87, 1982-83, 1976-77, 1972-73, 1969, 1965-66, 1963, 1957-58, 1953, 1951

La Niña

2008-09, 2007, 2000-01, 1998-2000, 1995-96, 1988-89, 1973-75, 1970-71,1964, 1955-56, 1949-50

El Niño and La Niña are not recent phenomena. Documentary, archeological, and geological evidence indicates that ENSO has operated over at least tens of thousands of years. For more on this topic, refer to this chapter's second Essay.

HISTORICAL EPISODES

In 1982-83, the weather seemed to go wild in many parts of the world, with the total worldwide impacts including thousands of deaths and an estimated $13 billion in property damage. From mid-November 1982 until late January 1983, excessive rains caused the worst flooding of the century in usually arid Ecuador. Strong winds and torrential rains produced by six tropical cyclones lashed the islands of French Polynesia in a span of three months in a region where on average, only one tropical cyclone strikes about every 5 years. At the other extreme, drought parched eastern Australia, Indonesia, and southern Africa, with huge drought-related wildfires breaking out in Australia and Borneo. Australia's drought was that nation's worst in 200 years, causing $2 billion in crop losses and the deaths of millions of sheep and cattle. Meanwhile, persistent drought in sub-Saharan Africa grew worse. Over North America, the winter storm track shifted hundreds of kilometers south, bringing episodes of destructive high winds and heavy rains to portions of California. Flooding rains also caused havoc across the southeastern United States. In the northern U.S., ski resorts experienced a snow drought and considerable economic loss as a consequence.

Just prior to these worldwide weather extremes, the ocean circulation off the northwest coast of South America changed drastically with dire implications for marine food webs. Along the coast of Ecuador and Peru, plankton populations plunged to about 5% of their normal level. The decline in plankton reduced the numbers of anchovy, which feed on plankton, to a record low. Other fish dependent on plankton, such as jack mackerel, suffered a similar fate and the commercial fisheries off the coast of Ecuador and Peru collapsed. With the decline in fish populations, marine birds (e.g., frigate birds and terns) and marine mammals (e.g., fur seals and sea lions) also experienced major population declines, food scarcity causing breeding failure as well as migration or starvation.

At first, some scientists attributed the weather extremes to the violent eruption of the Mexican volcano, El Chichón in March-April 1982. But it quickly became apparent that the worldwide weather extremes were linked to large-scale atmosphere/ocean circulation changes in the tropical Pacific and soon a new scientific term was added to the public's vocabulary: El Niño. The El Niño of 1982-83 spurred further research on atmosphere/ocean interactions and the deployment of an array of in situ and remote sensing instrument platforms in the tropical Pacific to provide early warning of the development of El Niño. In the late 1990s, when El Niño returned in its full fury, the global community was better prepared thereby lessening the impact.

The 1997-98 El Niño rivaled the 1982-83 El Niño as the most intense of the 20th century (Figure 8.15). This El Niño evolved rapidly, the trade winds weakening and eventually reversing direction in the western tropical Pacific in early 1997. Equatorial upwelling ceased during the Northern Hemisphere summer of 1997 and a pool of exceptionally warm surface waters (SST greater than 29 °C or 84 °F) migrated eastward from the western tropical Pacific. During the Northern Hemisphere fall of 1997, SST over the eastern tropical Pacific climbed at least 5 Celsius degrees (9 Fahrenheit degrees) above the long-term average. Warming was rapid in the eastern tropical Pacific with SST setting new record highs each successive month from June through December 1997. By late 1997, the thermocline essentially flattened across the tropical Pacific basin, rising some 20 to 40 m (65 to 130 ft) in the west and falling more than 90 m (295 ft) in the east. Somewhat lower sea-surface temperatures in the west and much higher than usual sea-surface temperatures in the east weakened the usual east-

west SST gradient. By further reducing the east-west air pressure gradient, the weaker SST gradient contributed to additional slackening of the trade winds.

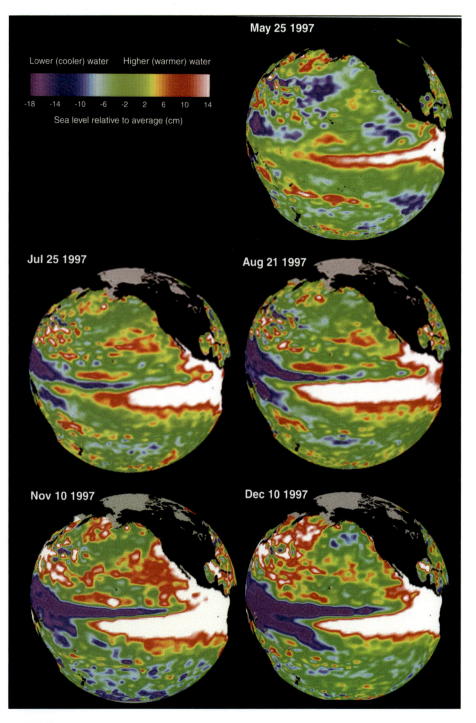

FIGURE 8.15
Evolution of the 1997-98 El Niño as derived from changes in ocean surface height (compared to the long-term average) as measured by altimeter sensors on board the TOPEX/Poseidon satellite. On the color scale, whites and reds indicate elevated areas (warmer than normal water). In the white areas, the sea surface is 14 to 32 cm (6 to 13 in.) above normal; in the red areas it is about 10 cm (4 in.) above normal. Green indicates normal sea level whereas purple corresponds to areas that are at least 18 cm (7 in.) below normal sea level (colder than normal water). [NASA Goddard Space Flight Center]

The 1997-98 El Niño came to an abrupt end in mid-May 1998. Trade winds strengthened rapidly, upwelling resumed along the equator and off the northwest coast of South America, and SST over the eastern tropical Pacific plummeted in response to upwelling of very cold water. At one location along the equator (125 degrees W), the SST fell 8 Celsius degrees (14 Fahrenheit degrees) in only four weeks.

In various parts of the world, weather extremes and climatic anomalies attributed to the 1997-98 El Niño were responsible for considerable human mortality, crop damage, and economic loss. Outbreaks of water-borne diseases such as malaria and cholera related to rainfall and standing water claimed thousands of lives. Storms and floods took 389 lives in the United States. Worldwide economic losses were estimated at $36 billion.

In the United States, however, a study undertaken by S.A. Changnon, K.E. Kunkel, and D. Changnon of the Illinois State Water Survey found that the 1997-98 El Niño had mixed impacts on the nation's economy, some negative and some beneficial. Their report, published in June 2007, concluded that the economic benefits significantly outweighed the losses.

During the 1997-98 El Niño, a succession of powerful winter storms from the Pacific struck the West Coast bringing floods, landslides, and considerable crop damage. Heavy winter precipitation also produced widespread flooding and crop loss in the Southeast. According to representatives of the property insurance industry, 15 weather-related catastrophes, each in excess of $25 million in losses, occurred from October 1997 to May 1998. The estimated economic loss in these events totaled $1.7 billion. In addition, the nation's tourism industry suffered a 30% drop in income. As listed in Table 8.2, national economic losses associated with the El Niño amounted to $10.735 billion (in 1998 dollars).

On the other hand, northern sections of the nation benefited from a mild, almost snow-free winter that was also attributed to El Niño. Heating costs declined with reduced demand for heating oil and natural gas, saving consumers about $6.7 billion. Unusually long episodes of mild, dry weather had a positive economic impact on construction activity, retail shopping, and home sales. Major savings came from a lack of spring snowmelt floods and the absence of major Atlantic hurricanes during 1997. As listed in Table 8.3, national economic benefits attributed to the 1997-98 El Niño totaled $19.825 billion (in 1998 dollars). Hence, comparing Tables 8.2 and 8.3, weather extremes and climatic anomalies associated with the 1997-98 El Niño had a net positive impact on the nation's economy amounting to $9.09 billion.

TABLE 8.2

Losses and Costs from Climate and Weather Conditions in the coterminous U.S. attributed to the 1997-98 El Niño (1998 dollars)[a]

Activity/sector	*Losses/costs ($ billions)*
Property	2.800
Federal relief	0.610
States	0.125
Agriculture	5.760
Environmental damage	1.180
Sales lost (snow equipment)	0.060
Tourism	0.200
Total	10.735

[a]From Changnon, S.A., K.E. Kunkel, D. Changnon, June 2007. *Impacts of Recent Climate Anomalies: Losers and Winners*, Illinois State Water Survey, Champaign, IL, p. 55.

TABLE 8.3

Economic Gains from Climate and Weather Conditions in the coterminous U.S. attributed to the 1997-98 El Niño (1998 dollars)[a]

Activity/sector	*Gains ($ billions)*
Heating costs down	6.7
Increased sales (homes and retail)	5.6
Reduced costs (snow removal)	0.4
Reduced costs (few floods and no hurricanes)	6.3
Construction	0.650
Reduced costs for transportation	0.175
Total	19.825

[a]From Changnon, S.A., K.E. Kunkel, D. Changnon, June 2007. *Impacts of Recent Climate Anomalies: Losers and Winners*, Illinois State Water Survey, Champaign, IL, p. 57.

PREDICTING AND MONITORING ENSO

Scientists have developed numerical models that simulate El Niño and La Niña. These models approximate oceanic processes that alter sea-surface temperatures, and the atmospheric response, including convection, clouds, and winds. Forecasters rely on two basic types of numerical models to predict the onset, evolution, and decay of El Niño (or La Niña): empirical (or statistical) models and dynamical models. An *empirical model* compares current and evolving oceanic and atmospheric conditions with comparable observational data for periods preceding El Niño (or La Niña) episodes over the prior 40 years. A match or at least some resemblance between past and present conditions is the basis for a prediction. A *dynamical model* consists of mathematical equations that simulate interactions or coupling among the atmosphere, ocean, and land. These so-called *coupled models* are more sophisticated than empirical models and run on supercomputers.

Assessment of the performance of 12 models (6 empirical and 6 dynamical) in predicting the onset, evolution, and demise of the 1997-98 El Niño was disappointing. Only the dynamical model operated by the Climate Prediction Center (CPC) of NOAA's National Centers for Environmental Prediction (NCEP) predicted the onset of the 1997-98 El Niño. But the NCEP and all other models greatly underestimated (by about 50%) the magnitude of sea-surface warming in the eastern tropical Pacific, and demonstrated little skill in forecasting the decay of El Niño. The NCEP model predicted gradual cooling in the eastern tropical Pacific through 1998 when sea-surface temperatures plunged in May 1998.

Official forecasts of an impending El Niño appeared in the spring of 1997, after warming of the eastern tropical Pacific had already begun. Model output strongly influenced the Climate Prediction Center's winter outlook issued in November 1997, for December 1997 through February 1998. That outlook correctly predicted heavier than usual precipitation across the southern U.S. and anomalous warmth over the northern third of the nation. Predictions of unusually dry conditions in Indonesia, northern South America, and southern Africa were verified. Above average rainfall predicted for Peru, east Africa, and northern Argentina was also correct. But the extreme drought expected for northeast Australia did not materialize.

Reliable observational data from the tropical Pacific and atmosphere are essential for detecting a developing El Niño or La Niña, as well as initializing numerical models. The accuracy of dynamical models also depends on how well their component equations simulate the coupled atmosphere/ocean/land system, and reliable observa-

tional data. Numerical models use those initial conditions as a starting point for predicting future states of the ocean and atmosphere, and for verifying the model predictions and results. Data are also assimilated into model runs to correct or "nudge" the model as events evolve.

Atmosphere/ocean observational data from the tropical Pacific were obtained by monitoring systems deployed as part of the 10-year (1985-1994) international *Tropical Ocean Global Atmosphere (TOGA)* study. TOGA was aimed at improving understanding, detection, and prediction of ENSO variability. An important product of TOGA, the **ENSO Observing System**, was fully operational by December 1994. The ENSO Observing System consists of an array of moored and drifting instrumented buoys, island and coastal tide gauges, ship-based measurements, and satellites (Figure 8.16).

One component of the ENSO Observing System is the *TAO (Tropical Atmosphere/Ocean)* array of moored buoys (small, unmanned, instrumented platforms) in the tropical Pacific Ocean (Figure 8.17). Buoys are strategically placed over an area bounded by 8 degrees N, 8 degrees S, 95 degrees W, and 137 degrees E. This instrument array, renamed TAO/TRITON in 2000, consists of approximately 70 deep-sea moorings that measure several atmospheric variables, including air temperature, wind, relative humidity, as well as oceanic parameters, sea-surface and subsurface temperatures at 10 depths in the upper 500 m, (1640 ft). Several newer moorings also have salinity sensors, along with additional meteorological sensors. Five moorings along the equator also measure ocean current velocity using Subsurface Acoustic Doppler Current Profilers. Observational data are transmitted to NOAA's Pacific Marine Environmental Laboratory (PMEL) in Seattle, WA, via a NOAA polar-orbiting satellite. Those data are available on the Internet in near real-time.

Remote sensing by Earth-orbiting satellites plays an important role in providing early warning of an evolving El Niño or La Niña. Sensors onboard NOAA and NASA satellites monitor cloud cover and map sea-surface temperatures. The *TOPEX/Poseidon satellite*, a joint mission of NASA and the *Centre National d'Etudes Spatiales (CNES)* in France, provided the first continuous global coverage of ocean surface topography (sea level) at 10-day intervals (Figure 8.18). Launched in 1992, it ceased operating in January 2006 after almost 62,000 orbits of the planet. Radar (microwave) altimeters on board the satellite (at an altitude of 1336 km or 830 mi) bounced microwaves off the ocean surface to obtain precise measurements of the distance between the satellite and the sea surface. These data, combined with data from

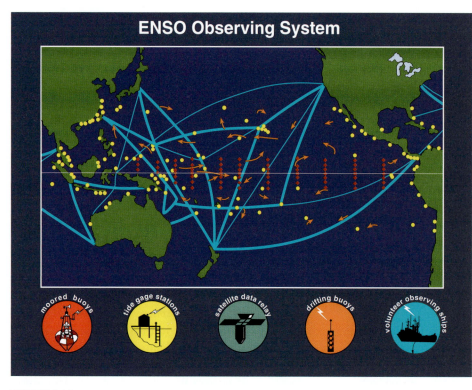

FIGURE 8.16
Components of the ENSO Observing System provide advance warning and monitor the development and decay of El Niño and La Niña events. [NOAA, Pacific Marine Environmental Laboratory (PMEL), Seattle, WA]

In December 2001, NASA and its French counterpart CNES, launched *Jason 1*, successor to TOPEX/Poseidon, that continues to gather data on how the ocean circulation affects Earth's climate. Sensors on Jason 1 map ocean surface topography (to an accuracy of 3.3 cm), wind speed, and wave height of 95% of Earth's ice-free ocean every 10 days. These data are also used in weather and climate forecast models. On 15 June 2008, *Jason 2*, the successor to Jason 1, was launched into the same orbit as Jason 1. This latest Ocean Surface Topography Mission (OSTM) is an international collaborative effort involving NASA, NOAA, CNES, and the *European Organisation for the Exploitation of Meteorological Satellites (EUMETSAT)*. Designed to operate for at least three years, Jason 2 will extend the continuous climate record of precise sea surface height measurements into the decade of the 2010s using the next generation of more accurate instruments.

the Global Positioning System (GPS), generated images of sea surface height. Elevated topography (hills) indicates warmer than usual water whereas areas of low topography (valleys) indicate colder than usual water. Such images can be used to calculate ocean surface currents based on the geostrophic flow model and to identify and track El Niño and La Niña.

FIGURE 8.17
An instrumented TAO moored buoy photographed at NOAA's Pacific Marine Environmental Laboratory in Seattle, WA. An array of similar moored buoys gathers oceanic and atmospheric data from the tropical Pacific as part of the ENSO Observing System. [Photo by J.M. Moran]

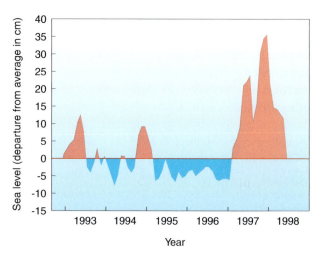

FIGURE 8.18
Sea level record at a location along the equator in the eastern tropical Pacific derived from measurements made by the TOPEX/Poseidon satellite. Sea level is expressed in cm as departure from the long-term average. Relatively high sea levels correspond to El Niño episodes. [From NOAA PMEL]

Another satellite mission that is helping scientists detect the onset and follow the evolution of El Niño and La Niña is the joint U.S.-Japanese *Tropical Rainfall Measuring Mission (TRMM)* launched in November 1997 (Figure 8.9). The TRMM satellite uses active radar and passive microwave energy sensors to monitor clouds, precipitation, and radiation over the area of the Pacific between 40 degrees N and 40 degrees S. During the 1997-98 El Niño, sensors on the TRMM satellite detected reduced precipitation in the western tropical Pacific and increased precipitation east of 150 degrees W longitude. An important component of the satellite is its TRMM Microwave Imager (TMI), an instrument that "sees through" clouds and measures sea-surface temperatures. The TMI uses the microwave energy emitted by the sea surface to characterize its IR emission spectrum and determine its radiation temperature. Microwaves pass through clouds with little attenuation but are strongly absorbed and reflected by rainfall so that TMI can measure SST only during fair weather.

More accurate predictions of El Niño and La Niña will enable humankind to better cope with the impacts of inter-annual climate variability. Such forecasts allow for informed strategic planning in agriculture, fisheries, and water resource management. Consider an example: Peru's economy is very sensitive to climate variability. In Peru, El Niño is bad for fishing and often accompanied by destructive flooding while La Niña benefits fishing but may bring drought and crop failure. Warning of an impending El Niño or La Niña prior to the start of the growing season allows agricultural interests and government officials to consult on what crops to plant for optimal food production. If El Niño is predicted, rice would be favored over cotton because rice thrives during a wet growing season whereas cotton is drought-tolerant and therefore suitable during La Niña. Also, in anticipation of the heavy rainfall, which often accompanies a full-blown El Niño, water resource

managers can direct the gradual draw down of reservoirs to reduce flooding.

North Atlantic Oscillation

Other regular climate-impacting oscillations that involve the interaction of the atmosphere and ocean include the North Atlantic Oscillation (NAO), the Arctic Oscillation (AO), and Pacific Decadal Oscillation (PDO). In general, NAO, AO, and PDO affect more restricted geographical areas and operate over longer time periods than either El Niño or La Niña.

Over the North Atlantic Ocean, the time-averaged planetary-scale atmospheric circulation features a subpolar low pressure system near Iceland (the *Icelandic low*) and a massive subtropical anticyclone centered near 30 degrees N that stretches from Bermuda to near the Azores (Chapter 6). The **North Atlantic Oscillation (NAO)** refers to a seesaw variation in air pressure between Iceland and the Azores. When air pressure is higher than the long-term average over the Azores, it is lower than the long-term average over Iceland and vice versa. The air pressure gradient between the Bermuda-Azores subtropical anticyclone and the Icelandic low drives winds that steer storms from west to east across the North Atlantic.

The North Atlantic Oscillation influences precipitation and temperatures primarily in winter (December to March) over eastern North America and much of Europe and North Africa. The **NAO Index** is directly proportional to the strength of the North Atlantic air pressure gradient, i.e., the difference in sea level air pressure between the Bermuda-Azores high and the Icelandic low. The NAO Index time series in Figure 8.19 is based on the difference in sea level pressure over Gibraltar and the sea level pressure over southwest Iceland. When the NAO Index is relatively high, stronger

FIGURE 8.19
Record of the North Atlantic Oscillation (NAO) during winter (December to March) through 2007-2008, based on the difference between the normalized sea-level air pressure at Gibraltar and the normalized sea-level air pressure over southwest Iceland. The solid black line is a running mean. [From Tim Osborn, Climatic Research Unit, University of East Anglia, Norwich, UK]

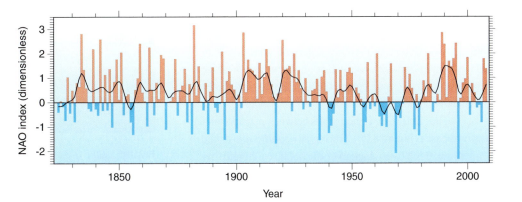

than usual winter winds blow across the North Atlantic moving cold air masses over eastern Canada and the U.S. so that winters tend to be colder than usual in that region. But cold air masses modify considerably as they move over the relatively mild ocean surface, warming and becoming more humid so that winters are milder and wetter downstream over Europe. Simultaneously, dry winter weather prevails in the Mediterranean region. On the other hand, when the NAO Index is low, steering winds over the North Atlantic shift southward so that winters are colder than usual over northern Europe, and wet and mild conditions prevail from the Mediterranean eastward into the Middle East. The eastern U.S. and Canada tend to experience relatively mild winters while winters in the southeast U.S. are colder than usual.

The NAO Index varies significantly from one year to the next and from decade to decade, and is much less regular than the ENSO cycle. The NAO Index was generally low in the 1950s, trended upward from the 1960s through the early 1990s, and then generally downward into the winter of 2007-08.

Changes in winter moisture supply associated with NAO have had varied impacts in Europe and North Africa. During recent decades of relatively high NAO-Index, wetter winters have increased hydroelectric power potential in the Scandinavian nations, and lengthened the growing season over northern Eurasia, but diminished the snow cover for winter recreation. Meanwhile, a moisture deficit has been the problem in the Iberian Peninsula, the watershed of the Tigris and Euphrates Rivers, and the Sahel of North Africa.

Arctic Oscillation

Related to the North Atlantic Oscillation, the **Arctic Oscillation (AO)** is a seesaw variation in air pressure between the North Pole and middle latitudes (about 45 degrees N). Associated changes in the horizontal air pressure gradient alter the speed of the band of winds aloft (the *polar vortex*) that blows counterclockwise (viewed from above) around the Arctic. Alternate strengthening and weakening of these polar winds impact winter weather in middle latitudes and contribute to climate variability and changes in ocean circulation.

The Arctic Oscillation shifts between negative and positive phases (Figure 8.20). During its *negative phase*, the air pressure is relatively high over the polar region and low at middle latitudes, and the polar vortex circulation is not as strong as usual. This allows bitterly cold arctic air masses

to more frequently move out of their source regions in the far north, and plunge southeastward into middle latitudes. This brings colder than usual winter weather to most of the U.S., Northern Europe, Russia, China, and Japan. Heavy lake-effect snows downwind from the Great Lakes are more frequent and nor'easters are more likely along the U.S. Eastern Seaboard (Chapter 7).

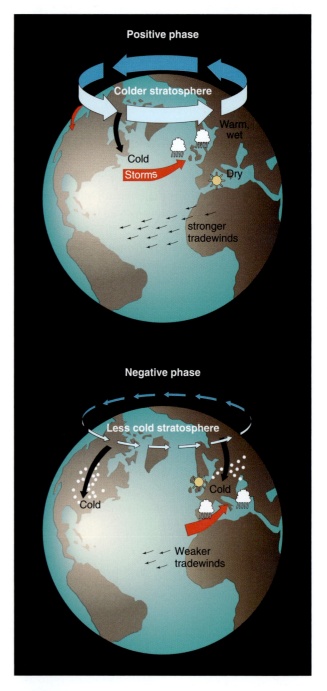

FIGURE 8.20
Atmospheric circulation changes between the positive phase (top) and negative phase (bottom) of the Arctic Oscillation. [From J. Wallace, University of Washington]

But when the Arctic Oscillation is in its *positive phase*, the pressure pattern is reversed and the air pressure gradient is greater so that winds encircling the Arctic are stronger. These stronger winds act as a dam that impedes the southeastward flow of arctic air so that ocean storms are shifted farther north. The middle latitude westerlies also strengthen and blow more directly from west to east, flooding much of the U.S. with relatively mild air from the Pacific Ocean. Winter temperatures are milder than usual especially east of the Rocky Mountains, and major snowstorms are less likely. Meanwhile, Alaska, Scotland, and Scandinavia are snowier while the western U.S. and the Mediterranean are drier.

Although the Arctic Oscillation shifts frequently between its negative and positive phases, extended periods occur when either the negative or positive phase dominates the winter season. In the 1960s, the negative phase of the Arctic Oscillation dominated. Since then, the positive phase has been more frequent, a trend that is consistent with observed climate variability in middle latitudes (e.g., less frequent episodes of extreme cold and fewer major snowstorms). Furthermore, during this recent episode of dominantly positive AO phase, winds have been delivering warmer than usual air and ocean water into the Arctic. As discussed in Chapter 11, this may at least help explain the recent shrinkage of Arctic ice cover.

Pacific Decadal Oscillation

The **Pacific Decadal Oscillation (PDO)** is a long-lived variation in climate over the North Pacific Ocean and North America. Sea-surface temperatures fluctuate between the north central Pacific Ocean and the west coast of North America. During a PDO *warm phase*, SST are lower than usual over the broad central interior of the North Pacific Ocean and above average in a narrow strip along the coasts of Alaska, the Pacific Northwest, and western Canada. In an interesting parallel to what happens off the coast of Ecuador and Chile during El Niño, the layer of relatively warm surface waters off the Pacific Northwest Coast significantly reduces upwelling of nutrient-rich bottom water. Populations of phytoplankton and zooplankton plummet and juvenile salmon migrating to coastal waters from streams and rivers starve. On the other hand, during a PDO *cool phase*, SST are higher in the North Pacific interior and lower along the coast associated with the return of nutrients, primary production, and salmon.

Key to the climatic impact of PDO is the strength of the subpolar *Aleutian low*, which persists through the winter off the Alaskan coast. During a PDO *warm phase*, the Aleutian low is well developed and its strong counterclockwise winds steer mild and relatively dry air masses into the Pacific Northwest. Winters tend to be mild and dry and water supplies suffer from reduced mountain snow pack. But during a PDO *cool phase*, the Aleutian cyclone is weaker so that cold, moist air masses more frequently invade the Pacific Northwest. Winters are colder and wetter, and the mountain snowpack is thicker.

The PDO Index, plotted in Figure 8.21), is based on monthly SST anomalies in the North Pacific Ocean, poleward of 20 degrees N. PDO phases tend to persist for 20 to 30 years. Cool phases prevailed from 1890 to 1924, from 1947 to 1976, and from 2007 through the present (mid 2009) whereas warm phases dominated from 1925 to 1946 and 1977 through 2006.

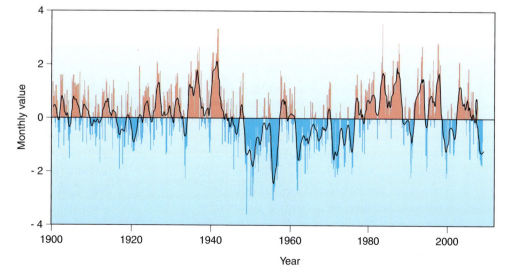

FIGURE 8.21
Variations in the Pacific Decadal Oscillation (PDO) Index between warm phases (positive values) and cool phases (negative values), 1900-2008, based on monthly SST anomalies in the North Pacific Ocean poleward of 20 degrees N. [University of Washington]

Conclusions

El Niño and La Niña involve interactions between the tropical Pacific Ocean and atmosphere. These phenomena underscore the importance of the flux of heat energy and moisture between the ocean and atmosphere. Changes in these fluxes during El Niño and La Niña have relatively short-term (one to two year) impacts on ocean circulation, and the weather and climate in various parts of the world, with implications for marine life, fisheries, and hydrologic budgets on land (e.g., drought, flooding rains). Short-term climate variability induced by El Niño and La Niña, as well as the longer-term NAO, AO, and PDO, are superimposed on much longer period variations in climate that have longer lasting impacts on the Earth-atmosphere-land-ocean system. The ocean also plays a central role in these long-term climate changes.

ENSO occurs in the tropics where the ocean can drastically alter its structure on time scales of months to years without restraint from Earth's rotation, which confines the higher latitudes. Therefore, any large-scale atmosphere/ocean interactions at higher latitudes will necessarily have a time scale of years to decades. Scientists are just beginning to understand the character of the higher latitude variations, in part because the long time-scale requires decades of records which they have only begun to collect.

Basic Understandings

- The ocean is a major player in Earth's climate system, operating on time scales from days to millennia, and spatial scales from local to global. The ocean influences the radiational heating and cooling of the planet, is an active mechanism for the global redistribution of heat energy, helps regulate atmospheric CO_2, and is the primary source for water vapor, the principal greenhouse gas.

- The ocean influences the planetary energy budget not only by affecting radiational heating and cooling of the entire planet but also by contributing to non-radiative latent heat and sensible heat fluxes at the air/sea interface.

- The ocean and atmosphere are closely coupled. Cold ocean surface currents are heat sinks; they chill and stabilize the overlying air, thereby increasing the frequency of sea fogs and reducing the likelihood of thunderstorms. Relatively warm ocean surface currents are heat sources; they supply heat and moisture to the overlying air, destabilizing the air, thereby energizing storm systems. Furthermore, the global thermohaline circulation contributes to the poleward transport of heat energy.

- Wind-driven currents are maintained by kinetic energy transferred from the horizontal winds to ocean surface waters. Once the wind sets surface waters in motion as a current, the Coriolis Effect, Ekman transport, and the configuration of the ocean basin modify the speed and direction of the current.

- The Coriolis Effect, combined with the frictional coupling of successively deeper layers of water, cause the horizontal movement of water to change direction and decrease in magnitude with increasing depth, producing the Ekman spiral. In an ideal case, a steady wind would cause surface waters to move at an angle of 45 degrees to the right of the wind in the Northern Hemisphere (to the left in the Southern Hemisphere). The Ekman spiral brings about net water movement through a depth of about 100 m (330 ft) at 90 degrees to the right of the surface wind direction in the Northern Hemisphere (to the left in the Southern Hemisphere). This net transport of water due to the coupling between surface wind and water is known as Ekman transport.

- Because of the wind circulation in semi-permanent subtropical highs, Ekman transport causes surface waters to converge toward the central region of subtropical gyres, producing a mounding of surface water near the center of the gyre. In response to this horizontal gradient in water pressure, water moves from higher to lower pressure, that is, downhill and directly opposite Ekman transport.

- Surface water parcels flow outward and down slope, away from the center of the gyre, while simultaneously the Coriolis Effect causes water parcels to shift direction to the right in the Northern Hemisphere (to the left in the Southern Hemisphere). Eventually, the outward-directed pressure gradient force balances the Coriolis Effect and water parcels flow around the subtropical gyre and parallel to contours of elevation of sea level. This horizontal movement of surface water is known as geostrophic flow.

- The pattern of ocean surface currents resembles the time-averaged pattern of Earth's planetary-scale winds. Surface currents form subtropical gyres roughly centered in each ocean basin near 30 degrees latitude. Viewed from above, currents in these gyres flow in a clockwise direction in the Northern Hemisphere and a counterclockwise direction in the Southern Hemisphere.

- In the sub-polar gyres of the far north, ocean surface current directions are opposite that of the Northern Hemisphere subtropical gyres.

- Winds associated with passing storm systems disturb the ocean surface and cause the flow of local ocean surface currents to deviate temporarily from long-term average patterns.

- In coastal areas, the combination of persistent winds blowing parallel to the coast, the Coriolis Effect, and restrictions on lateral movements of water caused by shorelines and shallow bottoms induces upward and downward movements of water. Where winds generate Ekman transport of surface waters away from the coast, colder nutrient-rich water wells up from below, a process called coastal upwelling. Upwelling supplies nutrient-rich waters to the sunlit, surface zone of the ocean spurring biological productivity.

- Where winds generate Ekman transport of surface waters toward a coast, water piles up and sinks; this is called coastal downwelling.

- Upwelling and downwelling also occur in the open ocean where winds cause surface waters to diverge away from a region (causing upwelling) or to converge toward some region (causing downwelling). For example, upwelling takes place along much of the equator in response to Ekman transport associated with convergence of the trade winds of the two hemispheres.

- Deep-ocean currents are driven primarily by slight differences in seawater density caused by variations in temperature and salinity with cold salty water being the densest combination. The deep-ocean circulation driven by variations in density is termed the thermohaline circulation, a major contributor to planetary heat transport.

- In 1924, Sir Gilbert Walker discovered the Southern Oscillation, a seesaw variation in air pressure across the tropical Indian and Pacific Oceans. In the mid-1960s, Jacob Bjerknes found that an El Niño episode begins when the air pressure gradient across the tropical Pacific begins to weaken, perhaps in response to changes in sea-surface temperatures.

- Long-term, neutral conditions of southerly and southeasterly winds along the west coast of South America drive warm surface waters westward (via Ekman transport), away from the coast. Departing warm surface waters are replaced by cold, nutrient-rich water that wells up from below the pycnocline fueling high biological production. Upwelling also occurs along the equator, east of the international dateline.

- During ENSO neutral conditions, relatively cool surface waters in the central and eastern tropical Pacific chill the overlying air and suppress convection so that rainfall is light in that region and along the adjacent western coastal plain of South America. Meanwhile, over the western tropical Pacific, relatively warm surface waters heat the overlying air, strengthening convection and giving rise to heavy rainfall. Higher air pressure in the east and lower air pressure in the west result in trade winds blowing from the east, piling up a thick layer of warm surface water in the western tropical Pacific.

- With the onset of El Niño, air pressure falls over the eastern tropical Pacific and rises over the western tropical Pacific as part of the Southern Oscillation. The air pressure gradient across the tropical Pacific weakens and the trade winds slacken, and may even reverse direction west of the international dateline. This allows the thick layer of warm surface water in the western tropical Pacific to slosh eastward along the equator into the eastern tropical Pacific, decreasing coastal upwelling.

- As El Niño evolves, the western tropical Pacific sea-surface temperatures decline, sea level falls, and depth of the thermocline decreases. Meanwhile, in the eastern tropical Pacific sea-surface temperatures rise, sea level climbs, and depth of the thermocline increases. Conditions are drier than usual in the western tropical Pacific and wetter than usual in the central tropical Pacific.

- A linkage between changes in atmospheric circulation occurring in widely separated regions of the world is known as a teleconnection. Through teleconnections, El Niño has a ripple effect on the weather and climate of middle latitudes, especially in winter. During El Niño, prevailing westerlies tend to blow more directly from west to east so that winters are milder than normal in western Canada and across parts of the northern U.S., and wetter and cooler along the Gulf Coast.

- Although some weather extremes almost always accompany El Niño, no two events are exactly the same because El Niño is only one of many factors that influence short-term climate variability.

- La Niña is a period of unusually strong trade winds and exceptionally vigorous upwelling in the eastern tropical Pacific. During La Niña, sea-surface temperature anomalies are essentially opposite those observed during El Niño, and so worldwide weather extremes are also opposite those observed during El Niño.

- El Niño occurs about once every 3 to 7 years. La Niña appears more likely to follow a strong El Niño than a weak one.

- Scientists employ two types of numerical models to predict the evolution of El Niño or La Niña: empirical (or statistical) models and dynamical models. An empirical model bases predictions on records of past atmosphere/ocean conditions. A dynamical model consists of mathematical equations that simulate atmosphere/ocean coupling.

- The accuracy of dynamical models in predicting the evolution of El Niño or La Niña depends not only on how well their constituent equations simulate the coupled ocean/atmosphere system, but also the reliability of observational data used to initialize, correct, and verify the model.

- The ENSO Observing System in the tropical Pacific consists of an array of instrumented moored buoys, current meters, and satellites. Radar (microwave) altimeters on board the *Jason* series of satellites accurately measure ocean surface elevation (sea level) changes that accompany El Niño or La Niña.

- Other quasi-regular oscillations involving the ocean and atmosphere that impact climate variability include the North Atlantic Oscillation, Arctic Oscillation, and the Pacific Decadal Oscillation.

- The North Atlantic Oscillation (NAO) refers to a seesaw variation in air pressure between Iceland and the Azores. When air pressure is higher than the long-term average over the Azores, it is lower than the long-term average over Iceland and vice versa. The North Atlantic Oscillation influences precipitation patterns and winter temperatures over eastern North America and much of Europe and North Africa.

- The Arctic Oscillation (AO) is a seesaw variation in air pressure between the North Pole and middle latitudes (about 45 degrees N). Associated changes in the horizontal air pressure gradient alter the speed of the band of winds aloft that encircle the Arctic. Strengthening and weakening of these polar winds impact winter weather in middle latitudes and contribute to climate variability and changes in ocean circulation.

- The Pacific Decadal Oscillation (PDO) is a long-lived variation in climate over the North Pacific and North America. Sea-surface temperatures fluctuate between the north central Pacific and along the west coast of North America.

Enduring Ideas

- The ocean, closely coupled with the atmosphere, is a major player in Earth's climate system, operating on time scales from days to millennia and spatial scales from local to regional, to global.
- Winds drive surface ocean waters such that the long-term average pattern of ocean surface currents and gyres resemble the long-term average pattern of the planetary-scale atmospheric circulation.
- In the tropical Pacific Ocean, during neutral conditions, trade winds transport surface ocean water westward causing upwelling, lower SST, and dry weather to the east and downwelling, higher SST, and wetter weather to the west.
- With regular changes in the east-west air pressure gradient across the tropical Pacific Ocean (the Southern Oscillation), the climate varies as the ENSO cycle with El Niño conditions at one extreme and La Niña conditions at the other extreme.
- The 1997-98 El Niño demonstrates that a mix of economic losses and benefits can result from a pattern of climatic anomalies that are geographically non-uniform in sign and magnitude.

Review

1. Describe how the ocean affects Earth's radiation budget.
2. Provide three examples of the role played by the ocean in Earth's climate system.
3. Describe Ekman transport and its significance for coastal upwelling.
4. Discuss the role of the ocean's thermohaline circulation in Earth's climate system.
5. What is the Southern Oscillation? Describe the link between the Southern Oscillation and El Niño.
6. During an El Niño event, how does sea-surface temperature and sea level change over the eastern tropical Pacific?
7. Describe the sea-surface temperature anomalies over the tropical Pacific during a La Niña event.
8. Distinguish between the two principal types of numerical models that forecasters rely upon to predict the evolution of El Niño or La Niña.
9. What is the North Atlantic Oscillation and what is its significance?
10. How do negative and positive phases of the Arctic Oscillation relate to winter weather in the U.S.?

Critical Thinking

1. What is the role of the ocean in Earth's greenhouse effect?
2. In what sense are the ocean and atmosphere coupled during an El Niño event?
3. Explain how changes in patterns of sea-surface temperature (SST) can alter the circulation of the atmosphere.
4. What are the early signs of a developing El Niño in the tropical Pacific Ocean?
5. What are the early signs of a developing La Niña in the tropical Pacific Ocean?
6. Why is La Niña more likely to follow a strong El Niño than a weak one?
7. What is meant by a teleconnection and what is its significance in forecasting the impact of ENSO?
8. No two El Niño events are the same. Explain the significance of this statement.
9. Identify and describe several ways whereby atmosphere/ocean interactions contribute to climate variability.
10. How does the SST influence the temperature and humidity of the overlying air mass?

ESSAY: Heinrich Events

Cores extracted from sediment deposited on the deep floor of the North Atlantic Ocean contain layers of pebbles and other coarse rock fragments, called *Heinrich layers*, named for the German geologist Hartmut Heinrich who first described them in 1988. Heinrich identified six layers dating from the past 100,000 years, deposited during episodes of exceptionally low sea-surface temperatures. The size of the Heinrich sediments is in marked contrast to the typically fine-grained pelagic sediments that blanket most of the deep-ocean floor. Heinrich sediments apparently were released by unusually large numbers of melting icebergs in the North Atlantic Ocean and settled to the seafloor (Figure 1). Heinrich layers are associated with massive discharges of icebergs from northeastern Canada (Hudson Bay and the St. Lawrence River) during the Pleistocene Ice Age.

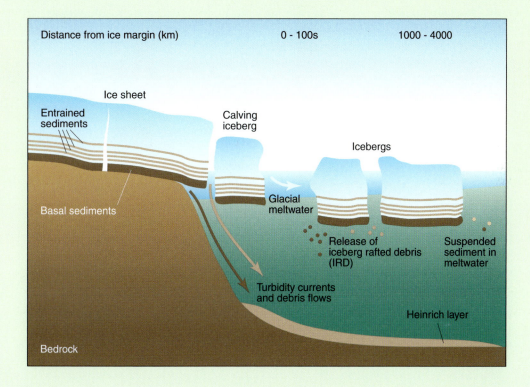

FIGURE 1
During Heinrich Events, ice rafting delivered relatively coarse sediments to the deep ocean floor. [From John T. Andrews and Thomas G. Andrews, NOAA Paleoclimatology Program and INSTAAR, University of Colorado, Boulder, CO]

During that time, Earth's mean surface temperature may have fluctuated as much as 2 to 5 Celsius degrees (5 to 10 Fahrenheit degrees) every 2000 to 3000 years, with deposition of Heinrich layers coinciding with the close of a cold episode. In addition to producing Heinrich layers, the massive influx of melting icebergs chilled the surface waters and the addition of large quantities of fresh water may have altered the ocean's thermohaline circulation that exerts a major influence on the climate of northern Europe. Another effect of injections of melting icebergs was a rapid rise in sea level that drowned Caribbean coral reefs.

The cause of Heinrich events is still disputed but Doug MacAyeal of the University of Chicago has proposed an intriguing explanation. Scientists know that major fluctuations in the planet's glacial ice cover arise from regular variations in Earth's orbital parameters, which control the seasonal and latitudinal distribution of incoming solar radiation. These regular changes in Earth-Sun geometry are known as *Milankovitch cycles* (Chapter 11). During episodes when Earth's orbital parameters favored warmer winters and cooler summers in central and northern Canada, some of the winter snows persisted year-round. In time, these climatic conditions gave rise to the Laurentide ice sheet which thickened and spread over much of Canada and the northern tier of the U.S. According to MacAyeal, over Hudson Bay the ice sheet initially was frozen to the bedrock but as the ice sheet thickened it acted as an insulating blanket over Earth's surface. This trapped enough geothermal heat from Earth's interior so that the bottom layer of the ice sheet thawed. Loaded with rock debris, this lubricated ice flowed rapidly into the North Atlantic releasing a massive surge of icebergs responsible for the Heinrich layers, changes in ocean circulation, and sea level rise. The now thinner ice sheet then re-froze to the bedrock and the cycle began anew as the ice sheet again thickened.

ESSAY: Evidence of El Niño in the Past

Lengthy records of El Niño episodes of the past would provide a useful perspective on more recent events, and perhaps help determine whether a connection exists between global climate change and the frequency and intensity of El Niño. However, in most areas of the world, reliable instrument-based weather records that could signal past El Niño events do not extend much earlier than the mid 19th century. For information on prior occurrences of El Niño, scientists must rely on proxy climate data sources, that is, information inferred or reconstructed from documentary (e.g., ship's logs, diaries), geological (e.g., ocean or lake bottom sediment cores, glacial ice cores), or biological (e.g., tree growth rings, tropical corals) indicators of climate (Chapter 9). Suppose, for example, that written records from India and Southeast Asia document agricultural losses due to prolonged drought while a sediment core extracted from a lake bottom contains evidence of heavy rainfall in normally arid coastal Peru for the same period. These simultaneous events are consistent with a strong El Niño.

Geological evidence indicates that El Niño episodes were occurring at least as far back as late in the Pleistocene Ice Age (1.7 million to 10,500 years ago). In 1999, based on his analysis of ancient corals from Sulawesi Island, Indonesia, Harvard researcher Daniel Schrag concluded that El Niño was occurring at 3- to 7-year intervals about 124,000 years ago, during the previous interglacial epoch. Schrag used oxygen isotope analysis to distinguish between El Niño and neutral episodes. Living corals build their external skeletons of limestone ($CaCO_3$) that incorporates two isotopes of oxygen: O^{16} and O^{18}. The lighter oxygen isotope evaporates more readily than does the heavier oxygen isotope. Hence, during relatively dry weather, evaporation rates are high, and the coral skeleton is enriched in O^{18} relative to O^{16}. On the other hand, during rainy episodes, evaporation rates are low, and the coral skeleton is enriched in O^{16} relative to O^{18}. Corals build their skeletons in annual growth rings that can be dated so that oxygen isotope analysis permits reconstruction of a detailed chronology of general weather conditions that can be tied to El Niño.

David Lea of the University of California, Santa Barbara, and Sandy Tudhope of the University of Edinburgh, Scotland studied coral reef terraces in New Guinea dating back some 130,000 years. In 2001, they reported on their isotopic and chemical analyses of these corals. They determined that El Niño events were more intense over the past 100 years than at any time during the past 130,000 years. Furthermore, they concluded that El Niño was about 50% weaker during the Ice Age and intensified during warm episodes, spurring speculation that the recent upturn in El Niño intensity is linked to global climate change.

An El Niño signal shows up in a 4000-year record of lake sediments extracted from Glacial Lake Hitchcock, which occupied the Connecticut River valley during the waning phase of the last Ice Age. Layers of lake sediments chronicle short-term climate variability spanning the period from 17,500 to 13,500 years ago (during the final retreat of glacial ice from New England), and apparently resolve both intense and weak El Niño events (having periods of 2.5 to 5 years).

In 2002, Geoffrey Seltzer of Syracuse University and his colleagues reported on their analysis of two 8-m (26-ft) sediment cores extracted from the bottom of Laguna (Lake) Pallcacocha high in the Andes Mountains of southern Ecuador. Rainfall governs the amount of sediment delivered to the lake but rainfall associated with a weak El Niño is not likely to reach the 4200-m (13,800-ft) high lake. Hence, the researchers interpreted anomalously thick layers of silt in the cores as indicating heavy rainfall associated with an intense El Niño. The lake sediment cores span the past 12,000 years and suggest that between 12,000 and 7000 years ago, strong El Niño episodes occurred five or fewer times per century. The frequency of intense El Niño episodes then increased and peaked during the 9th century CE when they occurred about every three years.

Using documentary evidence from explorers and early settlers along the coast of northwest South America, W.H. Quinn and colleagues reconstructed the chronology of El Niño episodes back to the year CE 1525. For example, El Niño conditions prevailed during 1531-32, when Francisco Pizarro (1478-1541) conquered the Incas. Although heavy rains impeded his advance, his horses were well fed by the vegetation made unusually lush by the moist conditions. Records of the maximum discharge of the Nile River at Cairo extend back to CE 622 and are another valuable source of information on past El Niño episodes. Year to year changes in discharge are linked to variations in the summer monsoon rains in the Ethiopian highlands, near the headwaters of the Nile, and decrease during El Niño.

The long-term reconstructed climate record suggests that El Niño has occurred regularly since at least the late Pleistocene—albeit with at least one significant lull from 12,000 to 7000 years ago. However, El Niño intensity has varied over thousands of years (i.e., the tempo is roughly the same but the beat has alternately intensified and weakened). We now appear to be in a period of particularly intense El Niño events.

CHAPTER 9

THE CLIMATE RECORD: PALEOCLIMATES

Clues such as ancient river beds help scientists reconstruct what a region's climate must have been in the past. The Safsaf Oasis in the Sahara was clearly once a thriving river valley. An ancient river's widest channel runs from the lower left corner towards the center of the image. The image was taken by the Spaceborne Imaging Radar-C/X-band Synthetic Aperture Radar (SIR-C/X-SAR), which uses radar to penetrate the thin sand cover. The sensor flew aboard the space shuttle Endeavour, and took this image on April 16, 1994. [Courtesy NASA]

Case-in-Point

Late in the 13th century CE, the Anasazi people (also known as the Ancestral Pueblo) abandoned their homelands near the present-day Four Corners area of the American Southwest (where Arizona, New Mexico, Colorado, and Utah adjoin). At about the same time the Anasazi's neighbors to the north and northwest (in present-day Utah and western Colorado), the Fremont people, permanently left the region after 1000 years of habitation (Figure 9.1).

Why did these people leave their ancestral lands? The first archaeological studies of the ruins of the Anasazi

FIGURE 9.1
Map of the American Southwest showing the ancestral lands of the Anasazi and Fremont peoples.

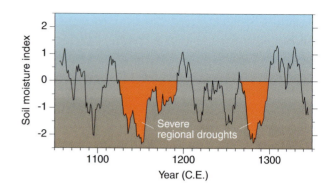

FIGURE 9.2
Drought record from the Four Corners Area of the Southwest derived from tree growth ring data. [Modified after Larry Benson of the U.S. Geological Survey and Edward R. Cook of Lamont-Doherty Earth Observatory]

cliff dwellings attributed abandonment to a single factor, either climate change or conflict. In the March-April 2008 issue of *American Scientist*, Timothy A. Kohler, an anthropologist at Washington State University, and his colleagues proposed a combination of causes: climate change, population growth, competition for resources, and conflict. This proposed explanation came from the *Village Ecodynamics Project* which focused on how the Anasazi interacted with their environment. The southwest Colorado study area covered 1816 km², and among the data gathered were proxy climate records from tree growth rings, pollen profiles, and macrofossils. Organic materials were dated using the radiocarbon method.

Kohler and colleagues noted that the Anasazi experienced two major cycles of population growth and decline: CE 600-920 and CE 920-1280. In both cycles, the drop in population was preceded by aggregation of the population in relatively dense clusters. Abandonment took place by the end of the second cycle.

The Anasazi were farmers and hunters who, in the 12th and 13th centuries, depended heavily on maize (corn). It was their principal source of carbohydrate calories and was fed to domesticated turkeys, their primary protein source. Droughts in the 1100s and 1200s caused shortfalls in the maize harvest and serious nutritional deficiencies. Some of these episodes constituted **megadroughts**, which are droughts that persist for multiple decades (Figure 9.2). Readily recognized in the tree growth ring records of the period, megadroughts occurred from CE 1135-1180 and

CE 1276-1299. The first of these (perhaps the most severe of the past two millennia) forced the Anasazi to leave Chaco Canyon (New Mexico) and move to the cliffs of Mesa Verde, CO (Figure 9.3).

Between droughts, during the relatively wet periods, the population soared, making them all the more vulnerable when drought and food shortages returned. In response to food scarcity, people stored food in granaries hidden away in places difficult to access such as high cliffs and narrow ledges. And the repeated failures of the maize harvest forced the Anasazi to depend increasingly on hunting and gathering. Inevitably, competition for declining food sources led to conflict among neighbors; some sought shelter in cliff dwellings, but there is evidence that conflict culminated in violent confrontations and death.

To the north and northwest, similar stresses were impacting the Fremont people as megadroughts caused food shortages and civil strife during the 1200s. The

FIGURE 9.3
Cliff dwellings of Mesa Verde in southwestern Colorado.

Fremont people were somewhat more versatile than the Anasazi; depending on environmental conditions, they readily shifted between hunting, farming, and foraging. Nonetheless, individuals were forced to hide what little food was available and to build defensive structures on ridge tops. Eventually the Fremont people also had no alternative but to abandon the Southwest.

In summary, many factors combined to force the Anasazi and Fremont peoples to abandon their ancestral homelands. A climate change involving more frequent megadroughts was one of those factors. In fact, often climate is only one of many factors that in combination result in societal upheaval.

Driving Question:
How and why do scientists reconstruct the climate record prior to the instrument era?

Extending the climate record as far back in time as possible aids our understanding of Earth's climate system, provides a valuable perspective on the present climate, and with an eye to the future, gives insight as to the nature of climate change. For example, reconstruction of the long-term mean annual Northern Hemisphere temperature indicates that the observed warming trend of the late 20th century was not equaled in magnitude during the past millennium. However, the reliable instrument-based climate record taken under standardized conditions extends back to only the 1870s. Climate information prior to the instrument era is drawn from a variety of climate-sensitive sources, including historical documents, tree growth rings, pollen profiles, coral, and deep-sea sediment cores.

Climate reconstruction often requires calibration of modern climate forcing with modern environmental response. That calibration is applied to ancient environmental response records (e.g., fossil flora and fauna in bedrock) to unlock information on the climate past. Reconstruction of past climates is the principal realm of the subfield of climatology known as **paleoclimatology**. The accuracy of climate reconstructions generally decreases with increasing time before present because of limitations, such as declining resolution of information, increasingly fragmented data, and problems with time control and correlation of data. Ideally, proxy climate data can be tested for validity by comparing the data with the instrument record for overlapping periods. The problem here is that instrument-derived data typically offer much finer resolution. Consequently, the climate of geologic time, spanning hundreds of millions of years, or longer,

can be described in only very general terms. Resolution improves for the latter portion of the Pleistocene Ice Age and into the Holocene Epoch.

The primary purpose of this chapter is to survey the techniques and findings of paleoclimatology, beginning with a general overview of why and how past climates are reconstructed, followed by a brief summary of the major proxy climate data sources. We then summarize the climate record prior to the instrument era.

Reconstructing Past Climates: Why and How?

Why reconstruct climates of the past, that is, climates prior to the era of reliable instruments and standardized methods of observation? As with any mystery, the climatic past appeals to human curiosity and offers a challenge. But beyond satisfying an inquisitive nature, many practical reasons justify our delving into the climate past. For one, the process of climate reconstruction improves our understanding of environmental response to climate variability and climate change. A lengthier climate database encompasses a broader range of fluctuations in Earth's climate than is represented in the reliable instrument-based climate record, thereby aiding our study of the possible causes of climate variability and climate change. In addition, coupling the instrument-based and reconstructed climate records provides a valuable perspective on the present climate. And it appears reasonable to assume that *what has happened in the past could happen again in the future.*

The validity of climate reconstruction rests on a systematic analysis and synthesis of a complex and diverse array of data. Information on climate prior to the era of instrument records consists of a montage of contributions from many disciplines including palynology, paleontology, glaciology, and geology. This information consists of biotic and abiotic response records, that is, climate-sensitive time series. To be most useful in climate reconstruction, these heterogeneous data sets must be converted into records of compatible climate elements, such as mean annual temperature and precipitation.

Climate reconstruction requires identification of a link between quantitative climate forcing and environmental response. This link is formulated by calibrating the record of modern environmental response against modern instrument-derived climate data and then applying it to the ancient (fossil) record. This procedure assumes *methodological uniformitarianism*; that is, factors that are ecologically limiting today were limiting in the same way in the past. For example, Bergthórsson developed a calibration between historic records of the duration and extent of drift ice off the coast of Iceland with climatic data for a recent 150-year period. Applying this modern calibration to documentary records of drift ice conditions in the past, Bergthórsson was able to reconstruct the decadal mean annual temperature in Iceland dating back to almost CE 900 (Figure 9.4). The credibility of Bergthórsson's model is supported by the close association between relatively cold episodes and the occurrence of famine years in Iceland as documented by Thoroddsen.

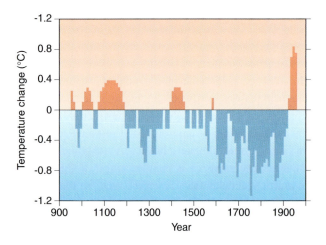

FIGURE 9.4
Reconstructed record of the decadal mean annual temperature for Iceland over the recent millennium based on historical accounts of the duration and extent of sea ice along the shore. [Data from Bergthórsson]

Proxy Climate Data Sources

For times and places where no instrument-derived record of climate exists, past climate information may be inferred from various sensors that substitute for actual weather instruments. These sensors of climate, known as **proxy climate data sources**, include historical documents, tree growth rings, pollen profiles, deep-sea sediment cores, speleothems, corals, and glacial ice cores. No one type of proxy alone is sufficient to enable scientists to reconstruct broad scale patterns of climate. For a summary of the overall limitations of proxy climate data sources, refer to this chapter's first Essay.

HISTORICAL DOCUMENTS

Under cautious scrutiny, certain historical documents archived in libraries and museums can yield a wealth of information on past climates. Personal diaries, almanacs, old newspapers, and mariner's log books may yield qualitative and some quantitative references to weather and climate. Other types of documents refer only indirectly to weather and climate but can be useful nonetheless. Records of success of grain harvests, quality of wine, or various phenological phenomena (such as dates of blooming of plants in spring) provide indirect indications of growing season weather. For example, in his book *Times of Feast, Times of Famine* (1971), Emmanuel Le Roy Ladurie relies heavily on vineyard records to reconstruct the climate of Western Europe during the Middle Ages. More recently, researchers at the University of Bern reconstructed summer weather/climate patterns for parts of Switzerland based on records of grape harvests dating as far back as the late 15th century. The growth of grapevines and the ripening of the fruit strongly respond to average April through August temperatures.

Caution must be exercised in inferring climate information from historical documents because many factors, in addition to weather and climate, usually influence such records. For example, aside from growing season weather, harvest dates in vineyards are affected by fluctuations in the wine market. Usually, the authors of such documents had no intention of chronicling the weather or climate and it is also important to bear in mind that people have long applied ingenuity to moderate the impact of climate—particularly extremes in climate. (Refer, for example, to the second Essay in Chapter 4 on agricultural freeze protection strategies.) Hence, climate information derived from written records of human activity is not always reliable and corroborating data from other independent sources are necessary to support these climatic inferences.

TREE GROWTH RINGS

Analysis of variations in the thickness and density of annual growth rings of certain tree species can yield detailed information on past climates. The study of tree growth rings for climate data is known as **dendroclimatology**. Andrew E. Douglass (1867-1962), a solar astronomer, pioneered this work in the American Southwest between 1894 and 1901 while at Lowell Observatory, a private non-profit research institution in Flagstaff, AZ. In 1937, he founded the Laboratory of Tree-Ring Research at the University of Arizona. Today, his successors at the Laboratory are reconstructing past climates using computers programmed with special statistical techniques.

At the onset of the growing season in spring, plant tissue located immediately beneath tree bark produces relatively large thin-walled wood cells, which give the wood a relatively light appearance. Wood cells produced in summer, however, are thick-walled, giving the wood a darker appearance. A year's growth of spring wood plus summer wood constitutes an annual growth ring, so counting the number of growth rings gives the age of the tree in years. Because the width of growth rings normally decreases as the tree ages, widths are usually expressed in terms of a *tree-growth index*, defined as the ratio of the actual tree growth-ring width to the width expected based on the tree's age. The index is relatively low in stressful growing seasons and high in favorable growing seasons.

Only trees living in subpolar terrestrial regions are useful in dendroclimatic reconstructions and primarily record conditions during the warm season. (Tropical species do not have well-defined seasonal growth rings.)Trees growing near the limits of their range are the most sensitive to climate variability so that their growth rings are the most reliable sensors of climate. A simple hollow drill is used to extract cores from living trees or cut timber (Figure 9.5). Usually cores are taken from many trees at one site, and tree-ring indexes are averaged. In western and southwestern North America, the primary locale for dendroclimatic research, scientists sample ponderosa pine, Douglas fir, and three closely related species of Bristlecone pine. Some of the longest tree ring records are obtained from Great Basin Bristlecone pine (*Pinus longaeva*) that grows at or near the tree line in mountains of Utah, Nevada, and eastern California (Figure 9.6). At up to almost 5000 years old, the Great Basin Bristlecone pine is thought to be one of the oldest known living tree species in the world. Bristlecone pine is characterized by heavy gnarled limbs, a spiked top, and in the case of the oldest specimens, only a narrow ribbon of living bark (cambium).

FIGURE 9.5
A simple hollow drill is used to extract a core from a living tree or cut timber. [Frome Hannes Grobe]

Typically, tree ring chronologies date back some 500 to 700 years but range up to 11,000 years in a few cases. By assiduous matching of tree growth ring records from living trees with those from timbers in prehistoric dwellings, detailed tree ring chronologies are extended back in time thousands of years. This matching technique is known as *cross-dating*.

Although other environmental factors (e.g., soil type, drainage conditions which is accounted for by sampling trees in the same area) can be important, the thickness and density of tree growth rings are especially sensitive to moisture stress and have been used to reconstruct lengthy drought chronologies.

FIGURE 9.6
The long-lived Bristlecone pine growing near the tree line in the White Mountains of California is a valuable source of tree growth ring data for dendroclimatic reconstructions. [Photograph courtesy of Mark A. Wilson, Department of Geology, The College of Wooster, Wooster, OH.]

Tree growth ring analysis has enabled researchers to extend the drought record across large portions of North America and many centuries into the past. In 1998, Edward R. Cook of Columbia University's Lamont-Doherty Earth Observatory and colleagues from Arizona and Arkansas reconstructed drought chronologies across the nation based upon annual tree ring data obtained from a network of 388 climatically sensitive tree ring sites. From these data, time series of summer (June through August) Palmer Drought Severity Index (PDSI) values were determined stretching back to 1700 at 155 grid points. These gridded tree ring chronologies were calibrated with PDSI instrument-based records from selected Historical Climatology Network stations commencing in the late 19th century. Researchers found that the 1930s drought (discussed in the Case-in-Point of Chapter 6) was the most severe drought to impact the nation since 1700.

By 2004, the drought record had been expanded to include 835 tree ring sites, primarily in the West, where precisely dated annual tree ring chronologies were obtained. The new grid covered most of North America with a latitude/longitude spacing of 2.5 degrees. In addition to the 286 grid point PDSI time series, annual contour maps of PDSI were constructed that span much of the continent. This work permitted extension of the spatial and temporal coverage of drought reconstruction not only into Canada and Mexico, but also back 2000 years. From this more extensive data set, researchers produced an online *North American Drought Atlas*. They identified several North American droughts that were even more severe than the 1930s drought. In addition to being more severe, some droughts persisted through several decades, considerably longer than those of the 20th century. A megadrought that occurred in the 16th century, along with another megadrought extending into the early 17th century, may have contributed to the disappearance of the Roanoke Colony on Roanoke Island in Dare County in what is now North Carolina. Intended to be the first English colony in America, the *Lost Colony* vanished in the 1580s.

Tree growth ring records have been used to reconstruct past variations in the flow of the Colorado River prior to 1896 when the gauge-based record began (Chapter 7). The flow of the Colorado River correlates with the mean annual precipitation in the river basin. In addition, the width of tree rings from living trees from many sites on the Colorado Plateau correlates with mean annual precipitation. Hence, the flow of the Colorado River directly correlates with the width of tree rings. Application of this correlation to the prehistoric tree ring record enables scientists to reconstruct the prehistoric flow of the Colorado River.

In 2006, Connie A. Woodhouse, a climatologist at the University of Arizona, and her colleagues used tree ring records to estimate the long-term natural mean annual flow of the Colorado River to be 14.6 MAF. They found that drought can occur suddenly and may persist from several years to a few decades. For example, during the drought of 1844-1848, the mean annual flow was only 9.6 MAF. Evidence was found that a 60-year *megadrought* impacted the Colorado Plateau during the 12th century. Analysis of tree growth rings enabled these scientists to extend the region's drought chronology back to CE 762. They relied on data from both living trees and cross-dating of ring patterns in tree trunks scattered throughout the upper Colorado River drainage basin, where dry conditions preserved them. About half way through the megadrought that lasted from CE 1118 to 1179, flow of the Colorado River remained below the long-term average for 13 consecutive years. By contrast, over the past century, consecutive years of below average flow numbered no more than 5 years.

POLLEN PROFILES

Ponds, peat bogs, marshes, and swamps are favorable sites for the accumulation and preservation of wind-borne pollen. *Pollen* is the tiny dust-like fertilizing component of a seed plant that is dispersed by the wind. Mixing with other sediments (clay, silt and organic particles), pollen grains settle and accumulate in low-lying depositional areas. Upward of 20,000 pollen grains may be mixed in a single cubic centimeter of pond mud. Assuming that the pollen is the product of nearby vegetation and that climate largely governs vegetation types, climate may be inferred from pollen. When climate changes, vegetation changes, and so too do the types of pollen delivered to depositional sites. Hence, changes in the abundance of pollen of different species at various depths within accumulated sediment may provide a record of past climatic regimes and climate change.

Scientists use a corer to extract a sediment column (core), then separate pollen from its host sediments, identify pollen species and frequency of occurrence of each species, and reconstruct the sequence of past changes in vegetation. From the climate requirements of the reconstructed vegetation (based on modern species distribution and modern climate), scientists decipher the sequence of past climate changes.

Pollen is a valuable source of information on the vegetation and climate of the late Pleistocene Ice Age and subsequent Holocene Epoch, especially over the past 15,000 years. Using sophisticated statistical techniques to calibrate climate and pollen, scientists have reconstructed remarkably detailed quantitative climate data. For example, a pollen record (profile) from Kirchner Marsh near Minneapolis, MN, yielded a reconstructed record of variations in July mean temperature and annual precipitation back to 12,000 years ago. Unfortunately, relatively few sites favor the accumulation and preservation of a continuous long-term pollen/climate record. In North America, most such records come from a few geographical areas, including western mountain valleys, the Great Lakes region, and interior New England.

DEEP-SEA SEDIMENT CORES

Cores extracted from sediments that blanket the ocean floor yield a continuous record of sedimentation dating back many hundreds of thousands of years and in some places, millions of years (Figure 9.7). Much of what we know about the climate of the Pleistocene Ice Age is based on analysis of the shell and skeletal remains of microscopic marine organisms found in deep-sea sediment cores. Identification of the environmental requirements of these organisms, plus oxygen isotope analysis of their remains, enables scientists to distinguish between relatively cold and warm climatic episodes of the past.

With **oxygen isotope analysis**, scientists use a special property of water to reconstruct large-scale climate fluctuations of the Pleistocene Ice Age. A water molecule (H_2O) is composed of one of two stable isotopes of oxygen, ^{16}O or ^{18}O. (*Isotopes* consist of atoms that are chemically identical, with the same number of protons but different numbers of neutrons in the nucleus.) In the Earth system, the lighter isotope (^{16}O) is much more abundant than the heavier isotope (^{18}O); only one ^{18}O exists for every thousand or so ^{16}O. Nonetheless, small but significant variations occur in the ratio of light oxygen to heavy oxygen circulating in the global water cycle, which have important implications for past fluctuations in glacial ice volume.

On average at a particular temperature, water molecules containing the lighter ^{16}O isotope move slightly faster than water molecules containing the heavier ^{18}O isotope and therefore evaporate more readily. Hence, water vapor is enriched with light oxygen compared to heavy oxygen and the amount of ^{16}O compared to ^{18}O is greater in cloud particles and precipitation versus liquid water on Earth's surface. When it rains or snows, the ^{16}O returns to the ocean, replenishing the ocean's supply of light oxygen and maintaining a relatively constant average ratio of light to heavy oxygen. However, geographical variations in the oxygen isotope ratio of seawater arise because of differences in precipitation amounts and evaporation rates. Seawater has more ^{18}O at subtropical latitudes, where evaporation exceeds precipitation, and less in middle latitudes, where rainfall is greater.

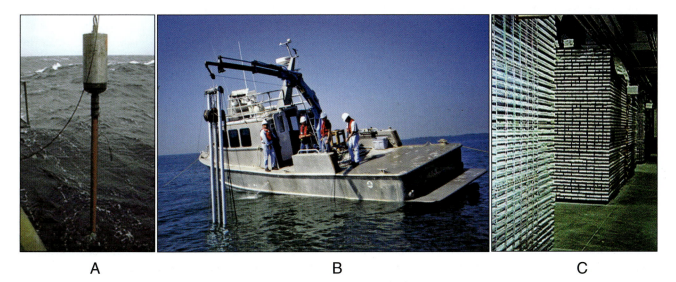

A **B** **C**

FIGURE 9.7
Sediment cores extracted from beneath the ocean floor provide valuable information on the geologic and climatic past. (A) A hollow pipe lined with plastic tubing and coupled to a weight at the top is lowered over the side of a ship. When within about 8 m of the bottom, the corer free-falls into the sediment as a piston suctions sediment into the tube. The coring device is recovered and the sediment core is removed and split lengthwise for analysis. (B) USGS crew on the research vessel *G.K. Gilbert* collect a 20-ft sediment core, using an electric coring system and hydraulic crane. [USGS] (C) Core racks holding a total of 72,000 m of sediment cores at the Deep-Sea Sample Repository of Lamont-Doherty Earth Observatory in Palisades, NY. [Courtesy of the Lamont-Doherty Earth Observatory]

During a climatic episode that favors the formation or growth of a glacier, snow that accumulates on land converts to ice. Heavy water molecules condense and precipitate slightly more readily than light water molecules. Moisture plumes moving from the tropics to high latitudes lose heavy oxygen along the way, so snow falling at high latitudes has less ^{18}O than rain falling in the tropics. The result is that growing glacial ice sheets sequester more and more light oxygen, while ocean water has less and less. With a shift to an interglacial climate, ice sheets shrink and meltwater rich in ^{16}O drains back into the ocean, increasing the ratio of ^{16}O to ^{18}O.

Organic sediments that accumulate on the ocean floor record fluctuations in the oxygen isotope ratio of seawater. Marine organisms, such as foraminifera living in the sunlit surface waters, build their shells from calcium carbonate ($CaCO_3$) that is dissolved in seawater. Shells formed during warmer interglacial climatic episodes contain more light oxygen than those formed during colder glacial climatic episodes. When these organisms die, their shells settle to the ocean floor and mix with other marine sediments. With specially outfitted deep-sea drilling ships, scientists extract cores from an undisturbed sequence of ocean bottom sediments. In the laboratory, the core is split open, and shells are extracted and analyzed for their oxygen isotope ratio. The youngest sediments are at the top of the core and the oldest sediments at the bottom. Variations in oxygen isotope ratio document changes in the planet's glacial ice volume, a measure of past changes in temperature. The proportion of light to heavy oxygen in ocean water decreases with increasing glacial ice volume.

Oxygen isotope analysis of deep-sea sediment cores indicates that the Pleistocene Ice Age (1.7 million to 10,500 years ago) was punctuated by numerous abrupt changes between glacial and interglacial climatic episodes. Oxygen isotope analysis has also been applied to ice layers within cores extracted from the Greenland and Antarctic ice sheets. These analyses confirm the abrupt change behavior of climate dating back hundreds of thousands of years.

SPELEOTHEMS

A **speleothem**, also called *dripstone*, is a calcite ($CaCO_3$) deposit in a limestone cave or cavern that can yield high-resolution records of past temperature and rainfall. A speleothem forms when calcite precipitates from groundwater that seeps into a cave and can either build downward from the roof of the cave creating a stalactite, or grow upward from the floor of the cave to form a stalagmite.

Climate is reconstructed using oxygen isotope analysis from samples of calcite extracted from a speleothem. As noted above, scientists measure the ratio of two isotopes of oxygen from the sample: ^{16}O (light oxygen) and ^{18}O (heavy oxygen). The ratio of these isotopes is sensitive to temperature and rainfall. The age of the sample is derived independently using the ratio of uranium-234 to thorium-230, a very precise radiometric technique that determines the age of a speleothem sample with a resolution ranging from a year to decades. A potential radiometric dating technique, involving the decay of uranium to lead, may extend speleothem-derived climate records back 600,000 years. The Case-in-Point that opens Chapter 7 summarizes how a speleothem from a cave in northern China was used to reconstruct monsoon rainfall over a recent 1800-year period.

CORALS

A **coral reef** is among the most spectacular biological features of the ocean. Primarily built by carbonate-secreting colonial animals, these slow growing structures can be centuries old, and may be so large that they are visible from space (e.g., Great Barrier Reef along Australia's East Coast). In many parts of the tropical ocean, reefs stand hundreds of meters above the sea bottom and can extend hundreds of kilometers along the shoreline or cap extinct undersea volcanoes. They consist of thin veneers of living organisms growing on older layers of dead coral (limestone) or volcanic rock, and are bound together by layers of calcareous algae.

Worldwide, coral reefs provide shelter and food for up to 9 million species of marine life (about one-third of all known marine species) including fish, invertebrates, and plants. Coral reefs are among Earth's most productive habitats, ranking second to rain forests in biodiversity; most are in the tropical Pacific and Indian Oceans between about 30 degrees N and 30 degrees S. Reef-building corals prefer waters with an average annual temperature of 23 °C to 25 °C (73 °F to 77 °F) and most corals cannot tolerate prolonged exposure to high or low temperatures or to large fluctuations in temperature. For this reason, even small changes in sea-surface temperatures—such as associated with large-scale climate change—threaten coral reefs (Chapter 12). Corals also require clear water and are endangered by sediment runoff from land, oil spills, and other forms of water pollution. Excess nutrient input washed from the land can stimulate the growth of algae on the surface of coral reefs, smothering coral polyps (a process known as *eutrophication*).

Each type of coral animal builds a characteristic structure that is conspicuous on reef surfaces. Some corals (e.g., brain corals) form robust compact structures; others build delicate, complex branching forms. Many corals have growth rings, much like trees, that can be used to reconstruct past variations in maritime conditions and climate in the tropics and subtropics. Through oxygen isotope analysis and measurement of chemical species ratios, scientists acquire climate data including information on SST and past El Niño and La Niña events. Corals provide a resolution typically measured in months (weeks in rare cases) and the time range is up to about 400 years.

Edward R. Cook and colleagues at the Lamont-Doherty Earth Observatory relied on climate data inferred from coral and tree growth rings to demonstrate a link between exceptionally long-lived La Niña episodes in the tropical Pacific Ocean and persistent droughts in the American Southwest. Analysis of coral from the central Pacific revealed that SST were much lower than normal from 1855 to 1863. This SST anomaly was interpreted to be a manifestation of an exceptionally long-lived La Niña and also coincided with the driest decade in Texas since 1700 based on tree growth ring studies. The drought that began in 1855 along the edge of the Great Plains was more intense than the Dust Bowl drought of the 1930s. Other widespread droughts that corresponded to persistent La Niña conditions occurred in the Southwest from 1703 to 1709 and 1818 to 1824.

GLACIAL ICE CORES

To better understand Earth's climate system and climate change as a basis for predicting the climate future, scientists are collecting and analyzing proxy climate data from the Pleistocene Ice Age. Ice cores extracted from the Antarctic and Greenland ice sheets are important sources of information on climate change, as well as the chemical composition of air during the Pleistocene Ice Age (Figure 9.8).

In 1988, Soviet and French scientists reported on their analysis of a 2200 m (7200 ft) ice core extracted at Vostok station on the East Antarctic ice sheet. The ice core spanned 160,000 years. Oxygen isotope analysis yielded a temperature record, and chemical analysis of trapped air bubbles revealed trends in the greenhouse gases carbon dioxide and methane. During the summers of 1991-93, two independent scientific teams, one American and the other European, drilled into the thickest portion of the Greenland ice sheet. The two drill sites were located within 30 km (19 mi) of each other, about 650 km (404 mi) north of the Arctic Circle. Both cores were about 3000 m

(9840 ft) in length and spanned a time interval of roughly 200,000 years. In the mid-1990s, drilling at Vostok recovered a 3100 m (10,170 ft) ice core spanning the past 425,000 years. In 2004, the *European Project for Ice Coring in Antarctica (EPICA)* extracted an ice core from East Antarctica representing a time interval of 740,000 years. By 2008, the recovered ice core encompassed about 800,000 years.

To derive air temperature from glacial ice cores, scientists use either oxygen isotope analysis (described above) or analysis of the ratio of deuterium to hydrogen in ice. The water molecule is composed of two different isotopes of hydrogen: ^1H and ^2H. ^1H consists of one proton and no neutrons whereas ^2H (also called deuterium or D) has one proton and one neutron. Isotopic ratios are compared to that of *standard mean ocean water (SMOW)*. Compared to SMOW, glacial ice cores contain slightly less of the heavier oxygen and deuterium isotopes. As the temperature falls, there is comparatively less and less ^{18}O and D.

In addition to reconstructing temperature, glacial ice cores are also analyzed for changes in the composition of the atmosphere, especially greenhouse gases. The present level of atmospheric carbon dioxide is about 27% higher than the highest levels detected in air bubbles trapped in glacial ice cores dating back 650,000 years. In addition, the reconstructed temperature record closely parallels the concentration of the greenhouse gases carbon dioxide and methane. This prompts the question as to whether greenhouse gas concentration drives climate or climate drives greenhouse gas concentration. It appears that with a slow change in atmospheric CO_2 concentration

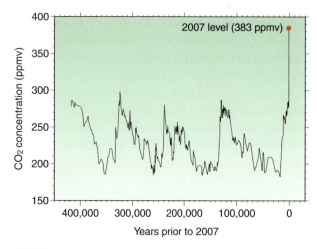

FIGURE 9.8
The concentration of atmospheric carbon dioxide in parts per million by volume (ppmv) from about 425 thousand years ago to 2007 based on glacial ice core analysis from the Vostok station in Antarctica. [Scripps Institution of Oceanography, University of California, San Diego]

over a long period, temperature drives CO_2 concentration by altering the global carbon cycle. On the other hand, with a rapid change in atmospheric CO_2 concentration over a short period (e.g., since the beginning of the Industrial Revolution), CO_2 appears to drive temperature.

Where snow accumulation rates are relatively high, glacial ice cores can be used to resolve short-term climate fluctuations of decades or less. As noted later in this chapter, cores extracted from the Greenland ice sheet reveal major abrupt oscillations in climate over periods ranging from millennia down to decades that are superimposed on much longer-term gradual climate cycles (i.e., Milankovitch cycles, Chapter 11).

STRATIGRAPHY AND GEOMORPHOLOGY

Bedrock, particularly sedimentary strata, is a widespread source of generalized information on environmental conditions of the geologic past, including climate. As noted in Chapter 1, internal and surface geological processes continually shape and reshape Earth's lithosphere. These same processes produce a multitude of rock types that compose the crust. Rocks, in turn, are made up of one or more minerals. (A *mineral* is a naturally occurring inorganic solid characterized by an orderly internal arrangement of atoms with fixed physical and chemical properties). Rocks composing the crust are classified as igneous, sedimentary, or metamorphic based on the general environmental conditions in which the rock formed.

Cooling and crystallization of hot molten *magma* produces **igneous rock**. Magma originates in the lower portion of the lithosphere or upper mantle, and migrates upward towards the Earth's surface (Figure 1.12). Magma may remain within the crust and cool slowly, forming coarse-grained igneous rock such as granite, or it may spew onto Earth's surface as *lava* through vents or fractures in bedrock and solidify rapidly, forming fine-grained igneous rock such as basalt or glassy material such as obsidian. Igneous rock-forming processes (e.g., volcanism) affect the chemical composition of the atmosphere and may contribute to climate change (Chapter 11).

Sedimentary rock may be composed of any one or a combination of compacted and cemented fragments of rock and mineral grains, partially decomposed remains of plants and animals (e.g. shells, skeletons), or minerals precipitated from solution. Sediments form as rocks undergo weathering (physical disintegration and chemical decomposition) when exposed to rain, atmospheric gases, and fluctuating temperatures at or near Earth's surface. Sediments are washed into rivers that transport them to the sea, and other standing bodies of water, where they settle out of suspension, accumulate on the bottom, and eventually compact into layers of solid sedimentary rock. Sediments are also transported and deposited by wind, glaciers, and icebergs at sea. Most sedimentary rocks have a granular texture; they are composed of individual grains that are compressed or cemented together (e.g., sandstone, shale), although some consist of precipitated minerals and are crystalline (e.g., limestone, rock salt).

Unless disturbed by tectonic forces, sedimentary rock occurs as a sequence of layers (strata) having a horizontal orientation (at least initially) with the youngest layer at the top of the sequence and the oldest layer at the bottom (Figure 9.9). By interpreting the environmental conditions represented by each layer (based on factors such as mineral composition, grain size distribution, fossil fauna or flora), geoscientists can infer environmental conditions (including climate) and change in general terms through hundreds of millions of years. In many cases, however, sediment deposition was not continuous, and portions of the environmental record are missing because of erosion or non-deposition of sediment. A hiatus in the sedimentary record may span many millions of years. So, like a history book with missing pages, this environment (and climate) record is incomplete.

Consider a few examples of climate inferences drawn from sedimentary strata. Layers of rock salt indicate a warm, dry climate that favored net evaporation of sea water and crystallization of salt. *Till* is the general term for sediments deposited by the direct action of advancing glacial ice without the influence of running water. Sediments are angular and occur in a broad size range (poorly sorted). Through lithification, till becomes *tillite*, a sedimentary rock that is evidence of ancient ice ages and glacial climates.

FIGURE 9.9
Exposure of fossil-rich sedimentary bedrock (limestone) around 300 million years old at a road cut in northwestern Missouri.

Like many sedimentary rocks, **metamorphic rock** is derived from other rocks. A rock is metamorphosed (changed in form) when exposed to high pressure, intense heat, and chemically active fluids—conditions that exist in geologically active mountain belts. Like igneous rocks, metamorphic rocks are crystalline; that is, they are composed of crystals that interlock like the pieces of a jigsaw puzzle. Marble is a common metamorphic rock formed by the metamorphism of limestone ($CaCO_3$) and quartzite is a very durable metamorphic rock formed by metamorphism of sandstone (mostly SiO_2).

Most of the bedrock composing Earth's crust is igneous with some metamorphic rock locally or regionally. In many places, thick layers of sedimentary rock and unconsolidated sediments overlie crystalline igneous and metamorphic rocks. Unconsolidated sediments include soils (on land) and clay, silt, sand, or gravel (on land and ocean bottom). Deposits of sediment vary widely in thickness, from a thin veneer to thousands of meters.

Geomorphology is the scientific study of landforms on Earth's surface and the processes responsible for their formation. In some cases, climate plays a major role in these processes such that some generalized information can be inferred from landforms regarding past climates. As noted later in this chapter, many different landforms are the consequence of glacial activity such that a distinction can be made between climate regimes that favor the thickening and expansion of glaciers (*glacial climates*) and climate regimes that favor the melting of glaciers (*interglacial climates*).

In another example of climate reconstruction based on geomorphology, large fields of active (mobilized) sand dunes usually occur where the climate is arid. According to the *United States Geological Survey (USGS)*, moisture balance and the amount of vegetation cover are more important than wind in controlling whether dunes are stable or active. Aerial photographs from 1936 and 1938 identified mobilized dunes in parts of North Dakota, Colorado, Kansas, Oklahoma, Texas, and New Mexico. From analysis of photos and climate data, scientists concluded that loss of vegetation cover (and its anchoring ability) because of higher temperatures and reduced rainfall was responsible for dune mobilization during the 1930s Dust Bowl.

VARVES

A **varve** is a thin layer (lamina) of sediment deposited annually in a body of still water, usually a lake fed by a stream. The stream may or may not drain a melting glacier, but the climate must be sufficiently cold

for ice to cover the lake surface during winter. In the case of a glacial meltwater stream, rapid melting of glacial ice during the warm summer months delivers to the water body relatively coarse-grained, light-colored sediment (sand or silt). This sediment settles to the bottom and constitutes the summer band of a varve. In winter, lower temperatures slow the discharge of meltwater and ice covers the lake. Suspended sediment slowly settles out of the still water to the lake bottom. Consequently, the summer layer grades upward to the thinner winter layer consisting of very fine sediment (clay), often dark in color because of the presence of organic particles. Hence, a single varve consists of one light band and one dark band.

If undisturbed, a sequence of varves may provide a high-resolution record of variations in the annual mass budget of a glacier. A coring device is used to extract a sediment core from the lake bottom. From a reconstructed mass budget record, past climate information may be inferred.

Climates of Geologic Time

For convenience of study, the geologic past plus its climate record is subdivided based on the **geologic time scale**, a standard division of Earth history into eons, eras, periods, and epochs based on large-scale geological events (Figure 9.10). Throughout most of the approximately 4.5 billion years that constitute *geologic time*, information on climate is unreliable and descriptions of climate are highly generalized. Accounts of early climates often are suspect because of lengthy gaps in the proxy climate records and difficulties in determining the timing of events and correlating events that occurred in widely separated locations. As pointed out earlier in this chapter, in many areas lengthy episodes of erosion or non-deposition of sediment, sometimes spanning hundreds of millions of years, mean no bedrock for reconstruction of past climates. Furthermore, *plate tectonics* complicates climate reconstruction efforts that focus on periods of hundreds of millions of years (Figure 9.11). Nonetheless, the available evidence supports some conclusions regarding the climate over geologic time.

Geologic evidence points to an interval of extreme climate fluctuations about 570 million years ago, corresponding to the transition between Proterozoic and Phanerozoic Eons. In southern Africa, along Namibia's Skeleton Coast (South Atlantic Ocean), rock layers that formed in tropical seas directly overlie glacial deposits. Geoscientists interpret these rock sequences as indicating abrupt changes in climate between extreme cold and

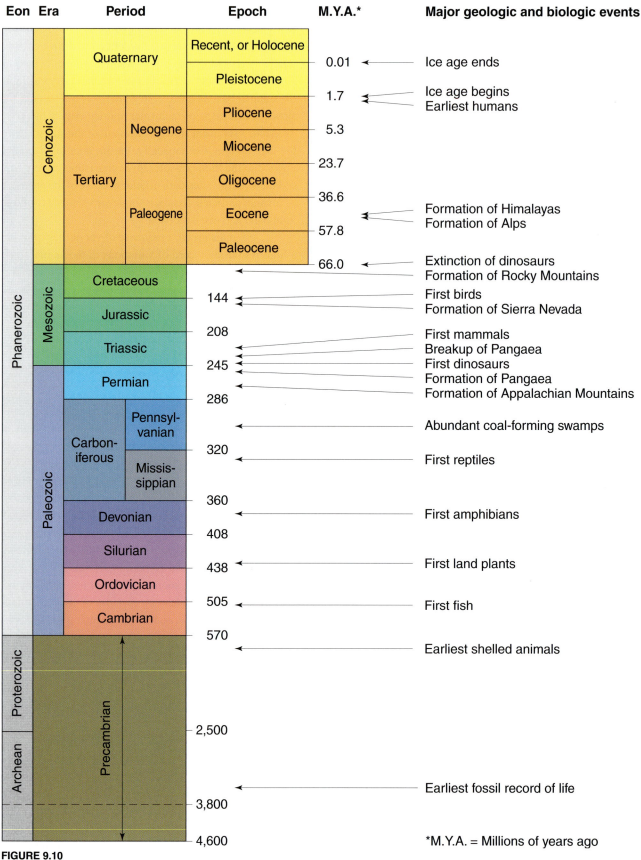

FIGURE 9.10
Geologic time scale.

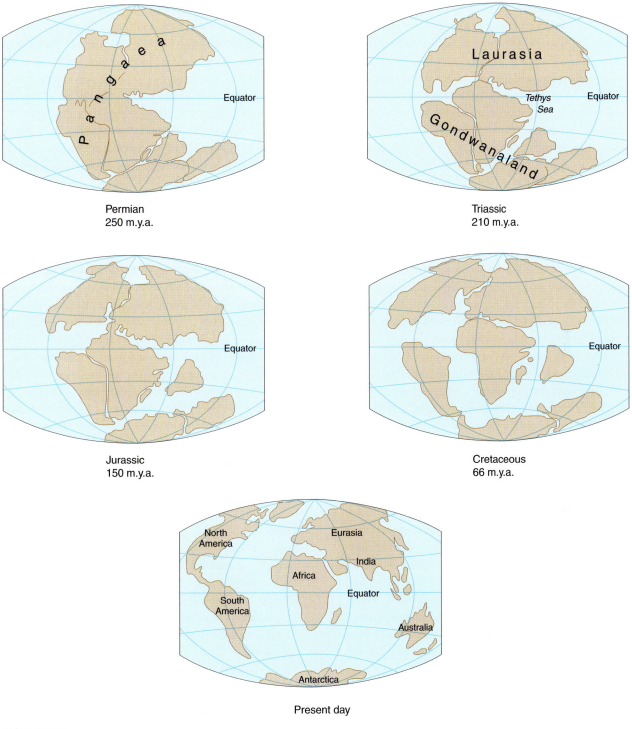

Permian
250 m.y.a.

Triassic
210 m.y.a.

Jurassic
150 m.y.a.

Cretaceous
66 m.y.a.

Present day

FIGURE 9.11
About 200 million years ago, the super-continent Pangaea began to split apart into separate continents that slowly drifted apart. Ocean basins opened and eventually the continents reached their present positions. Continental drift, a manifestation of plate tectonics, is responsible for climate changes operating over hundreds of millions of years. [U.S. Geological Survey]

tropical heat. During as many as four cold episodes, each lasting perhaps 10 million years, the continents were encased in glacial ice and the ocean froze to a depth of more than 1000 m (3300 ft). At the close of each cold episode, temperatures rose rapidly, and within only a few centuries, all the ice melted.

As noted in the second Essay of Chapter 1, five major mass extinctions occurred over the past 550 million years. Drastic changes in Earth's environment during each of these episodes eliminated up to 50% or more of all plant and animal species then in existence. Four of the five mass extinctions (at the close of the Ordovician, Devonian, Permian, and Triassic periods) were linked to a combination of chemical and circulation changes in the ocean, coupled with global warming due to an enhanced greenhouse effect.

The Mesozoic Era, from about 245 million to 70 million years ago, was characterized by a generally warm Earth free of large glacial ice sheets. At the boundary between the Triassic and Jurassic periods, the global mean temperature rose perhaps 3 to 4 Celsius degrees (5.5 to 7 Fahrenheit degrees), contributing to a major extinction or displacement of many animal and plant species. At peak warming during the Cretaceous Period, the global mean temperature was perhaps 6 to 8 Celsius degrees (11 to 14 Fahrenheit degrees) higher than now. Subtropical plants and animals lived as far north as 60 degrees N and dinosaurs roamed what is now the North Slope of Alaska.

A fifth mass extinction at the end of the Cretaceous period, about 65 million years ago, was likely caused by an asteroid impact that threw huge quantities of dust into the atmosphere, blocking sunlight, and causing cooling that contributed to the demise of the dinosaurs. An explosion in the population of mammals followed the extinction of the dinosaurs.

Great fluctuations in climate have characterized the present Cenozoic Era. About 55 million years ago, near the transition between the Paleocene and Eocene Epochs, the concentration of atmospheric greenhouse gases (either methane or carbon dioxide) increased dramatically. Scientists propose that these gases were released to ocean waters during massive intrusions of basaltic magma into carbonaceous marine sediments during tectonic events related to the opening of the North Atlantic Ocean. Some scientists propose that methane was released from gas hydrates. The gases escaped the ocean waters to the atmosphere, enhancing the greenhouse effect, and causing an already warm planet to become even warmer. Over a period of only a few thousand years, global temperatures climbed 5 to 10 Celsius degrees (9 to 18 Fahrenheit degrees).

By 40 million years ago, however, Earth's climate began shifting toward colder, drier, and more variable conditions, setting the stage for the Pleistocene Ice Age—an epoch of numerous major glacial advances and recessions that commenced approximately 1.7 million years ago. According to W.F. Ruddiman of the University of Virginia and J.E. Kutzbach of the University of Wisconsin-Madison, mountain building may explain this change in Earth's climate, specifically the rise of the Colorado Plateau, Tibetan Plateau, and Himalayan Mountains. Prominent mountain ranges influence the geographical distribution of clouds and precipitation and can alter the planetary-scale circulation (Chapters 5 and 6). Furthermore, mountain building may alter the global carbon cycle. Enhanced weathering of bedrock exposed in mountain ranges sequesters more atmospheric carbon dioxide in sediments thereby weakening the natural greenhouse effect.

In the American West, the region from the California Sierras to the Rockies, known as the Colorado Plateau, has an average elevation of 1500 to 2500 m (5000 to 8200 ft). Although mountain building began about 40 million years ago, about half of the total uplift took place between 10 and 5 million years ago. The Tibetan Plateau and Himalayan Mountains of southern Asia cover an area of more than 2 million km^2 (0.8 million mi^2) and have an average elevation of more than 4500 m (14,700 ft). About half of total Himalayan uplift took place over the past 10 million years. The plateaus diverted the planetary-scale westerlies into a more meridional pattern, increasing the north-south exchange of air masses and altering the climate over a broad region of the globe. Also, seasonal heating and cooling of the plateaus cause low pressure to develop in summer and high pressure in winter, enhancing the monsoon circulation over southern Asia (Chapter 7).

In 2008, Dennis V. Kent, a geoscientist at Rutgers University, and colleagues proposed an alternate tectonics-based explanation for the cooling trend that ultimately led to the formation of Antarctica's ice sheets and the Pleistocene Ice Age. Following the break-up of the supercontinent Pangaea, the tectonic plate carrying the Indian subcontinent moved northward and about 50 million years ago slammed into Asia. Prior to this collision, a million-year period of volcanic activity along Asia's southern border generated some 4 million km^3 of basaltic lava derived from carbonate-rich sediments on the sea floor. As a consequence, the level of atmospheric CO_2 increased to more than 1000 ppmv and temperatures rose worldwide. The collision, however, cut off the supply of carbonate sediments and, at the same time, weathering and erosion of rock exposed on the Indian subcontinent caused a gradual decline in atmospheric CO_2 to perhaps 300 ppmv by 30 million years ago. According to Kent and colleagues, the consequent weakening of the greenhouse effect would have been responsible for long-term cooling.

Climates of the Pleistocene Ice Age

Climate varies over a broad spectrum of time scales so that viewing the climate record of the past two million years in progressively narrower time frames is useful. Such an approach helps to resolve climate change into more detailed fluctuations, especially over the recent past. During the past two million years, plate tectonics has not been a major factor in climate change. For example, assuming a spreading rate of 10 cm (3.9 in.) per year, in two million years the Atlantic basin would spread a total distance of 200 km (124 mi), not very significant in a broad climatic sense. For all practical purposes, over the past two million years mountain ranges, continents, and ocean basins were essentially as they are today.

Compared to the climate that prevailed through most of geologic time, the climate of the last two million years was also unusual in favoring the development of huge glacial ice sheets (although evidence also exists of ice ages earlier in geologic time). During much of Earth's history, the average global temperature may have been 10 Celsius degrees (18 Fahrenheit degrees) higher than it was over the past two million years.

HISTORICAL PERSPECTIVE

The Swiss naturalist Louis Agassiz (1807-1873) is credited with championing the **glacial theory** during the mid-1800s. Previously, the prevailing view—as espoused by the Scottish geologist Charles Lyell (1797-1875)—attributed erratic boulders observed in Europe to rafting by icebergs during a massive flood. *Erratics* are large boulders that differ geologically from the local bedrock, implying transport from elsewhere—in some cases over distances of hundreds of kilometers. Interestingly, Swiss peasants long thought that alpine glaciers were once more extensive and interpreted erratics, polished, striated (scratched), and grooved rock surfaces, and other landscape features as indicators of ancient glaciation. At the invitation of Jean de Charpentier (1786-1855), a German-Swiss geologist and early proponent of the glacial theory, Agassiz had an opportunity to examine such evidence near Bex, Switzerland, in the summer of 1836. Agassiz quickly became convinced of the glacial origins of the landscape features and set out to develop a glacial theory.

At scientific meetings and in his writings, Agassiz argued that some time in the past (a period he called an *ice age*), large glacial ice sheets spread over Europe as far south as the Mediterranean and over much of northern Asia and North America. At first, his ideas met with considerable skepticism by the scientific establishment. However, after 1840, as more geologists had an opportunity to examine the field evidence, an increasing number of prominent scientists converted to the glacial theory. By 1865, the theory had gained widespread acceptance on both sides of the Atlantic. A major reason for the initial reluctance to accept the glacial theory was the fact that most geologists were unfamiliar with glaciers, having never observed them or their impacts in the field. Furthermore, scientific expeditions did not discover that Greenland's glaciers formed an ice sheet until 1852 and the size of the Antarctic ice sheets was not determined until the late 19th century.

During the 20th century much field and laboratory work was directed at working out the chronology and causes of the climate and glacial fluctuations of the Pleistocene Ice Age. In the early 1880s, the eminent geologist Thomas C. Chamberlin (1843-1928) proposed that the Ice Age had involved multiple advances and recessions of glacial ice sheets. Chamberlin's insight on the Ice Age came from his many years of field work in formerly glaciated southeastern Wisconsin. Believing that an ice sheet had advanced and receded over North America four times, geologists subdivided the Pleistocene Epoch into four glacial stages: from oldest to youngest, designated the Nebraskan, Kansan, Illinoian, and Wisconsinan, named for the states where the best evidence for glacial advance is present. European geologists had identified corresponding glaciations in the Alps that they called the Günz, Mindel, Riss, and Würm. Analysis of deep-sea sediment cores, however, reveals perhaps 12 to 15 major advances and recessions of glacial ice at periodic intervals. Nonetheless, division of the Pleistocene into the four glacial stages is still used today.

More than three decades ago, the *Climate Long-Range Investigation, Mapping and Prediction (CLIMAP)* project completed a global-scale reconstruction of sea-surface temperatures during the last glacial maximum. Although lacking many of the climate reconstruction techniques available today, this international, multidisciplinary enterprise was the first to reconstruct changes in SST by latitude and note the greater sensitivity of high latitudes to climate change. Today, an expanded and improved version of CLIMAP, known as *Multiproxy Approach for the Reconstruction of the Glacial Ocean Surface (MARGO)* is underway. Using a variety of temperature proxies and greater spatial resolution of data, an international team intends to develop an updated reconstruction of SST and sea-ice extent during the last glacial maximum.

GLACIERS

A **glacier** is a mass of ice that flows internally under the influence of gravity (Figure 9.12). Glacial ice is the second largest reservoir in the global water cycle and the single largest freshwater reservoir on Earth (Chapter 5). Like rivers of liquid water, glaciers flow toward the ocean as part of the runoff component of the global water cycle. Some glaciers partially or completely melt prior to reaching the sea and the meltwater supplies rivers, lakes, and groundwater. Other glaciers enter the ocean directly where they break up into icebergs that float out to sea and eventually melt. At present, glacial ice covers about 10% of Earth's land area. Major ice sheets, up to 3 km (1.5 mi) thick, cover most of Antarctica and Greenland while much smaller glaciers occupy some mountain valleys at lower latitudes and even at the equator. At times during the Pleistocene Ice Age, glacial ice may have covered as much as 30% of the planet's land area, but through most of Earth's 4.6 billion years, little glacial ice existed.

Epochs of extensive glaciation occurred prior to the Pleistocene Ice Age although supporting evidence is scarce because of subsequent erosion. Late in the Paleozoic Era, during the Caboniferous and Permian periods (360 to 250 million years ago), widespread glaciation occurred in what is now southern Africa, South America, India, Australia, and Antarctica. During the Ordovician and Silurian periods of the Paleozoic Era (450 to 420 million years ago), glaciation affected part of South America. During the early Paleozoic (500 million years ago), glaciers developed in Africa. At the end of the Precambrian (570 million years ago), glaciation occurred

FIGURE 9.12
A glacier is a mass of ice that flows internally under the influence of gravity. Climate change can affect the energy (heat) and mass balance causing the glacier to expand or shrink. This glacial ice stream is located on Fitz Roy Mountain in southern Patagonia, near the Chilean border. [Photo courtesy of Paul Sager, Professor Emeritus, University of Wisconsin-Green Bay.]

in various continents of the Northern Hemisphere. Glaciation may have occurred several times during the Precambrian, including the Huronian Ice Age, about 2.4 to 2.1 billion years ago.

Ultimately, large-scale climate change is responsible for variations in Earth's glacial ice cover. And now changes in glacial ice or sea ice cover, in turn, may be influencing or hastening climate change.

Snow is the raw material for the formation of glacial ice. A glacier only develops where the climate is such that snowfall exceeds snowmelt over a succession of many years. Accumulating snow compacts under its own weight and gradually transforms into tiny granular spheres of ice (called *firn*) which eventually re-crystallize into solid ice. As snow converts to firn and then ice, its physical properties change. For one, density increases from an average value of about 0.1 g per cm^3 for fresh-fallen snow to about 0.9 g per cm^3 for ice. Spherical granules of ice pack more closely than plate-like snowflakes, increasing the density. As more and more snow accumulates, confining pressure increases and firn granules re-crystallize into fewer but larger crystals that interlock like the pieces of a jigsaw puzzle. The pressure at a depth of about 80 m (260 ft) is sufficient to convert densely packed snow and firn into essentially impermeable ice thereby trapping air bubbles within.

A distinction is made between temperate and polar glaciers. In a **temperate glacier**, the internal ice temperature rises to the *pressure-melting point* sometime during the year (most likely by late summer); the confining pressure slightly depresses the melting point of ice. Meltwater produced in a temperate glacier accelerates the transition of snow to ice and increases the flow rate of the glacier by reducing frictional resistance. This water accelerates compaction of the ice mass and seeps into open spaces where it refreezes, further increasing the density. Most mountain glaciers such as those in the Rockies are temperate. On the other hand, ice temperature within a **polar glacier** remains below the pressure melting point throughout the year so that little or no meltwater is generated. Conversion of snow to ice and the internal motion of ice are extremely slow processes in a polar glacier. The Antarctic ice sheets are polar glaciers.

The greater supply of meltwater in temperate versus polar glaciers largely explains why temperate glaciers form more rapidly than polar glaciers, but also have a shorter life expectancy. The greater time constant of a polar glacier means that it responds more slowly to climate change.

CLIMATE AND GLACIERS

The climate system governs the exchange of both heat energy and ice mass between a glacier and its environment. For a glacier to exist, the relationship between heat input and heat output must maintain water in the solid phase for an extended period of time. Heat flows between a glacier and its surroundings chiefly through radiation (solar and terrestrial infrared), conduction, and phase changes of water at the glacier-atmosphere interface. If radiational heating exceeds radiational cooling, then ice temperature rises; if radiational cooling exceeds radiational heating, ice temperature falls. If the air temperature were higher than the ice surface temperature, then heat would be conducted to the ice; if the air temperature were lower than the ice surface temperature, then heat would be conducted from the ice to the atmosphere. Deposition or condensation of water vapor onto the ice surface and freezing of water are phase changes that release latent heat to the ice (warming). Evaporation of water, sublimation of ice, and melting of ice absorb latent heat from the ice (cooling). In addition, geothermal heat may be conducted to the ice from the underlying bedrock.

Ice mass is added to a glacier principally through precipitation and avalanches (in mountain valleys). Ice mass is removed from a glacier primarily via melting (including drainage of meltwater away from the glacier), sublimation, and wind erosion of snow. As a glacier enters the ocean or a large lake, ice bergs break off at its leading edge; this process of ice loss is known as *calving*. Recall from the first Essay in Chapter 3 that shrinkage of the ice cap on Mount Kilimanjaro is primarily due to enhanced sublimation coupled with dry conditions and reduced snowfall rather than a warming trend. The general term for all processes that result in a loss of glacial ice mass is **ablation**.

The difference between ice mass gain (accumulation) and ice mass loss (ablation) over the course of a year (usually measured at the end of the melt season in late summer) is the glacier's **mass balance**, that is,

Mass balance = Mass gain − Mass loss.

If a glacier gains more ice mass than it loses, its mass balance is positive and the glacier thickens and expands. If a glacier loses more ice than it gains, its mass balance is negative and the glacier thins and shrinks. A **glacial climate** favors a positive mass balance so that new glaciers form and existing glaciers advance. An **interglacial climate** favors a negative mass balance so that glaciers fail to form or existing glaciers retreat and eventually may completely waste away.

During the Pleistocene Ice Age, the climate shifted numerous times between glacial climates and interglacial climates. The square wave in Figure 9.13A models the behavior of climate during the Pleistocene; typically, the climate shifts abruptly between interglacial and glacial climatic episodes. The sinusoidal wave in Figure 9.13B models the response of glacial ice volume to large-scale shifts between glacial and interglacial climates. During a glacial climatic episode, ice volume increases slowly at first. But after an ice sheet reaches a critical size, it begins to influence atmospheric conditions in ways that promote preservation of ice or growth of the ice sheet (an example of *positive feedback*). The surface of a snow covered ice sheet strongly reflects sunlight. Less solar radiation is absorbed (converted to heat) lowering the temperature and reducing the amount of melting, especially in summer. But as the growing ice sheet expands into lower latitudes, its growth rate slows in response to the more intense solar radiation. With a shift to an interglacial climatic episode, the ice sheet slowly begins to shrink. As the ice warms to its pressure melting point, melting accelerates. The rate of shrinkage slows, however, as the ice sheet melts back into higher latitudes where solar radiation is less intense and air temperatures are lower.

Comparison of Figure 9.13A with Figure 9.13B reveals that abundant glacial ice may exist during an interglacial climatic episode. The unusually great amount of latent heat of fusion of ice is another reason for the stability of a glacial ice sheet. The geologic record confirms

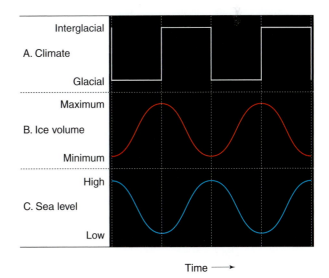

FIGURE 9.13
A graphical model showing (A) the variation of climate between glacial and interglacial episodes, (B) the response of glacial ice volume, and (C) the response of mean sea level. [Modified after R.A. Bryson, W.M. Wendland, and J.M. Moran, University of Wisconsin-Madison]

that climate change tends to be abrupt whereas ice sheets respond sluggishly to changes in climate; that is, it takes time for huge masses of ice to grow or melt away. On the other hand, small glaciers have relatively short time constants, are less stable, and melt faster.

Variations in the volume of glacial ice on Earth have important implications for mean sea level. As noted in Chapter 5, the total amount of water on the planet is essentially constant, a reasonable assumption throughout at least recent geologic time. Almost all the water locked in glaciers is drawn from the ocean as part of the global water cycle, so as the volume of glacial ice increases, the volume of water in the ocean basins decreases and sea level falls. Conversely, as glaciers melt, the volume of water in the ocean basins increases and sea level rises. This is an example of **eustatic sea-level change**, a worldwide fluctuation in sea level. Figure 9.13C models the response of mean sea level to changes in glacial ice volume. Note that the model eustatic sea-level curve is 180 degrees out of phase with the glacial ice volume curve.

Ocean water temperature also affects sea level, because water expands when heated and contracts when cooled. During the last glacial maximum (about 20,000 to 18,000 years ago) when Earth's glacial ice cover was about as extensive as anytime during the Pleistocene, mean sea level was 113-135 m (370-443 ft) lower than it is today. Roughly 10 m (33 ft) of this decline in sea level was due to greater water density in response to lower temperatures. The drop in sea level exposed portions of the continental shelf, including a land bridge linking Siberia and North America. If all the glacial ice currently on the planet were to melt, sea level would likely rise some 70 m (230 ft).

GLACIERS AND LANDSCAPES

The varied landscapes of the formerly glaciated regions of Canada, northern United States, and northwestern Europe are part of the legacy of the Pleistocene Ice Age. Lakes (including the Great Lakes), marshes, and swamps developed in lowlands excavated by advancing lobes of glacial ice. Meltwater streams draining shrinking glaciers deposited extensive layers of sand and gravel (called *outwash*) over broad areas. The fertile soils of the grain belt developed from dusty sediment, called *loess*, derived from sediment deposited first by glacial meltwater, and then transported and deposited by Ice Age winds. For more on loess and climate, refer to this chapter's second Essay.

During major glacial climatic episodes of the Pleistocene Epoch, the Laurentide ice sheet formed over what is now east-central Canada and spread westward to the Rocky Mountains, eastward to the Atlantic, and southward over the northern tier states of the United States (Figure 9.14). At about the same time, mountain

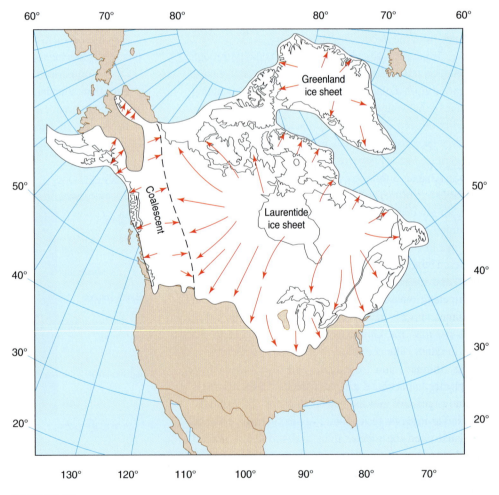

FIGURE 9.14
Glacial ice cover over North America about 20,000 to 18,000 years ago, the time of the last glacial maximum.

FIGURE 9.15
This glacial erratic near Washburn Observatory on the University of Wisconsin-Madison campus is called Chamberlin Rock, named for Thomas C. Chamberlin (1843-1928), a highly regarded glacial geologist.

glaciers in the Rockies coalesced into the Cordilleran ice sheet, and a relatively thin ice sheet spread over the Arctic Archipelago. In northwest Europe, the Fennoscandian ice sheet, much smaller than the Laurentide, covered what are now the United Kingdom, Norway, Sweden, and Finland. The Laurentide and Finnoscandian ice sheets thinned and retreated, and may even have disappeared entirely, during relatively mild interglacial climatic episodes, which typically lasted about 10,000 years. Throughout these interglacials, glacial ice cover persisted over most of Antarctica and Greenland as it still does today.

The way a glacier impacts a landscape depends to a large extent on whether the glacier is advancing (positive mass balance) or retreating (negative mass balance). During a glacial climatic episode, advancing glaciers

can be formidable agents of erosion. Glaciers are more competent than rivers of water, and can incorporate and transport much larger sediments. Glacial ice transported and dumped huge angular boulders, some the size of cars, which today are strewn over parts of Wisconsin, Michigan, New York, and New England (Figure 9.15). Armed with rock debris at their base, advancing lobes of glacial ice scratch and scrape underlying surfaces to sculpture the landscape. Broad U-shaped mountain valleys and grooves excavated in bedrock attest to the great erosive power of glacial ice (Figure 9.16). Elsewhere, glaciers plastered broad areas with poorly sorted sediment (known as *ground moraines*).

During an interglacial climatic episode, glaciers stagnate, melt, and recede. Sediments are released from ice during melting and may be transported and deposited by meltwater streams. Many streams that drain melting glaciers are choked with sediment. Sediment deposition creates various landforms, including deltas, eskers, and kames, all of which are composed of mostly sand and gravel. An *esker* is a long steep-sided sinuous ridge composed of sediment deposited by a meltwater stream flowing through a tunnel within the ice (Figure 9.17). A *kame* is a cone-shaped hill of sediment formed when a meltwater stream flowing on the surface of a mass of stagnant ice plunges into a crevasse. The sediment accumulates at the base of the crevasse similar to the sand at the bottom of an hourglass (Figure 9.18). The presence of any of these glacial landforms today points to a time in the past when the climate was quite different from what it is today.

FIGURE 9.16
Armed with rock debris at its base, an advancing lobe of glacial ice sculptured this exposure of igneous bedrock into a streamlined form.

FIGURE 9.17
Vertical cross-section through a tunnel in an accumulation of sand and gravel deposited by a meltwater stream flowing through a mass of stagnant glacial ice. Viewed from above, the landform, known as an esker, is a long narrow steep-sided sinuous ridge that winds tens of kilometers through the countryside.

FIGURE 9.18
A cone-shaped hill of sand and gravel, known as a kame, formed when a meltwater stream plunged into a crevasse in a mass of stagnant glacial ice.

CHRONOLOGY AND TEMPERATURE TRENDS

During the Pleistocene, glacial ice sheets advanced and receded numerous times. The leading edge of a glacial ice sheet consisted of a series of lobes that sometimes moved independently of its neighboring lobes. Glacial lobes act like huge erasers as they advance across the landscape, modifying, and even eradicating, much of the evidence from prior glacial advances and recessions. Hence, geoscientists know the most about the more recent major fluctuation of Earth's glacial ice cover because it left the last imprint. To find out more about earlier glaciations, scientists rely on other techniques.

Oxygen isotope analysis of deep-sea sediment cores shows numerous fluctuations between major glacial and interglacial climatic episodes over the past 600,000 years (Figure 9.19A). Shifting focus to the past 160,000 years, resolution of the climate record improves. The temperature curve in Figure 9.19B is based on analysis of an ice core extracted from the Antarctic ice sheet at Vostok. A relatively mild interglacial episode, referred to as the *Eemian*, began about 127,000 years ago and persisted for nearly 7000 years. In some localities, temperatures may have been 1 to 2 Celsius degrees (2 to 4 Fahrenheit degrees) higher than during the warmest portion of the present interglacial. The Eemian interglacial was followed by

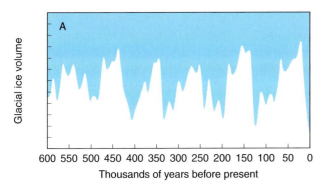

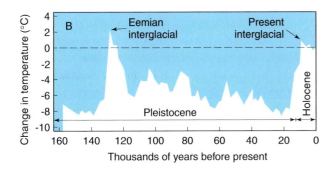

FIGURE 9.19
Reconstructed records of (A) the variation in global glacial ice volume over the past 600,000 years based on analysis of oxygen isotope ratio of shells in deep-sea sediment cores, and (B) temperature variation over the past 160,000 years based on oxygen isotope analysis of an ice core extracted from the Antarctic ice sheet at Vostok and expressed as a departure in Celsius degrees from the 1900 mean global temperature. [Compiled by R.S. Bradley and J.A. Eddy from J. Jousel et al., *Nature* 329(1987):403-408 and reported in *EarthQuest* 5, No. 1 (1991).]

numerous fluctuations between glacial and interglacial climatic episodes.

The last major glacial climatic episode began about 27,000 years ago and reached its peak about 20,000 to 18,000 years ago (Figure 9.14). At that time, glacial ice 3 km (5.1 mi) thick covered most of New England and the Great Lakes states, reaching as far south as the Ohio River Valley and Long Island, NY. About 18,000 years ago, the large-scale climate shifted to dominantly interglacial. Although occasionally interrupted by relatively brief shifts back to glacial climatic regimes, the Laurentide ice sheet gradually melted back, until it finally withdrew from what is now the coterminous northern United States about 10,500 years ago (Figure 9.20). Not until 5500 years ago had most of the residual Laurentide ice in northern Canada melted.

During glacial climatic episodes, temperatures were lower than today, but the magnitude of cooling was not the same everywhere. A variety of geologic evidence indicates that during the Pleistocene, temperature fluctuations between major glacial and interglacial

climatic episodes typically amounted to as much as 5 Celsius degrees (9 Fahrenheit degrees) in the tropics, 6 to 8 Celsius degrees (11 to 14 Fahrenheit degrees) at middle latitudes, and 10 Celsius degrees (18 Fahrenheit degrees) or more at high latitudes. An increase in the magnitude of a temperature change with increasing latitude is known as **polar amplification**, indicating that polar areas are more sensitive to climate change.

Glacial ice cores from both Greenland and Antarctica clearly reveal an approximately 100,000 year Ice Age cycle consisting of cold glacial climatic episodes (e.g., the Wisconsinan stage) sandwiched between mild interglacial climatic episodes of approximately 10,800-year average duration. Perhaps 16 of these long-term cycles operated over the span of the Pleistocene Epoch. Evidence from deep-sea sediment cores indicates that regular variations in Earth-Sun geometry (the Milankovitch cycles) drive this approximately 100,000-year glacial/ interglacial cycle (Chapter 11).

Greenland and Antarctic ice-core records correlate well both in terms of magnitude of temperature change and the timing of events suggesting that the 100,000-year Ice Age cycles were globally synchronous. However, comparison of the Greenland and Antarctic ice core data over the most recent Ice Age cycle (i.e., from about 142,000 years ago to 10,500 years ago) reveals marked differences between the Southern and Northern Hemispheres. Whereas the Antarctic record is reasonably smooth and "calm," the Greenland record shows numerous abrupt and drastic flip-flops between glacial and interglacial climatic episodes. Temperatures changed as much as 7 Celsius degrees (12.6 Fahrenheit degrees) over periods of decades or less (in some cases in only 3 years.) These abrupt temperature

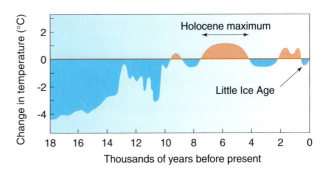

FIGURE 9.20
Reconstructed temperature variation over the past 18,000 years based on analysis of a variety of proxy climate indicators and expressed as a departure in Celsius degrees from the 1900 mean global temperature. [Compiled by R.S. Bradley and J.A. Eddy based on J.T. Houghton et al. (eds.), *Climate Change: The IPCC Assessment*, Cambridge University Press, U.K., 1990, and reported in *EarthQuest* 5, No. 1 (1991).]

changes, having two basic periodic components of 2000 to 3000 years and 7000 to 12,000 years, occurred during the Wisconsinan stage of the Pleistocene Epoch but not during the subsequent Holocene Epoch. The periods of these temperature fluctuations are much shorter than those of the Milankovitch cycles and hence are probably unrelated to changes in Earth-Sun geometry.

The most likely explanation for these short-term changes in temperature is the alternate weakening and strengthening of the ocean's meridional overturning circulation (Chapter 4). This may explain, for example, the occurrence of the relatively cool episode from 11,000 to 10,000 years ago, known as the **Younger Dryas** (named for the polar wildflower *Dryas octopetala* that reappeared in portions of Europe at the time). The return of glacial climatic conditions triggered short-lived re-advances of remnant ice sheets in North America, Scotland, and Scandinavia.

The Younger Dryas began abruptly when glacial ice lobes disrupted drainage patterns, diverting meltwater from the Mississippi River into the St. Lawrence River and North Atlantic. With this input of fresh water, North Atlantic surface waters became less saline and eventually were not sufficiently dense to sink and form North Atlantic Deep Water (NADW). This weakened the meridional overturning circulation, which in turn diminished the warm water flowing into the central and northern North Atlantic, causing a marked cooling of the surrounding lands. The Younger Dryas ended just as abruptly as it began when the input of fresh water into the North Atlantic decreased and formation of NADW resumed. The geographic pattern of the Younger Dryas climatic impacts (e.g., little in western North America and only a muted response in the Antarctic ice core record) suggests that the Younger Dryas was not part of the larger ice age variability driven by the Milankovitch cycles. Rather, the Younger Dryas was a regional short-term climate fluctuation linked to changes in the Atlantic meridional overturning circulation.

Atlantic Ocean circulation changes that may have triggered the Younger Dryas likely also occurred at other times. Scientists have interpreted certain layers of lithogenous sediment in cores extracted from the floor of the North Atlantic as materials released during melting of fleets of icebergs. These icebergs surged or slid off glaciated North America and floated out onto the North Atlantic every 2000 to 3000 years as the climate flip-flopped between warm and cold episodes. Melting icebergs freshened North Atlantic surface waters thereby weakening the meridional overturning circulation. With the return of colder conditions and fewer icebergs, freshening of the surface waters ceased, the water became salty again due to

wind-driven evaporation, and the thermohaline circulation strengthened. After two or three of these events, an even greater discharge of icebergs occurred at intervals of 7000 to 12,000 years.

The smaller, shorter, more frequent occurrences are called *Dansgaard-Oeschger* or *D-O events*, named for the Danish paleoclimatologist Willi Dansgaard and the Swiss climatologist Hans Oeschger (1927-1998). They are also referred to as "flickers" because of their relatively short period. Each D-O event consisted of abrupt warming (almost to interglacial levels) followed by gradual cooling. The Greenland ice core record contains evidence of some 23 Dansgaard-Oeschger events between 110,000 and 15,000 years ago. The larger, longer, and less frequent episodes, known as *Heinrich events*, are discussed in the second Essay of Chapter 8. Flickers and Heinrich events occurred during both glacial and interglacial regimes and are evident in the temperature record reconstructed from Greenland ice cores.

Climates of the Holocene

Glacial ice finally withdrew from the North American Great Lakes region about 10,500 years ago ushering in the present interglacial, the **Holocene Epoch**, when civilization and agriculture developed. Although the Laurentide ice sheet was melting and disappeared almost entirely 5500 years ago, the Holocene was an epoch of spatially and temporally variable temperature and precipitation. Cores extracted from the Greenland ice sheet and sediment cores taken from the bottom of the North Atlantic Ocean reveal that the overall post-glacial warming trend was interrupted by abrupt millennial-scale fluctuations in climate.

Post-glacial warming during the Holocene gave way to a notably cold episode about 8200 years ago. The likely reason for this abrupt climate change was a sudden influx of fresh water into the North Atlantic when the ice dam holding back Lake Agassiz burst. Lake Agassiz was centered over south-central Canada and contained the meltwaters of the shrinking Laurentide ice sheet (a volume of water greater than the present-day Great Lakes combined). The arrival of fresh water in the North Atlantic suppressed the meridonal overturning circulaton causing regional cooling. A similar mechanism was responsible for the Younger Dryas described earlier.

Significant cooling also occurred from about 3100 to 2400 years ago. On the other hand, at times during the mid-Holocene (classically known as the *Hypsithermal*), mean annual global temperature was perhaps 1 Celsius degree (2 Fahrenheit degrees) higher than it was in 1900, the warmest in more than 110,000 years, that is, since the Eemian interglacial. A pollen-based climate reconstruction indicates that 6000 years ago, July mean temperatures were about 2 Celsius degrees (3.6 Fahrenheit degrees) higher than now over most of Europe.

In 2001, V.J. Polyak and Y. Asmerom of the University of New Mexico reported on their analysis of speleothems (stalagmites) recovered from Carlsbad Cavern and Hidden Cave in the Guadalupe Mountains of southeast New Mexico. The speleothems span a 4000-year interval through the late Holocene and provide annual resolution. In this area, the mid-Holocene (about 5600 years ago) was characterized by drier conditions than at present. From 4000 to 3000 years ago, the climate was similar to present, but from 3000 to 800 years ago, it became cooler and wetter. Subsequently, the present-day climate prevailed with the exception of a wetter period from 440 to 290 years ago.

Across North Africa, one of the most dramatic environmental changes of the past 10,000 years began during the mid-Holocene. Recall from the Case-in-Point in Chapter 5 that a long-term decline in rainfall caused North Africa to transition from a green savanna to the world's largest warm desert, the Sahara.

Climates of the Recent Millennium

A generalized Northern Hemisphere temperature curve for a recent 1000-year period, derived mostly from historical documents, is shown as Figure 9.21. The most notable features of this record are (1) the Medieval Warm

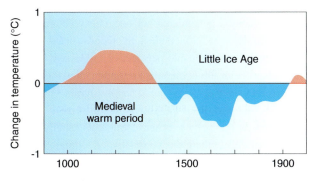

FIGURE 9.21
Reconstructed temperature variation of the Northern Hemisphere over the past 1000 years based on analysis of historical documents and expressed as a departure in Celsius degrees from the 1900 mean global temperature. [Adapted from J.T. Houghton et al. (eds.), *Climate Change: The IPCC Assessment*, Cambridge University Press, U.K., 1990, p. 202.]

Period from about CE 950 to 1250 and (2) the cooling that followed, from about CE 1400 to 1900, known as the Little Ice Age. The Medieval Warm Period and the Little Ice Age were not episodes of sustained warming and cooling, respectively. On the contrary, climate reconstructions based on tree growth rings, sediment cores, and glacial ice cores plus historical documents point to significant decadal-scale fluctuations in temperature and precipitation. The first Norse settlements in Greenland were established during the Medieval Warm Period and failed during the Little Ice Age. (For details on the Norse settlements in Greenland, refer to the second Essay in Chapter 2.)

MEDIEVAL WARM PERIOD

In 1965, the British climatologist Hubert H. Lamb (1913-1997) was the first to characterize the High Medieval (CE 1100 to 1200) as an episode of relatively mild winters and warm dry summers in Western Europe. During this relatively stable climate, harvests were generally bountiful and vineyards thrived in the British Isles. Apparently, a westerly (zonal) circulation pattern prevailed during winter whereas high pressure systems dominated during summer. Lamb estimated that average temperatures were 1 to 2 Celsius degrees (2 to 4 Fahrenheit degrees) higher than normal, based on anecdotal information plus proxy climate data from Western Europe. However, he had no data that would support a global-scale warming trend during Medieval time.

According to paleoclimatologist R.S. Bradley of the University of Massachusetts and colleagues, establishing the magnitude and spatial and temporal extent of warming during the Medieval Warm Period is problematic partially because of scarce observational data (especially from the tropics and Southern Hemisphere) and because Medieval time can encompass a relatively broad interval from CE 500 to 1500. Bradley notes that during the High Medieval, temperatures were higher than during the subsequent Little Ice Age. Large-scale climate reconstructions show that High Medieval mean temperatures were about the same as in the 1901 to 1970 period. During the following three decades, the mean temperature rose by about 0.35 Celsius degree (0.63 Fahrenheit degree). Bradley concludes that the available evidence does not support the notion that the High Medieval was equally as warm or warmer than the late 20th century.

While the Medieval Warm Period is usually defined in terms of positive temperature anomalies, drought was more notable in some regions of the globe. In 2004, Edward R. Cook of Lamont-Doherty Earth Observatory and colleagues reported evidence of long-term "elevated aridity" in the American West during the period CE 900 to 1300. This overlaps with the Medieval Warm Period and is based on a drought index derived from tree growth ring records spanning 600 to 1200 years. As discussed in the Case-in-Point of this chapter, a succession of megadroughts in the 12th and 13th centuries was likely a major reason why the Anasazi and Fremont peoples abandoned their homelands in the American Southwest.

In a 2006 report, V. Sridhar of the University of Nebraska and colleagues found that sand dunes, covering 7500 km[2] of the Nebraska Sand Hills, were last mobilized (indicating arid conditions) about 800 to 1000 years ago, during the early phase of the Medieval Warm Period. Today, these dunes are anchored in place (stabilized) by grasses that are supported by moisture transported in spring and early summer from the Gulf of Mexico on south to southeast winds. Reconstruction of prevailing winds based on dune characteristics indicates that the dunes were last mobilized when the prevailing spring-summer winds had shifted to a southwest direction. This flow descended from the Rockies and Colorado Plateau, undergoing adiabatic compressional warming and drying.

LITTLE ICE AGE

Cooling that heralded the Little Ice Age may have begun as early as CE 1200 in Greenland and the Arctic but closer to CE 1300 at lower latitudes. (François Matthes (1874-1948), a Dutch-American glacial geologist, was first to use the term *Little Ice Age* in 1939.) Based on climatic inferences drawn from tree growth rings, glacial ice cores, and historical documents, scientists characterize the Little Ice Age as a multi-century interval when the climate was more variable than the climate before (the Medieval Warm Period) or since. The climates of Europe and many other regions of the world were volatile, abruptly shifting from one extreme to another. In Europe, bitter cold winters, hot summers, drought, and torrential rains punctuated decades of mild winters and warm summers. Overall, the mean annual global temperature during the Little Ice Age was perhaps 0.5 Celsius degree (0.9 Fahrenheit degree) lower than it was in 1900.

Sea-ice cover expanded, mountain glaciers advanced, the growing season became shorter, and erratic harvests caused food scarcity and considerable hardship for many people. The unstable climate of the Little Ice Age contributed to economic, political, and social unrest in Europe. Interestingly, the Little Ice Age also coincided with the Renaissance, the Age of Discovery, the Enlightenment, the American and French Revolutions, and the Industrial Revolution.

Severe winters characterized the 1430s, culminating in the European-wide famine of 1433-1438. By 1440, almost all wine growing had ceased in England. An unusually cold interval from 1590 to 1610 was apparently synchronous at many localities around the globe. Perhaps the coldest periods of the Little Ice Age prevailed from 1670-1710 and during the 1810s. From about 1680 to 1730, the growing season in England was about five weeks shorter than at present and the average number of days with snow on the ground was 20 to 30, compared to today's 2 to 10 days. Recall the discussion of 1816, *the year without a summer*, in the Case-in-Point of Chapter 2. From the 17th century into the mid 19th century, mountain glaciers advanced well beyond their present-day limits in the Alps, Scandinavia, Iceland, Alaska, China, the southern Andes, and New Zealand. In the 19th century, glaciers also thickened and expanded in the Himalayas and Caucasus Mountains.

Conclusions

Our understanding of Earth's climate system benefits from in depth analysis of the climate record. Unfortunately, the reliable instrument-based climate record worldwide is limited to not much more than 140 years. To improve on this situation, scientists have devised many techniques that rely on climate proxy data to reconstruct the climate prior to the instrument era. Sources of climate proxy data range from low-resolution sedimentary strata to high-resolution tree growth rings. Central to this approach is our ability to calibrate climate forcing with environmental response.

A climate record based on proxy climate data may be compromised by many factors. Proxy climate data sources vary in resolution, time control, and spatial continuity. Consequently, our description of climate history becomes more fragmented, generalized, and uncertain with increasing time before present. Nonetheless, the reconstructed climate record provides a valuable perspective on present climate and prospects for future climate, keeping in mind the axiom, *if it has happened before, it could happen again*.

The portion of the climate record that is derived from measurements by weather instruments usually merits more confidence than the reconstructed climate record. However, many factors influence the integrity of the instrument-based climate record. For example, upgrades in instrument technology, relocation of weather stations from urban to rural sites, and the use of instrument shelters impact the climate record. The next chapter takes a closer look at the instrument-based climate record with special emphasis on what it reveals about recent trends in climate.

Basic Understandings

- Scientists reconstruct climates of the past for many reasons. It appeals to curiosity, adds to our understanding of how the environment responds to climate variability and climate change, aids our study of the nature and causes of climate variability and climate change, and provides a unique perspective on the present climate.

- For information on climate prior to the era of reliable instrument-based records, scientists rely on climate inferences drawn from historical documents, and geological/biological evidence such as bedrock, fossils, pollen, tree growth rings, glacial ice cores, and deep-sea sediment cores. These are known as proxy climate data sources.

- Climate reconstruction requires identification of a link between quantitative climate forcing and environmental response. Such modern calibrations are then applied to the ancient or fossil environmental response record to reveal the past climate.

- Caution is advised when attempting to infer climate information from historical documents because many factors in addition to weather and climate can influence the content of such documents.

- The width of tree growth rings normally decreases as the tree ages. Hence, widths are usually expressed in terms of a tree-growth index, that is, the ratio of the actual tree growth ring width to the width expected based on the tree's age. The index is relatively low in stressful growing seasons and relatively high in favorable growing seasons.

- Assuming that the fossil pollen recovered from a sediment core is the product of nearby vegetation and that climate largely governs vegetation types, climate and climate change can be inferred from analysis of pollen profiles.

- Cores extracted from sediments that blanket almost all the ocean floor yield a continuous record of sedimentation dating back hundreds of thousands to millions of years.

- Oxygen isotope analysis of calcium carbonate shells extracted from deep-sea sediment cores is used to reconstruct large-scale climate fluctuations of the Pleistocene Ice Age. Variations in oxygen isotope ratio ($^{16}O/^{18}O$) correspond to changes in Earth's glacial ice volume and hence, past changes in climate. The proportion of light to heavy oxygen in ocean water decreases with increasing glacial ice volume.

- The ratio of light oxygen to heavy oxygen in a speleothem, a cave deposit precipitated from groundwater, is sensitive to temperature and rainfall.

- Most corals cannot tolerate sea-surface temperatures outside of a relatively narrow range. They can be used as high resolution sensors of SST, El Niño, and La Niña.

- Scientists derive air temperature from glacial ice cores using either oxygen isotope analysis or analysis of the ratio of deuterium (D) to hydrogen in ice. As the temperature falls, there is less and less ^{18}O or D.

- For convenience of study, the geologic past and its climate record is subdivided using the geologic time scale, a standard division of Earth history into eons, eras, periods, and epochs based on large-scale geological events.

- Plate tectonics complicate climate reconstruction of periods spanning hundreds of millions of years. In the context of geologic time, topography and the geographical distribution of continents and the ocean are controls of climate change.

- Geologic evidence points to an interval of extreme climate fluctuations beginning about 570 million years ago, corresponding to the transition between the Precambrian and Phanerozoic Eons. During as many as four cold episodes, each lasting perhaps 10 million years, Earth was encased in glacial ice. At the close of each cold episode, temperatures rose rapidly, and within only a few centuries, all the ice melted. Relatively warm conditions appear to have persisted through much of the Mesozoic Era, from about 245 to 70 million years ago.

- By 40 million years ago, Earth's climate began shifting toward colder, drier, and more variable conditions. Scientists have implicated tectonic forces and the building of the Colorado Plateau, Tibetan Plateau, and Himalayan Mountains as the principal causes of this climate change.

- During the Pleistocene Ice Age that began about 1.7 million years ago, the climate shifted numerous times between glacial climatic episodes (favoring expansion of glaciers) and interglacial climatic episodes (favoring shrinkage of glaciers).

- Climate change during the Pleistocene was geographically non-uniform in magnitude; cooling was greatest at high latitudes and least in the tropics. This latitudinal variation in the magnitude of temperature change is known as polar amplification.
- A glacier is a mass of ice that flows internally under the influence of gravity. Presently, glacial ice covers about 10% of Earth's land area; at times during the Pleistocene Ice Age, ice cover appears to have increased to about 30% of the planet's land area.
- Earth's climate system governs the exchange of both heat energy and ice mass between a glacier and its surroundings. Heat flows between a glacier and its environment via radiation, conduction, and phase changes of water. Ice mass is added to a glacier via precipitation and avalanches. Ice mass is removed from a glacier primarily through melting, sublimation, and wind erosion of snow.
- A glacial climate favors a positive mass balance so that new glaciers form and existing ones advance. An interglacial climate favors a negative mass balance so that glaciers do not form and existing glaciers retreat.
- During glacial climatic episodes, glaciers advance and can be formidable agents of erosion, resulting in features such as bedrock grooves and striations, drumlins, and erratics. However, during an interglacial climatic episode, glaciers stagnate, melt, and recede. Sediment liberated during ice melt is transported and deposited by meltwater streams, building a variety of landforms such as deltas, eskers, and kames.
- Weakening of the meridional overturning circulation in the Atlantic Ocean caused an episode of cooling in the lands surrounding the North Atlantic from about 10,000 to 11,000 years ago. This cool period is known as the Younger Dryas.
- The last major glacial climatic episode began about 27,000 years ago and reached its peak 20,000 to 18,000 years ago when the glacial ice cover over North America was as extensive as it had ever been.
- The Holocene was an epoch of spatially and temporally variable temperature and precipitation.
- A generalized temperature curve for the past 1000 years, derived mostly from historical documents, reveals that the Medieval Warm Period lasted from about CE 950 to 1250 and was followed by the Little Ice Age, from about CE 1400 to 1900. The Medieval Warm Period and the Little Ice Age were not episodes of sustained warming or cooling, respectively. On the contrary, sediment cores and glacial ice cores, as well as historical documents point to significant decadal-scale fluctuations in temperature and precipitation.

Enduring Ideas

- One of the principal reasons why scientists attempt to reconstruct the climate record prior to the instrument era is to gain some perspective on the present climate and how climate might change in the future.
- Proxy climate data sources include historical documents, bedrock, fossil plants and animals, pollen, tree growth rings, glacial ice cores, deep-sea sediment cores, and varves.
- In the context of the hundreds of millions of years that constitute geological time, topographic relief and the geographical distribution of continents and the ocean are important controls of climate change.
- During the Pleistocene Ice Age, the climate shifted numerous times between glacial climatic episodes and interglacial climatic episodes with major implications for the landscape.
- The Holocene was an epoch of spatially and temporally variable temperature and precipitation.

Review

1. Of what value is reconstructing the climate record for the time prior to the era of weather instruments?
2. What is the significance of methodological uniformitarianism in reconstructing the climate past?
3. What two basic assumptions are made when analyzing fossil pollen for the purpose of reconstructing past climates?
4. How does the proportion of light oxygen (^{16}O) to heavy oxygen (^{18}O) in ocean water change as the volume of glacial ice on Earth increases?
5. How are past changes in the chemical composition of air determined from analysis of glacial ice cores?
6. Summarize the principal reasons why the climate during the Paleozoic Era is described in generalized terms.
7. What was a major reason for the slow pace of widespread acceptance of the glacial theory during the 19th century?
8. Compare the physical properties of temperate glaciers to those of polar glaciers.
9. Distinguish between glacial climates and interglacial climates in terms of glacial mass balance.
10. What evidence exists for the abrupt change behavior of glacial/interglacial climates?

Critical Thinking

1. Explain why trees growing near the limits of their range are the most sensitive to climate change and variability.
2. What is the advantage of cross-dating in climate reconstructions based on analysis of tree growth rings?
3. How does plate tectonics affect the reconstruction of climates of geologic time?
4. By about 40 million years ago, Earth's climate began to shift toward colder, drier, and more variable conditions. What might explain this climate change?
5. What is the relationship between large-scale shifts between glacial and interglacial climatic episodes and mean sea level?
6. Explain how it is possible for a great amount of glacial ice to cover the land even though the climate is interglacial.
7. How is the impact of a glacier on the landscape dependent on the sign of the glacier's mass balance?
8. Speculate on the significance of polar amplification for future climate change.
9. How might ocean circulation have played a key role in the occurrence of the relatively cool Younger Dryas?
10. Speculate on any possible parallels between what happened in the American West during the Medieval Warm Period and future warming in the American West.

ESSAY: Limitations of Proxy Climate Data

Reconstructing past climates can be challenging. Because the relationship between climate forcing mechanisms and environmental response can be complex, derivation of a reliable modern calibration between the two can be problematic. Even if the modern climate/environment calibration were reasonable, its application to the documentary or other proxy records may yield erroneous or misleading information. The older the period under study, the more fragmented the proxy record becomes, the sampling interval usually lengthens, resolution of events becomes poorer, and correlating events in widely separated regions becomes more challenging.

The diversity of proxy climate data is a major reason why climate reconstruction becomes less reliable, especially as the record becomes more fragmented. Although climate is the principal ecological control, it is not the only factor affecting ancient biotic or abiotic records. For example, local topography or drainage may influence the tree growth ring record. Furthermore, the relationship between climate forcing mechanisms and environmental response (even when non-climate factors are insignificant), is not always straight forward. Combinations of many climatic elements may be responsible for the environmental response.

Another complication that arises in efforts to reconstruct past climates is *lag response time*. According to *LeChatelier's principle*, a system responds to a change in its environment by shifting in such a way as to relieve the new stress, but different components of ecosystems may take different lengths of time to come into the new equilibrium. Hence, a modern calibration between climate forcing mechanisms and biotic/abiotic response is appropriately applied to the fossil record only when it is a reasonable assumption that an equilibrium condition actually existed.

Problems may also arise from an incomplete understanding of the frequency response mechanism of a fossil climate sensor. For example, some sensors, such as tree growth rings, do not respond significantly to high frequency climate variability but do respond to gradual systematic climate shifts. That is, tree growth rings do not record a single dry summer, but will contain the signature of a succession of several dry summers (drought). A distinction must be drawn between climate data reconstructed from high-resolution sensors (e.g., annually laminated ice cores) that can delineate dominant atmospheric circulation patterns versus low resolution sensors (e.g, sedimentary strata) that primarily mirror long-term average conditions.

If data are available from a high-resolution climate sensor, caution must then be exercised in extrapolating reconstructed climate data from one region to another. Recall from Chapter 2 that climate anomalies are geographically non-uniform in both sign (direction) and magnitude. In response to dominant atmospheric circulation patterns, average values of climatic elements (e.g., temperature, precipitation) calculated for specific weeks, months, or years over broad geographical areas typically show regions of positive anomalies and regions of negative anomalies. Hence, for example, unless corroborated by independent evidence, extrapolating a speleothem-derived temperature record from Missouri to Montana cannot be justified. Even if data are available to generate broad spatial patterns of climate anomalies, the data may not be synchronous. Radiometric dating techniques used to estimate the age of some material or object have many sources of error and generally magnify their effects the further back in time they are applied.

With increasing time before present, the decline in time control and resolution of climate detail means that the most reasonable large-scale climate reconstruction patterns represent lengthy periods of time. During the Holocene, available resolution permits a description of climate in increments of 1000 years. During the Pleistocene Ice Age, climate is generally described over periods of 10,000-yr increments. And in the Paleozoic Era, climate is described in highly generalized terms over hundreds of millions of years.

ESSAY: Pleistocene Climate and Loess Deposition

One of the most important legacies of the Pleistocene Ice Age is *loess* (Figure 1), unconsolidated fine sediment deposited by the wind across broad areas of the central United States, where it is the parent material for some of the most productive soils in the world, helping make possible the grain belt of North America. In the central U.S., loess owes its origin primarily to wind erosion of glacial outwash floodplain deposits. Hence, loess is intimately associated in time and place with continental-scale glacial ice sheets. Meltwater generation is key to floodplain sediment deposition so that loess was deposited primarily during interglacial climate episodes. Loess is also a valuable source of information on past climates including prevailing wind direction and strength (based on trends in loess thickness and particle size) and moisture balance.

FIGURE 1
An exposure of late Pleistocene loess near the Missouri River in Kansas City, MO. Loess consists of unconsolidated sediment, mostly silt-sized, transported by winds from the floodplains of rivers and streams that drained melting glacial ice.

Loess consists of accumulations of unconsolidated, moderately well sorted, and locally thinly layered blankets of sediment. Although the dominant (40% to 50%) grain size is silt, ranging from 0.015 mm to 0.05 mm in diameter, deposits generally include significant clay (5% to 30%) and sand (5% to 30%) fractions. Loess has a texture similar to flour or talcum powder, is porous and permeable, typically yellow to buff in color, and exposed in nearly vertical cliffs (up to 70 degrees). Ancient buried soils (*paleosols*) are often found between layers of loess representing periods of non-loess deposition. Layers of glacial *till* (sediments deposited by the direct action of glacial ice) alternating with layers of loess indicate alternating advances and recession of glacial ice during a rapidly changing climate regime.

Besides the central U.S., loess is widespread in Alaska, China, Russia, northern Europe, and Argentina. In the central U.S., late Pleistocene loess occurs in two convergent belts (Figure 2): one deposit thins eastward from eastern Colorado, Nebraska and Kansas through Iowa, Missouri, Wisconsin, Illinois, and Indiana into southern Ohio. The second belt is east of the Mississippi River and south of the Ohio River from Mississippi through western Tennessee and Kentucky.

FIGURE 2
Major areas of loess distribution (orange) across North America. [Modified after the U.S. Geological Survey]

Loess is a wind (*eolian*) deposit. In the Great Plains and South, the bulk of loess was derived via wind erosion of glacial outwash floodplains and channel bar deposits. Loess thickness decreases systematically with distance from source floodplains, suggesting that loess was a river deposit. However, loess is just as thick on hilltops as in valley bottoms, which confirms deposition by wind. In Nebraska and adjacent regions of Colorado and Kansas, wind erosion of floodplain deposits may have been secondary in importance to wind erosion of the Sand Hills of west-central Nebraska. According to the *United States Geological Survey (USGS)*, erosion of silt-rich poorly consolidated bedrock may also have contributed to loess deposition in the Great Plains. Furthermore, in the narrow periglacial zone along the southern edge of the Laurentide ice sheet, frost heaving and churning of soils probably were minor sources of silt for subsequent re-deposition as loess. The periglacial zone consisted of tree-less tundra underlain sporadically by permanently frozen ground (*permafrost*).

The relationship between loess deposition and Pleistocene climate fluctuations follows from the model presented in Figure 9.13. Elsewhere in this chapter, we noted that a lag time exists between climate forcing and the response of glacial ice volume so that considerable amounts of glacial ice persist long after the climate has shifted from glacial to interglacial. Therefore, the discharge of meltwater also peaks sometime after such a shift. Oxygen isotope analysis of planktonic foraminifera in deep-sea sediment cores extracted from the Gulf of Mexico confirms this lag time for the final disintegration of the Laurentide ice sheet. The influx of Laurentide meltwater reached dramatic proportions by about 15,000 years ago—at least 3000 years after the major shift from a glacial climate episode to a dominantly interglacial climate episode. The discharge of meltwater into the Gulf of Mexico then tapered off rapidly after 13,500 years ago, when the principal drainage shifted away from the Mississippi River to the St. Lawrence River. Furthermore, the most rapid rate of eustatic sea level rise apparently began thousands of years after the last Laurentide glacial maximum.

Fluctuations in climate affect the volume of glacial ice and thereby the discharge of meltwater into rivers and streams draining the ice sheet. Meltwater discharge, in turn, controls floodplain sedimentation and the amount of sediment (alluvium) available for wind erosion and re-deposition elsewhere as loess. (Typically, streams that drain melting glaciers are choked with fine sediment.) If, as suggested by our climate/glacier model (Figure 9.13), meltwater discharge peaks some time after the shift from a glacial to an interglacial climate episode, more alluvium would be available for loess generation during interglacial climate episodes. Regardless of climate, however, as long as glacial ice was present on the North American continent, there was some seasonal ablation so that some loess deposition took place during glacial climate episodes as well. This conclusion is supported by radiocarbon dating of organic matter in loess deposits and *optically stimulated luminescence (OSL)*, a technique that dates the last time the sediment was exposed to sunlight (Chapter 7).

CHAPTER 10

INSTRUMENT-BASED CLIMATE RECORD AND CLIMATOLOGY OF SEVERE WEATHER

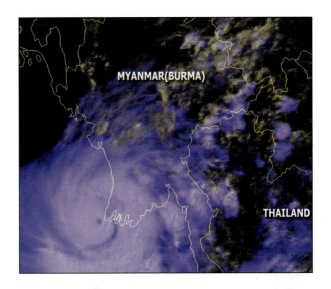

Satellite image of Tropical Cyclone Nargis, a category 4 storm which devastated the low-lying coastal plains of Myanmar (Burma) including the capital city Yangon (Rangoon) on 2 May 2008. Extensive flooding from the associated storm surge and heavy rains, and winds killed tens of thousands, made millions homeless, and destroyed an extensive rice growing region. [Courtesy NOAA]

Case-in-Point

The consensus of scientific opinion is that the present global warming trend is largely anthropogenic in origin. Human activity, principally the combustion of fossil fuels, is responsible for the build-up of carbon dioxide and other infrared-absorbing gases in the atmosphere, enhancing Earth's greenhouse effect. One of the most pervasive effects of global warming is the slow, inexorable rise in mean sea level (msl) due to the combination of thermal expansion of seawater and the melting of land-based glaciers and ice sheets. Rising sea level threatens to submerge islands and inundate low-lying coastal areas, forcing the relocation of human populations.

Even if the international community agrees to sharp cuts in greenhouse gas emissions within this century, the *tipping point* may already have been reached for many small island nations. Simply put, the amount of anthropogenic CO_2 already emitted into the atmosphere ensures a magnitude of warming that will cause an unacceptable rise in sea level in some localities. Natural removal mechanisms operating in biogeochemical cycles are insufficient to head off this warming. In the most conservative scenario, the 2007 *IPCC Fourth Assessment Report* estimates an msl rise of 0.2 to 0.6 m by the year 2100. However, this estimate is based primarily on thermal expansion of seawater and ignores the potential contribution of melting glacial ice sheets. At the other extreme, W.T. Pfeffer and colleagues of the Institute of Arctic and Alpine Research at the University of Colorado in Boulder include estimates of the input of glacier melt in their projection of a sea level rise of 0.8 to 2 m by 2100.

An example of an island nation that is particularly vulnerable to rising sea level is the Maldives, a chain of 1200 small flat islands near the middle of the Indian Ocean with a population of about 360,000. Another example is Kiribati, consisting of three island groups located in the Pacific Ocean about half way between Hawaii and Fiji. Kiribati is home to 100,000 people. The islands of both nations are 2 m (6.5 ft) or less above msl.

Adaptation to higher sea level is not likely to be a viable option for the people of the Maldives, Kiribati, and other low lying island nations. Migration may be their only alternative. Mohamed Nasheed, president of the Maldives since November 2008, proposes to raise funds to help purchase land abroad for a new homeland for his people. Relocation would take place during the present century. Anote Tong, president of Kiribati since 2003, has a somewhat different plan for his threatened island nation. He is asking for help from nearby Australia and New Zealand to train Kiribati's younger people in skilled professions so that they might find jobs and new homes abroad.

Driving Question:
What does the instrument-based climate record tell us about climate variability and climate change?

In the previous chapter, we examined how climate scientists use proxy climate data to extend the climate record as far back in time and with as much detail as possible. The focus of the first part of this chapter is the more recent instrument-based climate record, beginning with a general overview of global temperature and precipitation patterns. We then turn to the post-1880 trend in global mean annual air temperature and some recent changes in the global water cycle. This climate record is consistent with the view that the warming of the past several decades is largely due to enhancement of the greenhouse effect by human activity. Furthermore, some scientists propose that changes in the frequency or intensity of severe weather events will accompany (or already is accompanying) global warming. We frame this issue in the second part of this chapter by summarizing the basic characteristics and climatology of thunderstorms, tornadoes, and hurricanes—weather systems that can take lives and cause considerable property damage.

Global Climate Patterns

Instrument-based climate records enable climatologists to represent how the various boundary conditions in Earth's climate system interact with external forcing agents and mechanisms to shape the climates of the continents. This section summarizes the basic global patterns of climate on Earth. As we examine these patterns, keep in mind that they may be significantly modified by local and regional influences.

TEMPERATURE

Ignoring the effect of mountainous terrain on air temperature, mean annual isotherms roughly parallel latitude circles, underscoring the influence of solar radiation (solar altitude and day length) on climate. The latitude of highest mean annual surface temperature, the **heat equator**, is located about 10 degrees north of the geographical equator. Mean annual isotherms are symmetrical with respect to the heat equator, decreasing in magnitude toward the poles.

The heat equator is north of the equator because the Northern Hemisphere is warmer than the Southern Hemisphere. The polar regions of the two hemispheres have different radiational characteristics; most of the Antarctic continent is submerged under massive glacial ice sheets, so the surface has a very high albedo for solar radiation and is the site of intense radiational cooling, especially during the long polar night. By contrast, the Northern Hemisphere polar region is mostly ocean. Although the Arctic Ocean is usually ice covered, patches of open water develop in summer and lower the overall surface albedo so that the Arctic is warmer than the Antarctic. A second factor is that the Northern Hemisphere has a greater fraction of land in the tropical latitudes. Because land surfaces warm more than water surfaces from the same incoming solar radiation, tropical latitudes are warmer north of the equator instead of south. A third contributing factor is ocean circulation that transports more warm water to the Northern Hemisphere than to the Southern Hemisphere (Chapter 6).

Systematic patterns also appear when we consider the worldwide distribution of mean January temperature (Figure 10.1) and mean July temperature (Figure 10.2); January and July are usually the coldest/warmest months of the year in the respective hemisphere. As previously stated, neglecting the influence of topography on air temperature, isotherms tend to parallel latitude circles. However, monthly isotherms exhibit notable north-to-south bends, primarily because of land/sea contrasts and the influence of ocean currents. As the year progresses from January to July and to January again, isotherms in both hemispheres follow the Sun, shifting north and south in tandem. The latitudinal (north-south) shift in isotherms is greater over the continents than over the ocean; that is, the annual range in air temperature is greater over land than over the sea. Furthermore, the north-south mean temperature gradient is greater in the winter hemisphere than in the summer hemisphere. This is the consequence of greater north-south differences in incident solar radiation in which polar latitudes are in darkness during winter, while the tropics continue to receive relatively large amounts of incoming solar radiation. A steeper temperature gradient means a more vigorous atmospheric circulation and stormier weather in the winter hemisphere.

PRECIPITATION

The global pattern of mean annual precipitation (rain plus melted snow) exhibits considerable spatial variability (Figure 10.3). Some of this variability can be attributed to topography and land/sea distribution, but the planetary-scale circulation is also important. The intertropical convergence zone (ITCZ), subtropical anticyclones, and prevailing wind belts impose a roughly zonal signature on precipitation distribution. In addition, regular shifts in these circulation features through the year are responsible for the seasonality of precipitation that characterizes regional climates.

In tropical latitudes, convective activity associated with intense solar heating and the trade wind convergence triggers abundant rainfall for most of the year. In the adjacent belt poleward to about 20 degrees latitude, rainfall depends on seasonal shifts of the ITCZ and the subtropical anticyclones. Shift of the ITCZ toward the pole causes summer rains, whereas a shift of the subtropical highs toward the equator brings a dry winter. This climatic zone includes the belt of tropical monsoon circulation, described in Chapter 7.

Poleward of this belt, from about 20 to 35 degrees N and S, subtropical anticyclones, centered over the ocean basins, dominate the climate all year. Sinking dry air on the anticyclones' eastern flanks is responsible for Earth's major subtropical deserts (e.g., the Sahara). On the other hand, unstable humid air on the western flanks of subtropical anticyclones causes relatively moist conditions (e.g., over the southeastern U.S.). Between about 35 and 40 degrees latitude, the prevailing westerlies and subtropical anticyclones govern precipitation. Typically, on the western side of continents, winter cyclones migrating with the westerlies bring moist weather, but then in summer, westerlies shift poleward and the area lies under the dry eastern flank of a subtropical anticyclone. Hence, summers are dry. At the same latitudes, but on the eastern side of continents, the climate is dominated by westerlies in winter and the moist airflow on the western flank of a subtropical anticyclone in summer. Thus, rainfall is triggered by cyclonic activity in winter and by convection in summer, resulting in little seasonal variability in monthly precipitation totals.

Precipitation generally declines poleward of 40 degrees latitude, where lower temperatures reduce the amount of precipitable water (Chapter 5). The tendency in the continental interiors for more precipitation in summer is due to higher air temperatures, greater precipitable water, and more vigorous convection at that time of year.

Our description of the global pattern of annual precipitation is somewhat idealistic and requires some qualification. Land/sea distribution and topography complicate the generally zonal distribution of precipitation. While more rain falls over the ocean than over the continents, as pointed out in Chapter 5, more

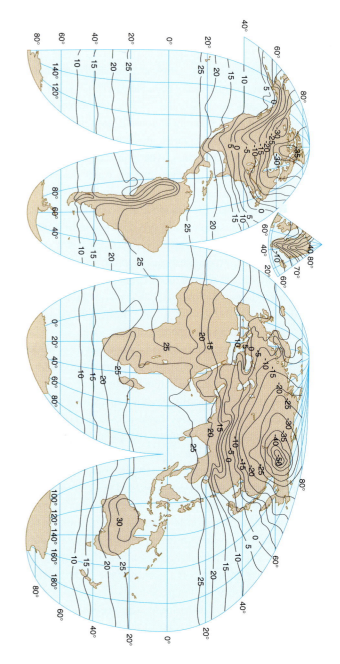

FIGURE 10.1
Mean sea level air temperature for January in degrees Celsius.

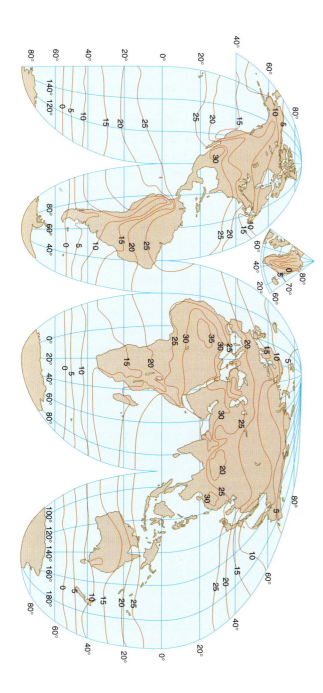

FIGURE 10.2
Mean sea level air temperature for July in degrees Celsius.

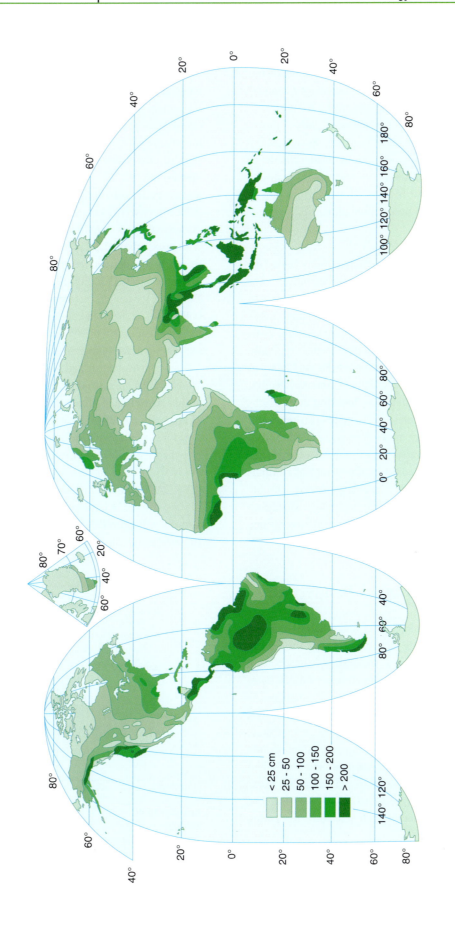

FIGURE 10.3
Mean annual precipitation (rain plus melted snow) in centimeters (cm).

precipitation falls on land than evaporates from it whereas less precipitation falls on the ocean than evaporates from it. Mountain belts induce wet windward slopes and extensive leeward rain shadows. Furthermore, annual precipitation totals fail to convey other important aspects of precipitation, including the average daily rainfall and the season-to-season and year-to-year reliability of precipitation. As a rule, rainfall is most reliable in maritime climates, less reliable in continental localities, and least reliable in arid regions. However, drought is possible anywhere, even in maritime climates.

CLIMATE CLASSIFICATION

In response to the interaction of many boundary conditions, Earth's climates form a complex mosaic. For more than a century, climatologists have attempted to organize the myriad of climate types by devising classification schemes to group together climates having common characteristics. Classification schemes typically group climates according to the meteorological basis of climate, or the environmental effects of climate. The first is a *genetic climate classification*, focusing on how climates form in certain locales. The second is an *empirical climate classification* that infers the type of climate from environmental indicators such as the distribution of indigenous vegetation. In addition, the advent of computers and databases has made possible numerical climate classification schemes that utilize sophisticated statistical techniques. Chapter 13 provides information on one of the more popular climate classifications.

Trends in Mean Annual Temperature

Invention of weather instruments and establishment of weather/climate observing networks around the world, combined with standardized methods of observation and record-keeping, made the climate record much more detailed, representative, and reliable. Confidence is greatest in temperature records dating from the late 1800s when the first national weather services were established, including those of the U.S. and Canada. At about the same time, in 1878, the *International Meteorological Organization (IMO)*, predecessor to today's *World Meteorological Organization (WMO)*, was founded. The WMO sets uniform international standards for taking observations.

In the late 1990s, climatologist Michael Mann and his colleagues at Pennsylvania State University combined a dozen Northern Hemisphere temperature

records covering the past 1000 years. Most of the combined records were derived from proxy climate data sources including tree growth rings and corals (Chapter 9), whereas the most recent segment of the record is largely instrument-based. Because of its shape, the temperature curve was popularly described as the "hockey stick" model. For most of the 1000-year period, the temperature trended gently downward (the handle) and then sharply upward into the 21st century (the blade). From this, Mann and his colleagues concluded that the final decades of the 20th century were likely warmer than any prior comparable period in the past millennium. This conclusion is consistent with proxy climate data from elsewhere in the world such as an ice-core derived temperature record extracted from a glacier at 7163 m (23,500 ft) in the Himalayan Mountains. Based on its review of research conducted by Mann and colleagues, a National Research Council (NRC) panel in 2006 endorsed the "hockey stick" model, although expressing less confidence in the early part of the record.

Instrument-based global average surface temperature records are essential for understanding climate variability and climate change and for identifying trends in climate. Today, scientists at three climate centers compute the global average surface temperature each month and add it to the time series (Figure 10.4). The climate centers are: (1) NASA's *Goddard Institute for Space Studies (GISS)*, (2) NOAA's *National Climatic Data Center (NCDC)*, and (3) the United Kingdom's Met Office in collaboration with the Climatic Research Unit (CRU) at the University of East Anglia (the *HadCRUT3* record). Working indepen-

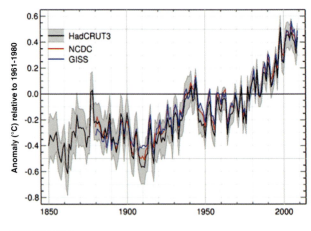

FIGURE 10.4
A comparison of three global mean surface temperature records, computed independently, showing a warming trend particularly over the past three decades. The grey area associated with the HadCRUT3 curve represents uncertainty. [Data provided by NCDC/NESDIS/NOAA and NASA; plot provided by UK Met Office]

dently, the three climate centers use different methods of collecting and processing data (including quality control by hand and computer), to arrive at a final calculation of the global mean monthly surface temperature. For more ready comparison, temperature readings are converted to anomalies, that is, departures from the long-term average. Although computed independently, the time series from each climate center are closely correlated. Note, however, that the HadCRUT3 record begins in 1850 whereas the NASA and NOAA records commence in 1880.

Enormous amounts of observational data from over the land and sea go into computing the global mean surface temperature. For example, the HadCRUT3 record uses daily observations from about 2000 land stations and sea-surface temperatures from about 1200 drifting buoys deployed across the ocean plus about 4000 ships in the *Voluntary Observing Ship* program. In addition, numerous instrumented moored buoys, located in the tropics and coastal regions (Chapter 8), are additional sources of sea-surface data.

Plotted in Figure 10.5 is NOAA's NCDC 1880 to 2008 instrument-derived record of variations in (1) global (land plus sea surface) mean annual temperature, (2) mean sea-surface temperature, and (3) mean land surface temperature. Temperature is expressed as a departure (in Celsius degrees and Fahrenheit degrees) from the long-term (1901-2000) average. For reasons given in Chapter 4, these temperature series indicate greater year-to-year variability over land than ocean. The trend in global mean temperature was generally upward from 1880 until about 1940, downward or steady from 1940 to about 1970, and upward again through the 1990s and early 2000s. The overall variation in temperature is mostly in the range of about ±0.4 Celsius degree (±0.7 Fahrenheit degree) about the period average.

In spring 2000, NCDC reported that global warming accelerated during the final quarter of the 20th century. From 1906 to 2005, the global mean annual temperature climbed 0.74 ± 0.18 Celsius degrees. Most of this warming (0.65 ± 0.15 Celsius degrees) took place from 1956 to 2005. In terms of global mean annual surface temperature, 11 of the 12 years prior to 2007 were the warmest since reliable instrument-based records began in the late 1800s.

The temperature record for the globe as a whole is not representative (in direction or magnitude) of all locations worldwide; that is, the trend was amplified or reversed, or both, in specific places. Figure 10.6 presents the temperature record for nine NCDC regions of the United States plus the coterminous U.S. as a whole. For

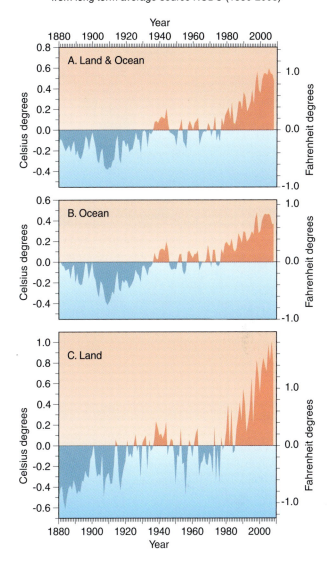

Annual average temperature departures from long term average source NCDC (1880-2008)

FIGURE 10.5

Instrument-derived NCDC mean annual global (land plus ocean), sea-surface, and land temperatures, expressed as departures in Celsius degrees and Fahrenheit degrees from the 20th century (1901-2000) average. [NOAA, National Climatic Data Center]

each region, the mean annual temperature in degrees Fahrenheit is plotted for the period 1895 to the fall of 2009. The five-year running mean, in red, enables us to follow trends. Temperature anomalies represent the departure of mean annual temperature in Fahrenheit degrees from the 1901-2000 average. Whereas an overall warming trend is evident through much of the period of record in all regions, considerable year-to-year variation occurs in mean annual temperature. Also, the magnitude of overall temperature change during the 20th century varies among regions.

U.S. standard climatological regions

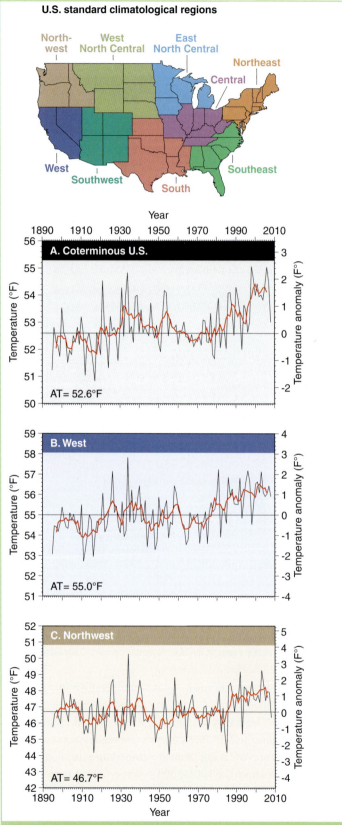

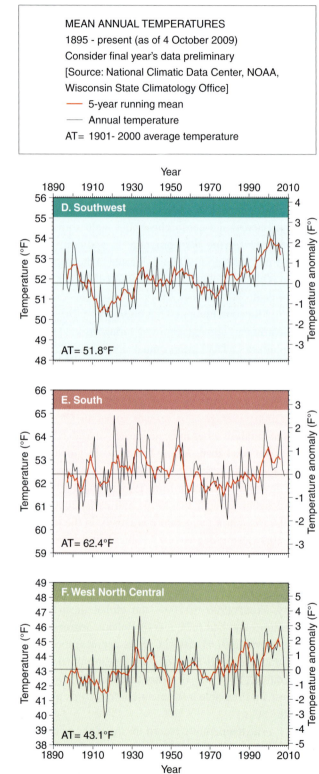

MEAN ANNUAL TEMPERATURES
1895 - present (as of 4 October 2009)
Consider final year's data preliminary
[Source: National Climatic Data Center, NOAA,
Wisconsin State Climatology Office]
— 5-year running mean
— Annual temperature
AT= 1901- 2000 average temperature

FIGURE 10.6
Mean annual temperature (in °F) for nine regions of the coterminous United States plus the coterminous U.S. as a whole for the period 1895-2009. The red curve is the 5-year running mean temperature. Also indicated is the departure of the temperature (in Fahrenheit degrees) from the 1901-2000 average. [Based on

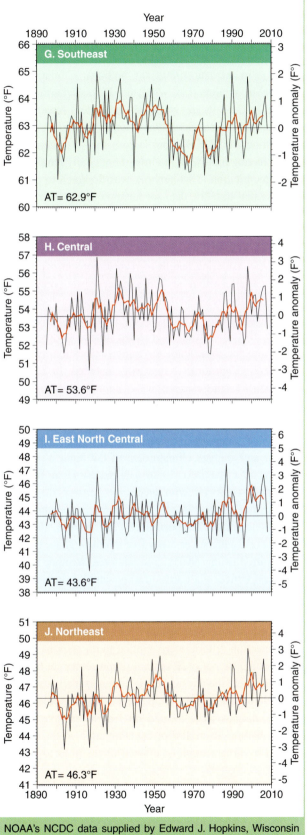

NOAA's NCDC data supplied by Edward J. Hopkins, Wisconsin State Climatology Office, University of Wisconsin-Madison]

INTEGRITY OF INSTRUMENT DATA

Systematic temperature and precipitation observations have been made at various locations across the nation for upwards of 140 years. Weather and climate observations grew from modest beginnings in the early 19th century when only a few tens of stations operated to today when hundreds of automatic weather stations operated by NOAA's National Weather Service (NWS) and the Federal Aviation Administration (FAA) plus the thousands of member stations of the NWS Cooperative Observer Network acquire a continuous stream of observational data (Chapter 1). Similar networks have been established in many other nations around the world. Data from these networks are used to quantitatively assess climate variability and climate change during the instrument era with an eye toward visualizing the climate future. However, some critics question the integrity of the instrument-based climate record, citing many potential sources of error.

The sensitivity and reliability of weather instruments have improved via technological advances achieved during the period of record. In recent decades, for example, electronic sensors have replaced liquid-in-glass thermometers at many NWS, FAA, and Cooperative Observer stations. At most long-term weather/climate stations, the location of instruments and the station itself has changed, perhaps many times. Relocations often involved changes in the elevation of instruments. With increasing altitude, air temperature normally decreases and wind speed increases. Hence, vertical displacement of instruments alters the climate record. A change in the type of surface underlying the instruments (e.g., grass instead of asphalt) is likely to affect the local radiation budget and temperature readings.

According to W.E.K. Middleton, author of *A History of the Thermometer and Its Use in Meteorology* (Johns Hopkins University Press, 1966), instrument shelters were not in widespread use in North America until the 1870s. Beginning with the U.S. Army Signal Corps weather observations (1870-91) and continuing well into the 20th century, instruments typically were mounted on the rooftop of the tallest building in a city (Figure 10.7). The goal was to monitor atmospheric conditions at an elevation of 30 m (100 ft) above the ground. In fact, the height of the anemometer was not standardized (at a height of 10 m or 33 ft) until the 1990s. Relocation of a weather/climate station may be accompanied by changes in topographic setting and exposure, or proximity to the moderating influence of large bodies of water or urban centers (Chapter 4).

FIGURE 10.7
The U.S. Weather Bureau office located at Johns Hopkins University about 1899. The instrument shelter (louvered box), rain gauge, and wind instruments are located on the roof of the building. [From *Maryland Weather Service*, 1899, Baltimore, MD: Johns Hopkins Press, Vol. I, p.422; NOAA's National Weather Service Collection, archival photograph by Sean Linehan]

At sea, huge gaps in monitoring networks are a broader source of error in computing sea-surface temperatures (SST). Also, prior to the mid 1940s, it was common practice for SST to be measured from a ship by lowering a bucket overboard to obtain a sample of water at the sea surface. The bucket was then raised to the deck where eventually a water temperature reading was made. In the meantime, as the bucket of water sat on the deck, the Sun may heat the water while evaporation may cool the water. After the mid 1940s, SST was measured by taking the temperature of surface ocean water that was drawn into a ship's engine room to cool the engine. It turns out that engine-room measurements are closer to the actual SST than the bucket measurements.

When dealing with huge amounts of observational data, some uncertainty is inevitable so that global mean temperature is expressed as a range rather than one definite figure. However, even when careful statistical analysis accounts for potential sources of error and a range of uncertainty is assigned, a significant global warming trend is evident over the past century and especially since the 1970s.

In the late 1980s, the National Climatic Data Center in conjunction with the U.S. Department of Energy's Oak Ridge National Laboratory created the *United States Historical Climatology Network (USHCN)* consisting of 1218 stations across the 48 coterminous states having long-term records of both daily temperature and precipitation. This network was designed to provide an essential baseline data set for monitoring the nation's climate commencing in the late 19th century. These stations were created from a subset of the Cooperative Observer Network stations, chosen based upon long-term data quality, considering length of record, percent of missing data, spatial distribution, and number of station changes. Many of the selected USHCN stations are located in rural areas to reduce the influence of urbanization. Using statistical techniques, data from these stations have been adjusted to account for station relocation, or when a different thermometer type was installed. In addition, an urban warming correction was applied based upon the population of the surrounding area.

More recently, NOAA began the *U.S. Climate Reference Network (USCRN)*, a project designed to collect and analyze climate data of the highest possible quality over the next 50 to 100 years. Each USCRN station is equipped with electronic sensors that make routine measurements of air temperature, precipitation, IR ground surface temperature, solar radiation, and wind speed at a frequency of every five minutes. These data are transmitted to the NCDC and National Weather Service Forecast Offices via Earth-orbiting satellites in near real-time. In addition to these measurements, additional sensors may be added to the USCRN stations for monitoring soil temperature and soil moisture. Conscientious and detailed site selection criteria were applied to all stations so that they would be not only spatially representative, but also located where the physical environment is likely to remain essentially unchanged over the next 50 to 100 years. Many of the stations were located on federal or state owned lands, helping to minimize contamination of the climate record by urbanization or other changes in local ground cover.

GLOBAL WARMING

A general consensus in the scientific community holds that an overall global-scale warming trend has prevailed since the end of the Little Ice Age, that is, since the end of the first decade of the 20th century. The three independently-derived global-average temperature records plotted in Figure 10.4 concur that Earth's climate has warmed over the past century and that warming has accelerated since the 1970s; that is, each decade (including the decade since 2000) has been warmer than the prior decade.

The most reasonable scientific explanation for the observed warming trend in the latter part of the 20th century and into the early 21st century is the steady build-up of carbon dioxide (CO_2) and other infrared-absorbing gases in the atmosphere primarily because of combustion of fossil fuels and the consequent enhancement of the natural greenhouse effect (Chapter 3). Continuation of the upward trend in the concentration of greenhouse gases inevitably will lead to global warming throughout this century and perhaps well beyond. We have more to say on this topic in Chapter 12.

Changes in the Water Cycle

When it comes to climate variability and climate change, the primary focus usually is on temperature. The prospect of continued global warming takes center stage in discussions regarding the climate future; after all, this anthropogenic influence on Earth's climate system is directly experienced by people. But just as important as temperature trends (if not more so) are concurrent changes in the global water cycle, including the amount and patterns of precipitation. For example, changes in the frequency and/or persistence of drought or episodes of exceptionally heavy rainfall are likely to seriously impact agricultural systems, community infrastructure, and hydroelectric facilities. This underscores the need to be just as cognizant of instrument-based precipitation and land-based runoff records as temperature records.

Recent analyses of instrument-based records reveal trends in river discharge, rainfall patterns, and surface humidity in various parts of the world. For example, according to a 2007 National Academy of Sciences report, the amount of atmospheric water vapor over the ocean (precipitable water) has increased by 0.41 kg per m² per decade since 1988, a change that is consistent with an enhanced greenhouse effect.

Global climate models predict that continuation of the current warming trend will be accompanied by an overall increase in precipitation. But climate change is geographically non-uniform in direction and magnitude so that rainfall is expected to increase in some areas and decrease in other areas. Already the instrument-based record is indicating that some regions are experiencing a greater frequency of extreme rainfall events and more days with rainfall. In 2008, P. Groisman and R. Knight of NOAA's National Climatic Data Center reported on their analysis of rainfall trends over the past 40 years at 4000 stations across the coterminous U.S. They found a significant increase in the mean duration of prolonged dry episodes during the warm season. Over the eastern U.S., the return period for a dry episode lasting one month decreased from every 15 years to 6-7 years.

Seung-Ki Min and colleagues at Environment Canada's Climate Research Division in 2008 reported on their analysis of trends in Arctic land precipitation over the period 1950-1999. Observational data, from NOAA's NCDC, was drawn from stations in North America and Eurasia north of 55 degrees N. They found an overall upward trend in Arctic precipitation, albeit with a decreasing trend over easternmost Eurasia. During the same period, river discharge into the Arctic Ocean increased, which weakens the meridional overturning circulation (MOC) in the Atlantic Ocean, with potential implications for global climate. These findings are consistent with climate model projections of future precipitation changes accompanying further enhancement of Earth's greenhouse effect.

Lessons of the Climate Record

Now that we have examined the reconstructed and instrument-based climate record, it is appropriate to revisit a topic first introduced in Chapter 2, that is, the lessons of the climate record. What does the climate record tell us about climate variability and climate change? The following lessons of the climatic past are useful in assessing prospects for the climate future and the possible impacts of climate change.

- *Climate is inherently variable over a broad spectrum of time scales ranging from years to decades, to centuries, through millennia. Variability is an endemic characteristic of climate. The question for the future is not whether the climate will change, but how climate will change and by how much.*
- *Climate variability and climate change are geographically non-uniform in both sign (e.g., warmer or cooler, wetter or drier) and magnitude. Some regions may experience warming while*

other regions experience cooling over the same period. Also, some regions may become wetter (with more frequent heavy rain events and potentially disastrous flooding) while other regions experience drying (and more frequent drought).

- *Global- and hemispheric-scale trends in climate are not necessarily duplicated at particular locations although the magnitude of temperature change tends to amplify with increasing latitude (known as polar amplification).* Partially for this reason, a large-scale change in climate is not likely to have the same impact everywhere.

- *Climate change may consist of a long-term trend in various elements of climate (e.g., mean temperature or average precipitation) and/or a change in the frequency of extreme weather events (e.g., drought, excessive cold).* Climate encompasses mean values plus extremes in climatic elements. A trend toward warmer or cooler, wetter or drier conditions may or may not be accompanied by a change in frequency of weather extremes. On the other hand, a climatic regime featuring relatively little change in mean temperature or precipitation through time could be accompanied by an increase or decrease in frequency of weather extremes.

- *Climate change tends to be more abrupt than gradual.* In the context of the climate record, "abrupt" is a relative term. If the time of transition between climatic episodes (e.g., glacial climate versus interglacial climate) were much shorter than the duration of the individual episodes, then the transition would be considered relatively abrupt. Analysis of glacial ice cores from Greenland indicates that cold and warm climatic episodes, each lasting about 1000 years, were punctuated by abrupt change over intervals as brief as a single decade or even several years. Abrupt climate change would test the resilience of society.

- *Only a few cyclical variations can be discerned from the long-term climate record, which are associated with several known periodic cycles in climate forcing.* Regular cycles include diurnal and seasonal changes in incoming solar radiation (the forcing) and temperature (the response). This means simply that days are usually warmer than nights and summers are warmer than winters. Quasi-regular variations

in climate include El Niño (occurring about every 3 to 7 years), Holocene millennial-scale fluctuations identified in glacial ice cores, and the major glacial-interglacial climate shifts of the Pleistocene Ice Age. These were unlocked from analysis of deep-sea sediment cores and operate over tens of thousands to hundreds of thousands of years (i.e., the Milankovitch cycles). Study of climate cycles adds to our understanding of the present climate and anthropogenic influences on climate.

- *Climate change impacts society.* History recounts numerous instances when climate change significantly impacted society. Examples discussed in this book thus far include the Norse settlements in Greenland (Chapter 2), the Garamantian civilization of North Africa (Chapter 5), the Tang, Yuan, and Ming Dynasties in China (Chapter 7), and the Anasazi and Fremont peoples in the American Southwest (Chapter 9). While climate played a role (perhaps a major one) in these and many other well-documented cases of social upheaval, other factors also contributed. Modern societies may be more capable of dealing with climate change than early peoples, but a rapid and significant change in climate would seriously impact all sectors of modern society. Recall, for example, the likely fate of inhabitants of low-lying island nations as described in this chapter's Case-in-Point.

In addition to these lessons of the climate record, it is interesting to note the occurrence of **singularities**, a weather episode that occurs on or about a certain time of year more frequently than chance would dictate. For a few examples, refer to this chapter's first Essay.

Climatology of Severe Weather

One of the major concerns regarding the climate future is the possibility that a continued rise in global mean surface temperature would somehow alter the frequency and/or intensity of severe weather events. Severe thunderstorms, tornadoes, and tropical cyclones (tropical storms and hurricanes) can take lives, cause considerable property damage, and disrupt commerce and industry. Assessing these potential hazards requires that we examine the atmospheric conditions that favor the development of these systems and their climatology.

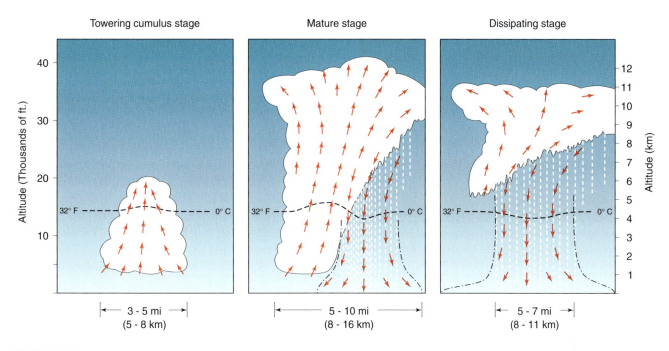

| Towering cumulus stage | Mature stage | Dissipating stage |

FIGURE 10.8
The life cycle of a thunderstorm cell consists of towering cumulus, mature, and dissipating stages.

Thunderstorms

A **thunderstorm** is a mesoscale weather system that is accompanied by lightning and thunder, affects a relatively small area, and is short-lived. A thunderstorm is the product of vigorous convection that extends high into the troposphere, sometimes to the tropopause or even the lower stratosphere. Upward surging air currents are made visible by billowing cumuliform clouds. A cumuliform cloud that shows significant vertical growth and resembles a huge cauliflower is known as a **cumulus congestus cloud**. If vertical growth continues, a cumulus congestus cloud builds into a **cumulonimbus cloud**, a thunderstorm cloud with characteristic anvil top, producing precipitation, lightning, and thunder.

A thunderstorm consists of one or more convection cells, each of which progresses through a three-stage life cycle: towering cumulus, mature, and dissipating (Figure 10.8). With favorable conditions in the troposphere, cumulus clouds build vertically and laterally during the initial **towering cumulus stage** (Figure 10.9). In about 10 to 15 minutes, cumulus cloud tops surge upward to altitudes of 8000 to 10,000 m (26,000 to 33,000 ft). At the same time, neighboring cumulus clouds merge so that by the end of the towering cumulus stage, the storm's lateral dimension may be 10 to 15 km (6 to 9 mi).

During the initial stage of the thunderstorm life cycle, saturated air streams upward throughout the cell as an **updraft**. The updraft is strong enough to keep water droplets and ice crystals suspended in the upper reaches of the cloud. By convention, the towering cumulus stage ends and the **mature stage** begins when precipitation first reaches Earth's surface. Typically, this stage lasts about 10 to 20 minutes. The cumulative weight of water droplets and ice crystals eventually becomes so great that they are no longer supported by the updraft. Rain, ice pellets, or snow descend through the cloud and drag the adjacent air downward, creating a strong **downdraft** alongside the updraft.

The downdraft leaves the base of the cloud and spreads out along Earth's surface, well in advance of the

FIGURE 10.9
Towering cumulus clouds may signal the initial stage in the development of a thunderstorm.

FIGURE 10.10
When the upward billowing cumulonimbus cloud reaches the tropopause, it spreads out forming a flat anvil top.

SINGLE CELL	MULTICELL CLUSTER	MULTICELL LINE	SUPERCELL
Weak updraft (non-severe) or Strong updraft (severe?)	Weak updraft (non-severe) or Strong updraft (severe)	Weak updraft (non-severe) or Strong updraft (severe)	Intense updraft (Almost always severe) Mesocyclone present
Slight threat	Moderate threat	Moderate threat	High threat

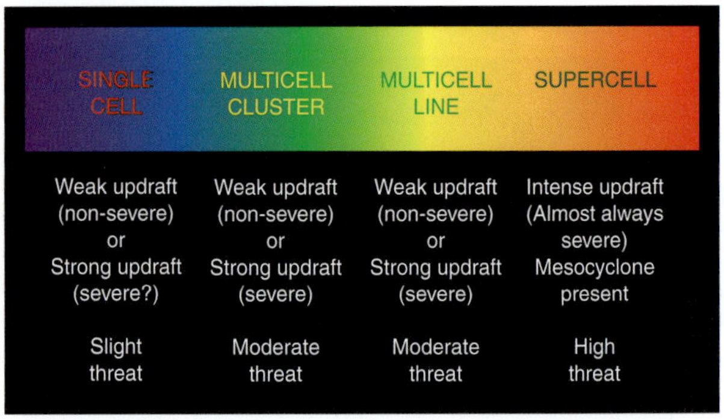

FIGURE 10.11
Classification of thunderstorms and the likelihood of severe weather. [Adapted from NOAA, *Basic Spotters' Field Guide*].

parent thunderstorm cell, as a mass of cool, gusty air. At the surface, the arc-shaped leading edge of downdraft air resembles a miniature cold front and is called a **gust front**. Uplift along the advancing gust front sometimes produces additional cumuliform clouds that may evolve into secondary thunderstorm cells tens of kilometers ahead of the parent cell.

A thunderstorm cell attains maximum intensity during its mature stage. Rain is heaviest, lightning is most frequent, and hail, strong surface winds, and even tornadoes may develop. Cloud tops can build to altitudes in excess of 18,000 m (about 60,000 ft). Strong winds at such great altitudes distort the cloud top into an anvil shape (Figure 10.10). The flat top of the anvil indicates that convection currents have reached the extremely stable air of the tropopause. Only in severe thunderstorms will convection currents overshoot this altitude, causing clouds to billow into the lower stratosphere before collapsing back into the troposphere.

As precipitation spreads throughout the thunderstorm cell, so does the downdraft, heralding the cell's demise. During the **dissipating stage**, subsiding air replaces the updraft throughout the cloud, effectively cutting off the supply of moisture provided by the updraft. Adiabatic compression warms the subsiding air, the relative humidity drops, precipitation tapers off and ends, and clouds gradually vaporize.

CLASSIFICATION OF THUNDERSTORMS

Thunderstorms are **mesoscale convective systems (MCS)** and are classified based on the number, organization, and intensity of their constituent cells. Thunderstorms occur as single cells, multi-cellular clusters, and supercells (Figure 10.11).

A *single-cell thunderstorm* is usually a relatively weak weather system that appears to pop up randomly almost anywhere within a warm, humid air mass. In reality, a single-cell thunderstorm is not a random phenomenon and almost always develops along a boundary within an air mass. The boundary, for example, may be the leading edge of the rain-cooled outflow from a distant thunderstorm cell. Outside of the tropics, solar heating of Earth's surface usually is insufficient for a thunderstorm cell to develop via convection alone. Frontal or orographic lifting or converging surface winds are needed to strengthen vertical motion.

Typically, a thunderstorm cell completes its life cycle in 30 minutes or so, but sometimes lightning, thunder, and bursts of heavy rain persist for many hours. This is an indication of the more common *multi-cellular thunderstorm*, which consists of more than one cell. Each cell may be at a different stage of its life cycle, and new cells form and old cells dissipate continually. A succession of many cells is thus responsible for a prolonged period of thunderstorm weather. Although a locality may be in the direct path of a distant, intense thunderstorm cell, the relatively brief life span of an individual cell means that severe weather may dissipate before reaching that locality. The multi-cellular nature of most thunderstorms complicates the motion of the weather system because a thunderstorm may track at some angle to the paths of its constituent cells (Figure 10.12). In terms of organization, two types of multi-cellular thunderstorms are the squall line and the mesoscale convective complex. Cells in either of these systems can produce severe weather.

A **squall line** is an elongated cluster of thunderstorm cells that is accompanied by a continuous gust front at the line's leading edge (Figure 10.13). A squall line is most likely to develop in the warm southeast sector of a mature extratropical cyclone, ahead of and parallel to

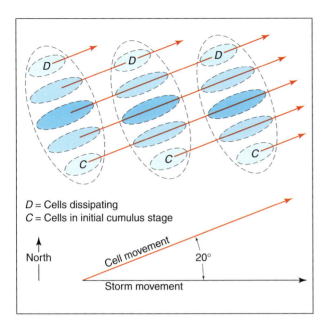

FIGURE 10.12
In this idealized situation viewed from above, the component cells of a multicellular thunderstorm travel at about 20 degrees to the eastward moving thunderstorm. As they travel toward the northeast, the individual cells progress through their life cycle.

the cold front (Chapter 7). Squall line thunderstorm cells are usually more intense than single-cell thunderstorms because the circulation in the associated low pressure system strengthens the updraft. As the squall-line gust front surges forward, warm and humid air is lifted and

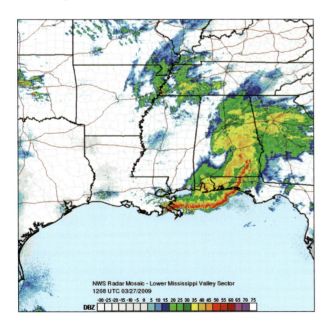

FIGURE 10.13
Radar image showing a squall line, an elongated cluster of thunderstorm cells, situated ahead of and parallel to a cold front. [NOAA]

forced into the updraft located at the leading edge of the squall line. The heaviest rain or hail typically occurs just behind (to the west of) this updraft.

A **mesoscale convective complex (MCC)** is a quasi-circular cluster of many interacting thunderstorm cells covering an area perhaps a thousand times greater than a single isolated thunderstorm cell. In fact, it is not unusual for one MCC to cover an area the size of the state of Iowa (Figure 10.14). W.S. Ashley and colleagues at the University of Georgia analyzed 15 years of observational data from MCC climate studies conducted between 1978 and 1999 to develop a comprehensive climatology of MCCs. Researchers examined a total of 527 MCCs, averaging more than 35 per year, and drew the following conclusions: An MCC is primarily a warm-season phenomenon (86% taking place from May through August with peak frequency in June) that generally develops at night and occurs chiefly over the eastern two-thirds of the United States. In an average year, MCCs are most likely in the lower Missouri River Valley. They are concentrated along a north-south axis from south Texas to the Canadian border and an east-west axis from the High Plains to the Ohio River Valley. MCCs are less common in the Southeast and along the East Coast. They account for upwards of 18% of growing season rainfall across the Great Plains and Midwest.

An MCC usually develops during weak synoptic-scale flow, often near an upper-level ridge of high pressure and on the cool side of a stationary front. A low-level jet

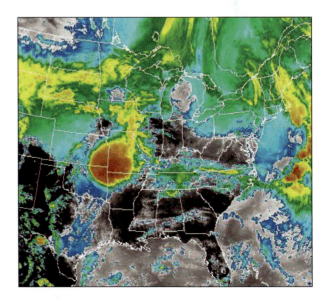

FIGURE 10.14
Infrared satellite image showing a mesoscale convective complex (MCC) over southwest Missouri, southeast Kansas, and northeast Oklahoma. [NOAA].

stream feeds warm humid air into a developing MCC while a low pressure system forms at middle levels of the troposphere. Rising temperatures at low levels and radiational cooling at upper levels destabilize the troposphere. New cells form while old cells dissipate continually within an MCC, so the life expectancy of the system is at least 6 hrs and often 12 to 24 hrs. The longevity and typically slow movement (15 to 30 km per hr, or 9 to 18 mph) of an MCC result in widespread and substantial rainfall. MCCs also can produce severe weather, including weak tornadoes, moderate-sized hail, and flash flooding. Ashley et al. point out that climate change that substantially reduces the frequency of MCCs would have major impacts on the moisture balance of the central and eastern U.S.

A **supercell thunderstorm** is a relatively long-lived, large, and intense system. It consists of a single cell with an exceptionally strong updraft, in some cases estimated at 240 to 280 km per hr (150 to 175 mph). A distinguishing characteristic of a supercell is the tendency for the updraft to develop a rotational circulation that may evolve into a tornado.

WHERE AND WHEN

Atmospheric conditions that favor thunderstorm development vary with latitude, season, and time of day. Key to understanding the climatology of thunderstorms is the conditions required for their formation: humid air in the low- to mid-troposphere, atmospheric instability, and a source of uplift.

Most thunderstorms develop within a maritime tropical (*mT*) air mass, which is destabilized by uplift. Usually, maritime tropical air is conditionally stable and becomes unstable (buoyant) only when lifted to the condensation level. Recall from Chapter 5 that *conditional stability* means that the ambient air is stable for unsaturated (clear) air parcels but unstable for saturated (cloudy) air parcels. The more humid the air, the less the ascent (expansional cooling) needed for destabilization. Most thunderstorms develop when maritime tropical air is lifted along fronts, up mountain slopes, or via converging surface winds. Furthermore, cold air advection aloft and/or warm air advection at the surface enhances the potential instability of *mT* air. Either of these processes increases the air temperature lapse rate and thereby reduces ambient air stability.

Solar heating drives atmospheric convection, so it is not surprising that thunderstorms tend to be most frequent during the warmest hours of the day. However, there are many exceptions to this rule; mesoscale

convective complexes and squall lines develop both night and day. In the Missouri River Valley and adjacent portions of the upper Mississippi River Valley, even single-cell thunderstorms are more frequent at night than during the day. One possible explanation for the nocturnal thunderstorm maximum in the upper Missouri/Mississippi Valleys is based on the role of a *low-level jet stream* of maritime tropical air that flows from the Gulf of Mexico northward up the Mississippi River Valley. This jet stream strengthens at night and causes warm air advection at low levels that destabilizes the air and spurs the buildup of cumuliform clouds.

Usually, *thunderstorm frequency* is expressed as the number of thunderstorm days per year, where a **thunderstorm day** is defined as a day when thunder is heard. This conventional method likely underestimates the actual number of thunderstorms, particularly if more than one line of thunderstorms passes over a weather station on the same day. With this limitation in mind, thunderstorms occur with greatest frequency over the continental interiors of tropical latitudes. The steamy Amazon Basin of Brazil, the Congo Basin of equatorial Africa, and the islands of Indonesia have the highest frequency of thunderstorms in the world, experiencing at least 100 thunderstorm days per year. Because the surfaces of large bodies of water do not warm as much as land surfaces in response to the same intensity of solar radiation, thunderstorms are less frequent over adjacent bodies of water. In the subtropics and tropics, intense solar heating may combine with converging surface winds to trigger thunderstorm development. As noted in Chapter 6, this combination characterizes the intertropical convergence zone (ITCZ).

In North America, thunderstorm frequency generally increases from north to south with the highest frequency over central Florida where some interior localities average 100 thunderstorm days per year (Figure 10.15). The Florida thunderstorm maximum near Orlando is due to convergence of sea breezes that develop along both the east and west coasts of the Florida peninsula (Figure 10.16). Sea breeze convergence over the interior induces ascent of maritime tropical air and formation of cumulonimbus clouds.

Portions of the Rocky Mountain Front Range rank second to interior Florida in thunderstorm frequency over North America. On average, more than 60 thunderstorm days occur per year in a band from southeastern Wyoming southward through central Colorado and into north central New Mexico. This high thunderstorm frequency is linked to topographically related differences in heating. Mountain slopes facing the Sun absorb direct solar radiation and

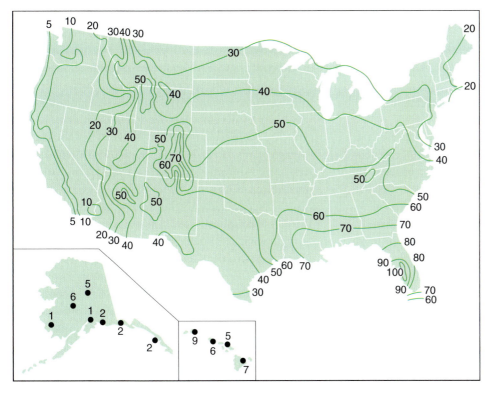

FIGURE 10.15
Average annual number of thunderstorm-days in the United States.

by cooler air sweeping westward from over the Plains. Updrafts over the mountain slopes produce cumuliform clouds that often evolve into thunderstorm cells, particularly during the warmest hours of the day. Thunderstorm development is enhanced whenever the synoptic-scale air pressure pattern favors winds blowing from the east over the western Great Plains.

Under some conditions, convection and ascent of air are inhibited and thunderstorms are rare or do not form. This is the case when air masses reside or travel over relatively cold surfaces and are thereby stabilized. Snow-covered ground is an example of such a surface. Convection is suppressed, so thunderstorms are relatively rare at middle and high latitudes in winter. Nonetheless, thunderstorms can be associated with winter extratropical cyclones, usually developing ahead of the surface warm front as lifting over the front destabilizes the near surface air. In that case, *thundersnow* can be locally heavy.

Thunderstorms are unusual over coastal areas that are downwind from relatively cold ocean waters. For example, thunderstorms are infrequent along coastal California, where prevailing winds blow onshore and a shallow layer of maritime polar air often flows inland from the relatively cold California Current. Cool *mP* air at low levels suppresses deep convection and thunderstorm development. Thus, the average annual number of thunderstorm days is only 2 at San Francisco, 6 at Los Angeles, and 5 at San Diego.

Thunderstorms are also relatively rare in Hawaii, primarily because of the trade wind inversion (Chapter 6). At NWS Forecast Offices at Hilo and Honolulu, the average annual number of thunderstorm days is under 10. The trade wind inversion, at an average altitude of about 2000 m (6600 ft), restricts vertical development of towering cumulus clouds. Cumulus clouds normally do not attain the altitude needed to develop into cumulonimbus clouds.

become relatively warm. The warm slopes, in turn, heat the air in immediate contact with the slopes, and that air rises. At the same time, the air at the same altitude, but located to the east of the mountains out over the relatively flat terrain of the western Great Plains, is much cooler. As warm air rises over the mountain slopes, it is replaced

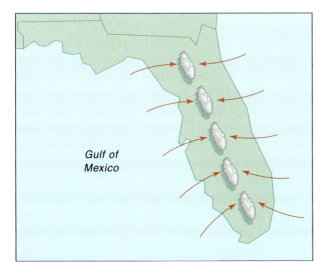

FIGURE 10.16
The relatively high frequency of thunderstorms over the interior of the Florida peninsula is linked to the convergence of sea breezes blowing inland from both the Gulf and East Coasts.

SEVERE THUNDERSTORMS

According to the official National Weather Service (NWS) criterion, a **severe thunderstorm** has surface winds stronger than 50 knots (93 km per hr or 58 mph) and/or produces hailstones 0.75 in. (1.9 cm) or larger in diameter (penny-size). Such thunderstorms may also produce flash floods or tornadoes.

As a general rule, the greater the altitude of the top of a thunderstorm, the more likely it is that the system will produce severe weather. Why do some thunderstorm cells surge to great altitudes and trigger severe weather, whereas others do not? The critical factor appears to be vertical **wind shear**, the change in horizontal wind speed or direction with increasing altitude. Weak vertical wind shear (little change in wind with altitude) favors short-lived updrafts, low cloud tops, and weak thunderstorms, whereas strong vertical wind shear favors vigorous updrafts, great vertical cloud development, and severe thunderstorms.

In the United States and Canada, most severe thunderstorms break out over the Great Plains and are associated with mature extratropical cyclones. Severe thunderstorm cells usually form as part of a squall line within the cyclone's warm sector, ahead of and parallel to a fast-moving and well-defined cold front. Severe cells can produce large hail, heavy rain, and downbursts. Most tornadoes that do form are weak, although some squall line cells evolve into supercells that can spawn strong to violent tornadoes.

A typical synoptic-scale circulation pattern conducive to severe thunderstorm development is shown schematically in Figure 10.17. From the center of a mature extratropical cyclone over western Kansas, a well-defined

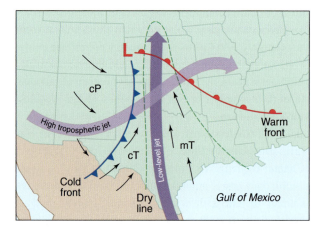

FIGURE 10.17
A map view of a synoptic-scale weather pattern that favors development of severe thunderstorms.

cold front trails across the Oklahoma and Texas panhandles into West Texas. At the same time, a warm front stretches southeastward from the low center. The polar front jet stream and a low-level jet stream of maritime tropical air, which are also plotted on the map, cross to the southeast of the cyclone center. The most intense thunderstorm cells are likely to develop near the intersection of these two jets. The *subtropical jet stream* is sometimes also present at high levels in the troposphere, blowing from west to east over the warm sector of the cyclone. In that event, intense squall lines are likely to form between the polar front jet stream and the subtropical jet stream.

The polar front jet stream produces strong vertical wind shear that maintains a vigorous updraft and favors great vertical development of thunderstorm cells. In addition, the jet contributes to a stratification of air that increases the potential instability of the troposphere. As described in Chapter 7, a *jet streak* induces both horizontal divergence and horizontal convergence of air in the upper troposphere. Recall that diverging horizontal winds trigger ascent of air and cyclone development under the left-front quadrant of a jet streak. Meanwhile, air converges in the right-front quadrant of a jet streak, causing weak subsidence of air over the warm sector of the cyclone. The subsiding air warms by compression and its relative humidity decreases. The subsiding, warming air is prevented from reaching Earth's surface by a shallow layer of maritime tropical air that surges northward as a low-level jet from the Gulf of Mexico. The jet of maritime tropical air occurs on the western flank of the Bermuda-Azores subtropical high and is particularly strong at an altitude of about 3000 m (9800 ft).

Compressional warming causes air subsiding from aloft to become warmer than the underlying layer of maritime tropical air. A zone of transition, that is, a temperature inversion, develops between the two layers of air (Figure 10.18). An air layer characterized by a temperature inversion is extremely stable, so the two air layers do not mix and convection is confined to the surface layer of *mT* air. In this circumstance, the inversion is known as a **capping inversion**. For as long as this stratification persists, the contrast between air layers increases; that is, subsiding air becomes drier while the underlying *mT* air becomes more humid. The potential for severe weather continues to grow; all that is needed is a trigger that causes updrafts to penetrate the capping inversion.

The necessary upward impetus may be supplied by a combination of the intense solar heating of mid-afternoon and the lifting of air caused by an approaching cold front or jet streak. Updrafts eventually break through the

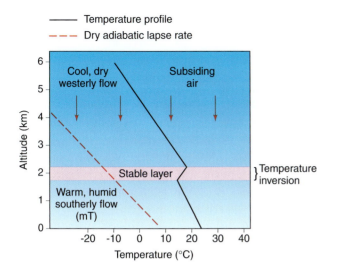

FIGURE 10.18
A temperature sounding that favors the development of severe thunderstorm cells. A capping temperature inversion separates subsiding dry air aloft from warm, humid air near the surface.

capping inversion, and cumulus clouds billow upward at explosive speeds that may exceed 100 km per hr (62 mph). A severe thunderstorm is born.

THUNDERSTORM HAZARDS

Thunderstorm hazards include lightning, downbursts, flash floods, hail, and tornadoes. A convective rain or snow shower is a thunderstorm if accompanied by lightning. **Lightning** is a brilliant flash of light produced by an electrical discharge within a cumulonimbus cloud or between the cloud and Earth's surface (Figure 10.19). According to NOAA's National Weather Service, an

FIGURE 10.19
Lightning is a brilliant flash of light associated with an electrical discharge between clouds and Earth's surface, within a cloud, or between clouds. [NOAA photograph]

estimated 25 million cloud-to-ground lightning flashes occur in the U.S. each year. Lightning is a weather phenomenon that can be directly hazardous to human life. Over a recent 30-year period (1971-2000), lightning killed an average of 73 people annually in the United States. According to NOAA statistics, in 2003 more people were killed by lightning than by tornadoes and hurricanes combined.

Lightning not only kills and injures people, but also ignites forest and brush fires. According to Stephen J. Pyne, an ecologist and fire expert at Arizona State University, the American Southwest leads the nation in the average number of forest fires ignited by lightning (60% to 70% of the total) and the number of hectares burned. A person might assume that heavy rain falling from a thunderstorm would quickly quench a lightning-ignited fire. In the western basins, however, the base of a cumulonimbus cloud is usually so far above the ground (perhaps 2400 m or 8000 ft), and the air below the cloud is so dry, that much if not all the rain evaporates before reaching a wildfire.

Where there is lightning, there is thunder, although sometimes we see distant lightning but cannot hear the thunder. Lightning heats the air along a narrow conducting path to temperatures that may exceed 25,000 °C (45,000 °F). Such intense heating occurs so rapidly that air density cannot initially respond. The rapid rise in air temperature is accompanied by a tremendous increase in air pressure locally, which generates a shock wave. The shock wave propagates outward, producing sound waves that are heard as **thunder**.

Severe, and sometimes not so severe, thunderstorms can produce a **downburst**, an exceptionally strong downdraft that, upon striking Earth's surface, diverges horizontally as a surge of potentially destructive winds. Downbursts occur with or without rain. T. Theodore Fujita (1920-1998) of The University of Chicago is credited with discovering downbursts, and he coined the term. Fujita's discovery stemmed from his airborne survey of property damage near Beckley, WV, shortly after the tornado *super-outbreak* of 3-4 April 1974. He observed that storm debris was spread over the countryside in a starburst pattern—distinct from the swirling pattern that is typical of tornado damage. Downbursts blow down trees, flatten crops, and wreck buildings.

A squall line sometimes produces a family of straight-line downburst winds that impacts a path that may be hundreds of kilometers long. This windstorm is known as a **derecho** (Spanish for *straight* as contrasted with tornado, Spanish for *turning*). The founding director of the Iowa weather service Gustavus Hinrichs (1836-1923) coined the term in 1888. The criterion for a derecho (pronounced deh-RAY-cho), as opposed to ordinary gusty thunderstorm winds, is sustained winds in excess of 94 km per hr (58 mph). At any location, the period of strongest winds typically lasts only 10 to 20 minutes or so and the system generally tracks from northwest to southeast. Derechos develop mostly from May through August with peak frequency in July. Infrequently, a mesoscale convective complex (MCC) produces a derecho.

On 4 July 1977, a derecho consisting of 25 individual downbursts struck several counties in northern Wisconsin, felling trees and destroying buildings along a path 268 km (166 mi) long, 27 km (17 mi) wide. One person was killed, 35 were injured, and damage to buildings and timber totaled in the millions of dollars. Based on the extent of damage, surface winds may have been as high as 250 km per hr (155 mph). Twenty-two years later, on 4 July 1999, a derecho struck Minnesota's Boundary Waters Canoe Area Wilderness. In about 20 minutes, winds as high as 150 km per hr (93 mph) blew down more than 20 million trees along a path measuring 60 km (37 mi) long and 20 km (12 mi) wide. Fortunately, there were no fatalities.

A **flash flood** is a short-term, localized, and often unexpected rise in stream level above bank full usually in response to torrential rain falling over a relatively small geographical area. Typically, the stream level rises and falls within 6 hrs of the rain event. Excessive rainfall may occur when successive thunderstorm cells (in a squall line or an MCC) mature over the same area. Alternatively, a stationary or slow-moving intense thunderstorm cell, which is embedded in weak steering winds aloft and/or maintained by a persistent flow of humid air up the slopes of a mountain range, may produce flooding rains.

Atmospheric conditions that favor flash floods differ somewhat from those that give rise to other types of severe weather (e.g., hail and tornadoes). Flash-flood producing thunderstorms are more common at night and form in an atmosphere with weak vertical wind shear and abundant moisture through great depths. Flash flooding is most likely in an atmosphere that is *precipitation efficient* with high values of precipitable water and relative humidity, and a thunderstorm cloud base having temperatures above freezing. The combination of high precipitable water and relative humidity reduces the amount of falling precipitation that vaporizes. A relatively warm cloud base favors the collision-coalescence precipitation process that leads to exceptionally heavy rainfall.

FIGURE 10.20
Road sign in the Colorado Rockies warns visitors to climb to higher ground in the event of a flash flood.

Flash flooding is especially hazardous in mountainous terrain, and motorists and campers are well advised to head for higher ground in the event of a flash flood warning (Figure 10.20). But even where the topography is relatively flat, a prolonged period of heavy rain (more than 7.6 mm, or 0.3 in., per hr) can greatly exceed the infiltration capacity of the ground. Simply put, the ground cannot absorb all the rainwater and becomes saturated. Excess water runs off to creeks, streams, rivers, or sewers or collects in other low-lying areas. If the drainage system cannot accommodate the sudden input of huge quantities of water, a flash flood is the consequence.

Because of their design and composition, urban areas are prone to flash floods during intense downpours. Concrete and asphalt make the surfaces of a city virtually impervious to water, so elaborate storm sewer systems are required to transport runoff to nearby natural drainageways. Storm sewer systems have a limited capacity for water, however, and may be unable to accommodate the excess runoff produced during a torrential rainfall. Water collects in underpasses, at intersections, in dips in the roadway, and in other low-lying areas. Sometimes water levels rise so quickly in these areas that motorists are trapped in their vehicles. In many cases, motorists misjudge the depth of the water and do not realize the power of even shallow water in motion. They unwittingly drive into water that sweeps their vehicle downstream and puts their lives in extreme danger.

Hail is frozen precipitation in the form of balls or lumps of ice more than 5 mm (0.2 in.) in diameter, called *hailstones* (Figure 10.21). The size of hailstones is usually reported to the National Weather Service by comparison to the size of an everyday object. Hailstones range in size

FIGURE 10.21
Hailstones are lumps of ice that fall from intense thunderstorm cells.

from a pea to an orange or larger. Hail is described as *severe* if its diameter is equal to or greater than 19 mm (0.75 in.), the size of a penny. *Extreme hail* has a diameter equal to or greater than 51 mm (2 in.) and accounts for less than 10% of all hail reports. In the United States, the largest hailstone on record fell at Aurora, NE, on 22 June 2003. It weighed just under 455 g (1.0 lb) and measured 47.6 cm (18.75 in.) in circumference and 17.8 cm (7 in.) in diameter, about the size of a cantaloupe.

Hail almost always falls from cumulonimbus clouds characterized by strong updrafts, great vertical development, and abundant supercooled water droplets. A hailstone develops when an ice pellet is transported vertically through portions of a cumulonimbus cloud containing varying concentrations of supercooled water droplets. The ice pellet may descend slowly through the entire cloud, or it may follow a more complex pattern of ascent and descent as it is caught alternately in updrafts and downdrafts. Along the way, the ice pellet grows by accretion (addition) of freezing water droplets. In general, the stronger the updraft, the greater is the size of the ice pellet. For example, an updraft of about 60 km per hr (37 mph) is required to produce a hailstone 1.9 cm (0.75 in.) in diameter

whereas an updraft of roughly 160 km per hr (100 mph) is needed for a 7.6 cm (3.0 in.) diameter hailstone. Eventually, when it becomes too large and heavy to be supported by updrafts, the ice pellet descends and falls out of the cloud base. If the ice does not melt completely during its journey through the above freezing air beneath the cloud, it reaches Earth's surface as a hailstone.

In the U.S. each year, hail causes an average $1 billion in damage. Large hailstones smash windows, dent automobiles, and damage the roof and siding of buildings. Small hailstones are often more costly to crops than large hailstones because there are more of them spread over a larger area. Hail usually falls during the growing season and in a matter of minutes can wipe out the fruits of a farmer's year of labor. On 11 July 1990, softball-sized hailstones in Denver caused $625 million in property damage, mostly to automobiles and roofs. On 5 May 1995, the most costly hailstorm in U.S. history struck the Fort Worth, TX area. Golf ball- to baseball-size hailstones injured scores of people who were caught in the open during the storm. Total property damage was estimated at more than $2 billion.

According to NOAA's Storm Prediction Center (SPC) in Norman, OK, the annual number of severe hail reports in the contiguous United States increased exponentially from less than 350 in 1955 to more than 17,760 in 2008 (Figure 10.22). The upward trend likely can be attributed to greater public awareness of the hail hazard, proliferation of cell phones making it easier to file reports, improved capabilities of weather radar (especially the *WSR-88D*), and the National Weather Service's more aggressive verification of hail reports.

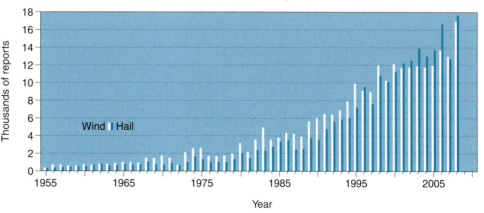

Severe wind and hail reports in U.S.

FIGURE 10.22
Annual reports of severe hail and wind occurrences in the contiguous United States have trended rapidly upward since 1955. [NOAA, NCEP, Storm Prediction Center]

In the contiguous U.S., severe hail is most likely in a belt stretching northward from south-central Texas to the Valley of the Red River in North Dakota and Minnesota, roughly corresponding to *tornado alley* (described later in this chapter). This locale of large hail is consistent with the region's relatively high frequency of supercell thunderstorms. Other areas of enhanced hail activity include a band from northeast Texas eastward to North Carolina plus northeast Colorado. The principal site of hail activity shifts seasonally from the Southeast in winter to Oklahoma by April and then to eastern Colorado and northern Minnesota by July.

Tornadoes

Of the roughly 10,000 severe thunderstorms that occur in the United States in an average year, about 10% produce tornadoes. A **tornado** is a violent rotating column of air in contact with the ground, usually associated with a thunderstorm. The system often (but not always) is made visible by water droplets formed by condensation and/or by dust and debris that are drawn into the tornado (Figure 10.23).

Tornadoes are the most violent of all weather systems. Fortunately, most are small and short-lived and often strike sparsely populated regions. Occasionally, however, a major tornado outbreak causes incredible

FIGURE 10.23
A tornado is a small-scale weather system that has the potential of taking many lives and causing considerable property damage. Photo shows the famous Elkhart, IN "double tornado" which killed 36 people on Palm Sunday 1965. It was an F5 tornado.

devastation, death, and injury. Over a 16-hr period on 3-4 April 1974, 148 tornadoes struck 13 states in the east central United States, the most widespread and costly outbreak of tornadoes in the nation's history. This super outbreak left 315 people dead, 6142 people injured, and caused property damage in excess of $600 million. The most prolific tornado outbreak on record occurred over the Great Plains and Midwest on 29-30 May 2004. More than 180 tornadoes were reported in a 24-hr period. In early May 1999, an outbreak of particularly intense tornadoes hit portions of Oklahoma, Kansas, Texas, and Tennessee. Oklahoma City, OK, was hardest hit. Fifty-five people lost their lives and property damage topped $1.1 billion.

TORNADO CHARACTERISTICS

Probably the most striking characteristic of a tornado is a localized lowering of cloud base into a tapered column composed of tiny water droplets. If this vortex remains aloft, it is called a **funnel cloud**, but if it reaches the ground, the system is classified as a tornado. Although often funnel-shaped, in fact, tornadoes take a variety of forms, ranging from cylindrical masses of roughly uniform dimensions in cross-section to long and slender, ropelike pendants. A funnel cloud forms in response to the steep air pressure gradient directed from the tornado's outer edge toward its center. Humid air expands and cools as it is drawn inward toward the center of the system and water vapor condenses into cloud droplets. But if the air were exceptionally dry, a funnel cloud may not form, and the tornado is made visible by a whirl of dust and debris lifted off the ground.

A funnel cloud may or may not develop into a tornado. In fact, in some cases, a whirl of dust or debris that appears on the ground signals the formation of a tornado prior to the appearance of a funnel cloud. Furthermore, the actual tornadic circulation covers a much wider area than is suggested by the funnel cloud. Typically, the diameter of a funnel cloud is only about one-tenth the diameter of the associated tornadic circulation in the storm.

A weak tornado's path on the ground typically is less than 1.6 km (1 mi) long and 100 m (330 ft) wide, and the system has a life expectancy of only a few minutes. Wind speeds are less than 180 km per hr (110 mph). Weak tornadoes account for about three-quarters of all tornadoes, but were responsible for only about 7% of all tornado fatalities from 1999-2008. At the other extreme, a violent tornado can cause damage along a path more than 160 km (100 mi) long and 1.0 km (3000 ft) wide, and the lifetime of the system may be 10 minutes to more than 2 hrs. One of the broadest tornadoes on record struck Hallam, NE, on 22 May 2004; its diameter was about 4 km (2.5 mi). Based on indirect measurements, wind speeds in violent tornadoes range up to 500 km per hr (300 mph).

The deadliest tornado in North American history was the Tri-State tornado of 18 March 1925. Traveling at a maximum forward speed of 118 km per hr (73 mph), the system persisted for 3.5 hrs and produced a 353-km (219-mi) path of devastation from southeastern Missouri through the southern tip of Illinois into southwest Indiana. The tornado caused 695 fatalities and 2000 injuries, and 11,000 people were made homeless. (Actually, more than one tornado may have contributed to the Tri-State disaster.) That same day, seven other tornadoes claimed an additional 97 lives across Kentucky and Tennessee.

Most tornadoes are spawned by and travel with intense thunderstorm cells. Tornadoes and their parent cells usually track from southwest to northeast, but any direction is possible. Tornado trajectories are often erratic; many tornadoes causing a hopscotch pattern of destruction as they alternately touch down and lift off the ground. Tornadoes have been known to move in circles and even to create figure eights. Average forward speed is around 48 km per hr (30 mph), although they can reach nearly 120 km per hr (75 mph), as happened in the 1925 Tri-State tornado.

An exceptionally great horizontal air pressure gradient is responsible for a tornado's vigorous circulation. The air pressure drop over a horizontal distance of only 100 m (330 ft) may equal the normal air pressure drop between sea level and an altitude of 1000 m (3300 ft), that is, a reduction of about 10%. The continually changing direction of the inward-directed pressure gradient force produces the centripetal force that maintains the rotation of the air column about a vertical axis. Most Northern Hemisphere tornadoes rotate in a counterclockwise direction (viewed from above); only about 5% rotate clockwise. While the Coriolis Effect has a negligible influence at the small scale of a funnel cloud, the counterclockwise bias is inherited from the larger parent thunderstorm cell.

TORNADO HAZARDS

Tornadoes threaten people and property because of extremely strong winds, a vigorous updraft, subsidiary vortices, and an abrupt drop in air pressure. Winds that may reach hundreds of kilometers per hour blow down trees, power poles, buildings, and other structures. Flying debris causes much of the death and injury associated with tornadoes. Shards of broken glass, splintered lumber, and even vehicles become lethal projectiles. In violent tornadoes, the updraft near the center of the storm's funnel is sometimes

strong enough to lift a house from its foundation. Some tornadoes consist of two or more *subsidiary vortices* orbiting about each other or a common center within a massive tornado. These **multi-vortex tornadoes** are usually the most destructive of all tornadoes.

In 1971, T.T. Fujita devised a six-level intensity scale for rating tornado strength and damage to structures. The **F-scale**, ranging from F0 to F5, was based on rotational wind speeds estimated from property damage (not measured). To more closely align estimated wind speeds with associated storm damage, the F-scale was revised and became operational in February 2007 as the **Enhanced F-scale**, ranging from EF0 to EF5 (Table 10.1). An EF0 tornado produces minor damage, snapping twigs and small branches and breaking some windows. EF1 and EF2 tornadoes can cause moderate to considerable property damage and even take lives. EF1 tornadoes can down trees and shift mobile homes off their foundations, and an EF2 tornado can rip roofs off frame houses, demolish mobile homes, and uproot large trees. An EF3 tornado can partially destroy even well constructed buildings and lift motor vehicles off the ground. At the violent end of the EF-scale, destruction is described as "devastating" to "incredible," with the potential for many fatalities. An EF4 tornado can level sturdy buildings and other structures and toss automobiles about like toys. In an EF5 tornado, sturdy frame houses are lifted and transported some distance before disintegrating.

Fortunately, EF5 tornadoes are rare. Of the nearly 1300 tornadoes that strike the United States in an average year, perhaps only one will be rated EF5. Since 1950 in the contiguous U.S., only 52 tornadoes rated F5 or EF5. When one does occur, however, the impact can be catastrophic. In about 1 minute, an EF5 tornado leveled the village of Barneveld, WI, on 8 June 1984; 100 homes

TABLE 10.1
The Enhanced Fujita Tornado Intensity Scale

EF-Scale	Damage	3 Second Wind Gust km/hr	mph
0	light	105-137	65-85
1	moderate	138-178	86-110
2	considerable	179-218	111-135
3	severe	219-266	136-165
4	devastating	267-322	166-200
5	incredible	> 322	> 200

were totally destroyed and 9 lives were lost. The Xenia, OH, tornado, which was part of the super outbreak of 3-4 April 1974, rated F5 over a portion of its 51-km (32-mi) path and claimed 34 lives.

According to NOAA's National Climatic Data Center, about 77% of tornadoes in the U.S. are considered weak (EF0 or EF1) and about 95% are below EF3 intensity. Only 0.1% of all tornadoes reach EF5 status. The few violent systems are responsible for the majority of all fatalities, however. Between 1999 and 2008, the average annual number of fatalities from tornadoes was 63, with 42% reported in April and May. In an average year, tornadoes injure about 1000 people.

WHERE AND WHEN

The central United States is one of only a few places in the world where synoptic weather conditions and terrain are ideal for tornado development; interior Australia is another. Although tornadoes have been reported in all 50 states and throughout southern Canada, most occur in **tornado alley**, a north-south corridor stretching from eastern Texas and the Texas Panhandle northward through Oklahoma, Kansas, Nebraska, and into southeastern South Dakota. Kansas and Oklahoma have the highest annual incidence of significant tornadoes (EF2 or higher) per unit

area, whereas local tornado frequency maxima occur from central Iowa eastward to central Indiana, and along the Gulf Coast. In terms of average annual number of tornadoes per 10,000 square miles by state, Kansas is the highest with 13.3 for the period 1999-2008 (Figure 10.24).

Concentration of tornadoes over the central Plains is sometimes taken to imply that tornadoes do not occur in regions of great topographical relief. Actually, strong to violent tornadoes are largely unaffected by terrain; there are numerous examples of such tornadoes traversing rugged and even mountainous topography. On the other hand, weak tornadoes are more common over flat rather than rough terrain.

A national record was set in 2004 with 1817 tornadoes reported but fatalities totaled only 36. Slightly more than half of all tornadoes develop during the warmest hours of the day (10 a.m. to 6 p.m., local time), and almost three-quarters of tornadoes in the United States occur from March to July. The months of peak tornado activity are May and June when atmospheric conditions are optimal for vigorous convection and the severe thunderstorms that spawn tornadoes. As of this writing, the monthly record for greatest number of tornadoes was set in May 2003 with 543 reported; the previous monthly record was 399 in June 1992.

One factor that contributes to the spring peak in tornado frequency is the relative instability of the lower atmosphere at that time of year. During the transition from winter to summer, daylight lengthens and solar radiation incident on Earth's surface becomes more intense. Heat is transported from the relatively warm ground into the cold troposphere, but it takes time for the temperature of the entire troposphere to adjust to heating from below. The upper troposphere, in fact, usually retains its winter-like cold well into spring. The steep air temperature lapse rate (warm at low levels and cold aloft) favors deep convection and severe thunderstorms.

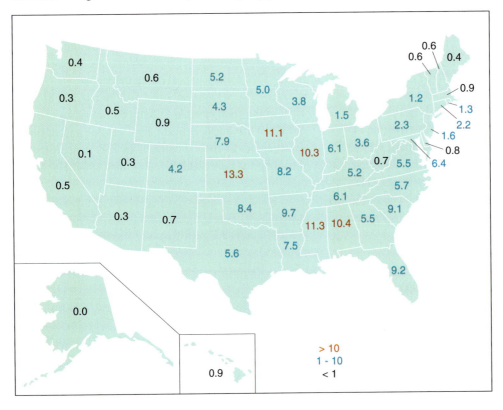

FIGURE 10.24
Average annual number of tornadoes per 10,000 square miles by state, 1999-2008. [NOAA/NWS/SPC]

Another factor that contributes to the spring tornado maximum is the greater likelihood that favorable synoptic weather conditions will occur at that time of year. Recall from earlier in this chapter that severe thunderstorms typically develop in the warm southeast sector of a strong extratropical cyclone. Such cyclones achieve their greatest intensity when sharp north-south temperature contrasts develop across the nation, that is, in spring when the polar front is well defined.

From late winter through spring, tornado occurrences progress northward. In effect, the center of maximum tornado frequency follows the Sun, as do the mid-latitude jet stream, the principal storm tracks, and northward incursions of maritime tropical air. By late February, maximum tornado frequency, on average, is along the central Gulf States. In April, the maximum frequency shifts to the southeast Atlantic States; in May and June, the highest tornado incidence is usually over the southern Plains, and by early summer, it moves into the northern Plains, the Prairie Provinces east of the Rockies, and the Great Lakes region. From late summer through autumn, tornado occurrence shifts southward. During spring, most tornadoes travel from southwest to northeast, steered by southwesterly winds in the middle and upper troposphere, but in summer and into fall, many tornadoes travel from northwest toward the southeast as the steering winds shift.

Tropical Storms and Hurricanes

Hurricane is likely derived from *Haracan*, the name of the storm god of the Taino people who inhabited Caribbean islands at the time of Spanish exploration of the New World. A **hurricane** is an intense tropical cyclone that originates over warm ocean waters, usually in late summer or early fall (when sea-surface temperatures are highest), and has a maximum sustained wind speed of at least 119 km per hr (74 mph). By convention in the United States, *sustained wind speed* is a one-minute average measured at the standard anemometer height of 10 m (33 ft).

A hurricane (Figure 10.25A) develops in a uniformly warm and humid air mass. Typically, the central pressure at sea level is considerably lower and the horizontal air pressure gradient is much greater in a hurricane than an extratropical cyclone (Chapter 7). A hurricane is usually a much smaller system, averaging a third the diameter of a typical extratropical cyclone. Rarely do hurricane-force winds extend much more than 120 km (75 mi) beyond the system's center. The circulation in a hurricane weakens rapidly with increasing altitude and usually becomes anticyclonic (clockwise in the Northern Hemisphere) in the upper troposphere at altitudes above about 12,000 m (40,000 ft).

At the center of a hurricane is an area called the **eye** of the storm (Figure 10.25B) with almost cloudless skies, subsiding air, and light winds (less than 25 km per

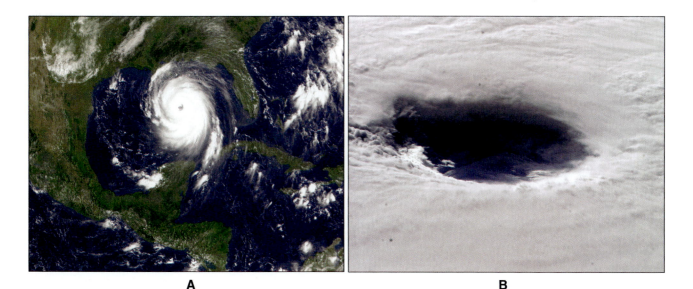

| A | B |

FIGURE 10.25
NOAA satellite image of (A) Hurricane Katrina on 28 August 2005, at 11:45 a.m., EDT, as the powerful storm churned in the Gulf of Mexico as a category 5 storm with sustained winds near 175 mph, a day before the storm made landfall on the U.S. Gulf Coast. [Photo courtesy of NOAA]. NASA photo shows (B) the eye of Hurricane Ivan, one of the strongest hurricanes on record, as the storm approached landfall on the central Gulf coast on September 15, 2004. The hurricane was photographed by astronaut Edward M. (Mike) Fincke from aboard the International Space Station. [Photo courtesy NASA]

hr or 16 mph). . The eye generally ranges from 10 to 65 km (6 to 40 mi) across, shrinking in diameter as the hurricane intensifies and winds strengthen. At a hurricane's typical rate of forward motion, the eye may take up to an hour to pass over a given locality. People are sometimes lulled into thinking the storm has ended when skies clear and winds abruptly weaken following a hurricane's initial blow. They may be experiencing passage of the hurricane's eye; heavy rains and ferocious winds soon will resume but blow from the opposite direction.

Bordering the eye of a hurricane is the **eyewall**, a ring of thunderstorm (cumulonimbus) clouds that produce heavy rains and very strong winds. The most dangerous and potentially most destructive part of a hurricane is the portion of the eyewall on the side of the advancing system where the wind blows in the same direction as the storm's forward motion. On that side, hurricane winds combine with the storm's forward motion to produce the system's strongest surface winds. In the Northern Hemisphere, this dangerous semicircle of high winds and high ocean waves is on the right side of the hurricane when facing in the direction of the system's forward movement. Cloud bands, producing heavy convective showers and hurricane-force winds, spiral inward toward the eyewall.

HURRICANE HAZARDS

Potentially the most devastating impact of tropical cyclones on the coastal zone is ocean water driven ashore by strong onshore winds (blowing from sea to land) associated with an intense storm system centered over or near the ocean. Strong winds (coupled with low

air pressure) pile up a dome of seawater 80 to 160 km (50 to 100 mi) wide that sweeps across the coastline bringing floodwaters inland that can take lives and cause considerable property damage (Figure 10.26). A **storm surge** can erode beaches, wash over barrier islands, wash out roads and railway beds, and demolish marinas, piers, cottages, and other coastal structures.

Other hazards of hurricanes are heavy rain and inland flooding, strong winds, and tornadoes. Hurricanes and tropical storms can produce very heavy rainfall with amounts often in the range of 13 to 25 cm (5 to 10 in.). Even if the system tracks well inland, heavy rains typically persist and may trigger costly flooding. According to research conducted by scientists at NOAA's Tropical Prediction Center/National Hurricane Center in Miami, FL, from 1970 to 1999, 60% of the 600 U.S. deaths attributed to tropical cyclones or their remnants was from fresh water flooding. In those three decades, far more people (351) died from inland flooding than from coastal storm surge flooding (only 6 fatalities). On the other hand, the majority of the more than 1300 fatalities associated with Hurricane Katrina in August 2005, were caused by the storm surge. The storm surge remains the most serious potential impact of a landfalling hurricane and is the primary reason people are evacuated from low-lying coastal areas.

From 1970 to 1999, about 12% of tropical cyclone fatalities were wind related. Strengthening winds exert a dramatically increasing pressure on building walls and infrastructure. Debris carried by the wind is hurled against these structures, adding to the potential damage. Fortunate-

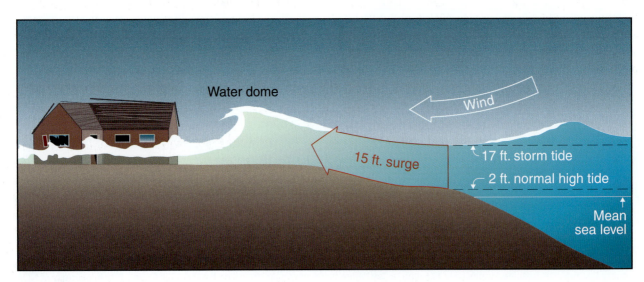

FIGURE 10.26
A storm surge is a dome of seawater 80 to 160 km (50 to 100 mi) wide that is driven onshore by strong winds associated with a tropical cyclone. The storm tide is the sum of the storm surge and the prevailing astronomical tide. [Adapted from NOAA]

ly, hurricane winds diminish rapidly once the system makes landfall, so most wind damage is confined to within 200 km (125 mi) of the coastline. Two factors account for the abrupt drop in wind speed: a hurricane over land is no longer in contact with its energy source, warm ocean water, and the frictional resistance of the rougher land surface slows the wind and shifts it toward the low-pressure center of the system. This wind shift causes the storm to begin to fill; that is, its central pressure rises, the horizontal air pressure gradient weakens, and winds slacken.

Although wind speed decreases once a hurricane makes landfall, the system may produce tornadoes. Typically, only a few tornadoes occur with a hurricane and most of these are weak, but in 1967, Hurricane Beulah reportedly spawned as many as 115 tornadoes across southern Texas. Tornadoes are most probable after the hurricane enters the westerly steering current and curves towards the north and northeast; they form mostly to the northeast of the storm center and often outside the region of hurricane-force winds.

In the early 1970s, H.S. Saffir (1917-2007), a consulting engineer, and R.H. Simpson, former director of the National Hurricane Center, designed a rating system for hurricanes known as the **Saffir-Simpson Hurricane Intensity Scale** (Table 10.2). Hurricanes are rated from category 1 to 5 corresponding to increasing intensity. The scale, first included in hurricane advisories in 1975, provides an estimate of potential coastal flooding and property damage from a hurricane landfall. Wind speed is the primary determining factor for a hurricane's rating on the Saffir-Simpson scale as storm surge heights are highly dependent on bathymetry and other factors in the region of landfall. Each intensity category specifies a range of central air pressure, a range of maximum sustained wind speed, storm surge potential, and the potential for property damage.

Of the 185 hurricanes that struck the U.S. Atlantic or Gulf Coasts between 1901 and 2008, 69 (about 37.3%) were classified as major; that is, they rated category 3 or higher on the Saffir-Simpson scale. Property damage potential rises rapidly with ranking on the Saffir-Simpson scale. In fact, destruction from a category 4 or 5 hurricane can be 100 to 300 times greater than that caused by a category 1 hurricane. The 25 major hurricanes that made landfall along the Gulf or Atlantic Coasts between 1949 and 1990 accounted for three-quarters of all property damage from all landfalling tropical storms and hurricanes during the same period.

WHERE AND WHEN

Three conditions are necessary for a tropical cyclone to form: high sea-surface temperature, adequate Coriolis Effect, and weak winds aloft. To a large extent, these requirements dictate the climatology of tropical cyclones (tropical storms and hurricanes), that is, where and when tropical cyclones are most likely to develop.

Tropical cyclone formation requires a sea-surface temperature (SST) of at least 26.5 °C (80 °F) through an ocean depth of about 45 m (150 ft) or more. Such exceptionally warm ocean waters sustain the system's circulation by the latent heat released when water vapor, evaporated from the ocean surface, is conveyed upward and condenses within the storm system. Temperature largely governs the rate of evaporation of water, so the higher the SST, the greater the supply of latent heat for the storm system. Furthermore, the spray from breaking ocean waves readily evaporates, adding to the supply of water vapor that condenses and releases latent heat in the developing tropical cyclone.

As a tropical cyclone makes landfall or moves over colder water, however, it loses its warm-water energy source and weakens. The strong winds of a tropical

TABLE 10.2
Saffir-Simpson Hurricane Intensity Scale

Category	Central pressure mb (in.)	Wind speed km/hr (mph)	Storm surge m (ft)	Damage potential
1	≥ 980 (≥ 28.94)	119-154 (74-95)	1-2 (4-5)	*Minimal*
2	965-979 (28.50-28.91)	155-178 (96-110)	2-3 (6-8)	*Moderate*
3	945-964 (27.91-28.47)	179-210 (111-130)	3-4 (9-12)	*Extensive*
4	920-944 (27.17-27.88)	211-250 (131-155)	4-6 (13-18)	*Extreme*
5	< 920 (< 27.17)	> 250 (> 155)	> 6 (> 18)	*Catastrophic*

cyclone stir up surface ocean waters, which induce Ekman transport and divergence of surface waters under the storm system, bringing cold water to the surface. Until the normal thermal structure of the ocean is restored, lower than usual sea-surface temperatures can inhibit development of subsequent tropical cyclones over the same region of the ocean.

The second condition required for tropical cyclone development is a significant Coriolis Effect; that is, the influence of Earth's rotation must be sufficiently strong to initiate a cyclonic circulation. As discussed in Chapter 6, the Coriolis Effect weakens toward lower latitudes and is zero at the equator. With very rare exceptions, tropical cyclones do not form within 5 degrees of the equator (a distance of about 480 km or 300 mi).

The first two conditions that favor tropical cyclone formation (i.e., high SST and sufficient Coriolis Effect) occur only over certain portions of the ocean. The main ocean breeding grounds for tropical cyclones along with average storm trajectories are plotted in Figure 10.27. Most hurricanes form in the 8- to 20-degree latitude belts. Major breeding grounds are: (1) the tropical North Atlantic west of Africa (including the Caribbean Sea and Gulf of Mexico), (2) the North Pacific Ocean west of Mexico, (3) the western tropical North Pacific and China Sea, where a hurricane is called a *typhoon* (from the Cantonese *tai-fung*, meaning great wind), (4) the South Indian Ocean east of Madagascar, (5) the North Indian Ocean (including the Arabian Sea and Bay of Bengal), and (6) the South Pacific Ocean from the east coast of Australia eastward to about 140 degrees W. In the Indian Ocean and near Australia, hurricanes are called *cyclones*.

The requirement of high sea-surface temperatures explains the seasonal occurrence of tropical cyclones. For reasons presented in Chapter 4, the temperature of surface ocean waters lags the regular seasonal variation in incoming solar radiation. Sea-surface temperatures do not reach their seasonal maximum until roughly 6 to 8 weeks after the date of most intense solar radiation. Most Atlantic hurricanes develop when surface waters are warmest, that is, in late summer and early autumn; the official hurricane season runs from 1 June to 30 November, with the peak hurricane threat for the U.S. coastline between mid-August and late October.

The third condition for tropical cyclone development is relatively weak winds in the middle and upper troposphere over oceanic breeding grounds. Weak winds aloft allow a cluster of cumulonimbus clouds to organize over tropical seas, the initial stage in the evolution of a hurricane. By contrast, strong west-to-east winds aloft shear off the tops of westward tracking thunderstorms, preventing the systems from building vertically and organizing. Strong vertical *wind shear* is the principal reason why hurricanes rarely form off the east or west coasts of South America (although Caribbean hurricanes occasionally impact the north coast of Venezuela).

According to NOAA's National Climatic Data Center's *International Best Track Archive for Climate Stewardship (IBTrACS)*, a worldwide average of about 80 named tropical cyclones develops each year and approximately one-half reach hurricane strength (based on the period 1958-2007). The western Pacific Ocean, with its vast expanse of warm surface waters, is the most

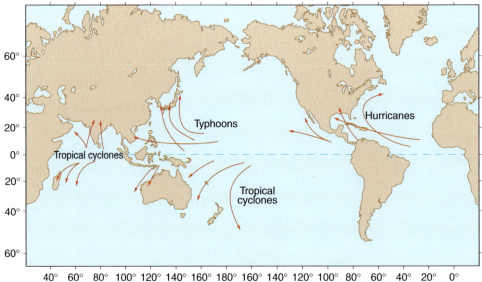

FIGURE 10.27
Tropical cyclone breeding grounds are located only over certain regions of the world ocean. Arrows indicate average hurricane trajectories.

active region for tropical cyclones, with an average of roughly 27 systems forming each season of which about 17 intensify into typhoons.

Hurricanes spawned over the tropical Atlantic, Caribbean Sea, and Gulf of Mexico pose the most serious threat to coastal North America. According to the Tropical Prediction Center/National Hurricane Center, based on the period 1931-2008, a seasonal average 10.6 named tropical storms form over these waters. Of these systems, on average about 6 intensify into hurricanes and of these 2.5 become major hurricanes. On average, two hurricanes strike the U.S. coast each year. The annual distribution of the 1245 tropical cyclones that reached at least tropical storm strength and the 738 reaching hurricane strength during the period 1870-2006 in the tropical North Atlantic are plotted in Figure 10.28.

The 2005 Atlantic hurricane season set a new record with 27 named tropical cyclones (and one unnamed subtropical cyclone); the previous record was 21 set in 1933. In 2005, 15 Atlantic tropical cyclones became hurricanes (a new record) and 4 major hurricanes struck the U.S. (also a record). The 1995 Atlantic hurricane season was the third most active in recorded history with 19 tropical storms, 11 of which attained hurricane strength. However, the annual number of hurricanes may have little bearing on the number of landfalling hurricanes and their impact. Whereas only 4 hurricanes occurred during the 1992 hurricane season, one of them,

Hurricane Andrew, was the second most costly in terms of property damage in U.S. history, amounting to $34.3 billion (in year 2000 dollars).

Hurricanes have hit every Atlantic and Gulf Coast state from Texas to Maine. Florida is the most hurricane-prone of all the states, with 66 hurricanes crossing its coastline between 1900 and 2008. In the same period, Texas was second with 41 hurricanes, while Louisiana had 31, and North Carolina had 29.

Some scientists attribute the recent upswing in Atlantic hurricane activity to global climate change and higher SST. However, reliable records of Atlantic hurricane frequency date back only about 130 years. To obtain a longer term perspective on the level of Atlantic hurricane activity, researchers have turned to a variety of proxy climate data. For more on this, refer to this chapter's second Essay.

The primary breeding ground for hurricanes in the Atlantic shifts east and west with the seasons. Early in the hurricane season (May and June), hurricanes form mostly over the Gulf of Mexico and the western Caribbean. By July, the main area of hurricane development begins to shift eastward across the tropical North Atlantic. By mid-September, most hurricanes form in a belt stretching from the Lesser Antilles (in the eastern Caribbean Sea) eastward to south of the Cape Verde Islands (off Africa's West Coast). After mid-September, hurricanes again originate mostly over the Gulf of Mexico and the western Caribbean.

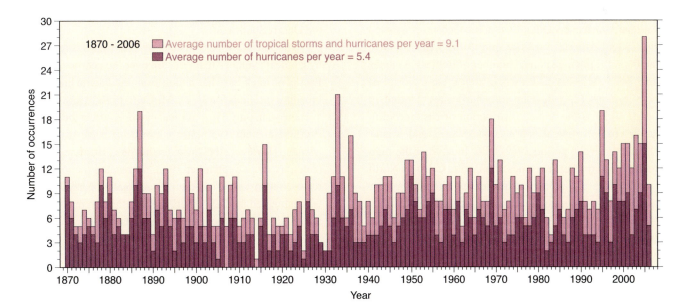

FIGURE 10.28
Annual distribution of the 1245 North Atlantic tropical cyclones reaching at least tropical storm strength (blue) and the 738 reaching hurricane strength (red), during the period 1870-2006. [From Historical Climatology Series 6-2, *Tropical Cyclones of the North Atlantic Ocean, 1851-2006*, C.J. McAdie et al., National Hurricane Center, and G.R. Hammer, National Climatic Data Center, Asheville, NC, July 2009, p. 19]

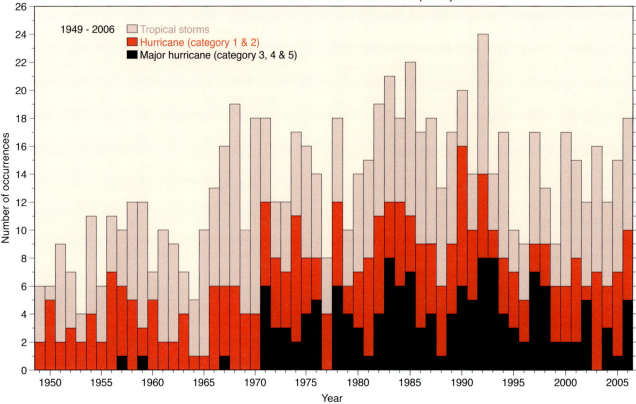

FIGURE 10.29
Annual distribution of eastern North Pacific basin tropical storms, hurricanes, and major hurricanes, during the period 1949-2006. [From Historical Climatology Series 6-5, *Tropical Cyclones of the Eastern North Pacific Basin, 1949-2006*, E.S. Blake et al., National Hurricane Center, and G.R. Hammer, National Climatic Data Center, Asheville, NC, June 2009, p. 15]

According to the Tropical Prediction Center/ National Hurricane Center, the Pacific Ocean off Mexico and Central America ranks second to the western North Pacific in the average annual number of tropical cyclones (15.2 per year between 1966 and 2008); the majority of these systems (8.2) develop into hurricanes. Plotted in Figure 10.29 is the annual distribution of eastern North Pacific basin tropical storms, hurricanes and major hurricanes for the period 1949-2006. In spite of the high frequency of tropical cyclones, the Pacific Coast of North America is rarely a target of hurricanes, although one or two tropical cyclones typically make landfall on Mexico's Pacific Coast each year. Prevailing winds (northeast trades) are directed offshore and usually steer tropical cyclones that form west of Central America away from the coast. Also, the southward flowing cold California Current plus upwelling just off the Southern California (and Baja California) coast produce sea-surface temperatures that normally are too low to sustain tropical cyclones that travel toward the northeast. However, during unusual atmospheric/oceanic circulation regimes, hurricanes have

struck coastal Southern California and even tracked over the Desert Southwest.

The Hawaiian Islands are sometimes threatened by tropical cyclones that develop over the central tropical Pacific or track into that region from the Pacific hurricane breeding grounds west of Mexico. Fortunately, in an average year, only 3 to 4 tropical cyclones affect the central Pacific and since 1957 only 4 hurricanes have struck the islands. However, the most recent one, Hurricane Iniki (category 4) in September 1992, devastated the island of Kauai. Seven people lost their lives. Total property damage was estimated at $2.3 billion, making this the most costly natural disaster in the history of the State of Hawaii.

HURRICANE LIFE CYCLE

The first sign of a developing tropical cyclone is the appearance of an organized cluster of thunderstorm clouds over tropical seas. This region of convective activity is labeled a **tropical disturbance** if a center of low pressure is detected at the surface. If atmospheric/ oceanic conditions favor hurricane development and if those conditions persist, the surface air pressure falls and a cyclonic circulation develops. Water vapor condenses within the storm, releasing latent heat of vaporization, and the heated buoyant air rises. Expansional cooling of the ascending air triggers more condensation, release of even more latent heat, and additional buoyancy. Rising temperatures in the core of the storm, coupled with an anticyclonic outflow of air aloft, cause a sharp drop in surface air pressure, which in turn, induces more rapid convergence of humid air at the surface. The consequent uplift surrounding the developing eye leads to additional condensation and release of latent heat.

Through this process, a tropical disturbance intensifies and its winds strengthen. When maximum sustained wind speeds reach 37 km per hr (23 mph) or higher, the developing system is called a **tropical depression**. When maximum sustained wind speeds reach at least 63 km per hr (39 mph), the system is classified as a **tropical storm** and assigned a name, such as Alberto or Beryl. Once maximum sustained winds reach 119 km per hr (74 mph) or higher, the storm is officially designated a hurricane. As a hurricane weakens and decays, the system is downgraded by reversing this classification scheme.

Hurricanes that form over the Atlantic Ocean near the Cape Verde Islands usually drift slowly westward with the trade winds (along the southern flank of the Bermuda-Azores subtropical high) across the tropical North Atlantic and into the Caribbean. At this stage in the storm's trajectory, it is not unusual for the system to travel at a mere 10 to 20 km per hr (6 to 12 mph) and take a week to cross the Atlantic. Once over the western Atlantic, however, the storm's forward speed usually increases and the system begins curving northward along the western flank of the subtropical high and then northeastward as the system enters the middle latitude westerlies. Precisely where this curvature takes place determines whether the hurricane enters the Gulf of Mexico (perhaps then tracking up the lower Mississippi River Valley or over the Southeastern States), moves up the Eastern Seaboard, or curves northeastward into the Atlantic.

Upon reaching about 30 degrees N, an Atlantic hurricane may begin to acquire extratropical characteristics as colder air circulates into the system and fronts develop. From then on the storm resembles an extratropical cyclone and completes its life cycle usually over the North Atlantic. Many hurricanes, however, depart significantly from the track just described. Some hurricane tracks, such as plotted in Figure 10.30, are very erratic. A hurricane can describe

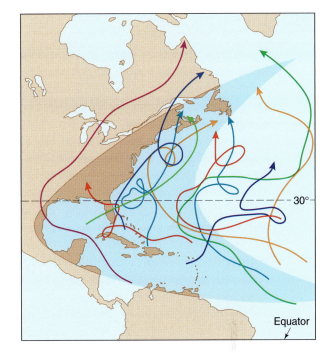

FIGURE 10.30
Tropical cyclone trajectories are sometimes erratic as shown by these samples. As indicated by the blue shaded area, however, most Atlantic tropical cyclones initially drift westward and then curve toward the north and northeast when they reach the western Atlantic. [From NOAA, *Hurricane*, Washington, DC: Superintendent of Documents, 1977.]

a complete circle or reverse direction. In addition, some hurricanes, fueled by warm Gulf Stream waters, maintain tropical characteristics far north along the Atlantic Coast. Coastal New England and Atlantic Canada, for example, have been the targets of hurricanes.

Conclusions

The climate record (both reconstructed and instrument-based) indicates that climate is inherently variable over a broad range of spatial and temporal scales. That is, climate variability and climate change are not everywhere the same in terms of direction (warmer, cooler, wetter, drier) or magnitude. Furthermore, conditions required for the development of specific weather systems such as an Arctic high, hurricane, or tornado occur in only certain locations. Boundary conditions within Earth's climate system ultimately govern climate variability and the geographical and seasonal biases of weather systems. In the next chapter, we examine the various forcing mechanisms responsible for natural climate variability and climate change.

Basic Understandings

- The latitude of highest mean annual surface temperature, the heat equator, is located in the Northern Hemisphere. Mean annual isotherms are symmetrical with respect to the heat equator, decreasing toward the poles.

- As the year progresses from January to July to January, isotherms in both hemispheres follow the Sun and shift north and south in tandem.

- The global pattern of mean annual precipitation exhibits considerable spatial variability due to topography, distribution of land and sea, and the planetary-scale circulation. In the tropics and subtropics, seasonal shifts of the ITCZ and subtropical anticyclones are responsible for rainy season/dry season climates.

- The most reliable instrument-based temperature records date from the late 1800s, with the worldwide establishment of standardized methods of observation and record keeping.

- The trend in global mean air temperature is generally upward from 1880 until about 1940, downward or steady from 1940 to about 1970, and upward again through the 1990s and early 2000s.

- Potential sources of error in global mean temperature records include improved reliability of weather instruments during the period of record, changes in location and exposure of instruments, gaps in monitoring networks (especially over the ocean), and the warming effect of urbanization.

- Global climate change encompasses not only trends in temperature but also changes in the water cycle. Recent analyses of instrument-based records reveal trends in river discharge, precipitation patterns, and surface humidity in various parts of the world.

- A major concern regarding the climate future is the possibility of changes in the frequency and/or intensity of severe weather.

- A thunderstorm is a mesoscale weather system produced by convection currents that surge to great altitudes in the troposphere. The life cycle of a thunderstorm cell consists of a three-stage sequence: towering cumulus, mature, and dissipating.

- A thunderstorm usually consists of more than one cell, each of which may be at a different stage in its life cycle.

- Most thunderstorms develop in a maritime tropical air mass as a consequence of uplift along fronts, on mountain slopes, via converging surface winds, or through intense solar heating of Earth's surface that triggers convection.

- Thunderstorms develop along boundaries within a maritime tropical air mass, as a squall line along or ahead of a cold front, or in a mesoscale convective complex (MCC). MCCs are warm season phenomena, occurring mostly from May through August, and account for upwards of 18% of growing season rainfall over the Great Plains and Midwest.

- A supercell thunderstorm is a relatively long-lived and intense system that consists of a single cell with an exceptionally strong updraft and may produce a tornado.

- Worldwide, thunderstorms are most common over the continental interiors of tropical latitudes. In North America, thunderstorm days are most frequent in central Florida where converging sea breezes induce uplift of maritime tropical air; they are also frequent on the eastern slopes of the central and southern Rockies, where they are associated with orographic lifting. Convection is inhibited and thunderstorms are unlikely to develop in air masses that reside or travel over relatively cold surfaces and are thereby stabilized.

- Severe thunderstorm cells typically form along a squall line ahead of a fast-moving, well-defined cold front associated with a mature extratropical cyclone. The polar front jet stream causes dry air to subside over a surface layer of maritime tropical air. This produces a layering of air that sets the stage for explosive convection and development of severe thunderstorms.

- Lightning is a brilliant flash of light produced by an electrical discharge within a cloud, between clouds, or between clouds and the ground.

- Some thunderstorm cells produce downbursts, which are intense downdrafts that spread out (diverge) at Earth's surface as potentially destructive winds. A squall line sometimes produces a family of potentially destructive straight-line downburst winds, known as a derecho.

- A flash flood is a short-term, localized, and often unexpected rise in stream level causing the stream to flow over its banks. Flash flooding is a special hazard in mountainous terrain, where steep slopes channel excess runoff into narrow stream and river valleys, and in urban areas, where impervious surfaces cause excess runoff to collect in low-lying areas.

- Hail develops in intense thunderstorm cells characterized by strong updrafts, great vertical development, and an abundant supply of supercooled water droplets.

- A tornado is a small mass of air that whirls rapidly about a nearly vertical axis and is usually made visible by condensed water vapor (funnel cloud), and dust and debris drawn into the system.

- An exceptionally steep horizontal air pressure gradient between the tornado center and outer edge is the force ultimately responsible for the violence of a tornado.

- When a tornado strikes, very high winds, a strong updraft, subsidiary vortices, and an abrupt air pressure drop are responsible for considerable property damage. The EF-Scale classifies tornadoes based on winds estimated from structural damage. Most tornadoes cause light or moderate damage (EF0 or EF1), but most fatalities are due to rare violent tornadoes that cause devastating or incredible damage (EF4 or EF5).

- Most tornadoes occur in spring within a corridor stretching from eastern Texas and the Texas Panhandle northward through Oklahoma, Kansas, Nebraska, and into southeastern South Dakota, from central Iowa eastward to central Indiana, and along the Gulf Coast states.

- Synoptic weather conditions that favor the outbreak of tornadoes progress northward (with the Sun), from the Gulf Coast in early spring to the Great Lakes region and southern Canada by early summer.

- A hurricane is an intense tropical cyclone that originates in a uniformly warm and humid air mass over tropical waters, usually in late summer or early fall, with a maximum sustained wind speed of at least 119 km per hr (74 mph).

- Potentially the most devastating impact of tropical cyclones in the coastal zone is a storm surge, a dome of ocean water driven ashore by strong onshore winds associated with an intense storm system centered over or near the ocean. Other hurricane hazards are flooding rains, strong winds, and tornadoes.

- The Saffir-Simpson Hurricane Intensity Scale rates hurricanes from category 1 to 5 corresponding to increasing intensity and provides an estimate of potential coastal flooding and property damage from a hurricane landfall.

- The three conditions required for a tropical cyclone to form are relatively high sea-surface temperature, adequate Coriolis Effect, and weak winds aloft (weak vertical wind shear). To a large extent, these requirements dictate the climatology of tropical cyclones, that is, where and when they occur.

- The energy source for tropical cyclones is latent heat released when water vapor evaporated from the ocean condenses in the system. As a tropical cyclone intensifies, the successive stages in its life cycle are designated tropical disturbance, tropical depression, tropical storm, and hurricane.

Enduring Ideas

- Mean annual surface isotherms decrease with distance north and south of the heat equator. For several reasons, the heat equator is located in the Northern Hemisphere.
- The spatial and seasonal variability of precipitation on Earth can be explained mostly by topography, the distribution of land and sea, and the planetary-scale atmospheric circulation.
- The instrument-based global mean temperature record indicates that Earth is warming. This temperature trend is consistent with enhancement of the greenhouse effect due to a build-up of CO_2 and other greenhouse gases in the atmosphere as a consequence of human activity.
- Global climate change may be accompanied by changes in the frequency and/or intensity of severe weather systems (e.g., severe thunderstorms, tornadoes, tropical cyclones).
- Understanding the geographical and seasonal biases of severe weather systems is key to human adaptation to climate change.

Review

1. Why is the *heat equator* located in the Northern Hemisphere?
2. Why does the instrument-based surface air temperature record indicate greater year-to-year variability over land than over the ocean?
3. Identify several factors that may influence the integrity of the long-term instrument-based climate record.
4. Briefly describe the characteristics of each of the three stages in the life cycle of a thunderstorm cell.
5. What atmospheric conditions favor the development of a thunderstorm?
6. Why are tornadoes so frequent in the region known as *tornado alley*?
7. What is the principal source of energy for a hurricane?
8. What hazards are associated with a hurricane that makes landfall along a low-lying coastal plain?
9. What three conditions are required for a tropical cyclone to form?
10. Hurricanes tend not to form within 5 degrees latitude of the equator. Explain why.

Critical Thinking

1. How do orographic features affect the global pattern of precipitation?
2. What are the sources of air temperature readings obtained from near the ocean surface?
3. How do the properties of Earth's surface affect the probability of thunderstorm development?
4. Identify the combination of atmospheric conditions required for hail development.
5. Explain why tornadoes are more frequent in spring.
6. From late winter through spring, tornado occurrences on average progress northward across the central U.S. Explain why.
7. How does a hurricane differ from an extratropical cyclone?
8. Why are Atlantic hurricanes most frequent in late summer and early autumn?
9. How does the magnitude of vertical wind shear affect the development of tropical cyclones?
10. If global warming means higher SST in the Atlantic hurricane breeding ground, would you expect an increase in the intensity of hurricanes? Explain your response.

ESSAY: Singularities in the Climate Record

A *singularity* is a weather episode that tends to occur on or about a certain date more frequently than chance would dictate. Most singularities are linked to regular (seasonal) changes in components of the planetary-scale atmospheric circulation. For example, seasonal north-south shifts in the intertropical convergence zone (ITCZ) and the subtropical anticyclones determine the onset and ending of the rainy season and dry seasons that are characteristic of monsoon climates and Mediterranean climates.

The *January thaw* is a widely recognized and perhaps the only real singularity in a statistically rigorous sense. It is an episode of relatively mild weather around 20 to 23 January and is most evident in the New England states and less so in the Midwest. The 20-23 January temperature peak shows up in climate records from Boston, MA (1873-1952), New York, NY, and Washington, D.C. The mild episode (thaw) is attributed to a southerly flow of warm air on the western flank of the Bermuda-Azores subtropical anticyclone, which for some reason, temporarily shifts north of its usual midwinter location.

A weather episode that does not fit precisely the definition of singularity, but nonetheless is quite regular in occurrence is the summer rainfall season in Arizona, New Mexico, and southern Colorado. As described in some detail in Chapter 7, the Southwest Monsoon is underway in early July and ends by early September. On average, some 40% to 70% of the annual rainfall in the American Southwest occurs during July and August.

Indian summer has no precise dates but tends to occur in October or November in the Midwest and Northeast, following the autumn's first killing frost. Typical Indian summer weather consists of an episode of abnormally warm conditions, sunny but hazy days, and frosty mornings. This weather episode is caused by a large warm anticyclone that stalls over the eastern United States, displacing the storm track northward along the St. Lawrence River Valley. An Indian summer does not occur every year and some years have two or three Indian summers. It is most commonly observed in the northeastern U.S. but is noted in many English-speaking nations. In Europe, the comparable period is known as *Old Wives' Summer* and in England, the episode may be either *St. Martin's Summer* or *St. Luke's Summer* depending on the date of occurrence.

ESSAY: Variability in Atlantic Hurricane Activity

Climate scientists are wrestling with the question of whether global climate change has or will affect the frequency and intensity of hurricanes. Following the lull in hurricane activity during the 1970s and 1980s, the frequency of major Atlantic hurricanes (category 3 or higher on the Saffir-Simpson Scale) increased substantially. From 1971-94, the annual average of major hurricanes was 1.5, but from 1995-2005, that annual average increased to 4.1. Is this change within the normal cyclic variability of hurricane frequency or is it the consequence of global climate change?

The detailed reliable instrument-based record of Atlantic hurricanes is limited to the past 130 years or so. Geoscientists have employed many methods to extend this record as far back in time as possible. An obvious advantage of a more lengthy record is possible insight on the variability of hurricane frequency and a historical perspective on the current active phase.

As we saw elsewhere in this chapter, two keys to hurricane formation are sea-surface temperatures (SST) and vertical wind shear. Even when SST are well above average, hurricane frequency depends more on strength of wind shear than SST. In 2007, K. Halimeda Kilbourne, a NOAA paleoclimatologist, and colleagues reconstructed the wind shear record back to 1730 by exposing coral to UV radiation to reveal the luminescence of its growth rings. Luminescence, in turn, is a measure of the amount of organic materials washed by thunderstorms from land to sea. Strong wind shear inhibits the development of thunderstorms and hurricanes. Hence, the luminescence of coral growth rings served as a proxy for Atlantic hurricane frequency.

Kilbourne and colleagues concluded that large variations in hurricane frequency are the norm in the Atlantic. From 1730 to 2005, the annual average of major hurricanes was 3.25, somewhat less than it was in the recent active period. Researchers identified at least 6 intervals since 1730 when hurricane activity was comparable in frequency to the present time.

Jeffrey P. Donnelly of Woods Hole (MA) Oceanographic Institution and colleagues demonstrated a relationship between Atlantic hurricane frequency and El Niño. Analysis of a 5000-year record of sedimentation from a lake in Ecuador and a lagoon in eastern Puerto Rico identify episodes of strong and frequent El Niño. Strong wind shear associated with El Niño favored a lower annual average frequency of Atlantic hurricanes.

In the March/April 2007 issue of *American Scientist*, Kam-biu Liu of Louisiana State University reported on his reconstruction of the frequency of major hurricanes along the Gulf Coast over a time frame of thousands of years. His working hypothesis was that the waves and storm surge associated with an intense hurricane would wash sand into coastal lakes located behind sandy beaches or dunes. In vertical cross-section, a sharply bounded sand layer situated between layers of fine mud served as a signature of hurricane impact. Radiocarbon or other radiometric dating technique was used to establish a hurricane chronology.

Liu demonstrated that along the Gulf Coast, major hurricane activity varied on time scales of centuries to millennia. At four coastal sites between Louisiana and Florida, catastrophic hurricanes (category 4 or 5) occurred 10 to 12 times every 3800 years or once every 300-350 years. Hurricane activity along the Gulf Coast was relatively low from 5000 to 3800 years before present and again during the past 1000 years when an average of only one catastrophic hurricane occurred per millennium. From 3800 to 1000 years ago, hurricane activity was relatively high with a catastrophic hurricane striking as frequently as once every two centuries.

CHAPTER 11

NATURAL CAUSES OF CLIMATE CHANGE

On the high peaks of Hawaii's Mauna Loa (south) and Mauna Kea (north) a cap of brilliant white snow covers the summits. The third volcano that makes up the island is Kilauea, where ever-present lava flows give off a heat signature detectable by NASA's MODIS (area outlined in red). [Courtesy NASA]

Case-in-Point

A violent volcanic eruption can have severe impacts locally and regionally. Mudflows (*lahars*) and ash falls take lives, destroy crops and livestock, and cause considerable property damage. In some cases, the eruption's impact spreads worldwide, altering the climate for a year or two and, in the extreme, brings hardship for millions of people.

Beginning on 19 February 1600 and lasting for several days, the volcano Huaynaputina erupted catastrophically. Huaynaputina is one of many volcanoes of the Andes Mountain Range that forms the backbone of South America. Located in southern Peru at 16.61

degrees S and 70.85 degrees W, with a summit elevation of 4850 m (15,912 ft), it had not erupted in historical time, nor has it since 1600. But that eruption was the largest of any South American volcano.

Regionally, the Huaynaputina eruption killed an estimated 1500 people, destroyed several villages, and ruined agriculture for two years; mudflows reportedly reached the ocean 120 km (74 mi) away. But of greater significance was the worldwide reach of the fine volcanic materials (ash and other aerosols) blasted into the atmosphere by the explosive eruption. As reported in 2008 by geophysicist Kenneth Verosub of the University of California, Davis, the eruption produced up to 12 km³ of ash. This means that the Huaynaputina eruption rated a category 6 on the *Volcanic Explosivity Index*, based on volume of ejecta and plume height. The Mt. Mazama eruption in BCE 5677 that formed Crater Lake, OR, had a VEI of 7. The Yellowstone eruption about 2.2 million years ago and the Lake Taupo, New Zealand, volcanic eruption 26,000 years ago had a VEI of 8. Since 1600, there have been only five category 6 volcanic eruptions (including Laki in 1783, Krakatau in 1883, and Pinatubo in 1991). The more violent Tambora eruption of 1815 was a category 7.

Verosub proposed that the Huaynaputina eruption may have impacted global climate because its ejecta was exceptionally rich in sulfur dioxide (SO_2). Within the atmosphere, SO_2 combines with water vapor to produce tiny sulfurous aerosols (sulfuric acid droplets and sulfate particles) that are transported by stratospheric winds around the world. Sulfurous aerosols reflect solar radiation causing cooling near Earth's surface. In fact, historical documents indicate that a relatively cool episode, lasting one or two years, followed the eruption in many areas of the globe. In some places, tree growth ring records indicate that 1601 was the coldest of any year in the past six centuries. Evidence of anomalous cold weather comes from Germany, France, Switzerland, Estonia and elsewhere, but the impact of this anomalous weather was probably most extreme in Russia.

In 1601, Russia experienced widespread crop failure and food scarcity, leading to that nation's worst famines in 1602 and 1603. The death toll was estimated at 2 million, about one-third of the total population; societal upheaval and failure of the government to alleviate the calamity at least contributed to the overthrow of Czar Boris Godunov (*ca.* 1551-1605).

To put this anomalous weather episode in context, Huaynaputina erupted during the Little Ice Age (Chapter 9), when poor growing conditions were a frequent occurrence, especially at northern latitudes. Although subsistence farming was practiced and food was stockpiled in an effort to offset crop failure, Russia's agricultural system was vulnerable to heavy rains and anomalous cold. According to Russian historian Chester Dunning of Texas A&M University, even prior to 1600 food production failed to keep pace with population growth, making society vulnerable to adverse growing conditions.

Driving Question:
What forcing mechanisms can bring about climate change and climate variability over a broad range of time and spatial scales?

No one simple explanation can account for climate variability and climate change. The complex spectrum of climate variability and climate change is a response to the interactions of many forcing agents and mechanisms operating both internally and externally relative to the Earth-atmosphere-land-ocean system (Figure 11.1).

One way to organize our thinking on the many possible causes of climate change is to match a possible cause (forcing agent or mechanism) with a specific climate fluctuation (response) based on similar periods (Figure 11.2). For example, with plate tectonics, atmospheric and oceanic circulation patterns change in response to continental drift, the opening and closing of ocean basins, and mountain building. Hence, plate tectonics might explain long-term climate changes operating over hundreds of millions of years. Systematic changes in Earth's orbit about the Sun affecting the latitudinal and seasonal distribution of incoming solar energy may account for climate shifts of the order of 10,000 to 100,000 years. Variations in sunspot number and the Sun's energy output may be associated with climate fluctuations of decades to centuries. Explosive volcanic eruptions, El Niño, or La Niña may account for inter-annual climate variability (Chapter 8). But matching some forcing mechanism with a climate response based on a similar periodicity is no guarantee of a real physical relationship.

This chapter covers the possible natural causes of climate change and climate variability. These forcing agents and mechanisms operate on Earth's climate system along with anthropogenic influences on climate which is explored in Chapter 12. We open with a look at climate change in the context of global radiative equilibrium.

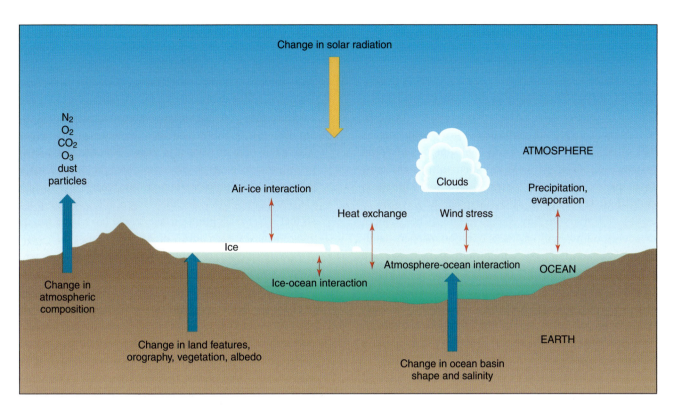

FIGURE 11.1
The complex spectrum of climate variability and climate change is a response to the interactions of many forcing mechanisms that operate both internal and external to the Earth-atmosphere-ocean system. [Modified from W.L. Gates and Y. Mintz, *Understanding Climatic Change: A Program for Action*, 1975, National Academy of Sciences, National Academy Press, Washington, DC]

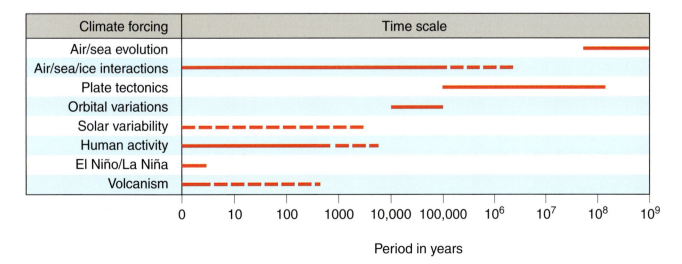

FIGURE 11.2
The various causes of climate change operate over a range of time scales.

Global Radiative Equilibrium and Climate Change

We can think about the possible causes of climate change in terms of the global energy budget. As noted in Chapter 3, **global radiative equilibrium** means that energy entering the Earth-atmosphere-land-ocean system (i.e., absorbed solar radiation) ultimately must equal energy leaving the system (i.e., infrared radiation emitted to space). This is a statement of the *law of energy conservation*. Any change in either energy input or energy output will shift the Earth-atmosphere-land-ocean system to a new equilibrium and change the planet's climate. A number of factors can alter the global radiative equilibrium, including fluctuations in solar energy output, changes in Earth's orbit about the Sun, volcanic eruptions, variations in atmospheric chemistry, alterations in Earth's surface properties, and certain human activities.

For more on global radiative equilibrium and climate change, see this chapter's first Essay.

Solar Variability and Climate Change

Fluctuations in the Sun's energy output, sunspots, or regular variations in Earth's orbital parameters are external factors in Earth's climate system that can bring about climate change. Satellite measurements taken in the 1980s and 1990s confirmed long-held suspicions in the scientific community that the Sun's total energy output at all wavelengths (total solar irradiance) is not constant (Figure 11.3). Furthermore, numerical global climate models predict that only a 1% change in the Sun's energy output could significantly alter the mean temperature of the Earth-atmosphere-land-ocean system.

FAINT YOUNG SUN PARADOX

According to standard models of stellar evolution, stars such as the Sun gradually become brighter (increasing luminosity) during their lifetime. Changes in nuclear fusion reactions within the Sun's interior are responsible for this steady increase in solar energy output (Chapter 3). On this basis, 4 billion years ago the Sun's energy output was about 70% of what it is today. With the same level of greenhouse gases in the atmosphere as today, Earth's average surface air temperature would have been about 25 Celsius degrees (45 Fahrenheit degrees) lower than now, so that water on Earth's surface would have been frozen and remained frozen until 1 or 2 billion years ago.

Evidence is convincing, however, that liquid water flowed on Earth's surface from very early in the planet's history. Some of the oldest rocks on Earth, dating to about 4 billion years ago, exhibit features indicative of running water. These rocks are sedimentary and contain pebbles that were rounded by the action of moving water. Other features preserved in rocks dating from 3 to 4 billion years ago include ripple marks, mud cracks, and fossil algae, all associated with liquid water. In 1972, astronomers Carl Sagan (1934-1996) and George Mullen were first to explain this paradox of a fainter Sun and warmer Earth.

The higher than expected temperature at Earth's surface is explained by a greenhouse effect that was stronger than at present. Early on, Earth's atmosphere

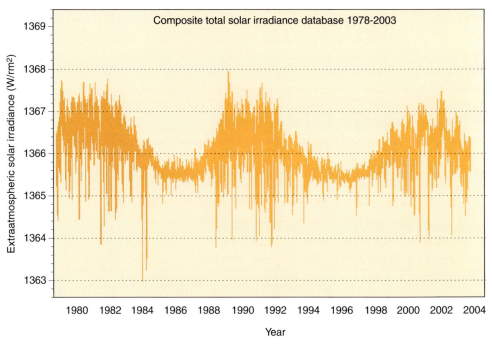

FIGURE 11.3
Satellite measured variations in the total solar irradiance 1978-2003. [ESA/NASA Mission SoHO]

contained very little free oxygen (O_2). In the presence of oxygen, methane (CH_4) breaks down to carbon dioxide (CO_2), so that we can assume that the atmospheric concentration of methane, a very efficient greenhouse gas, was much higher than at present. (For more on the evolution of Earth's atmosphere, refer to the first Essay in Chapter 1.) Furthermore, volcanic activity elevated the atmosphere's concentration of carbon dioxide, also a greenhouse gas. Hence, warming due to a stronger greenhouse effect partially offset the climatic impact of a Sun that was 30% fainter than now so Earth's surface was warm enough for water to exist in the liquid phase.

SUNSPOTS

Changes in solar energy output apparently are related to sunspot number. A **sunspot** is a dark blotch of irregular shape on the face of the Sun, typically thousands of kilometers across, that develops where an intense magnetic field suppresses the flow of gases transporting heat energy from the Sun's interior (Figure 11.4). A sunspot consists of a dark central region, called an *umbra*, ringed by a brighter, filamentary region, termed a *penumbra*. A sunspot appears dark because its temperature is about 400 to 1800 Celsius degrees (720 to 3240 Fahrenheit degrees) lower than the temperature of the surrounding surface of the Sun, the *photosphere*.

At least as early as BCE 28, Chinese astronomers observed sunspots with the unaided eye by viewing the Sun's reflection on the surface of a quiet pond at sunrise and sunset. Galileo Galilei (1564-1642) is credited with being among the first to study sunspots telescopically in 1610, and thereafter sunspots became objects of considerable scientific interest. In 1843, the German astronomer Samuel Heinrich Schwabe (1789-1875) reported regular variations in sunspot activity. A sunspot typically lasts only a few days, but the rate of sunspot generation is such that the number of sunspots varies systematically (Figure 11.5). The time between successive sunspot

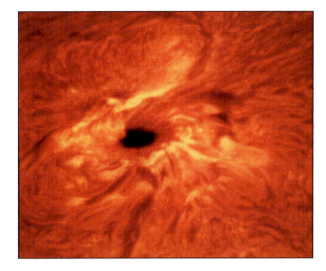

FIGURE 11.4
A sunspot is a dark blotch that appears on the Sun's photosphere, typically thousands of kilometers across, that develops where an intense magnetic field suppresses the flow of gases transporting heat from the Sun's interior. [Courtesy of Dr. Donat G. Wentzel, University of Maryland and the National Optical Astronomical Observatories.]

maxima or minima averages about 11 years with a range of 10 to 12 years. Also, an approximate 22-year oscillation (*double sunspot cycle*) characterizes the intense magnetic fields associated with sunspots. The strength of these magnetic fields is several thousand gauss (compared to Earth's magnetic field strength that averages about half a gauss).

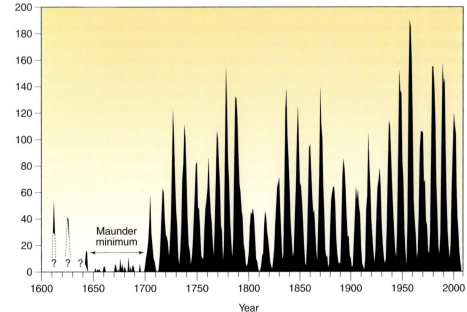

FIGURE 11.5
Variation in mean annual sunspot number since early in the 17th century. [National Geophysical Data Center]

In recent time, the sunspot number reached a minimum in 1996, a maximum in 2001, and a minimum in late 2008. As of mid 2009, the Sun remained extremely quiet with the next solar maximum expected in 2013.

Satellite monitoring reveals that the Sun's energy output varies directly with sunspot number; that is, a slightly brighter Sun has more sunspots whereas a slightly dimmer Sun has fewer sunspots (Figure 11.3). The variation in total solar energy output through one 11-year sunspot cycle is less than 0.1%, with much of this taking place in the ultraviolet (UV) portion of the solar spectrum. A brighter Sun is associated with more sunspots because of a concurrent increase in bright areas, known as *faculae*, which appear near sunspots on the photosphere. Faculae dominate sunspots and the Sun brightens. More sunspots may contribute to a warmer global climate and fewer sunspots may translate into a colder global climate.

MAUNDER MINIMUM AND THE LITTLE ICE AGE

How reasonable is the proposed link between global climate and sunspot number? In the late 1880s, the German astronomer F.W. Gustav Spörer (1822-1895) and the English solar astronomer E. Walter Maunder (1851-1928) reported that sunspot activity greatly diminished during the 70-year period from 1645 to 1715 and coincided with a cold episode in Europe. For reasons not yet understood, during the late 1600s, sunspot activity declined by a factor of 10 to 20 from its usual value during normal sunspot cycles. This period of greatly reduced solar activity is now referred to as the **Maunder minimum**. For the most part, the scientific community ignored the Maunder minimum until the American astronomer John A. Eddy (1931-2009) reinvestigated the phenomenon and published his findings in a landmark paper, "The Maunder Minimum," in the 18 June 1976 issue of *Science*. Eddy pointed out that the Maunder minimum plus a prior 90-year period of reduced sunspot number, called the **Spörer minimum** (1460 to 1550), occurred about the same time as relatively cold phases of the Little Ice Age in Western Europe (Chapter 9). Another period of reduced solar activity, known as the **Dalton minimum**, occurred from 1790 to 1830 and coincided with an interval of lower global temperature. Named for the English meteorologist John Dalton (1766-1844), the Dalton minimum includes the *year without a summer (1816)*. Furthermore, the Medieval Warm Period coincided with an interval of heightened sunspot activity between about 1100 and 1250.

Skeptics have dismissed the significance of the proposed match between the Maunder minimum and a cooler climate. They argue that relatively cold episodes occurred in Europe just prior to and after the Maunder minimum, and relatively cool conditions did not persist throughout the Maunder minimum and were not global in extent.

Support for a connection between variations in solar activity and climate comes from chemical analysis of tree growth rings. ^{14}C in tree growth rings is a proxy indicator of solar activity. ^{14}C is an isotope of carbon generated when cosmic rays from the Sun impact nitrogen in the upper atmosphere. When the Sun is relatively active (many sunspots), strong magnetic fields in the solar wind shield Earth from cosmic rays so that less ^{14}C is produced. With a lull in solar activity (few sunspots), cosmic rays generate more ^{14}C. Trees take up ^{14}C via photosynthesis so that tree growth rings provide a chronology of solar activity. According to John A. Eddy, tree growth ring records show that ^{14}C increased during the Maunder minimum, arguing for a link between reduced solar activity and climate change. Using the same method, the Spörer minimum can be identified in the tree growth ring record.

L.A. Scuderi, a researcher at Boston University, analyzed a 2000-year tree growth ring record extracted from foxtail pines in California's Sierra Nevada Mountains. Using growth rings as an index of solar activity (based on ^{14}C content) and air temperature (from growth ring width), Scuderi's year-by-year chronology reveals a close correlation between solar activity and temperature with a period of 125 years. Mild episodes of the Medieval Warm Period and cold episodes of the Little Ice Age are evident in the record.

Indications of the 11-year sunspot cycle show up in sea-surface temperature records. D. Cayan and W. White of Scripps Institution of Oceanography report that tropical and subtropical waters warm and cool about 0.1 Celsius degree (0.2 Fahrenheit degree) in phase with the 11-year sunspot cycle.

The 22-year double sunspot cycle closely corresponds to the frequency of drought on the western High Plains. Charles Stockton of the University of Arizona's Laboratory of Tree-Ring Research reconstructed the drought chronology of the western U.S. back to the 17th century. His analysis of tree growth rings from 40 sites revealed that droughts occurred every 20 to 22 years.

Do these statistical correlations actually indicate a cause-effect relationship between sunspots and climate or are they merely coincidental? If the relationship is real, is it global or indicative of some more local response mechanism? Some scientists argue that the variation in the solar radiation output during an 11-year solar cycle is too weak to significantly alter Earth's climate. During an 11-year solar cycle, the Sun's radiative output varies by only 0.1% from solar maximum to solar minimum, equivalent to a global average variation of only 0.2 W/m². However, the change in global SST of 0.1 Celsius degree recorded during a solar cycle requires more than 0.5 W/m². Hence, scientists have searched for mechanisms whereby the weak solar signal might be amplified within Earth's climate system.

Joanna Haigh of Imperial College London proposed a top-down mechanism for amplifying the solar signal. As noted earlier, a significant amount of the variation in solar output during an 11-year solar cycle is in the ultraviolet (UV) wavelengths. Much of the UV is absorbed in the natural generation and destruction of ozone (O_3) (Chapter 3). The accompanying changes in the temperature of the stratosphere would induce changes in the circulation of the troposphere, affecting the climate in the lower atmosphere.

In 2001, Drew Shindell of NASA's Goddard Institute for Space Studies (GISS) and his colleagues reported on their use of a global climate model to simulate what happened during the Maunder minimum. With reduced sunspot activity, less solar ultraviolet radiation reached Earth, decreasing the formation of ozone (O_3) in the stratosphere. Less ozone in the stratosphere was accompanied by changes in the upper atmospheric circulation (the planetary long-waves) that propagated into the troposphere and weakened the polar front jet stream. These changes, in turn, affected the Arctic Oscillation (AO) and North Atlantic Oscillation (NAO) (Chapter 8). A weaker jet stream reduced the transport of mild Pacific air into North America and mild Atlantic air into Europe. Although the change in global mean temperature during the Maunder minimum was quite small, temperatures were significantly lower in certain regions.

A recent reanalysis of the Shindell et al. study revealed that the assumed changes in values of solar energy output are considerably greater than what can be justified by current knowledge. With weaker forcing, the effect noted in the study becomes insignificant. However, the important point is that changes in incoming UV that affect the stratosphere may alter the dynamics of the troposphere in such a way as to augment regional climate change.

Another possible amplification mechanism works from the bottom-up. Increased solar energy output during the peak of a solar cycle increases the evaporation rate in the equatorial Pacific Ocean, triggering changes in atmospheric and oceanic circulation. Reduced cloud cover in the subtropics means that more solar radiation is incident on the ocean surface leading to more evaporation in a positive feedback cycle, amplifying the climate effect of the solar cycle.

Neither the top-down nor bottom-up mechanism acting alone is sufficient to explain the magnitude of climate changes associated with a solar cycle. However, in 2009, researcher Gerald Meehl and colleagues at the National Center for Atmospheric Research (NCAR) in Boulder, CO, found that the two mechanisms reinforced each other when combined in a climate model. The "dual-pronged solar amplifier" produces a climate response that is similar to that actually observed in the Pacific during recent solar maxima.

Earth's Orbit and Climate Change

In 1842, the French mathematician Joseph Alphonse Adhémar (1797-1862) proposed that regular variations in the shape (eccentricity) Earth's orbit about the Sun could explain the climate fluctuations that paced the glacial fluctuations of the Ice Age. Subsequently, Adhémar's ideas were expanded upon by James Croll and later by Milutin Milankovitch.

In 1864, James Croll (1821-1890), a self-educated Scottish scientist, attributed the climate changes responsible for the Ice Age to fluctuations in incoming solar radiation that accompanied regular changes in Earth's orbital parameters (eccentricity of the orbit, tilt of the rotational axis, and precession of the axis). On this basis, Croll worked out an Ice Age chronology, arguing that less incoming solar radiation in winter favored greater accumulations of snow. Furthermore, a more extensive snow cover would cool the atmosphere (a positive feedback) culminating in glaciation. Croll predicted multiple glaciations, an idea that was later confirmed through field work by the American geologist Thomas C. Chamberlain (1843-1928). However, by the close of the 19th century, Croll's theory fell into disfavor when European geologists discovered a serious discrepancy between his astronomically-based Ice Age chronology and field evidence. Croll's chronology showed the Ice Age ending about 80,000 years ago, whereas field evidence indicated that the date was closer to 10,000 years ago.

In the second decade of the 20th century, the Serbian astronomer Milutin Milankovitch (1879-1958) revived Croll's idea and set out to calculate the latitudinal and seasonal variations in solar radiation striking Earth's surface arising from long-term regular changes in Earth's three orbital parameters using computational procedures provided by astronomers. For more than 25 years, Milankovitch relentlessly pursued his goal. His collaboration with the German climatologist Wladimir

Köppen (1846-1940) and meteorologist Alfred Wegener (1880-1930) convinced Milankovitch that reduced solar radiation during summer (and not winter as Croll had proposed) at northern latitudes was key to initiating glaciation. In 1938, he published radiation curves for different latitudes (55 to 65 degrees N) extending back 600,000 years. In the 1930s and 1940s, support for Milankovitch's astronomical theory of the Ice Age was considerable, especially in Europe. However, in a dramatic turnaround, by the mid-1950s Milankovitch's views were rejected by most geologists. It was not until the mid-1970s that independent corroborative evidence from deep-sea sediment cores firmly established the Milankovitch cycles as the pacemaker of the major climatic fluctuations of the Pleistocene Epoch.

MILANKOVITCH CYCLES

Milankovitch cycles are regular variations in the precession and tilt of Earth's rotational axis and the eccentricity of its orbit about the Sun (Figure 11.6). These changes in Earth-Sun geometry are caused by gravitational influences exerted on Earth by other large planets, the Moon, and the Sun. Combined, Milankovitch cycles drive climate fluctuations operating over tens of thousands to hundreds of thousands of years (Figure 11.7).

Over a period of about 23,000 years, Earth's spin axis describes a complete circle, much like the wobble of a spinning top. This precession cycle changes the dates of *perihelion*, when Earth is closest to the Sun, and *aphelion*, when Earth is farthest from the Sun, increasing the summer-to-winter seasonal contrast in one hemisphere while decreasing the seasonal contrast in the other. At present, perihelion is in early January and aphelion is in early July. In about 11,000 years, those dates will be reversed (perihelion in July and aphelion in January) and the seasonal contrast will be greater than it is now in the Northern Hemisphere (i.e., colder winters and warmer summers) but less in the Southern Hemisphere (i.e., milder winters and cooler summers). A similar situation occurred 11,000 years ago.

The tilt of Earth's spin axis changes from 22.1 degrees to 24.5 degrees and then back to 22.1 degrees over a period of about 41,100 years, the consequence of long-period changes in orientation of Earth's orbital plane with respect to its spin axis. (Presently, the tilt is 23.5 degrees.) As the axial tilt increases, winters become colder and summers become warmer in both hemispheres. Earth's axial tilt has been slowly decreasing for about 10,000 years and will continue to do so for the next 10,000 years.

The eccentricity of Earth's elliptical orbit about the Sun changes from relatively high to low (nearly circular) during an irregular cycle of 90,000 to 100,000 years. Variations in the orbital eccentricity alter the distance between Earth and Sun at aphelion and perihelion, thereby changing the amount of solar radiation received by the planet at those times of the year. When Earth's orbit is highly elliptical, the difference in the amount of radiation received at perihelion versus aphelion is greater than when the orbit is less elliptical. Also, changes in eccentricity along with the precession cycle significantly influence the lengths of the individual astronomical seasons.

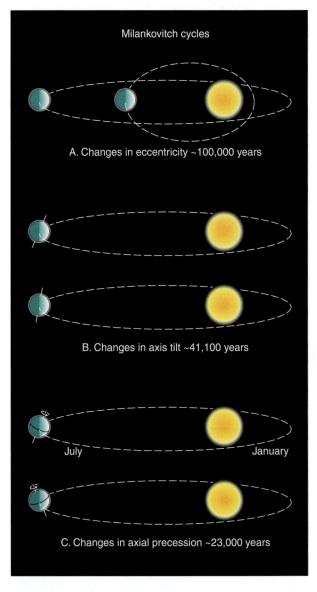

FIGURE 11.6
Milankovitch cycles likely explain the large-scale fluctuations of glacial ice cover during the Pleistocene Epoch. Note that diagrams greatly exaggerate changes in Earth-Sun geometry.

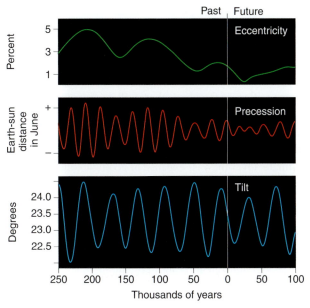

FIGURE 11.7
Past and future variations in the Milankovitch cycles.

Milankovitch cycles do not appreciably alter the total annual amount of solar energy received by Earth's planetary system, but these orbital variations change significantly the distribution of incoming solar radiation by latitude and season. The amplitude of glacial/interglacial climate cycles is not explained completely by the relatively small variations in solar radiation associated with the Milankovitch cycles. Feedback processes operating within Earth's climate system amplify the climatic forcing due to orbital variations. A major player in this regard is the greenhouse gas carbon dioxide. Analysis of glacial ice cores from Antarctica spanning the past 650,000 years reveals that the concentration of atmospheric CO_2 varies inversely with changes in the volume of glacial ice on Earth; that is, CO_2 concentration is higher during interglacial episodes (260 to 280 ppmv) and lower during glacial episodes (200 ppmv). During interglacial episodes, rising levels of CO_2 adds to the warming caused by orbital variations whereas during glacial episodes, falling levels of CO_2 contributes to the cooling caused by orbital variations.

Milankovitch developed a numerical model based on the three orbital cycles that calculated incoming solar radiation and the corresponding surface temperature by latitude for the time span of 600,000 years prior to the year 1800. He proposed that glacial climatic episodes began when Earth-Sun geometry favored an extended period of increased solar radiation in winter and decreased solar radiation in summer at 65 degrees N. More intense winter radiation in eastern Canada translated

into somewhat higher temperatures, higher humidity, and more snowfall. Weaker solar radiation in summer meant that some of the winter snow cover, especially north of 65 degrees N, survived summer and a succession of many such cool summers would favor formation of a glacier. Once formed, a glacier feeds back on its environment by reflecting away solar radiation. This process, according to Milankovitch, was the origin of the Laurentide ice sheet. At other times, Earth-Sun geometry favored enhanced solar radiation in summer at 65 degrees N, triggering higher temperatures, an interglacial climatic episode, and shrinkage of the Laurentide ice sheet.

EVIDENCE FROM DEEP-SEA SEDIMENT CORES

Willard F. Libby (1908-1980) of the University of Chicago developed the radiocarbon method of age-determination of the remains of formerly living organisms. By 1951, geologists were applying the new dating technique to organic materials associated with the last major glacial advance. (The range of radiocarbon dating was limited to about 40,000 years.) Geologists discovered that the astronomical chronology did not match the radiocarbon chronology. However, the radiocarbon chronology was derived mostly from terrestrial sources and some scientists were convinced that a more reliable chronology could be obtained from analysis of deep-sea sediment cores extracted from the bottom of the open ocean.

Sedimentation rates in the open ocean are extremely slow, ranging from 1 to 3 mm per century. Hence, a relatively short undisturbed core could contain information on climate fluctuations through much of the Pleistocene Epoch with a resolution in the order of 10,000 years. With the development of new age-dating techniques (based on reversals in Earth's magnetic field) and oxygen isotope analysis, scientists soon possessed the means to reconstruct past climate oscillations from analysis of deep-sea sediment cores (Chapter 9). An international team of scientists analyzed two cores, spanning 450,000 years, extracted from the bottom of the Indian Ocean. In December 1976, scientists announced that major changes in climate occurred at essentially the same frequency as variations in the eccentricity, obliquity, and precession of Earth's orbit, even labeling the cycles as the "pacemakers of the ice ages."

Today Milankovitch's astronomical theory is widely accepted as the explanation for the major climate fluctuations of the Pleistocene Epoch, but some reservations persist on the part of many climate scientists. The 100,000 year cycle is much weaker than the other two orbital cycles. While positive feedback may well have amplified the climatic response to the 100,000 year eccentricity cycle, it is also conceivable that the 100,000 year climate cycle (the primary Ice Age cycle over the past 1 million years) is due to entirely different mechanisms so that the proposed relationship to the eccentricity cycle may be coincidental.

Milankovitch cycles do not account for the period of climate quiescence prior to the Pleistocene Epoch. Apparently, Milankovitch cycles (which likely operated throughout Earth history) were not effective in initiating continental-scale glaciation without the appropriate arrangement of ocean basins, continents, and mountain ranges—boundary conditions not fully achieved until onset of the Pleistocene Epoch.

Discovery of a likely connection between systematic changes in Earth's orbital parameters and major glacial-interglacial climatic shifts prompted climate scientists to search for other influences of Milankovitch cycles on climate variability and climate change. For example, in the early 1980s, J.E. Kutzbach and B.L. Otto-Bliesner of the University of Wisconsin-Madison investigated the possible link between regular variations in Earth's orbital parameters and the strength of the African-Asian monsoon circulation during the early Holocene, some 9000 years ago. At that time, the average annual incoming solar radiation was about the same as now but the obliquity was 24.23 degrees (compared to 23.45 today), perihelion was 30 July (compared to 3 January today), and the orbital eccentricity was 0.01928 (compared to the present 0.01676). These differences in orbital parameters altered the solar radiation at the solstices by approximately 7% (compared to the present), that is, more radiation in June-July-August and less radiation in December-January-February. The consequence was greater heating of the land surface in summer and reduced heating of the land surface in winter. At the same time, because of the ocean's great thermal inertia, sea-surface temperatures remained at present values. Using a low-resolution general circulation model, Kutzbach and Otto-Bliesner predicted an intensification of the African-Asian monsoon circulation—a prediction that was consistent with paleoclimatic evidence.

Plate Tectonics and Climate Change

Recall from Chapter 1 that Earth's solid outer skin is divided into a dozen gigantic rigid plates (and many smaller ones) drifting very slowly over the face of the planet. The spreading rates of plates range between about 20 and 200 mm per year. As plates move, the continents they carry also move (changing latitude), and ocean basins open and close (altering ocean currents). Tectonic stresses build mountain ranges (changing elevation and atmospheric circulation) and trigger volcanic eruptions (altering the composition of the atmosphere). Plate movements are so slow compared to the span of human existence that we can consider topography and the geographical distribution of the ocean and continents as essentially fixed controls of climate. Over the vast expanse of geologic time, however, plate tectonics has been a major player in large-scale climate change.

Changes in the location of continents (*continental drift*) altered the local and regional radiation budget and the response of air temperature. Continental drift explains such seemingly anomalous discoveries as glacial deposits in the Sahara Desert, fossil tropical plants in Greenland, and coal in Antarctica. These finds reflect climate conditions millions to hundreds of millions of years ago when landmasses were situated at different latitudes than they are today (Figure 11.8). Furthermore, colliding plates cause mountain ranges to rise and weathering causes them to erode away, altering atmospheric circulation, the patterns of clouds and precipitation, and the concentration of atmospheric CO_2 and other greenhouse gases.

Recall from Chapter 9 that about 55 million years ago, near the boundary between the Paleocene and Eocene Epochs, massive basaltic intrusions released greenhouse gases that caused considerable global

FIGURE 11.8
This bedrock exposed in northeastern Wisconsin contains fossil coral that dates from nearly 400 million years ago. Based on the environmental requirements of modern coral, geoscientists conclude that 400 million years ago, the climate of this region was tropical marine. Plate tectonics can explain such a drastic change between ancient and modern conditions.

warming. About 50 million years ago, the plate carrying the Indian subcontinent collided with Asia, cutting off a major source of atmospheric CO_2, thereby contributing to cooling. About 40 million years ago, uplift of the Colorado Plateau, Tibetan Plateau, and Himalayan Mountains caused a climate shift to colder and more variable conditions.

Opening and closing of ocean basins changed the course of heat-transporting ocean currents and altered the thermohaline circulation. About 100 million years ago, the *Tethys Sea* separated Africa from Eurasia, and Central America was submerged so that warm water currents flowed around the equator connecting what are now the Pacific, Atlantic, and Indian Ocean basins. About 40 million years ago, diverging plates pulled what is now Antarctica away from Australia, and about 10 million years later, the Drake Passage opened between Antarctica and the southern tip of South America. The Drake Passage permitted an ocean current to flow around Antarctica. In addition, these plate movements eventually left the Antarctic continent situated over the South Pole. By blocking the transport of heat from the tropics to the Southern Ocean, the Antarctic circumpolar current probably led to the formation of the Antarctic ice sheets about 17 million years ago.

About 20 million years ago, the movement of Saudi Arabia northward against Asia sealed off the Tethys Sea, forming the Mediterranean Sea. About 3 million years ago, volcanic eruptions in what is now Central America produced a narrow isthmus of land, blocking the equatorial currents that previously flowed from the Atlantic through the Caribbean and into the equatorial Pacific Ocean.

Another example of a link between tectonics and climate change comes from East Africa some 8 to 2 million years ago. In the 8 September 2006 issue of *Science*, French researcher Pierre Sepulchre and colleagues reported on their numerical simulation of the climate forcing that caused a major transition from woodlands to open grasslands. Prior to 8 million years ago, the prevailing atmospheric circulation pattern across East Africa was zonal with considerable moisture advection and heavy rainfall that supported woodlands. However, tectonic uplift from 8 to 2 million years ago altered the circulation pattern bringing increasing aridity and spread of grasslands. Uplift was associated with the East African Rift System.

For another example of how plate tectonics may have played a role in climate change, refer to this chapter's second Essay.

Volcanoes and Climate Change

The idea that volcanic eruptions influence climate has been around for more than two centuries. Benjamin Franklin (1706-1790) proposed that the Laki (Skaftareldar or Skaftar Fires) fissure eruption in southern Iceland during the summer of 1783 was responsible for the severe winter of 1783-1784 in Europe. In a *fissure eruption*, hot molten lava and gases from Earth's interior flow from fractures in the ground. (However, Franklin was unaware that other larger volcanic eruptions had also occurred in eastern Asia that same year, adding to the aerosol loading of the atmosphere.) The Laki eruption was one of the two largest fissure eruptions to occur in Iceland over the past 11 centuries. The other was the Eldgjá (Fire Chasim) eruption in 934 which lasted for about six years and emitted about twice as much sulfur oxides to the atmosphere as the Laki eruption. Evidence of both eruptions appears in glacial ice cores as sulfuric acid fallout and in the stunted growth rings of northern temperate trees.

The Laki eruption began in June 1783 and lasted for 8 months, effusively emitting about 14.7 km^3 of lava plus sulfur dioxide, hydrochloric acid, and hydrofluoric acid. The column of eruptive materials reached to altitudes as great as 13 km (8 mi). The eruption killed an estimated 10,000 Icelanders, roughly 20% of the population. The principal cause of death was likely contamination of food and drinking water supplies by fluoride produced by hydrofluoric acid that spread across the countryside during the eruption. The Laki eruption likely also contributed to a spike in mortality in Britain and the European continent from weather extremes and famine.

Richard B. Stothers of NASA's Goddard Institute for Space Studies (GISS) reconstructed the impacts of the Eldgíá and Laki eruptions. In both cases, winds transported sulfurous aerosol clouds of volcanic origin eastward across northern Europe, dimming the Sun and producing red sunrises and sunsets. The Laki aerosol veil persisted over the Northern Hemisphere for more than 5 months. King Henry of Saxony noted a thick dry fog in 934 and Benjamin Franklin made a similar observation in France in 1783. Both eruptions were followed by several cold winters, poor harvests, and famine. Stothers estimated that adverse impacts from the two fissure eruptions persisted for 5 to 8 years after Eldgjá and 2 to 3 years following Laki.

In violent eruptions, volcanoes discharge ash particles and sulfur dioxide (SO_2) into the stratosphere either through fissures in the ground or from vents in volcanic mountains. The larger ash particles settle to Earth's surface in only a few days. Within the stratosphere, SO_2 combines with water vapor to form tiny droplets of sulfuric acid (H_2SO_4) and sulfate particles, collectively called **sulfurous aerosols**. The small size of sulfurous aerosols (averaging about 0.1 micrometer in diameter) coupled with the absence of precipitation in the stratosphere, allow sulfurous aerosols to remain suspended in the stratosphere for many months to perhaps a year or longer before they cycle out of the atmosphere to Earth's surface. Successive volcanic eruptions have produced a sulfurous aerosol veil in the stratosphere at altitudes of about 15 to 25 km (9 to 16 mi). Sulfur dioxide emissions from clusters of volcanic eruptions temporarily thicken the stratospheric aerosol veil impacting the climate.

While sulfurous aerosols are suspended in the stratosphere, winds transport them around the globe in either an easterly or westerly direction. North or south movement of aerosols ejected by a high latitude volcano is more limited so that these aerosols tend to be confined to a belt surrounding the polar cap. Only explosive tropical and subtropical volcanic eruptions rich in sulfur dioxide are likely to impact global or hemispheric climate. The unusually cool summer of 1816, described in the Case-in-Point of Chapter 2, occurred after the violent eruption of Tambora, an Indonesian volcano, in the spring of 1815. Several relatively cold years followed on the heels of the 1883 eruption of Krakatau, also an Indonesian volcano. In addition, as described in this chapter's Case-in-Point, the climatic impact of the violent eruption of the Peruvian volcano Huaynaputina in 1600 likely contributed to one of Russia's worst famines. Is the relationship between volcanic eruptions and cooling real or coincidental?

Although the eruption of Mount St. Helens, Washington, on 18 May 1980 was spectacular, ejecta were low in sulfur oxides and the eruption had no detectable influence on hemispheric- or global-scale climate. Some short-term localized effects on surface temperatures occurred immediately downwind of the volcano over eastern Washington, Idaho, and western Montana. Over those areas, the ash plume was sufficiently thick to alter the local radiation budget for 12 to 24 hrs following the eruption. C. Mass and A. Robock of the University of Maryland reported that on the day of the eruption, blockage of the Sun by volcanic ash lowered surface air temperatures over eastern Washington by up to 8 Celsius degrees (14.4 Fahrenheit degrees). That night, over Idaho and western Montana, low-level ash impeded the escape of infrared radiation to space, thereby elevating temperatures by as much as 8 Celsius degrees.

Research conducted following the 1991 eruption of Mount Pinatubo provided new insights on the relationship between sulfurous aerosols of volcanic origin and large-scale climate fluctuations. On 15-16 June 1991, Mount Pinatubo, located on Luzon Island in the Philippines, erupted violently injecting an estimated 20 megatons of sulfur dioxide into the stratosphere (Figure 11.9). This was the most massive stratospheric volcanic aerosol cloud of the 20th century. By altering both incoming and outgoing radiation, sulfurous aerosols affected temperatures in the stratosphere and troposphere, altered atmospheric circulation, and impacted surface air temperatures around the globe.

Sulfurous aerosols absorb both incoming solar radiation and outgoing infrared radiation. This absorption

FIGURE 11.9
The June 1991 explosive eruption of Mount Pinatubo (on Luzon Island in the Philippines) was rich in sulfur dioxide. The resulting sulfurous aerosol veil in the stratosphere was responsible for cooling at Earth's surface, interrupting the post-1970s global warming trend for a few years. [U.S. Geological Survey photo]

warms the lower stratosphere, especially in the tropics. Sulfurous aerosols also reflect solar radiation to space. NASA scientists reported that in the months following the Mount Pinatubo eruption, satellite sensors measured a 3.8% increase in solar radiation reflected to space. Also, in the presence of chlorine, sulfurous aerosols destroy ozone (O_3), allowing more solar UV radiation to reach Earth's surface. (This is essentially the same mechanism responsible for the Antarctic ozone hole described in Chapter 3.) In the two years following the Pinatubo eruption, ozone levels at middle latitudes declined by about 5% to 8%. The combination of stratospheric warming at low latitudes and ozone depletion at high latitudes strengthened the circumpolar vortex. Associated with this large-scale circulation change was a non-uniform change in surface temperatures; that is, some places cooled while other places warmed.

By influencing the flux of radiation and attendant changes in global circulation patterns, the Pinatubo eruption was likely responsible for the cool summer of 1992 over continental areas of the Northern Hemisphere (with surface temperatures up to 2.0 Celsius degrees or 3.6 Fahrenheit degrees lower than the long-term average) (Figure 11.10). The Pinatubo eruption was also implicated in the temperature anomalies of the winters of 1991-92 and 1992-93, featuring higher than average temperatures over most of North America, Europe, and Siberia and lower than average temperatures over Alaska, Greenland, the Middle East, and China.

A violent sulfur-rich volcanic eruption is unlikely to lower the mean hemispheric or global surface temperature by more than about 1.0 Celsius degree (1.8 Fahrenheit degrees) although the magnitude of local and regional temperature change may be greater (Figure 11.11). The 1963 eruption of Agung on the island of Bali lowered the mean temperature of the Northern Hemisphere an estimated 0.3 Celsius degree (0.5 Fahrenheit degree) for a year or two. The violent eruption of the Mexican volcano El Chichón in March-April 1982 may have produced hemispheric cooling of about 0.2 Celsius degrees (0.4 Fahrenheit degrees). Cooling associated with the Mount Pinatubo eruption temporarily interrupted the post-1970s global warming trend; from 1991 to 1992, the global mean annual temperature dropped 0.4 Celsius degrees (0.7 Fahrenheit degrees).

John A. Church, an Australian oceanographer, and his colleagues found that the cooling of the lower troposphere associated with the Mount Pinatubo eruption affected sea level. Ocean water contracts when cooled causing sea level to drop. Analyzing tide gauge, temperature, and salinity data from around the world plus the output from climate models, Church and colleagues estimated that mean sea level dropped a total of about 5 mm during the 18-month cool episode following the eruption. Then, as sulfurous aerosols settled out of the atmosphere and air temperatures returned to pre-eruption levels, SST increased causing sea level to recover gradually at the rate of about 0.5 mm per year. By comparison, measurements by satellite sensors indicate that sea level rose at the rate of 3.2 mm per year between 1993 and 2000.

Climatologist Reid A. Bryson and colleagues at the University of Wisconsin-Madison developed a chronology of volcanic eruptions extending back to approximately 40,000 years ago based upon radiometric

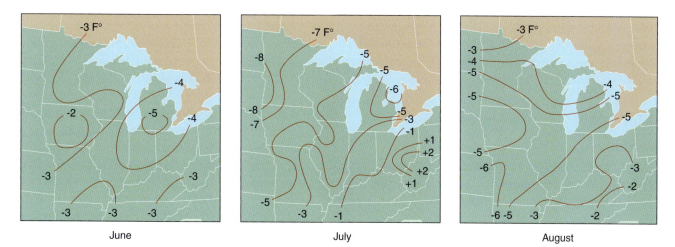

FIGURE 11.10
Temperature anomalies in Fahrenheit degrees in the Midwest during June, July, and August 1992. [From W.M. Wendland and J. Dennison, *Weather and Climate Impacts in the Midwest*, Midwestern Climate Center, Illinois State Water Survey, Champaign, IL]

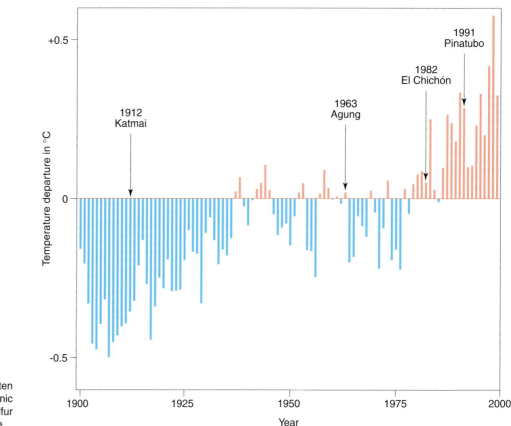

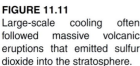

FIGURE 11.11
Large-scale cooling often followed massive volcanic eruptions that emitted sulfur dioxide into the stratosphere.

dates of volcanic ash layers worldwide. They found that the frequency of major volcanic eruptions was reasonably periodic and often numerous volcanic eruptions occurred during the same year, rather than a single isolated eruption. Bryson developed a simple interactive climate model that included the changes in the intercepted solar radiation due to the effects of volcanic aerosols along with those associated with the Milankovitch cycles. Using this model, he was able to develop climate chronologies for numerous locations from around the world through the late Pleistocene and Holocene that agreed with geologic and archeological field data.

Atmospheric Composition and Climate Change

Variations in the atmospheric concentration of radiatively active gases can cause global climate change. The concentrations of infrared-absorbing gases (e.g., carbon dioxide, methane, nitrous oxide) have fluctuated throughout Earth history with significant impact on Earth's surface temperature. In recent decades, considerable attention has been directed at the contribution of human activity to greenhouse gas concentrations in the atmosphere, but other natural processes have played important roles in regulating levels of greenhouse gases and global climate. In this section, we consider a few examples of natural processes that altered the level of certain greenhouse gases. In Chapter 12, we focus on the influence of human activity.

As described in some detail in the second Essay of Chapter 1, Earth's fossil record indicates five major mass extinctions of plants and animals over the past 550 million years. Four of the five extinctions likely are the consequence of increases in atmospheric CO_2 levels associated with massive volcanic eruptions. The largest eruptions of flood basalts in Earth history closely correspond in age to the times of most mass extinctions. Flood basalts consist of successive lava flows erupted from fissures in Earth's crust, accompanied by the release of gases including CO_2 and methane (CH_4) thereby enhancing the greenhouse effect and triggering global warming.

More recently in Earth history, analysis of air bubbles trapped in glacial ice cores indicates an abrupt warming following the last glacial maximum 20,000 to 18,000 years ago. Several centuries later, atmospheric methane concentration began a more gradual increase that continued into the early Holocene Epoch. Initially, scientists identified two possible sources for the methane: release from methane hydrates on the seafloor (Chapter 12), and emission from northern wetlands as temperatures rose. In 2007, K.M. Walter of the University of Alaska at Fairbanks and colleagues proposed a third substantial source of methane: thermokarst lakes.

A **thermokarst lake** occupies a depression in the ground formed when warming causes melting of ice-rich **permafrost** (permanently frozen ground) and water drains into the depression. (For more on permafrost, see the second Essay of Chapter 12.) Anaerobic decay of organic sediments on the lake bottom generates methane gas that bubbles to the lake surface and enters the atmosphere. Thermokarst lakes are most common at high northern latitudes in the unglaciated regions of Alaska, northern Canada, and Siberia. Methane emission from thermokarst lakes began about 14,000 years ago, continued through the Younger Dryas, accelerated during the early Holocene, and then declined.

Methane, a very efficient greenhouse gas, was an important contributor to post-glacial warming at high latitudes through the positive feedback mechanism described here. Higher levels of atmospheric methane meant higher air temperatures and higher air temperatures meant more methane released from northern wetlands and thermokarst lakes. These same sources of methane plus positive feedback may play a role in future climate change with continuation of the current warming trend.

Earth's Surface Properties and Climate Change

Earth's surface, which is mostly ocean water, is the principal absorber of solar radiation within the planetary system (Chapter 3). Any change in the physical properties of Earth's water or land surfaces or in the relative distribution of ocean, land, and ice may affect Earth's radiation budget and climate.

SNOW AND ICE COVER

Changes in mean regional snow cover may contribute to climate variability and climate change because an extensive snow cover has a refrigerating effect on the atmosphere. Fresh-fallen snow typically reflects 80% or more of incident solar radiation, thereby substantially reducing the amount of solar heating and lowering the daily maximum air temperature (Figure 11.12). Snow is also an excellent emitter of infrared radiation, so heat is

FIGURE 11.12
Snow has a cooling effect on the atmosphere by efficiently reflecting solar radiation during daylight hours and emitting infrared radiation to space at night.

efficiently radiated to space, especially on nights when the sky is clear, lowering the daily minimum air temperature. A snow cover tends to chill the near surface air, often increasing atmospheric stability and suppressing vertical mixing of air, which reduces convective heat flux. Because of this feedback, an extensive snow cover tends to be self-sustaining.

Persistence of a snow cover may be further enhanced by the tracks of extratropical cyclones that tend to follow the periphery of a regional snow cover, where horizontal air temperature gradients are relatively great. This places the snow-covered region on the cold, snowy side of migrating winter storms, thereby adding to the snow cover and reinforcing the chill (Chapter 7). An unusually widespread winter snow cover favors persistence of an episode of cold weather. On the other hand, less than the usual extent of winter snow cover raises average air temperatures.

In October 2005, a team of scientists reported in *Science* magazine on how land-surface changes were contributing to a major summer warming trend in the Alaskan Arctic. In that region, summers (June-August) are now warmer than at any time in the past 400 years. The summer warming trend accelerated from about 0.15-0.17 Celsius degree per decade in 1961-1990 and 1966-1995 to about 0.3-0.4 Celsius degree per decade in 1961-2004. Based on data acquired by in situ and remote sensing methods, scientists attributed the summer warming principally to a longer snow-free season. A significant lowering of land-surface albedo in late spring means that sensible heating of the lower atmosphere also begins earlier.

With an average snowmelt advance of about 2.5 days per decade in the Alaskan Arctic, the amount of additional energy absorbed and transferred to the atmosphere is approximately 3.3 W/m² per decade. Interestingly, this is comparable in magnitude to the amount of energy predicted by climate models to accompany a doubling of atmospheric CO_2 concentration over a period of many decades.

Also contributing to a reduction in land-surface albedo and summer warming in the Alaskan Arctic—albeit to a much lesser extent than snow cover changes—are shifts in vegetation. With warmer summers, tall shrubs have expanded their coverage and white spruce forests are invading the tundra. So far, shrub expansion and tree line advances have affected a relatively small area so that their combined contribution to summer warming caused by land-surface change is estimated at only about 5%. However, continuation of current trends in vegetation changes could significantly amplify this contribution to summer warming.

Whereas changes in regional snow cover might impact climate variability over the short-term (seasonal), changes in Earth's sea ice or glacial ice coverage are likely to have longer-lasting effects on climate. Worldwide, sea ice (formed from the freezing of seawater) covers an average area of about 25 million km² (9.6 million mi²), about the area of the North American continent. Terrestrial ice sheets, ice caps, and mountain glaciers cover a total area of about 15 million km² (5.8 million mi²), roughly 10% of the land area of the planet. Ice (especially snow-covered ice) reflects much more incident solar radiation than either the ocean or snow-free land so that any change in glacial or sea ice cover would affect climate.

SHRINKAGE OF ARCTIC SEA-ICE COVER

A major concern associated with the current global warming trend is shrinkage of Arctic sea-ice cover. Although melting of floating sea ice does not raise sea level, it can alter climate significantly. Shrinkage of Arctic sea ice is likely to trigger an **ice-albedo feedback** mechanism that would accelerate melting of sea ice and amplify warming (Figure 11.13). Sea ice insulates the overlying air from warmer seawater and reflects much more incident solar radiation than ocean water. The albedo of snow-covered sea ice is about 85% whereas ice-free Arctic Ocean water has an average albedo of about 40%. As sea ice cover shrinks, the greater area of ice-free ocean waters absorbs more solar radiation, sea-surface temperatures rise, and more ice melts. Furthermore,

warmer water slows the formation of ice in autumn. This positive feedback mechanism could rapidly reduce Arctic sea ice cover and greatly alter the flux of heat energy and moisture between the ocean and atmosphere with possible ramifications for global climate.

Less sea ice cover on the Arctic Ocean is likely to increase the humidity of the overlying air leading to more cloudiness. Clouds cause both cooling (by reflecting sunlight to space) and warming (by absorbing and emitting to Earth's surface outgoing infrared radiation). During the long dark polar winter, clouds would have a warming effect. In summer, the impact of a greater cloud cover depends on the altitude of the clouds. Cooling would prevail with an increase in low cloud cover whereas warming would likely accompany an increase in high cloud cover.

Recent research indicates that Arctic sea ice first formed about 47.5 million years ago. In 2009, Norwegian scientist Catherine Stickley and colleagues reported finding fossil algae in 47.5 million-year-old sediments extracted from a submarine ridge about 250 km (155 mi) from the North Pole. The fossil algae are related to algae that live in Arctic sea ice today. In modern time, Arctic sea ice cover varies seasonally (Figure 11.14) and exhibits some long term trends. A variety of sources provides information on the extent of Arctic sea ice cover since the early part of the 20th century. Ship and aircraft

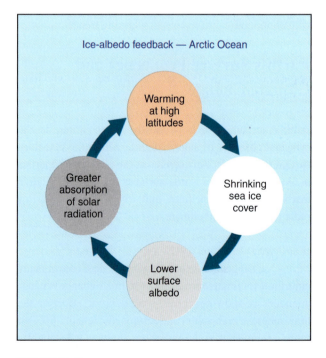

FIGURE 11.13
Positive ice-albedo feedback in the Arctic is likely to accelerate warming of surface waters and shrinkage of sea ice cover.

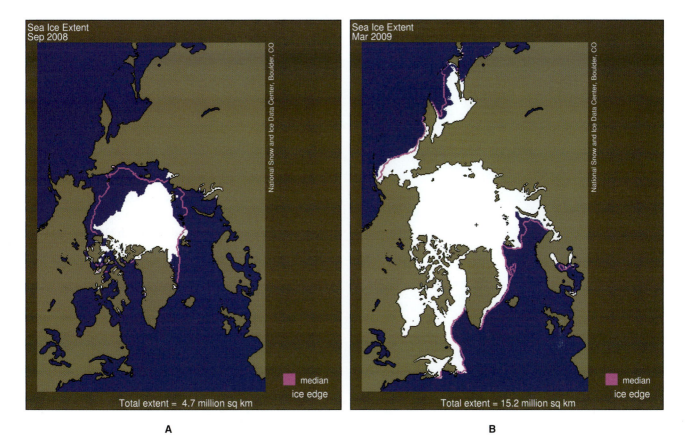

FIGURE 11.14
Total area covered by sea ice (not including the open water within the pack ice) in the Northern Hemisphere polar region in (A) September 2008 and (B) March 2009. These two months represent the times of minimum and maximum ice extent respectively over a 12-month period. Average ice conditions are estimated using passive microwave sensors on Earth-orbiting satellites. [NOAA, National Snow and Ice Data Center]

observations indicate that the multi-year Arctic ice cover remained essentially constant in all seasons through the first half of the 20th century. But beginning in the 1950s, observations by ships and aircraft detected shrinkage in the summer minimum extent of ice while the winter maximum remained nearly constant. By the mid-1970s, surveillance by satellites and submarines plus ice-core measurements found a decline in the winter maximum extent of ice as well.

Norwegian researchers reported that the area covered by multi-year ice in the Arctic decreased by about 14% between 1978 and 1998. Their findings were based on satellite monitoring of the spectrum of microwave energy emitted by the ice. Comparing ice-thickness measurements made by U.S. Navy submarines from 1958 to 1976 with those made during the *Scientific Ice Expedition* program in 1993, 1996, and 1997, University of Washington scientists discovered that ice had thinned from an average thickness of 3.1 m (10 ft) to an average thickness of 1.8 m (6 ft). Thinning at the

rate of 15% per decade translates into a total volumetric ice loss of about 40% in three decades. (Upward-looking acoustic sounders mounted on submarines were used to map ice depth.)

After 2000, the rate of reduction of Arctic sea ice cover accelerated. From analysis of satellite data, scientists from the Cooperative Institute for Research in Environmental Sciences (CIRES) at the University of Colorado reported that the extent of Arctic sea ice in 2002 was the lowest in the satellite record, likely the lowest since the early 1950s, and perhaps the lowest in several centuries. In September 2002, sea ice covered about 5.3 million km^2 (2 million mi^2) compared to the long-term average of about 6.3 million km^2 (2.4 million mi^2). According to the IPCC, from 1979 through 2006, the extent of Arctic sea ice decreased by 2.7 ± 0.6 percent per decade in the annual average and 7.4 ± 2.4 percent per decade for the end-of-summer. Between 1981 and 2000, average ice thickness decreased by about 1.13 m (3.7 ft) or 22%.

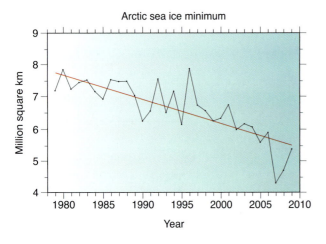

FIGURE 11.15
Arctic sea ice has been shrinking at least since the time satellite monitoring began in 1979, reaching a new end-of-season record low in 2007. [Adapted from NOAA's National Snow and Ice Data Center, Boulder, CO]

According to scientists at NOAA's National Snow and Ice Data Center, the extent of end-of-summer (September) sea ice cover in the Arctic in 2005 was the lowest since satellite monitoring of the polar region began in 1979 (Figure 11.15). That record was broken in 2007 when the summer minimum decreased to 4.2 million km^2 (1.6 million mi^2), which was 38% below the long-term average and 23% below the previous record low. This sharp decline in Arctic ice cover was attributed to a combination of factors including the long-term trend in ice thinning and shrinkage, unusually strong summer winds that pushed large amounts of ice out of the Arctic basin creating a broad area of open water and thin ice in the Arctic Ocean, and less than the usual cloud cover that allowed more solar radiation to reach the ocean surface. The ice-albedo feedback accelerated the warming and melting during the summer of 2007. In September 2008, the extent of Arctic sea ice cover was 4.67 million km^2, second to 2007 as the lowest since the beginning of satellite monitoring. However, going into the melt season, much of winter ice was thin (first-year ice) so that the volume of floating ice may have set a new record low. At the end of the 2009 Arctic summer, more ice cover remained than during the prior two record-setting years of 2007 and 2008, but the sea ice extent was the third lowest since 1979.

With the decline of Arctic sea-ice cover to record and near record summer minima, some scientists speculate that Arctic ice may have reached its tipping point; that is, complete loss of summer ice may be imminent. A team of scientists tested this hypothesis using 6 climate models (out of a field of 23) that best simulated the slow historical decline and summer-to-winter variation in Arctic sea ice cover. Although the models predict an Arctic Ocean free of summer ice by 2037, the loss is unlikely to occur as a single abrupt event (i.e., no tipping point). According to ice physicists Ian Eisenman of California Institute of Technology and John Wettlaufer of Yale University, the absence of a tipping point can be explained by the presence of an ice-thickness feedback that partially counters the ice-albedo feedback. Warmer conditions in the Arctic have caused the sea ice to thin. Thin ice conducts heat from the ocean to the atmosphere more readily than thick ice so that the ice cover grows back faster in winter.

Shrinkage of the Arctic sea ice may be the direct consequence of higher air temperatures or indirectly the result of changes in ocean circulation (i.e., greater input of warmer Atlantic water under the Arctic sea ice). Some scientists argue that higher air and ocean temperatures in the Arctic are natural variations in climate associated with the *Arctic Oscillation (AO)* and that the ice-cover will return to normal after the AO changes phase (Chapter 8). However the phase of the AO has not been consistently positive over the past decade when the Arctic sea-ice cover declined to record low levels. Instead, several other sea level pressure patterns have, perhaps coincidentally, provided wind patterns that exported sea ice out of the Arctic. It may be that thinner ice is easier to advect. Furthermore, the latest research shows that the warm water advected into the Arctic from the Atlantic has remained below the halocline, and so has not obviously contributed to sea ice decline.

Mean annual air temperatures in the Arctic have climbed about 0.5 Celsius degree (0.9 Fahrenheit degree) per decade over the past thirty years. In 2007, the mean annual temperatures in the Arctic were more than 3 Celsius degrees (5.4 Fahrenheit degrees) above the 1951-1980 mean. Proxy climate data indicate that present temperatures (especially in winter and spring) may be at their highest level in four centuries. In response, mountain glaciers in Alaska are shrinking at historically unprecedented rates (Chapter 12), permafrost is beginning to thaw, and freshwater runoff into the ocean has increased. More winter precipitation combined with an increased flow of groundwater (due to melting permafrost) is likely responsible for a 7% increase in the discharge of six major Eurasian rivers into the Arctic Ocean since the 1930s.

Input of more fresh water from rivers and melting glaciers would impact the ocean thermohaline circulation by reducing the salinity and thus the density of surface ocean waters. Recall from Chapter 6 that relatively dense cold, salty water sinks at high latitudes of the Atlantic Ocean, so the ocean may become more stratified between less dense surface waters and cold denser water below.

SEA-SURFACE TEMPERATURE PATTERN

Changes in ocean circulation and sea-surface temperatures (SST) contribute to large-scale climate change and climate variability. As described in detail in Chapter 8, changes in SST patterns accompanying El Niño and La Niña significantly influence inter-annual climate variability not only in the tropical Pacific but through teleconnections around the globe. Ocean circulation includes warm and cold surface currents and the deep-ocean thermohaline circulation that transports heat energy throughout the world. Regular changes in the strength of this circulation may explain millennial-scale (1400- to 1500-year) climate cycles over the past 10,000 years. A strong thermohaline circulation brings a relatively mild climate (for the latitude) to Western Europe whereas a weakening of the thermohaline circulation triggers cooling. Such climate shifts can be abrupt, occurring in a decade or less.

Conclusions

This chapter identifies and describes many potential causes of climate change, but we have treated the various forcing agents and mechanisms as if each one acted independently of all the others. While instructive, this simplification is not realistic. Within the Earth-atmosphere-land-ocean system, many factors are linked in complex cause-effect chains. Many forcing agents and mechanisms, both internal and external to Earth's climate system, work together in setting the boundary conditions that bring about climate variability and climate change. These factor interactions involve feedback loops that may at one extreme amplify (*positive feedback*) and at the other extreme weaken (*negative feedback*) fluctuations in climate (Chapter 2).

The next chapter continues our examination of the causes of climate variability and climate change, focusing on the influence of human activity. The anthropogenic impact on the climate system, particularly enhancement of the greenhouse effect, is the principal contributor to the post-1970s global warming trend. Continuation of this warming trend coupled with other changes in climate is likely to impact virtually every sector of society.

Basic Understandings

- Matching a particular forcing mechanism with a climate response based on similar periodicity is no guarantee of a real cause-effect relationship.
- Factors that could alter the global radiative equilibrium and change Earth's climate include fluctuations in solar energy output, volcanic eruptions, variations in atmospheric chemistry, changes in properties of Earth's surface, and certain human activities.
- According to standard models of stellar evolution, about 4 billion years ago, the Sun's energy output was about 70% of what it is today. On Earth, the cooling effect of a fainter Sun was offset by warming caused by elevated levels of greenhouse gases.
- A sunspot typically lasts only a few days, but the rate of sunspot generation is such that the number of sunspots varies systematically with the time between successive sunspot maxima or minima, averaging about 11 years.
- Changes in the Sun's total energy output are apparently related to sunspot activity. Solar output varies directly and minutely with sunspot number; that is, a slightly brighter Sun emitting more radiation has more sunspots (because of a concurrent increase in faculae, bright areas), and a slightly dimmer Sun emitting less radiation exhibits fewer sunspots.
- The Maunder minimum, a 70-year period of greatly diminished sunspot activity between 1645 and 1715, occurred about the same time as a relatively cold phase of the Little Ice Age, prompting some scientists to propose a cause-effect relationship.
- The magnitude of the variation in solar energy output during an 11-year solar cycle is so small that some amplification mechanism is required for sunspots to influence Earth's climate.
- Milankovitch cycles consist of regular changes in the precession of Earth's spin axis (period of 23,000 years), the tilt of Earth's spin axis (period of 41,000 years), and the eccentricity of Earth's orbit about the Sun (period of 90,000 to 100,000 years).
- Milankovitch cycles drive climatic oscillations operating over tens of thousands to hundreds of thousands of years and were likely responsible for the major advances and recessions of the

Laurentide ice sheet over North America during the Pleistocene Epoch. These same cycles also show up in deep-sea sediment cores that date from the Pleistocene Ice Age.

- Milankovitch cycles do not appreciably alter the total amount of solar energy received by Earth's planetary system annually, but they do change significantly the latitudinal and seasonal distribution of incoming solar radiation.

- Over the vast expanse of geologic time, plate tectonics was a major player in large-scale climate change. As tectonic plates move, the continents they carry also move (changing latitude), ocean basins open and close (altering ocean currents), and tectonic stresses build mountain ranges (shifting atmospheric circulation), and produce volcanic eruptions (changing the chemical composition of the atmosphere).

- Only violent volcanic eruptions in low latitudes rich in sulfur dioxide gases are likely to impact hemispheric or global-scale climate. Such eruptions are unlikely to lower the mean annual global surface temperature by more than about 1.0 Celsius degree (1.8 Fahrenheit degrees) for a year or two.

- Sulfurous aerosols (sulfuric acid droplets and sulfate particles) absorb both incoming solar radiation and outgoing infrared radiation, warming the lower stratosphere. Sulfurous aerosols also reflect solar radiation to space, thereby cooling the lower troposphere.

- Variations in the atmospheric concentration of radiatively active gases can cause global climate change. The concentrations of infrared-absorbing gases (e.g., carbon dioxide, methane, nitrous oxide) have fluctuated throughout Earth history with significant impact on Earth's surface temperature.

- A thermokarst lake occupies a depression in the ground formed when warming causes melting of ice-rich permafrost (permanently frozen ground) and water drains into the depression. Anaerobic decay of organic sediments on the lake bottom generates methane gas that bubbles to the lake surface and escapes to the atmosphere.

- Earth's surface, which is mostly ocean water, is the prime absorber of solar radiation so that any change in the physical properties of water or land surfaces or in the relative distribution of ocean and land may impact the global radiation balance and climate.

- Changes in mean regional snow cover may contribute to climate variability because an extensive snow cover has a refrigerating effect (positive feedback) on the atmosphere.

- Shrinkage of Arctic sea ice cover is likely to trigger an ice-albedo feedback mechanism that would accelerate the melting of sea ice and amplify warming. Sea ice insulates the overlying air from warmer seawater and reflects more incident solar radiation than ocean water.

- After 2000, the rate of reduction of Arctic sea ice cover accelerated and the end-of-summer ice extent reached a record low in 2007. If current trends continue, by 2037 the Arctic Ocean may be free of sea ice in summer.

Enduring Ideas

- Agents or mechanism that could alter global radiative equilibrium and thereby change the climate include fluctuations in solar energy output, regular variations in Earth-Sun geometry, plate tectonics, volcanic eruptions, variations in atmospheric chemistry, changes in Earth's surface properties, and some human activities.
- The Sun's total energy output varies directly with sunspot number and amounts to a fluctuation of only 0.1% between sunspot maximum and sunspot minimum. Hence, for sunspots to significantly impact climate requires an amplification mechanism within Earth's climate system.
- The Milankovitch Earth-Sun cycles, operating over tens of thousands to hundreds of thousands of years, are the principal forcing mechanism for the major fluctuations in climate and glacial ice cover during the Pleistocene Epoch.
- Continental drift, mountain building, and volcanic activity are climate change forcing mechanisms that operate over a broad range of time frames.
- Changes in Earth's surface properties can bring about change in both climate variability and climate change. Especially significant are changes in snow and ice cover on land and at sea because of the radiative properties of snow and ice.

Review

1. What is meant by global radiative equilibrium?
2. How does global radiative equilibrium illustrate the law of energy conservation?
3. Identify and describe three forcing agents or mechanisms that could alter the global radiative equilibrium and cause global climate change.
4. What is a sunspot and how does it develop?
5. How does the Sun's energy output vary with sunspot number?
6. How did Croll's explanation for the initiation of the major Pleistocene glaciations differ from the explanation put forth by Milankovitch?
7. Identify and describe the three Milankovitch Earth-Sun cycles.
8. What discovery convinced geoscientists that Milankovitch's astronomical theory explained the major climate fluctuations of the Pleistocene Epoch?
9. Describe the role of sulfurous aerosols in climate change induced by a volcanic eruption.
10. Why does an extensive winter snow cover tend to be self-sustaining?

Critical Thinking

1. Describe the Maunder minimum and its possible connection to the climate of the Little Ice Age.
2. Explain how chemical analysis of tree growth rings supports a linkage between variations in solar activity and climate.
3. In general, how do the Milankovitch cycles affect incoming solar radiation?
4. Describe how geoscientists reconstructed the Pleistocene Ice Age chronology from deep-sea sediment cores.
5. Identify the various ways whereby plate tectonics may cause large-scale climate change.
6. How does mountain building affect the atmospheric concentration of the greenhouse gas CO_2?
7. How do violent volcanic eruptions rich in sulfur dioxide (SO_2) affect the stratospheric ozone shield?
8. How does cloud cover affect surface air temperatures?
9. Explain how ice-albedo feedback affects the extent of Arctic sea ice cover.
10. Besides shrinkage of Arctic sea ice, what are some of the other impacts of warming in the Arctic region?

ESSAY: Radiative Equilibrium Climate Model

We can express the condition of radiative equilibrium in Earth's planetary (atmosphere, land, ocean) system using a simple numerical climate model that equates energy input to energy output. By perturbing the variables in the model, we can simulate climate change.

Solar radiation that is incident on Earth's planetary system is given by the solar constant, S (approximately 1370 W/m^2). A portion of this energy is scattered or reflected back into space, the amount depending on the planetary albedo, α. The energy actually available to drive the atmospheric circulation is

$$S(1-\alpha)$$

As noted in Chapter 3, Earth intercepts this energy as a disk having an area of πR^2, where R is the radius of the Earth. Hence, the net energy input is given by

$$S(1-\alpha)(\pi R^2)$$

Energy is emitted by Earth's planetary system in the form of infrared radiation. Assuming that Earth's planetary system were a perfect radiator (*blackbody*), then we can describe the energy output by the *Stefan-Boltzmann law*. According to this law, the total radiational energy output of a blackbody at all wavelengths is directly proportional to the radiating temperature, T (in kelvins) raised to the fourth power, that is, T^4. In this relationship, the constant of proportionality is σ, the Stefan-Boltzmann constant. However, Earth's planetary system is not quite a blackbody, so we must introduce a correction factor to account for its actual radiating properties. The correction factor, ε, is called the *effective emissivity*. The energy output of Earth's planetary system is therefore

$$\varepsilon\sigma T^4$$

This energy is emitted from the entire surface area ($4\pi R^2$) of the nearly spherical Earth. The total energy output is thus given by

$$(4\pi R^2)\,\varepsilon\sigma T^4$$

At radiative equilibrium, what enters the system (energy input) must equal what exits the system (energy output), so

$$S(1-\alpha)(\pi R^2)=(4\pi R^2)\,\varepsilon\sigma T^4$$

This relationship simplifies to

$$S(1-\alpha)=4\,\varepsilon\sigma T^4$$

Solving for T, the temperature of Earth's planetary system at radiative equilibrium, we have

$$T=[S(1-\alpha)/4\,\varepsilon\sigma]^{1/4}$$

Earth's radiative equilibrium temperature thus depends on the solar constant, planetary albedo, and the effective emissivity. A change in any one or combination of these variables will change the value of T and hence, the climate. Consider some illustrations.

Keeping α and ε constant, by mathematical manipulation, we can demonstrate that a 1% change in the solar constant translates into a 0.6 Celsius degree (1.1 Fahrenheit degrees) change in the planetary radiative equilibrium temperature. While this variation does not appear to be a major temperature change, recall that trends in climate are geographically non-uniform in magnitude (and direction). In some localities, this temperature change will be amplified considerably.

If we keep S and ε constant and increase the planetary albedo from its present value of 31% up to 36% (perhaps by increasing the ocean's ice cover), the radiative equilibrium temperature will drop by about 4.5 Celsius degrees (8.1 Fahrenheit degrees).

As a third illustration, we could hold S and α constant and vary the effective emissivity. As noted earlier, the effective emissivity depends on the radiative properties of Earth's planetary system, so any change in the composition (chemistry) of that system could alter ε. Such chemical changes might involve, for example, variations in levels of the greenhouse gases.

ESSAY: Climate Change and the Drying of the Mediterranean Sea

Late in the Miocene Epoch, about 5.6 million years ago, the Mediterranean Sea became isolated from the Atlantic Ocean and almost completely dried up. Then, abruptly, perhaps in less than two years, the Mediterranean refilled. Accompanying these drastic environmental changes were changes in climate.

The continents of Africa, Europe, and Asia surround the Mediterranean Sea, which connects to the Atlantic Ocean via the narrow Strait of Gibraltar. The Mediterranean is a remnant of the once vast *Tethys Sea* but was nearly squeezed shut during the Oligocene Epoch, 23 to 33 million years ago. The Mediterranean basin continues to be tectonically active and is slowly shrinking in size as the African plate moves northward, pushing (subducting) under the Eurasian plate. This subduction is responsible for large-scale bending and uplift of marine sedimentary rocks that formed the Alps along with volcanic activity at the northern edge of the Mediterranean.

The world's largest inland sea with an area of 2,499,350 km² (965,000 mi²), the Mediterranean is about 3900 km (2400 mi) long and its maximum width is about 1600 km (1000 mi) (Figure 1). The average water depth is 1500 m (4900 ft) with a maximum depth of 5150 m (16,900 ft) off the south coast of Greece. A bathymetric sill in the Strait of Gibraltar separates the West Mediterranean basin from the Atlantic Ocean basin; the maximum water depth over the sill is only 284 m (932 ft) at a point where the Strait is about 30 km (18.6 mi) wide. At its narrowest point, the Strait of Gibraltar is about 14 km (9 mi) wide. The Sicily sill separates the West Mediterranean basin from the deeper East Mediterranean basin.

FIGURE 1
Clouds and dust are visible over the East Mediterranean basin on 18 February 2010 in this image acquired by the Moderate Resolution Imaging Spectroradiometer (MODIS) on NASA's Terra satellite. Thick plumes of dust are transported from the Sahara to the Nile Delta, Crete, and southern Greece. [NASA image courtesy of Jeff Schmaltz, MODIS Rapid Response Team at NASA GSFC]

In the 1960s, William Ryan of the Lamont-Doherty Earth Observatory at Columbia University made a remarkable discovery regarding the composition of sediments on the floor of the Mediterranean. While sailing in the Mediterranean on the *R/V Chain* from Woods Hole (MA) Oceanographic Institution, Ryan used a new continuous seismic profiler that could penetrate sea bottom sediments. The acoustic return located a reflecting layer 100 to 200 m (325 to 650 ft) beneath the sea bottom, whimsically labeled the "mysterious layer," or M-layer. Through subsequent years of collecting M-layer data, it soon became obvious that the layer was ubiquitous in the Mediterranean and was deposited after the deep basin of the Mediterranean had already formed and had almost the same bathymetry as today.

In 1972, Ryan and Kenneth Hsü (a Swiss geologist) onboard the drill ship *Glomar Challenger* in the western Mediterranean brought up the first cores of the M-layer. The cores resembled marble and were labeled "the pillars of Atlantis." The core material turned out to be anhydrite (calcium sulfate, $CaSO_4$) and stromatolites (mats of the remains of sediment-trapping cyanobacteria) dating from the late Miocene Epoch. These two types of sediment core material were extremely unusual in that anhydrites form only in hot, dry deserts where salty groundwater close to the desert surface evaporates causing calcium sulfate to precipitate as a solid. How is it possible for evaporites to form beneath 200 m (650 ft) of marine sediments on the bottom of the Mediterranean Sea? Stromatolites also appear to be out of place on the bottom of the present Mediterranean. Today, stromatolites form in broad, intertidal mud flats in the Bahamas and in salty bays in Western Australia and require light for photosynthesis. What are they doing on the bottom of the Mediterranean, and why are they associated with desert evaporites?

Determination of the M-layer composition points to one of the most extraordinary events in the history of planet Earth. Plate tectonics, glaciation, and sea level fluctuation combined to drastically alter the climate of the region now occupied by the Mediterranean Sea. About 5.6 million years ago, the Mediterranean Sea became isolated from the Atlantic Ocean and almost completely dried up. For hundreds of thousands of years, the basin bottom consisted of deserts, salty lakes, and salt marshes. Then sea level rose and Atlantic water poured over the sill at the Strait of Gibraltar refilling the Mediterranean basins.

Several hypotheses seek to explain the drying up and refilling of the Mediterranean, most of which invoke tectonic plate movement, large-scale glaciation, and evaporation. The tectonic hypothesis suggests that the entire basin was uplifted and the Mediterranean Sea emptied into the Atlantic, and later dropped back down and refilled with water. This tectonic cycle may have repeated several times. The driving force for uplift may have been subduction of the African plate under the Eurasian plate, which caused similar uplift in the Alps. Another hypothesis attributes the drying to large-scale glaciation that caused sea level to drop below the sill at Gibraltar, cutting off input of water from the Atlantic Ocean. This is a reasonable hypothesis considering that the maximum depth of the sill at Gibraltar is only 284 m (932 ft). When the ice sheets melted, sea level rose above the sill and refilled the Mediterranean basin. Other hypotheses propose that either the Strait of Gibraltar was squeezed closed or parts of the Atlantic Ocean floor were deformed and uplifted forming a gate that alternately closed and opened the Strait of Gibraltar.

Evaporation of the isolated Mediterranean Sea is readily explained by reconstructions of the late Miocene climate. The rock record shows that at the time, the climate of the Mediterranean region probably was as hot and arid as it is today with little input of water by rivers and streams—the type of climate that would readily evaporate Mediterranean waters leaving behind anhydrite deposits.

The most reasonable explanation for the drying of the Mediterranean Sea is likely a combination of factors. In the late Miocene, about 5.6 million years ago, tectonic stresses (causing moderate regional uplift) combined with glaciation to drop sea level below the sill at Gibraltar, cutting off the Mediterranean Sea from the Atlantic Ocean. Over the subsequent 1000 years, the waters of the Mediterranean evaporated, eventually exposing the bottom of the basins where deserts formed among shallow hypersaline lakes, salt marshes, and salt flats. With the loss of the Mediterranean Sea, the climate of the region probably became cooler and even more arid. In the deeper East Mediterranean basin, a system of salty lakes, similar to North America's Great Salt Lake, was fed by salty overflow from the Black Sea located to the east. Such conditions can produce both anhydrites and stromatolites that together accumulate on the bottom of the Mediterranean. Alternating layers of biogenous sediments and evaporites suggest that the Mediterranean probably partial refilled and dried up between 8 and 40 times. Finally about 5.3 million years ago, tectonic forces relaxed dropping Gibraltar and/or sea level rose causing Atlantic waters to spill over the Gibraltar sill and refill the basins.

Hsü calculated that the flow of water over the Gibraltar sill was about 1000 times greater than Niagara Falls and took a century to fill the Mediterranean basins. However, recent field study supports an alternate view that refilling occurred in a much shorter time span. Key information came from sediment cores and seismic analysis associated with a train tunnel project planned for under the Strait of Gibraltar linking Europe and Africa. Exploration of the sea floor for the proposed tunnel revealed a deep sediment-filled channel with a U-shaped profile measuring 200 km (124 mi) long, 6-11 km (4-7 mi) wide, and 300-650 m (980-2100 ft) deep. D. Garcia-Castellanos of the Spanish National Research Council and colleagues interpreted the buried channel as excavated by a megaflood. Writing in the 10 December 2009 issue of *Nature*, the researchers proposed that the flow of Atlantic water into the dry Mediterranean rapidly eroded the sill at Gibralter so that the discharge of water increased exponentially. At its peak, the discharge of the megaflood was about 1000 times greater than that of today's Amazon River. With the water level of the Mediterranean rising more than 10 m (33 ft) per day, the basin refilled in only a few months to no more than two years.

CHAPTER 12

ANTHROPOGENIC CLIMATE CHANGE AND THE FUTURE

Smokestack emissions. [Courtesy U.S. Fish and Wildlife Service]

Case-in-Point

An estimated 1.34 billion people (20% of the world's population) live in China, a developing nation undergoing rapid urbanization and industrialization. As one indication of this growth, the number of cars sold in China increased from 1.2 million in 1999 to 13.6 million in 2009. China is now the world's largest auto market surpassing the U.S. in new vehicle production and sales volume. But with economic growth has come increasing demand for energy and, as a consequence, rising carbon emissions (i.e., emissions of the greenhouse gas CO_2). Coal burning, the chief source of carbon emissions, supplies about two-thirds of China's primary energy; the nation's coal reserves are among the largest in the world.

According to Ning Zeng of the University of Maryland and colleagues, a key economic consideration in assessing the potential climate impact of China's industrialization is *carbon intensity*, that is, the carbon emission per unit Gross Domestic Product (GDP). Over a recent 27-year period, China's GDP increased at an annual rate of 9.5% while CO_2 emissions increased at the rate of 5.4% per year so that carbon intensity actually decreased. This was attributed to higher energy efficiencies but may also reflect uncertainties in emission data bases. However, accelerated economic development coupled with China's large population will lead to a major increase in energy demand. If future carbon intensity were to keep pace with a GDP growth rate of 7% per year, by 2030, China's total carbon emissions would match today's total worldwide carbon emissions. With the downturn in economic conditions worldwide in 2008, China's GDP growth rate has slowed but only temporarily.

Compounding the challenge of maintaining economic growth and managing greenhouse gas emissions is China's vulnerability to climate change. China's climate is extremely diverse, ranging from subtropical in the south to subarctic in the north. Over most of the nation, monsoon circulation dominates the climate, determining the timing of the rainy season and the amount of precipitation. A varied topography accounts for considerable regional variation in precipitation and temperature. In general,

average annual rainfall decreases and becomes less reliable toward the north and northwest. Furthermore, on average, about 5 typhoons strike China's southern and eastern coasts each year and many areas of the nation are prone to devastating droughts and floods.

Climate models predict that climate change will severely impact portions of China. The nation's three main industrial areas are situated in lowlands which are within the 92,000 km² of the country that would be inundated if sea level rose 1 m. In northwest China, one-fifth of the mountain glaciers have melted over the past 50 years. An additional temperature rise of 3 to 6 Celsius degrees (5 to 11 Fahrenheit degrees) is predicted for Tibet by 2100. This warming will cause the permafrost to thaw and is likely to affect water resources for China and its neighbors. The headwaters of the Yangtze, Yellow, Ganges, and Mekong Rivers are in the Tibetan plateau. Over southern and eastern China, a greater frequency of floods and drought is expected. Additional drying of the nation's arid and semi-arid lands may cut agricultural output by 5-10% by 2030.

Decreasing China's carbon intensity will require increased reliance on alternative energy sources such as solar power and wind power. In addition, greater energy efficiency is needed

Driving Question:
How do human activities affect global climate and how significant are those influences compared to natural causes of climate change?

The previous chapter examined how various natural forcing agents and mechanisms contribute to climate variability and climate change. The principal focus of this chapter is the anthropogenic influence on climate. Human activity affects large-scale climate via emission of infrared-absorbing gases that enhance the greenhouse effect and aerosols that interact with incoming solar and outgoing terrestrial radiation. In addition, human transformations of the land surface impact climate. We compare the magnitude of anthropogenic versus natural influences on climate in terms of radiative forcing units, a means of scaling component effects upon the climate system in terms of radiative energy per unit time and unit area (i.e., W/m²).

Climate scientists rely primarily on numerical global climate models in experiments to predict the climate future. In addition, the instrument-based and reconstructed

climate record is subject to careful scrutiny in efforts to identify statistically significant cycles or trends that can be extrapolated into the future. Since economic and societal impacts are local or regional, scientists search for analogs that may help in assessing regional response to large-scale climate change. Regional climate models are also linked to global climate models. Because of increasing greenhouse gas emissions, global warming appears likely to continue through the remainder of this century and probably well beyond. Even if greenhouse gas emissions were reduced in the next several decades, given the long lifetime of CO_2 in the atmosphere, it will take some time for Earth's climate system to recover from the greenhouse gases already emitted to the atmosphere by anthropogenic activities. In closing, this chapter summarizes the actual and potential consequences of a continuation of the current global warming trend.

Human Activity and Climate Change

In 2007, the **Intergovernmental Panel on Climate Change (IPCC)** concluded that global warming since the mid-20th century *very likely* (estimated probability greater than 90%) was caused mostly by human activities. In a report issued six years earlier, the IPCC described the human role in global warming as *likely* (estimated probability higher than 66%).

Many human activities affect climate over broad ranges of spatial and temporal scales. Humans modify the landscape (e.g., urbanization, clear-cutting of forests) thereby altering radiation properties of Earth's surface. For reasons presented in Chapter 4, cities are slightly warmer than the surrounding countryside (the *urban heat island effect*). Combustion of fossil fuels (i.e., coal, oil, and natural gas) alters concentrations of certain key gaseous and aerosol components of the atmosphere. Of these human impacts on Earth's climate system, the last one is most likely influencing climate on a hemispheric or global scale.

TRENDS IN GREENHOUSE GASES

Many scientists, public policymakers, and the general public are concerned about how the steadily rising concentration of carbon dioxide (CO_2) and other infrared-absorbing gases in the atmosphere is affecting the climate. Higher levels of these gases enhance the greenhouse effect, contributing to warming on a global scale. The *Synthesis Report of the IPCC Fourth Assessment Report*, issued in November 2007, concluded that: "Warming of the climate system is unequivocal as is now evident from observations of increases in global average air and ocean temperatures, widespread melting of snow and ice, and rising global average sea level." The Report goes on to state: "Most of the increase in globally-averaged temperatures since the mid-20th century is *very likely* due to the observed increase in anthropogenic GHG (greenhouse gas) concentrations." This is the *Callendar effect* described in Chapter 3.

Recall from Chapter 3 that systematic monitoring of atmospheric carbon dioxide levels began in 1957 at NOAA's Mauna Loa Observatory in Hawaii under the direction of Charles D. Keeling (1928-2005) of Scripps Institution of Oceanography. The observatory, situated on the northern slope of Earth's largest volcano 3397 m (11,140 ft) above sea level in the middle of the Pacific Ocean, is sufficiently distant from major sources of air pollution that carbon dioxide levels are considered representative of at least the Northern Hemisphere. Also since 1957, atmospheric CO_2 has been monitored at the South Pole station of the U.S. Antarctic Program and that record closely parallels the one at Mauna Loa. The Mauna Loa record (the *Keeling curve*) shows a sustained increase in average annual atmospheric carbon dioxide concentration from about 316 ppmv (parts per million by volume) in 1959 to 386 ppmv near the end of 2008 (Figure 3.29).

Carbon dioxide is a byproduct of the burning of coal and other fossil fuels. The upward trend in atmospheric carbon dioxide was underway long before Keeling's monitoring and appears likely to continue well into the future. As pointed out in the first Essay of Chapter 2, the anthropogenic contribution to the buildup of atmospheric CO_2 may have begun thousands of years ago with land clearing for agriculture and settlement. By the middle of the 19th century, growing dependency on coal burning associated with the beginnings of the Industrial Revolution triggered a more rapid rise in CO_2 concentration. The concentration of atmospheric CO_2 is now about 35% higher than it was in the pre-industrial era. Fossil fuel combustion accounts for roughly 75% of the increase in atmospheric carbon dioxide while deforestation (and other land clearing) is likely responsible for the balance. With continued growth in fossil fuel combustion, the atmospheric carbon dioxide concentration could top 550 ppmv (double the pre-industrial level) by the end of the present century.

The ocean is a major reservoir in the global carbon cycle and as such plays an important role in governing the amount of carbon dioxide in the atmosphere. From the beginning of the Industrial Revolution, scientists were able to estimate the amount of CO_2 released to the atmosphere by human activity. But when the actual amount of carbon dioxide in the atmosphere was measured, about half the estimated amount was missing. The "missing" carbon is distributed between sinks in the ocean and on land. According to IPCC estimates in 2007, the ocean takes up 56.2% of the carbon dioxide of anthropogenic origin (via photosynthesis and cold surface waters absorbing CO_2 and sequestering it when it sinks below the surface) while terrestrial biomass is a sink for 13.7%. Ultimately, the rising concentration of atmospheric CO_2 is driving the net flux of carbon dioxide across the air-sea interface. Some anthropogenic carbon dioxide is transported with the ocean's thermohaline circulation and may be sequestered for about 1000 years before returning to the ocean-atmosphere interface. However, uptake of carbon dioxide by ocean waters is likely to decline in the future as the surface ocean warms in response to higher global air temperatures.

Besides carbon dioxide, rising levels of other infrared-absorbing gases associated with human activity (e.g., methane, nitrous oxide, halocarbons, and ozone) enhance the greenhouse effect. Based on analysis of air bubbles trapped in glacial ice cores, the concentration of methane in the atmosphere is now greater than at any time in the past 400,000 years. Methane is a product of the decay of organic matter in the absence of oxygen (*anaerobic decay*). Anthropogenic sources of methane are direct (e.g., rice cultivation, cattle, and landfills) and indirect (e.g., Arctic warming and release of methane from thermokarst lakes as described in Chapter 11). Anthropogenic sources of nitrous oxide (N_2O) include industrial air pollution, biomass burning, and fertilizer use. Although occurring in extremely low concentrations (typically measured in parts per billion), methane and nitrous oxide are very efficient absorbers of infrared radiation. In response to international efforts to protect the stratospheric ozone shield, the atmospheric concentration of halocarbons began leveling off in the early 1990s.

AEROSOLS

Aerosols are tiny (nanometer to micrometer in size) solid and liquid particles suspended in the atmosphere. Aerosols vary in size, shape, and chemical composition. Larger aerosols have short residence times in the atmosphere and tend to settle out quickly whereas smaller aerosols may remain suspended for many days to weeks and can be transported thousands of kilometers by the wind, possibly impacting climate on a planetary scale. Approximately 90% of anthropogenic aerosols are byproducts of fossil fuel burning in the Northern Hemisphere. Atmospheric aerosols of anthropogenic origin cause either cooling or warming of the atmosphere.

Sulfur oxides emitted to the atmosphere from coal-burning electric power plant smokestacks and boiler vent pipes combine with water vapor in the air to produce tiny droplets of sulfuric acid and sulfate particles, collectively called **sulfurous aerosols**. The effect of these aerosols on temperatures at Earth's surface is opposite that of greenhouse gases, causing cooling rather than warming. Sulfurous aerosols raise the atmosphere's albedo directly by scattering sunlight to space and indirectly by spurring cloud development. Aerosols of anthropogenic origin function as cloud condensation nuclei (CCN) that favor formation of more numerous and smaller cloud droplets and brighter clouds that reflect more solar radiation to space. Greater reflectivity cools the lower atmosphere.

Sulfurous aerosols in the troposphere have a shorter-term impact on climate than carbon dioxide and other greenhouse gases. Rain and snow wash sulfurous aerosols from the atmosphere so that the residence time of these substances in the atmosphere is typically only a few days. On the other hand, the lifetime of a CO_2 molecule in the atmosphere is much longer before being cycled out by natural processes. Approximately one-fifth of the carbon we release now will still be in the atmosphere after 200 years; after several thousand years about one-twentieth will remain.

On the other hand, aerosols that enter the atmosphere as soot (black carbon) strongly absorb solar radiation, causing warming. Soot consists of carbon particles produced during the incomplete combustion of fossil fuels. The global net effect of all anthropogenic aerosols is cooling, although the amount of cooling is uncertain.

Variations in the atmosphere's aerosol load likely contributed to observed trends in the transmission of sunlight through the atmosphere and the amount of solar radiation striking Earth's surface. From about 1960 to the late 1980s, a gradual dimming of sunlight incident on the land surface accompanied a decline in atmospheric transmission and an increase in cloud cover in many areas. According to one estimate, the solar energy flux incident at Earth's surface decreased 4% to 6% during the 30-year period. Since the late 1980s, however, the trend reversed and brightening has prevailed. The atmosphere became more transparent because of the reduction in aerosol emissions imposed by enforcement of air quality regulations (especially in the developed nations) and the downturn in the economy of Eastern Europe, as well as the decline in the Mount Pinatubo aerosol loading during the early 1990s (Chapter 11).

The net cooling effect of aerosols on Earth's climate partially offset the warming effect accompanying the build-up of greenhouse gases in the atmosphere. Consequently, scientists may have underestimated the magnitude of potential warming from greenhouse gases. This aerosol masking began to decline in the 1990s perhaps explaining why the global warming trend was greater than expected over the last few decades. Nonetheless, according to the *2007 IPCC Fourth Assessment Report*, the direct radiative forcing of aerosols is estimated to offset the CO_2 warming by almost one-third (Figure 12.1).

In recent years, researchers have been taking a closer look at how aerosols of anthropogenic origin are influencing cloudiness, precipitation, and the water cycle. Aerosols can alter the number, mean size, and size distribution of cloud droplets. In some cases, aerosols suppress precipitation by producing a larger number of

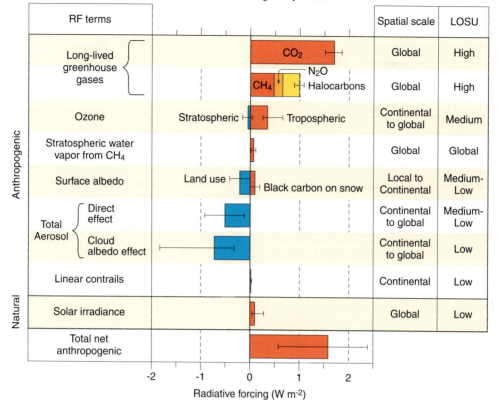

FIGURE 12.1
Estimates of global average radiative forcing (RF) for greenhouse gases and other important climate change agents and mechanisms, along with typical geographical extent (spatial scale) of the forcing and level of scientific understanding (LOSU). [After *IPCC, Climate Change 2007: The Physical Science Basis*]

smaller cloud droplets that less readily coalesce to form precipitation (Chapter 5). A reduction in orographic precipitation of up to 25% has been reported in some mountainous locales, especially downwind from urban centers. Over a recent 50-year period, a summit observatory on Mount Hua in China recorded a decline in visibility (presumably due to more aerosols in the air) accompanied by a 17% reduction in precipitation compared to nearby lowland sites.

In contrast, more aerosol pollution over Asia (due to growing industrial activity in China and India) may explain a trend toward increasing precipitation over northwest Australia. Apparently, aerosols that are transported downwind to the Pacific cause warming aloft and cooling at the surface. This temperature change alters convective activity so that more moist air (and precipitable water) is delivered to northwest Australia. Aerosols produced by biomass burning and fossil fuel combustion form brown clouds or brown haze that absorb solar radiation. The associated warming may be at least partially responsible for the shrinkage of the Hindu Kush and Himalayan glaciers.

In 2009, David Hoffman of NOAA in Boulder, CO, and colleagues reported that China's growing dependence on coal burning was responsible for a steady increase in stratospheric haze at altitudes of 20 to 30 km (12 to 19 mi) over that nation since 2000. With the rapid growth of China's economy, its sulfur emissions increased more than 60% between 2000 and 2005, causing stratospheric haze to thicken by 4% to 7% per year.

CHANGES IN LAND USE AND LAND COVER

As noted in the Case-in-Point of Chapter 1, the idea that anthropogenic changes in land use and land cover contribute to climate change has deep historical roots. During much of the 18th and 19th centuries, natural scientists debated the question of whether deforestation and cultivation practices were changing the climate of new settlements in America. Today, much of the research on the climatic impacts of alterations in land use and land cover focuses on deforestation in the tropics (e.g., the Amazon Basin). But as pointed out in a 2005 NASA news item, human development has had a much more extensive impact on land use and land cover having transformed an estimated one-third to one-half of Earth's land surfaces. The possible implications of this transformation on the climate system may be far reaching.

The spatial and temporal patterns of thunderstorms largely depend on the heat and moisture fluxes directed from the land to the atmosphere (Chapter 10). Most thunderstorms develop over land and the thousands of thunderstorms occurring at any time help transport heat and moisture to higher latitudes. Changes in land use or land cover can affect the vertical flux of heat and moisture, which in turn influences where and when thunderstorms occur and may modify atmospheric and oceanic circulation, with implications for global climate. In fact, as pointed out by Johannes Feddema of the University of Kansas and colleagues, the spatial scale of land use and land cover change is similar to SST anomaly patterns associated with ENSO, an important climate forcing mechanism (Chapter 8).

In 2001, Gordon Bonan of the National Center for Atmospheric Research (NCAR) found that at middle latitudes, clearing of forests for agriculture and reforestation of abandoned farm land caused regional climate change. Bonan compared the *diurnal temperature range (DTR)* at cropland sites in the Midwest with forested sites in the Northeast. He examined temperature records from 1986-1995 for 65 climate stations located away from cities and water bodies. Bonan found that the DTR was lower in the Midwest than the Northeast primarily because cropland has a higher albedo than forested land. Daytime heating was less in the Midwest than Northeast even though the Northeast had more cloudiness. Midwest cooling was greater in late spring and summer and less in fall after the crops were harvested. In expanding his study to a 100-year record of U.S. climate, Bonan found that prior to 1940, when the percent of cultivated land was similar in both the Midwest and the Northeast, the difference in DTR between the two regions was less than in recent decades. That is, the difference in DTR between the Midwest and Northeast has increased.

Anthropogenic versus Natural Forcing of Climate

The diagram in Figure 12.1, provided by the IPCC, summarizes the contributions of the various climatic forcing agents or mechanisms (excluding volcanic aerosols). These are global average radiative forcings expressed in units of watts per square meter (W/m^2) and

grouped by anthropogenic and natural sources. Positive forcings (e.g., greenhouse gases) cause the climate to become warmer whereas negative forcings (e.g., aerosols) cause the climate to become cooler. Black error bars represent the level of certainty of each forcing agent or mechanism; that is, the probability that values lie within the error bar is 90%. Also included is the typical spatial scale and assessed level of scientific understanding (LOSU) for each forcing agent or mechanism.

According to the diagram, the net radiative forcing is anthropogenic in origin. Specifically, the build-up of the greenhouse gases CO_2, CH_4, N_2O, halocarbons, and O_3 is primarily responsible for a net positive radiative forcing. Note, however, that the error bar is very large, ranging from 0.33 to about 1.5 times the net value of $1.6 \ W/m^2$. Furthermore, the global climate models used to predict the climate future may not adequately simulate the impact of aerosols and clouds.

The Climate Future

What does the climate future hold? Atmospheric scientists attempt to answer this question primarily by relying on global climate models that run on supercomputers.

GLOBAL CLIMATE MODELS

A **global climate model (GCM)** is a simulation of Earth's climate system. One type of global climate model consists of mathematical equations that describe the physical interactions among the various components of the climate system, that is, the atmosphere, ocean, land, ice-cover, and biosphere.

The focus of climate models differs from that of numerical weather prediction (NWP) models in that climate forecasting is a boundary value problem whereas weather forecasting is an initial value problem. NWP models start off with conditions that represent the state of the atmosphere at the present time and apply the laws of physics to work forward iteratively. The initial state of the atmosphere is specified quantitatively by simultaneous or near-simultaneous instrument-based measurements. The product is weather forecasts for the next 12, 24, 36, 48 hours or longer. A climate model is designed so that certain known fundamental variables can be changed (e.g., the forcing of the Sun or the boundary conditions of greenhouse gas concentration or Earth's surface properties). The climate model then predicts how the climate adjusts to these new conditions, specifying broad regions of expected positive and negative temperature and precipitation anomalies (departures from long-term averages) and the mean location of atmospheric circulation features such as jet streams and principal storm tracks over much longer time scales.

Global climate models are used to predict the potential climatic impacts of rising levels of atmospheric carbon dioxide (or other greenhouse gases). Using current boundary conditions, a global climate model simulates the present climate. Then, holding constant all other variables in the model, the concentration of atmospheric CO_2 (or other greenhouse gas) is elevated and the model is run to a new equilibrium state. Two different approaches are taken in adding CO_2 to the atmosphere: In a *transient run*, CO_2 is slowly added to the model and the effects are evaluated from moment to moment whereas with an *equilibrium run*, CO_2 is added all at once and the model is run until it achieves a new equilibrium. By comparing the new climate state with the present climate, scientists deduce the impact of an enhanced greenhouse effect on patterns of temperature and precipitation. Using boundary conditions derived from proxy climate data sources (Chapter 9), global climate models are also used to simulate climates that prevailed in the geologic past such as the last glacial maximum (20,000 to 18,000 years ago).

Most modelers agree that global climate models are in need of considerable refinement. Today's models may not adequately simulate the role of small-scale weather systems (e.g., thunderstorms) or accurately portray local and regional conditions and may miss important feedback processes. A major uncertainty is the net feedback of clouds. Clouds cause both cooling (by reflecting sunlight to space) and warming (by absorbing and emitting to Earth's surface outgoing infrared radiation). The cooling effect prevails with an increase in low cloud cover whereas the warming effect prevails with an increase in high cloud cover.

Problems with global climate models stem in part from the limited spatial resolution of the models. Today's models partition the global atmosphere into a three-dimensional grid of boxes with each box perhaps having an area of 250 km^2 (155 mi^2) and a thickness of 1 km (0.6 mi). Limited spatial resolution in climate models is caused by limited computational speed. Although today's supercomputers can perform trillions of operations per second, the complexity of the climate system means that simulating climate change over a century requires months of computing time. Much greater resolution of global climate models evolves with development of faster supercomputers.

SEARCH FOR CYCLES AND ANALOGS

Another approach to predicting future climate is empirical in nature and seeks to identify the various factors that may have contributed to past fluctuations in climate and to extrapolate their influence into the future. Atmospheric scientists have probed the instrument-based and reconstructed climate record in search of regular cycles that might be extended into the future, and analogs that might provide clues as to how the climate in specific regions responds to global-scale climate change. One formidable challenge in this search is to separate the signal from the noise. Time series of some climatic elements are so variable (noisy) that detection of any cycles or trends (the signal) requires close scrutiny of the climate record. Use of computers programmed with sophisticated statistical routines has greatly facilitated the search for regular rhythms and trends in the climate. The motivation behind this effort is obvious: Identification of any statistically significant periodicities or trends in the climate record would be a powerful tool in climate forecasting.

Few of the quasi-regular oscillations that appear in the climate record have much practical value for climate forecasting, at least over the next century or so. Cycles established as significant in a rigorous statistical sense are the familiar annual and diurnal radiation-temperature cycles and a less familiar quasi-biennial cycle (about every two years) in various climatic elements. The first merely means that summers are warmer than winters and days are warmer than nights. Examples of the quasi-biennial cycle include an approximate two-year fluctuation in Midwestern rainfall, a 25.5-month oscillation in a lengthy temperature record (1659 to present) from central England, and an approximately two-year cycle in the strength of the trade winds over the western Pacific and eastern Indian Oceans.

That same 335-year long central England temperature record reveals statistically significant oscillations of 7 to 8 years, 15 years, and 25 years. Scientists attribute these temperature fluctuations to regular changes in the atmospheric and thermohaline circulation of the North Atlantic Ocean. Trends may be visible in the climate record, but unless a trend is demonstrated to be part of a statistically significant cycle, there is no guarantee that the trend will not end abruptly or reverse direction at any time.

Although the climate record yields much useful information on how climate behaves through time, the search for realistic analogs of future global warming has been unsuccessful. Proposed analogs include relatively warm episodes of the mid Holocene Epoch and the Eemian interglacial about 127,000 years ago (Chapter 9). But those analogs are inappropriate because the mid Holocene and Eemian warming episodes primarily affected seasonal temperatures, with only a slight rise in global mean temperature. Furthermore, boundary conditions were different then. During the mid Holocene, sea level was lower, ice sheets were more extensive, and the seasonal and latitudinal receipt of solar radiation (due to different dates of perihelion and aphelion, the points when Earth is closest and farthest from the Sun) were not the same as they are now or will be for the next several centuries. Although the level of atmospheric CO_2 trended upward during the mid Holocene, the rate of increase (about 0.5 ppmv per century) was much slower than it is at present (more than 60 ppmv per century).

Proposed pre-Pleistocene analogs are also inappropriate because of the absence of ice sheets and significant differences in topography and land-ocean distribution. One proposed pre-Pleistocene analog is the **Paleocene/Eocene Thermal Maximum (PETM)**. Much of what is known about the PETM is based on analysis of deep-sea sediment cores. According to *IPCC Climate Change 2007: The Physical Science Basis Report*, the PETM occurred about 55 million years ago, during the Cenozoic Era, and near the transition from the Paleocene to Eocene Epochs. The PETM was a geologically brief interval (lasting 100,000 years) of widespread warming associated with a massive buildup of the greenhouse gases carbon dioxide and methane in the atmosphere. Over a period of 1000 to 10,000 years during the PETM, global temperatures rose several Celsius degrees.

Possible sources of greenhouse gases during the PETM include release of methane from the decomposition of hydrate deposits in seafloor sediments, carbon dioxide from volcanic activity, and oxidation of organic-rich sediments. (For more on methane hydrates, refer to this chapter's first Essay.) Interestingly, the estimated magnitude and rate of carbon release during PETM were comparable to modern and anticipated anthropogenic sources.

ENHANCED GREENHOUSE EFFECT AND GLOBAL WARMING

In the long run, the Milankovitch Earth-Sun orbital cycles favor an eventual return to Ice Age conditions (Chapter 11). In fact, when the link between the Pleistocene Ice Age and Milankovitch cycles was confirmed in the 1970s, some in the popular press warned of a coming ice age. How would cooling associated with the Milankovitch cycles compare to the warming projected to accompany the build-up of greenhouse gases and how

might the interaction of these climate forcing agents influence the onset of the next glaciation? Recall from Chapter 11 that continental-scale glaciation likely began when Milankovitch cycles favored reduced incoming solar radiation in summer at 65 degrees N so that some of the winter snows would persist through the summer. To trigger glaciation during a period of greenhouse warming would require an additional reduction in incoming summer solar radiation. According to David Archer, a geophysicist at the University of Chicago, the result could be a 50,000-year delay in the onset of the next glaciation.

What about climate in the near term? If all other boundary conditions remain fixed, rising concentrations of atmospheric CO_2 and other greenhouse gases would cause global warming to persist throughout this century and well beyond. The magnitude of warming depends on future emissions of greenhouse gases. According to the *2007 IPCC Assessment Report*, climate models predict that over the subsequent 20 years, the global mean annual temperature will rise at an average rate of about 0.2 Celsius degrees per decade. Again depending on the future greenhouse gas emission scenario, over the present century, climate models project that the global average surface temperature will rise by 1.8 to 4.0 Celsius degrees (3.2 to 7.2 Fahrenheit degrees).

Recall, however, that climate change is geographically non-uniform (in both magnitude and sign) so that this projected rise in global mean annual temperature is not necessarily representative of what might happen everywhere. Climate models predict *polar amplification*, i.e., warming will be greater at higher latitudes. In any event, enhancement of the greenhouse effect could cause a climate change that would be greater in magnitude than any previous climate change over the past 10,000 years.

Even if greenhouse gas emissions were to stabilize at present levels, global warming would likely continue well beyond the 21st century. The highest priority in curbing greenhouse warming is to cut emissions of CO_2 that accounts for about 77% of all anthropogenic greenhouse gases. Intuitively we would expect the atmospheric concentration of a greenhouse gas to decline as its emissions are reduced. However, the situation is not quite that straight forward.

How the atmospheric concentration of a gas responds to a decrease in emissions depends on the competition between the rate of emission of the gas into the atmosphere and the rates of various physical, chemical, and biological processes that remove the gas from the atmosphere. These interactions determine the lifetime of the gas in the atmosphere. Here, the **lifetime**

of a gas in the atmosphere is defined as the time it takes for a perturbation of the gas to be reduced to 37% of its original amount. The atmospheric concentration of some greenhouse gases decreases shortly after a decline in emissions whereas the atmospheric concentration of other greenhouse gases may continue to climb hundreds of years after emissions begin to decline. If emissions increase with time, the atmospheric concentration also increases regardless of the lifetime of the gas.

Consider how the atmospheric concentration of the greenhouse gases carbon dioxide (CO_2), nitrous oxide (N_2O), and methane (CH_4) adjust to future emissions scenarios. N_2O has a lifetime of 110 years; with an emission reduction of more than 50%, its atmospheric concentration would stabilize near present-day values. CH_4 has a lifetime of only 12 years; reducing emissions by less than 30% would stabilize its atmospheric concentration within a few decades. The situation is much more complex for CO_2.

More than 50% of CO_2 emitted into the atmosphere cycles out within a century but about 20% remains for millennia. Because of this slow rate of removal, even if emissions are reduced substantially from present levels, the concentration of CO_2 in the atmosphere will continue to increase over the long term. If emissions of CO_2 were to stabilize at current levels, the concentration of CO_2 in the atmosphere would continue to increase through the remainder of this century and beyond. Only the complete elimination of anthropogenic CO_2 emissions would stabilize the atmospheric concentration of CO_2 at current levels. The lengthy lifetime of CO_2 implies that greenhouse warming is inevitable over the long-term and that we are already committed to such impacts as shrinking glaciers, rising sea level, and ocean acidification. We have more on this topic in Chapter 14.

In predicting the climate future, another important consideration is the great inertia of Earth's climate system, primarily due to the dominating role of the ocean. As discussed in Chapter 4, ocean water has a high specific heat so that unusually great quantities of heat energy are required to bring about relatively small changes in temperature. Hence, the ocean has a tremendous capacity for storing heat energy. Furthermore, the transport and mixing of heat energy and carbon dioxide throughout the deep ocean operate over time scales of millennia. Hence, a considerable length of time is required for Earth's climate system to achieve equilibrium under new environmental conditions.

Further complicating efforts to predict the climate future are indications that the global radiation budget currently is not in equilibrium. In 2005, scientists reported that the planet is now absorbing about 0.85 W/ m^2 more energy from the Sun than it is emitting as infrared radiation to space; the imbalance is approximately 0.06% of the total incoming solar radiation at the top of the atmosphere. This conclusion is supported by measurements made by sensors onboard Earth-orbiting satellites and instrumented buoys at sea indicating an increase in the heat content of the ocean over the previous decade. The imbalance represents the delay in the response of Earth's climate (i.e., surface temperature) to some agent or forcing mechanism (i.e., build up of greenhouse gases). The delay, in turn, is the consequence of considerable thermal inertia in Earth's climate system, mostly due to the ocean. On this basis, global climate models predict a yet unrealized additional warming of 0.6 Celsius degree (1.1 Fahrenheit degrees) with no change in atmospheric composition as Earth shifts to a new state of global radiative equilibrium.

In many experiments involving global climate models, CO_2 concentration is elevated while all other boundary conditions are kept constant. This procedure may not be realistic. Comparing the post-1957 trend in atmospheric carbon dioxide to the trend in mean annual global temperature strongly suggests that recent climate has been shaped by many interacting forcing agents and mechanisms. The rapid rise in atmospheric CO_2 concentration was not accompanied by a consistent rise in global mean temperature over the same period. Recall, for example, that sulfurous aerosols from the June 1991 eruption of Mount Pinatubo apparently were responsible for significant global-scale cooling the following year (Chapter 11). Also, El Niño and La Niña influence inter-annual climate variability in most areas of the globe (Chapter 8).

Furthermore, analysis of the composition of tiny air bubbles in cores extracted from the Greenland and Antarctic ice sheets indicates that during the Pleistocene Ice Age, atmospheric CO_2 varied between about 260 and 280 ppmv. Although CO_2 was consistently higher during milder interglacial climatic episodes than during colder glacial climatic episodes, fluctuations in carbon dioxide levels lag reconstructed variations in temperature. For example, at the beginning of the last major glacial climatic episode, the decline in CO_2 concentration significantly lagged cooling in the Antarctic. Hence, fluctuations in atmospheric CO_2 during the Pleistocene Epoch may have been a response to large-scale climate oscillations rather than a cause of those oscillations.

On the other hand, the current rise in CO_2 due to anthropogenic activities is proceeding at a much more rapid pace than anytime during the Pleistocene Epoch.

There is reason to believe that in the present case of rapid build-up of atmospheric CO_2, the warming trend is a response to more CO_2 rather than the cause.

Potential Impacts of Global Climate Change

As we have seen, since the Little Ice Age ended in the late 19th century, the global mean temperature has generally trended upward. Since the 1970s, the warming trend has accelerated—probably because of the build up of carbon dioxide and other greenhouse gases in the atmosphere and the decline in aerosol loading. What are the potential impacts of a continued rise in global mean temperature? While global-scale trends in climate do not impact all regions in the same way, if global warming persists, the consequences for society are likely to be extensive and highly disruptive. Rising sea level is one of the most serious potential consequences because a third of the human population lives within 91 m (300 ft) of sea level.

RISING SEA LEVEL

Climate changes responsible for the waxing and waning of Earth's glacial ice cover during the Pleistocene Epoch also caused sea level to alternately fall and rise. Geological evidence such as offshore drowned beaches, submerged river valleys, and submarine canyons attests to periods when sea level was much lower than at present. Scientists estimate that during the last glacial maximum, about 20,000 to 18,000 years ago, mean sea level was 113 to 135 m (370 to 443 ft) lower than it is today. More than 90% of this drop in sea level was due to a change in the global water cycle brought about by a colder climate. As noted in Chapter 5, the total amount of water on the planet is essentially constant. Practically all the water locked in glaciers came from the ocean via the global water cycle. During glacial climatic episodes, glaciers on land thickened and expanded, and the volume of water in the ocean basins decreased (i.e., sea level fell). Conversely, during interglacial climatic episodes, glaciers on land thinned and retreated, and the volume of water in the ocean basins increased (i.e., sea level rose). Furthermore, perhaps 7% or 8% of the drop in sea level during the last glacial maximum was due to lower ocean temperature resulting in contraction of the ocean water and an increase in water density. Seawater always contracts when its temperature drops and expands when its temperature rises.

Waxing and waning of land-based glaciers plus ocean temperature fluctuations are two factors that govern

eustasy, the global variation in sea level brought about by changes in the volume of water occupying the ocean basins. Tectonic processes that alter the size of the ocean basins also contribute to sea-level change. Persistence of the current global warming trend appears likely to cause sea level to rise in response to melting of land-based polar ice sheets and mountain glaciers, coupled with thermal expansion of seawater.

When glacial ice flows from land to sea, the less dense ice floats on the sea surface and sea level rises immediately. An analogous situation is shown in Figure 12.2. The glass on the left is half filled with water. When ice cubes are added to the water, simulating a glacier moving from land to sea (center), the water level immediately rises to near the top of the glass. But as the floating ice cubes melt (right), the water level is unchanged remaining near the top of the glass. Hence, melting of floating ice shelves does not alter sea level. When glacial ice leaves the land and enters the ocean, it begins to float, displacing a volume of seawater equal to its own weight and causing a rise in sea level. When the ice melts, the smaller volume of meltwater occupies the same volume of sea water that the ice originally displaced.

How did sea level respond to the global warming trend of the 20th century? For most of the century, coastal tide gauges were the principal source of data on sea level

FIGURE 12.2
What happens to mean sea level when glacial ice enters the ocean? In this analogous situation, a glass initially is half filled with water (left). Ice cubes are added to the glass simulating a glacier moving from land to sea and the water level rises to near the top of the glass (center). As the floating ice melts (right), the water level remains unchanged. Hence, glacial ice has an immediate impact on mean sea level but melting of floating ice (e.g., an ice shelf) has no effect on mean sea level. [Modified after Robin E. Bell, "The Unquiet Ice," *Scientific American*, February 2008, page 61.]

change. Care must be taken to exclude (or adjust) those tide gauge records that might be influenced by geological processes (i.e., tectonic uplift or subsidence, or post-glacial rebound). Based on adjusted tide gauge records, mean sea level from 1870 to 1993 is estimated to have risen at the rate of 1.7 mm per year (Figure 12.3A). Beginning in 1993, microwave altimeters on Earth-orbiting satellites have permitted more precise measurement of the rise in global mean sea level (Figure 12.3B). (See the discussion of TOPEX/Poseidon and its successors in Chapters 2 and 8.) From 1993 to 2009 global mean sea level increased an estimated 3.32 mm per year. In total, mean sea level is estimated to have risen about 18 cm (7.1 in.) during the 20th century. How much of the recent rise in sea level was due to melting of glacial ice and how much was due to thermal expansion of warming ocean waters is not known. Most mountain glaciers have been shrinking since the mid 20th century, portions of the Greenland ice sheet have shown recent signs of accelerated melting, and portions of the ocean are warming.

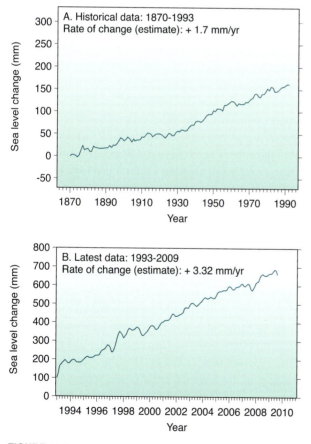

FIGURE 12.3
Upward trend in mean sea level as indicated by (A) coastal tide gauge records, 1870-1993, and (B) measurements by microwave altimeters flown on board Earth-orbiting satellites from 1993 to 2009. [From CSIRO (A) and CLS/Cnes/Legos (B)]

SHRINKING GLACIERS

Amplification of the global warming trend at higher latitudes threatens the ice sheets of Antarctica and Greenland. About 90% of the planet's glacial ice blankets Antarctica and melting could cause a considerable rise in sea level. How likely is this to happen? Two ice sheets cover Antarctica, separated by the Transantarctic Mountains. The larger of the two (accounting for about two-thirds of the ice), the East Antarctic ice sheet is situated on a continent about the size of Australia, averages a little more than 2 km (1.2 mi) thick, and is well above sea level. Complete melting of the East Antarctic ice sheet would raise mean sea level (msl) by about 60 m (197 ft). The West Antarctic ice sheet sits on a series of islands and the floor of the Southern Ocean with parts of the ice sheet more than 1.7 km (1 mi) below msl. Complete melting of the West Antarctic ice sheet would raise msl an estimated 5.8 m (19 ft). While geological evidence suggests that the East Antarctic ice sheet has been stable for the past 30 million years and remains fairly stable today, the West Antarctic ice sheet has undergone episodes of rapid disintegration and may have completely melted at least once in the past 600,000 years.

An important source of data on the Antarctic and Greenland ice sheets is remote sensing by satellite. NASA's *Ice, Cloud and Land Elevation Satellite (ICESat)* was the world's first laser-altimeter satellite (Figure 12.4). Launched in January 2003, ICESat monitored variations

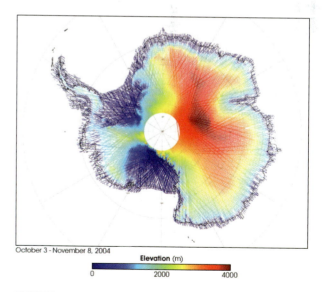

FIGURE 12.4
Topography of Antarctica from measurements by NASA's ICESat mission based on data compiled from 3 October – 8 November 2004. Red indicates the highest elevations (up to 4000 m above msl); yellow, green, and turquoise represent progressively lower elevations with green being 2000 m above msl. Dark blue signifies sea level. [NASA image courtesy of Christopher Shuman, ICESat Deputy Project Scientist, Goddard Space Flight Center]

FIGURE 12.5
View of a steep valley in the Antarctic Peninsula. Small caps of stagnant ice cover the summits while glacial ice in the valley moves quickly toward the coast. [Michael Studinger, Lamont-Doherty Earth Observatory]

in the thickness and mass of the Antarctic and Greenland ice sheets as well as changes in polar sea ice thickness until October 2009 when the satellite's Geoscience Laser Altimeter System (GLAS) failed. NASA continued monitoring the polar regions with *Operation Ice Bridge*, a 5-year airborne survey of ice sheets, ice shelves, and sea ice initiated in 2009 (Figure 12.5). Operation Ice Bridge is bridging the gap in satellite data until the launch of ICESat-2 scheduled for 2015. Additional information on changes in the mass of the Antarctic and Greenland ice sheets is provided by the NASA/German Aerospace Center's *Gravity Recovery and Climate Experiment (GRACE)*. Launched in March 2002, GRACE consists of twin satellites flying in formation and designed to obtain detailed measurements of Earth's gravity field.

The behavior of ice streams is an important consideration in predicting how the Antarctic and Greenland ice sheets might affect sea level. An **ice stream** is a zone of relatively fast flowing ice within an ice sheet. Just as excess water on land flows via rivers and streams to the ocean (Chapter 5), most of the ice (perhaps 90%) that flows from Antarctica and Greenland to the surrounding ocean occurs via ice streams and *outlet glaciers* (ice streams bounded by mountains). For perspective, the Rutford Ice Stream in West Antarctica is 150 km (93 mi) long, 25 km (16 mi) wide, up to 3 km (2 mi) thick, and moves at an average speed of 1.0 m per day.

Many of Antarctica's ice streams feed **ice shelves**, extensive areas of floating ice attached to land (at the *grounding line*) and fringing about 44% of the coast. Although the disintegration and melting of ice shelves does not affect sea level, they dam (or buttress) the flow of land-based glaciers that slowly feed the ice shelf. Ice shelves are vulnerable to warming seas and when an ice shelf disintegrates, the dam is gone, and the glacier surges forward into the ocean.

Although the average speed of ice streams is greater than that of the surrounding ice, ice stream speed can be highly variable depending on the frictional resistance of the terrain under the ice and the temperature of the ice. Recently, scientists confirmed the existence of large lakes underlying portions of some ice streams. Lake Vostok, under Russia's Vostok Station in East Antarctica, is about the size of Lake Ontario and the largest known *subglacial lake*; it was discovered in 1973 by scientists of the Scott Polar Research Institute. The ice floats in the water and the lake surface offers essentially no frictional resistance to the moving ice. In addition, the ice is warmed by the water thereby reducing its viscosity. Hence, an ice stream accelerates over a subglacial lake. Along the way, the ice stream erodes sediment and deposits it as a wedge where the ice stream enters the ocean, acting like a speed bump to slow the advancing ice.

More than 30 years ago, recognition of the relative instability of the West Antarctic ice sheet prompted speculation among some scientists that ice streams flowing from the interior to the Ross and Ronne ice shelves might cause a total collapse of the ice sheet in a few centuries or less. Such a catastrophic event would greatly accelerate the rate of sea level rise. As noted above, complete disintegration of the West Antarctic ice sheet would raise sea level by about 5.8 m or 19 ft.

In 2001, concerns about the possible disintegration of the West Antarctic ice sheet were alleviated with the discovery that new snowfall was keeping pace with the loss of ice from bergs breaking off the Ross Ice Shelf. In early 2002, scientists at the California Institute of Technology and the University of California-Santa Cruz reported that, based on satellite measurements of the flow of the Ross ice streams, the West Antarctic ice sheet appeared to be thickening. However, the region of the ice sheet that feeds the Thwaites and Pine Island glaciers is thinning. (Pine Island is the largest ice stream in the West Antarctic ice sheet.) These glaciers transport ice directly into the ocean (rather than adding to an ice shelf) and are responsible for about 10% of the average annual rise in mean sea level.

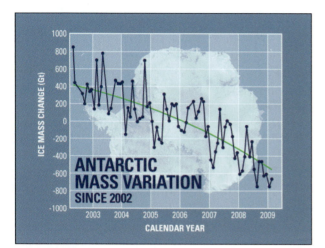

FIGURE 12.6
Based on measurements of changes in Earth's gravity field, Antarctica has been losing ice mass since at least 2002. [NASA Global Climate Change; GRACE mission]

More recently, oceanographer Eric Rignot of NASA's Jet Propulsion Laboratory in Pasadena, CA, and colleagues used satellite imagery to measure the thickness of ice fringing the coast of Antarctica. These data along with estimates of ice stream discharge and snowfall enabled them to determine the annual loss of ice from Antarctica. They concluded that the ice loss from Antarctica (mostly West Antarctica) was about 75% greater in 2006 than in 1996. Gravity data acquired by GRACE satellites indicate that Antarctica has been losing more than 100 km³ (24 mi³) of ice each year since 2002 and the rate of ice loss has accelerated in recent years (Figure 12.6). Nonetheless, the consensus of scientific opinion today is that the West Antarctic ice sheet is unlikely to catastrophically accelerate the current rise in sea level. According to some experts, long-term gradual shrinkage of the West Antarctic ice could raise sea level at a rate of about 1.0 m (3.3 ft) per 500 years.

Melting (and the rate of sea level rise) could accelerate if the Antarctic ice sheets begin to warm in response to global climate change. Over most of Antarctica, the mean air temperature has fluctuated very little over the past 50 years. The Antarctic Peninsula is the only part of the continent that has shown significant warming, with summer mean temperatures rising more than 2 Celsius degrees (3.6 Fahrenheit degrees) over the past half century. This warming has been accompanied by higher SST around Antarctica and breakup of ice shelves along the Antarctic Peninsula coast (Figure 12.7). A major review of the situation in 2009 found that the majority of glaciers of the Antarctic Peninsula were retreating at an accelerating rate.

The Greenland ice sheet presents a somewhat different situation because it is a temperate glacier whereas the Antarctic ice sheets are polar (cold) glaciers. That is, much of the Greenland ice is near the pressure-melting point whereas the Antarctic ice sheets have temperatures well below the pressure-melting point. (The melting point of ice decreases with increasing confining pressure or depth within the glacier.) Meltwater is more readily generated in temperate glaciers and they tend to move faster than polar ice sheets.

NASA research results released in 2000 concluded that while the central interior of the Greenland ice sheet showed no sign of thinning, about 70% of the margin was thinning substantially. Two research teams—one using a Global Positioning System (GPS) to monitor ice flow and the other relying on an airborne laser altimeter to measure ice thickness—reached the same conclusion based on observations made between 1993 and 1999. The maximum melting rate at the margin was about 1 m per year. An estimated 50 km³ (12 mi³) of Greenland's ice melts each year, enough to raise sea level by 0.13 mm annually.

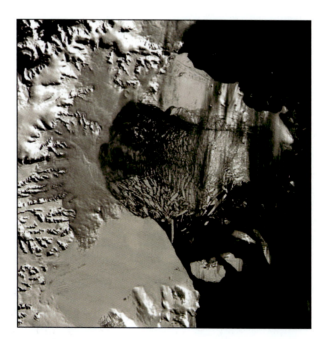

FIGURE 12.7
The northern section of Sector B of the Larsen ice shelf on the eastern side of the Antarctic Peninsula shattered and separated from the continent over a 35-day period beginning on 31 January 2002, sending thousands of icebergs adrift in the Weddell Sea. About 3250 km² of shelf (larger area than the state of Rhode Island) about 220 m thick disintegrated. This image was taken by the Moderate Resolution Imaging Spectroradiometer (MODIS) on board NASA's Terra satellite on 5 March 2002. [Courtesy of NASA]

Scientists at the University of Colorado reported that during the summer of 2002, surface melting on the Greenland ice sheet encompassed an area of about 695,000 km^2 (265,000 mi^2)—about 9% greater than observed during any summer since monitoring by satellite began 24 years previously. In addition, melting in the northern and northeastern portion of the ice sheet occurred at elevations as high as 2000 m (6550 ft) where normally temperatures are too low for any melting. The duration and extent of melting at elevations above 2000 m set a new record in 2007. Researchers at NASA's Goddard Space Flight Center monitored snow melt on the ice sheet surface using a satellite microwave sensor and computed a snow melt index by multiplying the area of snowmelt by the duration of melting. During 2007, the melting index above 2000-m elevation was about 2.5 times greater than the annual average from 1988 to 2006. It appears that winter snowfall is insufficient to offset the summer melt so that overall the Greenland ice sheet is shrinking (Figure 12.8).

The 2007 *IPCC Fourth Assessment Report* does not include a best estimate of the future contributions of the Antarctic and Greenland ice sheets to sea-level rise. In the past decade, flow of Antarctic and Greenland ice streams and outlet glaciers accelerated unexpectedly. However, much uncertainty surrounds the predictability of the processes that govern the speed of ice flow from land to sea.

In 2002, scientists discovered that meltwater lakes that form in summer on the surface of the Greenland ice sheet (known as *supraglacial lakes*) can force open a cre-

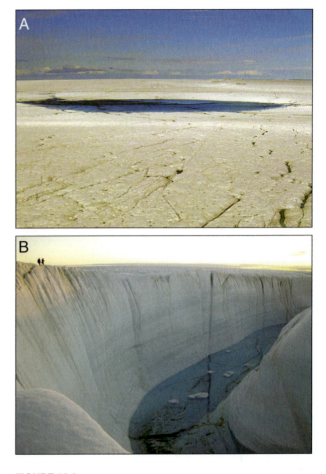

FIGURE 12.9
A meltwater lake on the surface of the Greenland ice sheet, one of thousands that form each summer (A). [Photo by Ian Joughin, University of Washington Polar Science Center] Over many years, a meltwater stream has excavated this large ice canyon in the Greenland ice sheet (B). [Photo by Sarah Das, Woods Hole Oceanographic Institution]

vasse allowing its waters to drain catastrophically to the base of the glacier. This phenomenon was documented by researchers from the Woods Hole (MA) Oceanographic Institution and the University of Washington in July 2006 (Figure 12.9). In only 90 minutes, a supraglacial lake covering 5.7 km^2 (2.2 mi^2) and containing 11.6 billion gallons of water drained through a crevasse, plunging some 980 m (3215 ft) to the base of the glacier. Reporting in the 14 November 2008 issue of *Science*, glaciologist Richard Alley of Pennsylvania State University and colleagues proposed that the lubricating effect of this water is not as important as the heat it delivers to the base of the ice sheet in accelerating glacier flow. Should the global warming trend lengthen the melt season and create meltwater lakes further inland on the Greenland ice sheet, the downward cascading meltwater could thaw areas where the glacier is now frozen to the ground. This is likely to significantly accelerate glacier flow

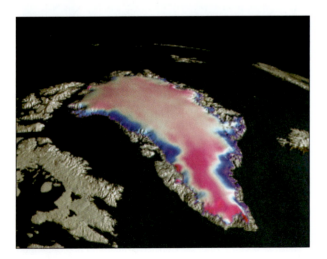

FIGURE 12.8
Measurements of the change in thickness of the Greenland ice sheet between 2003 and 2006 made by sensors on board NASA's ICESat satellite. Pink and red regions indicate slight thickening while shades of blue and purple signify thinning. [NASA/Goddard Space Flight Center Scientific Visualization Studio]

and perhaps destabilize the Greenland ice sheet. Complete melting of the Greenland ice sheet would raise mean sea level by an estimated 7.3 m (24 ft).

Alley and colleagues also point to another factor that influences the flow of outlet glaciers to the sea. At the glacier-ocean interface, outlet glaciers move faster when they do not encounter obstacles such as ice shelves and grounded ice. Ice shelves surround much of Antarctica but grounded ice, bedrock highs, and coastal landforms obstruct the flow of Greenland outlet glaciers. Melting of ice shelves or grounded ice would accelerate the flow of outlet glaciers.

What is happening to the much smaller glaciers that occupy mountain valleys? Roger G. Barry of NOAA's *National Snow and Ice Data Center* at the University of Colorado, Boulder, reports that the rate of melting of most of the world's mountain glaciers accelerated after the mid-1900s and especially since the mid-1970s (Figure 12.10). Some mountain glaciers have disappeared entirely. For example, today only 26 glaciers remain of the 150 glaciers that existed in Montana's Glacier National Park a century ago. Barry estimates that runoff from melting mountain glaciers contributes about 0.4 mm to the annual rise in mean sea level.

Alpine glaciers are also shrinking at accelerating rates. According to Frank Paul at the University of Zurich-Irchel in Zurich, Switzerland, the combined area covered by almost 940 Alpine glaciers decreased by about 18% from 1973 to 1999. Paul's conclusion was based on analysis of satellite images, aerial photographs, and land surveys. Swiss glaciers are now shrinking at an annual rate that is more than six times greater than the rate that prevailed from 1850 to 1973. At the present rate of shrinkage, Alpine glaciers at elevations below 2000 m (6500 ft) will likely disappear by the year 2070.

With the warming trend, the Rhône glacier in southern Switzerland could disappear by 2100. In 2008, Guillaume Jouvet of the Federal Polytechnic Institute in Lausanne, Switzerland, and colleagues used a glacier-climate model to simulate the retreat of the Rhône glacier since 1874. A 1 Celsius degree (1.8 Fahrenheit degree) rise in global mean temperature would reduce the glacier volume by about 35% by 2100. A temperature rise of 4 Celsius degrees (7.2 Fahrenheit degrees) would cause the glacier to disappear entirely.

A major concern is the shrinkage of mountain glaciers and ice fields where seasonal runoff is the principal source of fresh water for people, their crops and livestock. In India, 500 million people depend on runoff from the glaciers of the Himalayas for their fresh water supply. In early 2007, scientists reported the alarming news that this essential resource had diminished by 21% since 1962. Anil Kulkarni of the Indian Space Research Organization in Ahmedabad and colleagues compared glacial maps from 1962 with recent satellite images, estimating the retreat of 466 glaciers in three major river basins. They found that the extent of glacial ice had decreased from 2077 km² in the early 1960s to 1628 km² at present. Although many glaciers are shrinking, it is doubtful that the glaciers of the Himalayas will soon disappear. The largest glaciers are at high elevations, fed by copious monsoon moisture. Of more immediate concern are the many smaller glaciers

FIGURE 12.10
The present global warming trend has caused mountain glaciers to shrink with much of the meltwater eventually draining into the ocean. The dramatic recession of Grinnell Glacier in Montana's Glacier National Park is shown in photographs taken in 1938, 1981, 1998, and 2006. [Photos courtesy of Glacier National Park Archives and the USGS]

and ice fields at lower elevations; they are much more vulnerable to warming and monsoon failure.

In spring 2000, NOAA scientists reported that the combined Atlantic, Pacific, and Indian Oceans warmed significantly between 1955 and 1995. The greatest warming occurred in the upper 300 m (985 ft) of the ocean and amounted to 0.31 Celsius degree (0.56 Fahrenheit degree). Water in the upper 3000 m (9850 ft) warmed by an average 0.06 Celsius degree (0.11 Fahrenheit degree). In February 2002, researchers at Scripps Institution of Oceanography reported that temperatures at mid-depths (between 700 m and 1100 m) in the Southern Ocean rose 0.17 Celsius degrees between the 1950s and 1980s. Although these magnitudes of temperature change may sound trivial, recall that water has an unusually high specific heat so that even a very small change in temperature of such vast volumes of water represents a tremendous heat input into the ocean. Sequestering of vast quantities of heat energy in the ocean may help explain why global warming during the 20th century was less than some climate models predicted. That is, heating of the ocean partially offset warming of the lower atmosphere. This finding also underscores the importance of the ocean's moderating influence on global climate change.

According to the 2007 *IPCC Fourth Assessment Report*, thermal expansion of warming seawater will be a greater contributor to mean sea level rise than melting of land-based glaciers during the 21st century. Climate models predict that global warming will cause a rise in mean sea level in the range of 0.2 to 0.6 m (8 to 24 in.) during the century. Thermal expansion of ocean waters would account for more than 60% of the rise, the balance due to melting glaciers. Some scientists view the IPCC estimates as too conservative and argue that they underestimate the contribution of melting glaciers and ice sheets. For example, W.T. Pfeffer and colleagues at the Institute of Arctic and Alpine Research at the University of Colorado, Boulder, include estimates of glacier melt in their projection of a rise in sea level of 0.8 to 2.0 m (31 to 79 in.) by 2100.

Higher mean sea level would accelerate coastal erosion by wave action, inundate wetlands, estuaries and some islands, and make low-lying coastal plains more vulnerable to storm surges. Recall from the Case-in-Point of Chapter 10, the threat posed by rising sea level to inhabitants of the Maldives in the Indian Ocean and Kiribati in the Pacific Ocean. People may have to abandon these island nations as sea level rises. Globally, a 50-cm (20-in.) rise in sea level would double the number of people at risk from storm surges from about 45 million at present to more than 90 million, not counting any additional population growth in the coastal zone. Rising sea level

would disrupt coastal ecosystems, ruin agricultural lands, and could threaten historical, cultural, and recreational resources (Figure 12.11). In some coastal areas, higher sea level is likely to exacerbate saltwater intrusion into

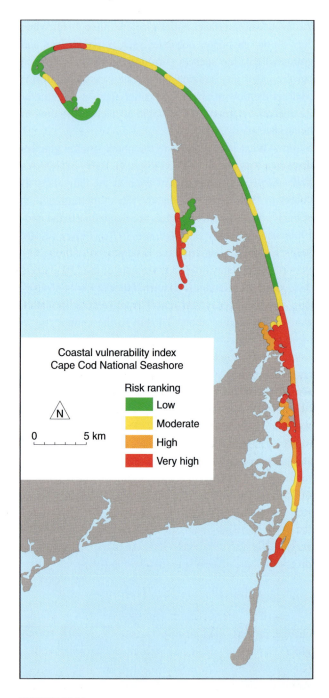

FIGURE 12.11
Map of the U.S. Geological Survey's *Coastal Vulnerability Index (CVI)* for Cape Cod National Sea Shore (MA) showing the vulnerability of the coast to changes in sea level. The CVI is based on tidal range, wave height, coastal slope, historic shoreline change rates, geomorphology, and historic rates of relative sea-level change due to eustatic sea level rise and tectonic uplift or subsidence. [USGA Fact Sheet FS-O95-02, September 2002]

groundwater. According to a 1997 report by the U.S. Office of Science and Technology Policy, a 50-cm (20-in.) rise in sea level would result in a substantial loss of coastal land, especially along the U.S. Gulf and southern Atlantic coasts. Particularly vulnerable is South Florida where one-third of the Everglades is less than 30 cm (12 in.) above sea level.

A combination of rising sea level and compaction of sediments is likely to mean further loss of coastal Louisiana. In 2009, Harry Roberts of Louisiana State University in Baton Rouge reported that perhaps 25% of the wetlands of the Mississippi River delta have been claimed by the ocean over the past few centuries. Sediments accumulate in reservoirs behind dams upriver, cutting off much of the sediment supply to the delta. Deprived of new sediments, the delta sediments compact, making the delta more vulnerable to erosion as sea level continues to rise. Roberts and Michael Blum of ExxonMobil reported on their use of numerical models to predict how these processes will impact the delta in the decades to come. At Grand Isle, LA, at the edge of the delta, the land is sinking up to 8 mm per year. With sea level rising at about 3 mm per year (and likely to accelerate with continued global warming), by the end of the century perhaps as much as 10% of the area of Louisiana will be submerged.

While higher temperatures would mean higher sea level, the level of North America's Great Lakes is likely to fall. Higher summer temperatures coupled with less winter ice cover on the Great Lakes are likely to translate into greater evaporation. And less winter snowfall would reduce spring runoff. Depending on the model used, forecasts call for a drop in mean water level of Lake Michigan of up to 2 m (6.5 ft) by the year 2070. Residents of the western Great Lakes may have previewed the impact of global warming from the late 1990s into the early 2000s when the levels of Lakes Michigan and Huron dropped to near historic record lows.

ARCTIC ENVIRONMENT

The Arctic is particularly sensitive to climate change. *Polar amplification* means that the current global warming trend is greater at higher latitudes (Chapter 9). Some perspective on recent warming in the Arctic is available from a recent 2000-year reconstruction of decadal mean temperature.

D.S. Kaufman of Northern Arizona University and colleagues inferred temperature data from a combination of Arctic lake sediments, ice cores, and tree rings from a total of 23 sites poleward of 60 degrees N. Reporting in the 4 September 2009 issue of *Science*, they found that a cooling trend in progress 2000 years ago persisted through the Little Ice Age culminating in the Arctic's lowest mean temperature of the past 8000 years. Researchers attributed the cooling trend to regular changes in Earth-Sun geometry that was responsible for a steady decline in incoming solar radiation during high latitude summers (Chapter 11). The long-term cooling trend was reversed during the 20th century so that 4 out of the 5 warmest decades of the 2000-year period occurred during the latter half of the 20th century.

Already significant environmental changes are taking place in the Arctic consistent with an enhanced greenhouse effect and higher temperatures. As noted in Chapter 11, through an ice-albedo feedback mechanism, Arctic sea-ice cover is shrinking at an accelerated rate so that the Arctic Ocean may be free of summer ice in the near future. At the same time, Arctic land precipitation and river discharge into the Arctic Ocean are increasing (Chapter 10). As described in Chapter 11, thawing of permafrost has produced thermokarst lakes, a source of the greenhouse gas methane. For more about permafrost and its vulnerability to warming, refer to this chapter's second Essay.

TROPICAL CYCLONES

Some scientists speculated that the record high Atlantic hurricane activity in 2005 was a consequence of global warming and a precursor of things to come. The prospect of continued warming and higher sea-surface temperatures (SST) prompted predictions of an overall upturn in both the number and intensity of tropical cyclones (i.e., hurricanes and tropical storms) through the remainder of this century. Recall from Chapter 10 that tropical cyclones ultimately derive their energy from evaporation of seawater, but relatively high SST is only one of several conditions required for tropical cyclone formation. In fact, some climate models predict that stronger winds aloft will accompany a warmer climate producing wind shear that would inhibit tropical cyclone development. According to long-held theory, higher SST and additional water vapor conveyed into the atmosphere can intensify existing storms and inhibit the formation of new storms.

In 2008, two research teams took somewhat different approaches in using global climate models to predict the impact of global warming on Atlantic tropical cyclones. Thomas Knutson and colleagues at NOAA's Geophysical Fluid Dynamics Laboratory (GFDL) in Princeton, NJ, applied enhanced computing power to simulate storm formation in the tropical Atlantic. Meanwhile, Kerry Emanuel and colleagues at the Massachusetts Institute of Technology (MIT) seeded

FIGURE 12.12
Output statistics (orange) from NOAA's GFDL model is a reasonably good simulation of the actual (green) year-to-year and long-term variation in Atlantic hurricane frequency since 1980. [Adapted from Knutson *et al.*, *Nature Geoscience*, June 2008]

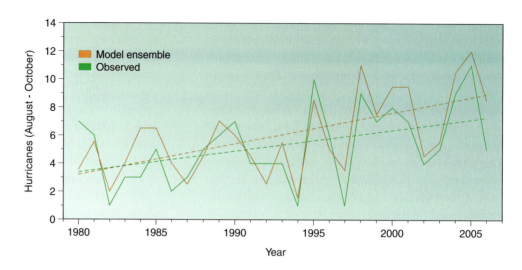

seven global climate models with early-stage tropical cyclones that either developed or died depending on ocean-atmosphere conditions. As shown in Figure 12.12, the GFDL model does a reasonably good job in simulating (orange) the actual (green) year-to-year and long-term variation in hurricane frequency. The GFDL model predicted an 18% decrease in frequency of Atlantic hurricanes by the end of this century with a few percent increase in storm intensity. The MIT model predicted an increase in hurricane frequency by a few percent and 7.5% increase in storm intensity.

In another study, James Elsner, a climatologist at Florida State University, and colleagues assembled wind speed data from tropical cyclones based on their analysis of infrared satellite imagery. Reporting in the 3 September 2008 issue of *Nature*, Elsner et al. found that the most intense tropical cyclones were becoming stronger mostly in the North Atlantic and northern Indian Oceans. That is, while the average frequency and intensity of all tropical cyclones had not changed significantly, category 4 and 5 storms were more frequent. Elsner et al. argue that intense tropical cyclones can overcome the weakening effect of wind shear. The researchers estimate that a rise in tropical SST by 1 Celsius degree (1.8 Fahrenheit degrees) could increase the frequency of category 4 and 5 storms by 31%. Since 1970, tropical SST have risen by an average of about 0.5 Celsius degree (0.9 Fahrenheit degree) and could warm an additional 2 Celsius degrees (3.6 Fahrenheit degrees) by the end of this century.

A major limitation for researchers seeking to predict long-term trends in hurricane activity is the relatively low resolution of global climate models. Recently, however, NOAA researchers developed a new prediction technique that combines climate models with high resolution models used by the U.S. National Weather

Service to routinely forecast hurricane development and tracks. The new findings are consistent with earlier more tentative studies that foresee Atlantic basin hurricanes becoming less frequent but more intense (more destructive) during this century.

Morris A. Bender and colleagues at NOAA's Geophysical Fluid Dynamics Laboratory reported on the findings of their hurricane prediction technique in the 22 January 2010 issue of *Science*. Researchers began with the output of 18 global climate models to predict average atmospheric and oceanic conditions for the end of the century. They then fed this prediction into a North Atlantic regional model that predicted hurricane frequency, finding an 18% reduction in the total number of Atlantic hurricanes by the close of the century.

Next, the NOAA researchers transferred the storms predicted by the regional model into a hurricane forecast model capable of identifying tropical cyclones that would evolve into a category 3, 4, or 5 system. They predict that by the end of the century, the frequency of category 4 and 5 hurricanes (maximum sustained wind speeds of 216 km per hr or 134 mph and higher) would about double while the frequency of the most intense hurricanes (maximum sustained winds of 234 km per hr or 145 mph or higher) would more than triple.

Applying this same technique to recent Atlantic hurricane activity, researchers find no support for the hypothesis that global warming has already impacted hurricane activity. The number of Atlantic basin hurricanes has doubled over the past 25 years. Based on output from the new modeling technique, global warming would cause a slight decrease in hurricane frequency during the same period. Natural variability rather than greenhouse enhancement more likely explains the recent upward trend in hurricane activity.

Many uncertainties surround the question of how global climate change might impact Atlantic hurricane activity. Development of more realistic, high-resolution models of Earth's climate system will aid understanding. Meanwhile, the prospect that the most intense tropical cyclones could become more frequent underscores the need for improved forecast and warning systems especially for the millions of people who live in low-lying coastal areas prone to a potentially catastrophic storm surge. Verification of the NOAA long-range prediction of more frequent intense Atlantic hurricanes could translate into a 30% increase in property damage.

MARINE LIFE

How might climate change impact marine life? Marine animals and plants are components of ecosystems. As noted in Chapter 1, an *ecosystem* is a fundamental subdivision of the Earth-atmosphere-land-ocean system in which organisms depend for their survival on other organisms and the physical and chemical constituents of their surroundings. Materials and energy flow from one organism to another via food webs within and between ecosystems. Climate change could alter the physical and chemical conditions in the ocean, perhaps exceeding the tolerance limits of organisms. If these organisms are unable to avoid these stressful conditions, they may perish.

Warming could raise the sea-surface temperature (SST) sufficiently to threaten coral reefs. As noted in Chapter 8, coral polyps are sensitive to small changes in water temperature and prolonged periods of exposure to excessively warm waters can lead to *coral bleaching*. A rise in SST of 1 to 2 Celsius degrees (2 to 4 Fahrenheit degrees) may cause temporary bleaching. However, if bleaching episodes persist or are unusually severe, coral polyps and the reef die. Because of the interdependency of organisms (e.g., for food, habitat), loss of one species can disrupt food webs and the integrity of the entire ecosystem.

Another potential threat to coral reefs is sea-level rise. If sea level rises slowly, healthy coral reefs can grow upward at a rate that will maintain them near the ocean surface and bright sunlight. However, if sea level rises too rapidly, reefs may be unable to grow fast enough. If they are too deeply submerged to receive sufficient sunlight for photosynthesis, they die. Many deeply submerged Pacific seamounts are capped by reef limestone. In these cases, coral growth could not keep pace with the rapid rise in sea level or sinking due to plate movement.

Organisms are particularly vulnerable to an environmental change that affects a limiting factor. A **limiting factor** is an essential resource that is in lowest supply compared to what is required by the organism. For example, climate change may indirectly affect the supply of an essential nutrient. Concurrent shifts in atmospheric circulation may suppress upwelling in some locations or reduce the transport of micronutrients to portions of the open ocean. Furthermore, rising sea level is likely to alter marine habitats, especially in the coastal zone. In some cases—where the coastline has been developed for roads and buildings—flooding may eliminate marine habitats entirely, causing the demise of organisms that are dependent on those habitats.

Consider, for example, the fate of salt marshes already threatened by development. Higher temperatures will increase the rate of evaporation from the soil surface thereby elevating the soil salinity. More saline soils will cut biological production and may exceed the tolerance limits of plants. Alternately, rising sea level threatens to drown salt marshes. If plants cannot shift inland to higher ground, the marsh may be lost. Loss of salt marshes, in turn, will make the coastline more vulnerable to flooding and erosion by storm waves. Furthermore the natural filtration function of salt marshes will be lost so that higher levels of pollution and nutrients will enter estuaries with runoff.

A key consideration in the potential impact of climate change on marine organisms is the rate of change. Marine ecosystems can more readily adjust to gradual rather than abrupt changes in the ocean environment brought on by global scale climate fluctuations. But as noted earlier in this chapter, the rate of climate change due to an enhanced greenhouse effect could be so rapid as to be without precedent over the past 10,000 years. If this accelerated global warming materializes, considerable disruption of marine ecosystems may occur.

GLOBAL WATER CYCLE

Higher global temperatures may translate into more severe drought in some locations and floods in others. Unfortunately, global climate models do not agree on the locations of precipitation extremes. An assessment of predictions generated by different coupled atmosphere-ocean climate models underscores the rudimentary state of climate prediction at the regional scale. For example, the Canadian Climate Centre model predicts that global warming will be accompanied by drier summer soil conditions over the eastern two-thirds of the coterminous U.S., whereas Great Britain's Hadley Centre for Climate Research and Prediction model projects wetter soils in the same area.

Whereas regional response to global climate change is uncertain, many scientists and public policymakers are confident that localities most vulnerable to water resource problems are places where the quality and quantity of fresh water are already a problem, that is, in semi-arid and arid regions. In the U.S., water supply problems are most likely in the drainage basins of the Missouri, Arkansas, Rio Grande, and lower Colorado Rivers. Water shortages in the Middle East and Africa may heighten political tensions especially in nations that depend on water supplies originating outside their borders.

FOOD SECURITY

Higher temperatures and more frequent drought may severely affect food production in certain regions of the world. Developed nations have the capacity to adapt their agricultural systems to climate change. While overall U.S. food production is expected to climb throughout the 21st century, there will be some losses. For example, greater demand for irrigation waters may further stress important reservoirs of groundwater (e.g., the High Plains aquifer). On the other hand, the agricultural systems of developing nations are much more vulnerable to the adverse effects of warming and are most likely to experience a significant decline in food production.

Ocean Acidification

Water in the various reservoirs of the Earth system can vary in acidity and alkalinity. An *acid* is a hydrogen-containing compound that releases hydrogen ions (H^{+1}) when dissolved in water. Strong acids more readily release hydrogen ions than weak acids. An *alkaline substance* releases hydroxyl ions (OH^{-1}) when dissolved in water and may also be weak or strong. Pure water has properties of both acids and alkaline materials as water molecules continually break up (into hydrogen and hydroxyl ions) and re-form. That is,

$$H_2O \leftrightarrows H^{+1} + OH^{-1}$$

The acidity of water (or any other substance) is expressed as **pH**, a measure of the hydrogen ion concentration. On the **pH scale**, the pH increases from 0 to 14 as the hydrogen ion concentration decreases (Figure 12.13). Pure water has a pH of 7, which is considered neutral; a pH above 7 is increasingly alkaline whereas a pH below 7 is increasingly acidic. The pH scale is logarithmic;

that is, each unit increment corresponds to a tenfold change in acidity or alkalinity. Hence, a two-unit drop in pH (e.g., from 5.6, the pH of natural rainwater, to 3.6) represents a hundred-fold (10×10) increase in acidity.

The pH of pristine seawater ranges between 8.0 and 8.3; that is, seawater is slightly alkaline. The

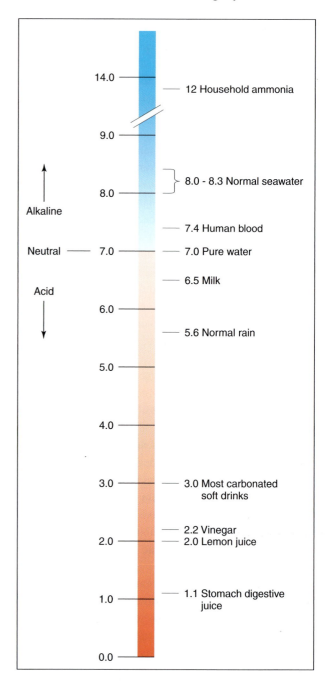

FIGURE 12.13
The acidity of water is expressed as pH, a measure of the hydrogen ion concentration. On this scale, pH increases from 0 to 14 as the hydrogen ion concentration decreases. Pure water has a pH of 7, which is considered neutral; a pH above 7 is increasingly alkaline and a pH below 7 is increasingly acidic.

relatively narrow variation in the pH of seawater is very important for marine organisms whose shells and skeletons are composed of calcium carbonate ($CaCO_3$). If ocean water were even slightly acidic, calcium carbonate would dissolve and would be unavailable for these organisms.

Carbon dioxide plays a key role in stabilizing the pH of seawater. A substance that stabilizes a chemical system in this way is known as a **buffer**. Atmospheric CO_2 dissolves in ocean water producing carbonic acid (H_2CO_3) that dissociates into hydrogen (H^{+1}), carbonate (CO_3^{-2}), and bicarbonate (HCO_3^{-1}) ions. A chemical equilibrium develops in which carbon dioxide, carbonic acid, hydrogen ions, carbonate ions, and bicarbonate ions co-exist. Adding acid to the ocean shifts the equilibrium so that there are fewer carbonate ions, which in turn reduces the hydrogen ion concentration and raises the pH. Adding alkaline materials to ocean water raises the pH but the equilibrium shifts in such a way as to return the pH to the normal range.

With continuation of the upward trend in the concentration of atmospheric carbon dioxide, ocean scientists are concerned about the potential effects on the ocean's buffering ability and the viability of certain marine organisms. Carbon dioxide that is absorbed by the ocean participates in chemical reactions that increase the acidity (lowers the pH) of ocean waters. This so-called **ocean acidification** is potentially harmful to those marine organisms that use carbonate ions (CO_3^{-2}) to build calcium carbonate ($CaCO_3$) shells or skeletons.

Today, about 40% of CO_2 from fossil fuel combustion stays in the atmosphere, 30% is taken up by terrestrial vegetation via photosynthesis, and 30% is absorbed by the ocean. This causes a decline in the pH of ocean waters and a reduction in the concentration of carbonate ions needed by certain marine organisms to build their shells and other hard parts. According to some studies, the current trend in the flux of CO_2 into the ocean will result in a 50% reduction in the ocean's carbonate concentration by the end of the century. According to Woods Hole (MA) Oceanographic Institution Senior Scientist Scott Doney, "the acidity of the ocean has increased by 26% from pre-industrial times to today." Over the past 200 years, the pH of surface ocean waters has decreased by about 0.1 unit and an additional drop in pH by 0.3 unit could occur by 2100, a level not experienced in the ocean for many millions of years.

Marine organisms that are particularly vulnerable to ocean acidification are coccolithophorids, foraminifera (phytoplanktonic organisms), and pteropods (small

FIGURE 12.14
These free-swimming planktonic molluscs form a calcium carbonate shell made of aragonite. They are an important food source for juvenile North Pacific salmon and also are eaten by mackerel, herring and cod. [Photo courtesy of Russ Hopcroft, UAF/NOAA.]

marine snails), which are an important food source in marine food webs (Figure 12.14). Also vulnerable are the corals which filter plankton from ocean water and secrete calcium carbonate. A reduction in carbonate ion concentration impairs the ability of reef-building coral to generate their calcium carbonate skeletons. Complicating matters is the fact that calcium carbonate shells occur in three different mineral forms that differ in solubility. Aragonite (a form of calcium carbonate) and magnesium calcite are more soluble than calcite; corals and pteropods produce aragonite shells that more readily dissolve in acidic waters. On the other hand, the shells of high latitude calcareous phytoplankton and zooplankton are composed of calcite and are less soluble.

FIGURE 12.15
Dead coral reef with skeletal remnants of reef-building corals. [NOAA Photo Library. Photograph by David Burdick]

Conclusions

Interaction of many factors is responsible for the inherent variability of climate. Although we can isolate specific climate forcing agents or mechanisms that operate internally or externally to the Earth-atmosphere-land-ocean system, our understanding of how these agents and mechanisms interact is far from complete. This limits the ability of climate scientists to forecast the climate future, so continued research on climate is needed. As scientists more fully comprehend Earth's climate system and the physical laws regulating climate variability and climate change, their ability to predict the climate future will improve, aided by more thorough global observational networks, faster supercomputers, and more realistic global climate models. Meanwhile, trends in climate must be monitored closely, especially in view of the strong dependence of our food and water supplies and energy demand on climate. In spite of, or perhaps because of, the scientific uncertainty, some experts argue that we should plan now for the worst possible future climate scenario.

Our study of climate science continues in the next chapter with a summary of how world climates are classified.

Basic Understandings

- In 2007, the IPCC concluded that observed increases in global temperature, often referred to as "global warming," since the mid 20th century very likely was caused mostly by human activities.

- Human activity may impact global-scale climate by elevating the concentration of greenhouse gases (causing warming) or sulfurous aerosols (causing cooling). The current upward trend in atmospheric carbon dioxide is primarily due to burning of fossil fuels and to a lesser extent the clearing of vegetation.

- The Keeling curve indicates a sustained increase in average annual atmospheric CO_2 level from about 316 ppmv in 1959 to 386 ppmv near the end of 2008.

- The ocean, a major reservoir in the global carbon cycle, plays an important role in governing the amount of carbon dioxide in the atmosphere. According to IPCC estimates in 2007, the ocean takes up 56.2% of the carbon dioxide of anthropogenic origin while terrestrial biomass is a sink for 13.7%. Ultimately, the rising concentration of atmospheric CO_2 is driving a net influx of carbon dioxide across the air-sea interface.

- Aerosols are tiny solid and liquid particles suspended in the atmosphere. Sulfurous aerosols that are byproducts of human activity (i.e., fossil fuel combustion) have the opposite effect on temperatures at Earth's surface as greenhouse gases; that is, they cause cooling rather than warming. By absorbing solar radiation, particles of soot (black carbon) have a warming effect.

- Sulfurous aerosols raise the atmosphere's albedo directly by reflecting sunlight to space and indirectly by spurring cloud development. These aerosols function as cloud condensation nuclei (CCN) that favor the formation of more numerous and smaller cloud droplets and brighter clouds that reflect more solar radiation to space.

- Changes in land use or land cover can affect the vertical flux of heat and moisture, which in turn influences where and when thunderstorms occur and atmospheric and oceanic circulation, with implications for global climate.

- The net radiative forcing of climate is anthropogenic in origin. Specifically, the build-up of the greenhouse gases CO_2, CH_4, N_2O, halocarbons, and O_3 is primarily responsible for a net positive radiative forcing.

- Global climate models consist of mathematical equations that describe the physical laws governing the interactions among the various components of Earth's climate system. The focus of climate models differs from that of numerical weather prediction (NWP) models in that climate forecasting is a boundary value problem whereas weather forecasting is an initial value problem.

- Global climate models are used to predict prevailing circulation patterns, jet streams, storm tracks, and broad regions of expected positive and negative temperature and precipitation anomalies, especially when certain forcing mechanisms (e.g., greenhouse gas concentrations) are changed in experiments designed to assess the anthropogenic impact on climate.

- The climate record does not contain cycles that are useful in forecasting the climate over a period of several centuries. Also, no satisfactory analogs exist in the climate record that would allow scientists to predict regional responses to future global-scale climate change.

- Current global climate models predict that significant global warming will accompany a doubling of atmospheric carbon dioxide, possible by the close of this century. Warming is likely to be greater in magnitude than any other prior climate change during the past 10,000 years.
- Projections of the magnitude of future global warming arising from an enhanced greenhouse effect assumes that all other boundary conditions of climate remain constant. To what extent this will happen is not known.
- Global warming is likely to cause a rise in mean sea level in response to melting of land-based glaciers and ice sheets, plus thermal expansion of seawater. Higher sea level would accelerate coastal erosion, inundate wetlands, estuaries and some islands, and make low-lying coastal plains more vulnerable to potentially catastrophic storm surges. Rising sea level would disrupt coastal ecosystems, destroy agriculture, and could threaten historical, cultural, and recreational resources.
- Amplification of the global warming trend at higher latitudes would threaten to melt the ice sheets of Antarctica and Greenland, reservoirs of about 90% of the planet's glacial ice, causing a considerable rise in sea level.
- The rate of melting of most of the world's mountain glaciers accelerated after the mid-1900s and especially since the mid-1970s. Some mountain glaciers have disappeared entirely. A major concern is the shrinkage of mountain glaciers and ice fields whose seasonal runoff is the principal source of fresh water for people, their crops and livestock.
- The Arctic region is particularly vulnerable to global climate change and already significant environmental changes are taking place.
- Some climate models predict that stronger winds aloft will accompany a warmer climate producing wind shear that would inhibit tropical cyclone development. However, recent studies suggest that while the total number of tropical cyclones has not increased, the number of major hurricanes (categories 4 and 5) appears to be increasing.
- Global warming could disrupt the functioning of marine ecosystems by exceeding tolerance limits or destroying habitats.
- CO_2 that is absorbed by the ocean participates in chemical reactions that increase the acidity (lowers the pH) of ocean waters. Ocean acidification is potentially harmful to marine organisms that use carbonate ions (CO_3^{-2}) to build calcium carbonate ($CaCO_3$) shells or skeletons.

Enduring Ideas

- The lengthy lifetime of the greenhouse gas carbon dioxide in the atmosphere commits us to continued global warming through this century and well beyond even if CO_2 emissions are stabilized. The consequences include melting glaciers, sea level rise, and ocean acidification.
- Some aerosols of anthropogenic origin cause cooling (e.g., sulfurous aerosols) whereas others cause warming (e.g., soot or black carbon).
- The net radiative forcing in Earth's climate system is anthropogenic in origin.
- Overall, global climate change has reduced the stability of the Antarctic and Greenland ice sheets. Ice streams, ice shelves and supraglacial lakes play important roles in the melting rates of these glaciers.
- Continued global warming is likely to impact the global water cycle, the intensity of tropical cyclones, habitation of the coastal zone, marine life, and food security.

Review

1. Why is the Keeling curve considered to be representative of CO_2 levels in the Northern Hemisphere?
2. How does the ocean influence the amount of CO_2 in the atmosphere?
3. How do sulfurous aerosols affect the climate?
4. What forcing agent or mechanism is primarily responsible for a net positive radiative forcing in Earth's climate system?
5. How does a global climate model differ from a numerical model used to forecast the weather?
6. Describe the relationship between the global water cycle, the volume of glacial ice on Earth, and mean sea level.
7. How do changes in temperature affect mean sea level?
8. Why is there uncertainty regarding the impact of global warming on tropical cyclones?
9. What causes coral reef bleaching and why is it significant for many forms of marine life?
10. What types of regions of the world are most vulnerable to food shortages caused by climate change?

Critical Thinking

1. Explain why scientists may have underestimated the magnitude of warming accompanying the build-up of greenhouse gases in the atmosphere.
2. Present an example of how changes in land use or land cover could affect the climate.
3. What is the most likely cause of the Paleocene/Eocene Thermal Maximum (PETM)?
4. Explain the considerable thermal inertia in Earth's climate system.
5. Identify and describe the various factors that govern the rise and fall of mean sea level.
6. Why is the behavior of an ice stream important in predicting how the Antarctic and Greenland ice sheets might influence sea level?
7. How have scientists determined that overall the Greenland ice sheet is shrinking?
8. Summarize some of the environmental changes that have already taken place in the Arctic region as a consequence of climate change.
9. What is the significance of the slight alkalinity of ocean water?
10. What causes ocean acidification and how is it likely to impact marine life?

ESSAY: Methane Hydrates: Greenhouse Gas and Energy Source

Methane hydrate (also called *methane ice*) is a clathrate, an ice-like solid formed from methane (CH$_4$) and water. The basic chemical structure of methane hydrate is a single molecule of methane enclosed in a rigid cage of water molecules. Cages are linked together forming a crystalline solid that is similar to water ice except that the crystal structure is stabilized by the caged gas molecule. Many gases besides methane can form hydrates, such as carbon dioxide, hydrogen sulfide, and other light hydrocarbons, but methane hydrates are the most common. Methane is the chief component of natural gas and methane hydrate could be a major source of natural gas in the future.

As early as 1811 the British chemist Sir Humphry Davy (1778-1829) made the first gas hydrates in his laboratory, but it was not until 1965 that a natural deposit was found (under permafrost in Siberia). Only recently have geoscientists started documenting the presence of methane hydrates in submarine and underground deposits and exploring their potential as an energy source. Ocean scientists first drilled through methane hydrates unintentionally on an expedition in 1970. Although that encounter was uneventful, research-drilling cruises avoided suspected methane hydrate deposits for two decades afterward, fearing they might encounter a pressurized pocket of methane that could blast away the drilling equipment. Concerns over pressurized gas gradually diminished, and mounting scientific curiosity emboldened researchers to try boring through gas hydrate fields. Starting in 1992, researchers with the international *Ocean Drilling Program (ODP)* intentionally breached methane hydrate deposits several times without incident. Through ongoing exploration and research, marine geologists have identified deposits at hundreds of sites around the globe along most continental margins of the world ocean (Figure 1).

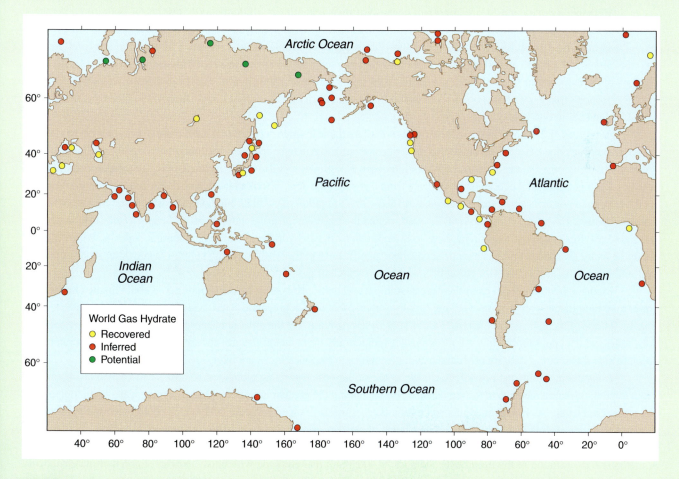

FIGURE 1
Worldwide locations of recovered, inferred, and potential methane hydrate deposits. [USGS]

Marine methane hydrates form in pore spaces within ocean sediments. Hydrates are not stable at typical surface temperatures and pressures but are found in sediments at ocean depths greater than 400 to 500 m (1300 to 1600 ft) where temperatures are sufficiently low and/or pressures are sufficiently high to squeeze water and methane into hydrates. When sediment cores containing methane hydrate, or solid chunks of methane hydrate, are brought up from the ocean floor, the reduction in pressure and rise in temperature cause the gas hydrates to become unstable and they decompose, much like ice melting when warmed. Without precautions, gas hydrate will melt and fizz away before reaching the ocean surface. On the other hand, because of its volatility, chunks of methane hydrate will burn at the touch of a match.

The precise origin of submarine methane hydrates is unknown, although scientists suspect that microbes, known as *archaea*, living within sediments generate methane from carbon and hydrogen extracted from rich organic materials. That is, biogenic methane is concentrated where organic detritus (from which archaea generate methane) and sediments (which protect detritus from oxidation) accumulate rapidly. At the appropriate temperatures and pressures, methane molecules are captured within the frozen hydrate structure. By contrast, conventional deposits of methane form through a different process. In that case, seafloor sediments are buried far deeper where temperatures are much higher, so that the organic material in the sediments simmers until it transforms into petroleum and eventually methane. Free methane then migrates upward through openings between sediments and is trapped below where hydrates might form from biological processes.

Terrestrial deposits of methane hydrates occur at shallow depths at high latitudes where low temperatures (rather than high pressure) keep them stable. For example, petroleum companies have encountered methane hydrates while drilling through permafrost (perennially frozen ground) in Alaska, Canada, and Siberia. In this environment, the flux of heat from Earth's interior generates methane from deeply buried organic matter. The gas migrates upward through openings in the sediments until it reaches the subsurface area where temperatures favor formation of methane hydrates. Methane hydrates formed below permafrost are much more concentrated than submarine deposits.

Methane hydrate deposits are important because of their potential as a future energy resource—both in the immense amounts of methane bound up in hydrates in sea-floor sediments as well as in the free methane trapped beneath hydrate ice layers. As a hydrate, methane is highly concentrated such that breakdown of one unit of methane hydrate ice at sea level pressure produces about 160 units of methane gas. Methane in submarine hydrates is conservatively estimated to contain at least 10,000 gigatons of carbon—about twice the amount of all other fossil fuel reserves on Earth. The United States Geological Survey (USGS) estimates that gas hydrates on the continental margin of North Carolina alone contain about 350 times the energy consumed by the U.S. in one year.

At present, the limiting factor in the exploitation of methane hydrate deposits is the difficulty of bringing the gas to market at competitive prices. Geologists report that most deposits are too thin to be recovered economically. Another major challenge is efficiently recovering the methane hydrate from the ocean bottom without losing the methane to hydrate decomposition.

The first successful extraction of methane from a gas hydrate deposit took place in March 2002 at a site in the Mackenzie River delta in northwestern Canada. Scientists drilled a 1200-m (3900–ft) well through permafrost into sediments containing methane hydrate. Hot water pumped down the well melted the crystals, releasing natural gas that came to the surface where it was flared off. Research continues on methods to mine methane in gas hydrates in continental margin deposits worldwide, but it may be many decades before gas hydrates contribute significantly to the world's energy supply. Meanwhile, Japan is drilling exploratory wells in the Nankai Trough subduction zone in search of gas hydrates.

Methane hydrates could exert their greatest impact on the Earth system by contributing to global climate change. Hydrates may function as a major source or sink for atmospheric methane, a greenhouse gas. Massive melting of hydrates and the ensuing release of methane gas could raise Earth's surface temperature. On the other hand, cooling could bind up more methane in hydrates, which could further lower Earth's surface temperature by reducing the greenhouse effect. As both temperature and pressure play a role in the stability of methane hydrates, multiple feedback loops are likely.

Consider the following possible feedback scenario: Suppose that rising global temperatures causes methane hydrate to begin to melt, releasing methane to the atmosphere and enhancing the greenhouse effect. More warming leads to melting of glacial ice and sea level rise. However, higher sea level increases pressure on ocean sediments, which could increase the formation of methane hydrates. Gas hydrates form at the expense of atmospheric methane, reducing the greenhouse effect and the Earth cools. Glaciers form and expand and sea level drops. While this scenario is pure speculation, evidence exists for dramatic shifts in methane concentration in the ocean over the past 70,000 years that coincide with episodes when Earth's climate suddenly warmed. As described elsewhere in this chapter, release of methane from gas hydrates may have contributed to global warming at the close of the Paleocene Epoch about 55 million years ago, known as the *Paleocene/Eocene Thermal Maximum (PETM)*.

ESSAY: Permafrost and Climate Change

Permafrost, a major component of Earth's cryosphere, refers to ground that is frozen year-round. Specifically, geologists define *permafrost* as soil and sediment at a temperature that remains below 0 °C (32 °F) for two or more consecutive years. At present, permafrost underlies about 25% of all land in the Northern Hemisphere. With the present global warming trend, permafrost is thawing in many areas surrounding the Arctic Ocean with ramifications for the climate system and the people of the region.

The presence or absence of permafrost depends on the subsurface temperature that, in turn, is governed by the relative flux of geothermal heat conducted from Earth's interior versus heat conducted from the soil surface to the overlying colder air. As shown in Figure 1, the amplitude of the annual temperature wave decreases with increasing soil depth and eventually levels out at the mean annual surface air temperature. If this temperature were -1 °C or lower, permafrost may be present; that is, the threshold for permafrost formation is a mean annual air temperature of -1 °C and permafrost becomes thicker and more extensive where air temperatures are lower. By convention, permafrost is described as *continuous* where it underlies 90% of the ground; *discontinuous* where more than 50% of the ground is frozen; and *sporadic* where less than 50% of the ground is frozen (Figure 2). In the continuous zone, the mean annual air temperature is lower than -6 °C and permafrost may have a thickness of 600 m (e.g. Alaska's North Slope) to 1400 m (Siberia)—although much of the thickest permafrost likely dates at least from the last glacial maximum 20,000 to 18,000 years ago.

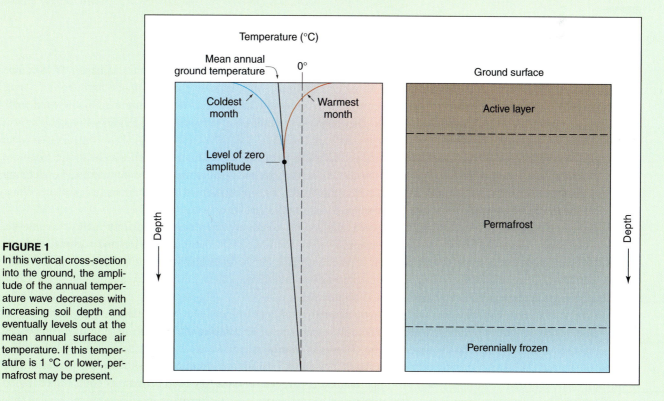

FIGURE 1
In this vertical cross-section into the ground, the amplitude of the annual temperature wave decreases with increasing soil depth and eventually levels out at the mean annual surface air temperature. If this temperature is 1 °C or lower, permafrost may be present.

At high latitudes and high elevations where permafrost is present, the ground is typically frozen from the surface down for much of the year. During the brief warm season, however, the surface layer of soil and sediment may thaw. The surface that divides this *active layer* from the underlying permafrost is known as the *permafrost table* and is analogous to the groundwater table. Because permafrost is essentially impervious to water, the active layer traps melt water and is transformed into a muddy quagmire in summer that is difficult to navigate, and a breeding ground for insects.

Local site conditions that influence the thickness and distribution of permafrost include aspect, snowfall, wildfires, and vegetation. Permafrost tends to be thicker under slopes that face away from the Sun. The insulating property of a snow

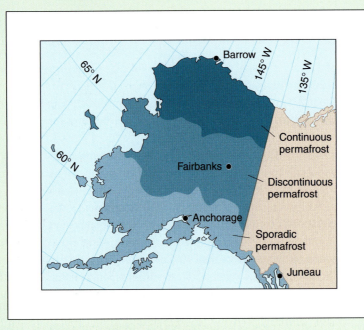

FIGURE 2
Zones of continuous, discontinuous, and sporadic permafrost in Alaska.

cover means that all other factors being equal, permafrost is thickest where the snow cover is thinnest (Chapter 4). Wildfires remove the forest litter which like snow insulates the underlying soil. Hence, following a wildfire and until the litter layer returns, permafrost thickens. Vegetation such as trees and shrubs slows the wind so that snow transported by the wind tends to accumulate downwind insulating the soil. On the other hand, in summer trees and shrubs cast shadows that cool the soil surface.

Gullies eroded by water flowing from thawing permafrost have been reported from many sites in the Arctic. Warming caused by anthropogenic enhancement of the greenhouse effect is likely responsible for thawing the permafrost. Accelerating the warming is the release of organic matter as the permafrost thaws. Decomposition of the organic matter produces either carbon dioxide (if aerobic decay) or methane (if anaerobic decay), both greenhouse gases.

Thawing of Arctic permafrost has many potentially serious consequences. Discharge of fresh melt water into the Arctic Ocean may weaken the thermohaline circulation, altering the large-scale ocean circulation and regional climates (Chapter 6). For people living in the Arctic, permafrost functions much like bedrock elsewhere, supporting buildings, roads, bridges, pipelines, and other structures. Thawing of permafrost is non-uniform, causing differential stresses that would cause considerable property damage and disruption of the infrastructure.

With continued warming, when will the Arctic permafrost disappear? Estimates vary widely ranging from decades to millennia. Much of Earth's permafrost occurs where the mean annual air temperature is at or close to 0 °C suggesting that the system is already close to its tipping point. However, warming permafrost is more readily accomplished than thawing permafrost. Whereas warming permafrost requires about 0.5 calorie per gram per Celsius degree, the latent heat of melting (fusion) is 80 calories per gram. Where permafrost is particularly thick, the deep permafrost chills the overlying permafrost inhibiting thawing. Hence, thin discontinuous or sporadic permafrost is much more vulnerable to thawing than thick continuous permafrost. For example, about 21% of Canada's land area is underlain by patchy thin permafrost that is particularly susceptible to thawing with continued warming.

In February 2010, researchers Serge Payette and Simon Thibault from Quebec's Université Laval reported that their field studies show the southern limit of permafrost in the James Bay region of central Canada to be about 130 km (80 mi) farther north than it was 50 years ago. They predict that if the present trend continues, permafrost will disappear from the region in the near future.

CHAPTER 13

CLIMATE CLASSIFICATION

Earth's many different biomes are named for the types of plants that dominate the landscape. [Photos courtesy of Sandia National Laboratory, NOAA, and USDA]

Case-in-Point

The tropical rain forests of the Amazon basin of South America and the Congo basin of equatorial Africa are the world's only remaining large tracts of undeveloped arable land where rainfall is sufficient to support intensive agriculture. The lush vegetation of tropical rain forests suggests high soil fertility and considerable potential for food production, but appearances are deceiving. In fact, soils in most tropical rain forests are low in nutrients and humus (the dark organic component); almost all the mineral nutrients are contained in the vegetation, not the soil.

Climate is a major contributing factor to low nutrient levels in tropical rain forests. Persistent high temperatures and frequent heavy rainfall accelerate the activity of decomposer organisms, so the remains of dead plants and animals (*detritus*) are broken down rapidly. However, in the decomposition process essentially no nutrients are released into the soil. For the most part, decomposition is carried out by fungi growing in living plant roots that extend into the soil. As these fungi decompose detritus, they take up mineral nutrients that are released and transport them directly into plant roots. The few nutrients that are released directly into the soil are quickly washed away in the runoff of heavy rain. Hence, soils are infertile.

The native peoples of the tropical rain forests adapted to this pattern of nutrient distribution by practicing **slash-and-burn agriculture** (Figure 13.1). Farmers clear trees and other vegetation from small plots of land, typically less than 1 hectare (2.5 acres) in area. Usually, some effort is made to harvest valuable tree species. Following a period of drying lasting from a week to a few months, they

FIGURE 13.1
In the Mato Grosso region of Brazil, fire (red dots) is a major cause of deforestation as revealed by this Moderate Resolution Imaging Spectroradiometer (MODIS) image. "Slash and burn" agriculture is used to clear tropical forests for farming and ranching. [Courtesy Jacques Descloitres, MODIS Rapid Response Team, NASA/GSFC] Inset shows children standing in a field recently cleared in slash-and-burn agriculture in the Amazon basin of South America. [Courtesy US Forest Service]

burn the vegetation and from the ashes of the burnt vegetation, a pulse of nutrients enters the soil. Ash also raises the soil pH, increasing the availability of certain nutrients (e.g., phosphorus). Within 3 to 5 years, successive harvesting of crops and their incorporated nutrients, coupled with the considerable runoff and leaching associated with heavy rainfall, depletes the soil of most of its nutrients. The soil becomes too infertile to produce adequate crops so that farmers must abandon their plots and move elsewhere to slash and burn. Within 20-25 years, vegetation overgrows the original plots, the farmers return, and the cycle of slash-and-burn agriculture begins anew.

Slash-and-burn agriculture is a human adaptation to the tropical rain forest ecosystem for the purpose of growing food, that is, to transform the natural ecosystem into productive agricultural lands. It is a low-tech farming method accessible to people even at the lowest socio-economic levels. Slash-and burn of woodlands and grasslands has been practiced throughout much of the world and may have had its origins in Neolithic time. Today, the technique is most often associated with tropical rain forests. The characteristics of the tropical rain forest ecosystem, in turn, are a response to a climate regime that features heavy rainfall and high temperatures.

Slash-and-burn farming has some serious drawbacks. For one, it is a labor-intensive, inefficient method for producing food in regions of rapid human population growth and large-scale industrial logging. These and other factors threaten the long-term sustainability of slash-and-burn agriculture. In addition, decay and burning of trees and other vegetation are sources of the greenhouse gas CO_2 for the atmosphere. These sources of CO_2 work counter to the photosynthetic removal of CO_2 in tropical rain forests thereby contributing to global climate change.

Driving Question:
How is Earth's mosaic of climate types classified?

Climatic elements such as temperature and precipitation vary both spatially and temporally over Earth's surface. This variability tends to be systematic and can be abrupt or gradual. Forcing agents and mechanisms operating in and on Earth's climate system are responsible for the planet's regular pattern of diverse climate types. Grouping together climate types having similar characteristics is the basis of climate classification systems.

This chapter opens by distinguishing between empirical and genetic climate classification schemes. Focus then turns to a version of the popular Köppen climate classification system, which is based on the strong correlation between climate and indigenous vegetation. After describing the major climate types, we summarize the relationship between climate and terrestrial ecosystems and the implications for climate change. This chapter closes with a brief look at several other climate classification systems.

Methods of Climate Classification

Climate scientists have attempted to simplify and organize the many climate types by devising classification schemes that group together climates having common characteristics. Classification schemes typically group climates according to:

(1) the meteorological basis of climate or

(2) the environmental effects of climate.

The first is a **genetic climate classification** that asks why climate types occur where they do. The second is an **empirical climate classification** that infers the type of climate from the environmental impacts of climate such as the geographical distribution of indigenous vegetation or soil type, or the degree of weathering of exposed bedrock. In addition, computers and databases have made possible *numerical* climate classification schemes that utilize objective statistical analysis techniques to group together climates having similar properties.

Traditionally, the climate of some locality or region is described quantitatively in terms of normals, means and extremes of various climatic elements. An alternative approach, known as **air mass climatology**, has some interesting implications. In this approach, the frequency with which various types of air masses develop over, or are advected into a locality is the basis for describing the climate of that locality. For example, during the month of January, a northern U.S. city on average might have cold, dry air 60% of the time; mild, humid air 30% of the time; and mild, dry air 10% of the time.

In Chapter 7, we identified and described the major air masses that regularly invade North America along with their source regions. Air masses originate in the diverging flow pattern that characterizes surface winds in anticyclones. As air streams outward and away from an anticyclone, the temperature, density and humidity of the air modify to some extent, depending on the properties of the surface over which the air travels (*air mass modification*). In this way, a single anticyclone can be the source of many different types of air masses, depending on the specific modifications that take place. As noted in Chapter 6, for example, surface winds spiraling outward from a semi-permanent subtropical anticyclone tend to be dry on the eastern flank and humid on the western flank. Hence, air masses to the east of a subtropical high are dry and those to the west are humid.

In a study of air mass climatology, W.M. Wendland, then at the Illinois State Water Survey, and R.A. Bryson at the University of Wisconsin-Madison identified the anticyclone sources for Northern Hemisphere air masses. They did this by tracing the average surface streamline pattern across the hemisphere for each month of the year. An average **streamline** represents the mean path of air moving horizontally. Anticyclonic flow is indicated by streamlines that turn into a clockwise and outward spiral. Of the 19 anticyclonic source regions identified by Wendland and Bryson, some were over the ocean, some were over the continents, and a few were in the Southern Hemisphere. Five sources persisted through the entire year and three persisted for about 11 months. The others were prominent for anywhere from 1 to 9 months.

Where streamlines from different anticyclones meet, a zone of confluence forms. A **confluence zone** is a boundary between air masses—equivalent to an average frontal boundary. As a rule, confluence zones separate distinctly different types of climate. Climate is quite uniform within airstreams but changes significantly across confluence zones.

Using air mass frequency as a basis for describing climate appears to be reasonable in view of the apparent air mass control of the location of certain biomes. In Chapter 1, for example, we described the correspondence between the region dominated by cold, dry arctic air and the location of Canada's boreal forest. Climate change may be manifested as changes in air mass frequency and the mean location of confluence zones with implications for both natural and anthropogenic ecosystems.

Köppen Climate Classification

One of the most widely used climate classification systems was designed by German climatologist and botanist Wladimir Köppen (1846-1940). The continued popularity of the Köppen climate classification system today is generally attributed to its simplicity. The Köppen system takes an empirical approach to organizing Earth's many climates. Recognizing that indigenous vegetation is a natural indicator of regional climate, Köppen and his students searched for patterns in mean annual and monthly temperature and precipitation, which closely correspond to the geographic limits of vegetative communities thereby revealing broad-scale climate boundaries throughout the world. Records of mean annual and monthly temperature and precipitation are sufficiently long and reliable in many parts of the world that they provide a good first approximation of large-scale patterns of climate.

Since its introduction in 1900, Köppen's climate classification has undergone numerous and substantial revisions by Köppen himself (notably in 1918 and 1936) and by other climate scientists and has had a variety of applications. The 1936 version was produced in collaboration with Rudolf Geiger (1894-1981), the founder of microclimatology, the study of the climate near the ground and its variability arising from topography and land use. The classification scheme used here is similar to a modification of the original Köppen version by Glenn T. Trewartha (1896-1984), a well regarded climate scientist and geographer at the University of Wisconsin-Madison.

In 2007, Murray C. Peel of the University of Melbourne, Australia, and colleagues published their revision of Köppen's climate classification. They drew on observational data from more than 4200 land weather/climate stations each having a continuous record of at least 30 years. Their high resolution analysis classifies climate in 1 degree by 1 degree squares in a grid covering Earth's entire land area. They found that the most widespread climates are hot desert (14.2%) and tropical savanna (11.5%).

Several criticisms have been leveled at Köppen-based climate classification systems. For one, the distribution of climate and vegetation do not always closely coincide. Although climate is the principal control of worldwide biomes, it is not the only control. (A *biome* is a community of plants and animals that usually occupies an extensive geographical area. Biomes are named for the types of plants that dominate the land-scape. Examples include temperate grassland, savanna, and boreal forest.) Soil type and drainage conditions, for example, also influence the geographical distribution of vegetation. Furthermore, in the Köppen scheme, climate and vegetation boundaries are portrayed as rigid whereas in fact, they shift location from one year to the next because of variability in temperature and precipitation. Nonetheless, the Köppen climate classification is quite useful.

As shown in Table 13.1, the Köppen climate classification system identifies six main climate groups; four are based on temperature, one is based on precipitation, and one applies to mountainous regions. Köppen's scheme uses upper case letters to symbolize major climatic groups: (A) Tropical humid, (B) Dry, (C) Subtropical (mesothermal), (D) Snow forest (microthermal), (E) Polar, and (H) Highland. Additional upper and lower case letters further differentiate climate types. Figure 13.2 is a global map of climate groups.

TABLE 13.1
Köppen-Based Climate Classification

Tropical humid (A)
Af	tropical wet
Am	tropical monsoon
Aw	tropical wet-and-dry

Dry (B)
BS	steppe or semiarid (BSh, BSk)
BW	arid or desert (BWh, BWk)
BWn	foggy desert

Subtropical (C)
Cs	subtropical dry summer (Csa, Csb)
Cw	subtropical dry winter
Cf	subtropical humid (Cfa, Cfb, Cfc)

Snow forest (D)
Dw	dry winter (Dwa, Dwb, Dwc, Dwd)
Df	year-round precipitation (Dfa, Dfb, Dfc, Dfd)
Ds	dry summer

Polar (E)
ET	tundra
EF	ice cap

Highland (H)

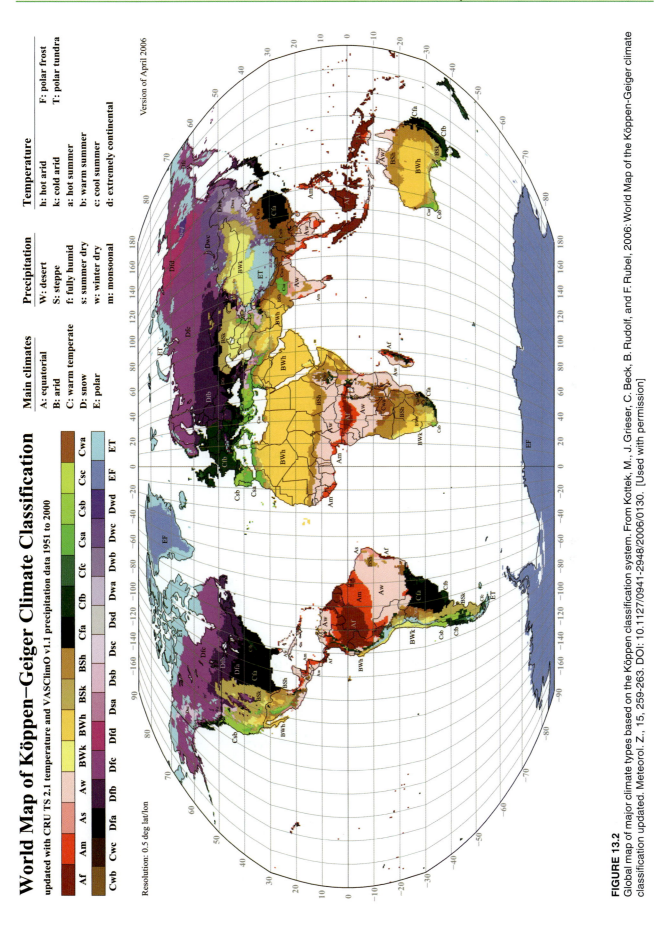

FIGURE 13.2
Global map of major climate types based on the Köppen classification system. From Kottek, M., J. Grieser, C. Beck, B. Rudolf, and F. Rubel, 2006: World Map of the Köppen-Geiger climate classification updated. Meteorol. Z., 15, 259-263. DOI: 10.1127/0941-2948/2006/0130. [Used with permission]

TROPICAL HUMID CLIMATES

Tropical humid climates (A) constitute a discontinuous belt straddling the equator and extending to higher latitudes to near the Tropic of Cancer in the Northern Hemisphere and the Tropic of Capricorn in the Southern Hemisphere. Mean monthly temperatures are high and exhibit little variability throughout the year. The mean temperature of the coolest month is no lower than 18 °C (64 °F), and there is no frost. The temperature contrast between the warmest and coolest month is typically less than 10 Celsius degrees (18 Fahrenheit degrees). In fact, the diurnal (day-to-night) temperature range generally exceeds the annual temperature range between the warmest and coolest months. This monotonous air temperature regime is the consequence of consistently intense incoming solar radiation associated with a high maximum solar altitude and little variation in the period of daylight throughout the year (Chapter 3).

Although tropical humid climate types are not readily distinguishable on the basis of temperature, important differences occur in precipitation regime. Tropical humid climates are subdivided into tropical wet (Af), tropical monsoon (Am), and tropical wet-and-dry (Aw). Although these climate types generally feature abundant annual rainfall, more than 100 cm (40 in.) on average, their rainy seasons differ in length and, in the case of Am and Aw, there is a pronounced dry season and wet season.

In tropical wet (Af) climates, the yearly average rainfall of 175 to 250 cm (70 to 100 in.) supports the world's most luxuriant vegetation. About 70% of all plant species growing in tropical rain forests are trees some of which grow to heights of 66 m (200 ft). The upper and middle levels of trees form a massive canopy that blocks much of the sunlight from reaching the ground so there is relatively little undergrowth. Tropical rain forests occupy the Amazon Basin of Brazil (Figure 13.3), the Congo Basin in Africa, the islands of Micronesia, and American Samoa (Figure 13.4). For the most part, rainfall is distributed uniformly throughout the year, although some areas experience a dry season. For example, July to November is the dry season for about two-thirds of the forested area of the Amazon basin. Nonetheless, photosynthesis persists and the rain forests remain green year-round.

Rainfall occurs as heavy downpours in frequent thunderstorms triggered by local con-

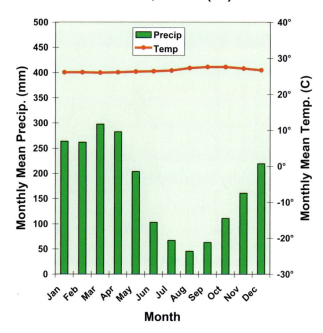

FIGURE 13.3
Mean monthly temperature and precipitation at Manaus, Brazil.

vection and the proximity of the intertropical convergence zone (ITCZ). Convection is largely controlled by solar radiation so that rainfall typically peaks in mid afternoon, the warmest time of day. Because the water vapor concentration is relatively high, even the slightest cooling at night leads to saturated air and the formation of dew or radiation fog, giving these regions a sultry, steamy appearance.

FIGURE 13.4
NOAA's Fagatele Bay National Marine Sanctuary in American Samoa, an example of a tropical humid Af climate. [NOAA Photo Library]

Tropical monsoon (Am) climates feature a seasonal rainfall regime with episodes of extremely heavy rainfall over a several month period followed by a lengthy dry season. The principal control for these climates involves seasonal shifts in wind direction between land and sea, a circulation that is best developed in Southeast Asia (Chapter 7). During the winter (*low-Sun* season), air pressure builds over the Asian continent causing dry air to flow southward into parts of Southeast Asia and India. During the summer (*high-Sun* season), low air pressure develops over the Tibetan Plateau and surface winds reverse direction, advecting moisture inland from the Indian Ocean. Local convection, orographic lifting, and shifts of the ITCZ combine to deluge the land with torrential rains. Am climates also occur in western Africa and northeastern Brazil. Compared to tropical rain forests, Am forests are less dense with individual trees more widely spaced and heavier ground cover (because of greater sunlight penetration). Tree species indigenous to Am climates include teak, mahogany, rubber, and banana.

For the most part, tropical wet-and-dry climates (Aw) border tropical wet climates (Af) on their higher latitude sides. Summers in Aw climates are wet and winters are dry. This marked seasonality of rainfall is linked to shifts of the intertropical convergence zone (ITCZ) and semipermanent subtropical anticyclones, which follow the seasonal excursions of the Sun. In the *high-Sun* season, surges of the ITCZ trigger convective rainfall; in the *low-Sun* season, the trade winds and dry eastern flank of the subtropical anticyclones dominate the weather. In a poleward direction, Aw climates transition to subtropical dry climates and the dry season lengthens. The drier Aw climate plus the bordering BS climate support the savanna, tropical grasslands with scattered deciduous trees. Savanna tree/shrub species include acacia and eucalyptus.

The mean annual temperature in Aw climates is only slightly lower, while the seasonal temperature range is only slightly greater, than in the tropical wet climates (Af). The diurnal temperature range varies seasonally, however. In summer, frequent cloudy skies and high humidity suppress the diurnal temperature range by reducing both solar heating of the ground during the day and radiational cooling at night. In winter, on the other hand, persistent fair skies have the opposite effect on the rates of radiational heating and cooling, causing an increase in the diurnal temperature range. Cloudy, rainy summers plus dry winters also mean that the year's highest temperatures typically occur toward the close of the dry season in late spring.

DRY CLIMATES

Dry climates (B) characterize those regions of the world where average annual potential evaporation exceeds average annual precipitation. *Potential evaporation* is the quantity of water that would vaporize into the atmosphere from a surface of fresh water during long-term average weather conditions. Air temperature largely governs the rate of evaporation so it is not possible to specify some maximum rainfall amount as the threshold for dry climates. Rainfall is not only limited in B climates but also is highly variable and unreliable. As a general rule, the lower the mean annual rainfall, the greater is its variability from one year to the next.

Earth's dry climates encompass a larger land area than any other single climate grouping. Perhaps 30% of the planet's land surface stretching from the tropics into middle latitudes, experience a moisture deficit of varying degree. These include the climates of the world's deserts and steppes, where vegetation is sparse and equipped with special adaptations that permit survival under conditions of severe moisture stress. Common features of arid or semiarid climate regions are stream or creek beds or water-carved gullies that are dry most of the time. However, an occasional thunderstorm over the watershed will produce a sudden surge of water that flows rapidly down the drainage way and may cause flash flooding. These normally dry waterways are called *arroyos* or *washes* and are common in the American Southwest, parts of India, and around the Mediterranean Sea.

Based on the degree of dryness, a distinction is made between two dry climate types: steppe or semiarid (BS) and arid or desert (BW). Steppe or semiarid climates are transitional between more humid climates and arid or desert climates. Mean annual temperature is latitude dependent, as is the range in variation of mean monthly temperatures through the year. Hence, a distinction is made between warm dry climates of tropical latitudes (BSh and BWh) and cold, dry climates of higher latitudes (BSk and BWk).

Dryness is the consequence of subtropical anticyclones, cold surface ocean currents, or the rain shadow effect of high mountain ranges. Subsiding stable air on the eastern flanks of subtropical anticyclones gives rise to tropical dry climates, designated as BSh and BWh (Figure 13.5). These huge semipermanent pressure systems, centered over the ocean basins, dominate the weather year-round in the 18 to 28 degree latitude belt of both hemispheres. Hence, dry climates characterize North Africa eastward to northwest India, the southwestern United States (Figure 13.6) and northern Mexico,

Khartoum, Sudan (BWh)

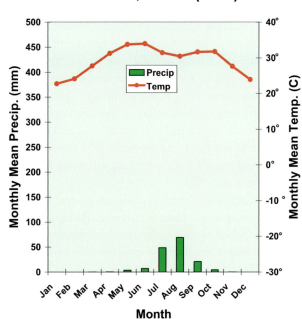

FIGURE 13.5
Mean monthly temperature and precipitation at Khartoum, Sudan.

coastal Chile and Peru, southwest Africa, and much of the interior of Australia. They include the world's great deserts: Sahara, Sonoran, Thar, Kalahari, and Great Australian.

Although persistent and abundant sunshine is generally the rule in dry tropical climates, there are some important exceptions. Where cold ocean water borders a

FIGURE 13.6
Desert landscape near Yuma, AZ, an example of a BWh climate. [Photo by Jason West]

coastal desert, a shallow layer of stable marine air drifts inland. The desert air thus features high relative humidity, persistent low stratus clouds and fog, and considerable dew formation. Examples are the Atacama Desert of Peru and Chile, the Namib Desert of southwest Africa, portions of the coastal Sonoran Desert of Baja California, and stretches of the coastal Sahara Desert of northwest Africa. These anomalous foggy desert climates are designated BWn. In fact, many desert plants have leaves and stems that are adapted to use dew for moisture.

Cold, dry climates of higher latitudes (BWk and BSk) are situated in the rain shadows of great mountain ranges. They occur primarily in the Northern Hemisphere, to the lee of the Sierra Nevada and Cascade ranges in North America, and the Himalayan chain in Asia. Because these dry climates are at higher latitudes than their tropical counterparts, mean annual temperatures are lower and the seasonal temperature contrast is greater. Anticyclones dominate winter weather, bringing cold and dry conditions, whereas summers are hot and generally dry. Scattered convective showers, mostly in summer, produce relatively meager precipitation.

SUBTROPICAL CLIMATES

Subtropical climates (C) are located just poleward of the Tropics of Cancer and Capricorn and are dominated by seasonal shifts of subtropical anticyclones. There are three basic climate types: subtropical dry summer (or *Mediterranean*) (Cs), subtropical dry winter (Cw), and subtropical humid (Cf), which receives precipitation throughout the year.

Mediterranean climates occur on the western side of continents between about 30 and 45 degrees latitude. In North America, mountain ranges confine this climate to a narrow coastal strip of California. Elsewhere, Cs climates rim the Mediterranean Sea and occur in portions of extreme southern Australia. Summers are dry because at that time of year Cs regions are under the influence of stable subsiding air on the eastern flanks of the semipermanent subtropical highs. Equatorward shift of subtropical highs in autumn allows extratropical cyclones to migrate inland from the ocean, bringing moderate winter rainfall. Mean annual precipitation varies greatly—ranging from 30 to 300 cm (12 to 80 in.) with the wettest winter month typically receiving at least three times the precipitation of the driest summer month.

Although Mediterranean climates exhibit a pronounced seasonality in precipitation (dry summers and wet winters), the temperature regime is quite variable. In coastal areas, cool onshore breezes prevail producing a strong maritime influence (Chapter 4), lowering the mean annual temperature and reducing the seasonal temperature contrast. Well inland, however, away from the ocean's moderating influence, summers are considerably warmer; hence, inland mean annual temperatures are higher and seasonal temperature contrasts are greater than in coastal Cs localities. The climate records of coastal San Francisco and inland Sacramento, CA, illustrate the contrast in temperature regime within Cs regions. Although the two cities are separated by only about 145 km (90 mi.), the climate of Sacramento is much more continental (much warmer summers and somewhat cooler winters) than the climate of San Francisco (Figure 13.7). The warm climate subtype is designated Csa and the cooler subtype is Csb.

Wildfires driven by hot, dry Santa Ana winds are a major hazard in portions of southern California especially in autumn. For details on conditions contributing to the California wildfire climate, refer to this chapter's first Essay.

Subtropical dry winter climates (Cw) are transitional between Aw and BS climates and located in South America and Africa between about 20 and 30 degrees S. Cw climates also occur between the Aw and H climates of the Himalayas and Tibetan plateau and between the BS and Cfa climates of Southeast and East Asia (Figure 13.8). Northward shift of the subtropical high pressure systems is responsible for the dry winter in South America and Africa. The narrowness of the two continents between 20 and 30 degrees S means a relatively strong maritime influence and dictates against extreme dryness. In spring, subtropical highs shift southward and rains return. In Asia, winter dryness is caused by winds radiating outward from the massive cold *Siberian high*. As the continent warms in spring, the Siberian high weakens and eventually is replaced by low pressure. Moist winds then flow inland bringing summer rains. Mean annual precipitation in Cw climates ranges from 75 to 150 cm (30 to 60 in.).

Subtropical humid climates (Cf) occur on the eastern side of continents between about 30 and 40 degrees latitude (and even more poleward where the maritime influence is strong). Cfa climates are the most important of

FIGURE 13.7
San Francisco, CA, an example of a subtropical Csb climate. [NOAA Photo Library]

the Cf climate subtypes in terms of land area and number of people impacted. Cfa climates are situated primarily in the southeastern United States, a portion of southeastern South America, eastern China, southern Japan, on the extreme southeastern coast of South Africa, and along much of the east coast of Australia. These climates fea-

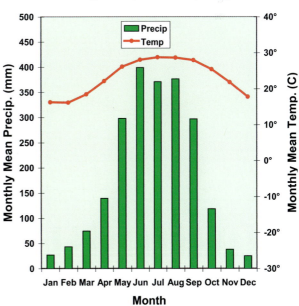

FIGURE 13.8
Mean monthly temperature and precipitation at Hong Kong, China.

ture abundant precipitation (75 to 200 cm, or 30 to 80 in., on average annually), which is distributed throughout the year. In summer, Cfa regions are dominated by a flow of sultry maritime tropical air on the western flanks of the subtropical anticyclones. Consequently, the long summers are hot and humid with frequent thunderstorms, which can produce brief periods of substantial rainfall. Hurricanes and tropical storms contribute significant rainfall (up to 15% to 20% of the annual total) to some North American and Asian Cfa regions, especially from summer through autumn. The short winters are cool and humid. After the subtropical highs shift toward the equator in autumn, Cfa regions come under the influence of the prevailing wester-lies and migrating extratropical cyclones and anticyclones. In Cfa localities, mean temperatures of the warmest month are typically in the range of 24 to 27 °C (75 to 81 °F). Average temperatures for the coolest months typically range from 4 to 13 °C (39 to 55 °F). Subfreezing temperatures and snowfalls are infrequent.

A strong maritime influence is responsible for the cool summers and mild winters of Cfb climates. These climates occur over much of northwestern Europe, New Zealand, and portions of southeastern South America, southern Africa, and Australia (Figure 13.9). The coldest subtype, the Cfc climate, is relegated to coastal areas of

southern Alaska, Norway, and the southern half of Iceland. Cfb and Cfc climates are relatively humid with mean annual precipitation ranging between 100 and 200 cm (40 and 80 in.).

SNOW FOREST CLIMATES

Snow forest climates (**D**) occur in the interior and on the leeward sides of large continents. The name underscores the link between biogeography and the Köppen climate classification system. These climates feature cold snowy winters (except for the Dw subtype in which the winter is dry) and occur only in the Northern Hemisphere. Snow forest climates are subdivided according to seasonal precipitation regimes with Df climates experiencing year-round precipitation whereas Dw climates have a dry winter. D climates with dry summers (Ds) are rare and small in extent. Additional distinction is made between warmer subtypes (Dwa, Dfa, Dwb, and Dfb) and colder subtypes (Dwc, Dfc, Dwd, and Dfd).

The warmer subtypes, sometimes termed *temperate continental climates*, feature warm summers (mean temperature of the warmest month greater than 22 °C or 71 °F) and cold winters. They are located in Eurasia, the northeastern third of the United States, southern Canada, and extreme eastern Asia. Continentality increases inland with maximum temperature contrasts between the coldest and warmest months as great as 25 to 35 Celsius degrees (45 to 63 Fahrenheit degrees). The southerly Dfa climates (Figure 13.10) have cool winters and warm summers and the more northerly Dfb climates have cold winters and mild summers. The length of the freeze-free period varies from 7 months in the south to only 3 months in the north. The weather is very changeable because these regions are swept by extratropical cyclones and anticyclones and by surges of contrasting air masses. Polar front cyclones dominate winter, bringing episodes of light to moderate frontal precipitation. These storms are followed by incursions of dry polar and arctic air masses. In summer, cyclones are weak and infrequent as the principal storm track shifts poleward. Summer rainfall is mostly convective, and locally amounts can be very heavy in severe thunderstorms, squall lines, and mesoscale convective complexes (MCCs). Although precipitation is distributed rather uniformly throughout the year, most places experience a summer maximum.

In northern portions, winter snowfall becomes an important factor in the climate. Mean annual snowfall and the persistence of a snow cover increase northward. Because of its high albedo for solar radiation and its

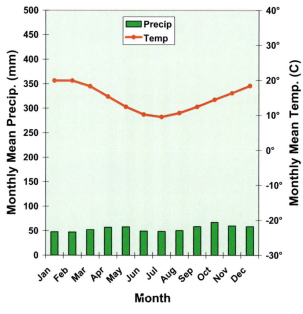

FIGURE 13.9
Mean monthly temperature and precipitation at Melbourne, Australia.

FIGURE 13.10
Autumn colors on Clopper Lake in Seneca Creek State Park, Maryland, an example of a Dfa climate. [NOAA Photo Library]

efficient emission of infrared, a snow cover chills and stabilizes the overlying air. For these reasons, a snow cover tends to be self-sustaining; once established in early winter, an extensive snow cover tends to persist.

Moving poleward, summers get colder and winters are bitterly cold. These so-called *boreal climates* (Dfc, Dfd, Dwc, Dwd) occur only in the Northern Hemisphere as an east-west band between 50 to 55 degrees N and 65 degrees N (Figure 13.11). It is a region of extreme continentality and very low mean annual temperature. Summers are short and cool, whereas winters are long and bitterly cold. Because midsummer freezes are possible, the growing season is precariously short. Both continental polar (cP) and arctic (A) air masses originate here, and this area is the site of an extensive coniferous (boreal) forest. As noted in Chapter 1, in summer, the mean position of the leading edge of arctic air (the arctic front) is located approximately along the northern border of the boreal forest. In winter, the mean position of the arctic front is situated near the southern border of the boreal forest.

Weak cyclonic activity occurs throughout the year and yields meager annual precipitation (typically less than 50 cm, or 20 in.). Convective activity is rare. A summer precipitation maximum is due to the winter dominance of cold, dry air masses. Snow cover persists throughout the winter and the range in mean temperature between winter and summer is among the greatest in the world.

POLAR CLIMATES

Polar climates (E) are situated poleward of the Arctic and Antarctic circles. These boundaries correspond roughly to localities where the mean temperature for the warmest month is 10 °C (50 °F) and approximate the tree line, the poleward limit of tree growth. Poleward are tundra and the Greenland and Antarctic ice sheets. A distinction is made between tundra (ET) and ice cap (EF) climates, with the dividing criterion being 0 °C (32 °F) for the mean temperature of the warmest month. Vegetation is sparse in ET regions and almost nonexistent in EF areas (Figures 13.12 and 13.13).

Polar climates are characterized by extreme cold and slight precipitation, which falls mostly in the form of snow (less than 25 cm, or 10 in., melted, per year). Greenland and Antarctica could be considered

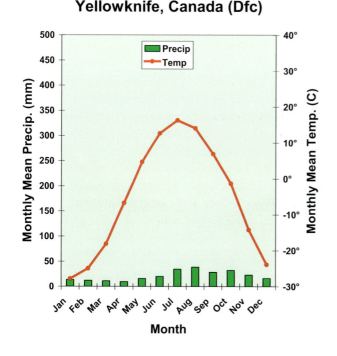

FIGURE 13.11
Mean monthly temperature and precipitation at Yellowknife, Canada, an example of a Dfc climate.

McMurdo, Antarctica (EF)

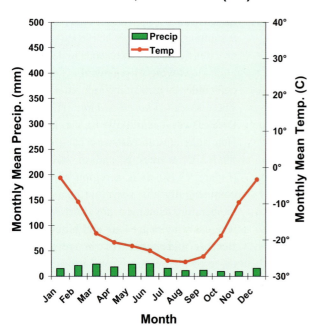

FIGURE 13.12
Mean monthly temperature and precipitation at McMurdo, Antarctica, an example of an EF climate.

FIGURE 13.13
Penguins explore snow dunes in Antarctica, an example of an EF climate. [NOAA Photo Library]

deserts for lack of significant precipitation, despite the presence of massive ice sheets. Although summers are cold, winters are so extremely cold that polar climates feature a marked seasonal temperature contrast. Mean annual temperatures are the lowest of any place in the world.

HIGHLAND CLIMATES

Highland climates (H) include a wide variety of climate types that characterize mountainous terrain (Figure 13.14). Altitude, latitude, and aspect are among the factors that differentiate a complexity of climate types. For example, temperature decreases rapidly with increasing altitude and windward slopes tend to be wetter than leeward slopes. Climate-ecological zones are telescoped in mountainous terrain. That is, in ascending several thousand meters of altitude, a mountain climber would encounter the same bioclimatic zones that he/she would experience in traveling several thousand kilometers of latitude. As a general rule, every 300 m (980 ft) of elevation corresponds roughly to a northward advance of 500 km (310 mi). However, in some cases, changes in climate-ecological zones with elevation are much more rapid. Consider for example, Mount Washington, New Hampshire, where a tree-less alpine zone occurs at elevations between 1524 m (5000 ft) above mean sea level and the summit at 1917 m (6288 ft). For more on Mount Washington's extreme climate, see this chapter's second Essay.

FIGURE 13.14
A variety of climate-ecological zones can be found in close proximity in mountainous terrain. [NOAA Photo Library]

Climate Change and Ecosystem Response

In our study of climate science, climate variability, and climate change, our principal focus has been the temporal fluctuations of climate (past, present, and future) with less attention paid to spatial shifts in climate types or climate zones. As we have seen, climate is the prime ecological control, shaping the fabric of Earth's mosaic of numerous terrestrial ecosystems and biomes. Hence, temporal changes in climate are likely to be accompanied by spatial shifts in both climate type and corresponding ecosystems.

With continued global warming, climate models predict a general poleward shift of climate zones. By 2100, these projections may mean that some polar climates will vanish, while previously unknown climates will appear in the tropics. In mountain ranges, climate zones will move upward with some high-altitude climates disappearing entirely off the summit along with snowfields and glaciers.

Plants and animals respond to climate change but reconstructions of past climates indicate that climate can change more rapidly than ecosystems. A major climate shift may exceed the tolerance limits of certain organisms, and the assemblage of species that live in the region will change. **Ecological succession** refers to the replacement of one ecosystem by another through a sequence of colonization and replacement of species until an assemblage of species is established that is able to maintain itself on a site. Organisms vary greatly in their ability to disperse and become established so that climate change may disrupt population interactions (e.g., predation, parasitism, and competition) for scores of years; some organisms will succumb, while others will migrate into or out of a region in search of more favorable habitat. Centuries may pass before the climax stage of an ecosystem is established that is adapted to the new climate regime if it ever does. (The *climax stage* is the relatively stable stage of ecological succession characterized by an assemblage of species that is able to maintain itself on a site.)

Extreme shifts in the location and species composition of ecosystems occurred during the Pleistocene Ice Age in response to global-scale climate change. A look at some of these extremes provides us with a perspective on the magnitude of possible climatically-induced ecosystem change. In fact, some scientists suggest that the forests of North America are still not in equilibrium with their environment thousands of years after deglaciation.

That is, today's pattern of ecosystem distribution reflects merely the current transient stage of species migration following glacial retreat.

As described in Chapter 9, cooling that set in about 27,000 years ago triggered the final major advance of the Laurentide ice sheet that spread over most of Canada and by 20,000 to 18,000 years ago engulfed the northern tier states of the coterminous United States. In non-glaciated areas south of the ice sheet, climate change disrupted ecosystems and Earth's mosaic of climate types was quite different than it is today. A narrow belt of treeless tundra, underlain by sporadic permafrost, developed along the edge of the ice sheet while moist conditions erased arid ecosystems in the American Southwest. In the East, deciduous forest species retreated to small refuges in Appalachian mountain valleys and along a portion of the continental shelf exposed by falling sea level.

During glacial climatic episodes, a forest community similar (but not identical) in species composition to the modern boreal forest of Canada spread over much of eastern North America. This boreal forest was made up of smaller percentages of pine and larger percentages of temperate deciduous trees than its modern counterpart. Based on reconstruction of past vegetation based on pollen profiles from sediment cores (Chapter 9), boreal species of the late Pleistocene were reported from as far south as Lee County in south central Texas, the Ozark Highlands of southwestern Missouri, and in several localities in the southeastern United States. The greater proportion of pine and deciduous trees in the Ice Age boreal forest may have been due to higher levels of solar radiation at the forest's lower latitude location. There, pine trees anchored in frozen soil would have been prone to desiccation during the more frequent warm sunny winter days. On the other hand, deciduous components of the forest would have weathered such conditions quite well.

In spite of the widespread presence of boreal species in late Pleistocene pollen profiles extracted from sites south of the Laurentide ice sheet, evidence exists that the Ice Age conifer forest was locally confined to limited stands, especially near the forest's southern border. This characteristic coupled with the difference in species composition between the modern and Ice Age boreal forest may imply that the properties of Arctic air of late Pleistocene differed from that of modern time.

With the advent of agriculture, humans greatly simplified ecosystems—in some cases cultivating only a few species of crops—so that agricultural systems tend to be more vulnerable than natural ecosystem to climate change and other stressors. Numerous examples could be cited of climate shifts that forced people to abandon their lands because of a shift to an inhospitable climate. More frequent megadroughts caused food shortages that forced the Anasazi and Fremont peoples to abandon the American Southwest in the 13th century. (Refer to the Case-in-Point in Chapter 9.) More recently, during the Dust Bowl era of the 1930s, farms were abandoned by the score. (Refer to the Case-in-Point in Chapter 6.) Furthermore, agriculture that is most vulnerable to climate change is located in areas where the climate is already marginal for food production; that is, where just enough rain falls or the growing season is barely long enough for crops to mature. Even a small unfavorable change in these critical parameters can spell disaster. (Refer, for example, to the second Essay of Chapter 2 for a discussion of the fate of the first European settlements in Greenland.)

Other Climate Classification Systems

In addition to the Köppen climate classification system are many other schemes designed to group together climates having similar characteristics. These include systems derived by C.W. Thornthwaite (based on soil moisture budget) and T. Bergeron (based on air mass frequency) plus the Holdridge life zones (identifies bioclimatic zones). These are summarized below.

THORNTHWAITE

The American climate scientist and geographer C.W. Thornthwaite (1892-1963) developed a climate classification system based on annual soil moisture. In 1931, the initial version of the **Thornthwaite classification** utilized a precipitation-evaporation index that was the basis for identifying humidity provinces as the major subdivision of climates. In a significant revision of his classification system in 1948, Thornthwaite added a measure of transpiration to direct evaporation (*evapotranspiration*). In climates where moisture is non-limiting, he used potential evapotranspiration. The revised classification depends on a *precipitation effectiveness index*, an indicator of net moisture supply, and a *temperature efficiency index*, a measure of heat energy relative to evaporation rate.

The precipitation effectiveness index is computed from the monthly ratio of precipitation to estimated evaporation and summed through the course of the year. Values are the basis for identifying nine moisture provinces. The temperature efficiency index is computed

in the same way as the precipitation effectiveness index except that temperature is used in place of evaporation. Values are the basis for identifying nine major temperature provinces.

The combination of mean temperature, mean precipitation, and vegetation type are used to generate a humidity index and aridity index that characterize a region's moisture regime. The lower the value of either index, the drier is the region. In order of declining moisture, the Thornthwaite classification system employs the descriptive terms *hyperhumid, humid, subhumid, subarid, semi-arid,* and *arid.* In a humid climate, annual precipitation exceeds annual evaporation whereas in an arid climate, annual potential evaporation is greater than annual precipitation.

Temperature is the principal distinction among microthermal, mesothermal, and megathermal climates. A *microthermal climate* features a mean annual temperature of 0 °C to 14 °C (32 °F to 57 °F), short summers, and potential evaporation of 14-43 cm (5.5-17 in.). In a *mesothermal climate*, neither heat nor cold is persistent and potential evaporation ranges from 57 to 114 cm (22-45 in.). A *megathermal climate* is characterized by persistent high temperature and abundant rainfall. Potential evaporation exceeds 114 cm (45 in.).

BERGERON

In 1928, Tor Bergeron (1891-1971) developed a climate classification system based on the frequency with which the various air masses occur in a particular location. The **Bergeron classification** uses three letters to identify the properties of an air mass in its source region. The initial letter stands for the moisture properties with c for continental (dry) air masses and m for maritime (moist) air masses. The second letter corresponds to thermal characteristics with T for tropical, P for polar, A for Arctic or Antarctic, M for monsoon, E for equatorial, and S for dry air produced when air descends within the atmosphere. The third letter symbol conveys the stability of the air mass. An air mass is labeled k (unstable) if it were colder than the ground beneath it or w (stable) if the air mass were warmer than the ground beneath it.

HOLDRIDGE LIFE ZONES

The **Holdridge life zones** constitute a global bioclimatic classification system, first published by Leslie Holdridge (1907-1999) in 1947 and revised in 1967. In his classification system, Holdridge, an American botanist and climate scientist, utilized mean annual biotemperature derived from growing season

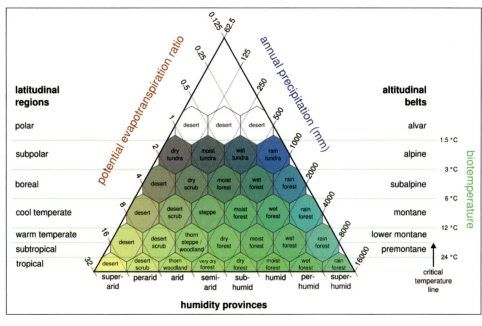

FIGURE 13.15
Holdridge life zones.

This chapter brings to a close our discussion of the basic aspects of climate science, climate variability, and climate change. The next chapter covers the various strategies relating to climate change mitigation, climate change adaptation, geoengineering, and climate-conscious architecture. While innovative technologies promise to reduce greenhouse gas emissions from anthropogenic sources, we are all part of the problem and there is much that each of us can do to reduce our carbon footprint.

length and temperature, the annual precipitation, and the ratio of potential evapotranspiration to mean annual precipitation. The system recognizes 38 bioclimate classes ranging from polar desert to tropical rainforest. These bioclimatic classes are displayed graphically as a two-dimensional array of hexagons in a triangular frame in Figure 13.15 with latitudinal regions on the left, altitudinal belts on the right, and humidity provinces across the base of the triangle.

Conclusions

Earth is a mosaic of numerous types of climate. Climates having similar characteristics are grouped together in classification systems that are empirical, genetic, or numerical. The Köppen system, the most popular empirical classification, matches the distribution of biomes to patterns of mean annual and monthly temperature and precipitation. On this basis, Earth's land areas are divided into six main climate groups; four are based on temperature, one is based on precipitation, and one applies to mountainous regions. Climate change may shift the boundaries of biomes and alter the species composition and interactions within natural ecosystems and agricultural systems. Because climate change is faster than ecological succession, considerable environmental disruption may result.

Basic Understandings

- The various forcing agents and mechanisms operating in and on Earth's climate system are responsible for producing the world's diverse mosaic of climates.
- Climate classification schemes typically group climates according to the meteorological basis of climate (genetic classification) or the environmental effects of climate (empirical classification).
- An alternative to the traditional method of describing the climate of some locality quantitatively in terms of normals, means and extremes of various climatic elements is air mass climatology. In this approach, the frequency with which various types of air masses develop over or are advected into a locality is the basis for describing the climate of that locality.
- The popular Köppen climate classification system takes an empirical approach to organizing Earth's many climates. It recognizes patterns in mean annual and monthly temperature and precipitation that closely correspond to the geographic limits of vegetative communities thereby revealing broad-scale climate boundaries throughout the world.
- The Köppen climate classification system identifies six main climate groups; four are

temperature-based, one is precipitation-based, and one applies to mountainous regions. They are (A) Tropical humid, (B) Dry, (C) Subtropical (mesothermal), (D) Snow forest (microthermal), (E) Polar, and (H) Highland.

- Although tropical humid climates are not readily distinguishable on the basis of temperature, important differences exist in precipitation regime. Tropical humid climates are divided into tropical wet (Af), tropical monsoon (Am), and tropical wet-and-dry (Aw).

- Dry climates (B) characterize those regions where the average annual potential evaporation is greater than the average annual precipitation. Because air temperature largely governs the rate of evaporation, it is not possible to specify some maximum rainfall as the threshold for dry climates.

- Dry climates encompass a larger land area than any other single climate grouping, stretching from the tropics into middle latitudes. Based on the degree of dryness, a distinction is made between steppe or semiarid (BS) and arid or desert (BW). A further distinction is made between warm dry climates of tropical latitudes (BSh and BWh) and cold, dry climates of higher latitudes (BSk and BWk).

- Subsiding stable air on the eastern flanks of subtropical anticyclones gives rise to tropical dry climates whereas cold, dry climates are located in the rain shadows of great mountain ranges.

- Subtropical climates (C) are dominated by seasonal shifts of subtropical anticyclones. Mediterranean climates (Cs) occur on the west side of continents between about 30 and 45 degrees latitude and feature dry summers and wet winters. In coastal areas, the strong maritime influence lowers the mean annual temperature and reduces the summer-winter seasonal temperature contrast.

- Subtropical humid climates (Cf) occur on the eastern side of continents between about 30 and 40 degrees latitude. Cfa climates occur in the southeastern U.S. and feature abundant precipitation which is distributed throughout the year. Summers are hot and humid with frequent thunderstorms. After the subtropical anticyclones shift toward the equator in autumn, Cfa climates come under the influence of migrating extratropical cyclones and anticyclones so the short winters are cool and humid.

- Snow forest climates (D) occur in the interior and on the leeward sides of large continents. Most of these climates feature cold snowy winters and occur only in the Northern Hemisphere. The warmer subtypes, temperate continental climates, feature warm summers and cold winters and occur in Eurasia, northeastern third of the U.S., southern Canada, and extreme eastern Asia. The southerly Dfa climates have cool winters and warm summers whereas the more northerly Dfb climates have cold winters and mild summers.

- Boreal climates occur only in the Northern Hemisphere as an east-west band between 50 to 55 degrees N and 65 degrees N. This is a region of extreme continentality and very low mean annual temperature. Summers are short and cool, and winters are long and bitterly cold.

- Polar climates (E), characterized by extreme cold and slight precipitation, are located poleward of the Arctic and Antarctic circles. In the Northern Hemisphere, it includes tree-less tundra and the Greenland ice sheet. In the Southern Hemisphere, it coincides with the Antarctic ice sheet.

- Highland climates (H) encompass a diversity of climate types that characterize mountainous terrain where climate-ecological zones are telescoped. Altitude, latitude, and aspect are among the factors that differentiate climates.

- With continued global warming, climate models predict a general poleward shift of climate zones. By 2100, this shift may mean that some polar and highland climates may vanish, while previously unknown climates may appear in the tropics.

- Ecological succession refers to the replacement of one ecosystem by another through a sequence of colonization and replacement of species until an assemblage of species is established that is able to maintain itself on a site.

- With the advent of agriculture, humans greatly simplified ecosystems so that agricultural systems tend to be more vulnerable than natural ecosystems to climate change and other stressors.

- In addition to the Köppen climate classification are many other schemes designed to group together climates having similar characteristics. These include systems derived by C.W. Thornthwaite (based on soil moisture budget) and T. Bergeron (based on air mass frequency) plus the Holdridge life zones (identifies bioclimatic zones).

Enduring Ideas

- Climate classification systems group together climates having common characteristics. These systems are genetic, empirical, and numerical.
- With air mass climatology, the climate of a locality or region is described in terms of the frequency with which various air masses develop over or are advected into that locality or region.
- As an expression of climate variability, boundaries between climate types are not necessarily rigid.
- Climate change is likely to be accompanied by spatial (e.g., latitude, altitude) shifts in both climate type and corresponding biome or ecosystem.

Review

1. Distinguish between a genetic climate classification and an empirical climate classification.
2. What is the basis of the Köppen climate classification?
3. In tropical humid climates, how does the typical diurnal temperature range compare to the annual temperature range?
4. In tropical wet-and-dry climates (Aw), the year's highest temperature typically is recorded toward the close of the dry season in late spring. Explain why.
5. Define dry climates (B) in terms of potential evaporation and annual precipitation.
6. Describe the atmospheric circulation characteristics that are responsible for tropical dry climates.
7. Describe the atmospheric circulation characteristics that are responsible for middle latitude dry climates.
8. Explain the development of persistent low stratus clouds and fog in the Atacama Desert of Peru and Chile.
9. Why do Mediterranean climates feature wet winters and dry summers?
10. In what way is the extent of ground that is snow covered an important factor in the northern portions of D climates?

Critical Thinking

1. What are the implications for the Köppen climate classification system if the climate and indigenous vegetation are not in equilibrium?
2. How might differences in drainage conditions influence the geographical distribution of vegetation independent of climate?
3. What role is played by regular seasonal shifts in the intertropical convergence zone and subtropical anticyclones in climates that feature a distinct dry season and wet season?
4. What is the connection between cold surface ocean currents and dry climates?
5. Why is the winter weather in snow forest climates so changeable?
6. Boreal climates feature a summer precipitation maximum. Explain why.
7. Why does climate change more rapidly than terrestrial ecosystems?
8. How might large-scale climate change alter the world mosaic of climate types?
9. Why are climate-ecological zones telescoped in mountainous regions?
10. How do large glacial ice sheets feed back on polar climates?

ESSAY: California Wildfire Climate

During autumn, *Santa Ana winds* frequently descend the mountain slopes of interior southern California and sweep over the coastal plain to the Pacific Ocean (Chapter 7). These hot dry winds, named for the Santa Ana Canyon or the Santa Ana Mountains southeast of Los Angeles, blow over a landscape parched by the long dry summer. Santa Ana winds further desiccate the vegetation, making the southern California landscape susceptible to an outbreak of wildfires. Also contributing to the wildfire hazard is a shrub community known as *chaparral* that dominates the hill slopes (Figure 1). The tissues of these plants contain oils that readily ignite; in fact, chaparral shrubs practically explode when exposed to flames. Periodic wildfires, whipped by Santa Ana winds roar through the canyons of southern California. Juan Rodriguez Cabrillo (ca. 1499-1543) provided one of the earliest accounts of such fires in October 1542 while onboard his ship anchored off the Los Angeles Basin. He observed hot desert winds blowing and fires burning.

FIGURE 1
Chaparral surrounds a California lake. [Courtesy of U.S. Department of the Interior/Bureau of Reclamation/Mid-Pacific Region/Lake Berryessa Recreation Division, Administrative Office.]

In southern California, winter rains usually follow the summer dry season. (Dry summers and wet winters characterize a *Mediterranean-type* climate. For example, in Los Angeles, on average, almost 95% of annual rainfall occurs from November through April.) Falling on the nutrient-rich ashes of burned vegetation, winter rains spur re-sprouting of shrubs. Furthermore, fire followed by rain also triggers germination of seeds that have been dormant in the soil. Renewal of vegetative cover helps to stabilize hillsides denuded by wildfires. If heavy rains arrive before the vegetation has a chance to stabilize burnt-over hill slopes, then soil and debris are washed from the bare ground and flushed into streambeds. Rainwater also percolates into the loosely consolidated upper soil layer. When this layer becomes saturated, it flows down steep hillsides as a river of mud, covering roadways and inundating homes. These mudflows are the most serious post-fire problem. Furthermore, even a single winter of below-average rainfall significantly elevates the potential for autumnal wildfires.

Native Americans adapted to southern California's climate/fire regime by building temporary shelters that they could readily move out of harm's way. Furthermore, these people routinely set small fires (*controlled burns*) that improved the habitat for game animals and thus made hunting easier. Arrival of European settlers in southern California marked the establishment of permanent dwellings and efforts to suppress wildfires. In subsequent centuries the population of southern California grew slowly, but since the end of World War II, southern California has been one of the fastest growing regions of the nation. Hundreds of thousands of homes were constructed in the chaparral-covered foothills. A combination of urban sprawl, summer drought, fire-prone vegetation, and autumnal Santa Ana winds set the stage for highly destructive wildfires.

Such a destructive event erupted on 27 October 1993 when fifteen major fires broke out on hillsides from Ventura County south to San Diego County. Firefighters were largely unable to contain the wildfires until Santa Ana winds finally died down three days later. After several days of relatively weak winds, the regional air pressure gradient again strengthened and Santa Ana winds resumed, whipping flames through previously untouched areas of Malibu and Topanga. In the aftermath, state and federal emergency services estimated total property damage at more than $1 billion. Fires scorched more than 80,000 hectares (almost 200,000 acres) and destroyed more than 1200 structures.

About ten years later, in October 2003, southern California experienced the most devastating wildfires in the state's history. Fourteen major fires broke out affecting Ventura, Los Angeles, San Bernardino, and San Diego counties (Figure 2). The fires claimed 24 lives, destroyed 3710 homes, and burned over 303, 542 hectares (750,043 acres). Thousands of residents

FIGURE 2
Several massive wildfires were raging across southern California over the weekend of 25 October 2003. Whipped by hot, dry Santa Ana winds that blow toward the coast from the interior deserts, at least one wildfire grew 10,000 acres in just 6 hours. The Moderate Resolution Imaging Spectroradiometer (MODUS) on the Terra satellite captured this image of the fires and clouds of smoke spreading over the region on 26 October 2003. [Courtesy of NASA]

were forced from their homes and fled to evacuation centers. At the peak of the fire siege, more than 14,000 firefighters from federal, state, and local agencies battled the flames from the ground and air. Once again Santa Ana winds fanned the flames that consumed the considerable fuel (dead, dying, and diseased vegetation) that had accumulated at the wildland/urban interface.

Massive wildfires erupted from Malibu, CA, to San Diego County in October and November of 2007 (Figure 3). Twenty-three fires took 10 lives, injured 292 people, burned 211,407 hectares (522,398 acres), and destroyed more than 3290 structures. In southern California, 7 counties were declared major disaster areas.

FIGURE 3
Helicopters drop water and fire retardant chemicals on the Harris fire, near the Mexican border, on 25 October 2007. [Andrea Booher/FEMA]

What can be done to reduce the Santa Ana wildfire hazard? Nothing can be done to alter the normal sequence of summer drought, autumnal Santa Ana winds, and variable winter rains. These are basic characteristics of the region's climate. People can change where they chose to live, however. Some people argue that zoning ordinances should be enacted and enforced that prohibit construction of houses in locales subject to natural hazards such as wildfires and mudflows. Alternatively, ordinances could require that dwellings be constructed of fireproof materials (particularly roofs and outside walls). Furthermore, chaparral vegetation could be cleared within a designated perimeter of buildings. At present, many people permit shrubby vegetation to grow up to the edge of their homes for privacy and a sense of living with nature.

If wildfire were suppressed, chaparral shrubs grow larger and denser and annually add more leaves and twigs to the ground litter layer. Because of the buildup of fuel, the chaparral, once ignited, burns even more intensely. Periodic setting of controlled burns would reduce the frequency and severity of wildfires. With such a strategy, people adapt their behavior to the environment and work with and not against the forces of nature.

ESSAY: The Extreme Climate of Mount Washington, NH

It may not be home to the world's worst weather, but the summit of Mount Washington, in Coos County, New Hampshire, comes close to justifying that dubious distinction[1]. Mount Washington is part of the Presidential Range of the White Mountains in the northern Appalachians (44.28 degrees N, 71.30 degrees W). At 1917 m (6288 ft) above mean sea level, Mount Washington is the highest peak in the northeastern United States and reaches above the tree line (at about 1524 m or 5000 ft above msl) into the alpine zone (Figure 1). At the summit, *permafrost* (permanently frozen ground) is found at a depth of about 6 m (19.7 ft).

FIGURE 1
Mount Washington, New Hampshire, viewed here in mid winter, is the highest peak in the northeastern United States (1917 m or 6288 ft above msl) and is home to some of the world's most extreme weather. [Photo courtesy of Marsha Rich]

Personnel of the U.S. Army Signal Service took weather observations on Mount Washington's summit operating year-round from December 1870 through the winter of 1886-1887 and then in summer only through 1892. A private, non-profit organization, known as the Mount Washington Observatory, has occupied the summit since 1932 and maintained a continuous climate record. The summit climate is subarctic with exceptionally strong winds, heavy winter snowfall, low temperatures, and frequent dense fog and clouds (averaging about 310 days per year). In winter, weather instruments are often heavily encrusted in ice (rime) and snow so that weather observers frequently must brave hurricane-force winds and the elements to de-ice the instruments (Figures 2 and 3).

---

[1]Under the leadership of meteorologist Charles F. Brooks (1891-1958), the Mount Washington Observatory became an important center for the investigation of mountain meteorology and climatology. Brooks is also credited with being the first to describe in print the summit of Mount Washington as possibly having the "worst weather in the world." Brooks authored "The Worst Weather in the World," in the December 1940 issue of *Appalachia* magazine, published by the Appalachian Mountain Club. In this article, Brooks concluded that the severity of the climate on Mount Washington is equaled or slightly exceeded on the highest mountains of middle or high latitudes and in Antarctica.

FIGURE 2
In spite of bitter cold, frequent hurricane-force winds, and heavy snow, the summit of Mount Washington, New Hampshire is occupied year-round by weather observers of the Mount Washington Observatory who have maintained a continuous climate record since 1932. [Photo courtesy of Marsha Rich]

FIGURE 3
During Mount Washington's long winter, buildings and outdoor weather instruments become encrusted in ice (rime) and snow so that weather observers must brave subarctic conditions to de-ice the instruments. [Photo courtesy of the Mount Washington Observatory]

Mount Washington has long been a popular destination for hikers, climbers, skiers, and tourists. Access to the summit observatory is by the old stagecoach (carriage) road (opened in 1861), now known as the Mount Washington Auto Road (12.3 km or 7.6 mi long), the Mount Washington Cog Railway (operating since 1869), and various hiking paths including the Appalachian Trail. Extreme weather, however, sometimes creates life-threatening conditions. Since 1849, more than 135 people have perished on and around Mount Washington in all seasons due to hypothermia (affecting hikers not adequately prepared for rapidly changing weather conditions), falls, ski accidents, falling ice, avalanches, and other causes. According to the Appalachian Mountain Club, more people have died climbing Mount Washington (and its neighboring peaks) than any other mountain in the U.S.

For almost 62 years, the Mount Washington Observatory held the world record for the fastest wind gust measured directly at the Earth's surface. On 12 April 1934, the wind gusted to 372 km per hr (231 mph) with a five-minute average wind speed of 303 km per hr (188 mph). However, on 22 January 2010, the *World Meteorological Organization (WMO)* reported that this wind record was officially toppled on 10 April 1996 at Barrow Island, Australia (20.7 degrees S, 115.4 degrees E, at 64 m or 210 ft above msl) during Typhoon Olivia. The new record is 408 km per hr (253 mph) and was discovered by a WMO evaluation panel (within the WMO Commission for Climatology) during a comprehensive review of global weather and climate extremes. The Mount Washington record still stands as the fastest surface wind ever measured at any staffed weather station in the Northern and Western Hemispheres. At the summit of Mount Washington, winds exceed hurricane-force more than 105 days per year—mostly from November through April. The mean annual wind speed at the summit is about 56 km per hr (35 mph).

On average, the coldest month of the year on the mountain top is January; the average high temperature is -10.9 °C (12.3 °F) and the average low temperature is -20.3 °C (-4.6 °F). The warmest month is July with an average high temperature of 12 °C (53.6 °F) and an average low temperature of 6.1 °C (46 °F). On average, freezing temperatures occur at the summit on 243 days of the year. The all-time record low temperature was thought to be -43.9 °C (-47 °F) set on 29 January 1934. However, recently discovered documents indicate a temperature reading of -45.6 °C (-50 °F) on 22 January 1885 and a reading of -50.6 °C (-59 °F) on 5 February 1871. At the other extreme, the all-time record high temperature at the summit is only 22.2 °C (72.0 °F), recorded on 26 June 2003 and 2 August 1975. (Note that there may be some question regarding the accuracy of the Signal Service thermometer.)

Average annual precipitation (rain plus melted snow and ice) at the summit is 252 cm (99.2 in.), ranging from an average of 7.1 cm (2.80 in.) in July to 10.4 cm (4.09 in.) in November. Snow can fall at any time of year on the summit with seasonal totals increasing with elevation from about 287 cm (113 in.) in the surrounding lowlands (at 610 m above msl) to an estimated 800 cm (254 in.) at the summit. The snowiest winter on record was 1968-69 when 1438 cm (566.1 in.) was recorded at the summit. The record high daily (24-hour) snowfall was 125.2 cm (49.3 in.) on 25 February 1969.

A combination of atmospheric and orographic factors explains the extreme climate of Mount Washington's summit. The rate of change of climatic/ecological zones with increasing elevation on Mount Washington is unusually great. As shown in Figure 7.21, the principal storm tracks affecting the coterminous U.S. converge toward New England, a preferred (climatological) location of the west-to-east flowing polar front jet stream. In winter and spring, the jet accelerates near the coastal boundary between cold, often snow-covered land and the milder ocean surface. Storm systems, especially those tracking northward along the Atlantic coast (nor'easters), bring copious moisture to the Northeast and orographic lifting adds to the mountain's snowfall. Mount Washington's north/south orientation forms a barrier to the prevailing westerlies forcing the wind to flow between the summit and the tropopause. Constricted, the wind accelerates over the summit creating exceptionally windy conditions especially above the tree line.

CHAPTER 14

RESPONDING TO CLIMATE CHANGE

In 1953, the North Sea surged onto the southwest coast of the Netherlands. with devastating results. The Dutch responded by fortifying and extending their system of dikes and seawalls. Global warming may require that similar measures be taken in other low-lying coastal areas around the world. [Courtesy of NASA/GSFC/METI/ERSDAC/JAROS, and U.S./Japan ASTER Science Team.]

Case-in-Point

Effective adaptation to climate change sometimes can be achieved via innovative low-tech strategies that are relatively inexpensive and require little maintenance. In the 30 October 2009 issue of *Science*, Gaia Vince reports on one such strategy successfully employed in a very remote part of the world experiencing a scarcity of water.

As discussed in Chapter 12, mountain glaciers are shrinking in various parts of the world in response to global climate change. The impact of vanishing glaciers is especially serious in dry climates where the local or regional agriculture depends on glacial melt water and

snowmelt for irrigation of crops. One such place is Ladakh, India, an ancient Buddhist kingdom situated in the state of Jammu and Kashmir high in the valleys between the Kunlun mountain range to the north and the Himalayas to the south, nestled between Pakistan, Afghanistan, and China (Figure 14.1). The population numbers about 270,000 (2001) and subsistence farming is practiced.

Much of Ladakh is at an elevation above 3000 m (9800 ft) making it the highest inhabited region of the world. Ladakh lies in the rain shadow of the Himalayas and receives an average annual rainfall of only about

FIGURE 14.1
The climate of Ladakh, India, located high in the valleys between the Kunlun mountain range to the north and the Himalayas to the south, is arid. Traditionally, subsistence farmers have depended on glacier melt and snow melt to irrigate crops. With climate change making that source of water less reliable, an innovative idea was developed to increase the supply of irrigation water.

5 cm (2 in.). Farmers have relied on melt water draining from mountain snows and glaciers to irrigate their crops during the precariously short growing season. However, beginning in the 1990s, climate change has caused the glaciers to shrink and some to disappear entirely leading to water scarcity and declining harvests. The glaciers that do remain are located at such high elevations that their melt water is not available until May or June, too late to benefit crops (e.g., wheat, barley) that must be planted in March.

In a moment of inspiration, a local civil engineer named Chewang Norphel devised a unique plan to help alleviate Ladakh's water scarcity problem and supply water when needed during the growing season. Norphel's idea was to build artificial glaciers, actually large masses of ice that would supply melt water for irrigation on a seasonal schedule similar to that of real glaciers. He observed that in winter swift mountain streams remained liquid as they flowed down slope, but the water froze where diverted into still ponds. Norphel directed construction of a shaded cul-de-sac (spoon) shaped enclosure bordered by rock walls and designed to collect some of the winter runoff from mountain streams. The water slows as it spreads over a large depression, loses heat by upward radiation, and freezes. In time the winter runoff forms a huge mass of ice. In early spring, the ice begins to melt, supplying runoff for irrigation from late March to late April. After that, melt water is available from snowmelt and glacier melt at higher elevations.

As of this writing, Norphel has built 10 artificial glaciers whose melt water supports crops that feed about 10,000 people. Each artificial glacier measures about 250 m (820 ft) by 100 m (330 ft) and holds an estimated 23,000 m³ (6 million gallons) of water. As a consequence of this simple low-tech adaptation to climate change, agricultural production has doubled in the region.

Driving Question:
How does humankind mitigate and adapt to climate change?

Human activity plays a major role in global climate change and is likely to continue to do so well into the future. Climate change imposes known and unknown risks on the orderly functioning of the natural world as well as all sectors of society (Chapter 12). Climate change is not necessarily gradual and linear, making it sometimes difficult for society to smoothly adjust to a new climatic regime. Climate change can be abrupt causing considerable environmental and societal upheaval and instability, even leading to civil conflict. But it should be noted that climate change can at times deliver beneficial impacts (e.g., milder winters in the northern United States and consequent reduced need for space heating).

Strategies that effectively and efficiently address the challenges of anthropogenic climate change must recognize the fact that our demand for energy lies at the heart of the challenge. About 80% of the primary energy used worldwide is derived from the burning of fossil fuels (coal, oil, and natural gas). A byproduct of fossil fuel combustion is carbon dioxide (CO_2), an important contributor to Earth's greenhouse effect (Chapters 3 and 12). Total global carbon emissions have trended upward since the middle of the 18th

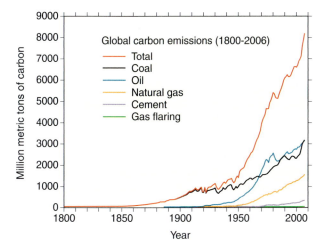

FIGURE 14.2
Global carbon emissions, 1800-2006. [Carbon Dioxide Information Analysis Center, DOE Oak Ridge National Laboratory, TN]

TABLE 14.1
Sources and sinks of carbon in North America in 2003.[a]

Sources	*gigatons per year*
Vegetation respiration	55.5
Ocean outgassing	90.5
Fossil fuels and cement emissions	6.4
Changes in land use	1.2

Sinks	*gigatons per year*
Vegetation net production	57.0
Ocean absorption	92.2
Land sinks	2.3

[a]Source: *The First State of the Carbon Cycle Report (SOCCR)*, U.S. Climate Change Science Program, Synthesis and Assessment Product 2.2, November 2007.

One gigaton equals one billion tons.

century, the beginnings of the industrial era (Figure 14.2). From 2000 to 2006, CO_2 was emitted to the atmosphere at almost three times the rate of the 1990s primarily due to the surging economies of China and India. The consequent enhancement of the greenhouse effect accelerated in recent decades with more than half of the global temperature rise since the beginning of the industrial era occurring after 1970 (Chapter 10). According to the United Nations Environmental Programme's *Climate Change Science Compendium 2009*, "the pace and scale of climate change may now be outstripping even the most sobering predictions of the *2007 Fourth Assessment Report* of the IPCC."

Human activities emit more than 7 gigatons of CO_2 into the atmosphere each year, about half of which remains there for an extended period while the balance is taken up by the ocean, northern forests, and soils. Table 14.1 summarizes the magnitudes of the various North American sources and sinks of carbon as CO_2 in 2003. (One *gigaton* equals one billion tons.) That year, North America's net contribution to the atmosphere was nearly 2 gigatons. North America's fossil fuel emissions in 2003 were 27% of total global emissions, 85% from the U.S., 9% from Canada, and 6% from Mexico.

This chapter considers how we can minimize the risks and potential disruptions threatened by anthropogenic climate change. We cover climate risk management strategies under the general headings of mitigation, adaptation, and geoengineering. This three-pronged division of risk management strategies is a relatively loose classification with some strategies overlapping more than one category. The chapter closes with a discussion of aspects of climate-conscious architecture.

Managing Anthropogenic Climate Change

How much is too much CO_2 in the atmosphere? In order to set realistic goals for reducing emissions of CO_2 and other greenhouse gases, climate scientists must determine the maximum allowable concentration that will avoid dangerous consequences. The consensus among European policymakers is that global average temperatures should rise no more than 2 Celsius degrees (3.6 Fahrenheit degrees) above preindustrial levels by 2100. This would correspond to a CO_2 concentration in the atmosphere of about 450 ppmv. NASA climate scientist James Hansen argues for a more conservative target of 350 ppmv and preferably less. Myles Allen of the University of Oxford states that a maximum warming of 2 Celsius degrees could be achieved even if human activity emits one trillion metric tons of CO_2 into the atmosphere by 2050. About half of that amount has already been emitted so that only about 25% of the remaining coal, oil, and natural gas can be burned to stay below the threshold of 2 Celsius degrees.

According to the 1992 United Nations Framework Convention on Climate Change (UNFCCC), **mitigation** encompasses action designed to stabilize greenhouse gas emissions at a level that avoids a dangerous disturbance of the climate system, by reducing the pace and magnitude of anthropogenic climate change. More specifically, the IPCC defines mitigation as a human action that (1) reduces any process, activity, or mechanism that releases a greenhouse gas, an aerosol or a precursor of a greenhouse gas or aerosol into the atmosphere, or (2) enhances any process, activity or mechanism that removes a greenhouse gas, an aerosol or a precursor of a greenhouse gas or aerosol from the atmosphere.

Adaptation refers to steps taken for the purpose of making communities more climate-resilient so that society can more effectively cope with the inevitable harmful impacts of climate change while taking advantage of the beneficial impacts. The IPCC defines adaptation as an adjustment in natural or human systems in response to actual or expected climate stimuli or their effects, which moderates harm or exploits beneficial opportunities. In essence, mitigation is concerned with reducing heat-trapping gas emissions to prevent dangerous climate change whereas adaptation involves coping with those impacts that cannot be avoided. Mitigation and adaptation are not alternative responses to anthropogenic climate change; both responses are necessary.

Climate models generate predictions that guide (inform) both mitigation and adaptation strategies, but with some important differences. Uncertainties in climate prediction vary with the time scale over which the climate is simulated and the climate prediction lead time. Adaptation strategies tend to be local or regional in scope and typically depend on climate predictions having lead times of less than a decade. With such predictions, anthropogenic influences are small compared to natural decadal-scale climate variability. In addition, errors or gaps in initial conditions are largely responsible for the uncertainty of these climate predictions. Mitigation strategies are more global in scope and depend on climate predictions having lead times of a century or so and are much less sensitive to initial conditions and more sensitive to boundary conditions and processes (e.g., feedbacks) in the climate system. Anthropogenic influences are much more important at this longer time scale. However, uncertainties in future anthropogenic emissions of greenhouse gases (and aerosols) are a major source of uncertainty in these climate predictions.

Geoengineering refers to large-scale, intentional human manipulation of components of Earth's climate system that is intended to offset the consequences of increasing greenhouse gas emissions.

Climate Mitigation

With the atmospheric concentration of the greenhouse gas carbon dioxide continuing to climb, global warming appears inevitable through the remainder of the present century and well beyond. Even in the unlikely event that CO_2 emissions were somehow slashed immediately, the amount of CO_2 already released to the atmosphere by human activities plus the lengthy lifetime of CO_2 in the atmosphere commit the planet to higher average global temperatures, shrinking glaciers, sea level rise, and ocean acidification (Chapter 12). By sometime near the middle of the present century, Earth's mean surface temperature will rise to levels that likely will pose serious consequences for society and the environment.

What can be done to mitigate anthropogenic climate change? Coal burning (e.g., for electric power generation) contributes about 25% of the world energy supply and is responsible for approximately 40% of all carbon emissions to the atmosphere. Combustion of coal produces more CO_2 per unit of energy than any other fossil fuel (about twice that of natural gas). The U.S., Russia, China, and India account for an estimated two-thirds of all coal reserves worldwide. In the U.S., primary energy consumption by source and sector shows that 91% of coal is used for electric power generation (Figure 14.3). Most energy experts agree that the coal-rich nations are unlikely anytime soon to abandon coal as a major energy source.

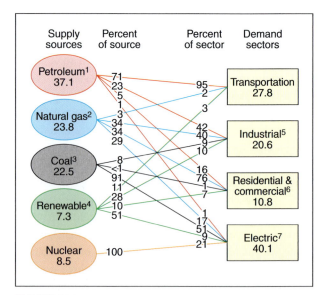

FIGURE 14.3
U.S. primary energy consumption by source and sector, 2008, in quadrillion Btu. [U.S. Department of Energy, Energy Information Administration

The immediate objective in curbing greenhouse warming is to cut emissions of CO_2 that accounts for about 77% of anthropogenic greenhouse gases. Intuitively we would expect the atmospheric concentration of a greenhouse gas to decline as its emissions are reduced. However, as noted in Chapter 12, the situation is not quite that straight forward. How the atmospheric concentration of a gas responds to decreasing emissions depends on the competition between the rates of emission of the gas into the atmosphere and the rates of various physical, chemical, and biological processes that remove the gas from the atmosphere and hence, the lifetime of the gas in the atmosphere. Here, the **lifetime of a gas** in the atmosphere is defined as the time it takes for a perturbation of the gas to be reduced to 37% of its original amount. The atmospheric concentration of some greenhouse gases decreases shortly after a decline in emissions whereas the atmospheric concentration of other greenhouse gases may continue to climb hundreds of years after emissions begin to decline. If emissions increase with time, the atmospheric concentration also increases regardless of the lifetime of the gas.

Consider how the atmospheric concentration of each of the greenhouse gases carbon dioxide (CO_2), nitrous oxide (N_2O), and methane (CH_4) adjust to five different future emissions scenarios: stabilization at present levels, and immediate reductions of 10%, 30%, 50%, and 100% (Figure 14.4). N_2O has a lifetime of 110 years; with an emission reduction of more than 50%, its atmospheric concentration would stabilize near present-day values (Figure 14.4B). CH_4 has a lifetime of only 12 years; reducing emissions by less than 30% would stabilize its atmospheric concentration within a few decades (Figure 14.4C). The situation is much more complex for CO_2.

The lifetime of CO_2 in the atmosphere is not as clearly determined as it is for N_2O and CH_4. More than 50% of CO_2 emitted into the atmosphere cycles out within a century but about 20% remains for millennia. Because of this slow rate of removal, even if emissions are reduced substantially from present levels, the concentration of CO_2 in the atmosphere will continue to increase over the long term (Figure 14.4A). If emissions of CO_2 were to stabilize at current levels, the concentration of CO_2 in the atmosphere would continue to increase through the remainder of this century and beyond. Only the complete elimination of anthropogenic CO_2 emissions would stabilize the atmospheric concentration of CO_2 at current levels.

Recent research conducted by NASA climate modeler Drew Shindell suggests that the focus on reducing CO_2 emissions should be broadened to include limits on emissions of certain short-lived air pollutants. In April 2009, Shindell reported that *black carbon*, a component of soot, absorbs solar radiation (causing warming) and may be responsible for half or more of the warming observed in the Arctic. Tropospheric ozone (O_3), a component of photochemical smog, also absorbs solar radiation. Meanwhile, methane (CH_4) and hydrofluorocarbons (HFCs) are very efficient greenhouse gases. Black carbon persists in the atmosphere for only a few weeks, the lifetime of methane is 15 years or less, whereas CO_2 can persist for millennia. Working to reduce emissions of short-lived air pollutants is likely to yield comparatively speedy results, buying time while governments overhaul global energy systems to reduce CO_2 emissions.

As described in this chapter's first Essay, one strategy to slow the rate of CO_2 emissions into the atmosphere is to increase energy efficiency. Other strategies include carbon trading, increased reliance on low-carbon or carbon-free alternate energy sources, carbon capture and storage, and changes to the transportation sector.

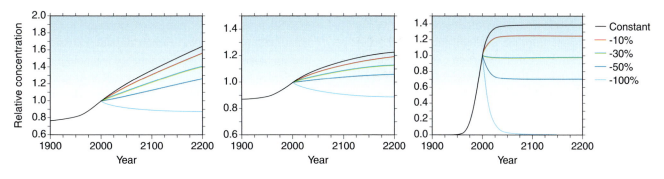

FIGURE 14.4
Changes in the relative atmospheric concentration of (A) carbon dioxide, (B) nitrous oxide, and (C) methane with changes in emission rates.

CARBON TRADING AND TAXATION

Economics and market forces play an important role in regulating the amount of CO_2 and other greenhouse gases emitted to the atmosphere. One approach, known as **cap-and-trade**, is similar to a successful system implemented in 1990 by the U.S. Environmental Protection Agency (USEPA) to reduce sulfur dioxide (SO_2) emissions from point sources, mostly coal-fired electric power plants. The objective was to reduce acid rain. Quotas or caps are imposed on each SO_2 source. Sources whose emissions are below the quota earn tradable credits that can be marketed to sources that have exceeded their quotas.

Cap-and-trade is more difficult to apply to CO_2 than to SO_2. Whereas the sources and sinks of SO_2 are relatively few, the carbon biogeochemical cycle is more complex with numerous and widely distributed sources and sinks of carbon (e.g., ocean, forests, soils, agriculture). Furthermore, the lifetime time of CO_2 in the atmosphere is much longer than that of SO_2. The simplest and most effective use of cap-and-trade for regulating CO_2 emission to the atmosphere is to focus exclusively on sources of CO_2 from fossil fuel combustion.

From a global climate perspective, the total amount of CO_2 emitted into the atmosphere is much more important than the contribution from individual sources. This is the rationale underlying cap-and-trade. Suppose, for example, that utility A emits more CO_2 than permitted while utility B emits less CO_2 than permitted. Utility A purchases carbon credits (unused permits) from utility B and uses them to comply with its environmental obligation.

In another market-based approach, known as **offset exchange**, carbon credits are earned for investment in projects that compensate for (offset) CO_2 emissions. For example, utility C, located in an industrial nation such as Germany, exceeds its CO_2 emission allocation. Utility C chooses to invest in a windmill farm being constructed in a developing nation such as China. Thereby, utility C earns carbon credits based on the difference between CO_2 emission levels from a coal-fired power plant versus the zero emissions from the windmill farm for the same amount of electricity generated. Both cap-and-trade and offset exchange benefit from the typically lower costs (e.g., for low-carbon or no-carbon energy alternatives) in developing nations.

Imposing a fee or tax on CO_2 emissions from fossil fuel sources spreads the burden of mitigation equitably among all sectors of society. Such a tax on gasoline, for example, would encourage drivers to choose more energy efficient motor vehicles (i.e., more kilometers per liter or miles per gallon). A similar tax on emissions from coal-fired power plants, passed on to consumers via their monthly electric bill, would encourage the further development and use of alternative energy sources and greater energy efficiency in the home and industry.

Politicians and industrial interests tend to favor carbon trading over taxation. Taxes represent a much more visible cost. We have more on cap-and-trade and the public policy aspects of carbon markets in Chapter 15.

ALTERNATIVE ENERGY SOURCES

Earlier in this book, we examined solar-thermal power (Chapter 3) and wind power (Chapter 6), two alternative sources of energy that one day are likely to play major roles in replacing fossil fuels. Other non-carbon renewable energy sources are hydropower, geothermal power, and tidal power. In addition, nuclear energy sources may play a greater role in the future. Scaling up these energy sources, solving associated environmental problems (e.g., nuclear waste disposal), and making them more cost competitive with fossil fuels, will require time (perhaps 50 years or so). Today, renewable energy supplies less than 7% of all power consumed in the U.S. and if hydroelectric power were excluded, this number drops to less than 4.5%. Worldwide, renewable energy sources provide only about 3.5% of electricity. According to the United Nations, the global investment in renewable energy in 2007 was $148.4 billion, up about 60% from the previous year. Nonetheless, coal-fired power plants continue to proliferate. To buy time until non-carbon energy sources emerge as feasible alternatives to fossil fuels, considerable interest is directed at the potential of carbon capture and storage (CCS) to reduce the amount of CO_2 that enters the atmosphere as a byproduct of fossil fuel combustion.

CARBON CAPTURE AND STORAGE

Carbon capture and storage (or sequestration) (CCS) is a climate mitigation strategy designed to minimize anthropogenic carbon dioxide emissions into the atmosphere using well-established technologies. In the 25 September 2009 issue of *Science*, R.S. Haszeldine of the University of Edinburgh, Scotland, reported that CCS potentially could reduce worldwide emissions of CO_2 from energy generation by 20%. For perspective, U.S. coal-fired power plants have a total capacity in excess of 300,000 megawatts (MW), accounting for about half of the electric power generated nationally and more than 30% of CO_2 emissions.

FIGURE 14.5
A traditional coal-fired electric power plant.

The most effective application of CCS promises to be large single-source facilities such as coal-fired electric power plants where carbon is captured at either the pre- or post-combustion stage. Plants that burn pulverized coal capture CO_2 before it reaches the smokestack whereas plants using *coal gasification*, separate out CO_2 following gasification but prior to combustion.

In a traditional coal-fired power plant, coal is burned in air (Figure 14.5). Heat converts water to high pressure steam that drives a turbine that, in turn, generates electricity. For air quality purposes, sulfur and most particulate matter are removed from the exhaust gases prior to release to the atmosphere via a tall smoke stack. CO_2 typically comprises about 15% of the exhaust gases; nitrogen and water vapor account for the rest. To capture CO_2, an absorption tower takes the place of the smokestack and droplets of chemicals known as amines selectively absorb CO_2. Heating the amines releases concentrated CO_2 that then enters the sequestration phase.

In a coal gasification combined-cycle power plant, coal is burned partially in the presence of oxygen producing a synthetic gas (*syngas*) composed of hydrogen and carbon monoxide (CO). Following removal of sulfur, the syngas burns in air in a gas turbine that generates electricity. In addition, heat from exhaust gases exiting that turbine turns water into steam that drives a steam turbine for generating additional electricity. To capture CO_2, steam is added to syngas converting most CO to CO_2 and hydrogen. The CO_2 is then filtered out prior to burning the syngas (mostly hydrogen) for electric power generation.

Following capture, the pure stream of CO_2 is compressed to its liquid phase and then piped to an underground storage site located either onshore or offshore beneath the ocean floor. CO_2 is injected into interstices (tiny openings)

in geologic formations at depths of at least 800-1000 m (2600-3300 ft) for long-term storage. A thick layer of impermeable rock (cap rock) must overlie the reservoir rock to prevent vertical escape of the sequestered CO_2.

The injection technology used with CCS is the same one employed successfully in oil fields for *enhanced oil recovery* for more than three decades. Liquid CO_2 is injected into pore spaces in the reservoir rock of depleted oil or natural gas fields or a subterranean saline aquifer (where pore space is saturated with salty water). Other potential deep storage sites include unmineable coal beds, organic-rich shale formations, and submarine layers of basalt.

In early 2010, D.S. Goldberg and colleagues at Columbia University's Lamont-Doherty Earth Observatory reported on the potential for certain submarine basalt layers for sequestration of CO_2. Basalt is a fine-grained igneous rock formed from the cooling and crystallization of lava (Chapter 1). CO_2 would be injected into fractures in the basalt where chemical reactions would produce carbonate minerals in a relatively short period of time. Goldberg identified suitable CO_2 basalt storage sites off the U.S. West Coast (from California to Washington) and East Coast (Georgia, South Carolina, New York, New Jersey, and Massachusetts). The potential storage capacity is enormous according to Goldberg; a basalt formation off the New Jersey coast could accommodate an estimated one billion metric tons of CO_2.

Worldwide, saline aquifers are capable of sequestering hundreds of years supply of the CO_2 emitted by power plants. The goal is to identify storage sites that are large enough to store tens of millions of metric tons of CO_2 for thousands of years. For perspective, one new large coal-fired power plant (e.g., rated at 1000 megawatts) emits about 6 million metric tons of CO_2 yearly to the atmosphere. The U.S. Department of Energy estimates that the total subsurface storage capacity in appropriate geologic formations nationwide would accommodate 3.5 trillion metric tons of CO_2, more than adequate for the 2.9 billion metric tons of CO_2 emitted every year. (Worldwide, over 30 billion metric tons of CO_2 are emitted each year.) In the U.S., perhaps 95% of all large point sources of CO_2 are located within 80 km (50 mi) of a suitable geologic storage site.

In spite of the fact that technologies for capture, transport, and storage of CO_2 are proven and readily available and that abundant subsurface storage space exists, as of this writing, only one commercial demonstration coal-fired power plant with CCS capability is operating in the U.S. More than 20 experimental and pilot plants are currently operating worldwide. According to R.S. Haszeldine, these plants capture and store some 3 megatons of CO_2 each year mostly from power plants and natural gas cleanup. Future prospects are encouraging with electric utilities planning to install many CCS facilities in the next few decades with the first full-scale CCS facility expected to be operational in about 2015.

American Electric Power's Mountaineer coal-fired power plant along the Ohio River near New Haven, WV, is equipped with a demonstration CCS unit that captures about 2% of the plant's CO_2 emissions. The Mountaineer facility is one of the largest in the nation, capable of generating 1300 MW of electricity, and also a leading source of CO_2. The demonstration began in September 2009 and is expected to operate for five years with plans to eventually scale up to capture 20% of total CO_2 emissions. Ammonium carbonate chemistry separates the CO_2 from the other exhaust gases. The captured CO_2 is compressed and liquefied, and then injected into the tiny pore spaces within layers of sandstone and dolostone formations at depths near 2400 m (8000 ft). Layers of impermeable rock overlie the reservoir rock. The plan is to sequester at least 0.45 million metric tons of CO_2 during the demonstration phase. Three wells (in addition to two injection wells) will monitor conditions in the subsurface storage reservoir. Safety concerns focus on the potential for either gradual or abrupt leakage. In 2007, the U.S. Department of Energy estimated that the added cost of CCS would nearly double the price of electricity from $63 per megawatt-hour without CCS to $114 per megawatt-hour with CCS.

Since 1996, Norway's state-controlled oil company (*StatoilHydro*) has operated the world's longest running, commercial-scale CCS facility associated with the Sleipner Vest natural gas field located in the North Sea about halfway between Norway and Scotland. About 9% of the extracted gas is CO_2 that is separated from methane, the chief component of natural gas, and injected into a porous submarine sandstone layer about 1000 m (3300 ft) below the sea surface. The CO_2 reservoir rock is topped by an impermeable rock layer about 700 m (2300 ft) thick and the rate of injection is about 1 million metric tons of CO_2 per year. The methane is piped ashore and marketed.

IMPROVING TRANSPORTATION SECTOR EFFICIENCY

Worldwide, the transportation sector is the source of about 23% of energy-related emissions of carbon dioxide (estimated at 31.5 billion metric tons in 2008). Oil-based fuels (e.g., gasoline, diesel, jet fuel) are favored because they are liquid at ordinary temperatures and pressures, easily stored and handled, have a relatively high energy content per unit volume, and historically have been inexpensive and in abundant supply. But when burned, they release copious amounts of the greenhouse gas CO_2 to the atmosphere.

As the population continues to climb and national economies grow, worldwide reliance on oil-based fuels and attendant CO_2 emissions are likely to increase. As pointed out by Andreas Schäfer and colleagues at the Massachusetts Institute of Technology in the November-December 2009 issue of *American Scientist*, the per capita passenger kilometers traveled (PKT) increases with the growth in per capita gross domestic product (GDP). By the middle of this century, the world population is expected to increase by 44% (U.N. projection), and the per capita world gross product is projected to multiply by a factor of 2.2 to 2.6. According to Schäfer et al., world travel could triple or quadruple in the same period. The increase in travel demand compared to 2005 could increase by a factor of 3 to 6 in developing nations and double or triple in the industrial nations by 2050.

With greater affluence, people demand faster means of transport, improved safety, and greater personal comfort. Although passenger vehicles are being built to be more fuel efficient (more kilometers per liter or miles per gallon), energy use per passenger kilometer has increased because the number of people occupying the same vehicle has declined. With an increase in energy use per-PKT of road and air vehicles, CO_2 emissions will increase. Schäfer et al., estimate that if per-PKT energy use by vehicles manufactured in 2005 were to stay constant and oil-based fuels continued as the primary source of energy for transportation, by 2050, energy use per PKT would increase 12 to 25% globally. Consequently, transportation's greenhouse gas contribution would triple to quintuple.

Although there is no known practical way to apply CCS technology to individual motor vehicles, a combination of other technological innovations plus a shift in consumer preference away from large, powerful vehicles could head off the anticipated increase in greenhouse emissions from the transportation sector. For example, hybrid-electric autos could cut fuel consumption in half, the new generation of biofuels, derived from cellulose rather than feedstock, emit less CO_2, and new types of batteries could extend the range and efficiency of electric vehicles. Under optimum conditions, the global average auto energy use per PKT could decline by 35% (compared to 2005) by the middle of the century. In the same period, technological advances in aviation could realize a 40% reduction in energy use per PKT for aircraft.

Climate Adaptation

From the time our ancient ancestors first mastered fire and through the ages, human ingenuity has developed technological innovations that have enabled us to travel and live in regions where the climate is hostile. In some locales, in fact, the climate is so inhospitable that human survival is marginal at best. Irrigation opened arid and semi-arid lands to agriculture; scientists developed drought- and frost-resistant varieties of crops; dams and reservoirs on rivers regulate water supply through the course of a year and provide flood control and hydroelectric power; central heating and air conditioning make it possible for people to live and thrive in places where the climate otherwise would be too hot or too cold; structures such as breakwaters and seawalls and improvements in infrastructure may protect against the erosive impact of storm surges along low-lying coastal areas.

Human intervention by way of technological innovations has enabled communities to adapt to some extent to local climate and climate variability—encompassing seasonal changes and weather extremes. While mitigation strategies (discussed in the previous section) aim to reduce emissions of greenhouse gases over the long-term, society must adapt to inevitable and sometimes irreversible changes in climate happening on the decadal scale. With rising sea level, a greater frequency of intense hurricanes, heat waves, drought, excessive rainfall and other extremes, cities and towns must be made more climate-resilient to minimize the risk to the population. Such action must include making the public better informed on climate change issues and implementing more effective and appropriate response strategies.

Writing in *Scientific American* in 2007, Jeffrey D. Sachs, director of the Earth Institute at Columbia University, cautions that climate change will not be the same everywhere. Some human populations are much more vulnerable than others to the adverse effects of climate change, especially as the change may reduce the potable water supply. For hundreds of millions of people, migration may be the only alternative. Sachs identifies these potential "climate change refugees" as people currently living in one of four zones. As noted in the Case-in-Point of Chapter 10, adaptation to rising sea level is not possible for people living in low-lying island nations. Many residents of densely populated low-lying coastal plains also will be forced to abandon their homes. Agricultural areas dependent on irrigation water supplied by rivers fed by snowmelt and glacier melt may be stressed by disappearing glaciers or spring snowmelt that becomes increasingly out of sync with the growing season. (Refer to this chapter's Case-in-Point.) Indications that the sub-humid and arid regions of Africa will dry further can only exacerbate an already desperate human situation (Chapter 7). In Southeast Asia, shifts in monsoon circulation patterns may make the water supply less dependable for hundreds of millions of people.

Adaptation to climate change will require modification of the infrastructure especially in urban areas. Consider as an example, an 80-km (50-mi) wide strip along the Gulf Coast from Houston, TX to Mobile, AL. This region, home to about 10 million people, was the subject of a 3-year study by the U.S. Department of Transportation on potential impacts of future climate change on the transportation system, that is, roads, airports, seaports, and railroads. Released in March 2008, the report was a collaborative effort of outside transportation experts and climate scientists.

Drawing on predictions generated by runs of 21 global climate models, investigators identified likely changes in climate, how those climate changes would impact the transportation infrastructure, and what might be done to alleviate disruptions. The mean temperature of the region is predicted to rise some 1 to 2 Celsius degrees (1.8 to 3.6 Fahrenheit degrees) by 2050. The warming trend is expected to be accompanied by a greater frequency of heat waves that will buckle highways and railroad tracks. More frequent heat waves call for construction and improved maintenance of more durable road beds. Rising sea level, ground subsidence, and perhaps more intense tropical cyclones (hurricanes and tropical storms) will result in more frequent and extensive flooding plus erosion of barrier islands. This threat calls for installation of more effective drainage systems, flood control barriers, altered traffic patterns, and swifter evacuation routes.

In designing standards for the transportation infrastructure, engineers traditionally consult historical records of temperature, precipitation, and other climate elements. Now and in the future, however, extreme weather events may occur with greater frequency than in the past, requiring new strategies to adapt to future climate change. Furthermore, urban and regional planners are beginning to question whether encouraging coastal development (Figure 14.6) is prudent in view of anticipated climate change.

Climate change adaptation must also be a high priority in densely-populated low-lying coastal developing nations such as Bangladesh. With an area slightly smaller than Wisconsin, Bangladesh's population is estimated at about 162 million (2009) and expected to climb to 205 million by 2050. The region's monsoon climate and topography, as well as its relatively frequent exposure to

FIGURE 14.6
This summer home, located very near the ocean along the U.S. East Coast, is being elevated so that the unoccupied first floor will allow flow-through of storm-surge floodwaters. The second and third floor living areas of the home are supported by wooden beams driven deep into the sand. Many people question whether this type of construction should be allowed in the coastal zone in view of rising sea level. Is this an example of climate change adaptation?

intense tropical cyclones, make Bangladesh vulnerable to deluges of rain and extensive flooding. The nation borders India and the Bay of Bengal and about 20% of Bangladesh is less than 1 m (3 ft) above mean sea level (Figure 14.7).

FIGURE 14.7
Adaptation to climate change is a high priority in Bangladesh, a low-lying developing nation that borders on India and the Bay of Bengal.

Climate change may cause mean sea level to rise up to 2 m (6 ft) by 2100 making vast areas of Bangladesh unsuitable for agriculture, increasing the salinity of rivers, and displacing millions of people. According to Sheikh Ghulam Hussian, a scientist with the Bangladesh Agricultural Research Council, a rise in the nation's mean annual temperature of 3 to 4 Celsius degrees (5.4 to 7.2 Fahrenheit degrees), possible by 2100, would cut rice production by about 25%. Also, climate scientist Noah Diffenbaugh of Purdue University predicts an average two-week delay in the onset of monsoon rains. His research also indicates that during the wet monsoon, rainy episodes will be less frequent whereas intense bursts of rainfall will be more frequent so that severe flooding will be more likely.

As reported by Mason Inman in the 30 October 2009 issue of *Science*, the government of Bangladesh is moving forward with many climate adaptation strategies to improve community resilience to climate change. To help control flooding and protect farmland from rising sea level, coastal polders will be extended, strengthened, and elevated. A **polder** is a pocket of land enclosed by earthen embankments that provide protection against unusually high tides and some storm surges. Other strategies aimed at reducing the impact of climate change on agriculture include improved crop varieties, that is, a shift to new crops such as maize, varieties of rice that can be submerged up to two weeks, and a variety of rice that can tolerate high salt levels.

A major obstacle to achieving adaptation to climate change is cost. According to estimates by the World Bank, perhaps as much as $100 billion a year will be required to help vulnerable areas worldwide to prepare for climate change.

Geoengineering the Climate System

Geoengineering involves large-scale, intentional manipulation of Earth's climate system for the purpose of reducing substantially the anthropogenic contribution to global climate change. Geoengineering schemes tend to be controversial and attract considerable media attention. They generally fall into one of two categories based on objective: (1) reduce the amount of solar energy reaching Earth's surface or (2) increase the flux of heat escaping from Earth to space by reducing the amount of CO_2 in the atmosphere. As pointed out previously, rising levels of atmospheric CO_2 and other greenhouse gases reduce the flux of outgoing infrared radiation thereby altering Earth's heat balance and causing surface warming. Some geoengineering schemes are designed to avoid the energy imbalance responsible for warming by reducing the flux of incoming solar radiation.

Carbon capture and sequestration (CCS) could be applied on a large-scale to capture CO_2 before it enters the atmosphere or CO_2 could be extracted from the ambient air for subsequent storage in geologic formations (described earlier in this chapter). The partial pressure of CO_2 in the atmosphere, however, is only about 1/300 of the concentration of CO_2 emitted by a coal-fired electric power plant. Hence, the CCS system for ambient air requires more expensive equipment for conveying high volumes of air through an absorbing medium. Other proposed geoengineering schemes include ecosystem sequestration, injecting sulfur dioxide (SO_2) into the stratosphere, raising the albedo of ocean clouds, and ocean fertilization.

ECOSYSTEM SEQUESTRATION

As part of the carbon cycle, ecosystems naturally sequester carbon. According to Anthony W. King, an ecologist with the Oak Ridge National Laboratory in Tennessee, on average, North American ecosystems temporarily sequester a total of about 505 million metric tons of carbon each year. Major carbon reservoirs include organic soils, wetlands, and sediments (84 million metric tons), woody plants or trees in shrub lands (120 million metric tons), and forests and wood products (301 million metric tons).

Replanting tropical forests has been proposed as a low-cost way to remove carbon dioxide from the atmosphere. Plantation costs are low in many developing countries in the tropics and rain forest vegetation grows rapidly, thus storing carbon for decades to centuries. Unfortunately, such projects do not provide enough carbon storage over the long term to be effective. Growth slows as trees mature, so the carbon uptake also diminishes. Furthermore, forests are either harvested or burned due to natural causes (e.g., lightning) every 30 to 50 years. As populations grow and more agricultural land is needed, forests will probably shrink, rather than expand.

Nonetheless, Ning Zeng, an atmospheric scientist at the University of Maryland, College Park, studied the feasibility of taking advantage of the carbon sequestering ability of forests. Worldwide, an estimated 65 billion metric tons of carbon has accumulated on forest floors. This forest litter (coarse wood) consists of mostly downed twigs, branches, and entire fallen trees. In time, decomposition (aerobic decay) and wildfires return the carbon in the forest litter to the atmosphere in the form of CO_2.

Zeng notes that if the carbon-rich litter material were buried in an oxygen-poor environment, the carbon could be sequestered for hundreds of years. According to Zeng's calculations, the litter of one km² of forest floor may contain 500 metric tons of carbon, enough to fill a trench 10 m wide, 10 m deep, and 25 m long. To sequester 5 billion metric tons of carbon each year would require about 10 million trenches having these dimensions. For perspective, burning of fossil fuels annually releases about 6.9 billion metric tons of carbon to the atmosphere. Perhaps the greatest downside to this plan is the potential for considerable environmental disruption.

SULFUROUS HAZE

As discussed in Chapter 11, an explosive volcanic eruption that is rich in sulfur dioxide (SO_2) can impact climate on a hemispheric or global scale (e.g., 1991 eruption of Mount Pinatubo). SO_2 of volcanic origin reaches the stratosphere where it combines with water vapor forming tiny droplets of sulfuric acid (H_2SO_4) and sulfate particles, collectively known as **sulfurous aerosols**. These droplets and particles form a haze that reflects a considerable amount of incoming solar radiation to space while absorbing both incoming solar radiation and outgoing infrared radiation. The net effect of these interactions is to lower Earth's mean surface temperature by as much as 1.0 Celsius degree (1.8 Fahrenheit degrees) for a year or two following the eruption before sulfurous aerosols cycle out of the atmosphere. Some scientists propose mimicking this process to counteract global warming caused by an enhanced greenhouse effect. According to Alan Robock and colleagues at Rutgers University, climate models predict that offsetting global warming over the next few decades would require injection of an amount of SO_2 equivalent to that emitted by one Pinatubo-type eruption every four years.

In 1974, the Russian climatologist Mikhail I. Budyko (1920-2001) was among the first to propose injecting SO_2 into the stratosphere to bring about large-scale cooling. The idea gained more widespread attention following publication of an essay on the subject in the August 2006 issue of *Climate Change* by Paul Crutzen of the Max Planck Institute for Chemistry and a winner of the Nobel Prize for his work on stratospheric ozone chemistry. Crutzen was responding to what he described as lack of "international political response" to the risk of global warming. With a scientist of Crutzen's stature expressing interest in the sulfurous haze idea, interest in this and other geoengineering schemes has increased in recent years but many scientists have pointed out the potential downsides of the scheme.

Maintaining a sulfurous sunshade would require an injection of millions of tons of SO_2 every year. Delivery could be by large balloons, aircraft powered by high-sulfur fuel, or missiles charged with SO_2 fired from ships at sea. Shading the planet with a **sulfurous haze** could have many adverse and perhaps unexpected consequences. Changes could occur in atmospheric circulation patterns, evaporation rates, and rainfall amounts and distribution. Furthermore, rains would become more acidic, recovery of the Antarctic ozone hole could be delayed by an estimated 30 to 70 years (Chapter 3), and the natural cycling of sulfurous aerosols out of the atmosphere would require continual renewal of the sulfurous haze layer.

Another drawback of the sulfurous aerosol scheme hinges on the difference in radiational properties of CO_2 and SO_2. CO_2 warms Earth's surface continually, night and day, throughout the year plus warming undergoes poleward amplification. SO_2, on the other hand, blocks sunlight only where and when the Sun is shining. No effect is noted during the polar winter and the tropics are cooled more than the poles.

BRIGHTER OCEAN CLOUDS

Another geoengineering scheme intended to make the planet more reflective, thereby heading off global warming, proposes to brighten the low clouds that are widespread over the ocean. Stratocumulus cloud layers, composed of tiny water droplets, cover on average almost one-quarter of the ocean surface. For the same amount of water in equal volumes of air, a large number of small cloud droplets has a greater total surface area than a small number of large cloud droplets. Hence, a cloud composed of small droplets reflects more sunlight than does a cloud composed of large droplets.

Suspended in the lower troposphere over the ocean is an abundance of tiny salt crystals originating from seawater sprayed into the atmosphere from the ocean surface. Below an altitude of about 300 m (1000 ft) above sea level, seawater spray evaporates leaving behind salt crystals. With ascent and expansional cooling of air, water vapor condenses on these sea salt crystals, producing a large number of very small cloud droplets. In fact, the addition of sea salt crystals quadruples the number of cloud droplets making the clouds brighter so that they reflect more sunlight to space. John Latham, an English cloud physicist, proposes to enhance this process by spraying microscopic seawater droplets upward into the atmosphere from special unmanned satellite-guided ships at sea.

On the down side, this cloud brightening scheme could add to the longevity of clouds and reduce rainfall. Also, changes may occur in regional temperature patterns along with greater variability of temperature with time.

OCEAN IRON FERTILIZATION

Ocean iron fertilization (OIF) has been proposed as a means of boosting the ocean's uptake of carbon dioxide from the atmosphere. OIF refers to the intentional introduction of iron into surface ocean waters to stimulate a phytoplankton (algal) bloom that will remove CO_2 via photosynthesis (Figure 14.8). Marine organisms feed on the bloom and produce carbon-containing waste particles that settle to the sea bottom where the carbon is sequestered in deep water for centuries. Less CO_2 in surface waters increases the flux of CO_2 from the atmosphere to the ocean.

Broad areas of the open ocean receive abundant sunshine, but are nearly devoid of one-celled plants (algae) that form the base of marine food webs. In some areas the reason is low concentrations of dissolved phosphorus and

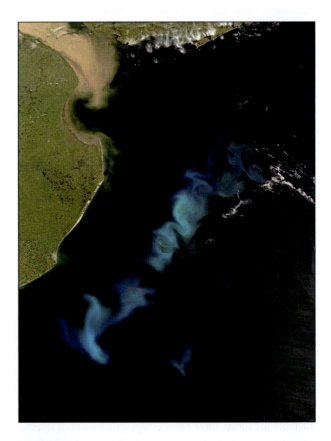

FIGURE 14.8
A phytoplankton bloom in the South Atlantic Ocean, off the coast of Argentina. Image obtained by the Moderate Resolution Imaging Spectroradiometer (MODIS) on NASA's Aqua satellite. [Courtesy of NASA]

nitrogen compounds, nutrients essential for most plant life. Other areas, however, have high concentrations of these nutrients but low concentrations of algae, as evidenced by lack of chlorophyll (plant pigments). Such waters occur in the Antarctic, subarctic, and equatorial Pacific Ocean. Apparently, something is keeping the algae populations from growing. This has implications for climate in that fewer algae mean less photosynthetic removal of CO_2 dissolved in surface ocean waters.

The low growth mystery was solved by John H. Martin (1935-1993), an oceanographer based at Moss Landing (CA) Marine Laboratories. Martin's story is interesting not only in terms of his climate change hypothesis but also in exemplifying the interdisciplinary nature of climate change research. Initially, Martin studied how the classic plant nutrients (nitrogen, phosphorus) limit algal growth. Later his research focused on the role of metals that occur in trace concentrations in seawater and algae. Martin's study of trace metals was made difficult by potential contamination, which usually occurred while sampling seawater onboard ship and during laboratory chemical analyses. To avoid contamination, Martin used special clean rooms with filtered air and ultra-clean plastics. He even made his own ultra-pure chemical reagents so that he could accurately measure concentrations of trace metals in uncontaminated seawater. This work showed that most trace-metal concentrations reported in the scientific literature were too high because of contamination.

Martin discovered that algal growth normally is inhibited by the presence of copper and zinc in sea water, but trace amounts of these metals are essential for their growth. He then proposed that the high-nutrient low-chlorophyll (HNLC) areas of the ocean were deficient in iron and other trace elements. Over about one-third of the ocean, nutrients are abundant but iron concentrations and productivity are low. To test his hypothesis, Martin proposed a field experiment in which a large quantity of soluble iron would be added to high nutrient waters in an attempt to stimulate algal blooms and thereby spur photosynthetic uptake of CO_2.

Building on his iron limitation hypothesis, Martin also turned his attention to the possible role of marine algae in the large-scale climate fluctuations of the Pleistocene Ice Age (Chapter 9). Martin proposed that during cold, dry glacial climatic episodes, strong planetary-scale winds transported iron and other trace metals in dust blown off mountains and deserts. Plumes of dust moved over the ocean and dust particles, containing the missing trace ingredient, settled into the ocean

stimulating blooms of phytoplankton. More phytoplankton meant more photosynthesis, less atmospheric CO_2, a weaker greenhouse effect, and a cooler planet. During mild, wet interglacial climatic episodes, less dust reached the ocean and phytoplankton populations declined. Less phytoplankton meant reduced photosynthesis, more CO_2 in the atmosphere, an enhanced greenhouse effect, and a warmer planet.

Lecturing at the Woods Hole (MA) Oceanographic Institution in July 1988, Martin said, "Give me a half tanker of iron, and I will give you an ice age." Martin's hypothesis came at a time of growing scientific interest in the prospects for global warming due to a carbon dioxide-enhanced greenhouse effect. Martin argued that the build-up of atmospheric CO_2 (due to fossil fuel combustion and deforestation) could be offset by increasing the ocean's uptake of CO_2. He proposed an experiment in which large amounts of dissolved iron would be released into open-ocean waters where nutrient concentrations are high and chlorophyll levels are low.

Ocean iron fertilization (OIF) attempts to boost the ocean's natural biological pump. For this to work as a climate change mitigation strategy, two additional consequences are required. The enhanced biological pump must lead to a lower dissolved CO_2 concentration in surface ocean waters thereby increasing the transfer of carbon dioxide from the atmosphere. Also the carbon transported to the deep ocean by the biological pump must stay there (sequestered) and not be quickly upwelled. Unfortunately, HNLC regions are typically associated with regions of upwelling (e.g., the tropical Pacific Ocean and Southern Ocean) where models suggest that most carbon pumped down would be upwelled and vented to the atmosphere.

Martin died before he could carry out his ocean iron fertilization experiment. In 1993, Kenneth Coale, a researcher at Moss Landing Marine Laboratories, and his colleagues did carry out Martin's experiment in a patch of HNLC water near the Galápagos Islands in the equatorial eastern Pacific. From their ship, the R.V. *Melville*, they released a half a ton of iron over an ocean area of 64 km² (25 mi²). Within a day, the clear water turned soupy green as phytoplankton levels increased three-fold, confirming Martin's hypothesis that at least some species of phytoplankton benefit from the addition of iron. However, the hoped for sequestering of carbon dioxide in the deep ocean did not materialize because zooplankton consumed much of the phytoplankton releasing CO_2 to surface waters. In 1995, Coale and colleagues conducted another OIF experiment in the eastern Pacific Ocean that increased phytoplankton biomass and reduced the carbon dioxide concentration of surface waters. The different result was attributed at least in part to changes in experimental procedures. Subsequently, three open-ocean iron fertilization experiments were conducted in the Southern Ocean. All three experiments increased phytoplankton biomass and reduced the amount of carbon dioxide dissolved in surface waters. However, evidence that particulate organic carbon (POC) sank into the deep ocean was limited.

As of this writing, the international oceanographic community has conducted at least a dozen OIF experiments since 1993. Although adding to our understanding of the role of iron in marine ecosystems, these experiments did not reveal conclusively whether OIF is effective in sequestering carbon in the deep ocean. The next generation of OIF experiments will try to answer some key questions including the fate of the algae blooms: how much is consumed by microorganisms and zooplankton and how much sinks to the ocean floor? The goal is to sequester carbon below the 100-year horizon, equivalent to a water depth of about 500 m (1600 ft). Water below this level does not come into contact with the atmosphere for at least a century. Numerical models predict that HNLC regions of the ocean are not likely to sequester more than several hundred million tons of carbon each year. Furthermore, the scale of these field experiments makes accurate assessment very challenging.

Some scientists and public policymakers vigorously oppose geoengineering schemes such as ocean iron fertilization aimed at combating human contributions to global climate change. They argue that such quick fixes deflect public attention away from the root cause of this global environmental issue, that is, humankind's dependence on fossil fuels. Besides, there are ethical considerations and many uncertainties in tinkering with the Earth system on such a grand scale. Not enough is known about the workings of the Earth system. Open-ocean and deep-ocean ecosystems are too poorly understood to accurately predict possible unintended consequences of OIF. How might the explosive growth in the populations of certain phytoplankton species impact marine ecology? Might OIF spur the growth of harmful algal blooms (HABs)? Others counter by pointing out that large-scale geoengineering projects may be the only feasible way to stem global warming.

Nonetheless, despite scientific uncertainties surrounding OIF and over the objections of scientists and conservationists, some private companies are going forward with plans for iron fertilization of the ocean. Their goal is to market credits for carbon emission reductions (carbon offsets) based on the amount of CO_2 estimated to be sequestered through phytoplankton blooms stimulated by OIF. In this way, a carbon dioxide emitting industry could reduce its carbon footprint by purchasing carbon credits.

POTENTIAL OF GEOENGINEERING

Geoengineering schemes are likely to have broad appeal because they promise a quick fix to global warming and would be an alternative to a major overhaul of the global energy system that would be required to cut greenhouse gas emissions. However, geoengineering fails to address the root cause of anthropogenic climate change, that is, our strong and growing dependence on carbon fuels for energy. In fact, geoengineering reduces the incentive to cut greenhouse gas emissions and does nothing to alleviate ocean acidification (Chapter 12). Furthermore, some critics argue that should a geoengineered sunshade fail, the equivalent of many decades of greenhouse warming could occur in only a few years.

Many uncertainties surround geoengineering schemes aimed at curbing anthropogenic climate change. In the 21 August 2009 issue of *Science*, G.C. Hegerl of Grant Institute, Edinburgh, Scotland, and S. Solomon of NOAA argue for a thorough feasibility study of geoengineering options prior to implementation. They point out that although a geoengineering scheme may provide a means of rapid response in the event of a potentially catastrophic climate change, limitations in our understanding of the climate system may entail considerable risk. For example, a scheme to reduce the flux of incoming solar radiation is likely to affect both temperature and precipitation. While the temperature response to changes in the flux of both incoming solar radiation and outgoing infrared radiation is well known, the precipitation response is uncertain. Climate model simulations and observations of precipitation often do not agree. Also, models tend to underestimate variations in precipitation extremes. Hegerl and Solomon raise the possibility that the proposed geoengineering scheme could lead to a greater frequency of drought and more conflict over water resources worldwide.

Growing interest in geoengineering schemes to offset global warming prompted the American Meteorological Society (AMS) in September 2009 to issue the policy statement, *Geoengineering the Climate System*. The statement cautions that any manipulation of Earth's climate system risks negative and unpredictable consequences. While geoengineering may reduce adverse impacts of climate change, it may also create new risks and/or redistribute risk among nations. The AMS recommends the following:

(1) Enhanced research on the scientific and technological potential for geoengineering the climate system, including research on intended and unintended environmental responses.

(2) Coordinated study of historical, ethical, legal, and social implications of geoengineering that integrates international, interdisciplinary, and intergenerational issues and perspectives and includes lessons from past efforts to modify weather and climate.

(3) Development and analysis of policy options to promote transparency and international cooperation in exploring geoengineering options along with restrictions on reckless efforts to manipulate the climate system.

Climate-Conscious Architecture

For millennia, humans have constructed houses and other buildings not only to protect themselves from the weather and other environmental conditions, but also to create a comfortable indoor environment that is energy efficient (at least in terms of time, effort, and cost) especially where the climate is extreme. For centuries, natural or primitive housing reflected an adaptation to the climate of the locale and the types of building materials available locally. Unfortunately, inexpensive energy during the last century meant that modern architecture in Western cultures strayed away from the energy conservation principles followed by our ancestors. However, recent concerns about increasing dependency on fossil fuels for space heating and cooling of buildings, along with the rising cost of energy, have turned attention toward improving the energy efficiency of a suitable "indoor climate." These efforts often address architecture.

INDOOR COMFORT

In order to maintain a tolerable level of comfort within a building, consideration must be given to thermal effects, ventilation, illumination, and humidity. The indoor thermal state ultimately depends on the building's energy budget involving incoming and outgoing radiation, latent and sensible heating, and interior heat sources or sinks. Whereas heat generated inside the building from appliances, lighting, and inhabitants is important, the indoor thermal level is mainly associated with the external energy load on the building. Space heating or artificial air conditioning is designed to maintain a desirable ambient indoor air temperature. In addition, the relative temperature of the interior walls also has been found to affect human comfort. The external energy load on the building, in turn, depends upon the latitude of the building, season of the year, and time of day.

In tropical climates and elsewhere during midday hours in summer, the Sun's path across the local sky increases the solar radiation incident upon the roof and walls of the building. In polar latitudes or during the winter or when the amount of available sunlight is significantly lower, the loss of infrared radiation causes a net cooling of the building. The color of the roof and the outer walls can affect the amount of incoming sunlight absorbed. Building orientation and the effective use of overhangs can also affect the amount of sunlight absorbed. Furthermore, the amount of insulation, often related to the thickness of the walls, reduces the conduction of heat into or out of the building. Thick adobe walls have been used effectively in the Southwest to moderate indoor air temperature. These walls reduce the heat flow into the building during the daytime and in summer and out from the building at night or during winter.

Size and placement of windows also affects the energy balance. Large windows on the side of the building facing the Sun's path tend to admit large amounts of sunlight into the building. However, large windows on the side facing away from the Sun can cause heat loss due to conduction, as window panes are not energy efficient compared to insulated walls.

Appropriate landscaping can reduce energy demands upon a dwelling. Deciduous trees planted on the south and west sides of the home provide cooling shade during the summer, keeping sunlight from entering the windows. These trees lose their foliage in fall and allow the Sun to shine through in winter, warming south facing rooms. Evergreen trees or dense shrubbery on the north and west sides of the building can serve as a windbreak, in winter protecting the house from cold northerly and westerly winds.

Adequate ventilation is desired to help moderate temperature and humidity levels inside the house while refreshing the air. Many buildings in the tropics have louvers or screens to allow the through-flow of air, reducing daytime temperatures and humidity levels, especially in humid tropical climates. In middle latitudes, ceiling fans can be used to help move air and reduce the need for air conditioning or space heating. During the summer, fans circulate cooler air near floor level upward, whereas in winter, fans direct warm air from near the ceiling downward.

Illumination levels are related to the available sunlight within a building. In addition to affecting indoor temperature, illumination also affects human activity and wellbeing. The amount of natural illumination within a building can be modified by appropriate placement and size of windows, as well as by outdoor landscaping.

Atmospheric humidity affects human comfort by helping to regulate the loss of heat from the body. Low humidity cools the body via evaporative cooling, whereas high humidity inhibits cooling and contributes to heat stress when accompanied by high air temperatures. Very humid indoor air also can contribute to other environmental problems, such as mold. Ventilation or air conditioning can be used to reduce humidity to more comfortable levels. At low relative humidity, some people experience discomfort caused by irritation of mucous membranes in the nose, sinus glands, and throat. House plants require more frequent watering, and wood furniture may crack and become unjointed. Hence, it is often desirable to raise indoor humidity during the winter heating season. Raising the humidity reduces evaporative cooling and makes people feel more comfortable at lower room temperature.

HEATING AND COOLING DEGREE DAYS

Television and newspaper weather summaries routinely report heating or cooling degree-day totals in addition to daily maximum and minimum temperature. Heating and cooling degree-days are indicators of household energy consumption for space heating and cooling, respectively.

In the United States, **heating degree-days** are based on the Fahrenheit temperature scale and are computed only for days when the average outdoor air temperature is lower than 65 °F (18 °C). Heating engineers who formulated this index early in the 20th century found that when the average outdoor air temperature drops below 65 °F, space heating is required in most buildings to maintain an average indoor air temperature of 70 °F (21 °C). The average daily temperature is the simple arithmetic average of the 24-hr maximum and minimum air temperatures. Each degree of average temperature below 65 °F is counted as *one heating degree-day*. Subtracting the average daily temperature from 65 °F yields the number of heating degree-days for that day. Suppose, for example, that this morning's low temperature was 28 °F, and this afternoon's high temperature was 46 °F. Today's average temperature would then be 37 °F, for a total of 28 heating degree-days (65 − 37 = 28). It is usual to keep a running total of heating degree-days, that is, to add degree-days for successive days through the heating season (actually from 1 July of one year through 30 June of the next).

Fuel distributors and power companies closely monitor heating degree-days. Fuel oil dealers base fuel use rates on cumulative degree-days and schedule deliveries accordingly. Natural gas and electrical utilities anticipate power demands on the basis of degree-day totals, and implement priority use policies on the same basis when capacity fails to keep pace with demand.

The map in Figure 14.9 is a plot of the average annual cumulative heating degree-day totals over the United States. Outside of mountainous areas, regions of equal heating degree-day totals roughly parallel latitude circles with degree-day totals increasing toward the pole. As an example, the average annual space-heating requirement in Chicago (about 6500 heating degree-days) is approximately four times that of New Orleans (about 1600 heating degree-days). If per unit fuel costs are the same in both cities, then, in an average winter, Chicago homeowners can expect to pay about four times as much for space heating as homeowners in New Orleans. This assumes that buildings in the two cities are comparable in structure and insulation.

Cooling degree-days are computed only for days when the average outdoor air temperature is higher than 65 °F (although higher base temperatures are sometimes used). Supplemental air conditioning may be needed on such days. Again, a cumulative total is maintained through the cooling season (1 January through 31 December). Across the United States, average annual cooling degree-day totals range from less than 500 in the northern tier states and along much of the Pacific coast to more than 4000 in South Texas, South Florida, and the Desert Southwest (Figure 14.10).

Indexes of indoor heating and cooling requirements are based on outdoor air temperatures and fail to account for other weather elements, such as air circulation and humidity, which also influence human comfort and demands for space heating and cooling. Heating and cooling degree-days are therefore only approximations of residential fuel demands for heating and cooling. Nonetheless, for the consumer, comparing energy use for space heating and cooling to heating and cooling degree-days is a convenient way of gauging the effectiveness of home or building energy conservation measures.

Extremes in outdoor air temperature and humidity affect human well-being. With climate change, heat waves and cold waves may become more frequent and intense. As discussed in this chapter's second Essay, special indexes are available to gauge the potential hazard of extremes in air temperature and humidity.

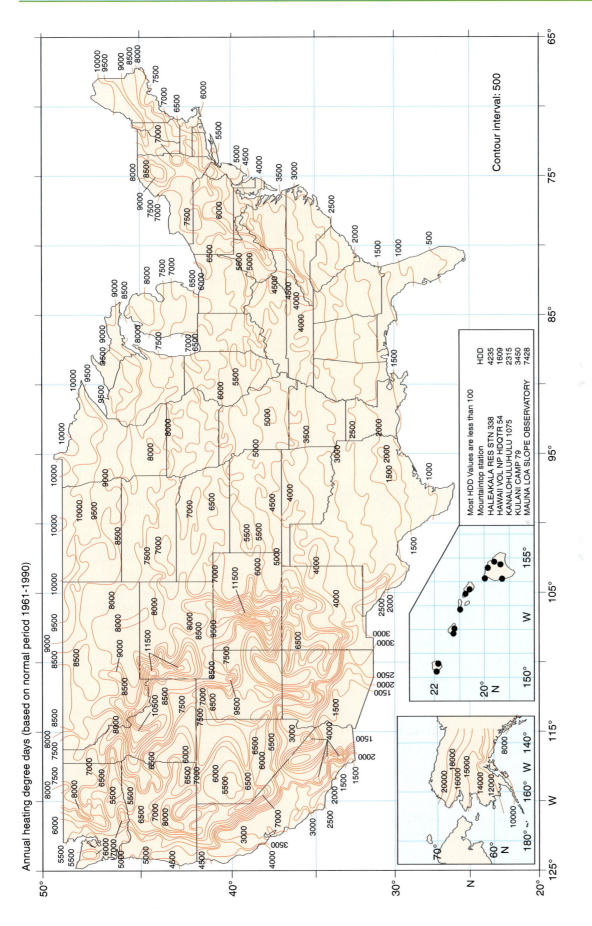

Annual heating degree days (based on normal period 1961–1990)

Contour interval: 500

Most HDD Values are less than 100	HDD
Mountaintop station	
HALEAKALA RES STN 338	4235
HAWAII VOL NP HDQTR 54	1609
KANALOHULUHULU 1075	2315
KULANI CAMP 79	3450
MAUNA LOA SLOPE OBSERVATORY	7428

FIGURE 14.9
Average annual heating degree-day totals over the U.S. using a base of 65 °F. [NOAA's National Climatic Data Center]

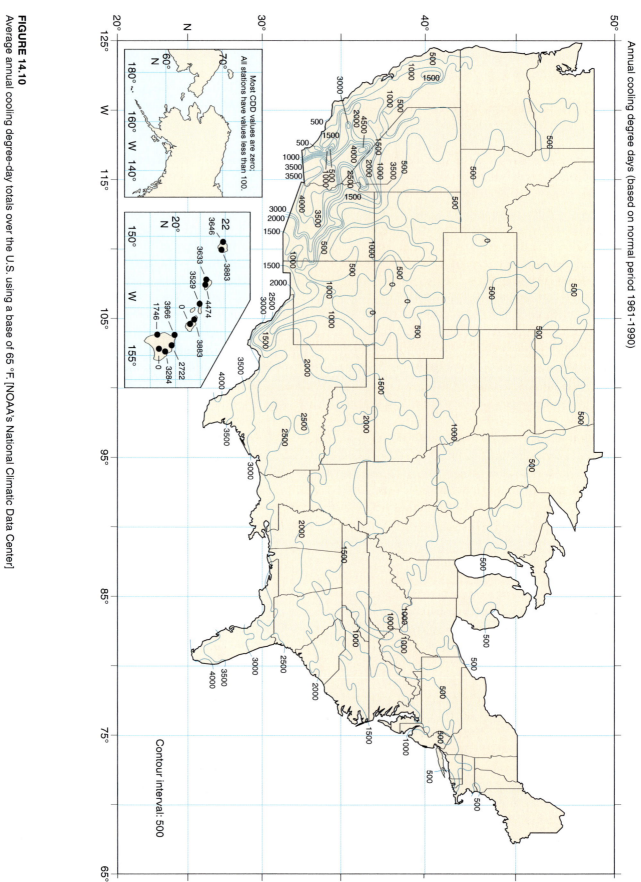

Conclusions

At the heart of anthropogenic climate change is our dependence on fossil fuels for energy. In view of the long lifetime of the greenhouse gas CO_2 in the atmosphere, we are already committed to major disruptions of the environment such as glacier melting, sea level rise, changes in the global water cycle, and ocean acidification.

Humankind is responding to the probability that anthropogenic climate change will persist at least through this century. Climate change mitigation seeks to cut emissions of CO_2 to the atmosphere via cap-and-trade, carbon capture and storage, and higher energy efficiency especially in the generation of electricity, the transportation sector, and the built environment (referring to the human-made surroundings where human activity takes place). Climate change adaptation has a more immediate objective of making communities more resilient to climate change having lead times of a decade or less. A variety of large-scale geoengineering projects have been proposed to alter Earth's climate system in ways that will cool the Earth's surface by either raising the planetary albedo or increasing the flux of infrared radiation from Earth's surface to space.

Our consideration of humankind's response to anthropogenic climate change continues in this book's final chapter with a focus on the design and application of appropriate and effective public policy and economic considerations.

Basic Understandings

- Mitigation encompasses action taken over the present century that is designed to stabilize greenhouse gas emissions at a level that avoids a dangerous disturbance of the climate system, reducing the pace and magnitude of anthropogenic climate change.
- Adaptation refers to steps taken to make communities more climate-resilient so that society can more effectively cope with those climate changes that appear inevitable over the next decade or so.
- Geoengineering refers to large-scale human manipulation of components of Earth's climate system that is intended to offset the environmental consequences of greenhouse gas emissions.
- Adaptation strategies typically depend on climate predictions having lead times of less than a decade

whereas mitigation strategies depend on climate predictions having lead times of a century or so.
- The amount of CO_2 already released to the atmosphere by human activities plus the long lifetime of atmospheric CO_2 commit the planet to global warming, sea level rise, shrinking glaciers, and ocean acidification.
- The most promising use of cap-and-trade for regulating CO_2 emission to the atmosphere is to focus exclusively on sources of CO_2 from fossil fuel combustion.
- In another market-based approach to reducing CO_2 emissions, known as offset exchange, carbon credits are earned for investment in low-carbon energy-generating projects (in developing nations) that compensate for (offset) CO_2 emissions (in developed nations).
- Carbon capture and storage (CCS) is most effectively applied to large single-source facilities such as coal-fired electric power plants where carbon is captured at either the pre- or post-combustion stage.
- Following capture, the pure stream of CO_2 is compressed to a liquid and then piped to an underground storage site located on land or offshore beneath the ocean floor. CO_2 is then injected into pore spaces in geologic formations at depths of at least 800-1000 m for long-term storage. A thick layer of impermeable rock (cap rock) must overlie the reservoir formation to prevent vertical escape of the sequestered CO_2.
- As the global population continues to climb and national economies grow, world-wide reliance on oil-based fuels for transportation and CO_2 emissions are likely to increase. With greater affluence, people demand faster means of transport, improved safety, and greater personal comfort.
- Human intervention by way of technological innovation has enabled communities to adapt to some extent to the local climate and climate variability. However, climate change is not the same everywhere and some human populations are much more vulnerable than others to the adverse effects of climate change especially as the change may involve a reduction in water supply.
- Geoengineering schemes fall into one of two general categories depending on objective: reduce the amount of solar radiation reaching Earth's

surface or increase the flux of heat escaping from Earth to space by reducing the amount of CO_2 in the atmosphere.

- One proposed geoengineering scheme would mimic the effect of a violent sulfur-rich volcanic eruption by injecting millions of tons of SO_2 each year into the stratosphere to bring about large-scale cooling at Earth's surface.

- Another geoengineering scheme would raise the reflectivity (albedo) of low stratocumulus clouds over the ocean by introducing sea salt crystals that would favor formation of relatively large numbers of small cloud droplets that are more reflective.

- Geoengineering fails to address the root cause of anthropogenic climate change, that is, our strong and growing dependence on carbon fuels for energy.

- Inexpensive energy available during the last century meant that modern architecture in Western cultures strayed from the energy conservation principles applied by our ancestors. However, recent concerns about increasing dependency on fossil fuels for energy, foreign sources of oil, and global climate change has spurred renewed interest in improving energy efficiency.

- In order to maintain a tolerable level of comfort within a building, attention must be paid to thermal effects, ventilation, illumination, and humidity.

- Heating and cooling degree-days are indicators of household energy consumption for space heating and cooling, respectively. Heating degree-days are computed only for days when the average outdoor air temperature is lower than 65 °F (18 °C). Cooling degree-days are computed only for days when the average outdoor air temperature is higher than 65 °F (18 °C), although higher base temperatures are sometimes used.

Enduring Ideas

- The lengthy lifetime of carbon dioxide in the atmosphere means that we are committed to continued global warming through this century and beyond.
- The principal climate risk management strategies are mitigation, adaptation, and geoengineering.
- Mitigation encompasses actions designed to stabilize greenhouse gas emissions at a level that avoids a dangerous disturbance of Earth's climate system.
- Adaptation refers to steps taken for the purpose of making communities more climate-resilient so that society can more effectively cope with the inevitable harmful impacts of climate change while taking advantage of the beneficial impacts.
- Geoengineering consists of large-scale, intentional human manipulation of components of Earth's climate system to counteract the consequences of rising levels of heat-trapping gases in the atmosphere.

Review

1. What is the principal objective of climate change mitigation?
2. How does the time scale of climate change mitigation differ from that of climate change adaptation?
3. What is the significance of coal burning for anthropogenic climate change?
4. With carbon capture and storage, the geologic formation intended for storage of CO_2 must be overlain by a thick layer of impermeable cap rock. Explain why.
5. Give several examples of how human intervention by way of technological innovations has enabled communities to adapt to climate variability.
6. Explain how injection of sulfurous aerosols into the stratosphere could help offset global warming caused by an enhanced greenhouse effect.
7. What is the purpose of iron fertilization of the ocean?
8. What is the appeal of geoengineering schemes with regard to reducing anthropogenic climate change?
9. What is the impetus for climate-conscious architecture?
10. What is the purpose of heating and cooling degree-days?

Critical Thinking

1. What is the advantage of applying carbon capture and storage to single-source facilities such as coal-fired power plants rather than individual motor vehicles?
2. How are geoengineering schemes intended to influence Earth's climate system?
3. How might we increase the amount of carbon sequestered in forests?
4. Identify the fundamental drawbacks to geoengineering schemes aimed at offsetting global climate change.
5. What are some of the limits to climate change adaptation?
6. Identify some of the limits to climate change mitigation.
7. How might predictions made by climate models guide the development of climate change mitigation and adaptation strategies?
8. What is a major shortcoming of ocean iron fertilization?
9. How are heating and cooling degree-days used to evaluate the energy efficiency of residential and commercial buildings?
10. Explain why we are already "committed" to global warming, shrinking glaciers, sea level rise, and ocean acidification.

ESSAY: Increasing Energy Efficiency and Decreasing CO_2[1]

A potentially very effective strategy aimed at slowing the anthropogenic contribution to atmospheric CO_2 is to increase energy efficiency. *Energy efficiency* refers to the portion of available energy that is converted to useful energy services. Between 1900 and 1997, energy efficiency in the U.S. grew very slowly from 2.2% to 13%. At 13% efficiency, for every unit of energy converted to useful work, about seven units of energy are lost to the environment, mostly as heat.

In a traditional coal-fired electric power plant, only about one-third of the energy in the fuel (coal) is converted to electricity. The rest is usually vented to the atmosphere as waste heat. Making such plants more energy efficient would reduce the demand for fuel and new power plants, and cut emissions of CO_2 with obvious implications for global climate change.

Much greater energy efficiency could be achieved in coal-fired power plants via energy recycling, that is, "rescuing" waste energy (i.e., heat). (Note that recycled energy is not the same as renewable energy such as solar-thermal or wind power.) At present, few plants use their waste heat. Heat energy is not readily conveyed long distances and most power plants are located too far from where there is a market for waste heat, say for space heating. Of all energy used in the U.S., only about 8% is recycled, mostly by *cogeneration facilities* that serve as combined sources of electricity and space heating for nearby customers.

According to the U.S. Environmental Protection Agency (USEPA) and Department of Energy (DOD), the greatest potential for recycled energy in the U.S. is electric power plants. The total potential recycled energy is equivalent to about 20% of all power generated in the nation. Recycling of this energy would cut CO_2 emissions by an estimated 20%.

Higher energy efficiency is needed wherever energy is used, and that need is greatest wherever the use of energy is greatest. Access to energy and energy services is strongly correlated to economic growth. As rapidly industrializing nations such as China and India continue to build more and more coal-fired power plants, the impact on climate will also increase unless concurrent steps are taken to increase energy efficiency.

Some scientists and engineers such as Leon R. Glicksman of the Massachusetts Institute of Technology, argue that by reducing CO_2 emissions, improved energy efficiency will help buy time for environmentally acceptable (low-carbon or no-carbon) energy sources to come online. Writing in the July 2008 issue of *Physics Today*, Glicksman points out that the largest U.S. consumer of primary energy (amounting to about 40% of the total) is buildings, residential and commercial combined, larger than either the transportation or industrial sectors (Figure 1). (*Primary energy* encompasses fossil fuels consumed in a building plus the energy that generates electricity used in the building.) Some 70% of U.S. electricity is used in the residential and commercial buildings sector.

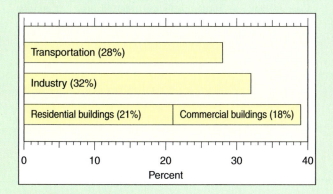

FIGURE 1
Consumers of primary energy by sector. [Modified after Glicksman, L.R., 2008. "Energy efficiency in the built environment," *Physics Today*, July, 35-40; Energy Information Administration, *Annual Energy Review 2005*, rep. no. DOE/EIA-0384(2005), U.S. Department of Energy, Washington, DC (July 2006).]

According to Glicksman, a 25% reduction in energy use by buildings is feasible. The American Institute of Architects set a goal of cutting in half the amount of fossil fuels used in the construction and operation of new and renovated buildings. A long-term goal in some quarters is for new construction to approach *zero net energy* whereby onsite renewable energy production each year will compensate for all primary energy consumed during the same period.

In residential buildings, space heating is the largest user of energy. Active or passive solar-energy systems, high efficiency windows, and heavily insulated walls will add to energy efficiency. Over the past 30 years, the energy efficiency of light bulbs and home appliances such as refrigerators has improved 2- to 4-fold.

In commercial buildings, the single largest consumer of primary energy is lighting. Use of occupancy sensors for lighting (Figure 2) will cut energy use but at present these

[1]Much of this Essay is based on:
Casten, T.R. and P.F. Schewe, 2009. "Getting the Most from Energy," *American Scientist* 97:26-33.
Glicksman, L.R., 2008. "Energy efficiency in the built environment," Physics Today, July, 35-40.

sensors are used in less than 10% of total floor space. Shading and spectrally selective window glazing reduce solar heating, highly reflective roof tops cut the heat load especially where the climate is hot (Figure 3), and natural ventilation may reduce the need for air conditioning.

FIGURE 2
A sensor measures the amount of natural light coming in and adjusts overhead fluorescent lights accordingly. [Courtesy Building Energy Codes Resource Center, Pacific Northwest National Laboratory, US Department of Energy]

FIGURE 3
View of high-rise building with high glazing area. [Courtesy Building Energy Codes Resource Center, Pacific Northwest National Laboratory, US Department of Energy]

ESSAY: Air Temperature Extremes and Human Wellbeing

Scientists have developed special environmental indexes to gauge the influence of temperature and humidity extremes on human well-being. These indexes are useful in evaluating the impact of climate change on humans (perhaps involving a change in frequency or intensity of heat waves or cold waves). These indexes account for the combined influence of heat and humidity and how wind speed affects cooling rates.

Statistics compiled by NOAA's National Weather Service reveal that heat (along with high humidity) was responsible for the greatest number of weather-related deaths across the U.S. during the 10-year period, 1998-2007, with an average of 170 fatalities (mostly the elderly) occurring per year. By comparison, 62 deaths per year were associated with tornadoes and 14 deaths were annually caused by the cold (low temperatures). Furthermore, concern has been raised that during this century, heat waves associated with global climate change could become more frequent and severe, leading to a greater risk of hyperthermia and, ultimately, to higher morbidity rates.

The relationship between the vapor pressure of ambient air and evaporative cooling helps explain why people usually are more uncomfortable when the weather is both hot and humid. At air temperatures above about 25 °C (77 °F), perspiring and consequent evaporative cooling enhance heat loss from the body. But if the air were also humid, the evaporation rate is reduced, hampering the body's ability to maintain a nearly constant core temperature of about 37 °C (98.6 °F). (The body's *core* refers to vital organs such as the heart and lungs.) In contrast, drier air promotes evaporative cooling. At air temperatures above about 25 °C (77 °F), most people feel more comfortable when the humidity is relatively low.

A combination of high temperature and high humidity adversely affects everyone to some extent. People generally are more irritable and less able to perform physical and mental tasks. During hot and humid weather, the efficiency of factory workers declines and students do not concentrate as well on their studies. Extreme heat and humidity are a potentially lethal combination when these weather conditions overtax the body's capacity to maintain a constant core temperature. Heat and humidity can also exacerbate other health problems leading to premature death.

In mid-July 1995, high air temperature combined with unusually high humidity pushed the *heat index* to record high levels over the Midwest and in many cities along the East Coast. More than 1000 people lost their lives. Chicago was particularly hard hit. Public safety officials estimate the number of deaths from heat and humidity based on the *excess death rate*, that is, the number of reported deaths minus the typical number of deaths over a specified period. During the week of 14-20 July 1995, 739 more Chicagoans died than the norm. According to the U.S. Centers for Disease Control and Prevention, individuals most vulnerable to the combined hazard of heat and humidity are elderly people who have no air conditioning, live alone, do not leave home daily, lack access to transportation, are sick or bedridden, and do not have social contacts nearby.

During summer 2003, extreme heat rather than a combination of heat and humidity was to blame for one of Europe's deadliest heat waves in a century. A 2006 report by the Earth Policy Institute placed the total heat-related death toll at more than 52,000. Based on the *excess death rate* method, the estimated mortality exceeded 18,000 in Italy and 14,000 in France. Persistent episodes of hot, dry weather began in late spring and in some parts of Europe persisted until early October. From Spain to Hungary, the June mean temperature was 3 to 5 Celsius degrees (5.4 to 9 Fahrenheit degrees) above the long-term average. The heat wave during the first half of August was particularly severe, with some localities establishing all-time maximum temperature records. During July-August, from southern Spain to central France, the daily maximum temperature topped 34 °C (93 °F) on 30 to 50 days, some 20 days more than normal. Most Europeans are not accustomed to such extreme heat. The elderly without air conditioning were especially hard hit. In addition, the accompanying dry weather contributed to wildfires and significantly cut crop yields.

Scientists have experimented with a variety of indexes that attempt to gauge the combined influence of temperature and humidity on human wellbeing and advise people of the potential hazards of heat stress. Since the summer of 1984, NOAA's National Weather Service has regularly reported the heat index, or apparent temperature index, which was developed by R.G. Steadman in 1979. The *heat index* accounts for the increasing inability of the human body to dissipate heat to the environment as the temperature and relative humidity rise. As illustrated in Figure 1A, at an air temperature of 90 °F (32 °C) and a relative humidity of 60%, the body loses heat to the environment at the same rate as if the air temperature were 105 °F (41 °C) with a relative humidity of 10%. In both cases, the *apparent temperature* is 100 °F (38 °C). As shown in Figure 1A, various combinations of air temperature and relative humidity produce the same apparent temperature, implying that all these temperature/relative-humidity combinations have the same impact on human comfort and wellbeing.

A

Relative humidity (%)

Air temperature (°F)

	0	5	10	15	20	25	30	35	40	45	50	55	60	65	70	75	80	85	90	95	100
120	107	111	116	123	130	139	148														
115	103	107	111	115	120	127	135	143	151												
110	99	102	105	108	112	117	123	130	137	143	150										
105	95	97	100	102	105	109	113	118	123	129	135	142	149								
100	91	93	95	97	99	101	104	107	110	115	120	126	132	138	144						
95	87	88	90	91	93	94	96	98	101	104	107	110	114	119	124	130	136				
90	83	84	85	86	87	88	90	91	93	95	96	98	100	102	106	109	113	117	122		
85	76	79	80	81	82	83	84	85	86	87	88	89	90	91	93	95	97	99	102	105	108
80	73	74	75	76	77	77	78	79	79	80	81	81	82	83	85	86	86	87	88	89	91
75	69	69	70	71	72	72	73	73	74	74	75	75	76	76	77	77	78	78	79	79	80
70	64	64	65	65	66	66	67	67	68	68	69	69	70	70	70	70	71	71	71	71	72

B

Category	Apparent temperature*	Heat syndrome
I	130 °F (54 °C) or higher	Extreme danger; heat stroke imminent.
II	105 to 130 °F (41 to 54 °C)	Danger; heat cramps or heat exhaustion likely. Heat stroke possible with prolonged exposure and physical activity.
III	90 to 105 °F (32 to 41 °C)	Heat cramps or heat exhaustion possible with prolonged exposure and physical activity.
IV	80 to 90 °F (27 to 32 °C)	Fatigue possible with prolonged exposure and physical activity.

Source: NOAA National Weather Service

*Apparent temperature combines the effects of heat and humidity on human comfort.

FIGURE 1
Heat index (A) and hazards posed by heat stress for ranges of apparent temperature (B). [Courtesy of NOAA]

The heat index is divided into four categories based on the severity of potential health impacts; these categories and associated symptoms of heat stress are listed in Figure 1B. Apparent temperatures in category I pose the greatest health risk. When the heat index is 130 °F (53 °C) or higher, the danger is extreme and heatstroke (*hyperthermia*) may be imminent. During the heat wave of 10-16 July 1995, the highest heat index values were in the range of 115 °F to 120 °F (46 °C to 49 °C) (Figure 2).

The significance of a specific apparent temperature for personal wellbeing depends upon (1) the assumptions that form the basis of Steadman's index, and (2) a person's age, general health, and body characteristics. Steadman's heat index assumes a person at rest in the shade with light winds. Personal experience tells us that exercise and exposure to the direct rays of the Sun add considerably to the heat load that the body must dissipate to maintain a

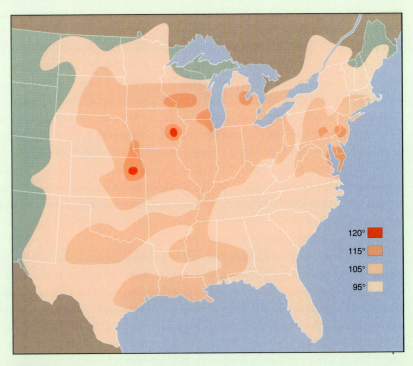

120°
115°
105°
95°

FIGURE 2
Maximum apparent temperature (in °F) recorded during the heat wave of 10-16 July 1995. [NOAA, NWS Climate Prediction Center]

normal core temperature. For example, the apparent temperature may be as much as 8 Celsius degrees (14 Fahrenheit degrees) higher in direct sunlight than the shade. Furthermore, if the ambient air temperature were higher than skin temperature, the wind will convey more heat to the body than away from the body. Any combination of physical activity, exposure to direct sunlight, or hot winds elevates the danger of high temperature and high humidity to levels higher than suggested by the apparent temperature.

Where do people experience the most uncomfortable (and potentially life-threatening) summer weather? In view of the adverse impact of high humidity on evaporative cooling, many of us would probably point to locales bordering the Gulf of Mexico where high temperature and high humidity are the norm. For example, during a summer heat wave, a typical temperature/relative humidity combination in New Orleans is 90 °F (32 °C)/60%. According to Figure 1, these conditions translate into an apparent temperature of 100 °F (38 °C), a category III hazard. During such weather conditions, most people would experience some discomfort and everyone is well advised to closely monitor his or her level of physical activity.

Perhaps surprisingly, North America's most stressful summer weather typically does not occur near the Gulf of Mexico, but instead in the deserts of Arizona and California. Although the relative humidity is considerably lower, much higher ambient air temperatures translate into greater heat stress than along the Gulf Coast. For example, residents of Phoenix, AZ, experience summer afternoon temperatures that often approach 110 °F (43 °C) with a relative humidity of 20%. This combination of heat and humidity translates into an apparent temperature of 112 °F (44 °C), a category II hazard. In general, residents of the eastern half of North America associate greater discomfort with the muggy days of summer whereas residents of the American Southwest ascribe their days of greatest discomfort to the very hot, albeit dry, conditions of the desert in summer.

Scientists with NOAA's National Climatic Data Center recently conducted research on the threshold apparent temperatures that cause an increase in morbidity rates across the nation. They used a slightly different way of calculating apparent temperatures than is used by the National Weather Service for computing the heat index. They determined the 85th percentiles of daily average apparent temperatures over the period 1961-1990 at many first-order U.S. weather stations and used these as threshold values for identifying extreme heat stress conditions (Figure 3). The 85th percentile values are closely correlated with weather related mortality. Threshold temperatures are higher in the mid-south probably due to acclimatization to higher temperatures in that part of the country. Additionally, their research showed that apparent temperatures across many areas of the nation have increased over the past half-century.

Low humidity also influences human comfort. As the relative humidity declines, the rate of evaporative cooling increases. The lower the relative humidity, the higher is the rate of heat loss from the body via evaporative cooling. As shown in Figure 1A, for a specified ambient air temperature and successively lower values of relative humidity, the difference between apparent temperature and ambient air temperature decreases. Below some threshold value of relative humidity, the apparent temperature is actually lower than the ambient air temperature. At low relative humidity, people experience the apparent temperature as lower than the ambient air temperature because the higher rate of evaporation causes excessive heat loss. That is, enhanced evaporative cooling of the skin at low relative humidity gives the sensation that the air is cooler than it actually is.

At low air temperature, the wind increases human discomfort outdoors and heightens the danger of *frostbite*, the freezing of body tissue, or *hypothermia,* a potentially lethal condition brought on by a drop in the temperature of the body's vital organs. Air in motion is more effective than still air in transporting heat away from the body. To account for the effect of wind on human comfort and wellbeing, weather reports during winter in northern and mountainous regions include the *wind-chill equivalent temperature (WET)* along with the actual air temperature.

A very thin layer of motionless air (thickness measured in millimeters) next to the skin helps insulate the body from heat loss to the environment. Within this *boundary layer*, heat is lost through the very slow process of conduction. (Recall that calm air is a poor conductor of heat.) The boundary layer is thickest and offers maximum insulation when the wind speed is under 0.5 mph (0.2 m per sec). Because water vapor molecules diffuse slowly through the skin's boundary layer, the rate of evaporative cooling is also relatively slow. Hence, even at very low air temperatures, people who are appropriately dressed are comfortable as long as winds are light. As winds strengthen, the boundary layer becomes thinner and the rate of heat and water vapor transport from the skin surface increases. Thinning of the boundary layer (and accelerated heat loss) is most pronounced at low to moderate wind speeds. As wind speed increases above about 35 mph (55 km per hr), incremental heat loss becomes relatively small as the boundary layer approaches its minimum thickness.

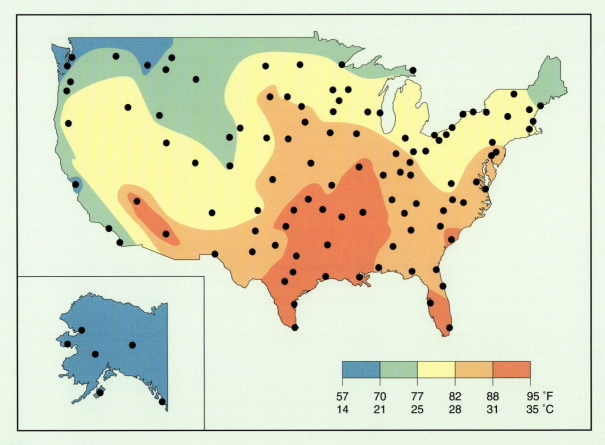

FIGURE 3
The color scale on this map indicates 85th percentile values of climatological daily-average apparent temperature in °C, derived from 3-hourly July and August data for the period 1961-1990. The range of values (14 to 35 °C) is indicated on the scale. These threshold values for extreme heat stress conditions are each exceeded about 12 days per year. Black squares give location of the first-order weather stations (mostly airports) used in this study. [NOAA's National Climatic Data Center]

Scale	57	70	77	82	88	95 °F
	14	21	25	28	31	35 °C

The *wind-chill equivalent temperature (WET)*, also called the *wind-chill index*, is a measure of the rate of body heat loss due to a combination of wind and low air temperature. The original index was based upon research conducted in Antarctica by P.A. Siple and C.F. Passel during the winter of 1941. Their original objective was to measure the time required for water to freeze under various weather conditions. Only later did the idea arise to develop a wind-chill index based upon this research. Their experiments using a small water-filled non-insulated plastic cylinder provide, at best, a rough approximation of the thermal response of a human body dressed appropriately for cold and windy weather. The wind-chill index also fails to account for the appreciable warming that occurs when the body absorbs solar radiation—Siple and Passel made their measurements in darkness.

In the winter of 2001-02, the National Weather Service (NWS) and the Meteorological Services of Canada (MSC) implemented a major revision of the wind-chill index (Figure 4). The new index incorporates recent advances in science, technology, and computer modeling to provide a

Wind (mph)

Calm	5	10	15	20	25	30	35	40	45	50	55	60
40	36	34	32	30	29	28	28	27	26	26	25	25
35	31	27	25	24	23	22	21	20	19	19	18	17
30	25	21	19	17	16	15	14	13	12	12	11	10
25	19	15	13	11	9	8	7	6	5	4	4	3
20	13	9	6	4	3	1	0	-1	-2	-3	-3	-4
15	7	3	0	-2	-4	-5	-7	-8	-9	-10	-11	-11
10	1	-4	-7	-9	-11	-12	-14	-15	-16	-17	-18	-19
5	-5	-10	-13	-15	-17	-19	-21	-22	-23	-24	-25	-26
0	-11	-16	-19	-22	-24	-26	-27	-29	-30	-31	-32	-33
-5	-16	-22	-26	-29	-31	-33	-34	-36	-37	-38	-39	-40
-10	-22	-28	-32	-35	-37	-39	-41	-43	-44	-45	-46	-48
-15	-28	-35	-39	-42	-44	-46	-48	-50	-51	-52	-54	-55
-20	-34	-41	-45	-48	-51	-53	-55	-57	-58	-60	-61	-62
-25	-40	-47	-51	-55	-58	-60	-62	-64	-65	-67	-68	-69
-30	-46	-53	-58	-61	-64	-67	-69	-71	-72	-74	-75	-76
-35	-52	-59	-64	-68	-71	-73	-76	-78	-79	-81	-82	-84
-40	-57	-66	-71	-74	-78	-80	-82	-84	-86	-88	-89	-91
-45	-63	-72	-77	-81	-84	-87	-89	-91	-93	-95	-97	-98

Temperature (°F)

Frostbite times (minutes)   30   10   5

FIGURE 4
Wind-chill. [NOAA, NWS]

more accurate, understandable, and useful formula for representing the combined hazard of wind and low air temperatures. The new wind-chill index uses wind speed adjusted for the average height (1.5 m or 5 ft) of a human's face instead of standard anemometer height (10 m or 33 ft), applies modern heat transfer theory, and assumes the worst case scenario for solar radiation (clear night sky).

Suppose, for example, the air temperature is 35 °F and the air is calm; then the wind-chill equivalent temperature (35 °F) is the same as the actual air temperature. If the wind strengthens to 25 mph with no change in actual air temperature, the WET drops to 23 °F. Contrary to popular opinion, this does not mean that the temperature of exposed skin actually drops to 23 °F. Skin temperature can drop no lower than the actual air temperature, which in this example is 35 °F. When skin temperature equals the air temperature, the temperature gradient between the skin surface and adjacent air is zero and no heat is exchanged between the skin and the air. Even though the WET may fall below the freezing point, skin will not freeze if the actual air temperature were above the freezing point.

On a cold and windy day, an exposed body part loses heat to the environment at the same rate that it would if the air were calm and the actual air temperature equaled the wind-chill equivalent temperature. In fact, many combinations of air temperature and wind speed produce about the same rate of heat loss. For instance, heat loss is essentially the same for air temperature/wind speed combinations of 15 °F/5 mph and 25 °F/35 mph. In both cases, the wind-chill equivalent temperature is 7 °F.

During cold and blustery weather, if the human body cannot supply heat to the skin at a rate sufficient to compensate for heat loss, skin temperature declines. Initially a person has a sense of discomfort. If the rate of heat loss from exposed skin causes skin temperature to fall to subfreezing levels, then frostbite may ensue. Particularly susceptible to frostbite are body parts that are usually exposed and have a relatively high surface-to-volume ratio, such as the ears, nose, and fingers. Frostbite may occur in 30 minutes or less at wind-chill values of −18 °F (−28 °C) or lower. In extreme circumstances, the rate of heat loss from the body may be great enough to cause life-threatening hypothermia.

CHAPTER 15

CLIMATE CHANGE AND PUBLIC POLICY

UN Secretary-General Ban Ki-moon (right) confers with Felipe Calderón Hinojosa (second from right), President of Mexico, and Jens Stoltenberg (center), Prime Minister of Norway, during the final hours of the UN Climate Change Conference in Copenhagen, Denmark, December 2009. [UN Photo]

Case-in-Point

The prospect of continued global warming, shrinking ice sheets, thermal expansion of sea water, and sea level rise spurred representatives of the *Small Island Developing States (SIDS)* to join forces in a coordinated response to anthropogenic climate change. Founded in 1990 during the Second World Climate Conference in Geneva, the **Alliance of Small Island States (AOSIS)** is an intergovernmental organization of low-lying coastal and small island nations (Figure 15.1). AOSIS operates as an ad hoc lobbying and negotiating presence in the United Nations (UN) engaging in consultation and consensus building. It has 42 member states (e.g., Bahamas, Maldives, Fiji) and observers (e.g., Guam) all told representing 28% of all developing nations. Some 37 are UN members, 20% of the UN's total membership. AOSIS represents about 5% of the global population.

     AOSIS reports that many member states are already experiencing serious adverse impacts from climate change. These impacts include coastal erosion, flooding, coral bleaching, and more frequent and intense weather extremes such as tropical cyclones with a heightened risk of storm surges. Scientific experts warn that many

FIGURE 15.1
A view of Fanning Island from a space craft at an altitude of 298 km (161 nautical miles). Fanning Island (also known as Tabuaeran) is one of 32 atolls in the Republic of Kiribati in the central Pacific at about 4 degrees N, 159.4 degrees W. The island covers an area of about 39 km² (15 mi²) with a maximum elevation of only about 3 m (10 ft) above high tide level. Many of the approximately 2500 residents make their living via tourism from cruise ships. Fanning Island is one of numerous islands that are threatened by rising sea level brought on by global climate change. [NASA, Earth from Space]

water and agriculture, (3) higher sea water temperatures plus ocean acidification that would destroy coral reefs and cause a drastic decline in fisheries. UN representatives warn that some of these locales are "very likely to become entirely uninhabitable." (Refer to the Case-in-Point in Chapter 10).

Underscoring the urgency of the situation for low-lying coastal and small island nations, global greenhouse gas emissions have risen faster than predicted in IPCC's most pessimistic scenario published in 2007. Also, glacial ice is melting faster than projected so that the rise in sea level by the end of this century may be a meter or more rather than the few tenths of a meter projected by IPCC several years ago.

To avoid the potentially devastating effects of global warming, AOSIS is urging the world community of nations to take action that would limit global warming to less than 1.5 Celsius degrees (2.7 Fahrenheit degrees) above preindustrial levels. In this recommendation, AOSIS was joined in 2009 by the 80-member *Group of Least Developed Countries (LDCs)* and others.

According to AOSIS estimates, limiting global warming to less than 1.5 Celsius degrees would require stabilization of the atmospheric concentration of greenhouse gases to the equivalent of less than 350 ppmv of CO_2. AOSIS proposes that global greenhouse gas emissions should peak no later than 2015 and then decline to at least 85% below 1990 levels by 2050. If emissions continue to fall after 2050, the projected global mean temperature increase would decline from less than 2 Celsius degrees (3.6 Fahrenheit degrees) to less than 1.5 Celsius degrees by 2100.

of these low-lying coastal and small island nations face the prospect of (1) extensive disruption of their socio-economic infrastructure, (2) saltwater intrusion that would contaminate coastal aquifers essential for drinking

Driving Question:
How does public policy on climate change evolve?

We have reached the stage in the scientific assessment of climate change where the cumulative evidence is now convincing that (1) global warming is real and unequivocal, and (2) human activity is primarily responsible for warming since the mid-20th century. Considering the magnitude of the likely impacts of climate change on society now and in the future, climate change mitigation and adaptation are high priorities

and receiving much attention worldwide (Chapter 14). Interest in geoengineering schemes is also growing. In 2009, James J. McCarthy, a climate expert at Harvard University, observed that "we may reach a point where we're going to be so desperate that we need to look critically at various geoengineering approaches …" Research, observation systems, scientific assessment, and technological development support and enhance these risk management tools. Nonetheless, greenhouse gas emissions continue to soar so that Earth is already committed to major disturbances including glacier melt, sea level rise, and other changes in the global water cycle.

According to most climate scientists, dangerous disruptions of the environment and society can be avoided if the rise in global mean temperature by 2100 were limited to less than 2 Celsius degrees (3.6 Fahrenheit degrees) above preindustrial levels or less than 1 Celsius degree (1.8 Fahrenheit degree) higher than the current value. For a realistic chance for this to happen, the developed nations must reduce greenhouse gas emissions by 50 to 80% of 1990 levels by 2050. For reasons presented in Chapter 14, action is needed immediately. However, major uncertainties surround certain key greenhouse gas abatement options making it difficult to accurately assess the costs and benefits of alternative strategies.

This final chapter describes how our scientific understanding of anthropogenic climate change and its societal implications translates into public policy enacted at the local, regional, national, and international levels. The chapter opens with a summary of the public policy lessons from another global environmental problem: stratospheric ozone depletion. Focus then turns to how the international community of nations is attempting to curb global warming by negotiating mandatory limits on greenhouse gas emissions into the atmosphere. The next section outlines the workings of the U.S. political system and explains the nation's relatively slow pace toward adopting remedial action legislation. The discussion then covers some of the economic dimensions of anthropogenic climate change. This chapter closes with a rationale for establishing a National Climate Service that would serve the needs of public policymakers, decision makers, and other stakeholders as they develop climate change mitigation and adaptation strategies tailored to specific regions.

Policy Lessons from Stratospheric Ozone Depletion

Starting in the late 1970s, recognition of threats to the stratospheric ozone shield and the likely adverse impacts on the health and wellbeing of humans and other living organisms spurred international action to eliminate use of ozone-depleting substances (ODSs) (Chapter 3). Humankind's successful intervention in this global-scale environmental problem encourages scientists and policymakers that similarly effective action can be taken to curb anthropogenic climate change.

Worldwide acceptance of the threat posed by chlorofluorocarbons (CFCs) and other ODSs to the stratospheric ozone shield prompted the United Nations Environment Programme (UNEP) in 1987 to draw up the *Montreal Protocol on Substances That Deplete the Ozone Layer*. This international treaty entered into full force on 1 January 1989 and, as of 16 September 2009, all 192 UN member states had ratified the original Montreal Protocol. The initial goal of the treaty was to cut CFC production in half by 1992 (compared to 1986 levels). However, the seriousness of the problem led to seven subsequent amendments (revisions) that included expanding the list of regulated substances (e.g., to include bromine-containing halons) and requiring the worldwide phase out of the manufacture and use of CFCs beginning in January 1996.

CFCs and halons have long lifetimes in the atmosphere so that these substances will continue to threaten the stratospheric ozone shield for some time to come in spite of the phase out. All ODSs contain either chlorine or bromine and although not as abundant as chlorine, bromine is much more effective per atom in destroying stratospheric ozone. The 2006 UNEP *Scientific Assessment of Ozone Depletion Concentrations* found that ozone-depleting chlorine levels in the stratosphere are on the decline (Figure 15.2). The *Assessment* concluded that the *Montreal Protocol* was successful in reducing concentrations of major ODSs. Recovery of the ozone layer to pre-1980 levels is not likely until the mid 21st century in middle latitudes and a decade or two later in polar regions. Some world leaders

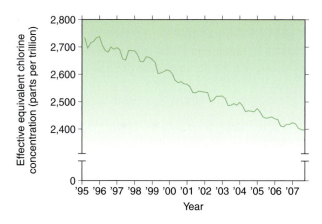

FIGURE 15.2
The post-1995 downward trend in the atmospheric concentration of effective equivalent chlorine (EECl) in parts per trillion by volume. EECl represents atmospheric concentrations of ozone depleting substances (ODSs), weighted by their potential to catalyze the destruction of stratospheric ozone. [Source: NOAA 2008]

such as Kofi Annan, UN Secretary-General from 1997 through 2006, characterize the Montreal Protocol as the most successful international environmental agreement to date.

Some important differences exist between protecting the stratospheric ozone shield and curbing anthropogenic climate change. The Montreal Protocol of 1987 affected only a relatively small segment of national economies whereas greenhouse gases are emitted from all sectors of national economies worldwide and, as emphasized in Chapter 14, the complex global energy system is at the heart of anthropogenic climate change. Nevertheless, the Montreal Protocol of 1987 initiated an international-scale process and established a framework for change that could be strengthened with time. International negotiations regarding limits on greenhouse gas emissions are following a similar model of incremental action.

While the Montreal Protocol halted production and use of CFCs and other ODSs, the agreement does not address the status of ODSs produced prior to the mandated phase-out date. Emissions of ODSs to the atmosphere are not regulated nor does the agreement require the destruction of existing stores of ODSs. In fact, significant amounts of ODSs are sequestered in older refrigeration and air conditioning units as well as building insulation and fire suppression systems. (Release of halons by fire extinguishers explains the continued increase in the concentration of halons in the atmosphere although the rate of increase is slowing.) ODSs are also powerful greenhouse gases and their eventual release to the atmosphere is likely to contribute in a major way to anthropogenic climate change. In terms of the *Global Warming Potential (GWP)* (previously discussed in Chapter 3), the current global store of ODSs is estimated by scientists to be the equivalent of 16 to 18 billion tons of CO_2.

In the 13 November 2009 issue of *Science*, Jeff Cohen of EOS Climate and colleagues summarized various options for preventing the release of ODSs into the atmosphere thereby curbing further exacerbation of global warming. Governments could require destruction of all accessible ODSs using proven technologies. However, ODSs are widely dispersed around the world and some financial incentive is probably necessary for this approach to succeed. One possible incentive is issuance of carbon credits in national and international markets in exchange for the removal and destruction of ODSs. The number of carbon credits awarded would depend on determination of the GWP.

Global Climate Change and International Response

Anthropogenic climate change is a global-scale problem whose solution requires international cooperation. For a variety of reasons, mainly economic and political, little progress has been made in the past two decades toward a coordinated international response to the problem of anthropogenic climate change. Instead, existing climate policy has evolved via initiatives taken by specific nations, regions, and individual firms and hence, is fragmented and decentralized.

Stewardship of the climatically-sensitive Arctic Ocean also requires international cooperation. For more on this topic, refer to this chapter's first Essay.

UN FRAMEWORK CONVENTION ON CLIMATE CHANGE

A global approach to solving the problem of anthropogenic climate change was advocated by the **United Nations Framework Convention on Climate Change (UNFCCC)** treaty that was produced at the UN Conference on Environment and Development convened in Rio de Janeiro, Brazil, in June 1992. The principal objective of the treaty was "stabilization of greenhouse gas concentrations in the atmosphere at a level that would prevent dangerous anthropogenic interference with the climate system."

The UNFCCC treaty entered into full force in March 1994 and, as of October 2009, had 192 signatories. Its initial action was to establish national greenhouse gas emissions and removal inventories. The treaty is not legally binding, that is, it does not require signatory nations to adhere to fixed limitations on greenhouse gas emissions and enforcement mechanisms are non-existent. The treaty calls for updates, known as *protocols* that would specify mandatory limits on greenhouse gas emissions.

KYOTO PROTOCOL

In the initial update to the UNFCCC treaty, representatives of the global community met in Kyoto, Japan, in December 1997 to formulate an acceptable and effective international treaty limiting greenhouse gas emissions worldwide. As a key provision of the **Kyoto Protocol**, 39 developed (industrial) nations (referred to as *Annex I* countries) agreed to reduce emissions of four greenhouse gases (CO_2, CH_4, N_2O, and sulfur hexafluoride) plus two groups of gases produced by them (hydrofluorocarbons and perfluorocarbons). Annex

I countries were responsible for almost two-thirds of all greenhouse gas emissions in 1990. As signatories to the Kyoto Protocol, these countries commit in principle to cutting emissions by a group average of 5.2% below 1990 levels. Excluded from caps are greenhouse gases emitted by international shipping and aviation. The Global Warming Potential (GWP) is the basis for weighing the relative climatic significance of emissions by specific greenhouse gases.

Signatories to the Kyoto Protocol must be parties to the UNFCCC treaty. Adopted in December 1997, the Kyoto Protocol entered into full force on 16 February 2005. As of November 2009, some 186 nations plus the European Union (EU), a regional economic organization, had ratified, accepted, or agreed to the treaty. Notably absent from the list of signatories to the Kyoto Protocol, however, was the United States, which had ratified the UNFCCC treaty and was responsible for 36.1% of greenhouse gas emissions by all Annex I countries in 1990.

The U.S. essentially ignored the Kyoto Protocol. The George W. Bush administration claimed that the high cost of compliance would harm the nation's economy and objected to the fact that the Kyoto Protocol exempted rapidly industrializing nations, such as China and India, where economic growth requires a continual increase in energy supply, most of which comes from fossil fuel combustion. (By the end of 2009, China had surpassed the United States as the largest single emitter of greenhouse gases worldwide.) The rationale for treating developing nations differently than the industrial nations is primarily economic. Cleaner alternatives to fossil fuels tend to be more technologically sophisticated and expensive. The poorer developing nations do not have the resources to immediately switch to low-carbon or no-carbon energy sources so that they were given more time to follow a less carbon intensive path to industrialization.

Nonetheless, some regional initiatives aimed at cutting greenhouse gas emissions are taking place within the U.S. For example, in 2005, nine northeastern states (the six New England states plus New York, New Jersey, and Delaware) formed the Regional Greenhouse Gas Initiative limiting CO_2 emissions from large power plants and allowing trading of carbon credits.

Some developed nations worked diligently in designing and implementing strategies in support of the Kyoto Protocol. The largest such entity was the European Union (EU), which in 2005 adopted a cap-and-trade system to regulate greenhouse gas emissions among its now twenty-seven member states. In spite of some start-up problems, the EU experience with cap-and-trade was viewed as successful in that it set a price on carbon and reduced emissions. The EU caps CO_2 effluent from about 12,000 industrial plants. Each government issues free of charge emission credits, each one of which permits release of one ton of CO_2. As described in Chapter 14, carbon credits are bought and sold in the marketplace. The EU limits the total number of carbon credits and thereby controls the concentration of the main greenhouse gases released to the atmosphere while market forces (supply and demand) regulate the price emitters pay for credits. Ultimately the goal of this approach is to create incentives for companies to employ innovative technologies that cut greenhouse gas emissions and encourage development and use of climate-neutral no-carbon or low-carbon energy sources.

To accommodate developing nations and their rejection of emission caps, the EU implemented the **clean development mechanism (CDM)**, a provision of the Kyoto Protocol that consists of carbon trading via *offset exchange*. As described in Chapter 14, with offset exchange, an investor from a developed nation funds a carbon-free energy project such as a windmill farm in a developing nation. In turn, the investor is awarded carbon credits in an amount corresponding to the emissions that would have been produced by a conventional coal-fired power plant for the same amount of electricity generated by the carbon-free energy source (windmill farm). The reduction in greenhouse gas emissions must be *in addition to* what would have happened without CDM. The investor cannot obtain or sell carbon offset credits if the carbon-free project was to be constructed anyway. In addition, through its Prototype Carbon Fund (PCF), the World Bank invests in carbon-cutting projects mostly in developing nations.

The original goal of the UNFCCC for the creation of a binding international treaty to significantly reduce greenhouse gas emissions was not achieved. Rather, individual nations designed their own unique strategies to control emissions. The different strategies reflect variations in governments and political systems, local business interests plus uncertainties regarding the optimum approach to greenhouse gas abatement. According to a UN report issued on 21 October 2009, by 2007 greenhouse gas emissions by industrial nations collectively had dropped to about 4% under their 1990 levels. However, 18 of the 40 industrial nations now emit more greenhouse gases than they did in 1990. The United States, for example, releases 16.8% more now than in 1990 (Table 15.1).

TABLE 15.1
Percent Change in Greenhouse Gas Emissions (in equivalent CO_2) for Developed Nations, 1990-2007.

Turkey	+119.1
Netherlands	-2.1
Spain	+ 53.5
France	-5.3
Australia	+30.0
United Kingdom	-17.3
United States	+16.8
Russian Federation	-33.9
Japan	+ 8.2
Latvia	-54.7
Slovenia	+ 1.9

Source: United Nations

With the growing consensus of scientific opinion that anthropogenic climate change represents a serious threat to society and human well-being, the U.S. appeared more ready and willing to set targets for reduction of greenhouse gas emissions. For example, in June 2005, the U.S. Senate passed a resolution recommending establishment of a "comprehensive and effective national program of mandatory, market-based limits and incentives on emissions of greenhouse gases that slow, stop, and reverse the growth of such emissions." Although President Barack Obama advocates targeted reductions in CO_2 emissions (primarily via cap-and-trade), his efforts so far have been stymied by continuing partisanship in Washington, DC, polls that say the public is less convinced of the threat of global warming and more concerned about the economy and possible adverse impacts of remediation on employment opportunities. As of this writing, the upcoming mid-term congressional elections make it less likely that the U.S. will agree any time soon to a legally-binding treaty to cut CO_2 levels to specific levels.

Researchers David G. Victor and Danny Cullenward of Stanford University's Program on Energy and Sustainable Development argue that the U.S. can learn much from the initial efforts at reducing the greenhouse gas concentration in the atmosphere, specifically the experiences of the European Union. Reporting in the December 2007 issue of *Scientific American*, Victor and Cullenward recommend that in moving toward an effective emissions reduction strategy, the U.S. follow these five steps:

(1) The federal government should impose a tax on CO_2 emitters instead of relying on a cap-and-trade system. While this approach would appear equitable for all parties concerned, Congress tends to look unfavorably on new taxes.

(2) If Congress prefers cap-and-trade over taxation, a ceiling should be placed on the price of carbon credits to avoid the price volatility of carbon that typically characterizes cap-and-trade and led to windfall profits for some participants in the EU cap-and-trade scheme. Also, credits should be auctioned in the public arena to eliminate charges of favoritism. The EU gave away most credits (permits or allowances) rather than selling them at auction. In one recent U.S. proposal, some credits would be provided free of charge (80% at first and then decreasing to 30% by 2030) while the balance would be sold at an annual auction. In addition, the overall total number of carbon credits issued would decrease with time as an incentive to reduce emissions.

(3) Efforts must be made to create innovative strategies that encourage developing nations to slow (and eventually reverse) their growth in greenhouse gas emissions. A reliable method is needed to validate use of the clean development mechanism and offset exchange.

(4) Economic incentives alone cannot solve the problem of anthropogenic climate change; hence, the government should encourage adoption of renewable and non-renewable no-carbon or low-carbon energy sources.

(5) The government should encourage the development and use of innovative technologies such as carbon capture and storage (CCS), described in Chapter 14.

In an editorial in *Science* magazine on 4 December 2009, Sir David King of the University of Oxford proposed a global cap-and-trade scheme to equitably cut CO_2 emissions. The goal is to reduce those emissions (or the equivalent in other greenhouse gases) so that the atmospheric concentration of CO_2 is no higher than about 450 ppmv by 2050. Reaching this goal would require reducing global annual emissions from today's 30 gigatons to 18 gigatons by mid century. Based on the projected growth in human population, this would limit each person to an allowance of 2 tons of CO_2 per year. Individual nations would come up with a plan to meet this target. King's idea is to permit China and India to temporarily increase their carbon emissions prior to decreasing to the target level. Also the poorest developing nations where CO_2 emissions currently are under 2 tons per person per year, could market their excess carbon emission credits and use the proceeds to install low-carbon or no-carbon alternate energy sources.

COPENHAGEN CLIMATE CHANGE CONFERENCE

In December 2009, climate scientists, policymakers, and others representing 193 nations convened at the 15th UN Climate Change Conference in Copenhagen, Denmark. As a follow-up to the Kyoto Protocol, attendees aimed to draft a new international treaty limiting global warming primarily by cutting greenhouse gas emissions. Negotiations focused on what would happen after the Kyoto agreement expires in 2012. The goal was an equitable agreement that would be effective for both developing and developed nations as well as for future generations. A major stumbling block in this and prior negotiations was the conflicting interests of developed versus less-developed nations. Developed nations are responsible for most of the increase in greenhouse gas emissions thus far, whereas the emission rates in less-developed nations have been low in the past but are expected to grow at a relatively rapid pace in the future.

Attendees agreed on the *Copenhagen Accord*, a framework for an international treaty to counter global warming, with a goal of limiting the rise in temperature to 2 Celsius degrees (3.6 Fahrenheit degrees) above the pre-industrial level by 2100. The Accord calls for each nation to report on their plans for voluntary emissions-reduction targets from 2010 to 2020 (with individual nations allowed to conduct their own auditing). Developed nations will provide about $30 billion during 2010-2012 for climate change mitigation and adaptation, with this figure increasing to $100 billion per year by 2020. Also, the developed nations will provide $3.5 billion to tropical nations to slow the destruction of forests. However, the Accord is not legally binding, does not specify emissions-reduction targets for developed nations for 2050, and fails to deal with major developing nations. The next step in the negotiating process will take place in Mexico in December 2010 at a meeting of the UN Framework Convention on Climate Change.

Climate Change Policy

Policy- and decision-making regarding anthropogenic climate change follow from the potential impacts on human society and affect societal development (Figure 15.3). Hence, climate policy choices require policymakers

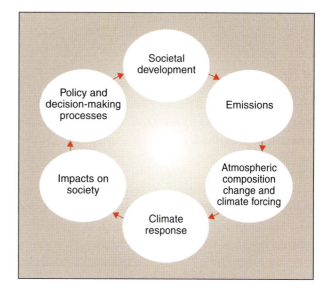

FIGURE 15.3
This framework portrays the essential relationships among climate science, policy- and decision-making, societal development, and impacts analyses. [After V. Ramaswamy, 2009. "Anthropogenic Climate Change in Asia: Key Challenges." *EOS, Transactions*, American Geophysical Union 90(49):469.]

to become more informed about the societal risks posed by greenhouse gas emissions. A key challenge is to translate existing knowledge into sound societal decisions directed at managing climate risks. For example, climate sensitivity is an essential understanding in formulating climate change policy. **Climate sensitivity**, first defined in the 1970s by Stephen Schneider of Stanford University, refers to the equilibrium change in global mean annual surface temperature caused by an increment in downward infrared radiative flux that would result from sustained doubling of atmospheric CO_2 concentration compared to its preindustrial level.

In formulating public policy aimed at curbing anthropogenic climate change, many people advocate adherence to the precautionary principle. Furthermore, the long-term goal of sustainability should underlie public policy on climate change.

PRECAUTIONARY PRINCIPLE

Development of an effective climate change policy is likely to involve adherence to the **precautionary principle**: "When an action causes a threat to human health or the environment, precautionary measures should be taken even if some cause-and-effect relationships are not fully established scientifically." The 1992 United Nations Conference on Environment and Development held in Rio de Janeiro, Brazil, adopted the precautionary principle. It was also included in the Rio Declaration's "Agenda 21," which was adopted by 178 nations in June 1992, and later ratified by the United States. The precautionary principle helps to overcome the enormous barriers to action posed by the inevitable scientific uncertainty about cause-effect relationships in complex systems. This is especially important in Earth's climate system where many sub-systems and processes, and their interactions are not completely understood.

CLIMATE AND SUSTAINABILITY

The United Nations World Commission on Environment and Development defines **sustainability** as "developments that meet the needs of the present without compromising the ability of future generations to meet their own needs." Achieving sustainability calls for balancing environmental issues with social and economic concerns. Sustainability requires intergenerational equity in managing resources and the environment; that is, we must consider the needs of future generations as well as our own.

There is an ethical dimension to sustainability as the concept relates to anthropogenic climate change. Every day each of us benefits from activities that emit greenhouse gases into the atmosphere. Driving an auto, typing at the computer, or using anything whose fabrication, transport, or use consumes energy produces greenhouse gases, mostly CO_2. The benefits we experience ultimately may contribute to climate change and more frequent weather extremes (e.g., floods, excessive heat, or drought) that will bring harm or even death to future generations. Most of us would probably agree that it is morally unacceptable for one person to benefit at someone else's expense. On the other hand, the costs of mitigation and adaptation that are borne by the present generation must be compared to the benefits of higher quality lives enjoyed by future generations.

Climate Policy Making at the National Level

Within the scientific community is a growing sense of ur-gency that we need to take substantive action if we are to curb anthropogenic climate change. However, this sense of urgency does not appear to extend to the general population. Most Americans and the majority of people in other nations are aware of climate change and favor action to deal with the problem. Yet polls indicate that among people's major concerns, climate change ranks well be-hind issues such as the economy, health care reform, ter-

rorism, and war. The majority oppose cuts in greenhouse gas emissions if it means higher energy prices. In a Gallup poll released in March 2009, 41% of Americans surveyed believed that the seriousness of global warming is exag-gerated. (This is a record high response in the 12 years that Gallup has been asking this question.) In September 2009, a poll conducted by the Pew Research Center for the People and Press found that the proportion of Americans who "think there is solid evidence that the average tem-perature on Earth has been getting warmer over the past few decades" was 57%, down from 71% in April 2008.

Many people argue that uncertainties associated with the risks of climate change do not justify the tremendous cost of remedial action. A 2007 survey of Americans revealed that 54% of respondents favored a cautious wait-and-see approach to reductions in greenhouse gas emissions. (Even larger majorities felt the same way in the developing nations of China and India.) However, this view ignores the nature of Earth's climate system. As we have seen from earlier in this book, this complex dynamical system is characterized by positive and negative feedback loops, nonlinear behavior, and inertia. Hence, for example, substantial delays exist between implementation of greenhouse gas abatement strategies and the response of ice sheets and sea level. As emphasized in Chapter 14, we have already committed to serious consequences of anthropogenic climate change. Geoscientist Michael Oppenheimer of Princeton University contends that human activity could emit a sufficient amount of greenhouse gases into the atmosphere during this century to "guarantee" the melting of the Greenland ice sheet only centuries from now.

Calls for action on anthropogenic climate change (or any other environmental problem) are usually directed at the government. These calls derive from the belief that only governments have the political authority and access to financial resources to formulate public policy that would effectively address the problem. (**Public policies** are defined to be those actions that governments take or decide not to take.)

POLITICAL RESPONSE

Under ideal circumstances, we would expect that public policy on climate change should address the root cause of the problem and reflect a holistic perspective. The root cause of anthropogenic climate change may be traced to the widely accepted practice of resource consumption, especially in the Western world. Ultimately, responsibility for anthropogenic climate change rests with a society that expects an ever rising standard of living that is powered mostly by fossil fuels. Policymakers who accept this explanation realize that fundamental shifts in life styles and existing public policy are necessary to solve the problem.

A **holistic perspective** considers a problem in its entirety. As discussed in Chapter 1, the various components of Earth's climate system are interrelated. Hence, when one component is disturbed (e.g., shrinking glaciers), other parts are affected (e.g., rising sea level, coastal flooding). The holistic perspective calls on government agencies that have climate and climate change in their purview to coordinate and integrate their activities for greater efficiency and synergy, and to avoid conflicting or contradictory policies.

Addressing the root cause of the problem and maintaining a holistic perspective may be reasonable expectations, but the government has difficulty doing this. Policymakers are continually subject to competing pressures brought by various interest groups as well as individual citizens who view the problem from the perspective of different values and goals. In democracies especially, formulation of public policy depends not only on the beliefs of scientific experts but also the public. The government cannot satisfy all interests, so that public policy inevitably favors some groups at the expense of others. Groups that enjoy certain advantages want to retain their privileges and oppose efforts to alter the status quo. It is up to the groups who want to change the status quo to persuade policymakers that existing policies are unsatisfactory and that change is necessary.

Although the climate science community overwhelmingly agrees that anthropogenic climate change poses a serious threat to society and should be remedied (Figure 15.4), this consensus does not necessarily extend to policymakers (e.g., elected officials, senior bureaucrats, and other public servants). For policymakers to address the problem, it must be recognized as a public problem and compete for a place on policymakers' agenda that is already overloaded with other issues that demand their full attention. Policymakers prefer to deal with old, familiar and recurring issues such as health care, foreign policy, national security, and the economy. Any new issue that gains a position on the government's already full agenda must replace another issue that another group had justified as deserving a place on that agenda. A change in agenda means that a group loses its advantageous position, which explains why some groups oppose any attempt to alter the agenda and why change occurs so slowly. Nonetheless, agendas do change. In 2009, the U.S. Congress began to seriously address anthropogenic climate change.

FIGURE 15.4
In a hearing on December 2, 2009, the House Select Committee on Energy Independence and Global Warming explored with climate scientists from the Obama administration the urgent, consensus view that global warming is real, and the science indicates that it is getting worse. In the picture, Dr. Jane Lubchenco, NOAA Administrator demonstrates the effects of carbon dioxide in water, which, on a global scale produces ocean acidification. [Courtesy of the US House of Representatives]

Issues that are most likely to be added to an agenda tend to have at least some of the following attributes: The issue is easy to understand, is the result of a crisis or catastrophe, has a relatively inexpensive solution, attracts widespread public attention, and spurs demands for action. Policymakers tend to avoid or neglect consideration of complex problems that are not readily understood, and are potentially costly to solve. This helps to explain the relatively slow pace at which U.S. policymakers have thus far responded to anthropogenic climate change.

What happens when a policymaker decides that a problem requires a response? As noted above, a holistic approach to anthropogenic climate change is desirable from a scientific perspective. However, policymakers (especially elected officials) rarely adopt a holistic perspective. Often decisions are based largely on considerations that have little to do with the issue at hand. These factors include partisan politics, elections, and the basis of government representation.

Terms of office dictate that most state legislators and all members of the U.S. House of Representatives stand for re-election every two years. The President and most governors have 4-year terms while U.S. senators serve 6-year terms. Politicians who want to be re-elected must continually demonstrate to their constituents those accomplishments that reflect favorably on them. Because terms of office are relatively short, many legislators throw their support behind laws that promise quick results. In addition, many politicians favor programs that benefit the present generation of voters and avoid programs that take many years to produce results or benefit future generations. As noted in Chapter 14, the benefits of climate change mitigation operate on a time scale of a century or so whereas the benefits of climate change adaptation may be realized on a decadal scale or even sooner. Hence, legislation directed at adaptation is probably more politically expedient than legislation directed at mitigation.

All state and federal legislators are elected to represent specific geographical districts. To be re-elected, a legislator must be responsive to the needs and interests of the majority of his or her voting constituents. Consequently, legislators can ignore the concerns of all other districts or even the concerns of the nation as a whole. For example, climate change legislation may result in net economic losses in some districts (e.g., closing of an outmoded coal mine) and net gains in other districts (e.g., new climate neutral energy technology). This parochial view of issues is often at the expense of the much more important regional, national, and international perspectives.

INCREMENTAL DECISION MAKING

Public policy is formulated through **incremental decision making**, a process whereby small changes are made in existing policies or programs. The basic assumption is that existing approaches to public problems are satisfactory. Most laws and public policies are actually little more than small revisions of existing practices implying that the best guide to the future can be found in what has already been done. When new problems arise, public officials usually assume that existing responses to similar problems in the past can be applied. Hence, as noted earlier in this chapter, our successful response to the stratospheric ozone depletion problem to some extent can serve as a model for the way we deal with anthropogenic climate change.

This approach appeals to policymakers because it makes fewer demands on them. It enables elected officials to be responsive to public opinion and to engage in bargaining and compromise. Compromise permits legislators to respond to competing demands and develop policies that are acceptable to as many powerful and well-organized interest groups as possible.

Climate Change: The Economic Perspective

In view of the alternative and its implications for life on Earth, most of us probably favor a stable climate. However, a stable climate does not appear likely unless we are willing to alter our lifestyles (reducing our carbon footprint) and pay for measures that directly or indirectly reduce emissions of greenhouse gases into the atmosphere. We must also decide how much protection we really need from the greater intensity and higher frequency of weather extremes (e.g., tropical cyclones, drought) expected to accompany global warming. The more protection we demand (or the less risk we are willing to tolerate), the higher is the cost. For example, along the Gulf and Southeast Atlantic coast, protection might take the form of shoring up infrastructure and constructing beach barriers at great expense. For another example, refer to this chapter's second Essay.

Solving the anthropogenic climate change problem has major economic dimensions. Today, economic growth depends on a continually increasing input of energy derived mostly from fossil fuels. Continued reliance on current technologies cannot substantially reduce greenhouse gas emissions and, at the same time, allow the global economy to expand. In fact, reducing emissions without innovative technologies will inhibit economic growth. Examples of these innovative low-emissions technologies include renewable energy resources (e.g., solar-thermal power, wind power), emissions control devices (e.g., carbon capture and storage), and more efficient motor vehicles (e.g., plug-in hybrid autos).

Economists are interested in the cost of curbing anthropogenic climate change in the context of their fundamental concern regarding society's use of resources. Economists assume that resources such as money and raw materials are scarce, that each of us acts to maximize our self-interest, and that we want to obtain the greatest possible benefit for ourselves. Hence, one goal of economists is to determine how society's resources can be allocated most efficiently among competing demands. An **efficient economy** is operating when resources are used in the best possible way to satisfy the largest number of consumer demands. An **inefficient economy** is operating when resources are not used to their greatest advantage, so that if they were more productively employed, they could be used to satisfy more demands.

A company that uses 1000 units of energy to manufacture a product is considered to be inefficient if a change in technology would enable the company to produce the same product using only 800 units of energy. If the energy source were fossil fuels (e.g., electricity from a coal-fired power plant), such inefficiency translates into more greenhouse gas emissions. No nation's economy is entirely efficient and hence, all societies have opportunities to upgrade their economic efficiency and their use of scarce resources.

Solving the anthropogenic climate change problem calls for improving economic efficiency. Economists are interested knowing how the people, corporations, or industries that contribute to the problem can be induced into changing their behavior to improve efficiency. The range of possible strategies that can be used to bring about change is broad, but the initial issue has to do with the relative roles of market forces versus government regulation.

FREE MARKETS

According to many economists, free markets are inherently efficient. In a **free market**, resources are allocated and prices established based upon individual voluntary exchange among producers and consumers. In a free market system, we obtain what we are willing and able to pay for, and markets adjust to accommodate changing consumer preferences. By this view, supply, demand, and price tend toward a state of equilibrium.

Some critics of the free market approach argue that many economists idealize the role played by markets in the efficient allocation of a nation's resources. For markets to operate efficiently, a large number of wise buyers and sellers must have complete information about the price, quality, and availability of all products. Such markets do not exist. For example, it would be desirable if we all knew the carbon footprint associated with the manufacture, use, and disposal of the products we use every day. Such information would better enable us to make climate neutral choices among the various products and services available in the marketplace.

Critics also point out that not all of the consequences of economic activity are reflected in market transactions. These un-reflected consequences are known as **externalities** and are either positive or negative. When we are advantaged by what someone else does, we reap external benefits and enjoy *positive externalities*. For example, if several people in the neighborhood landscape their yards and paint and refurbish their homes, all residents of the neighborhood benefit in terms of higher property values even though they did not help pay for the improvements. In another example of a positive externality, a local manufacturer replaces its fleet of gasoline-powered delivery vans with plug-in hybrid vehicles. By reducing emissions of greenhouse gases, benefits accrue to the entire community including those people who would never directly or indirectly use the services or products of the firm.

On the other hand, *negative externalities* refer to the cost of undesirable side effects of some activity that must be borne by people who are not directly involved in that activity. We bear the negative consequences of someone else's actions and the external cost. Negative externalities lead to what the late ecologist Garrett J. Hardin (1915-2003) described as the *tragedy of the commons*. The commons are the resources shared by all of us, such as the atmosphere, a river, or the ocean. These resources directly or indirectly contribute to our personal well-being. But when an individual degrades the commons for personal (or collective) gain, that gain accrues exclusively to the individual while the degradation affects all who use the commons. As noted by geophysicist David Archer in his book, *The Long Thaw*, "Individuals profit from releasing CO_2, but everyone collectively pays the price."

Market forces offer both advantages and disadvantages in solving environmental problems, but few economists favor relying on them exclusively. More popular are mixed-market economies. A **mixed-market economy** describes a system that combines private, competitive enterprise with some government involvement. This system is used in the U.S.

Cap-and-trade is a greenhouse gas abatement program that involves some government regulation to ensure integrity and the orderly operation of the free market where carbon credits are exchanged.

GOVERNMENT REGULATION

An alternative to a completely free market system is either partial or total government regulation. Instead of relying exclusively on market forces to affect behavior, government regulation imposes command-and-control techniques and specifies exactly what the private sector is obliged to do. For example, to help reduce CO_2 emissions from the transportation sector, the government could impose mandatory fuel efficiency standards for motor vehicles.

Government regulation is the most common method used to change the behavior of those whose actions adversely affect the quality of the environment. But despite widespread reliance on government regulation, many people oppose government regulation, including many economists. These critics argue that government regulation—because it does not rely on the profit motive—discourages efficiency, innovation, and improved productivity. Furthermore, according to many who are subject to it, government regulation on the national level too often ignores the cost of compliance and the variations that exist among different industries and regions of the country.

Given the drawbacks of free markets and government regulation, economists often recommend problem-solving approaches that combine government intervention with economic incentives. To cut greenhouse gas emissions, such an approach could include government-funded subsidies for development of low-carbon or no-carbon energy sources.

ANALYTIC TOOLS

As noted earlier, economists are interested in improving the allocation of a nation's resources. Hence, they want to know how much of a change in behavior is necessary or how large an expenditure is desirable to control greenhouse gas emissions in terms of economic efficiency. To answer these questions, economists use several analytical tools including cost-benefit analysis, cost-effectiveness analysis, and risk-benefit analysis.

Theoretically at least, **cost-benefit analysis** compares all the gains or benefits of a project with all the corresponding losses or costs of that project. If benefits exceed costs by a significant margin, a project is usually considered desirable and worth pursuing. A controversial aspect of cost-benefit analysis is the assumption that everything, such as a human life, has a monetary value that can be measured.

With cost-benefit analysis, care must be taken to be as inclusive as possible, that is, to inventory all costs and all benefits of an action. For example, replacing a traditional coal-fired electric power plant with a wind farm not only cuts greenhouse gas emissions but also benefits human health by reducing emissions of tiny particulates into the atmosphere. When inhaled, these particulates can contribute to respiratory distress in humans.

Cost-effectiveness analysis makes no judgment regarding the desirability of a project. Instead, given a particular goal, an economist tries to determine how the goal can be achieved for the lowest cost. Different technologies may produce the same results but one of those technologies may do so at the lowest price. With cap-and-trade, limits are set on the quantity of allowable carbon emissions but emitters are free to buy and sell permits (credits) so that the cap is achieved at the least cost.

Risk-benefit analysis weighs the risks of an activity against its benefits to determine whether the activity should be allowed. Risk-benefit analysis appeals to people who are involved in environmental decision-making. Most human activities that emit heat-trapping gases into the atmosphere also produce benefits that we enjoy daily. However, the risks associated with anthropogenic climate change can outweigh these benefits.

Although economists employ many other analytic tools, those described above provide an initial framework for understanding how economists analyze environmental issues including anthropogenic climate change. Decisions should be made rationally with the goal of making the best use of our resources. How likely we are to reach this goal depends on our choice of decision-making processes and how responsive the political system is to both scientific and economic concerns regarding climate change.

National Climate Service

In cities, towns and villages around the world, policymakers/decision-makers face the often daunting challenge of how to select and implement the most appropriate and effective climate change mitigation and adaptation strategies. This is also true for policymakers at the national level. Their goal is to write legislation that balances the environmental and economic impacts of specific greenhouse gas emissions reduction strategies. For that, they require credible information. In an editorial published in the 30 October 2009 issue of *Science*, Eric J. Barron, then Director of the National Center for Atmospheric Research (NCAR) in Boulder, CO, observed that the U.S. currently does not have "a deliberate approach to generating the environmental intelligence needed to support good decisions" regarding climate change. The same can be said of most other nations. Many decision-makers simply do not know what climate information is available, where it is located, how to access it, or its limitations.

Over the past 40 years or so, considerable effort has gone into developing predictive climate models with reasonably good success. However, little progress has been made in simulating how ecosystems, water resources, agriculture, human health, and energy respond to climate change. Such climate impact models are essential for shaping appropriate climate mitigation and adaptation plans. Barron notes that human vulnerability depends to a large extent on how climate change affects the occurrence of severe weather systems (e.g., hurricanes). Furthermore, climate information is most useful for water, forest, and agriculture resource management if made available on a regional scale.

Barron and many other climate scientists recommend creation of a **National Climate Service** in the U.S. that would function as an authoritative source of intelligence on climate change in support of decision-making regarding mitigation and adaptation. The National Climate Service (NCS) would parallel the National Weather Service (NWS) and, like the NWS, would be part of NOAA. The NCS would synthesize climate research from many federal agencies and oversee expanded research in high-resolution regional climate models. It would offer to decision-makers and other stakeholders many short-term products such as drought forecasts and seasonal climate outlooks.

Conclusions

There is now strong consensus in the scientific community that global warming is real, humankind's demand for energy is at the heart of anthropogenic climate change, future climate change will continue to have significant impacts on societies and ecosystems through the 21st century and beyond, and the need for remedial action is urgent. An international framework exists for the community of nations to enact protocols designed to stabilize concentrations of greenhouse gases in the atmosphere. But the process is slowed by persistent conflicts of interest between the developed nations such as the U.S. and the rapidly industrializing nations such as China and India. Anthropogenic climate change is a global problem whose solution calls for equitable treatment of all nations and all people.

The concluding paragraph of an Information Statement on Climate Change, issued by the American Meteorological Society on 1 February 2007, succinctly summarizes the role of public policy in climate change. "Policy choices in the near future will determine the extent of the impacts of climate change. Policy decisions are seldom made in a context of absolute certainty. Some continued climate change is inevitable, and the policy debate should also consider the best ways to adapt to climate change. Prudence dictates extreme care in managing our relationship with the only planet known to be capable of sustaining human life."

Basic Understandings

- We have reached the stage in the scientific assessment of climate change where the cumulative evidence is now convincing that global warming is real and unequivocal, and human activity is primarily responsible for warming since the mid-20th century.

- Humankind's successful intervention in the global-scale problem of stratospheric ozone depletion encourages scientists and policymakers that similarly effective action also can be taken to curb anthropogenic climate change.

- The Montreal Protocol of 1987 initiated an international-scale process and established a framework for action to deal with ozone depleting substances that could be strengthened with time. International negotiations regarding limits on greenhouse gas emissions are following a similar model of incremental evolution.

- The 1992 United Nations Framework Convention on Climate Change (UNFCCC) called for "stabilization of greenhouse gas concentrations in the atmosphere at a level that would prevent dangerous anthropogenic interference with the climate system."

- In the initial update to the UNFCCC treaty, representatives of the global community met in Kyoto, Japan, in December 1997 to formulate an acceptable and effective international treaty limiting greenhouse gas emissions worldwide. As a key provision of the Kyoto Protocol, 37 developed (industrial) nations (referred to as *Annex I* countries) agreed to reduce emissions of four greenhouse gases (CO_2, CH_4, N_2O, and sulfur hexafluoride) plus two groups of gases produced by them (hydrofluorocarbons and perfluorocarbons).

- Although a signatory to the UNFCCC treaty, the United States essentially ignored the Kyoto Protocol claiming that the cost of compliance would be too great or politically inconvenient. Opposition to cuts in emissions also came from developing nations such as China and India where rapid economic growth requires a continual increase in energy supply, most of which comes from fossil fuel combustion.

- To deal with the developing nations and their rejection of greenhouse gas emissions caps, the European Union implemented the clean development mechanism (CDM), that is, carbon trading via offset exchange.

- Of the two market mechanisms for reducing greenhouse gas emissions, cap-and-trade is generally favored over emission fees (taxes). With cap-and-trade, a limit is set on the quantity of allowable emissions but emitters are free to buy and sell permits (credits) so that the cap is achieved at the lowest cost. With emission fees, policymakers set the price sources must pay for every ton they emit.

- With the growing consensus of scientific opinion that anthropogenic climate change represents a serious threat to society and human well-being, the U.S. appeared more ready and willing to set targets for reduction of greenhouse gas emissions. Although President Barack Obama advocates targeted reductions in CO_2 emissions (primarily via cap-and-trade), his efforts so far have been stymied by continuing partisanship in Washington, DC, polls that say the public is less convinced of the threat of global warming and more concerned about the economy and possible adverse impacts of remediation on employment opportunities. As of this writing, the upcoming mid-term congressional elections make it less likely that the U.S. will agree any time soon to a legally-binding treaty to cut CO_2 levels to specific levels.

- The precautionary principle helps to overcome the barrier to action posed by scientific uncertainty, especially when dealing with complex systems such as Earth's climate system.

- Sustainability requires intergenerational equity when designing solutions to the problem of anthropogenic climate change.

- Many people argue that uncertainties associated with the risks of climate change do not justify the tremendous cost of remedial action. They advocate a wait-and-see approach. However, this view ignores the long lifetime of CO_2 in the atmosphere and the complex dynamical nature of Earth's climate system.

- Ideally, public policy on climate change should address the root cause of the problem and reflect a holistic perspective.

- For policymakers to address anthropogenic climate change, it must be recognized as a public problem and compete for a place on policymakers' agenda that is already overloaded with issues demanding their full legislative attention.

- Issues that are most likely to be added to an agenda tend to have at least some of the following attributes: easy to understand, the result of a crisis or catastrophe, has an inexpensive solution, attracts widespread public attention, and spurs calls for action.

- Often policymakers make decisions based on considerations that have little to do with the problem at hand, that is, partisan politics, elections, and the basis of representation in the government.

- Politicians tend to favor legislation that promises quick results for the present generation of voters. Hence, legislation directed at short-term climate change adaptation is probably more politically expedient than legislation directed at long-term climate change mitigation.

- Public policy is formulated through a process known as incremental decision making whereby small changes are made in existing policies or programs.

- We must decide on how much protection we really need from predicted increases in the intensity and frequency of weather extremes. The more protection we demand (or the less risk we are willing to tolerate), the higher is the cost.

- Given the drawbacks of free markets and government regulation, economists often recommend problem-solving approaches that combine government intervention with economic incentives.

- Economists want to know how much of a change in behavior is necessary or how large an expenditure is desirable to control greenhouse gas emissions in terms of economic efficiency. To answer these questions, they employ many analytical tools including cost-benefit analysis, cost-effectiveness analysis, and risk-benefit analysis.

- A proposed National Climate Service in the U.S. would serve as an authoritative source of intelligence on climate change in support of decision-making regarding mitigation and adaptation.

Enduring Ideas

- The cumulative evidence is now convincing that global warming is real, and human activity culminating in the enhancement of Earth's greenhouse effect is responsible for warming since about the mid-20th century.
- Although a signatory to the UNFCCC treaty, the U.S. essentially ignored the Kyoto Protocol underscoring the fundamental differences between the developed and developing nations.
- Sustainability calls for intergenerational equity when designing solutions to the problem of anthropogenic climate change.
- The long lifetime of CO_2 in the atmosphere argues for immediate efforts to cut CO_2 emissions.
- We must decide on how much protection we really need from anticipated increases in the intensity and frequency of weather extremes. The less risk we are willing to tolerate, the higher is the cost.
- Adding a price to greenhouse gas emissions is widely viewed as a key component for cutting emissions and reducing the threat of climate change. The two principal market approaches are cap-and-trade and emissions fees.

Review

1. What are two major differences between protecting the stratospheric ozone shield and curbing anthropogenic climate change?
2. Summarize the principal objective of the 1992 UN Framework Convention on Climate Change.
3. Why did the United States oppose the Kyoto Protocol?
4. How does the European Union (EU) seek to regulate greenhouse gas emissions by sources within its member states?
5. What is the significance of the precautionary principle as it applies to anthropogenic climate change?
6. For action on anthropogenic climate change or any other environmental problem, the public usually looks to the government. Explain why.
7. Identify the attributes of issues most likely to be added to the policymakers' agenda for action.
8. Terms of office for elected officials are relatively short. How does this influence the types of issues that legislators are most likely to address?
9. What is meant by incremental decision making?
10. What would be the primary purpose of a National Climate Service?

Critical Thinking

1. Explain how the clean development mechanism (CDM) is designed to reduce greenhouse gas emissions in developing nations.
2. Why are U.S. politicians more likely to favor cap-and-trade over taxation as a mechanism to regulate greenhouse gas emissions?
3. How does sustainability apply to future generations?
4. Surveys indicate that the sense of urgency in the scientific community regarding the need to cut greenhouse gas emissions does not appear to carry over into the general population. Explain why.
5. What is the value of adopting a holistic perspective in solving environmental problems such as anthropogenic climate change?
6. What is the "root cause" of anthropogenic climate change?
7. Explain why a legislator is more likely to focus his or her efforts on legislation that is directed at climate change adaptation rather than climate change mitigation.
8. Why are rapidly developing nations less likely than developed nations to agree to mandatory reductions in greenhouse gas emissions?
9. List the advantages and disadvantages if India were to invest in nuclear power plants rather than coal-fired power plants to meet future energy needs.
10. Why might different countries adopt different strategies to curb anthropogenic climate change?

ESSAY: Some Implications of an Open Arctic Ocean

As discussed in Chapter 11, the ice cover on the Arctic Ocean is shrinking in response to climate change. The situation is exacerbated by polar amplification and ice-albedo feedback. Recall that *polar amplification* refers to the increase in warming with increasing latitude. Warming in the Arctic region so far has been at roughly twice the rate of warming in middle latitudes. The *ice-albedo feedback* refers to the lower albedo of open water compared to ice and snow. With melting of pack ice, the area of open water increases, and positive feedback accelerates the melting. Through climate change, the Arctic Ocean is in the process of transitioning from a perpetual ice-covered state to a seasonally (summer) ice-free state.

Already the Arctic Ocean is sufficiently ice-free that its waters have become much more navigable in recent years. In 2009, the minimum ice cover was the third lowest on record. In September of that year, two German ships completed the first commercial crossing of the Arctic Ocean as a segment of their journey from South Korea to Siberia and then on to Rotterdam. The ships followed the Northern Sea Route past Russia and through Canada's Northwest Passage (Figure 1). By traversing the Arctic Ocean rather than the Suez Canal, the ships trimmed the distance by one-third at a considerable cost savings. Beluga Shipping of Bremen, Germany, announced plans for up to six vessels to make the same trip in 2010.

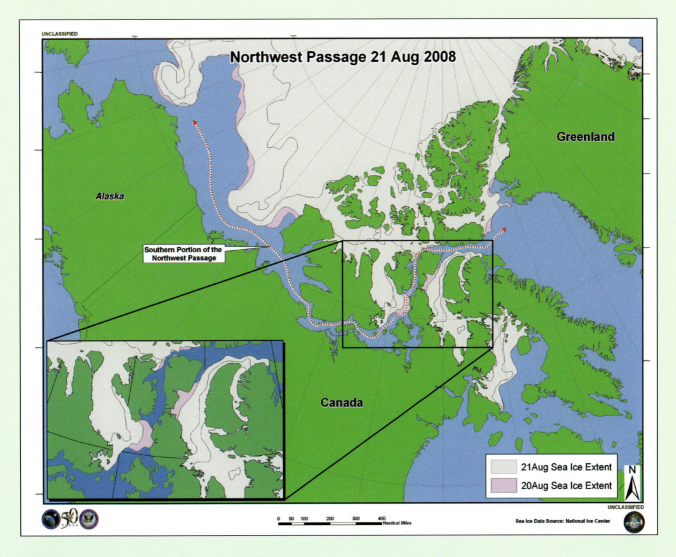

FIGURE 1
Canada's Northwest Passage was ice-free on 21 August 2008, the third year in a row. Shrinkage of the Arctic sea-ice cover has important ramifications for climate, weather, ecosystems, and navigation. [Courtesy National Ice Center/NOAA]

The opening of the Arctic Ocean to summer navigation not only is heightening prospects for increased shipping activity in the Arctic Region, it also has spurred interest from nations around the world in Arctic fishing, tourism, and extraction of energy resources. (The Arctic Region contains an estimated 20% of Earth's energy resources.) Potential jurisdictional conflicts, however, are also increasing military interest in the Arctic.

Unlike Antarctica, no international treaties at present protect the Arctic from completing demands. With regard to governance, the entire Arctic Basin is subject to the 1982 United Nations Convention on the Law of the Sea (LOSC). The LOSC applies to all Arctic rim states (Canada, Denmark, Norway, and Russia) except for the United States. The U.S. accepts specific provisions of the LOSC as "customary international law." All five rim states are reinforcing their jurisdiction seaward from the coastline to include the seabed beyond the Exclusive Economic Zone (EEZ) to the plate boundary. Representatives of all five rim states also serve on the intergovernmental forum of the Arctic Council. Although lacking any regulatory authority, the Arctic Council has produced policy-relevant scientific assessments of the Arctic. An example is the *2004 Arctic Climate Impact Assessment*. With increasing international activity in the Arctic Ocean, made possible by climate change, regulations or codes are needed that cover all types of shipping and marine pollution, regulates tourism to protect against ecological disturbance, and supports research on biophysical and socio-economic systems as a key part of ecosystem-based management (EBM).

Some scientists and policy-makers suggest treating the central Arctic (considered part of the high seas and beyond national jurisdiction) as *international space*, similar to Antarctica (since 1959), drawing a clear distinction between the sea floor and the overlying water column. Both Arctic and non-Arctic states would participate in governance of the central Arctic in a cooperative decision-making process under the auspices of the LOSC and customary international law and dedicated to peaceful uses. The successful stewardship of Antarctica has demonstrated that many nations can work together to maintain sound oversight of a large international region. The same needs to be done in the climatically sensitive Arctic Region.

ESSAY: Protecting Venice, Italy, from Sea-Level Rise

Many of the world's most populous cities are located along the coast, and efforts to protect them against flooding from storm-generated surges are becoming more challenging as mean sea level (msl) continues to rise in response to global climate change. In many instances, the problem is not new as demonstrated by what has happened to Venice, a city in northern Italy situated on a cluster of 120 salt-marsh islands in a large coastal lagoon at the head of the Adriatic Sea (Figure 1). Bridges connect the islands and a causeway and ferries link Venice to the mainland. A long narrow barrier island separates the lagoon from the sea (Gulf of Venice) except for three major tidal inlets. At one time, the sea and the lagoon protected the city from foreign invaders. In fact, tradition has it that Venice was founded in CE 452 by people from northern Italian cities who fled to the islands seeking protection from invading Teutonic tribes. Today, however, the encroaching sea threatens Venice's art and architectural treasures and perhaps the city's continued existence.

FIGURE 1
ASTER satellite image of Venice, Italy, located in the saltwater lagoon between the mainland and the barrier islands. This image was acquired on 9 December 2001. ASTER (Advanced Spaceborne Thermal Emission and Reflection Radiometer) is flown on NASA's Terra satellite orbiting Earth at an altitude of 705 km (430 mi). [Courtesy of NASA/GSFC/METI/ERSDAC/JAROS, and U.S./Japan ASTER Science Team]

Slow subsidence of Venice combined with rising sea level has increased the frequency of flooding. Venice has been sinking slowly under its own weight for centuries. From the 1930s to 1970s, withdrawals of large quantities of groundwater for industrial use exacerbated the problem. Groundwater occupies the tiny pore spaces in sediment and as the water is pumped out, the sediment compacts and the ground subsides. Land subsidence is also occurring because of oxidation of organic-rich soils in fertile farmland that was reclaimed from marshland at the southern end of the Venice Lagoon. A drainage system that keeps the soil relatively dry promotes oxidation and consequent loss of soil mass. Reclamation of marshland began in the late 19th century and continued into the late 1930s. Cumulative land subsidence since reclamation is up to 2 m (6.5 ft) and is currently taking place at a rate of 1.5 to 2 cm per year. According to estimates by the *Venice Organic Soil Subsidence (VOSS)* project, in 50 years oxidation could remove all the organic soil with additional land subsidence of 75 to 100 cm (30 to 40 in.).

Ten major floods have inundated Venice over the past 70 years and smaller floods are now much more frequent. During the often stormy months of November through January, some areas of the city are under water almost every day. Portions of Venice, including the historic Piazza San Marco, are flooded about 100 days per year; a century ago, the flood frequency was about 7 days per year. Flooding forces businesses to close and inconveniences residents and tourists alike but more significantly, the salt water corrodes the brick underpinnings of historic churches and other buildings along the Grand Canal and elsewhere in the city.

To protect Venice from future flooding, the Italian government embarked on a $7 billion project to construct 78 floodgates at the three tidal inlets linking the Venice lagoon to the Adriatic Sea. Individual gates measure up to 5 m (16 ft) thick, 20 m (66 ft) wide, and 27 m (89 ft) in length. Most of the time, the hollow gates will be filled with seawater and lie horizontally within a foundation (caisson) on the sea floor so as not to impede tidal currents or navigation. But when a storm surge is predicted to elevate the tide to a height of 110 cm (43 in.) or more, the floodgate system will be activated. (For perspective, the water level reached 194 cm, or 6.4 ft, during a disastrous flood in November 1966.) Compressed air will be pumped into the gates expelling the water. The gates are hinged at one end so that the free end can swing to the surface and block the flow of floodwaters through the inlets. This system is designed to protect Venice from a 60-cm (24-in.) rise in mean sea level and a 3-m (10-ft) flood. Locks (already completed) will allow passage of ocean-going container-cargo and cruise ships as well as the fishing fleet when the gates are up. Similar flood-control structures have been built on the Thames River in England to protect London and at the mouths of estuaries to protect the low-lying Dutch coastal plain.

Although the idea for the Venice floodgates dates back to at least the mid 1980s, it was not until May 2003 that a stone-laying ceremony signaled the beginning of construction on the *Moses Project* (known in Italian as *Mose* for Modulo Sperimentale Elettromeccanico). As of this writing, the project is slated for completion in 2012. However, environmental protests have slowed construction. Opponents of the Venice floodgate scheme argue that closing the inlets will cause substantial environmental damage, especially to the lagoon ecosystem. Most of Venice's sewage empties untreated into the many canals dissecting the city or directly into the lagoon. Normally tidal currents dilute and transport these wastes out to sea but if the gates are closed for extended periods, wastes will accumulate in the city's canals or in the lagoon. Some scientists speculate that the gates could be closed for as many as 50 days per year.

APPENDIX I

CONVERSION FACTORS

	Multiply	*By*	*To obtain*
LENGTH	inches (in.)	2.54	centimeters (cm)
	centimeters	0.3937	inches
	feet (ft)	0.3048	meters (m)
	meters	3.281	feet
	statute miles (mi)	1.6093	kilometers (km)
	statute miles (mi)	0.869	nautical miles
	kilometers	0.6214	statute miles
	kilometers	0.54	nautical miles
	nautical miles	1.852	kilometers
	kilometers	3281	feet
	feet	0.0003048	kilometers
SPEED	miles per hour (mph)	1.6093	kilometers per hour (kph)
	miles per hour	0.869	knots (kts)
	miles per hour	0.447	meters per second (m/s)
	knots	1.1508	miles per hour
	knots	0.5144	meters per second
	kilometers per hour	0.6214	miles per hour
	kilometers per hour	0.540	knots
	meters per second	1.938	knots
	meters per second	2.237	miles per hour
WEIGHTS AND MASS	ounces (oz)	28.35	grams (g)
	grams	0.0353	ounces
	pounds (lb)	0.4536	kilograms (kg)
	kilograms	2.205	pounds
	tons	0.9072	metric tons
	metric tons	1.102	tons
LIQUID MEASURE	fluid ounces (fl oz)	0.0296	liters (L)
	gallons (gal)	3.785	liters
	liters	0.2642	gallons
	liters	33.814	fluid ounces
AREA	acres (A)	0.4047	hectares (ha)
	square yards (yd^2)	0.8361	square meters (m^2)
	square miles (mi^2)	2.590	square kilometers (km^2)
	hectares	0.010	square kilometers
	hectares	2.471	acres

PRESSURE/ FORCE	pounds force (lb)	4.448	newtons (N)
	newtons	0.2248	pounds
	millimeters of mercury at 0 °C	133.32	pascals (Pa; N per m^2)
	pounds per square inch (psi)	6.895	kilopascals (kPa; 1000 pascals)
	pascals	0.0075	millimeters of mercury at 0 °C
	kilopascals	0.1450	pounds per square inch
	bars	1000	millibars (mb)
	bars	100000	pascals
	bars	0.9869	atmospheres (atm)
ENERGY	joules (J)	0.2389	calories (cal)
	kilocalories (kcal)	1000	calories
	joules	1.0	watt-seconds (W-sec)
	kilojoules (kJ)	1000	joules
	calories	0.00397	Btu (British thermal units)
	Btu	252	calories
POWER	joules per second	1.0	watts (W)
	kilowatts (kW)	1000	watts
	megawatts (MW)	1000	kilowatts
	kilocalories per minute	69.78	watts
	watts	0.00134	horsepower (hp)
	kilowatts	56.87	Btu per minute
	Btu per minute	0.236	horsepower

APPENDIX II

MILESTONES IN THE HISTORY OF CLIMATE SCIENCE

ca. 2950 BCE	Earliest portion of the astronomical calculator at Stonehenge in southern England.
ca. 525 BCE	Greek philosopher Anaximenes of Miletus proposed that winds, clouds, rain and hail are formed by thickening of air, the primary substance.
ca. 500 BCE	Greek philosopher Parmenides classified world climates by latitude as torrid, temperate, or frigid.
ca. 400 BCE	Rainfall measured in India using the first known rain gauge.
ca. 400 BCE	Hippocrates (ca. BCE 460-370) authored the first climatography, *On Airs, Waters, and Places,* a treatise on climate and medicine.
ca. 350 BCE	Aristotle produced his *Meteorologica,* the first work on the atmospheric sciences.
ca. 300 BCE	Theophrastus (ca. 371-287 BCE), a student of Aristotle, wrote about how agricultural activity could alter the climate.
ca. 28 BCE	Chinese astronomers first observed sunspots with the unaided eye by viewing the Sun's reflection on the surface of a quiet pond at sunrise and sunset.
1287	First snow gauges in China.
1442	Rain gauges used in Korea.
1450	Cardinal Nicholas of Cusa (1400-1464) constructed the first balance hygrometer.
1592	Galileo Galilei (1564-1642) invented the thermoscope, forerunner of the thermometer.
1600	Violent eruption of the volcano Huaynaputina in southern Peru; affected climate in various parts of the world for one or two years.
1610	Galileo Galilei was among the first to study sunspots using a telescope.
1639	First mention of a rain gauge in European literature.
1641	Ferdinand II of Tuscany (1610-1670) constructs a thermometer containing liquid.
1643	Evangelista Torricelli (1608-1647) invented the mercury barometer.
1644-45	Rev. John Campanius (1601-1683) made the first systematic weather observations in America, near the present site of Wilmington, DE.
1648	Florin Périer ascended the Puy-de-Dôme in France, and with Blaise Pascal (1623-1662) demonstrated that air pressure decreases with increasing altitude.
1666	Sir Isaac Newton (1643-1727) used a glass prism to demonstrate that sunlight is made up of a spectrum of colors.

ca. 1670 — Mercury was used in thermometers for the first time.

1683 — The English astronomer Edmund Halley (1656-1742) published the first comprehensive map of winds along with a partial explanation of the trade winds.

1686 — Edmund Halley proposed an explanation for monsoons.

1687 — Isaac Newton developed his three laws of motion.

1687 — French physician Guillaume Amontons (1663-1705) invented a hygrometer.

1709 — German physicist Gabriel Daniel Fahrenheit (1686-1736) constructed a thermometer with alcohol as the working fluid.

1714 — Gabriel Daniel Fahrenheit introduced the Fahrenheit temperature scale.

1730 — French entomologist and physicist René Réamur (1683-1757) introduced a thermometer using a water/alcohol mixture.

1735 — German explorer Johann G. Gmelin (1709-1755) discovered permafrost in Siberia.

1735 — George Hadley (1685-1768) proposed the Hadley cell circulation.

1738 — Long-term instrument-based temperature record began in Charleston, SC, in 1738.

1742 — Swedish astronomer Anders Celsius (1701-1744) introduced the Celsius temperature scale.

1743 — Jean Pierre Christin (1707-1778), inverted the fixed points on the Celsius temperature scale, producing the scale used today.

1743 — Benjamin Franklin (1706-1790) deduced the progressive movement of a storm along the U.S. East Coast.

1749 — In Glasgow, Scotland, Alexander Wilson used thermometers attached to a kite to obtain the first temperature profile of the lower atmosphere (up to perhaps 60 m or 200 ft).

1752 — Benjamin Franklin performed his famous kite experiment demonstrating the electrical nature of lightning.

1760 — Scottish chemist Joseph Black (1728-1799) formulated the concept of specific heat.

1778 — Thomas Jefferson and James Madison took the nation's first simultaneous weather observations.

1781 — Systematic weather observations began at New Haven, CT.

1783 — Laki fissure volcanic eruption in southern Iceland; proposed by Benjamin Franklin (1706-1790) to have caused the severe winter of 1783-84 in Europe.

1800 — Astronomer Sir William Herschel (1738-1822) discovered energy transfer in the infrared (IR).

1802-03 — British pharmacist and amateur meteorologist Luke Howard (1772-1864) developed his classification of cloud types.

1802 — British chemist John Dalton (1766-1844) defined *relative humidity*.

1804 — J. L. Gay-Lussac and Jean Biot conducted the first manned balloon exploration of the atmosphere, reaching a maximum altitude of 7000 m (23,000 ft).

1805 — British admiral Sir Francis Beaufort (1774-1857) proposed a scale of winds (now known as the *Beaufort scale*).

1811 — Italian chemist Amedeo Avogadro (1776-1856) first stated Avogadro's law.

1814 — U.S. Surgeon General James Tilton ordered the Army Medical Corps to begin a diary of weather conditions at army posts.

1815	Violent eruption of the Indonesian volcano Tambora; may have been responsible for the 1816 "year without a summer."
1819	German physicist Heinrich Wilhelm Brandes (1777-1834) drew the first weather map, depicting a storm over the English Channel.
1820	U.S. Army Major Stephen Long described the western plains as the *Great American Desert*.
1820	English chemist John Frederic Daniell (1790-1845) invented the dew-point hygrometer.
1821	Swiss geologist Ignatz Venetz (1788-1857) proposed that glaciers formerly occurred throughout Europe.
1822	French mathematician Jean Baptiste Joseph Fourier (1768-1830) made reference to the greenhouse effect in his book *Théorie analytique de la chaleur*.
1825	E.F. August developed the psychrometer.
1835	Gaspard-Gustav de Coriolis (1792-1843) demonstrated quantitatively how Earth's rotation affects large-scale motion in the atmosphere or ocean.
1837	Swiss naturalist Louis Agassiz (1807-1873) used the term *Eiszeit* (Ice Age) for the first time; he championed the glacial theory during the mid 1800s.
1837	James P. Espy (1785-1860) published the first U.S. weather map.
1838	British Navy adopted the Beaufort scale for estimating wind speed from the state of the sea.
1840	Agassiz publishes his *Etudes des glaciers* (*Studies on Glaciers*).
1842	Austrian physicist Johann Christian Doppler (1803-1853) first explained the Doppler Effect.
1842	French mathematician Joseph Alphonse Adhémar (1797-1862) proposed that regular variations in Earth's orbit explained climate fluctuations during the Ice Age.
1843	German astronomer Samuel Heinrich Schwabe (1789-1875) reported regular variations in sunspot activity (the sunspot cycle).
1844	Aneroid barometer invented.
1853	James Coffin suggested that there are three distinct wind zones in the Northern Hemisphere.
1853	First International Meteorological Conference, held in Brussels.
1856	American meteorologist William Ferrel (1817-1891) proposed a model of the general circulation of the atmosphere consisting of three cells.
1857	Lorin Blodget published *Climatology of the United States*.
1859	John Tyndall (1820-1893), an Irish physical scientist, conducted the first experiments on the radiative properties of gases and established the experimental basis for the greenhouse effect.
1862	James Glaisher and Henry Coxwell set manned balloon altitude record of 9000 m (29,500 ft) over Wolverhampton, England.
1864	Scottish scientist James Croll (1821-1890) advanced study on the astronomical theory of the Ice Age.
1869	In Cincinnati, OH Cleveland Abbe (1838-1916) prepared the first regular weather maps for part of the United States.
1870s	Beginning of widespread use of instrument-shelters in North America.
1870	President Ulysses S. Grant (1822-1885) signed a Congressional resolution into law establishing a national weather

service operated by the U.S. Army Signal Corps.

1871   Cleveland Abbe appointed chief meteorologist of the new national weather service.

1874   Based on field studies, American geologist Thomas Chrowder Chamberlin (1843-1928) suggested that there were several Ice Ages separated by nonglacial epochs.

1874   Hurricane plotted for the first time on a surface weather map (offshore near Savannah, GA).

1878   The International Meteorological Organization (IMO) was founded.

1880s   American geologist Thomas C. Chamberlin (1843-1928) proposed multiple glaciations during the Ice Age.

1880s   German astronomer F.W. Gustav Spörer (1822-1895) and English solar astronomer E. Walter Maunder (1851-1928) reported a 70-year period (1645-1715) of greatly reduced sunspot activity (now referred to as the Maunder minimum).

1881   Lord Rayleigh (1842-1919) demonstrated that scattering of visible light by gas molecules is inversely proportional to the fourth power of the wavelength.

1884   Ludwig Boltzmann (1844-1906) derived the Stefan-Boltzmann law.

1884   French chemist and engineer Henri-Louis Le Chatelier (1850-1936) first proposed what is now called Le Chatelier's Principle.

1888   Gustavus Hinrichs (1836-1923), an Iowa weather researcher, first used the word *derecho* for a straight-line windstorm.

1891   Luis Carranza, President of the Lima Geographical Society, describes a countercurrent flowing north to south from Paita to Pacasmayo and called El Niño by the local fishermen.

1891   The national weather service was transferred from military to civilian control in a new Weather Bureau within the U.S. Department of Agriculture, with a special mandate to provide weather and climate guidance for farmers.

1893   British astronomer E. Walter Maunder (1851-1928) identified a period of low solar activity between 1645 and 1715 (now known as the *Maunder minimum*).

1894   Wilhelm Wien (1864-1928) developed his *displacement law of radiation*.

1894   First sounding of the atmosphere using a self-recording thermometer attached to a kite at Blue Hill Observatory, Milton, MA.

1895   Svante A. Arrhenius (1859-1927), a Swedish electrochemist, proposed that increasing levels of atmospheric carbon dioxide could alter Earth's heat budget and temperature.

1900   Russian-born German climatologist Wladimir Peter Köppen (1846-1940) first published his original climate classification scheme.

1902   French meteorologist Léon Philippe Teisserenc de Bort (1855-1913) identified and named the troposphere and stratosphere as layers of the atmosphere.

1904   Vilhelm Bjerknes (1862-1951) laid the foundations of synoptic meteorology at the Bergen School of Meteorology in Norway.

1905   Swedish physicist V. Walfrid Ekman (1874-1954) first described mathematically the coupling of surface winds and surface ocean waters (the *Ekman spiral*).

1910   German meteorologist Alfred Wegener (1880-1930) and American geologist Frank B. Taylor (1860-1939) independently proposed the concept of continental drift.

1910   On 5 May, a train of 10 instrumented kites reached an altitude of 7265 m (23,835 ft).

1916   Pyranometer for measuring global radiation developed.

1917   Vilhelm Bjerknes formulated polar front theory.

1920s   Serbian astronomer Milutin Milankovitch (1879-1958) revived the theory of long-term cyclic climate change based on regular changes in Earth-sun geometry (*Milankovitch cycles*); he calculated latitudinal and seasonal changes in incoming solar radiation.

1920   Theory of atmospheric front was developed by V. Bjerknes, H. Solberg, and J. Bjerknes.

1924   Sir Gilbert Walker (1868-1958) discovered the Southern Oscillation, a key to understanding El Niño and La Niña.

1928   Tor Bergeron (1891-1971) developed a climate classification system based on air mass frequency.

1928   Chlorofluorocarbons (CFCs) first synthesized.

ca. 1928   First radiosonde developed.

1929   Robert H. Goddard (1882-1945) conducted the first rocket probe of the atmosphere.

1933   Tor Bergeron (1891-1977) published his paper on *Physics of Clouds and Precipitation*.

1933   The U.S. Weather Bureau closed its last operating weather kite station at Ellendale, ND.

1935   Radar invented.

1935   Selection of the 30-year period 1901-1930 as the basis for calculating climatic normals, at the International Meteorological Conference in Warsaw, Poland.

1937   First official U.S. Weather Bureau radiosonde launch, at East Boston, MA.

1937   Andrew E. Douglass (1867-1962) founded the Laboratory of Tree-Ring Research at the University of Arizona.

1938   The Callendar Effect, the theory that global climate change can be brought about by enhancement of the greenhouse effect caused by increasing levels of atmospheric carbon dioxide from anthropogenic sources, was articulated by the noted British steam engineer, Guy Stewart Callendar (1898-1964).

1938   Milutin Milankovitch (1879-1958) published radiation curves for different latitudes extending back 600,000 years based on regular changes in Earth-Sun geometry (the Milankovitch cycles).

1939   Work by Walter Findeisen supported Bergeron's theory of precipitation development in cold clouds (later known as the *Bergeron-Findeisen process*).

1941   Radar applied to weather systems for the first time; used to track a thunderstorm on the south coast of England.

1944   Hurd Curtis Willett produced cross sections of the atmosphere showing the jet stream.

1946   Vincent J. Schaefer (1906-1993) and Irving Langmuir (1881-1957) performed the first cloud-seeding experiments.

1946   John von Neumann (1903-1957) and colleagues began mathematical modeling of the atmosphere.

1946   U.S. Weather Bureau established the first River Forecast Centers at Cincinnati, OH, and Kansas City, MO.

1947   American chemist Willard F. Libby (1908-1980) invented the radiocarbon dating technique, a valuable tool in reconstructing late-glacial climates.

1947   First photographs of the Earth's cloud cover obtained by a V2 rocket at altitudes of 110 to 165 km (70 to 100 mi).

1950   A team led by the Hungarian-born American mathematician John von Neumann produced the first computer-generated weather forecasts using the ENIAC computer.

1951   The World Meteorological Organization (WMO) commenced operation.

1951   The American geographer and climatologist Arthur N. Strahler introduced a climate classification scheme based on air masses.

1955   Beginning of the era when electronic computers were routinely generating weather forecasts from surface and upper-air weather observations.

1956   Physicist Gilbert N. Plass (1920-2004) authored an article in *American Scientist* that helped revive interest in the carbon dioxide theory of climate change.

1957   Atmospheric CO_2 measurements began at Mauna Loa Observatory, Hawaii, and the South Pole station of the U.S. Antarctic Program.

1959   A temperature-humidity index first introduced by the U.S. Weather Bureau.

1959   Satellite surveillance of Earth's climate system began on 13 October with the U.S. launch of *Explorer7*. An on board radiometer provided the first measurements of Earth's radiation budget from space.

1960   The United States successfully orbited the first weather satellite, TIROS-I.

1961   American meteorologist Edward N. Lorenz (1917-2008) observed that computer predictions of weather are highly sensitive to small differences in initial conditions, an aspect of chaos theory.

1963   Edward N. Lorenz applied chaos theory to meteorology.

1964   U.S. government introduced the first Clean Air Act to set national air pollution standards.

1965   British climatologist Hubert H. Lamb (1913-1997) was the first to characterize the High Medieval (CE 1100 to 1200) as an episode of relatively mild winters and warm dry summers in Western Europe.

1966   Jacob Bjerknes (1897-1975) demonstrated a relationship between El Niño and the Southern Oscillation.

1967   Verner E. Suomi (1915-1995) of the University of Wisconsin-Madison processed the first geostationary satellite image.

1970s   American engineer Herbert Saffir (1917-2007) and meteorologist Robert Simpson (1912-) developed their hurricane intensity scale (the *Saffir-Simpson scale*).

1971   Japanese-born American meteorologist Tetsuya T. Fujita (1920-1998) with Allen Pearson introduced their tornado intensity scale (*Fujita* or *Fujita-Pearson scale*).

1972   Earth Resources Technology Satellite 1 (later called Landsat-1) was launched.

1974   The Global Atmospheric Research Program (GARP) conducted the GARP Atlantic Tropical Experiment (GATE) to improve understanding of the tropical atmosphere.

1974   M. J. Molina and F. S. Rowland warned of the CFC threat to the stratospheric ozone layer.

1974   T. T. Fujita discovered *downbursts* and coined the term.

1974   Russian climatologist Mikhail I. Budyko (1920-2001) was one of the first scientists to propose injecting sulfur dioxide (SO_2) into the stratosphere to bring about large-scale cooling at Earth's surface.

1975   First GOES (Geostationary Operational Environmental Satellite) was launched.

1975   The European Space Agency was formed.

1975   The Saffir-Simpson Hurricane Intensity Scale became operational.

1976   American astronomer John A. Eddy (1931-2009) linked periods of minimum sunspot activity to relatively cold phases of the Little Ice Age.

1976   In December, scientists announced that based on their analysis of deep-sea sediment cores, major changes in Ice Age climate occurred at essentially the same frequency as variations in the eccentricity, obliquity, and precession of Earth's orbit.

1977   The first Meteosat weather satellite was launched by the European Space Agency.

1979   R.G. Steadman developed the heat index (apparent temperature index).

1979   NOAA scientists developed the SLOSH model that predicts the location and height of a storm surge.

1980s   Creation of the United States Historical Climatology Network (USHCN).

1985   The British Antarctic Survey first reported a drastic decline in stratospheric ozone over Antarctica during the Southern Hemisphere spring.

1985   The Vienna Convention for the Protection of the Ozone Layer was convened.

1985   The Tropical Ocean-Global Atmosphere Program (TOGA) began.

1987   United Nations Environmental Programme (UNEP) drafted the *Montreal Protocol on Substances That Deplete the Ozone Layer*, subsequently ratified by all 192 UN member states.

1987   *The Joint Global Ocean Flux Study (JGOFS)* initiated to study the fluxes of carbon in the ocean.

1988   Soviet and French scientists reported on their analysis of a 2200 m (7200 ft) ice core extracted at Vostok station on the East Antarctic ice sheet, spanning 160,000 years.

1988   German geologist Hartmut Heinrich described layers of relatively coarse rock fragments (called Heinrich layers) on the floor of the deep North Atlantic Ocean; these sediments likely were released by melting icebergs.

1988   The Global Energy and Water Cycle Experiment (GEWEX) launched to examine the global hydrologic cycle.

1988   Formation of the Intergovernmental Panel on Climate Change (IPCC) by the World Meteorological Organization and the UN Environmental Programme.

1988   During a lecture at Woods Hole (MA) Oceanographic Institution, oceanographer John H. Martin (1935-1993) declared, "Give me a half tanker of iron, and I will give you an ice age."

1990   First Doppler radar was introduced into meteorological service.

1991   On 15-16 June, the violent eruption of Mount Pinatubo in the Philippines likely was responsible for cooling that temporarily interrupted the post-1970s global warming trend.

1992   *UN Framework Convention on Climate Change* was ratified.

1992   UN Conference on Environment and Development in Rio de Janeiro, Brazil, adopted the precautionary principle.

1992	Launch of the *TOPEX/Poseidon satellite*; provided the first continuous global coverage of ocean surface topography (sea level) at 10-day intervals.
1993	Beginning of a series of ocean iron fertilization (OIF) experiments.
1994	In December, the ENSO Observing System was fully operational.
1997	In December, representatives of the global community met in Kyoto, Japan, to formulate an acceptable and effective international treaty limiting greenhouse gas emissions worldwide. Notably absent from the list of signatories to the Kyoto Protocol was the United States.
1997	In November, launch of the joint U.S.-Japanese *Tropical Rainfall Measuring Mission (TRMM)* satellite for remote sensing of precipitation between 40 degrees S and 40 degrees N.
1999	Landsat-7 was launched.
2001	NWS and the Meteorological Services of Canada (MSC) implemented a major revision of the wind-chill index.
2001	Beginning of the *African Monsoon Multidisciplinary Analysis (AMMA)*, a 10-year field experiment to upgrade forecasting of the West African monsoon.
2003	Launch of NASA's *Ice, Cloud and Land Elevation Satellite (ICESat)*, the world's first laser-altimeter satellite, an important source of data on the Antarctic and Greenland ice sheets; operational until October 2009.
2004	New Ultraviolet (UV) Index introduced.
2005	The European Union (EU) adopted a cap-and-trade system to regulate greenhouse gas emissions among its member states.
2007	Enhanced F-Scale (EF-Scale) became operational for rating tornadoes.
2007	The *IPCC Fourth Assessment Report* concluded that global warming since the mid 20th century very likely was caused by human activity.
2007	By 1 November, some 3000 Argo profiling floats had been deployed over the world ocean for the purpose of monitoring the climate of the ocean's wind-driven surface layer and underlying pycnocline.
2007	The extent of end-of-summer (September) sea ice cover in the Arctic was the lowest since satellite monitoring of the polar region began in 1979.
2009	Beginning of NASA's *Operation Ice Bridge*, a 5-year airborne survey of ice sheets, ice shelves, and sea ice.
2009	Beginning of a five-year demonstration of carbon capture and storage (CCS) technology at the Mountaineer coal-fired electric power plant in West Virginia.
2009	In December, representatives of 193 nations convened at the 15th UN Climate Change Conference in Copenhagen, Denmark, as a follow-up to the Kyoto Protocol.

GLOSSARY

ablation—All processes that contribute to a loss of glacial ice mass (e.g., melting, sublimation).

absolute instability—If the sounding indicates that the *temperature* of the ambient air were dropping more rapidly with altitude than the dry adiabatic lapse rate (that is, more than 9.8 Celsius degrees per 1000 m), then the ambient air would be unstable for both saturated and unsaturated air parcels.

absolute stability—An air layer is stable for both saturated and unsaturated air parcels when the sounding indicates any of the following conditions: (1) the *temperature* of the ambient air drops more slowly with altitude than the moist adiabatic lapse rate; (2) the temperature does not change with altitude (isothermal); (3) the temperature increases with altitude (temperature inversion).

absolute zero—Theoretical *temperature* at which all molecular motion ceases and a body does not emit *electromagnetic radiation*.

absorption—An *energy* conversion process whereby some of the radiation striking an object is converted to *heat* energy.

acoustic thermometry—Technique using variations in the speed of sound between many transmitters and receivers combined to obtain a three-dimensional representation of seawater *temperature*.

adaptation—Steps taken over the next decade or so that make communities more *climate-resilient* so that society can more effectively cope with the inevitable harmful impacts of *climate change* while taking advantage of beneficial impacts.

adiabatic process—As an air parcel ascends or descends within the atmosphere, no *heat* is exchanged between the air parcel and the surrounding (ambient) air.

advection fog—Ground-level cloud generated by the cooling of a mild, humid *air mass* (to its *dewpoint*) as it travels over a relatively cool surface.

aerosols—Tiny (nanometer to micrometer in size) liquid or solid particles of various compositions that occur suspended in the *atmosphere*.

agroclimatic compensation—When poor growing *weather* and consequent reduced crop yields in one area are compensated to some extent by better growing weather and increased crop yields elsewhere. This generally applies to crops such as corn, soybeans, and other grains that are grown over broad geographical areas.

air mass—A huge volume of air covering thousands of square kilometers that is relatively uniform horizontally in *temperature* and *humidity*. The specific characteristics of an air mass depend on the type of surface over which the air mass forms (its *source region*) and travels.

air mass advection—Refers to the movement of an air mass from one locality to another. With advection, one air mass replaces another air mass having different *temperature* and/or *humidity* characteristics.

air mass climatology—Method of describing *climate* in which the basis is the frequency with which various types of *air masses* develop over, or are advected into a locality.

air mass modification—Changes in the *temperature, humidity,* and/or *stability* of an air mass as it travels away from its source region. Occurs primarily by exchange of *heat*, moisture, or both, with the surface over which the air mass travels; *radiational heating* or cooling; or adiabatic heating or cooling associated with large-scale vertical air motion.

air pressure—The weight per unit area of the column of air above any given location on Earth's surface.

air pressure gradient—Change in *air pressure* with distance.

albedo—The fraction or percent of radiation striking a surface (or interface) that is reflected by that surface (or interface); usually applied to the reflectivity of an object to visible radiation.

Alliance of Small Island States (AOSIS)—An intergovernmental organization of low-lying coastal and small island nations. AOSIS operates as an ad hoc lobbying and negotiating presence in the United Nations (UN) engaging in consultation and consensus building.

Antarctic Circle—The latitude circle of 66 degrees 33 minutes S.

Antarctic ozone hole—Thinning of the stratospheric ozone layer over the continent to levels significantly below pre-1979 levels; typically develops annually between late August and early October and generally ends in mid-November.

anticyclone—A uniform *air mass* that exerts relatively high surface pressure compared with surrounding air; same as a *High*. Viewed from above, surface *wind*s in an anticyclone blow clockwise and outward in the Northern Hemisphere but counterclockwise and outward in the Southern Hemisphere.

aphelion—Time of the year when Earth is farthest (152 million km or 94 million mi) from the Sun (currently about 4 July).

apparent temperature—A measure of the combined effect of high air *temperature* and high *relative humidity* on human comfort and well-being. It can be used to describe what hot *weather* "feels like" to the average person for various combinations of temperature and relative humidity.

arctic air (A)—Exceptionally cold and dry *air mass* that forms over the *snow* or ice covered regions of Siberia, the Arctic Basin, Greenland, and the northern interior of North America north of about 60 degrees N, in much the same way as *continental polar* air, but in a region that receives very little solar radiation in winter.

Arctic Circle—The latitude circle of 66 degrees 33 minutes N.

arctic high—A *cold-core anticyclone* that originates in a source region for *arctic air*; the product of extreme *radiational cooling* over the *snow*-covered continental interior of North America well north of the *polar front*.

Arctic Oscillation (AO)—A seesaw variation in *air pressure* that occurs between the North Pole and middle latitudes. Changes in the horizontal *air pressure gradient* alter the speed of horizontal *wind*s in the polar vortex.

aspect—The direction faced by a sloping surface and is important in the study of *microclimates* in that it affects the intensity of solar radiation incident on Earth's surface.

atmosphere—A relatively thin envelope of gases and tiny suspended particles surrounding the planet.

atmospheric boundary layer—The atmospheric zone to which frictional resistance (*eddy viscosity*) is essentially confined; on average, the zone extends from Earth's surface to an altitude of about 1000 m (3300 ft).

atmospheric stability—Determined by comparing the *temperature* change of an ascending or descending air parcel with the temperature profile, or sounding, of the ambient air through which the parcel ascends or descends.

atmospheric window—Range of *wavelength*s over which little or no radiation is absorbed.

Automated Surface Observing System (ASOS)—Consists of electronic meteorological sensors, computers, and fully automated communications ports for monitoring the state of the atmosphere.

B

back-door cold front—A *cold front* that propagates southward or southwestward along the North Atlantic coast east of the Appalachian Mountains. These *front*s occur most frequently in the summer and fall and usher in Canadian *continental polar air* or Atlantic *maritime polar air*.

Bergeron classification—Climate classification that uses three letters to identify the properties of an air mass in its source region, developed in 1928 by Tor Bergeron.

Bergeron process—*Precipitation* formation in *cold cloud*s whereby ice crystals grow at the expense of supercooled water droplets in response to differences in *vapor pressure* relative to water and ice surfaces; also known as the ice-crystal process.

biogeochemical cycles—Pathways along which solids, liquids, and gases move among the various reservoirs on Earth, often involving physical or chemical changes to these substances.

biosphere—The portion of the planet that includes all living plants and animals.

blackbody—A hypothetical "body" that absorbs all *electromagnetic radiation* that is incident on it at every *wavelength* and emits all radiation at every wavelength; no radiation is reflected or transmitted. The "body" must be large compared to the wavelength of incident radiation.

blocking system—A cutoff low or a cutoff high that prevents the usual west-to-east movement of *weather system*s. Because a blocking circulation pattern tends to persist for extended periods (often several weeks or longer), extremes of *weather* such as drought or flooding *rain*s or excessive *heat* or cold can result.

bora—A cold *katabatic wind* that originates on the high plateau region of Croatia and cascades onto the narrow Dalmatian coastal plain along the Adriatic Sea; a winter phenomenon.

Bowen ratio—For a moist surface, the ratio of *heat energy* used for *sensible heating* to the heat energy used for *latent heating*. The Bowen ratio varies from one locality to another depending on the amount of surface moisture, ranging from 0.1 for the ocean surface to about 5.0 for deserts; negative values are also possible.

brine rejection—Salts are excluded from the ice structure as seawater freezes and the remaining unfrozen water becomes saltier and therefore freezes at still lower temperatures.

British thermal unit (Btu)—Quantity of *heat* needed to raise the *temperature* of 1 pound (lb) of water 1 Fahrenheit degree (technically, from 62 °F to 63 °F).

buffer—Substance that stabilizes a chemical *system*.

C

Callendar effect—Theory that global *climate change* can be brought about by enhancement of Earth's natural *greenhouse effect* by increased levels of atmospheric carbon dioxide from anthropogenic sources, principally the burning of *fossil fuels*; named for the British steam engineer Guy Stewart Callendar (1898-1964) who investigated the link between global warming and fossil fuel combustion in the 1930s.

calorie (cal)—The amount of *heat* needed to raise the *temperature* of 1 gram of water 1 Celsius degree (technically, from 14.5 °C to 15.5 °C).

cap-and-trade—Quotas or caps are imposed on each CO_2 source. Sources whose emissions are below the quota earn tradable credits that can be marketed to sources that have exceeded their quotas.

capping inversion—An elevated *temperature* inversion layer that caps a convective boundary layer, keeping the convective elements from rising higher into the *atmosphere*.

carbon capture and storage (CCS)—Climate *mitigation* strategy designed to minimize anthropogenic carbon dioxide emissions into the *atmosphere* using well-established technologies. CO_2 emissions are captured, liquefied and then stored in subsurface geological formations.

cellular respiration—Process by which an organism processes food and liberates *energy* for maintenance, growth, and reproduction; also releasing carbon dioxide, water, and *heat* energy to the environment.

centripetal force—Inward-directed force that acts on an object moving in a curved path, confining the object to the curved path; the result of other forces.

chinook wind—Relatively warm and dry *wind*, which develops when air aloft is adiabatically compressed as it is drawn down the leeward slopes of a mountain range.

chromosphere—Outward from the *photosphere* of the Sun; consists of ions of hydrogen and helium at 4000 °C to 40,000 °C (7200 °F to 72,000 °F).

cirriform cloud—A thin wispy or fibrous *cloud* composed of ice crystals.

clean development mechanism (CDM)—A provision of the *Kyoto Protocol* implemented by the European Union to accommodate developing nations and their rejection of emission caps; consists of carbon trading via *offset exchange*.

climate—The state of the *atmosphere* at some locality averaged over a specified time period plus extremes in weather observed during the same period or during the entire period of record. Climate must be specified for a particular place and time interval because, like *weather*, climate varies both spatially and temporally.

climate change—Any systematic fluctuation in the long-term statistics of climatic elements (e.g., *temperature*, pressure, or *wind*) that is sustained over several decades or longer.

Climate Prediction Center (CPC)—Component of NOAA that "assesses and forecasts the impacts of short-term *climate variability*, emphasizing enhanced risks of *weather*-related extreme events for use in mitigating losses and maximizing economic gain."

climate science—Systematic study of the mean state of the *atmosphere* at a specified location and time period as governed by natural laws.

climate sensitivity—Refers to the equilibrium change in global mean annual surface temperature caused by an increment in downward infrared radiative flux that would result from sustained doubling of atmospheric CO_2 concentration compared to its preindustrial level.

climate system—The totality of Earth's *atmosphere*, *hydrosphere* (including the *cryosphere*), biosphere and geosphere and their interactions.

climate variability—The variations in *climate* around a mean state and other statistics (e.g., standard deviation, variance, or statistics of extremes) of the climate on all time and space scales beyond that of individual *weather* events.

climatic anomalies—Departures from long-term climatic averages.

climatic norm—Equated to the average value of some climatic element such as *temperature* or *precipitation*.

climatology—The study of *climate*, its controls and spatial and temporal variability.

cloud—A visible suspension of minute water droplets and/or ice crystals in the *atmosphere* above Earth's surface. Clouds form in the free atmosphere primarily as a result of *condensation* or *deposition* of water vapor in ascending air that nears saturation.

cloud condensation nuclei (CCN)—Tiny solid and liquid particles that promote the *condensation* of water vapor at *temperatures* both above and below the freezing point of water.

cloud forest—Tropical forests perpetually shrouded in clouds and mist.

coastal downwelling—Downward motion of warm surface waters along the coast caused by *Ekman transport* directed onshore.

coastal upwelling—Upward motion of cold, nutrient-rich deep water along the coast caused by *Ekman transport* directed offshore.

cold air advection—Occurs when the *wind* transports

colder air into a previously warmer area. On a weather map, cold air advection is indicated by winds blowing across regional *isotherms* from a colder area to a warmer area. Cold air advection occurs behind a *cold front*.

cold cloud—*Cloud* with *temperature* at or below 0 °C (32 °F).

cold front—*Front* that moves in such a way that relatively cold (more dense) air advances and replaces relatively warm (less dense) air.

cold-core anticyclone—A dome of continental polar (*cP*) or arctic (*A*) air; product of extreme *radiational cooling* over the often *snow*-covered continental interior of North America well north of the *polar front*. They are shallow *system*s in which the clockwise circulation weakens with altitude and frequently reverses direction aloft.

cold-core cyclone—An occluded *extratropical cyclone* in which the lowest *temperature*s occur throughout the column of air above the low-pressure center. Furthermore, a cyclonic circulation prevails throughout the *troposphere* and is most intense at high altitudes.

collision-coalescence process—Growth of *cloud* droplets into raindrops within a *warm cloud*; droplets merge upon impact. This process takes place in a cloud made up of droplets of different sizes; larger droplets with higher terminal velocity overtake, and then collide and coalesce with smaller droplets in their paths.

comma cloud—A large-scale comma-shaped *cloud* pattern associated with a mature *extratropical cyclone*; reflects the strengthening of the *system*'s circulation.

compressional warming—A rise in *temperature* rise that accompanies an increase in pressure on a gas or mixture of gases.

condensation—Process whereby more water molecules return to the water surface as liquid than enter the *atmosphere* as vapor; a net gain of liquid water mass results.

conditional stability—An air layer that is stable for unsaturated air parcels, but unstable for saturated air parcels.

conduction (of heat)—Transfer of kinetic *energy* of atoms or molecules via collisions between neighboring atoms or molecules.

confluence zone—A boundary between *air masses*; equivalent to an average frontal boundary. As a rule, confluence zones separate distinctly different types of *climate*.

consumers—An organism that is unable to manufacture its food from nonliving materials, but is dependent on the energy stored in other living things.—

continental climate—At the same latitude, summers are cooler and winters are milder in *maritime climates* than in continental climates.

continental polar air (cP)—Dry *air mass*es that develop over the northern interior of North America. In winter, it is typically very cold because the ground in its source region is often *snow* covered, daylight is short, solar radiation is weak, and *radiational cooling* is extreme. In summer, when the snow-free source region warms in response to extended hours of bright sunshine, *cP* air is mild and pleasant.

continental tropical air (cT)—A hot and dry air mass that develops over the subtropical deserts of Mexico and the American Southwest primarily in summer.

convection—The transport of *heat* within a fluid via motions of the fluid itself; generally occurs only in liquids or gases.

cooling degree-days—Computed only for days when the average outdoor air *temperature* is higher than 65 °F (although higher base temperatures are sometimes used). Each degree of average temperature above 65 °F is counted as one cooling degree-day.

coral bleaching—The whitening of coral colonies due to the loss of symbiotic zooxanthellae from the tissues of coral polyps; may be triggered by a rise in seawater *temperature*.

coral reef—A calcareous reef in relatively shallow, tropical seas composed of a thin veneer of living coral growing on older layers of dead coral or volcanic rock.

Coriolis Effect—A deflective force arising from the rotation of the Earth on its axis; affects principally synoptic-scale and planetary-scale *wind*s. Winds are deflected to the right of their initial direction in the Northern Hemisphere and to the left in the Southern Hemisphere. Magnitude depends on latitude and speed of the moving object.

cost-benefit analysis—All the gains or benefits of a project are compared with all the corresponding losses or costs of that project. If benefits exceed costs by a significant margin, a project is usually considered desirable and worth pursuing.

cost-effectiveness analysis—Given a particular goal, an economist tries to determine how the goal can be achieved for the lowest cost. Different technologies may produce the same results but one of those technologies may do so at the lowest price.

cryosphere—The frozen portion of the *hydrosphere*; encompasses massive continental (glacial) ice sheets, much smaller ice caps and mountain *glacier*s, *permafrost*, and the pack ice and icebergs floating at sea.

cumuliform cloud—Heaped or puffy *cloud* that exhibits significant vertical development due to convection.

cumulonimbus cloud—A *thunderstorm* cloud with considerable vertical development and characteristic anvil top; produces *precipitation*, *lightning*, and *thunder*.

cumulus congestus cloud—A *cumuliform cloud* showing significant vertical growth; resembles a huge cauliflower.

cup anemometer—An instrument used to monitor *wind* speed consisting of 3 or 4 open hemispheric cups mounted to spin horizontally on a vertical shaft. At least one open cup faces the wind at any time. The rotation rate of the cups is calibrated to read in m per sec, km per hr, or knots.

cycling rate—The amount of material transferred from one reservoir to another within a specified period of time.

cyclogenesis—The birth and development of an *extratropical cyclone*; usually occurs along the *polar front*, directly under an area of strong horizontal divergence in the upper *troposphere*.

cyclolysis—Process whereby an *extratropical cyclone* weakens as its central pressure rises, the horizontal pressure gradient weakens, and *wind*s diminish.

cyclone—A *weather* system characterized by relatively low surface *air* pressure compared to the surrounding area; same as a *Low*. Often brings stormy *weather*.

D

Dalton's law of partial pressures—States that the total pressure exerted by a mixture of gases is equal to the sum of the partial pressures of each constituent gas. Each gas species in the mixture acts independently of the other molecules. Stated another way, each gas exerts a pressure as if it were the only gas present.

dendroclimatology—The analysis of tree growth rings for information on the *climate* past.

deposition—The process whereby water vapor becomes ice without first becoming a liquid. *Frost* formation is an example of deposition.

derecho—A family of straight-line *downburst* winds associated with a *squall line* or *mesoscale convective complex* that impacts a path that may be hundreds of kilometers long.

dew—Tiny droplets of water formed when water vapor condenses on a cold surface such as blades of grass on a clear, calm night. Dew forms as a consequence of *radiational cooling* and is not a form of *precipitation*.

dewpoint—The *temperature* to which air must be cooled at constant pressure to achieve saturation of air relative to liquid water without the addition or removal of water vapor.

dewpoint hygrometer—Humidity measuring instrument in which air passes over the surface of a metallic mirror that is cooled electronically. An electronic sensor monitors the *temperature* of the mirror at the same time that an infrared beam is pointed at the mirror. With sufficient cooling, a thin film of water condenses on the mirror changing its reflectivity and altering the *reflection* of the infrared beam. The mirror temperature is automatically recorded as the *dewpoint*. Then the mirror is warmed electronically to evaporate the *dew* to prepare for the next measurement.

dissipating stage   Final phase in the life cycle of a *thunderstorm* where subsiding air replaces the *updraft* throughout the *cloud*, effectively cutting off the supply of moisture provided by the updraft. Adiabatic compression warms the subsiding air, the *relative humidity* drops, *precipitation* tapers off and ends, and clouds gradually vaporize.

distillation—Purification of water through phase changes (e.g., *evaporation* followed by *condensation*). When water vaporizes from Earth's surface, all suspended and dissolved substances such as sea salts and other contaminants are left behind.

doldrums—An east-west equatorial belt of light and variable surface *wind*s where *trade winds* of the two hemispheres converge.

downburst—An exceptionally strong *thunderstorm downdraft* that, upon striking Earth's surface, diverges horizontally as a surge of potentially destructive *wind*s.

downdraft—A strong, downward flowing air current within a thunderstorm, usually associated with precipitation.

drizzle—A form of liquid precipitation consisting of liquid water drops having diameters between 0.2 and 0.5 mm (0.01 and 0.02 in.); descends very slowly from low stratus clouds.

dropwindsonde—Similar to a *rawinsonde* except that instead of being launched by a balloon from a surface station, it is dropped from an aircraft.

Dry climates (B)—A climate group of the Köppen climate classification system; characterize those regions where average annual potential evaporation exceeds average annual precipitation.

dry slot—Dry air descending behind a *cold front* and drawn into the center of an *extratropical cyclone*.

dry-bulb temperature—Same as the ambient air *temperature*.

dust devil—A whirling mass of dust-laden air; a common sight over flat, dry terrain (not exclusively deserts). They develop on sunny days in response to local variations in surface characteristics (i.e., *albedo*, moisture supply, topography) that give rise to localized hot spots.

E

ecological succession—Replacement of one *ecosystem* by another through a sequence of colonization and replacement of species until an assemblage of species is established that is able to maintain itself on a site.

ecosystem—communities of plants and animals that interact with one another, together with the physical conditions and chemical substances in a specific geographical area.

ecotone—Boundary between *ecosystems* that is particularly vulnerable to excess soil erosion (by wind or water) or where barely enough *rain* falls or the growing season is hardly long enough to support crops and livestock.

eddy viscosity—Fluid *friction* that arises from large irregular motions, called eddies, which develop within fluids such as air or water.

efficient economy—Operating when resources are used in the best possible way to satisfy the largest number of consumer demands.

Ekman spiral—Simplified model of the three-dimensional water motion in the upper 100 m (330 ft) or so of the ocean produced by directional change and decreasing horizontal motion of successively lower layers of water; due to the combined influence of the *Coriolis Effect* and frictional drag of *wind* on surface water.

Ekman transport—Net horizontal movement of water in the top 100 m (330 ft) or so of the ocean induced by the coupling of surface *winds* and surface waters. The *Ekman spiral* causes this net transport to be about 90 degrees to the right of the surface wind direction in the Northern Hemisphere and to the left of the surface wind direction in the Southern Hemisphere.

El Niño—Anomalous warming of surface waters in the eastern tropical Pacific and along the equator; accompanied by suppression of upwelling off the coasts of Ecuador and northern Peru and along the equator east of the international dateline. Typically lasting for 12 to 18 months and occurring ever 3 to 7 years, El Niño is accompanied by changes in oceanic and atmospheric circulation plus *weather* extremes in various parts of the world.

electromagnetic radiation—Describes both a form of *energy* and a means of energy transfer. Energy in the form of waves that have both electrical and magnetic properties; these waves travel through gases, liquids, and solids, and can occur in a vacuum.

electromagnetic spectrum—Various forms of radiation arranged by *wavelength* or by *wave frequency* or by both. In order of increasing wavelength and decreasing frequency, forms of electromagnetic radiation include gamma rays, X-rays, ultraviolet (UV) radiation, visible light, infrared (IR) radiation, microwaves, and radio waves.

electronic hygrometer—Instrument for measuring the water vapor component of air that is based on changes in the electrical resistance of certain chemicals as they adsorb water vapor from the air. The adsorbing element may be a thin carbon coating on a glass or plastic strip. The more humid the air, the more water adsorbed, and the lower is the resistance to an electric current passing through the sensing element. Variations in electrical resistance are calibrated in terms of percent *relative humidity* or *dewpoint*. An electronic hygrometer is flown on board *radiosonde*s.

empirical climate classification—Climate classification scheme that infers the type of climate from the environmental impacts of climate such as the geographical distribution of indigenous vegetation or soil type, or the degree of weathering of exposed bedrock.

energy—The capacity for doing work; occurs in many different forms such as radiation and *heat*.

Enhanced F-scale—Updated version of T.T. Fujita's tornado intensity scale (*F-scale*) that became operational in February 2007. Revisions were made to more closely align estimated *wind* speeds with associated storm damage; ranges from EF0 to EF5.

ENSO (El Niño/Southern Oscillation)—The relationship between *El Niño* and the *Southern Oscillation*. An El Niño episode begins when the weakening of the surface *air pressure gradient* between the western and central tropical Pacific heralds the slackening of the *trade winds*.

ENSO Alert System—Launched by *NOAA*'s *Climate Prediction Center* in February 2009, an *El Niño* or *La Niña* watch is issued when conditions in the equatorial Pacific are favorable for the development of El Niño or La Niña within the next three months. An El Niño or La Niña advisory is issued when El Niño or La Niña conditions have developed and are expected to continue.

ENSO Observing System—Operational since December 1994, consists of an array of moored and drifting instrumented buoys, tide gauges, ship-based sensors, and satellites in the tropical Pacific Ocean.—

equatorial upwelling—Upward circulation of cold, nutrient-rich bottom water toward the ocean surface; the consequence of convergence of the *trade winds* of the Northern and Southern Hemispheres plus *Ekman transport*.

equinoxes—On or about 21 March and again on or about 23 September, the Sun's noon position (*solar altitude* of 90 degrees) is directly over the equator. At these times, day and night are approximately equal in length (12 hrs) everywhere, except at the poles.

erosion—Removal and transport of sediments (e.g., fragmented rock) by gravity, moving water, *glaciers*, or *wind*.

eustasy—Global variation in sea level brought about by a change in the volume of water occupying the ocean basins.

eustatic sea-level change—Worldwide fluctuations in mean sea level (msl) often because of changes in Earth's glacial ice cover.

evaporation—Process whereby more water molecules enter the *atmosphere* as vapor than return as liquid, a net loss occurs in liquid water mass.

evapotranspiration—Direct *evaporation* from Earth's surface plus the release of water vapor by vegetation (*transpiration*).

excess death rate—Number of reported human deaths minus the typical number of deaths over a specified period.

expansional cooling—When a gas or mixture of gases (such as air) expands, its *temperature* decreases.

externalities—Consequences of economic activity that are not reflected in market transactions; may be positive or negative.

extratropical cyclone—A synoptic-scale low pressure *system* that occurs in midlatitudes, often forming along the *polar front*. This low, characterized by fronts and a *comma cloud* pattern, becomes a *cold-core cyclone* especially in the later stage of its life cycle.

eyewall—Ring of *thunderstorm* (cumulonimbus) clouds surrounding the eye of a mature *hurricane*; produces heavy *rain*s and very strong *wind*s.

F

flash flood—A short-term, localized, and often unexpected rise in river or stream level above bankfull usually in response to torrential *rain* falling over a relatively small geographical area.

foehn—*Chinook*-like *wind* that blows into the Alpine valleys of Austria, Germany, and Switzerland.

fog—Visibility-restricting suspension of tiny water droplets or ice crystals (called ice fog) in an air layer next to Earth's surface.

food chain—Simple feeding relationships among organisms.

food web—Realistic feeding relationships among organisms.

force—A push or pull on an object computed as mass times acceleration.—

fossil fuels—Includes coal, oil, and natural gas.

free market—An economic system in which resources are allocated and prices established based upon individual voluntary exchanges among producers and consumers. In a free market system, we obtain what we are willing and able to pay for, and markets adjust to accommodate changing consumer preferences. By this view, supply, demand, and price tend toward a state of equilibrium.

freezing rain (or freezing *drizzle*)—Consists of *rain* (or drizzle) drops that become supercooled and at least partially freeze on contact with cold surfaces (at subfreezing temperatures), forming a coating of ice (glaze) on roads, tree branches, and other exposed surfaces.

friction—Resistance that an object or medium encounters as it moves in contact with another object or medium.

front—A narrow zone of transition between *air mass*es that differ in density. Density differences are usually due to *temperature* contrasts; for this reason we use the nomenclature *cold front*s and *warm front*s.

frontal fog—*Fog* that develops when *precipitation* falling from relatively warm air aloft into a wedge of relatively cool air near Earth's surface evaporates and increases the water vapor concentration in the cool air to saturation.

frontal uplift—Ascent of air along the surface of a boundary (*front*) between air masses of contrasting density; often accompanied by cloud and *precipitation* development.

frontogenesis—Formation or strengthening of a *front*.

frontolysis—Weakening of a *front*.

frost—Ice crystals that form on exposed surfaces such as vegetation due to cooling at constant pressure and producing saturation at an air *temperature* below freezing.

frost point—The *temperature* to which air must be cooled at constant *air pressure* to achieve saturation at or below the freezing point of water so that *frost* forms.

frostbite—Freezing of body tissue.

F-scale—A six-level intensity scale for rating *tornado* strength and damage to structures devised by T.T. Fujita in 1971; ranging from F0 to F5, was based on rotational *wind* speeds estimated from property damage. Replaced by the *Enhanced F-scale* in 2007.

funnel cloud—Tornadic circulation extending below *cloud* base but not reaching the ground; made visible by a cone-shaped cloud.

G

genetic climate classification—Climate classification scheme that asks why climate types occur where they do.

geoengineering—Large-scale human manipulation of components of Earth's *climate system* that is intended to offset the consequences of increasing greenhouse gas emissions.

geologic time scale—Standard division of Earth history into eons, eras, periods, and epochs based on large-scale geological events.

geosphere—The solid portion of the planet consisting of rocks, minerals, soil, and sediments.

Geostationary Operational Environmental Satellites (GOES)—Geostationary *weather* satellites for the United States operated by *NOAA* to provide a complete view of much of North America and adjacent portions of the Pacific and Atlantic Oceans. A geostationary satellite revolves around Earth at the same rate and in the same direction as the planet rotates so that the satellite is always positioned over the same spot on Earth's surface and its sensors have a consistent field of view.

geostrophic flow—The horizontal movement of water or air arising from a balance between the *pressure gradient force* and the *Coriolis Effect*.

geostrophic wind—A hypothetical unaccelerated, horizontal movement of air that follows a straight path parallel to *isobars* or height contours at altitudes above the *atmospheric boundary layer*; results from a balance between the horizontal *pressure gradient force* and the *Coriolis Effect*.

glacial climate—*Climate* that favors a positive *mass balance* so that new *glaciers* form and existing glaciers thicken or expand (advance).

glacial theory—Some time in the past, large glacial ice sheets spread over northwestern Europe and over much of what is now Canada and the northern United States.

glacier—Mass of ice that flows internally under the influence of gravity.

global climate model (GCM)—A numerical simulation of Earth's *climate system*.

global radiative equilibrium—*Energy* entering the Earth-*atmosphere*-land-ocean system (i.e., absorbed solar radiation) ultimately must equal energy leaving the system (i.e., *infrared radiation* emitted to space).

global warming potential (GWP)—The ratio of *heat* "trapped" by one unit mass of a greenhouse gas to that of one unit mass of CO_2 over a specified time period. Developed to compare the contributions of various gases to Earth's *greenhouse effect*.

global water budget—The balance sheet for inputs and outputs of water to and from the various global reservoirs; may be expressed in terms of water volume or depth of water.

global water cycle—The ceaseless movement of water among the atmospheric, oceanic, terrestrial, and biospheric reservoirs on a planetary scale; assumes an essentially fixed amount of water in the Earth-atmosphere-land-ocean system.

gradient wind—A hypothetical horizontal movement of air parallel to curved *isobars* or height contours above the *atmospheric boundary layer*. It differs from the *geostrophic wind* in that the path of the *gradient wind* is curved.

gravity—The force that accelerates air downward to Earth's surface.

greenhouse effect—Heating of Earth's surface and lower *atmosphere* caused by strong *absorption* and emission of *infrared radiation* by *greenhouse gases*.

greenhouse gases—Atmospheric gasses that absorb and emit *infrared radiation* that heats Earth's surface and lower *atmosphere*. The principal *greenhouse gas* is water vapor; others include carbon dioxide, ozone, methane, and nitrous oxide.

gust front—Leading edge of a mass of relatively cool gusty air that flows out of the base of a *thunderstorm cloud* (downdraft) and spreads along the ground well in advance of the parent thunderstorm cell; resembles a miniature *cold front*.

H

haboob—A dust- or sand-storm caused by the strong, gusty *downdraft* of a desert *thunderstorm*. The mass of dust or sand rolls along the ground as a huge ominous dark *cloud* that may be more than 100 km (60 mi) wide and may reach altitudes of several thousand meters.

Hadley cell—Thermally driven air circulation in tropical and subtropical latitudes of both hemispheres resembling a huge *convection* cell with rising air near the equator in the *intertropical convergence zone* and sinking air in the subtropical *anticyclones*. Equatorward blowing surface *wind*s and poleward directed upper-level winds complete the circulation.

hail—*Precipitation* in the form of jagged to nearly spherical chunks of ice that form within intense *thunderstorms* characterized by vigorous *updraft*s, an abundant supply of supercooled water droplets, and great vertical *cloud* development.

hair hygrometer—An instrument designed to monitor *relative humidity* by measuring the changes in the length of human hair that accompany variations in the *humidity*. Cells in the hair adsorb (collect on the surface) water and swell, causing the hair to lengthen slightly as the relative humidity increases.

heat—A form of *energy* transfer between systems or components of a system in response to differences in *temperature*. Heat energy is always transferred from a warmer system (or object) to a colder system (or object).

heat conductivity—Ratio of the rate of *heat* transport across an area to the *temperature gradient*.

heat equator—Latitude of highest mean annual surface *temperature*; located about 10 degrees north of the geographical equator.

heat index—Accounts for the increasing inability of the human body to dissipate *heat* to the environment as the *relative humidity* rises; a measure of human comfort and well-being during various combinations of heat and humidity.

heating degree-days—Based on the Fahrenheit *temperature* scale; computed only for days when the average outdoor air temperature is lower than 65 °F (18 °C). Each degree of average temperature below 65 °F is counted as one heating degree-day. Subtracting the average daily temperature from 65 °F yields the number of heating degree-days for that day.

Highland climates (H)—Climate group of the Köppen climate classification system; include a wide variety of climate types that characterize mountainous terrain. Altitude, latitude, and aspect are among the factors that differentiate a complexity of climate types.

Holdridge life zones—Climate classification system first published by Leslie Holdridge in 1947 and revised in 1967. It utilized mean annual biotemperature derived from growing season length and temperature, the annual precipitation, and the ratio of potential evapotranspiration to mean annual precipitation. The system recognizes 38 bioclimate classes ranging from polar desert to tropical rainforest.

holistic perspective—Consideration of a problem in its entirety. For example, a holistic perspective would call on government agencies that have *climate* and *climate change* in their purview to coordinate and integrate their activities for greater efficiency and synergy, and to avoid conflicting or contradictory policies.

Holocene Epoch—Interval of time since the end of the Pleistocene glaciation (approximately 10,500 years ago) that represents the present interglacial. This epoch has been characterized by spatially and temporally variable *temperature* and *precipitation*.

horse latitudes—Name applied to all latitudes between about 30 and 35 degrees N and S under the semi-permanent subtropical highs.

hot spot—A long-lived source of *magma* caused by rising plumes of hot material originating in the mantle (mantle plumes) of Earth's interior.

humidity—General term referring to any one of many ways of describing the amount of water vapor in the air.

hurricane—An intense tropical *cyclone* that originates over warm ocean waters, usually in late summer or early fall (when sea-surface *temperature*s are highest); has a maximum sustained *wind* speed of at least 119 km per hr (74 mph).

hydrosphere—The water component of the *climate system*

hygrometer—An instrument that measures the water vapor concentration of air.

hygroscopic nuclei—*Cloud condensation nuclei* that possess a chemical attraction for water molecules.

hypothermia—Potentially lethal condition brought on by a drop in the *temperature* of the body's vital organs (e.g., heart, lungs).

I

ice pellets—A form of *precipitation,* commonly called *sleet*, which consists of spherical or irregularly-shaped, transparent or translucent particles of ice that are 5 mm (0.2 in.) or less in diameter and bounce on impact with the ground.

ice stream—A zone of relatively rapidly flowing ice within a glacial ice sheet.

ice-albedo feedback—Warming that causes a reduction in Arctic sea-ice cover is enhanced by the lower *albedo* of the open ocean waters; an example of *positive feedback*.

ice-forming nuclei (IN) Tiny particles suspended in the atmosphere that promote formation of ice crystals only at *temperature*s well below freezing.

igneous rock—Type of rock produced by the cooling and crystallization of hot molten *magma.*

in situ measurement—A monitoring system in which the sensor is immersed in the medium that is being measured.

incremental decision making—Process whereby public policy evolves via small changes in existing policies or programs.

inefficient economy—Operating when resources are not used to their greatest advantage, so that if they were more productively employed, they could be used to satisfy more demands.

infrared radiation (IR)—*Electromagnetic radiation* at *wavelengths* ranging from 0.8 micrometer (near-infrared) to about 0.1 mm (far infrared). Most objects in the Earth-atmosphere-land-ocean system have their peak emission in the infrared.

infrared satellite image—Image or picture processed from radiometers on board an Earth-orbiting satellite that sense thermal (infrared) radiation (typically having wavelengths in the range of 8 to 12 micrometers) emitted by components of the Earth-atmosphere-land-ocean system.

interglacial climate—*Climate* that favors a negative *mass balance* so that *glaciers* fail to form or existing glaciers retreat and eventually may completely waste away.

Intergovernmental Panel on Climate Change (IPCC)—Established in 1988 by the World Meteorological Organization (WMO) and the United Nations Environmental Programme (UNEP). The IPCC is charged with evaluating the state of *climate science* as the basis for policy action and serving the interests of scientists, public policymakers, and through them the public at large.

internal energy—A measure of all the *energy* in a substance, that is, the kinetic energy of atoms and molecules plus the potential energy arising from forces between atoms or molecules.

intertropical convergence zone (ITCZ)—A discontinuous low-pressure belt with *thunderstorms* paralleling the equator. The average location of the ITCZ corresponds approximately to *heat equator*.

inverse square law—The radiation emanating in all directions from a nearly point source like the Sun spreads out as ever larger spheres, with the intensity decreasing as a function of the surface area of the sphere (where the area is a function of the square of the sphere's radius).

isobars—Lines on a *weather* map joining places having the same *air pressure* (adjusted to sea level). Usually isobars are drawn at 4-millibar intervals.

isotherms—Lines drawn on a *weather* map through localities having the same air *temperature*.

J

jet streak—An area of accelerated winds within a jet stream; the *wind* may strengthen by as much as an additional 100 km per hr (62 mph). Jet streaks occur where surface horizontal *temperature gradients* are particularly steep and they play an important role in the generation and maintenance of *extratropical cyclones*.

K

katabatic wind—Shallow layer of cold, dense air that flows downhill under the influence of gravity; usually originates in winter over an extensive *snow*-covered plateau or other highlands.

Kyoto Protocol—In the initial update to the *UNFCCC* treaty, representatives of the global community met in Kyoto, Japan, in December 1997 to formulate an acceptable and effective international treaty limiting greenhouse gas emissions worldwide.

L

La Niña—An episode of particularly strong *trade winds* and unusually low sea-surface *temperature*s (SST) in the central and eastern tropical Pacific; essentially the opposite of *El Niño*. La Niña is accompanied by *weather* extremes in various parts of the world.

lake breeze—A relatively cool surface *wind* directed from a large lake toward land in response to differential heating of land and lake; develops during daylight hours.

lake-effect snow—Highly localized fall of *snow* immediately downwind of an open lake at middle latitudes of the Northern Hemisphere; most common in autumn and early winter when lake surface *temperature*s are still relatively mild; can produce paralyzing accumulations of snow.

lake-enhanced snow—*Lake-effect snow* can develop on the western shores of the lakes which are normally upwind. For example, an early winter *extratropical cyclone* tracking through the lower Great Lakes region may produce strong northeast winds which blow onshore on the western shores. In this case, the lake's snow-generating mechanism adds to the snowfall produced by overrunning in the cyclone.

land breeze—Cool surface *wind* directed from land to sea or land to lake in response to differential cooling between land and a body of water; develops at night.

latent heat—Quantity of *heat* that is involved in the phase changes of water.

latent heat of fusion—*Heat* released to the environment when water changes phase from liquid to solid; 80 *calorie*s per gram.

latent heat of vaporization—*Heat* required to change the phase of water from liquid to vapor; 540 to 600 *calorie*s per gram, depending on the *temperature* of the water.

latent heating—Transport of *heat* from one location to another within the Earth-*atmosphere* system as a consequence of phase changes of water. For example, h*eat* is supplied for melting, *evaporation*, and *sublimation* of water at Earth's surface, and heat is released to the *atmosphere* during freezing, *condensation*, and *deposition*.

law of conservation of energy—See *law of energy conservation.*

law of energy conservation—Energy is neither created nor destroyed but can change from one form to another; same as the *first law of thermodynamics.*

Le Chatelier's principle—If a system in dynamic equilibrium experiences a change or perturbation (e.g., in *temperature*, concentration, volume, or stress), then the equilibrium shifts in such a way as to counteract or compensate for the change or perturbation.——

lifetime of a gas—The time it takes for a perturbation of a gas to be reduced to 37% of its original amount.

lightning—Brilliant flash of light produced by an electrical discharge within a *cumulonimbus cloud* or between the cloud and Earth's surface.

lithosphere—Rigid uppermost portion of Earth's mantle, plus the overlying crust, averaging 100 km (62mi) thick.

Little Ice Age—An interval of the late *Holocene Epoch* from about 1400 to 1900 when average global *temperatures* were lower and alpine *glaciers* increased in size and advanced down mountain valleys. The Little Ice Age followed the *Medieval Warm Period.*

local climate—Applies to a variety of *ecosystems* such as a city, cropland, marsh or forest. For example, a city tends to be somewhat warmer than the surrounding rural areas because of differences in composition (asphalt, brick, and concrete versus vegetation), water supply (more standing water in the countryside), and concentration of *heat* sources (motor vehicles, space heating and cooling).

macroclimate—Refers to climates in large regions up to continental in area.

Madden-Julian Oscillation (MJO)—Large-scale (1000 km) disturbance of the near-equatorial *troposphere* that slowly propagates from the Indian Ocean into the Western Pacific over the course of about 30 to 50 days; possible trigger for *ENSO.*

magma—Hot molten rock material that wells up from deep in Earth's crust or upper mantle and migrates along rock fractures.

maritime climate—Characteristic of places immediately downwind of the ocean (or large lake) that experience much less contrast between average winter and summer *temperatures* compared to places at the same latitude but well inland.

maritime polar air (mP)—Cool, humid *air mass* that forms over the cold ocean waters of the North Pacific and North Atlantic, especially north of 40 degrees N. Along the West Coast of North America, maritime polar air contributes to heavy winter *rains* (*snows* in the mountains), and persistent coastal *fog* in summer.

maritime tropical air (mT)—Warm, humid *air mass* that forms over tropical and subtropical ocean waters (e.g., Gulf of Mexico). Maritime tropical air is responsible for oppressive summer *heat* and *humidity* east of the Rocky Mountains.

mass balance—In a glacier, the difference between ice mass gain (accumulation) and ice mass loss (*ablation*) over the course of a year (usually measured at the end of the melt season in late summer).

mature stage (thunderstorm)—Middle phase in the life cycle of a *thunderstorm* which begins when *precipitation* first reaches Earth's surface. Typically, this stage lasts about 10 to 20 minutes. The cumulative weight of water droplets and ice crystals eventually becomes so great that they can no longer be supported by the *updraft.*

Maunder minimum—Seventy-year period between 1645 and 1715 when *sunspot* activity greatly diminished; corresponded to cold phases of the *Little Ice Age.*

Medieval Warm Period—Relatively mild episode of the *Holocene Epoch* between about CE 950 and 1250.

Mediterranean climate—Type of *subtropical climate* that occurs on the western side of continents between about 30 and 45 degrees latitude. In North America, mountain ranges confine this climate to a narrow coastal strip of California. Although *precipitation* is seasonal (dry summers and wet winters), the *temperature* regime is quite variable. Along the coast, cool onshore breezes prevail, lowering the mean annual temperature and reducing seasonal temperature contrasts. Well inland, away from the ocean's moderating influence, summers are considerably warmer; hence, inland mean annual temperatures are higher and seasonal temperature contrasts are greater than in coastal areas.

megadrought—Drought that persists for multiple decades.—

meridional flow pattern—Flow of the *planetary-scale* westerlies in a series of deep troughs and sharp ridges; westerlies exhibit considerable amplitude. In this pattern, cold *air mass*es surge southward and warm air masses stream northward, leading to strong *temperature gradient*s.

meridional overturning circulation (MOC)—*Thermohaline circulation* in the Atlantic basin; transports *heat*, salt, and dissolved gasses over great distances and to great depths in the world ocean and plays an important role in Earth's *climate system*.

mesoclimate—Characteristic of smaller and physically distinctive regions such as mountains or plains and is strongly influenced by regional atmospheric circulation patterns and systems (e.g., *mountain* and *valley breeze*s, *chinook wind*s).

mesoscale convective complex (MCC)—A nearly circular organized cluster of many interacting *thunderstorm* cells covering an area of many thousands of square kilometers.

mesoscale convective systems (MCS)—A regional *weather system* such as a *thunderstorm*.

mesoscale system—Atmospheric circulation *system* so small that it may influence the weather in only a portion of a large city or county. Examples include *thunderstorms*, *sea breezes, and land breezes*.

mesosphere—The atmospheric layer between the *stratosphere* and the *thermosphere*; the average

temperature generally falls with increasing.

metamorphic rock—A crystalline rock derived from other rocks that were subjected to high temperatures, high pressures, and chemically-active fluids, conditions that exist in geologically active mountain belts.

meteorological seasons—Successive three-month intervals centered on the typical date of occurrence of the warmest and coldest months of the year. Meteorological spring consists of the months of March, April, and May; summer encompasses June, July, and August; autumn is September, October, and November; and meteorological winter consists of December, January, and February.

meteorology—Scientific study of the *atmosphere*, processes that cause *weather*, and the life cycle of *weather system*s.

microclimate—Exchange of *energy* and matter in a small area such as an individual farm field, hillslope, or marsh.

microscale systems—*Weather* phenomena that represent the smallest spatial subdivision of atmospheric circulation, such as a weak *tornado*. These systems have dimensions of 1 m to 1 km (3 ft to 1 mi) and last from seconds to an hour or so.

midlatitude westerlies—Prevailing planetary-scale *winds* in the middle and upper *troposphere* between about 30 and 60 degrees of latitude; blow on average from the southwest in the Northern Hemisphere and from the northwest in the Southern Hemisphere.

Milankovitch cycles—Regular variations in three elements of Earth-Sun geometry: precession and tilt of Earth's rotational axis and the eccentricity of the planet's orbit about the Sun. These cycles affect the seasonal and latitudinal distribution of incoming solar radiation and influence climatic fluctuations over tens of thousands to hundreds of thousands of years.

mistral—A *katabatic wind* that descends from the *snow-*capped Alps down the Rhone River Valley of France and into the Gulf of Lyons along the Mediterranean coast; winter phenomenon.

mitigation—Actions taken over the present century that

are designed to stabilize greenhouse gas emissions at a level that avoids a dangerous disturbance of the *climate system*, reducing the pace and magnitude of anthropogenic *climate change*.

mixed-market economy—An economic system that combines private, competitive enterprise with some government involvement. This system is used in the U.S.

molecular viscosity—Fluid *friction* arising from the random motions and interactions of molecules composing a liquid or gas.

monsoon climate—Found in regions subject to a seasonally reversing monsoon circulation and is typically characterized by dry winters and wet summers.

mountain breeze—Shallow, gusty downslope flow of cool air that develops at night in some mountain valleys in response to differential heating between the air adjacent to the valley wall and air at the same altitude out over the valley floor.

multi-vortex tornadoes—*Tornadoes* that consist of two or more subsidiary vortices that orbit about each other or a common center within a massive tornado.

N

NAO Index—Measure of the strength of the horizontal *air pressure gradient* between Iceland and the Azores in the North Atlantic; influences the *climate* of eastern North America and much of Europe and North Africa; varies from year to year and decade to decade.

National Aeronautics and Space Administration (NASA)—Government agency with an Earth Science mission to "conduct aeronautical and space activities so as to contribute materially to . . . the expansion of human knowledge of the Earth and of phenomena in the atmosphere and space."

National Centers for Environmental Prediction (NCEP)—Government agency that aims to deliver "analyses, guidance, forecasts and warnings for *weather*, ocean, *climate*, water, land surface and space weather to the Nation and the world."

National Climate Service—Proposed government agency that would function as an authoritative source of intelligence on *climate change* in support of decision-making regarding *mitigation* and *adaptation*. The National Climate Service (NCS) would parallel the *National Weather Service* (NWS) and, like the NWS, would be part of *NOAA*.

National Climatic Data Center (NCDC)—An agency of *NOAA* that archives climatic data for the United States; located in Asheville, NC.

National Oceanic and Atmospheric Administration (NOAA)—Administrative unit within the U.S. Department of Commerce that oversees the *National Weather Service*.

National Weather Service (NWS)—Agency of *NOAA* that is responsible for *weather* observations, data analysis and interpretation, weather forecasting and dissemination, and storm watches and warnings. Also gathers the basic *weather* data used in generating the nation's climatological summaries.

neutral air layer—An ambient air layer in which an ascending or descending air parcel always has the same *temperature* (and density) as its surroundings. Hence, a neutral air layer neither impedes nor spurs upward or downward motion of air parcels.

Newton's first law of motion—An object at rest or in straight-line, unaccelerated motion remains that way unless acted upon by a net external force.

North American Monsoon System (NAMS)—Also called the *Southwest Monsoon*, a prominent feature of the *climate* of the American Southwest including Arizona, New Mexico, southern Nevada, and parts of southern Colorado. The monsoon brings a dramatic increase in rainfall to this region mainly during July and August.

North Atlantic Oscillation (NAO)—A seesaw variation in *air pressure* between Iceland and the Azores; influences the *climate* of eastern North America and much of Europe and North Africa over periods up to decades.

Norwegian cyclone model—Conceptual model derived primarily from surface *weather* observations to approximate the structure and life cycle of an extratropical low-pressure *system*; first proposed during World War I by researchers at the Norwegian School of Meteorology at Bergen.

nuclei—Tiny solid and liquid particles suspended in the *atmosphere* that provide surfaces on which water vapor condenses into droplets or deposits into ice crystals; essential for *cloud* formation.

NWS Cooperative Observer Network—Consists of more than 10,000 *weather* stations across the United State that record daily *precipitation* and maximum/minimum *temperatures* for climatic, hydrologic, and agricultural purposes.

O

occluded front—A *front* formed late in the life cycle of an *extratropical cyclone*; its behavior depends upon the characteristics of air behind the *cold front* and ahead of the *warm front*. Also known as an occlusion.

ocean acidification—Lowering of the pH of ocean water due to an increase in the amount of dissolved carbon dioxide.

offset exchange—Market-based approach in which carbon credits are earned for investment in projects that compensate for (offset) CO_2 emissions.

open ocean convection—The primary mechanism of deep water formation in the Northern Hemisphere. Cold *wind*s cool the surface water to the extent that its density becomes greater than that of the water beneath it, creating an unstable water column and driving overturning.

orographic lifting—The forced ascent of air up the slopes of a hill or mountain. Air that is forced to ascend the slopes facing the oncoming *wind* (windward slopes) expands and cools, which increases its *relative humidity*. With sufficient cooling, *clouds* and *precipitation* develop.

overrunning—Process whereby less dense air flows up and over denser air; occurs along a *warm front* and sometimes a *stationary front*. Ascending air cools by expansion, leading to the formation of *clouds* and

perhaps *precipitation* over a widespread area.

oxygen isotope analysis—A technique that enables scientists to identify *climate* fluctuations of the past by examining the ratio of light to heavy isotopes of oxygen (O^{16} and O^{18}) found, for example, in shells extracted from deep-sea sediment cores.

P

Pacific air—Cool, humid *maritime polar air* swept inland from the Pacific Ocean that undergoes *air mass modification* over the Rocky Mountains, emerging milder and drier to the east of the mountains. During a *zonal flow pattern*, Pacific air floods the eastern two-thirds of the United States and southern Canada, causing mild and generally dry *weather*.

Pacific Decadal Oscillation (PDO)—Long-lived variation in *climate* over the North Pacific Ocean and North America. Sea-surface *temperature*s fluctuate between the north central Pacific Ocean and the west coast of North America; linked to changes in strength of the Aleutian low.

Paleocene/Eocene Thermal Maximum (PETM) A geologically brief interval of widespread warming associated with a massive buildup of greenhouse gases; about 55 million years ago.

paleoclimatology—Subfield of *climatology* covering the reconstruction of past *climate*s.

perihelion—Day of the year when Earth is closest to the Sun (about 3 January).

permafrost—Permanently frozen ground.

photosphere—The intensely bright portion of the Sun visible to the unaided eye; this several-kilometer-thick layer is what we perceive as the surface of the Sun; radiating *temperature*s near 6000 K (11,000 °F).

photosynthesis—Process whereby green plants convert light *energy* from the Sun, carbon dioxide from the *atmosphere*, and water to sugars and oxygen (O_2).

planetary albedo—The fraction (or percent) of incident solar radiation that is scattered and reflected to space by the *Earth-atmosphere system*. Measurements by sensors on board Earth-orbiting satellites indicate that Earth's planetary albedo is about 31%.

planetary system—The Earth-*atmosphere*-land-ocean system.

planetary-scale systems—*Weather* phenomena operating at the largest spatial scale of atmospheric circulation; includes the global *wind* belts and *semipermanent pressure systems*. These systems have dimensions of 10,000 to 40,000 km (6000 to 24,000 mi) and exhibit patterns that persist from weeks to months.

planetary system—Equivalent to the *Earth-atmosphere-land-ocean sys*tem.—

plate tectonics—Concept that the outer 100 km (62 mi) of solid Earth is divided into a dozen or more gigantic rigid plates that move relative to one another slowly across the surface of the planet. The drift of these plates moves continents and opens and closes ocean basins over the vast expanse of *geologic time*. Mountain building and most volcanic activity occur at plate boundaries.

polar amplification—Tendency for polar areas to be subject to greater changes in temperature than localities at lower latitudes.

Polar climates (E)—Climate group of the Köppen climate classification system; situated poleward of the Arctic and Antarctic circles. These boundaries correspond roughly to localities where the mean temperature for the warmest month is 10 ºC (50 ºF). Polar climates are characterized by extreme cold and slight precipitation, which falls mostly in the form of snow (less than 25 cm, or 10 in., melted, per year).

polar front—Narrow transition zone where the relatively mild *midlatitude westerlies* meet and override the relatively cold polar easterlies. When the *temperature gradient* across the *front* is steep, the front is well defined and is a potential site for development of *extratropical cyclone*s.

polar front jet stream—A corridor of strong *westerlies* in the upper *troposphere* between the midlatitudes tropopause and the polar tropopause and directly over the *polar front*.

polar glacier—*Glacier* that remains below the pressure melting point throughout the year so that little or no meltwater is generated. Conversion of *snow* to ice and the internal motion of ice are extremely slow processes in a polar glacier. The Antarctic ice sheets are polar glaciers.

polar high—A *cold-core anticyclone* that originates in a source region for *continental polar air*; this shallow *system* is the product of intense *radiational cooling* over the snow-covered continental interior of North America well north of the *polar front*.

Polar-orbiting Operational Environmental Satellites (POES)—*NOAA* operated satellite that follows the Sun (Sun-synchronous). It passes over the same area twice during each 24-hr day. Other polar-orbiting satellites are positioned so that they require several days before passing over the same point on Earth's surface.

polder—Pocket of land enclosed by earthen embankments that provides protection against unusually high tides and some *storm surge*s.

poleward heat transport—Flow of *heat* from the tropics poleward into middle and high latitudes in response to latitudinal imbalances in net *radiational heating* and net *radiational cooling*. Brought about by *air mass* exchange, storm *system*s, and ocean circulation.

precautionary principle—When an action causes a threat to human health or the environment, precautionary measures should be taken even if some cause-and-effect relationships are not fully established scientifically.

precipitable water—Depth of liquid water that would be produced if all the water vapor in a vertical column of air were condensed; usually the column of air extends from Earth's surface to the top of the *troposphere*.

precipitation—Water in liquid, frozen or freezing form (i.e., *rain*, *drizzle*, *snow*, *ice pellets*, *hail*, and *freezing rain*) that falls from *clouds* under the influence of gravity to Earth's surface.

pressure gradient force—A force operating in the *atmosphere* that accelerates air parcels away from regions of relatively high *air pressure* directly across *isobars* toward regions of relatively low air pressure in response to an *air pressure gradient*.

producers—Plants that form the base of most *ecosystem*s, providing *energy*-rich carbohydrates; also called autotrophs for "self-nourishing."

profiling float—An instrument package that measures vertical profiles of ocean water *temperature*, pressure (a measure of depth), and conductivity (a measure of salinity).

proxy climate data sources—Various sensors that substitute for actual *weather* instruments; used for times and places when and where no instrument-derived record of *climate* exists.

psychrometer—An instrument that provides an indirect measure of *relative humidity*; consists of two identical liquid-in-glass *thermometer*s mounted side by side with the bulb of one thermometer wrapped in a muslin wick.

public policies—Those actions that governments take or decide not to take.

radiation fog—Forms when *radiational cooling* causes air near the ground to approach saturation. Usually occurs with a clear night sky, light *wind*s, and an *air mass* that is humid near the ground and relatively dry aloft.

radiational cooling—The drop in *temperature* of an object or a surface that occurs when the object or surface undergoes a net loss of *heat* due to a greater rate of emission of *infrared radiation* than *absorption*.

radiational heating—The rise in *temperature* of an object or a surface that occurs when the object or surface undergoes a net gain of *heat* due to a greater rate of *absorption* of *electromagnetic radiation* than emission.

radiosonde—Small balloon-borne instrument package equipped with a radio transmitter that takes altitude readings (*soundings*) of *temperature*, *air pressure*, and *humidity*.

rain—Form of *precipitation* consisting of liquid water drops with diameters generally in the range of 0.5 to 6 mm (0.02 to 0.2 in.) that falls mostly from *nimbostratus* and *cumulonimbus clouds*.

rain gauge—Device for collecting and measuring rainfall (or melted snowfall); a standard rain gauge consists of cylindrical container equipped with a cone-shaped funnel at the top.

rain shadow—Region situated downwind (often hundreds of kilometers) of a prominent mountain range and characterized by descending air and, as a consequence, a relatively dry *climate*.

rawinsonde— A *radiosonde* used by meteorologists to track the balloon's drift from ground stations using a radio direction-finding antenna or a global positioning system (GPS), thereby monitoring variations in *wind* direction and speed with altitude.

reflection—A special case of *scattering* that takes place at the interface between two different media, such as air and *cloud*, when some of the radiation striking that interface is redirected (backscattered).

relative humidity—Compares the actual amount of water vapor in the air with the amount of water vapor that would be present if that same air were saturated. Relative *humidity* (RH) is always expressed as a percentage and can be computed from the ratio of the *vapor pressure* to the *saturation vapor pressure*.

remote sensing—Acquisition of data on the properties of some object without the sensor being in direct physical contact with the object.

residence time—The average length of time for a substance in a reservoir to be replaced completely.

risk-benefit analysis—Weighs the risks of an activity against its benefits to determine whether the activity should be allowed; often used by people who are involved in environmental decision-making.

Rossby waves—Series of long-*wavelength* troughs and ridges that characterize the planetary-scale westerlies (above the 500-mb level) as they encircle the globe; also called *long waves* waves. At any time, typically between 2 and 5 waves encircle the hemisphere.

S

Saffir-Simpson Hurricane Intensity Scale—Rating system for *hurricanes* designed in the early 1970s by H.S. Saffir (1917-2007), a consulting engineer, and R.H. Simpson (1912-), former director of the National Hurricane Center. Hurricanes are rated from category 1 to 5 corresponding to increasing intensity.

Santa Ana wind—Chinook-type *wind* that typically occurs in autumn and winter. A strong high pressure system centered over the Great Basin sends northeast winds over the southwestern United States, driving air downslope from the desert plateaus of Utah and Nevada, around the Sierra Nevada, and as far west as coastal southern California. This *wind* desiccates vegetation and contributes to outbreaks of forest and brush fires.

saturation vapor pressure—The value of the *vapor pressure* when air is saturated with respect to water vapor; varies directly with *temperature*.

scattering—Process by which *aerosols* and molecules disperse radiation in all directions (forward, backward, and sideways).

sea breeze—Relatively cool mesoscale surface *wind* directed from the ocean toward land in response to differential heating of land and sea; develops during daylight hours.

second law of thermodynamics—All systems tend toward a state of disorder. A gradient in a system, such as a *temperature gradient*, signals order in the system. As a system tends towards disorder, gradients are eliminated.

sedimentary rock—Type of rock that may be composed of any one or a combination of compacted and cemented fragments of rock and mineral grains, partially decomposed remains of plants and animals (e.g. shells, skeletons), or minerals precipitated from solution.

semi-permanent pressure systems—Areas of relatively high and low *air pressure*. Although persistent features of the planetary-scale circulation, these *systems* undergo important seasonal changes in both location and surface *air pressure*, hence the modifier "semi-permanent." These systems include subtropical *anticyclones* and *subpolar lows*.

sensible heating—transport of *heat* from one location or object to another via *conduction*, *convection* or both.

severe thunderstorm—A *thunderstorm* producing surface *winds* stronger than 50 knots (93 km per hr or 58 mph) and/or hailstones 0.75 in. (1.9 cm) or larger in diameter (penny-size). Such thunderstorms may also produce heavy *rain*, *flash floods* or *tornadoes*.

short waves—Relatively small short-*wavelength* ripples (troughs and ridges) superimposed on *Rossby waves* in the planetary-scale westerlies. Although *Rossby waves* usually drift very slowly eastward, short waves propagate rapidly through the Rossby waves. Whereas five or fewer Rossby waves encircle the hemisphere, the number of short waves may be a dozen or more.

slash-and-burn agriculture—A human adaptation to the tropical rain forest *ecosystem* for the purpose of growing food, that is, transformation of the natural *ecosystem* into productive agricultural lands by cutting down and burning the indigenous vegetation.

sleet—See *ice pellets*.

Snow forest climates (D)—Climate group of the Köppen climate classification system; occur in the interior and on the leeward sides of large continents. These climates feature cold snowy winters (except for the Dw subtype in which the winter is dry) and occur only in the Northern Hemisphere.

snow—An agglomeration of ice crystals in the form of flakes.

societal resilience—The ability of a society to recover from *weather*- or *climate*-related extremes or other natural disasters

solar altitude—Angle of the Sun above the horizon that influences the intensity of solar radiation received at Earth's surface.

solar constant—Rate at which solar radiation falls on a unit area of a flat surface located at the outer edge of the *atmosphere* and oriented perpendicular to the incoming solar beam when Earth is at its mean distance from the Sun.

solar corona—Outermost portion of the solar atmosphere (above the chromosphere), a region of extremely hot (1 to 4 million °C) and highly rarefied ionized gases (predominantly hydrogen and helium) extending millions of kilometers into space, to the outer limits of the solar system. The solar wind originates in the solar corona and is intensified by solar flares that erupt from the *photosphere* into the corona.

solstice—Time during the year when the Sun is at its maximum poleward location relative to Earth (23.5 degrees N or S); the first day of astronomical summer and winter.

sonic anemometer—Instrument used to monitor *wind* speed and direction consisting of three arms that send and receive ultrasonic pulses. The travel times of sound waves with and against the wind are translated into wind speed and direction.

sounding—Continuous altitude measurements providing vertical profiles of environmental variables such as *temperature*, *humidity*, and *wind* speed.

Southern Oscillation—A seesaw variation in *air pressure* across the tropical Indian and Pacific Oceans.

Southwest Monsoon—See *North American Monsoon System*.

specific heat—The amount of *heat* that will raise the *temperature* of 1 gram of a substance by 1 Celsius degree.

speleothem—Also called dripstone, is a calcite ($CaCO_3$) deposit in a limestone cave or cavern that can yield high-resolution records of past *temperature* and rainfall.

split flow pattern—Flow of *westerlies* to the north have a wave configuration different from the westerlies to the south. For example, *winds* may be zonal across central Canada while winds are meridional over much of the coterminous United States.

squall line—Elongated cluster of intense *thunderstorm* cells, which is accompanied by a continuous *gust front* at the cluster's leading edge. A squall line is most likely to develop in the warm southeast sector of a mature *extratropical cyclone*, ahead of and parallel to the *cold front*.

stable air layer—An ambient air layer characterized by a vertical *temperature* profile such that air parcels return to their original altitudes following any upward or downward displacement. An ascending air parcel becomes cooler (denser) than the ambient air and a descending air parcel becomes warmer (les dense) than the ambient air.

stationary front—Narrow zone of transition between contrasting *air masses* that exhibits essentially no forward movement.

steam fog—Develops in late fall or winter when extremely cold and dry air flows over a large unfrozen body of water.

Stefan-Boltzmann law—A radiation law that states that the total energy radiated by a *blackbody* at all *wavelengths* is directly proportional to the fourth power of the *absolute temperature* (in kelvins) of the body.

storm surge—A dome of seawater perhaps 80 to 160 km (50 to 100 mi) wide that is driven onshore by strong *wind*s associated with a tropical or extratropical *cyclone*, often causing considerable coastal erosion and flooding.

stratiform cloud—Layered *cloud* often produced by *overrunning*.

stratosphere—The *atmosphere's* thermal subdivision situated between the *troposphere* below and the *mesosphere* above (10 to 50 km or 6 to 30 mi above Earth's surface); contains the *ozone shield*, which prevents organisms from exposure to potentially lethal levels of solar ultraviolet (UV) radiation.

stratospheric ozone shield—Ozone in the *stratosphere* that shields organisms at Earth's surface from exposure to potentially lethal intensities of solar *ultraviolet radiation*.

streamline—Represents the mean path of air moving horizontally.

sublimation—Process whereby *ice* or *snow* becomes vapor without first becoming a liquid.

subpolar gyres—Roughly circular pattern of surface ocean currents that occurs at high latitudes of the Northern Hemisphere; covers a smaller area than their subtropical counterparts.

subpolar lows—High-latitude, semipermanent *cyclones* marking the convergence of planetary-scale surface southwesterlies of midlatitudes with surface northeasterlies of polar latitudes in the Northern Hemisphere or midlatitude northwesterlies and polar southeasterlies in the Southern Hemisphere. The Icelandic low and Aleutian low are Northern Hemisphere examples.

Subtropical climates (C)—Climate group of the Köppen climate classification system; located just poleward of the Tropics of Cancer and Capricorn and dominated by seasonal shifts of subtropical anticyclones. There are three basic climate types: subtropical dry summer (or *Mediterranean*) (Cs), subtropical dry winter (Cw), and subtropical humid (Cf), which receive precipitation throughout the year.

subtropical gyres—Large-scale roughly circular pattern of surface ocean currents, centered near 30 degrees latitude in the North and South Atlantic, the North and South Pacific, and the Indian Ocean.

sulfurous aerosols—Tiny droplets of sulfuric acid (H_2SO_4) and sulfate particles created when a violent volcanic eruption sends sulfur dioxide (SO_2) high into the *stratosphere* where it combines with water vapor.

sulfurous haze—Atmospheric haze formed by *sulfurous aerosols* that reflects a considerable amount of incoming solar radiation to space while absorbing both incoming solar radiation and outgoing *infrared radiation*.

sunspot—Dark blotch of irregular shape on the face of the Sun, typically thousands of kilometers across, that develops where an intense magnetic field suppresses the flow of gases transporting heat *energy* from the Sun's interior.

supercell thunderstorm—Relatively long-lived, large, and intense *thunderstorm* cell characterized by an exceptionally strong *updraft*, in some cases estimated at 240 km per hr (150 mph) or higher. Tendency for the updraft to develop a rotational circulation that may evolve into a *tornado*.

sustainability—Developments that meet the needs of the present without compromising the ability of future generations to meet their own needs.

synoptic climatology A subfield of *climatology* that relates regional and local *climate* to atmospheric circulation.

synoptic-scale systems—Atmospheric circulation systems that are continental or oceanic in scale; *extratropical cyclones*, *hurricanes*, and *air masses* are examples.

system—An entity whose components interact in an orderly manner according to the laws of physics, chemistry, and biology.

T

teleconnection—Linkage between changes in atmospheric circulation occurring in widely separated regions of the globe, often over distances of thousands of kilometers.

temperate glacier—*Glacier* in which the internal ice *temperature* rises to the pressure-melting point sometime during the year (most likely by late summer); the confining pressure slightly depresses the melting point of ice. Meltwater produced accelerates the transition of *snow* to ice and increases the flow rate of the glacier by reducing frictional resistance.

temperature gradient—A change in *temperature* over a distance.

temperature—Measure of the average kinetic *energy* of the atoms or molecules composing a substance.

thermal inertia—Resistance to a change in *temperature*.

thermograph—Recording instrument that provides a continuous trace of *temperature* variations with time.

thermohaline circulation—Subsurface movement of

ocean water masses driven by variations in density arising from differences in *temperature* and salinity.

thermokarst lake—Occupies a depression in the ground formed when warming causes thawing of ice-rich *permafrost* and water drains into the depression. Anaerobic decay of organic sediments on the lake bottom generates methane that bubbles to the lake surface and enters the *atmosphere*.

thermometer—The usual instrument for measuring air *temperature*; an important element of *climate*.

thermosphere—Outermost thermal subdivision of the *atmosphere* where the average *temperature* is isothermal in the lower reaches and then increases with altitude, but is particularly sensitive to variations in the high *energy* portion of incoming solar radiation.

Thornthwaite classification—Climate classification system based on annual soil moisture. Developed by American climate scientist and geographer C.W. Thornthwaite (1892-1963), this system utilized a precipitation-evaporation index that was the basis for identifying humidity provinces as the major subdivision of climates.

thunder—Sound accompanying *lightning*; intense heating of air by a lightning discharge causes a tremendous increase in *air pressure* locally that generates a shock wave. The shock wave propagates outward producing sound waves heard as the rumble of thunder.

thunderstorm—Mesoscale *weather system*, accompanied by *lightning* and *thunder* and perhaps strong gusty surface winds, heavy rainfall, *hail*, or *tornadoes*. Produced by strong *convection* currents that surge to great altitudes within the *troposphere* or higher. Consists of one or more convective cells, each of which progresses through a three-stage life cycle of *towering cumulus, mature*, and *dissipating*. An individual thunderstorm cell affects a relatively small area and is short-lived.

tipping-bucket rain gauge—Instrument that consists of a free-swinging container partitioned into two compartments, each of which can collect the equivalent of 0.01 in. of rainfall. Each compartment alternately fills with water, tips and spills its contents, and trips an electric switch that either marks a chart

on a clock-driven drum or sends an electrical pulse to a computer for recording.

tipping point—A critical point or threshold in a dynamic *system* when a new and irreversible development, perhaps major, takes place.

tornado—Violently rotating column of air in contact with the ground, usually associated with a *thunderstorm*. The *system* often (but not always) is made visible by water droplets formed by *condensation* and/or by dust and debris that are drawn into the tornado.

tornado alley—Region of maximum tornado frequency in North America; a corridor stretching from eastern Texas and the Texas Panhandle northward into Oklahoma, Kansas, and Nebraska, and into southeastern South Dakota.

towering cumulus stage—Initial stage in the life cycle of a *thunderstorm* cell. Over a period of perhaps 10 to 15 minutes, cumulus *cloud* tops surge upward to altitudes of 8000 to 10,000 m (26,000 to 33,000 ft). At the same time, neighboring cumulus clouds merge so that by the end of this initial stage, the storm's lateral dimension may be 10 to 15 km (6 to 9 mi). The updraft throughout the cell is sufficiently strong to keep water droplets and ice crystals suspended in the *cloud* so that no *precipitation* occurs during this stage.

trade wind inversion—Elevated *stable air layer* that occurs on the eastern flank of subtropical *anticyclones* in the vicinity of the *trade winds*. Formed when the subsiding, compressionally warmed air in a subtropical *anticyclone* encounters the marine air layer, a layer of cool, humid, and stable air formed where sea-surface *temperatures* are relatively low. A temperature inversion develops at the altitude where air subsiding from above meets the top of the marine air layer.

trade winds—Prevailing planetary-scale surface *wind*s in tropical latitudes, blowing from the northeast in the Northern Hemisphere and from the southeast in the Southern Hemisphere, out of the equatorward flanks of the *subtropical anticyclones*.

transpiration—Process whereby water that is taken up from the soil by plant roots eventually escapes as vapor through tiny pores on the surface of green leaves.

triple point—Point of occlusion where cold, warm, and *occluded front*s intersect and where conditions can favor development of a new *cyclone*.

Tropic of Cancer—Latitude circle of 23 degrees 27 minutes N where the noon Sun is directly overhead (*solar altitude* of 90 degrees) on or about 21 June, the summer *solstice* in the Northern Hemisphere.

Tropic of Capricorn—Latitude circle of 23 degrees 27 minutes S where the noon Sun is directly overhead (*solar altitude* of 90 degrees) on or about 21 December, the winter *solstice* in the Northern Hemisphere.

tropical cyclone—Synoptic-scale low pressure system that originates over the tropical ocean; includes *tropical depression*, *tropical storm*, *hurricane*, and *typhoon*.

tropical depression—A *tropical cyclone* with sustained *wind* speeds of at least 37 km per hr (23 mph) but less than 63 km per hr (39 mph); an early stage in the development of a *hurricane*.

tropical disturbance—An organized cluster of *thunderstorm* clouds over tropical seas with a detectable center of low *air pressure* at the surface; the initial stage in the development of a *hurricane*.

Tropical humid climates (A)—Climate group of the Köppen climate classification system; constitute a discontinuous belt straddling the equator and extending poleward to near the Tropic of Cancer in the Northern Hemisphere and the Tropic of Capricorn in the Southern Hemisphere. Mean monthly temperatures are high and exhibit little variability throughout the year.

tropical storm—A *tropical cyclone* having sustained *wind* speeds of 63 to 118 km per hr (39 to 73 mph). A tropical cyclone that intensifies to tropical storm strength is assigned a name.

tropopause—Zone of transition between the *troposphere* below and the *stratosphere* above; the top of the troposphere. Average altitude ranges from about 6 km (3.7 mi) at the poles to about 20 km (12 mi) at the equator.

troposphere—Lowest thermal subdivision of the *atmosphere* (averaging about 10 km or 6 mi thick) where the *atmosphere* interfaces with the *hydrosphere*, *cryosphere*, *geosphere*, and *biosphere* and where most *weather* takes place. In the troposphere, the average air *temperature* drops with increasing altitude so that it is usually colder on mountaintops than in lowlands.

U

United Nations Framework Convention on Climate Change (UNFCCC)—Probably best known for its non-binding international treaty produced at the UN Conference on Environment and Development convened in Rio de Janeiro, Brazil, in June 1992. The UNFCCC treaty has as its principal objective the "stabilization of greenhouse gas concentrations in the *atmosphere* at a level that would prevent dangerous anthropogenic interference with the *climate system*."

unstable air layer—An ambient air layer characterized by a vertical *temperature* profile such that air parcels accelerate upward or downward and away from their original altitudes. An ascending air parcel remains warmer (less dense) than the ambient air and continues to ascend and a descending air parcel remains cooler (denser) than the ambient air and continues to descend.

updraft—Current of air that streams upward within a *thunderstorm* cell during the *towering cumulus* and *mature stages* of its life cycle. During the towering cumulus stage, the updraft is strong enough to keep water droplets and ice crystals suspended in the upper reaches of the *cloud*.

upslope fog—Formed as a consequence of the *expansional cooling* of humid air that is forced to ascend a mountain slope.

urban heat island—An area of higher air *temperatures* in a city setting compared to the air temperatures of the suburban and rural surroundings; shows up as an island of warmth in the pattern of *isotherms* on a surface map.

V

valley breeze—Shallow, upslope flow of relatively warm air that develops during daylight hours within *snow-free* mountain valleys in response to differential heating between the air adjacent to the valley wall and air at the same altitude out over the valley floor.

vapor pressure—Water vapor's contribution to the total *air pressure*.

varve—Thin layer (lamina) of sediment deposited annually in a body of still water, often a pond or lake fed by a glacial meltwater stream.

viscosity—*Friction* within fluids such as water and air.

visible radiation (light)—Electromagnetic radiation that is perceptible by the human eye; the *wavelength* of visible light ranges from about 0.40 micrometers (violet) to 0.70 micrometers (red).

visible satellite image—Image processed from radiometers on board a satellite that sense visible solar radiation reflected or backscattered from surfaces in the *Earth-atmosphere system*; essentially black and white photographs of the planet. From analysis of *cloud* patterns displayed on the image, atmospheric scientists can identify not only a specific type of *weather system* (such as a *hurricane*), but also the stage of its life cycle and its direction of movement when a sequence of images is animated.

W

Walker circulation—Direct zonal tropical circulation, thermally driven, in which air rises over the warm western Pacific Ocean and sinks over the cool eastern Pacific.

warm air advection—The flow of air across regional *isotherms* from a warmer locality toward a colder locality. Warm air advection occurs behind a *warm front* and ahead of a *cold front*.

warm cloud—Any *cloud* with *temperatures* above 0 °C (32 °F).

warm front—A narrow zone of transition between relatively warm and cold *air masses* that moves in such a way that the cold (more dense) air retreats, allowing the relatively warm (less dense) air to advance; may be associated with a broad band of cloudiness and light to moderate *precipitation*.

warm-core anticyclone—Massive high pressure *system* with a circulation extending from Earth's surface up to the tropical tropopause; forms south of the *polar front* and consists of extensive areas of subsiding warm, dry air. Like *cold-core cyclones*, warm-core *anticyclones* strengthen with altitude. *Subtropical anticyclones are examples*.

warm-core cyclone—A surface, synoptic-scale, stationary low-pressure system that develops as a consequence of intense solar heating of a broad expanse of arid or semiarid land such as the interior of Mexico and the American Southwest. The *system* is stationary, has no *fronts*, generally is associated with fair *weather*, and its circulation weakens rapidly with altitude, away from the *heat* source (the ground).

water vapor satellite imagery—Picture or image processed from a satellite radiometer that senses *infrared radiation* at those *wavelengths* (typically near 6.7 micrometers) emitted by *clouds* and water vapor in the *atmosphere*; enables scientists to track the movements of moisture plumes within the *atmosphere* over distances of thousands of kilometers.

wave frequency—The number of crests (or troughs) of a wave that passes a given point in a specified period of time, usually one second.

wavelength—The distance between successive wave crests (or equivalently, wave troughs).

weather—The state of the *atmosphere* at some place and time, described in terms of such variables as *temperature*, *humidity*, cloudiness, *precipitation*, and *wind* speed and direction.

weathering—The physical disintegration, chemical decomposition, or solution of exposed rock.

weighing-bucket rain gauge—Instrument that calibrates the weight of accumulating rainwater in terms of water depth. This instrument has a device that marks a chart on a clock-driven drum or sends an electronic signal to a computer for processing.

westerlies—Belt of planetary-scale *winds* that encircles middle latitudes in a wave-like pattern of clockwise turns (called ridges) and counterclockwise turns (called troughs).

wet-bulb depression—On a *psychrometer*, the difference between the *dry-bulb temperature* and the *wet-bulb temperature*; used to determine *relative humidity*.

wet-bulb temperature—The *temperature* an air parcel would have if cooled adiabatically to saturation at constant *air pressure* by *evaporation* of water into it; measured using a *psychrometer* in which the wick-covered bulb is first soaked in distilled water and the instrument is ventilated to promote evaporation.

Wien's displacement law—Hot objects (such as the Sun) emit radiation that peaks at relatively short *wavelength*s, whereas relatively cold objects (such as the Earth-*atmosphere*-land-ocean system) emit peak radiation at longer wavelengths (and lower frequencies).

wind—Local motion of air measured relative to the rotating Earth.

wind shear—Change in horizontal *wind* speed or direction with increasing altitude.

wind vane—Instrument used to monitor *wind* direction consisting of a free-swinging horizontal shaft with a vertical plate at one end and a counterweight (arrowhead) at the other end. The counterweight always points into the *wind*.

wind-chill equivalent temperature (WET)—See *wind-chill index*.

wind-chill index—A measure of the rate of body *heat* loss due to a combination of *wind* and low air *temperature*.

Younger Dryas—A relatively cool climatic episode from about 11,000 to 10,000 years ago; triggered short-lived re-advances of remnant ice sheets in North America, Scotland, and Scandinavia.

Z

zonal flow pattern—Flow of the midlatitude *westerlies* almost directly from west to east, nearly parallel to latitude circles, with only a weak meridional component. North-south exchange of *air mass*es is minimal.

zonda—Chinook-like *wind* that is drawn down the leeward slopes of the Andes in Argentina.

INDEX

N

O

U

X

Y

Z